Education and Child Development

Chess lessons and memory improvement: Ex 11.6
Childcare for preschoolers: Ex 4.15
College plans of high school seniors: Ex 7.4
Combining exam scores: Ex 7.16
Helping hands: Ex 2.7, 2.9
IQ scores: Ex 4.16, Ex 7.28
Predictors of writing competence: Ex 14.4
School enrollment in Northern and Central Africa: Ex 3.11
Standardized test scores: Ex 4.14, Ex 10.16
Students' knowledge of geography: Act 3.1
Television viewing habits of children: Ex 3.16

Environmental Science

Cosmic radiation: Ex 9.7
Lead in tap water: Ex 10.7
Rainfall frequency distributions for Albuquerque: Ex 3.19
River water velocity and distance from shore: Ex 5.17
Soil and sediment characteristics: Ex 14.12, Ex 14.14,
Ex 14.16 (online)
Water conservation: Ex 10.10
Water quality: Ex 1.2

Food Science

Calorie consumption at fast food restaurants: Ex 2.2
Effect of exercise on food intake: Ex. 4.1
Fat content of hot dogs: Ex 8.6
Pomegranate juice and tumor growth: Ex 5.4, Ex 5.5

Leisure and Popular Culture

Car preferences: Ex 6.1
Do U Txt?: Ex 1.6
Facebook and academic performance: Ex 11.1
iPod shuffles: Ex 7.7
Life insurance for cartoon characters: Ex 3.2
Number of trials required to complete game: Ex 7.2
Probability a Hershey's Kiss will land on its base: Act 6.1
Selecting cards: Ex 6.20
Selection of contest winners: Ex 6.7
Tossing a coin: Ex 6.8
Video game performance: Ex. 4.9, Ex 4.10, Ex 6.2, Ex 6.4

Manufacturing and Industry

Bottled soda volumes: Ex 8.5
Comprehensive strength of concrete: Ex 7.14
Computer configurations: Ex 6.19
Computer sales: Ex 7.19
Corrosion of underground pipe coatings: Ex 15.14 (online)
Durable press rating of cotton fabric: Ex 14.18 (online)
DVD player warranties: Ex 6.24
Engineering stress test: Ex 7.3
Ergonomic characteristics of stool designs: Ex 15.10,
Ex 15.11 (online)
Garbage truck processing times: Ex 7.30
GFI switches: Ex 6.12
Lifetime of compact florescent lightbulbs: Ex 10.2
On-time package delivery: Ex 10.18
Paint flaws: Ex 7.6
Testing for flaws: Ex 7.11, Ex 7.12

Marketing and Consumer Behavior

Car choices: Ex 6.10
Energy efficient refrige
High-pressure sales tac
Impact of food labels:
Online security: Ex 7.
Satisfaction with cell p

Medical Science

Apgar scores: Ex 7.10, Ex 7.13
Affect of long work hours on sleep: Ex 11.10
Births and the lunar cycle: Ex 12.1, Ex 12.2
Blood platelet volume: Ex 8.2
Blood pressure and kidney disease: Ex 16.5
Blood test for ovarian cancer: Ex 10.6
Cardiovascular fitness of teens: Ex 10.11
Chronic airflow obstruction: Ex 16.9
Contracting hepatitis from blood transfusion: Ex 8.8, Ex 8.9
Cooling treatment after oxygen deprivation in newborns: Ex 2.6
Diagnosing tuberculosis: Ex 6.15
Drive-through medicine: Ex 9.8
Effects of ethanol on sleep time: Ex 15.6
Evaluating disease treatments: Ex 10.3
Facial expression and self-reported pain level: Ex 12.7
Growth hormone levels and diabetes: Ex 16.10
Hip-to-waist ratio and risk of heart attack: Ex 14.2
Hormones and body fat: Ex 15.4, Ex 15.5
Lead exposure and brain volume: Ex 5.12
Lyme disease: Ex 6.27
Markers for kidney disease: Ex 7.33
Maternal age and baby's birth weight: Ex 13.2
Medical errors: Ex 6.9
Parental smoking and infant health: Ex 16.2, Ex 16.3
Pediatric tracheal tube: Ex 13.7
Platelet volume and heart attack risk: Ex 15.1, Ex 15.2,
Ex 15.3
Premature births: Ex 7.35
Sleep duration and blood leptin level: Ex 13.12 (online)
Slowing the growth rate of tumors: Ex 10.5
Stroke mortality and education: Ex 12.8
Surviving a heart attack: Ex 6.13
Time perception and nicotine withdrawal: Ex 10.13
Treating dyskinesia: Ex 16.8
Treatment for acute mountain sickness: Act 2.5
Ultrasound in treatment of soft-tissue injuries: Ex 11.5, Ex 11.8
Video games and pain management: Act 2.4
Vitamin B12 levels in human blood: Ex 16.7
Waiting time for cardiac procedures in Canada: Ex 9.9

Physical Sciences

Electromagnetic radiation: Ex 5.18
Rainfall data: Ex 7.32
Snow cover and temperature: Ex 13.9
Wind chill factor: Ex 14.3

Politics and Public Policy

Fair hiring practices: Ex 6.29
Predicting election outcomes: Ex 13.3, Ex 13.6
Recall petition signatures: Act 9.3
Requests for building permits: Ex 6.31
School board politics: Ex 14.10

Psychology, Sociology, and Social Issues

Benefits of acting out: Ex 1.3
Color and perceived taste: Act 12.2
Estimating sizes: Act 1.2
Extrasensory perception: Ex 6.33
Face-to-height ratio: Ex 5.6
Facial cues and trustworthiness: Ex 5.1
The "Freshman 15": Ex 11.4
Gender and salary: Ex 11.2
Golden rectangles: Ex 4.11
Hand-holding couples: Ex 6.30
Internet addiction: Ex 6.28
Motivation for revenge: Ex 2.4
One-boy family planning: Ex 6.32
Reading emotions: Ex 11.3
Stroop effect: Act 2.2
Subliminal messages: Ex 2.5
Weight regained proportions for three follow-up methods: Ex 12.6

Public Health and Safety

Careless or aggressive driving: Ex 9.5
Effect of cell phone distraction: Ex 2.8
Effects of McDonald's hamburger sales: Act 2.3
Crime scene investigators: Ex 5.14
Safety of bicycle helmets: Ex 5.2
Salmonella in restaurant eggs: Act 7.2
Teenage driver citations and traffic school: Ex 6.23

Sports

Age and marathon times: Ex 5.3
Calling a toss at a football game: Ex 6.6
Concussions in collegiate sports: Ex 12.4, Ex 12.5
Fairness of Euro coin-flipping in European sports: Act 6.2
Helium-filled footballs: Act 11.1
"Hot hand" in basketball: Act 6.3
NBA player salaries: Ex 4.5, Ex 4.13
Olympic figure skating: Ex 3.20
Racing starts in competitive swimming: Ex 16.6
Soccer goalie action bias: Ex 6.26
Tennis ball diameters: Ex 10.1
Time to first goal in hockey: Ex 8.3
Wrestlers' weight loss by headstand: Ex 13.1

Surveys and Opinion Polls

Are cell phone users different?: Ex 2.1
Cell phone usage: Ex 11.11
Collecting and summarizing numerical data: Act 2.2
Designing a sampling plan: Facebook friending: Act 2.1
Selecting a random sample: Ex 2.2

Transportation

Accidents by bus drivers: Ex 3.18
Airborne times for San Francisco to Washington, D.C. flight: Ex 9.3
Airline luggage weights: Ex 7.17
Airline passenger weights: Act 4.2
Comparing gasoline additives: Ex 2.10
Electric cars: Ex 11.9
Freeway traffic: Ex 7.15
Fuel efficiency of automobiles: Ex 16.1
Lost airline luggage: Ex 6.25
Motorcycle helmets: Ex 1.7, Ex 1.8
On-time airline flights: Ex 10.4
Predicting transit times: Ex 14.15 (online)
Turning directions on freeway off-ramp: Ex 6.3

Introduction to Statistics and Data Analysis

Rowan University Edition

Fifth Edition

Peck | Olsen | Devore

Australia • Brazil • Japan • Korea • Mexico • Singapore • Spain • United Kingdom • United States

Introduction to Statistics and Data Analysis: Rowan University Edition, Fifth Edition

Introduction to Statistics and Data Analysis, 5th Edition
Roxy Peck | Chris Olsen | Jay L. Devore

For product information and technology assistance, contact us at
Cengage Learning Customer & Sales Support, 1-800-354-9706
For permission to use material from this text or product,
submit all requests online at **cengage.com/permissions**
Further permissions questions can be emailed to
permissionrequest@cengage.com

This book contains select works from existing Cengage Learning resources and was produced by Cengage Learning Custom Solutions for collegiate use. As such, those adopting and/or contributing to this work are responsible for editorial content accuracy, continuity and completeness.

ISBN: 978-1-305-26596-7

WCN: 01-100-101

Cengage Learning
20 Channel Center Street
Boston, MA 02210
USA

Cengage Learning is a leading provider of customized learning solutions with office locations around the globe, including Singapore, the United Kingdom, Australia, Mexico, Brazil, and Japan. Locate your local office at:
www.international.cengage.com/region.

Cengage Learning products are represented in Canada by Nelson Education, Ltd.

For your lifelong learning solutions, visit **www.cengage.com/custom.**

Visit our corporate website at **www.cengage.com.**

Brief Contents

Chapter 1 The Role of Statistics and the Data Analysis Process 1
Chapter 2 Collecting Data Sensibly 29
Chapter 3 Graphical Methods for Describing Data 80
Chapter 4 Numerical Methods for Describing Data 152
Chapter 5 Summarizing Bivariate Data 202
Chapter 6 Probability 283
Chapter 7 Random Variables and Probability Distributions 352
Chapter 8 Sampling Variability and Sampling Distributions 437
Chapter 9 Estimation Using a Single Sample 461
Chapter 10 Hypothesis Testing Using a Single Sample 505
Appendices 759
Answers to Selected Exercises 783
Index 805

Preface

In a nutshell, statistics is about understanding the role that variability plays in drawing conclusions based on data. *Introduction to Statistics and Data Analysis,* Fifth Edition, develops this crucial understanding of variability through its focus on the data analysis process.

An Organization That Reflects the Data Analysis Process

Students are introduced early to the idea that data analysis is a process that begins with careful planning, followed by data collection, data description using graphical and numerical summaries, data analysis, and finally interpretation of results. This process is described in detail in Chapter 1, and the ordering of topics in the first ten chapters of the book mirrors this process: data collection, then data description, then statistical inference.

The logical order in the data analysis process can be pictured as shown in the following figure.

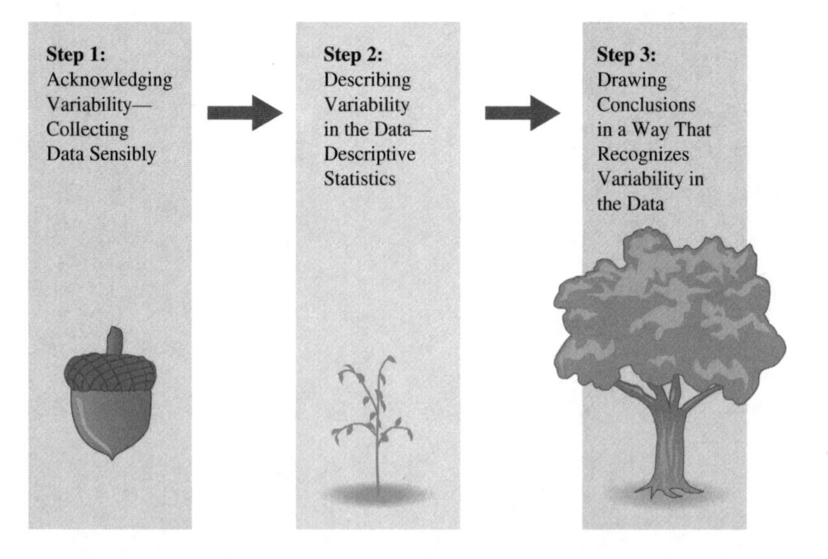

Unlike many introductory texts, *Introduction to Statistics and Data Analysis,* Fifth Edition, is organized in a manner consistent with the natural order of the data analysis process:

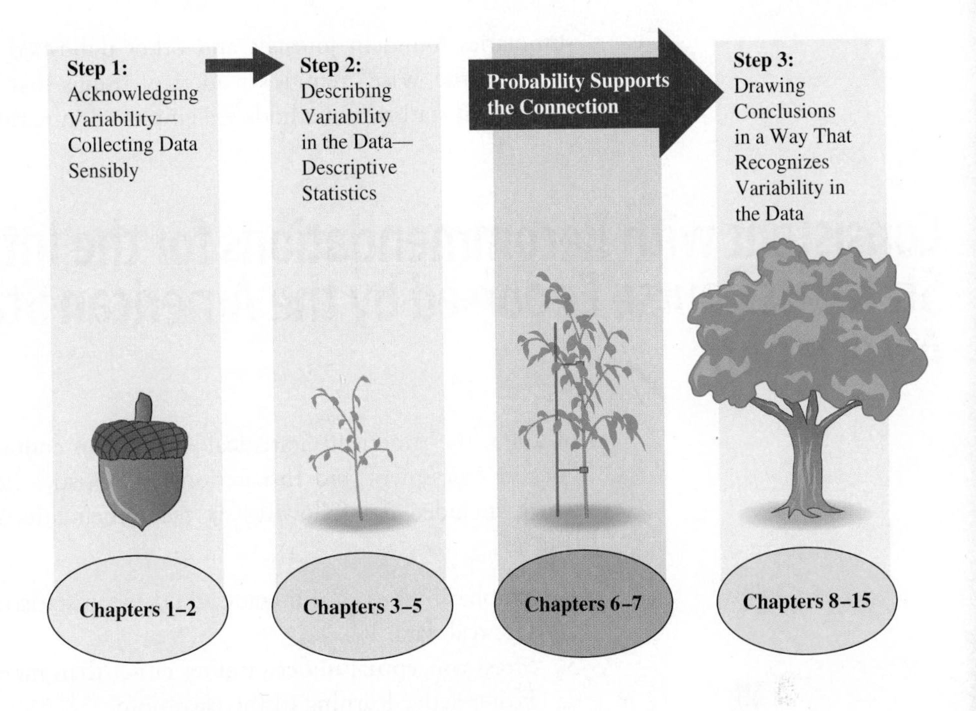

The Importance of Context and Real Data

Statistics is not about numbers; it is about data—numbers in context. It is the context that makes a problem meaningful and something worth considering. For example, exercises that ask students to compute the mean of 10 numbers or to construct a dotplot or boxplot of 20 numbers without context are arithmetic and graphing exercises. They become statistics problems only when a context gives them meaning and allows for interpretation. While this makes for a text that may appear "wordy" when compared to traditional mathematics texts, it is a critical and necessary component of a modern statistics text.

Examples and exercises with overly simple settings do not allow students to practice interpreting results in authentic situations or give students the experience necessary to be able to use statistical methods in real settings. We believe that the exercises and examples are a particular strength of this text, and we invite you to compare the examples and exercises with those in other introductory statistics texts.

Many students are skeptical of the relevance and importance of statistics. Contrived problem situations and artificial data often reinforce this skepticism. A strategy that we have employed successfully to motivate students is to present examples and exercises that involve data extracted from journal articles, newspapers, and other published sources. Most examples and exercises in the book are of this nature; they cover a very wide range of disciplines and subject areas. These include, but are not limited to, health and fitness, consumer research, psychology and aging, environmental research, law and criminal justice, and entertainment.

A Focus on Interpretation and Communication

Most chapters include a section titled "Interpreting and Communicating the Results of Statistical Analyses." These sections include advice on how to best communicate the results of a statistical analysis and also consider how to interpret statistical summaries found in journals and other published sources. A subsection titled "A Word to the Wise" reminds readers of things that must be considered in order to ensure that statistical methods are employed in reasonable and appropriate ways.

Consistent with Recommendations for the Introductory Statistics Course Endorsed by the American Statistical Association

In 2005, the American Statistical Association endorsed the report "College Guidelines in Assessment and Instruction for Statistics Education (GAISE Guidelines)," which included the following six recommendations for the introductory statistics course:

1. Emphasize statistical literacy and develop statistical thinking.
2. Use real data.
3. Stress conceptual understanding rather than mere knowledge of procedures.
4. Foster active learning in the classroom.
5. Use technology for developing conceptual understanding and analyzing data.
6. Use assessments to improve and evaluate student learning.

Introduction to Statistics and Data Analysis, Fifth Edition, is consistent with these recommendations and supports the GAISE guidelines in the following ways:

1. **Emphasize statistical literacy and develop statistical thinking.**
   Statistical literacy is promoted throughout the text in the many examples and exercises that are drawn from the popular press. In addition, a focus on the role of variability, consistent use of context, and an emphasis on interpreting and communicating results in context work together to help students develop skills in statistical thinking.
2. **Use real data.**
   The examples and exercises from *Introduction to Statistics and Data Analysis,* Fifth Edition, are context driven, and the reference sources include the popular press as well as journal articles.
3. **Stress conceptual understanding rather than mere knowledge of procedures.**
   Nearly all exercises in *Introduction to Statistics and Data Analysis,* Fifth Edition, are multipart and ask students to go beyond just computation. They focus on interpretation and communication, not just in the chapter sections specifically devoted to this topic, but throughout the text. The examples and explanations are designed to promote conceptual understanding. Hands-on activities in each chapter are also constructed to strengthen conceptual understanding. Which brings us to . . .
4. **Foster active learning in the classroom.**
   While this recommendation speaks more to pedagogy and classroom practice, *Introduction to Statistics and Data Analysis,* Fifth Edition, provides more than 30 hands-on activities in the text and additional activities in the accompanying instructor resources that can be used in class or assigned to be completed outside of class.
5. **Use technology for developing conceptual understanding and analyzing data.**
   The computer brings incredible statistical power to the desktop of every investigator. The wide availability of statistical computer packages such as Minitab,

JMP, and SPSS, and the graphical capabilities of the modern microcomputer have transformed both the teaching and learning of statistics. To highlight the role of the computer in contemporary statistics, we have included sample output throughout the book. In addition, numerous exercises contain data that can easily be analyzed using statistical software. However, access to a particular statistical package is not assumed. Technology manuals for specific software packages and for the graphing calculator are available in the online materials that accompany this text.

The appearance of handheld calculators with significant statistical and graphing capability has also changed statistics instruction in classrooms where access to computers is still limited. There is not universal or even wide agreement about the proper role for the graphing calculator in college statistics classes, where access to a computer is more common. At the same time, for tens of thousands of students in Advanced Placement Statistics in our high schools, the graphing calculator is the only dependable access to statistical technology.

This text allows the instructor to balance the use of computers and calculators in a manner consistent with his or her philosophy and presents the power of the calculator in a series of Graphing Calculator Explorations. These are part of the online materials that accompany this text. As with computer packages, our exposition avoids assuming the use of a particular calculator and presents the calculator capabilities in a generic format. For those using a TI graphing calculator, there is a technology manual available in the online materials that accompany this text. As much as possible, the calculator explorations are independent of each other, allowing instructors to pick and choose calculator topics that are most relevant to their particular courses.

6. **Use assessments to improve and evaluate student learning.**
   Assessment materials in the form of a test bank, quizzes, and chapter exams are available in the instructor resources that accompany this text. The items in the test bank reflect the data-in-context philosophy of the text's exercises and examples.

Advanced Placement Statistics

We have designed this book with a particular eye toward the syllabus of the Advanced Placement Statistics course and the needs of high school teachers and students. Concerns expressed and questions asked in teacher workshops and on the AP Statistics Teacher Community have strongly influenced our explanation of certain topics, especially in the area of experimental design and probability. We have taken great care to provide precise definitions and clear examples of concepts that Advanced Placement Statistics instructors have acknowledged as difficult for their students. We have also expanded the variety of examples and exercises, recognizing the diverse potential futures envisioned by very capable students who have not yet focused on a college major.

Topic Coverage

Our book can be used in courses as short as one quarter or as long as one year in duration. Particularly in shorter courses, an instructor will need to be selective in deciding which topics to include and which to set aside. The book divides naturally into four major sections: collecting data and descriptive methods (Chapters 1–5), probability

(Chapters 6–8), the basic one- and two-sample inferential techniques (Chapters 9–12), and more advanced inferential methodology (Chapters 13–16).

We include an early chapter (Chapter 5) on descriptive methods for bivariate numerical data. This early exposure raises questions and issues that should stimulate student interest in the subject. It is also advantageous for those teaching courses in which time constraints preclude covering advanced inferential material. However, this chapter can easily be postponed until the basics of inference have been covered, and then combined with Chapter 13 for a unified treatment of regression and correlation.

With the possible exception of Chapter 5, Chapters 1 through 10 should be covered in order. We anticipate that most instructors will then continue with two-sample inference (Chapter 11) and methods for categorical data analysis (Chapter 12), although regression could be covered before either of these topics. Optional portions of Chapter 14 (multiple regression) and Chapter 15 (analysis of variance) and Chapter 16 (nonparametric methods) are included in the online materials that accompany this text.

A Note on Probability

The content of the probability chapters is consistent with the Advanced Placement Statistics course description. It includes both a traditional treatment of probability and probability distributions at an introductory level, as well as a section on the use of simulation as a tool for estimating probabilities. For those who prefer a briefer and more informal treatment of probability, two chapters previously from the book *Statistics: The Exploration and Analysis of Data,* by Roxy Peck and Jay Devore are available as a custom option. Please contact your Cengage learning consultant for more information about this alternative and other alternative customized options available to you.

In This Edition

Look for the following in the Fifth Edition:

- **NEW A more informal writing style.** In this revision, an effort was made to accommodate a broader range of student reading levels.
- **NEW Helpful hints in exercises.** To help students who might be having trouble getting started, hints have been added to many of the exercises directing students to relevant examples in the text.
- **NEW Changes in exercise layout.** Many of the multistep exercises carried over from the previous edition have now been broken into parts to make them more manageable and to help students organize their thinking about the solution.
- **NEW Margin notes to highlight the importance of context and the process of data analysis.** Margin notations have been added in appropriate places in the examples. These include *Understanding the context, Consider the data, Formulate a plan, Do the work,* and *Interpret the results.* These notes are designed to increase student awareness of the steps in the data analysis process.
- **Updated examples and exercises that use data from current newspapers and journals are included.** In addition, more of the exercises specifically ask students to write (for example, by requiring students to explain their reasoning, interpret results, and comment on important features of an analysis).
- **Activities at the end of each chapter.** These activities can be used as a chapter capstone or can be integrated at appropriate places as the chapter material is covered in class.

- **Students can now go online with Aplia and CourseMate to further their understanding** of the material covered in each chapter.
- **Advanced topics** that are often omitted in a one-quarter or one-semester course, such as survey design (Section 2.6), logistic regression (Section 5.6), inference based on the estimated regression line (Sections 13.4 and 13.5), inference and variable selection methods in multiple regression (Sections 14.3 and 14.4), analysis of variance for randomized block and two-factor designs (Sections 15.3 and 15.4), and distribution-free procedures (Chapter 16) **are available in the online materials that accompany this text.**
- **Updated materials for instructors** are included. In addition to the usual instructor supplements such as a complete solutions manual and a test bank, the following are also available to instructors:
  - **An Instructor's Resource Binder,** which contains additional examples that can be incorporated into classroom presentations and cross-references to resources such as Fathom, Workshop Statistics, and Against All Odds. Of particular interest to those teaching Advanced Placement Statistics, the binder also includes additional data analysis questions of the type encountered on the free response portion of the Advanced Placement exam, as well as a collection of model responses.
  - For those who use student-response systems in class, **a set of "clicker" questions** (see JoinIn™ on TurningPoint® under Instructor Resources—Media) for assessing student understanding is available.

Instructor and Student Resources

MindTap™

MindTap™

New for the fifth edition, available via Aplia, is MindTap™ Reader, Cengage Learning's next-generation eBook. MindTap Reader provides robust opportunities for students to annotate, take notes, navigate, and interact with the text (e.g., ReadSpeaker). Annotations captured in MindTap Reader are automatically tied to the Notepad app, where they can be viewed chronologically and in a cogent, linear fashion. Instructors also can edit the text and assets in the Reader, as well as add videos or URLs.

Go to http://www.cengage.com/mindtap for more information.

aplia

Aplia™

Content

Aplia™ is an online interactive learning solution that improves comprehension and outcomes by increasing student effort and engagement. Founded by a professor to enhance his own courses, Aplia provides automatically graded assignments with detailed, immediate explanations for every question, along with innovative teaching materials. Our easy-to-use system has been used by more than 1,000,000 students at over 1,800 institutions. Exercises are taken directly from text.

Aplia homework engages students in critical thinking, requiring them to synthesize and apply knowledge, not simply recall it. The diverse types of questions reflect the types of exercises that help students learn. All homework is written by subject matter experts in the field who have taught the course before.

Aplia contains a robust course management system with powerful analytics, enabling professors to track student performance easily.

Service

Your adoption of Aplia® includes CourseCare, Cengage Learning's industry leading service and training program designed to ensure that you have everything that you need to make the most of your use of Aplia. CourseCare provides one-on-one service, from finding the right solutions for your course to training and support. A team of Cengage representatives, including Digital Solutions Managers and Coordinators as well as Service and Training Consultants assist you every step of the way. For additional information about CourseCare, please visit http://www.cengage.com/coursecare.

Our Aplia training program provides a comprehensive curriculum of beginner, intermediate, and advanced sessions, designed to get you started and effectively integrate Aplia into your course. We offer a flexible online and recorded training program designed to accommodate your busy schedule. Whether you are using Aplia for the first time or are an experienced user, there is a training option to meet your needs.

JMP Statistical Software

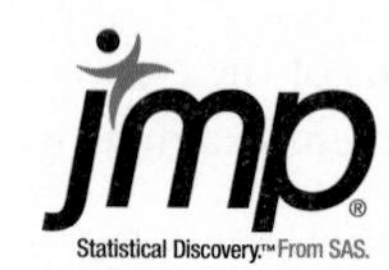

Access to JMP is free with the purchase of a new book.

JMP is a statistics software for Windows and Macintosh computers from SAS, the market leader in analytics software and services for industry. JMP Student Edition is a streamlined, easy-to-use version that provides all the statistical analysis and graphics covered in this textbook. Once data is imported, students will find that most procedures require just two or three mouse clicks. JMP can import data from a variety of formats, including Excel and other statistical packages, and you can easily copy and paste graphs and output into documents.

JMP also provides an interface to explore data visually and interactively, which will help your students develop a healthy relationship with their data, work more efficiently with data, and tackle difficult statistical problems more easily. Because its output provides both statistics and graphs together, the student will better see and understand the application of concepts covered in this book as well. JMP Student Edition also contains some unique platforms for student projects, such as mapping and scripting. JMP functions in the same way on both Windows and Mac platforms and instructions contained with this book apply to both platforms.

Access to this software is available for free with new copies of the book and available for purchase standalone at CengageBrain.com or http://www.jmp.com/getse. Find out more at www.jmp.com.

Student Resources

Digital

CENGAGE brain.com

To access additional course materials and companion resources, please visit www.cengagebrain.com. At the CengageBrain.com home page, search for the ISBN of your title (from the back cover of your book) using the search box at the top of the page. This will take you to the product page where free companion resources can be found.

- Complete step-by-step instructions for JMP, TI-84 Graphing Calculators, Excel, Minitab, and SPSS indicated by the icon throughout the text.

- Data sets in JMP, TI-84, Excel, Minitab, SPSS, SAS, and ASCII file formats indicated by the ● icon throughout the text.
- Applets used in the Activities found in the text.

Print

Student Solutions Manual *(ISBN: 978-1-3052-6582-0):* The Student Solutions Manual, prepared by Michael Allwood, contains fully worked-out solutions to all of the odd-numbered exercises in the text, giving students a way to check their answers and ensure that they took the correct steps to arrive at an answer.

Instructor Resources

Print

Annotated Instructor's Edition *(ISBN: 978-1-3052-5252-3):* The Annotated Instructor's Edition contains answers for all exercises, including those not found in the answer section of the student edition. There also are suggested assignments and teaching tips for each section in the book written by Kathy Fritz, an experienced AP Statistics teacher, along with an annotated table of contents with comments written by Roxy Peck.

Teacher's Resource Binder *(ISBN: 978-1-3052-6605-6):* The Teacher's Resource Binder, prepared by Chris Olsen, is full of wonderful resources for both college professors and AP Statistics teachers. These include

- Additional examples from published sources (with references), classified by chapter in the text. These examples can be used to enrich your classroom discussions.
- Model responses—examples of responses that can serve as a model for work that would be likely to receive a high mark on the AP exam.
- A collection of data explorations written by Chris Olsen that can be used throughout the year to help students prepare for the types of questions that they may encounter on the investigative task on the AP Statistics Exam.
- Advice to AP Statistics teachers on preparing students for the AP Exam, written by Brian Kotz.
- Activity worksheets, prepared by Carol Marchetti, that can be duplicated and used in class.
- A list of additional resources for activities, videos, and computer demonstrations, cross-referenced by chapter.
- A test bank that includes assessment items, quizzes, and chapter exams written by Chris Olsen, Josh Tabor, and Peter Flannigan-Hyde.

Online

- **Instructor Companion Site:** Everything you need for your course in one place! This collection of book-specific lecture and class tools is available online via www.cengage.com/login. Access and download PowerPoint presentations, images, instructor's manual, and more.
  - **JoinIn™ on TurningPoint®:** The easiest student classroom response system to use, JoinIn features instant classroom assessment and learning.

- **Cengage Learning Testing Powered by Cognero (ISBN: 978-1-3052-6589-9)** is a flexible, online system that allows you to author, edit, and manage test bank content, create multiple test versions in an instant, and deliver tests from your LMS, your classroom or wherever you want. This is available online via www .cengage.com/login.
- **Complete Solutions Manual** This manual contains solutions to all exercises from the text, including Chapter Review Exercises and Cumulative Review Exercises. This manual can be found on the Instructors Companion Site.

Acknowledgments

We are grateful for the thoughtful feedback from the following reviewers that has helped to shape this text over the last three editions:

Reviewers for the Fifth Edition

Kathleen Dale
Eastern Connecticut State University

Kendra Mhoon
University of Houston-Downtown

Mamunur Rashid
Indiana University-Purdue University Indianapolis

Leah Rathbun
Colorado State University

Cathleen ZuccoTeveloff
Rider University

Reviewers for the Fourth, Third, and Second Editions

Arun K. Agarwal, Jacob Amidon, Holly Ashton, Barb Barnet, Eddie Bevilacqua, Piotr Bialas, Kelly Black, Jim Bohan, Pat Buchanan, Gabriel Chandler, Andy Chang, Jerry Chen, Richard Chilcoat, Mary Christman, Marvin Creech, Ron Degged, Hemangini Deshmukh, Ann Evans, Guangxiong Fang, Sharon B. Finger, Donna Flint, Steven Garren, Mark Glickman, Rick Gumina, Debra Hall, Tyler Haynes, Sonja Hensler, Trish Hutchinson, John Imbrie, Bessie Kirkwood, Jeff Kollath, Christopher Lacke, Austin Lampros, Michael Leitner, Zia Mahmood, Art Mark, Pam Martin, David Mathiason, Bob Mattson, C. Mark Miller, Megan Mocko, Paul Myers, Kane Nashimoto, Helen Noble, Douglas Noe, Broderick Oluyede, Elaine Paris, Shelly Ray Parsons, Deanna Payton, Judy Pennington-Price, Michael Phelan, Alan Polansky, Michael Ratliff, David Rauth, Kevin J. Reeves, Lawrence D. Ries, Hazel Shedd, Robb Sinn, Greg Sliwa, Angela Stabley, Jeffery D. Sykes, Yolanda Tra, Joe Ward, Nathan Wetzel, Mark Wilson, Yong Yu, and Toshiyuki Yuasa, Cathleen Zucco-Teveloff.

We would also like to express our thanks and gratitude to those whose support made this fifth edition possible:

- Molly Taylor, our editor, for her support and guidance.
- Jay Campbell, our developmental editor, for his kindness and humor as he kept us on track and moving forward.

- Danielle Hallock for leading the development of all of the supporting ancillaries.
- Jill Quinn, our Cengage production editor.
- Edward Dionne, our project editor at MPS Limited.
- Christine Sabooni for her careful manuscript review and copyediting.
- Michael Allwood for his work in creating new student and instructor solutions manuals to accompany the text.
- Kathy Fritz for creating the new interactive PowerPoint presentations that accompany the text and also for sharing her insight in writing the Teaching Tips that accompany each chapter in the annotated instructor editions.
- Stephen Miller for checking the accuracy of examples and solutions.
- Josh Tabor and Peter-Flannagan Hyde for their contributions to the test bank that accompanies the book.
- Beth Chance and Francisco Garcia for producing the applet used in the confidence interval activities.
- Gary McClelland for producing the applets from Seeing Statistics used in the regression activities.
- Carolyn Crockett, our former editor at Cengage, for her support on the previous editions of this book.

And, as always, we thank our families, friends, and colleagues for their continued support.

Roxy Peck
Chris Olsen
Jay Devore

Peck, Olsen, Devore's

Introduction to Statistics and Data Analysis, Fifth Edition . . .

. . . Emphasizes Statistical Literacy and Statistical Thinking

Context-Driven Applications

Real data examples and exercises throughout the text are drawn from the popular press, as well as journal articles. Data sources are in a colored font for easy identification.

EXERCISES 2.1 - 2.12

2.1 The article **"How Dangerous Is a Day in the Hospital?"** (*Medical Care* **[2011]: 1068–1075**) describes a study to determine if the risk of an infection is related to the length of a hospital stay. The researchers looked at a large number of hospitalized patients and compared the proportion who got an infection for two groups of patients—those who were hospitalized overnight and those who were hospitalized for more than one night. Indicate whether the study is an observational study or an experiment. Give a brief explanation for your choice.

2.2 The authors of the paper **"Fudging the Numbers: Distributing Chocolate Influences Student Evaluations of an Undergraduate Course"** (*Teaching in Psychology* **[2007]: 245–247**) carried out a study to see if events unrelated to an undergraduate course could affect chocolate prior to having them fill out course evaluations. Students in the other three sections were not offered chocolate.

The researchers concluded that "Overall, students offered chocolate gave more positive evaluations than students not offered chocolate." Indicate whether the study is an observational study or an experiment. Give a brief explanation for your choice.

2.3 The article **"Why We Fall for This"** (*AARP Magazine*, **May/June 2011**) described a study in which a business professor divided his class into two groups. He showed students a mug and then asked students in one of the groups how much they would pay for the mug. Students in the other group were asked how much they would sell the mug for if it belonged to them. Surprisingly, the average value assigned to the

Page 33

EXAMPLE 2.2 Think Before You Order That Burger!

The article **"What People Buy from Fast-Food Restaurants: Caloric Content and Menu Item Selection"** (*Obesity* **[2009]: 1369–1374**) reported that the average number of calories consumed at lunch in New York City fast-food restaurants was 827. The researchers selected 267 fast-food locations at random. The paper states that at each of these locations "adult customers were approached as they entered the restaurant and asked to provide their food receipt when exiting and to complete a brief survey."

Approaching customers as they entered the restaurant and before they ordered may have influenced what they purchased. This introduces the potential for response bias. In addition, some people chose not to participate when approached. If those who chose not to participate differed from those who did participate, the researchers also need to be concerned about nonresponse bias. Both of these potential sources of bias limit the researchers' ability to generalize conclusions based on data from this study. ■

Page 37

Focus on Interpreting and Communicating

Chapter sections on interpreting and communicating results are designed to emphasize the importance of being able to interpret statistical output and communicate its meaning to non-statisticians. A subsection titled "A Word to the Wise" reminds students of things that must be considered in order to ensure that statistical methods are used in reasonable and appropriate ways.

3.5 Interpreting and Communicating the Results of Statistical Analyses

A graphical display, when used appropriately, can be a powerful tool for organizing and summarizing data. By sacrificing some of the detail of a complete listing of a data set, important features of the data distribution are more easily seen and more easily communicated to others.

Page 127

A Word to the Wise: Cautions and Limitations

When constructing and interpreting graphical displays, you need to keep in mind these things:

1. **Areas should be proportional to frequency, relative frequency, or magnitude of the number being represented.** The eye is naturally drawn to large areas in graphical displays, and it is natural for the observer to make informal comparisons based on area. Correctly constructed graphical displays, such as pie charts,

Page 131

What to Look for in Published Data

Here are some questions you might ask yourself when attempting to extract information from a graphical data display:

- Is the chosen display appropriate for the type of data collected?
- For graphical displays of univariate numerical data, how would you describe the shape of the distribution, and what does this say about the variable being summarized?
- Are there any outliers (noticeably unusual values) in the data set? Is there any plausible explanation for why these values differ from the rest of the data? (The presence of outliers often leads to further avenues of investigation.)
- Where do most of the data values fall? What is a typical value for the data set? What does this say about the variable being summarized?
- Is there much variability in the data values? What does this say about the variable being summarized?

Page 130

Interpreting the Results of Statistical Analyses

When someone uses a web search engine, do they rely on the ranking of the search results returned or do they first scan the results looking for the most relevant? The authors of the paper **"Learning User Interaction Models for Predicting Web Search Result Preferences"** (*Proceedings of the 29th Annual ACM Conference on Research and Development in Information Retrieval,* **2006**) attempted to answer this question by observing user behavior when they varied the position of the most relevant result in the list of resources returned in response to a web search.

They concluded that people clicked more often on results near the top of the list, even when they were not relevant. They supported this conclusion with the comparative bar graph in Figure 3.37.

Page 128

Peck, Olsen, Devore's

Introduction to Statistics and Data Analysis, Fifth Edition . . .

. . . Encourages Conceptual Understanding and Active Learning

Hands-on Activities in Every Chapter

More than 30 hands-on activities in the text, and additional activities in the accompanying instructor resources, can be used to encourage active learning inside or outside the classroom.

ACTIVITY 2.1 Facebook Friending

Background: The article **"Professors Prefer Face Time to Facebook"** appeared in the student newspaper at Cal Poly, San Luis Obispo (***Mustang Daily,* August 27, 2009**). The article examines how professors and students felt about using Facebook as a means of faculty-student communication. The student who wrote this article got mixed opinions when she interviewed students to ask whether they wanted to become Facebook friends with their professors. Two student comments included in the article were

"I think the younger the professor is, the more you can relate to them and the less awkward it would be if you were to become friends on Facebook. The older the professor, you just would have to wonder, 'Why are they friending me?'"

and

"I think becoming friends with professors on Facebook is really awkward. I don't want them being able to see into my personal

Page 69

Technology Notes

Technology Notes appear at the end of most chapters and give students helpful hints and guidance on completing tasks associated with a particular chapter. The following technologies are included in the notes: JMP, Minitab, SPSS, Microsoft Excel, and TI-83/84.

TECHNOLOGY NOTES

Confidence Intervals for Proportions

JMP

Summarized data

1. Enter the data table into the JMP data table with categories in the first column and counts in the second column

	Column 1	Column 2
1	Yes	130
2	No	45

2. Click **Analyze** and select **Distribution**
3. Click and drag the first column name from the box under **Select Columns** to the box next to **Y, Columns**
4. Click and drag the second column name from the box under **Select Columns** to the box next to **Freq**
5. Click **OK**
6. Click the red arrow next to the column name and click **Confidence Interval** then select the appropriate level or select **Other** to input a level that is not listed

3. Click and drag the first column na der **Select Columns** to the box ne
4. Click **OK**
5. Click the red arrow next to the co **Confidence Interval** then select th select **Other** to input a level that is

Minitab

Summarized data

1. Click **Stat** then click **Basic S** **1 Proportion...**
2. Click the radio button next to **Sum**
3. In the box next to **Number of Tria** n, the total number of trials
4. In the box next to **Number of even** the number of successes
5. Click **Options...**
6. Input the appropriate confidence le **Confidence Level**
7. Check the box next to **Use test an** **normal distribution**
8. Click **OK**
9. Click **OK**

Raw data

Page 502

Peck, Olsen, Devore's

Introduction to Statistics and Data Analysis, Fifth Edition . . .

. . . Uses Technology to Develop Conceptual Understanding

Applets Allow Students to See the Concepts

Within the Activities, applets are used to illustrate and promote a deeper understanding of the key statistical concepts.

▶

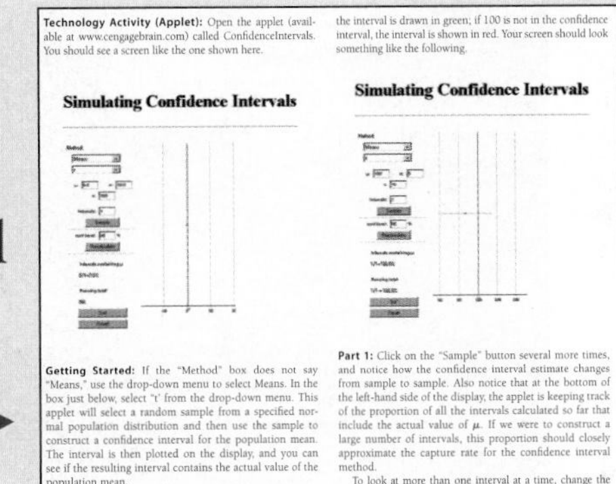

Technology Activity (Applet): Open the applet (available at www.cengagebrain.com) called ConfidenceIntervals. You should see a screen like the one shown here.

Simulating Confidence Intervals

Getting Started: If the "Method" box does not say "Means," use the drop-down menu to select Means. In the box just below, select "t" from the drop-down menu. This applet will select a random sample from a specified normal population distribution and then use the sample to construct a confidence interval for the population mean. The interval is then plotted on the display, and you can see if the resulting interval contains the actual value of the population mean.

For purposes of this activity, we will sample from a normal population with mean 100 and standard deviation 5. We will begin with a sample size of $n = 10$. In the applet window, set $\mu = 100$, $\sigma = 5$, and $n = 10$. Leave the conf-level box set at 95%. Click the "Recalculate" button to rescale the picture on the right. Now click on the sample button. You should see a confidence interval appear on the display on the right-hand side. If the interval contains the actual mean of 100, the interval is drawn in green; if 100 is not in the confidence interval, the interval is shown in red. Your screen should look something like the following.

Simulating Confidence Intervals

Part 1: Click on the "Sample" button several more times, and notice how the confidence interval estimate changes from sample to sample. Also notice that at the bottom of the left-hand side of the display, the applet is keeping track of the proportion of all the intervals calculated so far that include the actual value of μ. If we were to construct a large number of intervals, this proportion should closely approximate the capture rate for the confidence interval method.

To look at more than one interval at a time, change the "Intervals" box from 1 to 100, and then click the sample button. You should see a screen similar to the one at the top left of the next page, with 100 intervals in the display on the right-hand side. Again, intervals containing 100 (the value of μ in this case) will be green and those that do not contain 100 will be red. Also note that the capture proportion on the left-hand side has also been updated to reflect what happened with the 100 newly generated intervals.

Page 497

. . . And Analyze Data

Real Data Sets

Real data sets promote statistical analysis, as well as technology use. They are formatted for JMP, Minitab, SPSS, Microsoft Excel, TI-83/84, and ASCII and are indicated by the ● icon throughout the text.

EXAMPLE 3.11 Progress for Children

● The report **"Progress for Children" (UNICEF, April 2005)** included the accompanying data on the percentage of primary-school-age children who were enrolled in school for 19 countries in Northern Africa and for 23 countries in Central Africa.

Northern Africa

54.6	34.3	48.9	77.8	59.6	88.5	97.4	92.5	83.9	96.9	88.9
98.8	91.6	97.8	96.1	92.2	94.9	98.6	86.6			

Central Africa

58.3	34.6	35.5	45.4	38.6	63.8	53.9	61.9	69.9	43.0	85.0
63.4	58.4	61.9	40.9	73.9	34.8	74.4	97.4	61.0	66.7	79.6
98.9										

We will construct a comparative stem-and-leaf display using the first digit of each observation as the stem and the remaining two digits as the leaf. To keep the display simple the leaves will be truncated to one digit. For example, the observation 54.6 would be processed as

Page 95

3.21 ● The percentage of teens not in school or working in 2010 for the 50 states were given in the **2012 Kids Count Data Book (www.aecf.org)** and are shown in the following table:

State	Rate	State	Rate
Alabama	11%	Kansas	6%
Alaska	11%	Kentucky	11%
Arizona	12%	Louisiana	14%
Arkansas	12%	Maine	7%
California	8%	Maryland	8%
Colorado	7%	Massachusetts	5%
Connecticut	5%	Michigan	9%
Delaware	9%	Minnesota	5%
Florida	10%	Mississippi	13%
Georgia	12%	Missouri	9%
Hawaii	12%	Montana	9%

Page 98

Step-by-Step Technology Instructions

Complete online step-by-step instructions for JMP, Minitab, SPSS, Microsoft Excel, and TI-83/84 are indicated by the icon throughout the text.

Step-by-step technology instructions available online

Understand the data)

EXAMPLE 3.16 TV Viewing Habits of Children

The article **"Early Television Exposure and Subsequent Attention Problems in Children" (*Pediatrics*, April 2004)** investigated the television viewing habits of children in the United States. Table 3.5 gives approximate relative frequencies (read from graphs that appeared in the article) for the number of hours spent watching TV per day for a sample of children at age 1 year and a sample of children at age 3 years. The data summarized in the article were obtained as part of a large-scale national survey.

Page 105

Peck, Olsen, Devore's

Introduction to Statistics and Data Analysis, Fifth Edition . . .

. . . Evaluates Students' Understanding

Evaluate as You Teach Using Clickers

Using clicker content authored by Roxy Peck, evaluate your students' understanding immediately—in class—after teaching a concept. Whether it's a quick quiz, a poll to be used for in-class data, or just checking in to see if it is time to move on, our quality, tested content creates truly interactive classrooms with students' responses shaping the lecture as you teach.

The comparative bar chart shown illustrates how nonsmokers, previous smokers, and smokers differ with respect to their perceived risk of smoking. Which of the following is a correct conclusion based on this graph?

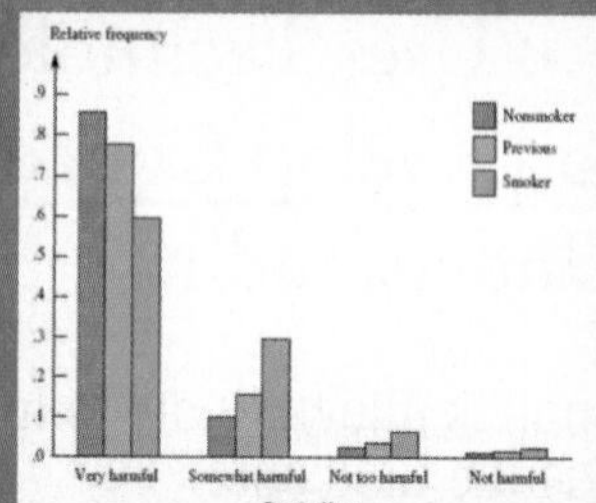

1. The proportion of smokers who perceive the risk of smoking to be very harmful is greater than the corresponding proportion of nonsmokers.
2. The biggest difference between smokers and previous smokers is in the not harmful category.
3. The proportion of smokers is higher than the proportion of nonsmokers and higher than the proportion of previous smokers in all but the very harmful category.
4. The proportion of nonsmokers is higher than the proportion of smokers in both the very harmful and somewhat harmful categories.

3.11 ▼ Poor fitness in adolescents and adults increases the risk of cardiovascular disease. In a study of 3110 adolescents and 2205 adults (*Journal of the American Medical Association*, December 21, 2005), researchers found 33.6% of adolescents and 13.9% of adults were unfit; the percentage was similar in adolescent males (32.9%) and females (34.4%), but was higher in adult females (16.2%) than in adult males (11.8%).

a. Summarize this information using a comparative bar graph that shows differences between males and females within the two different age groups.

b. Comment on the interesting features of your graphical display.

Page 90

Video Solutions Motivate Student Understanding

More than 90 exercises have video solutions, presented by Brian Kotz of Montgomery College, which can be viewed online or downloaded for later viewing. These exercises are designated by the ▼ in the text.

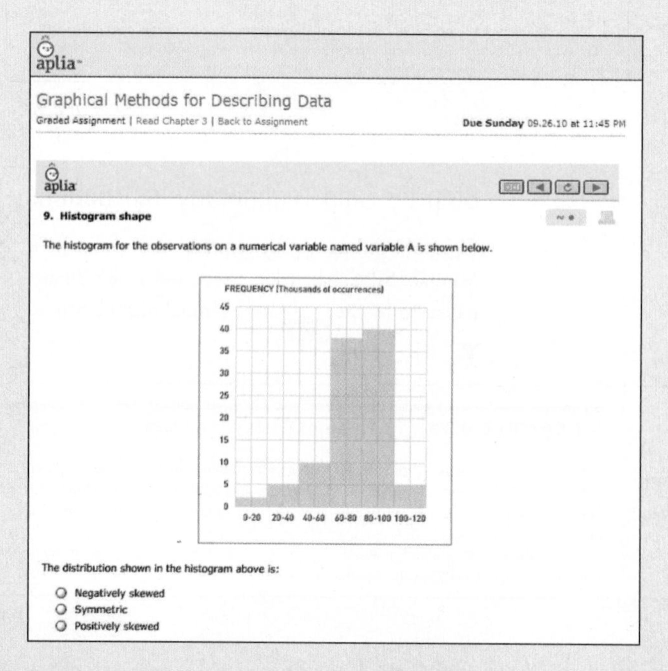

APLIA™ Online Interactive Learning Solution

Aplia provides automatically graded assignments with detailed, immediate explanations for every question, along with innovative teaching materials.

Resources for Students*

DIGITAL

MindTap™

Cengagebrain.com

- Step-by-step instructions for JMP, TI-84 Graphing Calculators, Excel, Minitab, and SPSS
- Data sets formatted for JMP, TI-84, Excel, Minitab, SPSS, SAS, and ASCII
- Video solutions
- Applets used in the Activities

Aplia™

Enhanced WebAssign

JMP statistical software

PRINT

Student Solutions Manual
(978-1-3052-6582-0)

Resources for Instructors*

DIGITAL

MindTap™

Aplia™

Companion website: cengage.com/login

- PowerPoint presentations
- Images
- Complete solutions manual
- Cengage Learning Testing by Cognero
- JoinIn on TurningPoint

JMP statistical software

PRINT

Annotated Instructor's Edition
(978-1-3052-5252-3)

Teacher's Resource Binder
(978-1-3052-6605-6)

* See the full preface for complete descriptions.

Options That **SAVE** Your Students Money

ONLINE LEARNING WITH INTERACTIVE EBOOK

Teach more, grade less with Aplia.

Aplia™ is an online interactive learning solution that improves comprehension and outcomes by increasing student effort and engagement.

Founded by a professor to enhance his own courses, **Aplia** provides automatically graded assignments with detailed, immediate explanations on every question, and innovative teaching materials. Our easy to use system has been used by more than 1,000,000 students at over 1800 institutions.

2-Semester Slim Pack
ISBN: 0-538-73475-2

Aplia can be bundled with the text. Contact your Cengage Learning representative to find out more about bundling options.

BOOK RENTALS UP TO 60% OFF

CENGAGE **brain**.com

Students have the CHOICE to purchase the eBook or rent the text at CengageBrain.com

OR

an eTextbook in PDF format is also available for instant access for your students at **www.coursesmart.com**.

CourseSmart

Students can rent Peck/Olsen/Devore's *Introduction to Statistics and Data Analysis*, Fifth Edition, for up to 60% off list price.

CUSTOM SOLUTIONS

Custom Solutions to Fit Every Need

Contact your Cengage Learning representative to learn more about what custom solutions are available to meet your course needs.

- Adapt existing Cengage Learning content by adding or removing chapters
- Incorporate your own materials

Source Code: 12M-MA0010

CHAPTER 1

The Role of Statistics and the Data Analysis Process

© Andresr/Shutterstock.com

We encounter data and make conclusions based on data every day. **Statistics** is the scientific discipline that provides methods to help us make sense of data. Statistical methods, used intelligently, offer a set of powerful tools for gaining insight into the world around us. The widespread use of statistical analyses in diverse fields such as business, medicine, agriculture, social sciences, natural sciences, and engineering has led to increased recognition that statistical literacy—a familiarity with the goals and methods of statistics—should be a basic component of a well-rounded educational program.

The field of statistics helps us to make intelligent judgments and informed decisions in the presence of uncertainty and variation. In this chapter, we consider the nature and role of variability in statistical settings, introduce some basic terminology, and look at some simple graphical displays for summarizing data.

Chapter 1: Learning Objectives

STUDENTS WILL UNDERSTAND:

- the steps in the data analysis process.

STUDENTS WILL BE ABLE TO:

- distinguish between a population and a sample.
- distinguish between categorical, discrete numerical, and continuous numerical data.
- construct a frequency distribution and a bar chart and describe the distribution of a categorical variable.
- construct a dotplot and describe the distribution of a numerical variable.

1.1 Why Study Statistics?

There is an old saying that "without data, you are just another person with an opinion." While anecdotes and coincidences may make for interesting stories, you wouldn't want to make important decisions on the basis of anecdotes alone. For example, just because a friend of a friend ate 16 apricots and then experienced relief from joint pain doesn't mean that this is all you need to know to help one of your parents choose a treatment for arthritis! Before recommending apricots, you would definitely want to consider relevant data—that is, data that would allow you to investigate the effectiveness of apricots as a treatment for arthritis.

It is difficult to function in today's world without a basic understanding of statistics. For example, here are just a few headlines from articles that draw conclusions based on data that all appeared over two days in ***USA Today* (December 19 and 20, 2013):**

- The article **"American Attitudes Toward Global Warming"** summarized data from a nationwide survey of adults. A variety of graphs and charts provide information on opinions regarding the existence of global warming, the impact of global warming, and what actions should be taken in response to global warming.
- **"Standardized Testing Fails College Students"** is the title of an article describing the use of standardized tests to place college students into appropriate college mathematics courses. The article concludes that many students are not aware of the importance of these exams and do not prepare for them. This results in many students being placed in developmental mathematics courses that slow their progress toward getting their degree.
- The article **"Shoppers Say Ho-Hum to Discounts"** describes conclusions drawn from a study of how consumers respond to e-mail advertising campaigns that offer discounts. The article concludes that this practice has become so widespread that shoppers largely ignore these e-mails and delete them without even reading them. This is information that retailers should consider when planning future advertising campaigns.
- **"Older Americans Could Opt Out of Blood Pressure Meds"** is the title of an article describing a study of the effect of high blood pressure on health for those over age 60. Data from this study lead to the conclusion that for older Americans, there is no further benefit to reducing blood pressure below 150/90. This is of interest to doctors because the previous recommendation was that blood pressure should be 140/90 or lower.
- The article **"College Coaching Gender Gap Persists"** reported data from a study of colleges in six large NCAA sports conferences. The study found that only 39.6% of the coaches of women's sports teams in 2013 were women, which is even lower than in previous years. The article concluded that although Title IX increased opportunities for participation of women in collegiate sports, it has not yet resulted in increased opportunities for women as coaches.

To be an informed consumer of reports such as those described above, you must be able to do the following:

1. Extract information from tables, charts, and graphs.
2. Follow numerical arguments.
3. Understand the basics of how data should be gathered, summarized, and analyzed to draw statistical conclusions.

Your statistics course will help prepare you to perform these tasks.

Studying statistics will also enable you to collect data in a sensible way and then use the data to answer questions of interest. In addition, studying statistics will allow you to critically evaluate the work of others by providing you with the tools you need to make informed judgments.

Throughout your personal and professional life, you will need to understand and use data to make decisions. To do this, you must be able to

1. Decide whether existing data is adequate or whether additional information is required.
2. If necessary, collect more information in a reasonable and thoughtful way.
3. Summarize the available data in a useful and informative manner.
4. Analyze the available data.
5. Draw conclusions, make decisions, and assess the risk of an incorrect decision.

These are the steps in the data analysis process. These steps will be considered in more detail in Section 1.3.

We hope that this textbook will help you to understand the logic behind statistical reasoning, prepare you to apply statistical methods appropriately, and enable you to recognize when statistical arguments are faulty.

1.2 The Nature and Role of Variability

Statistical methods allow us to collect, describe, analyze, and draw conclusions from data. If we lived in a world where all measurements were identical for every individual, these tasks would be simple. Imagine a population consisting of all students at a particular university. Suppose that *every* student was enrolled in the same number of courses, spent exactly the same amount of money on textbooks this semester, and favored increasing student fees to support expanding library services. For this population, there is *no* variability in number of courses, amount spent on books, or student opinion on the fee increase. A researcher studying students from this population in order to draw conclusions about these three variables would have a particularly easy task. It would not matter how many students the researcher studied or how the students were selected. In fact, the researcher could collect information on number of courses, amount spent on books, and opinion on the fee increase by just stopping the next student who happened to walk by the library. Because there is no variability in the population, this one individual would provide complete and accurate information about the population. The researcher could draw conclusions with no risk of error.

The situation just described is obviously unrealistic. Populations with no variability are rare. In fact, variability is almost universal. We need to understand variability to be able to collect, describe, analyze, and draw conclusions from data in a sensible way.

Examples 1.1 and 1.2 illustrate how describing and understanding variability are important.

EXAMPLE 1.1 If the Shoe Fits

Understand the context ⟩

The graphs in Figure 1.1 are examples of a type of graph called a histogram. (The construction and interpretation of such graphs is discussed in Chapter 3.) Figure 1.1(a) shows the distribution of the heights of female basketball players who played at a particular university between 2005 and 2013. The height of each bar in the graph indicates how many players' heights were in the corresponding interval. For example, 40 basketball players had heights between 72 inches and 74 inches, whereas only 2 players had heights between 66 inches and 68 inches. Figure 1.1(b) shows the distribution of heights for members of the women's gymnastics team. Both histograms are based on the heights of 100 women.

Consider the data ⟩

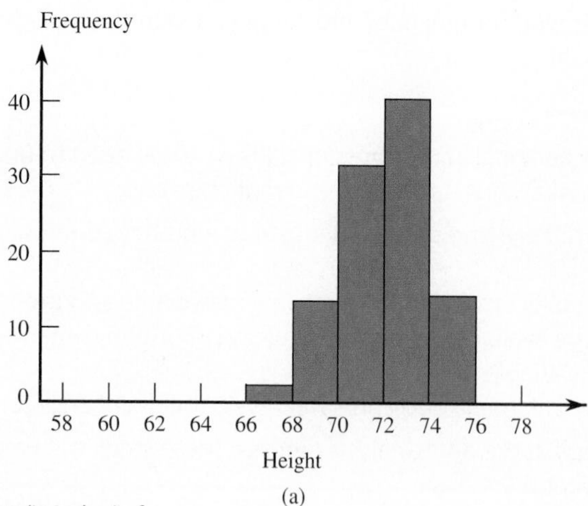

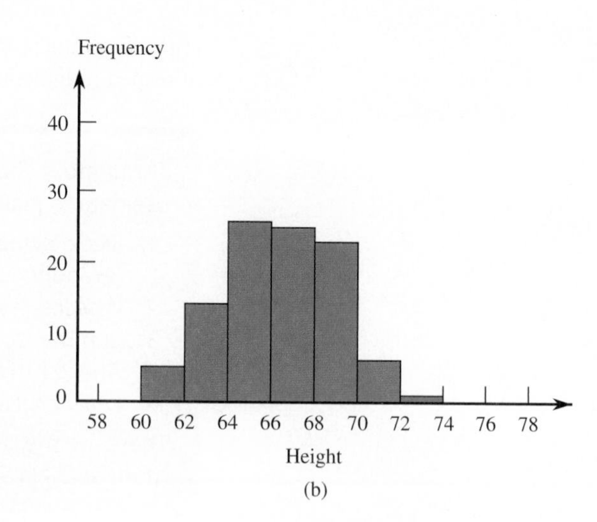

FIGURE 1.1
Histograms of heights (in inches) of female athletes:
(a) 100 basketball players;
(b) 100 gymnasts.

The first histogram shows that the heights of female basketball players varied, with most heights falling between 68 inches and 76 inches. In the second histogram we see that the heights of female gymnasts also varied, with most heights in the range of 60 inches to 72 inches. It is also clear that there is more variation in the heights of the gymnasts than in the heights of the basketball players, because the gymnast histogram spreads out more about its center than does the basketball histogram.

Interpret the results ⟩

Now suppose that a tall woman (5 feet 11 inches) tells you she is looking for her sister who is practicing with her team at the gym. Would you direct her to where the basketball team is practicing or to where the gymnastics team is practicing? What reasoning would you use to decide? If you found a pair of size 6 shoes left in the locker room, would you first try to return them by checking with members of the basketball team or the gymnastics team?

You probably answered that you would send the woman looking for her sister to the basketball practice and that you would try to return the shoes to a gymnastics team member. To reach these conclusions, you informally used statistical reasoning that combined your own knowledge of the relationship between heights of siblings and between shoe size and height with the information about the distributions of heights presented in Figure 1.1. You might have reasoned that heights of siblings tend to be similar and that a height as great as 5 feet 11 inches, although not impossible, would be unusual for a gymnast. On the other hand, a height as tall as 5 feet 11 inches would be a common occurrence for a basketball player.

Similarly, you might have reasoned that tall people tend to have bigger feet and that short people tend to have smaller feet. The shoes found were a small size, so it is more likely that they belong to a gymnast than to a basketball player, because small heights are usual for gymnasts and unusual for basketball players. ■

EXAMPLE 1.2 Monitoring Water Quality

David Chasey/Photodisc/Getty Images

Understand the context ⟩

As part of its regular water quality monitoring efforts, an environmental control board selects five water specimens from a particular well each day. The concentration of contaminants in parts per million (ppm) is measured for each of the five specimens, and then the average of the five measurements is calculated. The histogram in Figure 1.2 summarizes the average contamination values for 200 days.

Now suppose that a chemical spill has occurred at a manufacturing plant 1 mile from the well. It is not known whether a spill of this nature would contaminate groundwater in the area of the spill and, if so, whether a spill this distance from the well would affect the quality of well water.

Consider the data)

One month after the spill, five water specimens are collected from the well, and the average contamination is 15.5 ppm. Considering the variation before the spill, would you interpret this as convincing evidence that the well water was affected by the spill? What if the calculated average was 17.4 ppm? 22.0 ppm? How is your reasoning related to the histogram in Figure 1.2?

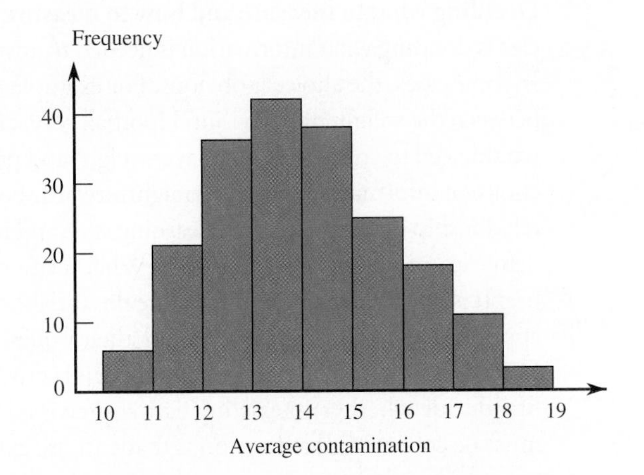

FIGURE 1.2
Average contamination concentration (in parts per million) measured each day for 200 days.

Interpret the results)

Before the spill, the average contaminant concentration varied from day to day. An average of 15.5 ppm would not have been an unusual value, so seeing an average of 15.5 ppm after the spill isn't necessarily an indication that contamination has increased. On the other hand, an average as large as 17.4 ppm is less common, and an average as large as 22.0 ppm is not at all typical of the pre-spill values. In this case, we would probably conclude that the well contamination level has increased. ■

In these two examples, reaching a conclusion required an understanding of variability. Understanding variability allows us to distinguish between usual and unusual values. The ability to recognize unusual values in the presence of variability is an important aspect of most statistical procedures. It also enables us to quantify the chance of being incorrect when a conclusion is based on data. These concepts will be developed further in subsequent chapters.

1.3 Statistics and the Data Analysis Process

Statistics involves collecting, summarizing, and analyzing data. All three tasks are critical. Without summarization and analysis, raw data are of little value. Even sophisticated analyses can't produce meaningful information from data that were not collected in a sensible way.

Statistical studies are undertaken to answer questions about our world. Is a new flu vaccine effective in preventing illness? Is the use of bicycle helmets on the rise? Are injuries that result from bicycle accidents less severe for riders who wear helmets than for those who do not? Do engineering students pay more for textbooks than psychology students? Data collection and analysis allow researchers to answer such questions.

The data analysis process can be viewed as a sequence of steps that lead from planning to data collection to making informed conclusions based on the resulting data. The process can be organized into six steps described in the following box.

The Data Analysis Process

1. **Understanding the nature of the problem.** Effective data analysis requires an understanding of the research problem. We must know the goal of the research and what questions we hope to answer. It is important to have a clear direction before gathering data to ensure that we will be able to answer the questions of interest using the data collected.
2. **Deciding what to measure and how to measure it.** The next step in the process is deciding what information is needed to answer the questions of interest. In some cases, the choice is obvious. For example, in a study of the relationship between the weight of a Division I football player and position played, you would need to collect data on player weight and position. In other cases the choice of information is not as straightforward. For example, in a study of the relationship between preferred learning style and intelligence, how would you define learning style and measure it? What measure of intelligence would you use? It is important to carefully define the variables to be studied and to develop appropriate methods for determining their values.
3. **Data collection.** The data collection step is crucial. The researcher must first decide whether an existing data source is adequate or whether new data must be collected. If a decision is made to use existing data, it is important to understand how the data were collected and for what purpose, so that any resulting limitations are also fully understood. If new data are to be collected, a careful plan must be developed, because the type of analysis that is appropriate and the subsequent conclusions that can be drawn depend on how the data are collected.
4. **Data summarization and preliminary analysis.** After the data are collected, the next step is usually a preliminary analysis that includes summarizing the data graphically and numerically. This initial analysis provides insight into important characteristics of the data and can provide guidance in selecting appropriate methods for further analysis.
5. **Formal data analysis.** The data analysis step requires the researcher to select and apply statistical methods. Much of this textbook is devoted to methods that can be used to carry out this step.
6. **Interpretation of results.** Several questions should be addressed in this final step. Some examples are: What can we learn from the data? What conclusions can be drawn from the analysis? How can our results guide future research? The interpretation step often leads to the formulation of new research questions. These new questions lead back to the first step. In this way, good data analysis is often an iterative process.

To illustrate these steps, consider the following example. The admissions director at a large university might be interested in learning why some applicants who were accepted for the fall 2014 term failed to enroll at the university. The **population** of interest to the director consists of all accepted applicants who did not enroll in the fall 2014 term. Because this population is large and it may be difficult to contact all the individuals, the director might decide to collect data from only 300 selected students. These 300 students constitute a **sample**.

DEFINITION

Population: The entire collection of individuals or objects about which information is desired is called the **population** of interest.

Sample: A **sample** is a subset of the population, selected for study.

Deciding how to select the 300 students and what data should be collected from each student are steps 2 and 3 in the data analysis process. Step 4 in the process involves organizing and summarizing data. Methods for organizing and summarizing data, such as the use of tables, graphs, or numerical summaries, make up the branch of statistics called **descriptive statistics**. The second major branch of statistics, **inferential statistics**, involves generalizing from a sample to the population from which it was selected. When we generalize in this way, we run the risk of an incorrect conclusion, because a conclusion about the population is based on incomplete information. An important aspect in the development of inferential techniques involves quantifying the chance of an incorrect conclusion.

DEFINITION

Descriptive statistics: The branch of statistics that includes methods for organizing and summarizing data.

Inferential statistics: The branch of statistics that involves generalizing from a sample to the population from which the sample was selected and assessing the reliability of such generalizations.

Example 1.3 illustrates the steps in the data analysis process.

EXAMPLE 1.3 The Benefits of Acting Out

Understand the context ⟩

A number of studies have reached the conclusion that stimulating mental activities can lead to improved memory and psychological wellness in older adults. The article **"A Short-Term Intervention to Enhance Cognitive and Affective Functioning in Older Adults"** (***Journal of Aging and Health*** **[2004]: 562–585**) describes a study to investigate whether training in acting has similar benefits. Acting requires a person to consider the goals of the characters in the story, to remember lines of dialogue, to move on stage as scripted, and to do all of this at the same time. The researchers conducting the study wanted to see if participation in this type of complex multitasking would lead to an improvement in the ability to function independently.

Participants in the study were assigned to one of three groups. One group took part in an acting class for 4 weeks. One group spent a similar amount of time in a class on visual arts. The third group was a comparison group (called the "no-treatment group") that did not take either class. A total of 124 adults age 60 to 86 participated in the study.

Interpret the results ⟩

At the beginning of the 4-week study period and again at the end of the 4-week study period, each participant took several tests designed to measure problem-solving ability, memory span, self-esteem, and psychological well-being. After analyzing the data from this study, the researchers concluded that those in the acting group showed greater gains than both the visual arts group and the no-treatment group in both problem solving and psychological well-being.

Several new areas of research were suggested in the discussion that followed the analysis. The researchers wondered whether the effect of studying writing or music would be similar to what was observed for acting and described plans to investigate this further. They also noted that the participants in this study were generally well educated and recommended study of a more diverse group before generalizing conclusions about the benefits of studying acting to the larger population of all older adults.

This study illustrates the nature of the data analysis process. A clearly defined research question and an appropriate choice of how to measure the variables of interest (the tests used to measure problem solving, memory span, self-esteem, and psychological well-being) preceded the data collection. Assuming that a reasonable method was used to collect the data (we will see how this can be evaluated in Chapter 2) and that appropriate methods of analysis were employed, the investigators reached the conclusion that the study of acting showed promise. However, they recognized the limitations of the study, which in turn led to plans for further research. As is often the case, the data analysis cycle led to new research questions, and the process began again. ■

EXERCISES 1.1 - 1.11

1.1 Give a brief definition of the terms *descriptive statistics* and *inferential statistics.*

1.2 Give a brief definition of the terms *population* and *sample.*

1.3 The following conclusion from a study appeared in the article **"Smartphone Nation" (*AARP Bulletin,* September 2009)**: "If you love your smart phone, you are not alone. Half of all boomers sleep with their cell phone within arm's length. Two of three people age 50 to 64 use a cell phone to take photos, according to a 2010 Pew Research Center report." Are the given proportions (half and two of three) population values, or were they calculated from a sample?

1.4 Based on a study of 2121 children between the ages of 1 and 4, researchers at the Medical College of Wisconsin concluded that there was an association between iron deficiency and the length of time that a child is bottle-fed **(*Milwaukee Journal Sentinel,* November 26, 2005)**. Describe the sample and the population of interest for this study.

1.5 The student senate at a university with 15,000 students is interested in the proportion of students who favor a change in the grading system to allow for plus and minus grades (e.g., B+, B, B−, rather than just B). Two hundred students are interviewed to determine their attitude toward this proposed change.

a. What is the population of interest?

b. What group of students constitutes the sample in this problem?

1.6 The increasing popularity of online shopping has many consumers using Internet access at work to browse and shop online. In fact, the Monday after Thanksgiving has been nicknamed "Cyber Monday" because of the large increase in online purchases that occurs on that day. Data from a large-scale survey by a market research firm **(*Detroit Free Press,* November 26, 2005)** was used to compute estimates of the percent of men and women who shop online while at work. The resulting estimates probably won't make most employers happy—42% of the men and 32% of the women in the sample were shopping online at work!

Are the estimates given computed using data from a sample or for the entire population?

1.7 The supervisors of a rural county are interested in the proportion of property owners who support the construction of a sewer system. Because it is too costly to contact all 7000 property owners, a survey of 500 owners is undertaken. Describe the population and sample for this problem.

1.8 A consumer group conducts crash tests of new model cars. To determine the severity of damage to 2014 Toyota Camrys resulting from a 10-mph crash into a concrete wall, the research group tests six cars of this type and assesses the amount of damage. Describe the population and sample for this problem.

1.9 A building contractor has a chance to buy an odd lot of 5000 used bricks at an auction. She is interested in determining the proportion of bricks in the lot that are cracked and therefore unusable for her current project, but she does not have enough time to inspect all 5000 bricks. Instead, she checks 100 bricks to determine whether each is cracked. Describe the population and sample for this problem.

1.10 The article **"Brain Shunt Tested to Treat Alzheimer's" (*San Francisco Chronicle,* October 23, 2002)** summarizes the findings of a study that appeared in the journal *Neurology*. Doctors at Stanford Medical Center were interested in determining whether a new surgical approach to treating Alzheimer's disease results in improved memory functioning. The surgical procedure involves implanting a thin tube, called a shunt, which is designed to drain toxins from the fluid-filled space that cushions the brain. Eleven patients had shunts implanted and were followed for a year, receiving quarterly tests of memory function. Another sample of Alzheimer's patients was used as a comparison group. Those in the comparison group received the standard care for Alzheimer's disease. After analyzing the data from this study, the investigators concluded that the "results suggested the treated patients essentially held their own in the cognitive tests while the patients in the control group steadily declined. However, the study was too small to produce conclusive statistical evidence."

a. What were the researchers trying to learn? What questions motivated their research?

b. Do you think that the study was conducted in a reasonable way? What additional information would you want in order to evaluate this study? (Hint: See Example 1.3.)

1.11 In a study of whether taking a garlic supplement reduces the risk of getting a cold, participants were assigned to either a garlic supplement group or to a group that did not take a garlic supplement

(**"Garlic for the Common Cold," Cochrane Database of Systematic Reviews, 2009**). Based on the study, it was concluded that the proportion of people taking a garlic supplement who get a cold is lower than the proportion of those not taking a garlic supplement who get a cold.

a. What were the researchers trying to learn? What questions motivated their research?

b. Do you think that the study was conducted in a reasonable way? What additional information would you want in order to evaluate this study?

1.4 Types of Data and Some Simple Graphical Displays

Every discipline has its own particular way of using common words, and statistics is no exception. You will recognize some of the terminology from previous math and science courses, but much of the language of statistics will be new to you. In this section, you will learn some of the terminology used to describe data.

Types of Data

The individuals or objects in any particular population typically possess many characteristics that might be studied. Consider a group of students currently enrolled in a statistics course at a particular college. One characteristic of the students in the population is the brand of calculator owned (Casio, Hewlett-Packard, Sharp, Texas Instruments, and so on). Another characteristic is the number of textbooks purchased that semester, and yet another is the distance from the college to each student's permanent residence. A **variable** is any characteristic whose value may change from one individual or object to another. For example, *calculator brand* is a variable, and so are *number of textbooks purchased* and *distance to the college*. **Data** result from making observations either on a single variable or simultaneously on two or more variables.

DEFINITION

Variable: A characteristic whose value may change from one observation to another.

Data: A collection of observations on one or more variables.

A **univariate data set** consists of observations on a single variable made on individuals in a sample or population. There are two types of univariate data sets: **categorical** and **numerical**. In the previous example, *calculator brand* is a categorical variable, because each student's response to the query, "What brand of calculator do you own?" is a category. The collection of responses from all these students forms a categorical data set. The other two variables, *number of textbooks purchased* and *distance to the college*, are both numerical in nature. Determining the value of such a numerical variable (by counting or measuring) for each student results in a numerical data set.

DEFINITION

Univariate data set: A data set consisting of observations on a single characteristic is a **univariate data set**.

Categorical data set: A univariate data set is **categorical** (or **qualitative**) if the individual observations are categorical responses.

Numerical data set: A univariate data set is **numerical** (or **quantitative**) if each observation is a number.

EXAMPLE 1.4 College Choice Do-Over?

Understand the context ⟩

The Higher Education Research Institute at UCLA surveys over 20,000 college seniors each year. One question on the 2008 survey asked seniors the following question: If you could make your college choice over, would you still choose to enroll at your current college? Possible responses were definitely yes (DY), probably yes (PY), probably no (PN), and definitely no (DN). Responses for 20 students were:

DY PN DN DY PY PY PN PY PY DY
DY PY DY DY PY PY DY DY PN DY

Consider the context ⟩

(These data are just a small subset of the data from the survey. For a description of the full data set, see Exercise 1.18). Because the response to the question about college choice is categorical, this is a univariate categorical data set. ■

In Example 1.4, the data set consisted of observations on a single variable (college choice response), so this is a univariate data set. In some studies, attention focuses simultaneously on two different characteristics. For example, both height (in inches) and weight (in pounds) might be recorded for each individual in a group. The resulting data set consists of pairs of numbers, such as (68, 146). This is called a **bivariate data set. Multivariate data** result from obtaining a category or value for each of two or more attributes (so bivariate data are a special case of multivariate data). For example, multivariate data would result from determining height, weight, pulse rate, and systolic blood pressure for each individual in a group. Example 1.5 illustrates a bivariate data set.

EXAMPLE 1.5 How Safe Are College Campuses?

Understand the context ⟩

● Consider the accompanying data on violent crime on college campuses in Florida during 2012. http://www.fbi.gov/

Consider the data ⟩

University/College	Student Enrollment	Number of Violent Crimes Reported in 2012
Edison State College	17,107	4
Florida A&M University	13,204	14
Florida Atlantic University	25,246	4
Florida Gulf Coast University	12,851	3
Florida International University	44,616	9
Florida State University	41,067	31
New College of Florida	845	1
Pensacola State College	11,531	3
Santa Fe College	15,493	1
Tallahassee Community College	15,090	2
University of Central Florida	58,465	26
University of Florida	49,589	18
University of North Florida	16,198	2
University of South Florida	4,310	2
University of West Florida	11,982	2

Here two variables—*student enrollment* and *number of violent crimes reported*—were recorded for each of the 15 schools. Because this data set consists of values of two variables for each school, it is a bivariate data set. Each of the two variables considered here is numerical (rather than categorical). ■

● Data set available online

Two Types of Numerical Data

There are two different types of numerical data: **discrete** and **continuous**. Consider a number line (Figure 1.3) for locating values of the numerical variable being studied. Each possible number (2, 3.125, 8.12976, etc.) corresponds to exactly one point on the number line.

Suppose that the variable of interest is the number of courses in which a student is enrolled. If no student is enrolled in more than eight courses, the possible values are 1, 2, 3, 4, 5, 6, 7, and 8. These values are identified in Figure 1.4(a) by the dots at the points marked 1, 2, 3, 4, 5, 6, 7, and 8. These possible values are isolated from one another on the number line. We can place an interval around any possible value that is small enough that no other possible value is included in the interval. On the other hand, the line segment in Figure 1.4(b) identifies a plausible set of possible values for the time (in seconds) it takes for the first kernel in a bag of microwave popcorn to pop. Here the possible values make up an entire interval on the number line, and no possible value is isolated from other possible values.

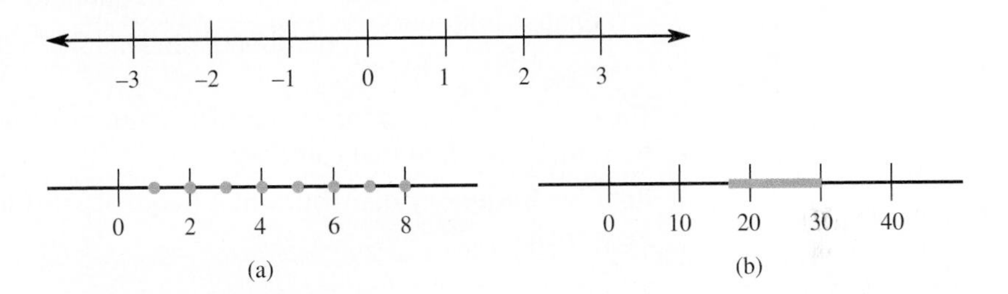

FIGURE 1.3
A number line.

FIGURE 1.4
Possible values of a variable:
(a) number of courses;
(b) popping time.

DEFINITION

Discrete numerical variable: A numerical variable results in **discrete** data if the possible values of the variable correspond to isolated points on the number line.

Continuous numerical variable: A numerical variable results in **continuous** data if the set of possible values forms an entire interval on the number line.

Discrete data usually arise when observations are determined by counting (for example, the number of roommates a student has or the number of petals on a flower).

EXAMPLE 1.6 Do U Txt?

● The number of text messages sent on a particular day is recorded for each of 12 students. The resulting data set is

23 0 14 13 15 0 60 82 0 40 41 22

Possible values for the variable *number of text messages sent* are 0, 1, 2, 3. . . . These are isolated points on the number line, so this data set consists of discrete numerical data.

Suppose that instead of the number of text messages sent, the *time spent texting* had been recorded. Even though time spent may have been reported rounded to the nearest minute, the actual time spent could have been 6 minutes, 6.2 minutes, 6.28 minutes, or any other value in an entire interval. So, recording values of *time spent texting* would result in continuous data. ■

In general, data are continuous when observations involve making measurements, as opposed to counting. In practice, measuring instruments do not have infinite accuracy, so possible measured values, strictly speaking, do not form a continuum on the number line. However, any number in the continuum *could* be a value of the variable. The distinction between discrete and continuous data will be important in our discussion of probability models in Chapter 6.

● Data set available online

Frequency Distributions and Bar Charts for Categorical Data

An appropriate graphical or tabular display of data can be an effective way to summarize and communicate information. When the data set is categorical, a common way to present the data is in the form of a table, called a **frequency distribution**.

DEFINITION

Frequency distribution for categorical data: A table that displays the possible categories along with the associated frequencies and/or relative frequencies.

Frequency: The **frequency** for a particular category is the number of times the category appears in the data set.

Relative frequency: The **relative frequency** for a particular category is calculated as

$$\text{relative frequency} = \frac{\text{frequency}}{\text{number of obervations in the data set}}$$

The relative frequency for a particular category is the proporton of the observations that belong to that category.

Relative frequency distribution: A frequency distribution that includes relative frequencies.

EXAMPLE 1.7 Motorcycle Helmets—Can You See Those Ears?

Understand the context ❯

The U.S. Department of Transportation establishes standards for motorcycle helmets. To ensure a certain degree of safety, helmets should reach the bottom of the motorcyclist's ears. The report **"Motorcycle Helmet Use in 2005—Overall Results" (National Highway Traffic Safety Administration, August 2005)** summarized data collected in June of 2005 by observing 1700 motorcyclists nationwide at selected roadway locations. Each time a motorcyclist passed by, the observer noted whether the rider was wearing no helmet, a noncompliant helmet, or a compliant helmet. Using the coding

NH = noncompliant helmet
CH = compliant helmet
N = no helmet

Consider the data ❯

a few of the observations were

CH N CH NH N CH CH CH N N

There were also 1690 additional observations, which we didn't reproduce here. In total, there were 731 riders who wore no helmet, 153 who wore a noncompliant helmet, and 816 who wore a compliant helmet.

The corresponding frequency distribution is given in Table 1.1.

Do the work ❯

TABLE 1.1 Frequency Distribution for Helmet Use

Helmet Use Category	Frequency	Relative Frequency
No helmet	731	0.430 ← 731/1700
Noncompliant helmet	153	0.090 ← 153/1700
Compliant helmet	816	0.480
	1700 ↖ Total number of observations	1.000 ↖ Should total 1, but in some cases may be slightly off due to rounding

Interpret the results ⟩

From the frequency distribution, we can see that a large percentage of riders (43%) were not wearing a helmet, but most of those who wore a helmet were wearing one that met the Department of Transportation safety standard. ■

A frequency distribution displays a data set in a table. It is also common to display categorical data graphically. A bar chart is one of the most widely used types of graphical displays for categorical data.

Bar Charts

A **bar chart** is a graph of a frequency distribution of categorical data. Each category in the frequency distribution is represented by a bar or rectangle, and the picture is constructed in such a way that the *area* of each bar is proportional to the corresponding frequency or relative frequency.

Bar Charts

When to Use Categorical data.

How to Construct

1. Draw a horizontal axis, and write the category names or labels below the line at regularly spaced intervals.
2. Draw a vertical axis, and label the scale using either frequency or relative frequency.
3. Place a rectangular bar above each category label. The height is determined by the category's frequency or relative frequency, and all bars should have the same width. With the same width, both the height and the area of the bar are proportional to frequency and relative frequency.

What to Look For

- Frequently and infrequently occurring categories.

EXAMPLE 1.8 Revisiting Motorcycle Helmets

Understand the context ⟩

Example 1.7 used data on helmet use from a sample of 1700 motorcyclists to construct a frequency distribution (Table 1.1). Figure 1.5 shows the bar chart corresponding to this frequency distribution.

Step-by-step technology instructions available online

Consider the data ⟩

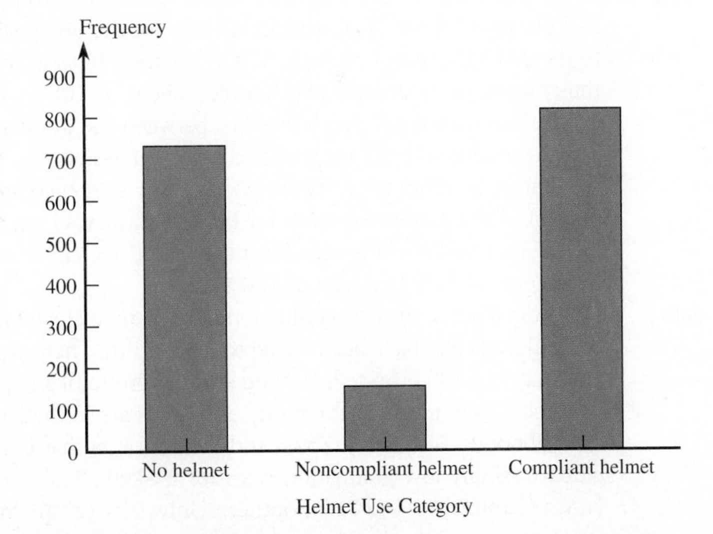

FIGURE 1.5
Bar chart of helmet use.

Interpret the results ⟩ The bar chart provides a visual representation of the information in the frequency distribution. From the bar chart, it is easy to see that the compliant helmet use category occurred most often in the data set. The bar for compliant helmets is about five times as tall (and therefore has five times the area) as the bar for noncompliant helmets because approximately five times as many motorcyclists wore compliant helmets than wore noncompliant helmets. ■

Dotplots for Numerical Data

A dotplot is a simple way to display numerical data when the data set is reasonably small. Each observation is represented by a dot above the location corresponding to its value on a horizontal measurement scale. When a value occurs more than once, there is a dot for each occurrence and these dots are stacked vertically.

Dotplots

When to Use Small numerical data sets.

How to Construct

1. Draw a horizontal line and mark it with an appropriate measurement scale.
2. Locate each value in the data set along the measurement scale, and represent it by a dot. If there are two or more observations with the same value, stack the dots vertically.

What to Look for

Dotplots convey information about:

- A representative or typical value in the data set.
- The extent to which the data values spread out.
- The nature of the distribution of values along the number line.
- The presence of unusual values in the data set.

EXAMPLE 1.9 Making It to Graduation . . .

Understand the context ⟩ ● The article **"Keeping Score When It Counts: Graduation Success and Academic Progress Rates for the 2013 NCAA Men's Division I Basketball Tournament Teams" (The Institute for Diversity and Ethics in Sport, University of Central Florida, March 2013)** compared graduation rates of basketball players to those of all student athletes for the universities and colleges that sent teams to the 2013 Division I playoffs. The graduation rates in the accompanying table represent the percentage of athletes who started college in 2006 who had graduated by the end of 2012. Also shown are the differences between the graduation rate for all student athletes and the graduation rate for basketball student athletes.

Minitab, a computer software package for statistical analysis, was used to construct a dotplot of the graduation rates for basketball players (see Figure 1.6). From this dotplot, we see that basketball graduation rates varied a great deal from school to school, ranging from a low of 17% to a high of 100%.

Consider the data ⟩ We can also see that the graduation rates seem to cluster in several groups, denoted by the colored ovals that have been added to the dotplot. There are quite a few schools with graduation rates of 100% (excellent!) and another group of 17 schools with graduation rates that are higher than most. The majority of schools are in the large cluster with graduation rates from about 40% to about 76%. And then there is that bottom group of four schools with embarrassingly low graduation rates for basketball players: University of Florida (17%), North Carolina A&T (25%), Southern University (27%), and New Mexico State (29%).

● Data set available online

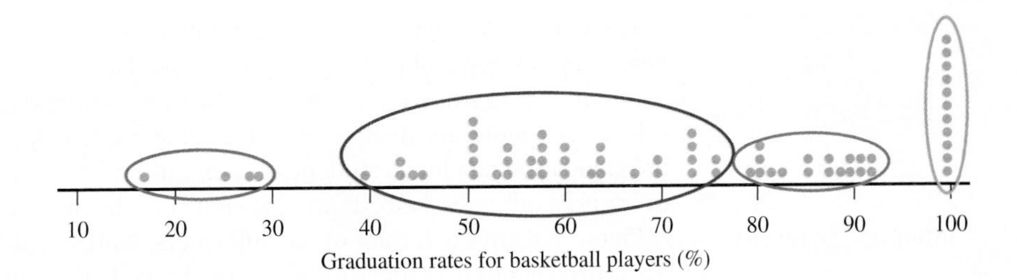

FIGURE 1.6
Minitab dotplot of graduation rates for basketball players.

GRADUATION RATES (%)

Basketball	All Athletes	Difference (All – BB)	Basketball	All Athletes	Difference (All – BB)	Basketball	All Athletes	Difference (All – BB)
100	92	−8	29	70	41	75	80	5
79	76	−3	25	55	30	50	81	31
100	99	−1	73	77	4	87	93	6
80	83	3	75	68	−7	64	84	20
53	82	29	50	77	27	54	83	29
91	94	3	64	87	23	56	76	20
100	97	−3	92	92	0	67	84	17
100	98	−2	62	73	11	73	80	7
73	73	0	44	83	39	92	76	−16
80	94	14	27	51	24	50	75	25
90	96	6	58	87	29	91	88	−3
100	98	−2	43	78	35	100	99	−1
43	80	37	45	85	40	70	72	2
60	81	21	60	75	15	85	80	−5
50	80	30	57	73	16	54	78	24
60	83	23	82	82	0	100	84	−16
58	77	19	54	68	14	40	83	43
64	91	27	70	84	14	80	94	14
58	74	16	50	80	30	100	94	−6
85	83	−2	56	78	22	73	80	7
87	91	4	17	82	65	100	79	−21
89	85	−4	100	89	−11	90	83	−7
83	78	−5	100	85	−15			

Figure 1.7 shows two dotplots of graduation rates—one for basketball players and one for all student athletes. There are some striking differences that are easy to see when the data is displayed in this way. The graduation rates for all student athletes tend to be higher and to vary less from school to school than the graduation rates for basketball players.

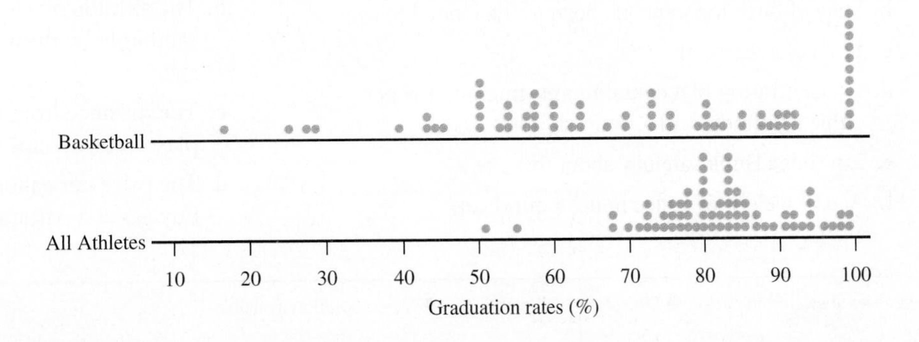

FIGURE 1.7
Minitab dotplot of graduation rates for basketball players and for all athletes.

The dotplots in Figure 1.7 are informative, but we can do even better. The data given here are an example of *paired data*. Each basketball graduation rate is paired with a graduation rate for all student athletes from the same school. When data are paired in this way, it is usually more informative to look at differences—in this case, the difference between the graduation rate for all student athletes and for basketball players for each school. These differences (all − basketball) are also shown in the data table.

Interpret the results)

Figure 1.8 gives a dotplot of the differences. Notice that three differences are equal to 0. This corresponds to schools for which the basketball graduation rate is equal to the graduation rate of all student athletes. There are 20 schools for which the difference is negative. Negative differences correspond to schools that have a graduation rate for basketball players that is higher than the graduation rate for all student athletes.

The most interesting features of the difference dotplot are the very large number of positive differences and the wide spread. Positive differences correspond to schools that have a lower graduation rate for basketball players. There is a lot of variability in the graduation rate difference from school to school, and one school has a difference that is noticeably higher than the rest. (In case you were wondering, this school is the University of Florida with a difference of 65%.)

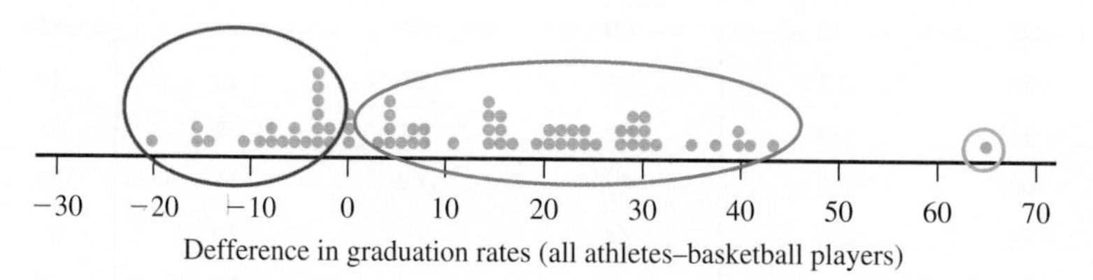

FIGURE 1.8 Dotplot of graduation rate differences (all athletes − basketball players).

■

EXERCISES 1.12 - 1.31

1.12 Classify each of the following variables as either categorical or numerical. For those that are numerical, determine whether they are discrete or continuous.

a. Number of students in a class of 35 who turn in a term paper before the due date

b. Gender of the next baby born at a particular hospital

c. Amount of fluid (in ounces) dispensed by a machine used to fill bottles with soda pop

d. Thickness of the gelatin coating of a vitamin E capsule

e. Birth order classification (only child, firstborn, middle child, lastborn) of a math major

1.13 Classify each of the following variables as either categorical or numerical. For those that are numerical, determine whether they are discrete or continuous.

a. Brand of computer purchased by a customer

b. State of birth for someone born in the United States

c. Price of a textbook

d. Concentration of a contaminant (micrograms per cubic centimeter) in a water sample

e. Zip code (Think carefully about this one.)

f. Actual weight of coffee in a 1-pound can

1.14 For the following numerical variables, state whether each is discrete or continuous.

a. The number of insufficient-funds checks received by a grocery store during a given month

b. The amount by which a 1-pound package of ground beef decreases in weight (because of moisture loss) before purchase

c. The number of New York Yankees during a given year who will not play for the Yankees the next year

d. The number of students in a class of 35 who have purchased a used copy of the textbook

1.15 For the following numerical variables, state whether each is discrete or continuous.

a. The length of a 1-year-old rattlesnake

b. The altitude of a location in California selected randomly by throwing a dart at a map of the state

c. The distance from the left edge at which a 12-inch plastic ruler snaps when bent sufficiently to break

d. The price per gallon paid by the next customer to buy gas at a particular station

1.16 For each of the following situations, give a set of possible data values that might arise from making the observations described.

a. The manufacturer for each of the next 10 automobiles to pass through a given intersection is noted.

b. The grade point average for each of the 15 seniors in a statistics class is determined.

c. The number of gas pumps in use at each of 20 gas stations at a particular time is determined.

d. The actual net weight of each of 12 bags of fertilizer having a labeled weight of 50 pounds is determined.

e. Fifteen different radio stations are monitored during a 1-hour period, and the amount of time devoted to commercials is determined for each.

1.17 In a survey of 100 people who had recently purchased motorcycles, data on the following variables were recorded:

Gender of purchaser
Brand of motorcycle purchased
Number of previous motorcycles owned by purchaser
Telephone area code of purchaser
Weight of motorcycle as equipped at purchase

a. Which of these variables are categorical?

b. Which of these variables are discrete numerical?

c. Which type of graphical display would be an appropriate choice for summarizing the gender data, a bar chart or a dotplot?

d. Which type of graphical display would be an appropriate choice for summarizing the weight data, a bar chart or a dotplot?

1.18 The report **"Findings from the 2008 Administration of the College Senior Survey" (Higher Education Research Institute, UCLA, June 2009)** gave the following relative frequency distribution summarizing student responses to the question "If you could make your college choice over, would you still choose to enroll at your current college?"

Response	Relative Frequency
Definitely yes	0.447
Probably yes	0.373
Probably no	0.134
Definitely no	0.046

a. Use this information to construct a bar chart for the response data.

b. If you were going to use the response data and the bar chart from Part (a) as the basis for an article for your student paper, what would be a good headline for your article?

1.19 ● The article **"Feasting on Protein" (*AARP Bulletin*, September 2009)** gave the cost (in cents per gram) of protein for 19 common food sources of protein.

Food	Cost	Food	Cost
Chicken	1.8	Yogurt	5.0
Salmon	5.8	Milk	2.5
Turkey	1.5	Peas	5.2
Soybeans	3.1	Tofu	6.9
Roast beef	2.7	Cheddar cheese	3.6
Cottage cheese	3.1	Nuts	5.2
Ground beef	2.3	Eggs	5.7
Ham	2.1	Peanut butter	1.8
Lentils	3.3	Ice cream	5.3
Beans	2.9		

a. Construct a dotplot of the cost data. (Hint: See Example 1.9.)

b. Locate the cost for meat and poultry items in your dotplot and highlight them in a different color. Based on the dotplot, do meat and poultry items appear to be a good value? That is, do they appear to be relatively low cost compared to other sources of protein?

1.20 ● Box Office Mojo (www.boxofficemojo.com) tracks movie ticket sales. Ticket sales (in millions of dollars) for each of the top 20 movies in 2007 and 2008 are shown in the accompanying table.

Movie (2007)	2007 Sales (millions of dollars)
Spider-Man 3	336.5
Shrek the Third	322.7
Transformers	319.2
Pirates of the Caribbean: At World's End	309.4
Harry Potter and the Order of the Phoenix	292.0
I Am Legend	256.4
The Bourne Ultimatum	227.5
National Treasure: Book of Secrets	220.0
Alvin and the Chipmunks	217.3
300	210.6
Ratatouille	206.4
The Simpsons Movie	183.1
Wild Hogs	168.3
Knocked Up	148.8
Juno	143.5
Rush Hour 3	140.1
Live Free or Die Hard	134.5

continued

Movie (2007)	2007 Sales (millions of dollars)
Fantastic Four: Rise of the Silver Surfer	131.9
American Gangster	130.2
Enchanted	127.8

Movie (2008)	2008 Sales (millions of dollars)
The Dark Knight	533.3
Iron Man	318.4
Indiana Jones and the Kingdom of the Crystal Skull	317.1
Hancock	227.9
WALL-E	223.8
Kung Fu Panda	215.4
Twilight	192.8
Madagascar: Escape 2 Africa	180.0
Quantum of Solace	168.4
Dr. Suess' Horton Hears a Who!	154.5
Sex and the City	152.6
Gran Torino	148.1
Mamma Mia!	144.1
Marley and Me	143.2
The Chronicles of Narnia: Prince Caspian	141.6
Slumdog Millionaire	141.3
The Incredible Hulk	134.8
Wanted	134.5
Get Smart	130.3
The Curious Case of Benjamin Button	127.5

a. Construct a dotplot of the 2008 ticket sales data. Comment on any interesting features of the dotplot. (Hint: See "What to Look For" in the Dotplots box on page 14.)

b. Construct a dotplot of the 2007 ticket sales data. Comment on any interesting features of the dotplot.

c. In what ways are the distributions of the 2007 and 2008 ticket sales observations similar? In what ways are they different?

1.21 ● About 38,000 students attend Grant MacEwan College in Edmonton, Canada. In 2004, the college surveyed non-returning students to find out why they did not complete their degree **(Grant MacEwan College Early Leaver Survey Report, 2004).** Sixty-three students gave a personal (rather than an academic) reason for leaving. The accompanying frequency distribution summarizes the primary reason for leaving for these 63 students.

Primary Reason for Leaving	Frequency
Financial	19
Health	12
Employment	8
Family issues	6
Wanted to take a break	4
Moving	2
Travel	2
Other personal reasons	10

a. Summarize the reason for leaving data using a bar chart.

b. Write a few sentences commenting on the most common reasons for leaving.

1.22 Figure EX-1.22 is a graph similar to one that appeared in ***USA Today* (June 29, 2009)**. This graph is meant to be a bar graph of responses to the question shown in the graph.

a. Is response to the question a categorical or numerical variable?

b. Explain why a bar chart rather than a dotplot was used to display the response data.

c. There must have been an error made in constructing this graph. How can you tell that the graph is not a correct representation of the response data?

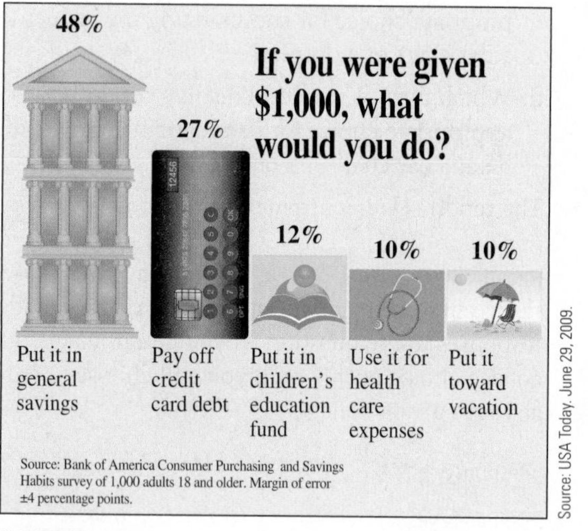

FIGURE EX-1.22

1.23 ● The online article **"Social Networks: Facebook Takes Over Top Spot, Twitter Climbs" (Compete.com, February 9, 2009)** included the accompanying data on number of unique visitors and total number of visits for January 2009 for the top 25 online social network sites. The data on total visits and unique visitors were used to compute the values in the final column of the data table, in which

$$\text{visits per unique visitor} = \frac{\text{total visits}}{\text{number of unique visitors}}$$

Site	Unique Visitors	Total Visits	Visits per Unique Visitor
facebook.com	68,557,534	1,191,373,339	17.3777
myspace.com	58,555,800	810,153,536	13.8356
twitter.com	5,979,052	54,218,731	9.0681
fixter.com	7,645,423	53,389,974	6.9833
linkedin.com	11,274,160	42,744,438	3.7914
tagged.com	4,448,915	39,630,927	8.9080
classmates.com	17,296,524	35,219,210	2.0362
myyearbook.com	3,312,898	33,121,821	9.9978
livejournal.com	4,720,720	25,221,354	5.3427
imeem.com	9,047,491	22,993,608	2.5414
reunion.com	13,704,990	20,278,100	1.4796
ning.com	5,673,549	19,511,682	3.4391
blackplanet.com	1,530,329	10,173,342	6.6478
bebo.com	2,997,929	9,849,137	3.2853
hi5.com	2,398,323	9,416,265	3.9262
yuku.com	1,317,551	9,358,966	7.1033
cafemom.com	1,647,336	8,586,261	5.2122
friendster.com	1,568,439	7,279,050	4.6410
xanga.com	1,831,376	7,009,577	3.8275
360.yahoo.com	1,499,057	5,199,702	3.4686
orkut.com	494,464	5,081,235	10.2762
urbanchat.com	329,041	2,961,250	8.9996
fubar.com	452,090	2,170,315	4.8006
asiantown.net	81,245	1,118,245	13.7639
tickle.com	96,155	109,492	1.1387

continued

a. A dotplot of the total visits data is shown in Figure EX-1.23a. What are the most obvious features of the dotplot? What does it tell you about the online social networking sites?

b. A dotplot for the number of unique visitors is shown in Figure EX-1.23b. In what way is this dotplot different from the dotplot for total visits in Part (a)? What does this tell you about the online social networking sites?

c. A dotplot for the visits per unique visitor data is shown in Figure EX-1.23c. What new information about the online social networks is provided by this dotplot?

1.24 Heal the Bay is an environmental organization that releases an annual beach report card based on water quality **(Heal the Bay Beach Report Card, May 2009).** The 2009 ratings for 14 beaches in San Francisco County during wet weather were:

A+ C B A A+ A+ A A+ B D C D F F

a. Would it be appropriate to display the ratings data using a dotplot? Explain why or why not.

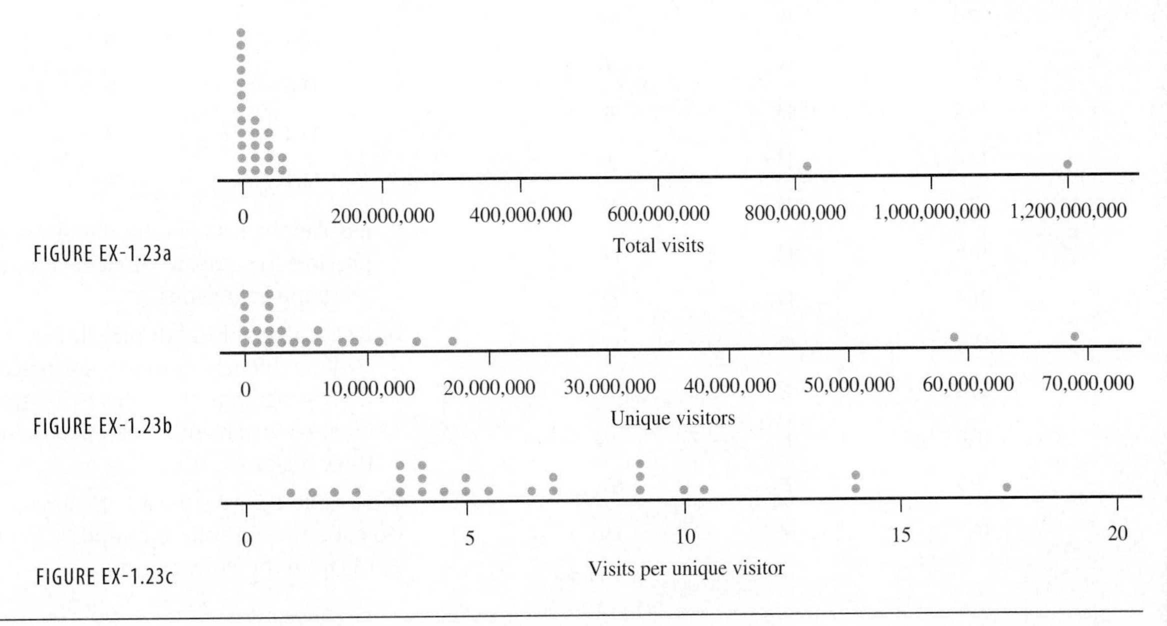

FIGURE EX-1.23a

FIGURE EX-1.23b

FIGURE EX-1.23c

b. Summarize the wet weather ratings by constructing a relative frequency distribution and a bar chart.

c. The dry weather ratings for these same beaches were:

A B B A+ A F A A A A A A B A

Construct a bar graph for the dry weather ratings.

d. Do the bar graphs from Parts (b) and (c) support the statement that beach water quality tends to be better in dry weather conditions? Explain.

1.25 ● The article **"Going Wireless" (*AARP Bulletin*, June 2009)** reported the estimated percentage of households with only wireless phone service (no landline) for the 50 states and the District of Columbia. In the accompanying data table, each state was also classified into one of three geographical regions—West (W), Middle states (M), and East (E).

Wireless %	Region	State
13.9	M	AL
11.7	W	AK
18.9	W	AZ
22.6	M	AR
9.0	W	CA
16.7	W	CO
5.6	E	CN
5.7	E	DE
20.0	E	DC
16.8	E	FL
16.5	E	GA
8.0	W	HI
22.1	W	ID
16.5	M	IL
13.8	M	IN
22.2	M	IA
16.8	M	KA
21.4	M	KY
15.0	M	LA
13.4	E	ME
10.8	E	MD
9.3	E	MA
16.3	M	MI

continued

Wireless %	Region	State
17.4	M	MN
19.1	M	MS
9.9	M	MO
9.2	W	MT
23.2	M	NE
10.8	W	NV
16.9	M	ND
11.6	E	NH
8.0	E	NJ
21.1	W	NM
11.4	E	NY
16.3	E	NC
14.0	E	OH
23.2	M	OK
17.7	W	OR
10.8	E	PA
7.9	E	RI
20.6	E	SC
6.4	M	SD
20.3	M	TN
20.9	M	TX
25.5	W	UT
10.8	E	VA
5.1	E	VT
16.3	W	WA
11.6	E	WV
15.2	M	WI
11.4	W	WY

a. Display the data graphically in a way that makes it possible to compare wireless percent for the three geographical regions.

b. Does the graphical display in Part (a) reveal any striking differences in wireless percent for the three geographical regions or are the distributions of wireless percent observations similar for the three regions?

1.26 ● Example 1.5 gave the accompanying data on violent crime on college campuses in Florida during 2012 (from the FBI web site):

University/College	Student Enrollment	Number of Violent Crimes Reported in 2012
Edison State College	17,107	4
Florida A&M University	13,204	14
Florida Atlantic University	25,246	4
Florida Gulf Coast University	12,851	3
Florida International University	44,616	9
Florida State University	41,067	31
New College of Florida	845	1
Pensacola State College	11,531	3
Santa Fe College	15,493	1
Tallahassee Community College	15,090	2
University of Central Florida	58,465	26
University of Florida	49,589	18
University of North Florida	16,198	2
University of South Florida	4,310	2
University of West Florida	11,982	2

a. Construct a dotplot using the 15 observations on number of violent crimes reported. Which schools stand out from the rest?

b. One of the Florida schools has only 845 students and a few of the schools are quite a bit larger than the rest. Because of this, it might make more sense to consider a crime rate by calculating the number of violent crimes reported per 1000 students. For example, for Florida A&M University the violent crime rate would be

$$\frac{14}{13,204}(1000) = (0.0011)(1000) = 1.1$$

Calculate the violent crime rate for the other 14 schools and then use those values to construct a dotplot. Do the same schools stand out as unusual in this dotplot?

c. Based on your answers from Parts (a) and (b), write a couple of sentences commenting on violent crimes reported at Florida universities and colleges in 2012.

1.27 ● The article **"Fliers Trapped on Tarmac Push for Rules on Release" (*USA Today*, July 28, 2009)** gave the following data for 17 airlines on number of flights that were delayed on the tarmac for at least 3 hours for the period from October 2008 to May 2009:

Airline	Number of Delays	Rate per 10,000 Flights
ExpressJet	93	4.9
Continental	72	4.1
Delta	81	2.8
Comair	29	2.7
American Eagle	44	1.6
US Airways	46	1.6
JetBlue	18	1.4
American	48	1.3
Northwest	24	1.2
Mesa	17	1.1
United	29	1.1
Frontier	5	0.9
SkyWest	29	0.8
Pinnacle	13	0.7
Atlantic Southeast	11	0.6
AirTran	7	0.4
Southwest	11	0.1

Figure EX-1.27 shows two dotplots: one displays the number of delays data, and one displays the rate per 10,000 flights data.

a. If you were going to rank airlines based on flights delayed on the tarmac for at least 3 hours, would you use the *total number of flights* data or the *rate per 10,000 flights* data? Explain the reason for your choice.

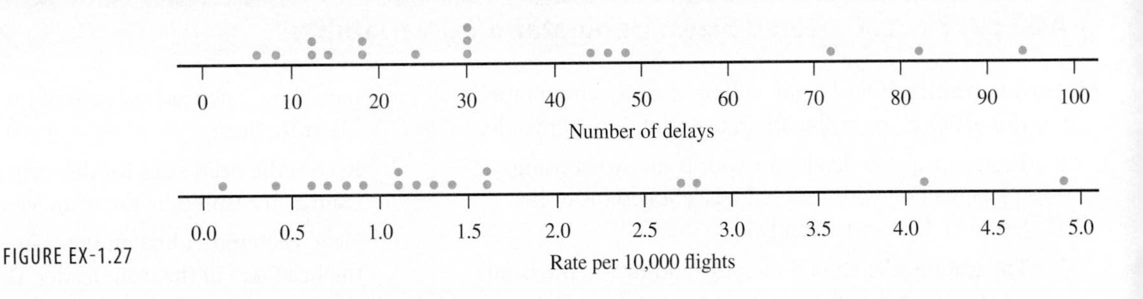

FIGURE EX-1.27

b. Write a short paragraph that could be used as part of a newspaper article on flight delays that could accompany the dotplot of the *rate per 10,000 flights* data.

1.28 The report **"Trends in Education 2010: Community Colleges" (www.collegeboard.com/trends)** included the accompanying information on student debt for students graduating with an AA degree from a public community college in 2008.

Debt	Relative Frequency
None	0.62
Less than $10,000	0.23
Between $10,000 and $20,000	0.10
More than $20,000	0.05

a. Use the given information to construct a bar chart.

b. Write a few sentences commenting on student debt for public community college graduates.

1.29 The article **"Where College Students Buy Textbooks" (*USA Today*, October 14, 2010)** gave data on where students purchased books. The accompanying frequency table summarizes data from a sample of 1152 full-time college students.

Where Books Purchased	Frequency
Campus bookstore	576
Campus bookstore web site	48
Online bookstore other than campus bookstore	240
Off-campus bookstore	168
Rented textbooks	36
Purchased mostly eBooks	12
Didn't buy any textbooks	72

a. Construct a bar chart to summarize the data distribution.

b. Write a few sentences commenting on where students are buying textbooks.

1.30 ▼ The article **"Americans Drowsy on the Job and the Road" (Associated Press, March 28, 2001)** summarized data from the 2001 Sleep in America poll. Each individual in a sample of 1004 adults was asked questions about his or her sleep habits. The article states that

> "40 percent of those surveyed say they get sleepy on the job and their work suffers at least a few days each month, while 22 percent said the problems occur a few days each week. And 7 percent say sleepiness on the job is a daily occurrence."

Assuming that everyone else reported that sleepiness on the job was not a problem, summarize the given information by constructing a relative frequency bar chart.

1.31 **"Ozzie and Harriet Don't Live Here Anymore" (*San Luis Obispo Tribune*, February 26, 2002)** is the title of an article that looked at the changing makeup of America's suburbs. The article states that nonfamily households (for example, homes headed by a single professional or an elderly widow) now outnumber married couples with children in suburbs of the nation's largest metropolitan areas. The article goes on to state:

> In the nation's 102 largest metropolitan areas, "nonfamilies" comprised 29 percent of households in 2000, up from 27 percent in 1990. While the number of married-with-children homes grew too, the share did not keep pace. It declined from 28 percent to 27 percent. Married couples without children at home live in another 29 percent of suburban households. The remaining 15 percent are single-parent homes.

Use the given information on type of household in 2000 to construct a frequency distribution and a bar chart. (Be careful to extract the 2000 percentages from the given information).

Bold exercises answered in back ● Data set available online ▼ Video Solution available

ACTIVITY 1.1 Head Sizes: Understanding Variability

Materials needed: Each team will need a measuring tape. For this activity, you will work in teams of 6 to 10 people.

1. Designate a team leader for your team by choosing the person on your team who celebrated his or her last birthday most recently.
2. The team leader should measure and record the head size (measured as the circumference at the widest part of the forehead) of each of the other members of his or her team.
3. Record the head sizes for the individuals on your team as measured by the team leader.
4. Next, each individual on the team should measure the head size of the team leader. Do not share your measurement with the other team members until

all team members have measured the team leader's head size.

5. After all team members have measured the team leader's head, record the different team leader head size measurements obtained by the individuals on your team.
6. Using the data from Step 3, construct a dotplot of the team leader's measurements of team head sizes. Then, using the same scale, construct a separate dotplot of the different measurements of the team leader's head size (from Step 5).

Now use the available information to answer the following questions:

7. Do you think the team leader's head size changed in between measurements? If not, explain why the measurements of the team leader's head size are not all the same.
8. Which data set was more variable—head size measurements of the different individuals on your team or the different measurements of the team leader's head size? Explain the basis for your choice.
9. Consider the following scheme (you don't actually have to carry this out): Suppose that a group of 10 people measured head sizes by first assigning each person in the group a number between 1 and 10. Then person 1 measured person 2's head size, person 2 measured person 3's head size, and so on, with person 10 finally measuring person 1's head size. Do you think that the resulting head size measurements would be more variable, less variable, or show about the same amount of variability as a set of 10 measurements resulting from a single individual measuring the head size of all 10 people in the group? Explain.

ACTIVITY 1.2 Estimating Sizes

1. Construct an activity sheet that consists of a table that has 6 columns and 10 rows. Label the columns of the table with the following six headings: (1) Shape, (2) Estimated Size, (3) Actual Size, (4) Difference (Estimated − Actual), (5) Absolute Difference, and (6) Squared Difference. Enter the numbers from 1 to 10 in the "Shape" column.
2. Next you will be visually estimating the sizes of the shapes in Figure 1.9. Size will be described as the number of squares of this size

   that would fit in the shape. For example, the shape

   would be size 3, as illustrated by

   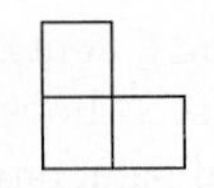

   You should now quickly *visually* estimate the sizes of the shapes in Figure 1.9. *Do not* draw on the figure—these are to be quick visual estimates. Record your estimates in the "Estimated Size" column of the activity sheet.
3. Your instructor will provide the actual sizes for the 10 shapes, which should be entered into the "Actual Size" column of the activity sheet. Now complete the "Difference" column by subtracting the actual value from your estimate for each of the 10 shapes.
4. What would cause a difference to be negative? What would cause a difference to be positive?
5. Would the sum of the differences tell you if the estimates and actual values were in close agreement? Does a sum of 0 for the differences indicate that all the estimates were equal to the actual value? Explain.
6. Compare your estimates with those of another person in the class by comparing the sum of the absolute values of the differences between estimates and corresponding actual values. Who was better at estimating shape sizes? How can you tell?
7. Use the last column of the activity sheet to record the squared differences (for example, if the difference for shape 1 was -3, the squared difference would be $(-3)^2 = 9$. Explain why the sum of the squared differences can also be used to assess how accurate your shape estimates were.
8. For this step, work with three or four other students from your class. For each of the 10 shapes, form a new size estimate by computing the average of the size estimates for that shape made by the individuals in your group. Is this new set of estimates more accurate than your own individual estimates were? How can you tell?
9. Does your answer from Step 8 surprise you? Explain why or why not.

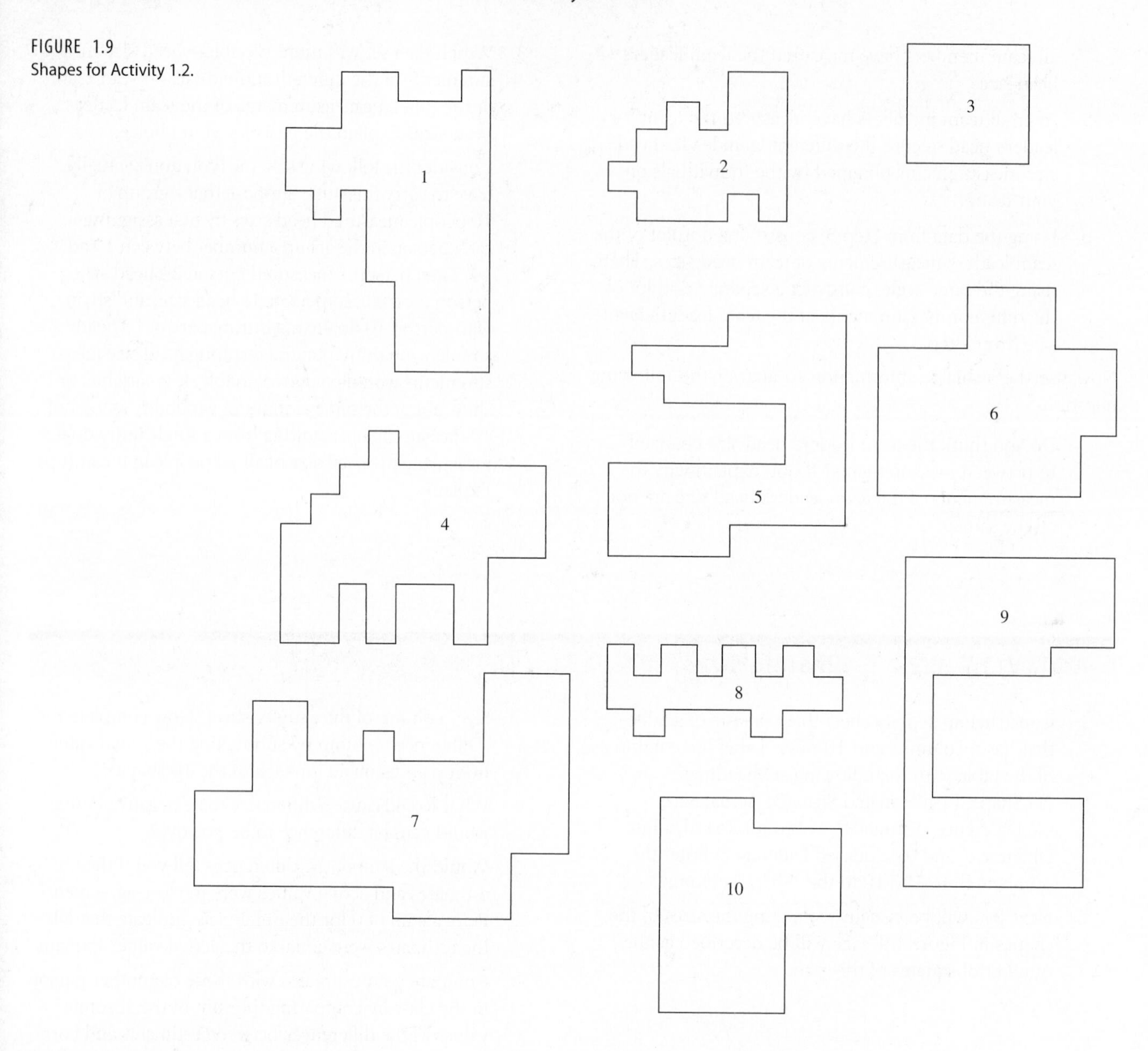

FIGURE 1.9
Shapes for Activity 1.2.

ACTIVITY 1.3 A Meaningful Paragraph

Write a meaningful paragraph that includes the following six terms: **sample, population, descriptive statistics, bar chart, numerical variable,** and **dotplot**.

A "meaningful paragraph" is a coherent piece of writing in an appropriate context that uses all of the listed words. The paragraph should show that you understand the meanings of the terms and their relationships to one another. A sequence of sentences that just defines the terms is *not* a meaningful paragraph. When choosing a context, think carefully about the terms you need to use. Choosing a good context will make writing a meaningful paragraph easier.

SUMMARY Key Concepts and Formulas

TERM OR FORMULA	COMMENT
Population	The entire collection of individuals or measurements about which information is desired.
Sample	A part of the population selected for study.
Descriptive statistics	Numerical, graphical, and tabular methods for organizing and summarizing data.
Inferential statistics	Methods for generalizing from a sample to a population.
Categorical data	Individual observations are categorical responses (nonnumerical).
Numerical data	Individual observations are numerical (quantitative) in nature.
Discrete numerical data	Possible values are isolated points along the number line.
Continuous numerical data	Possible values form an entire interval along then umber line.
Univariate, bivariate, and multivariate data	Each observation consists of one (univariate), two (bivariate), or two or more (multivariate) responses or values.
Frequency distribution for categorical data	A table that displays frequencies, and sometimes relative frequencies, for each of the possible values of a categorical variable.
Bar chart	A graph of a frequency distribution for a categorical data set. Each category is represented by a bar, and the area of the bar is proportional to the corresponding frequency or relative frequency.
Dotplot	A graph of numerical data in which each observation is represented by a dot on or above a horizontal measurement scale.

CHAPTER REVIEW Exercises 1.32 - 1.37

1.32 ● The report **"Testing the Waters 2009" (www.nrdc.org)** included information on the water quality at the 82 most popular swimming beaches in California. Thirty-eight of these beaches are in Los Angeles County. For each beach, water quality was tested weekly and the data below are the percent of the tests in 2008 that failed to meet water quality standards.

Los Angeles County

32	4	6	4	4	7	4	27	19	23
19	13	11	19	9	11	16	23	19	16
33	12	29	3	11	6	22	18	31	43
17	26	17	20	10	6	14	11		

Other Counties

0	0	0	2	3	7	5	11	5	7
15	8	1	5	0	5	4	1	0	1
1	0	2	7	0	2	2	3	5	3
0	8	8	8	0	0	17	4	3	7
10	40	3							

a. Construct a dotplot of the percent of tests failing to meet water quality standards for the Los Angeles County beaches. Write a few sentences describing any interesting features of the dotplot.

b. Construct a dotplot of the percent of tests failing to meet water quality standards for the beaches in other counties. Write a few sentences describing any interesting features of the dotplot.

c. Based on the two dotplots from Parts (a) and (b), describe how the percent of tests that fail to meet water quality standards for beaches in Los Angeles County differs from those of other counties.

1.33 The U.S. Department of Education reported that 14% of adults were classified as being below a basic literacy level, 29% were classified as being at a basic literacy level, 44% were classified as being at an intermediate literacy level, and 13% were classified as

Bold exercises answered in back ● Data set available online ▼ Video Solution available

being at a proficient level (**_2003 National Assessment of Adult Literacy_**).

a. Is the variable *literacy level* categorical or numerical?

b. Construct a bar chart to display the given data on literacy level.

c. Would it be appropriate to display the given information using a dotplot? Explain why or why not.

1.34 ● ▼ The Computer Assisted Assessment Center at the University of Luton published a report titled **"Technical Review of Plagiarism Detection Software."** The authors of this report asked faculty at academic institutions about the extent to which they agreed with the statement "Plagiarism is a significant problem in academic institutions." The responses are summarized in the accompanying table. Construct a bar chart for these data.

Response	Frequency
Strongly disagree	5
Disagree	48
Not sure	90
Agree	140
Strongly agree	39

1.35 ● The article **"Just How Safe Is That Jet?" (_USA Today_, March 13, 2000)** gave the following relative frequency distribution that summarized data on the type of violation for fines imposed on airlines by the Federal Aviation Administration:

Type of Violation	Relative Frequency
Security	0.43
Maintenance	0.39
Flight operations	0.06
Hazardous materials	0.03
Other	0.09

a. Use this information to construct a bar chart for type of violation.

b. Write a sentence or two commenting on the relative occurrence of the various types of violation.

1.36 ● Each year, **_U.S. News and World Report_** publishes a ranking of U.S. business schools. The following data give the acceptance rates (percentage of applicants admitted) for the best 25 programs in a recent survey:

16.3 12.0 25.1 20.3 31.9 20.7 30.1 19.5 36.2
46.9 25.8 36.7 33.8 24.2 21.5 35.1 37.6 23.9
17.0 38.4 31.2 43.8 28.9 31.4 48.9

a. Construct a dotplot.

b. Comment on the interesting features of the plot.

1.37 ● Many adolescent boys aspire to be professional athletes. The paper **"Why Adolescent Boys Dream of Becoming Professional Athletes" (_Psychological Reports_ [1999]:1075–1085)** examined some of the reasons. Each boy in a sample of teenage boys was asked the following question: "Previous studies have shown that more teenage boys say that they are considering becoming professional athletes than any other occupation. In your opinion, why do these boys want to become professional athletes?" The resulting data are shown in the following table:

Response	Frequency
Fame and celebrity	94
Money	56
Attract women	29
Like sports	27
Easy life	24
Don't need an education	19
Other	19

Construct a bar chart to display these data.

TECHNOLOGY NOTES

Bar Charts

JMP

1. Enter the raw data into a column (**Note:** To open a new data table, click **File** then select **New** then **Data Table**)
2. Click **Graph** and select **Chart**
3. Click and drag the column name containing your stored data from the box under **Select Columns** to the box next to **Categories, X, Levels**
4. Click **OK**

MINITAB

1. Enter the raw data into C1
2. Select **Graph** and choose **Bar Chart...**
3. Highlight Simple
4. Click **OK**
5. Double click C1 to add it to the **Categorical Variables** box
6. Click **OK**

Note: You may add or format titles, axis titles, legends, etc., by clicking on the **Labels...** button prior to performing step 6 above.

SPSS

1. Enter the raw data into a column
2. Select **Graph** and choose **Chart Builder...**
3. Under **Choose from** highlight **Bar**
4. Click and drag the first bar chart in the right box (Simple Bar) to the Chart preview area

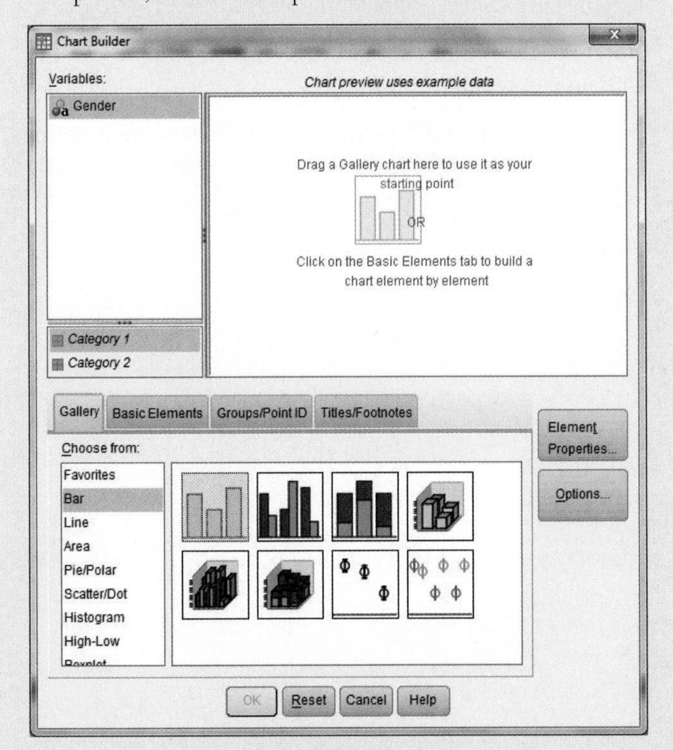

5. Click and drag the variable from the **Variables** box to the **X-Axis?** box in the chart preview area
6. Click **OK**

Note: By default, SPSS displays the count of each category in a bar chart. To display the percentage for each category, follow the above instructions to Step 5. Then, in the Element Properties dialog box, select Percentage from the drop-down box under Statistic. Click **Apply** then click **OK** on the Chart Builder dialog box.

Excel 2007

1. Enter the category names into column A (you may input the title for the variable in cell A1)
2. Enter the count or percent for each category into column B (you may put the title "Count" or "Percentage" in cell B1)
3. Select all data (including column titles if used)
4. Click on the **Insert** Ribbon

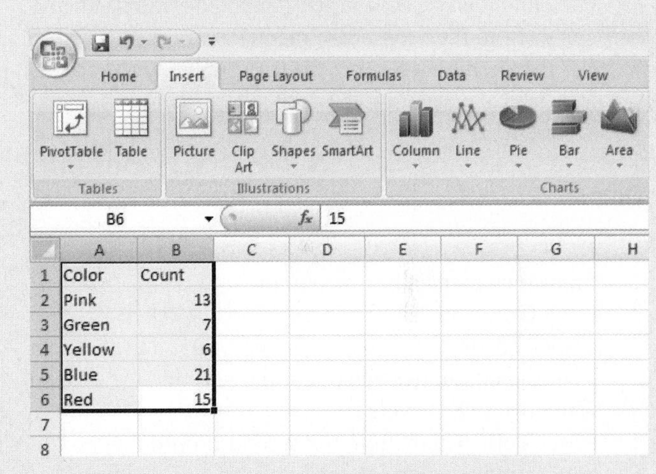

5. Choose **Column** and select the first chart under 2-D Column (Clustered Column)
6. The chart will appear on the same worksheet as your data

Note: You may add or format titles, axis titles, legends, and so on, by right-clicking on the appropriate piece of the chart.

Note: Using the Bar Option on the Insert Ribbon will produce a horizontal bar chart.

TI-83/84

The TI-83/84 does not have the functionality to produce bar charts.

TI-Nspire

1. Enter the category names into a list (to access data lists select the spreadsheet option and press **enter**)

Note: In order to correctly enter category names, you will input them as text. Begin by pressing ?!> and then select " and press **enter**. Now type the category name and press **enter**.

Note: Be sure to title the list by selecting the top row of the column and typing a title.

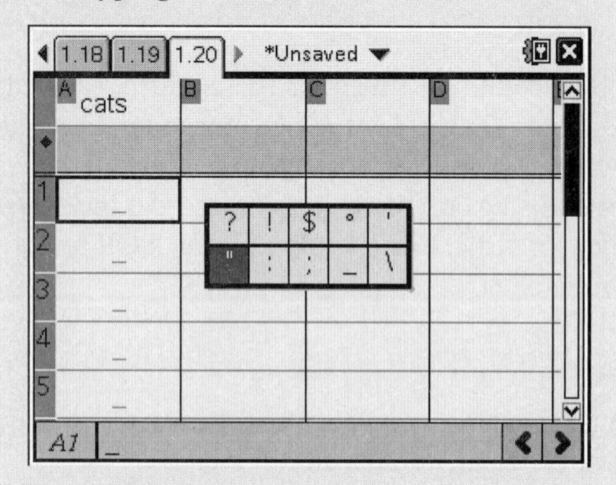

2. Enter the counts for each category into a second list (be sure to title this list as well!)
3. Press the **menu** key and navigate to **3:Data** and then **5:Frequency Plot** and select this option
4. For **Data List** select the list containing your category names from the drop-down menu
5. For **Frequency List** select the list containing the counts for each category from the drop-down menu
6. Press **OK**

Dotplots

JMP

1. Input the raw data into a column
2. Click **Graph** and select **Chart**
3. Click and drag the column name containing the data from the box under **Select columns** to the box next to **Categories, X, Levels**
4. Select **Point Chart** from the second drop-down box in the **Options** section

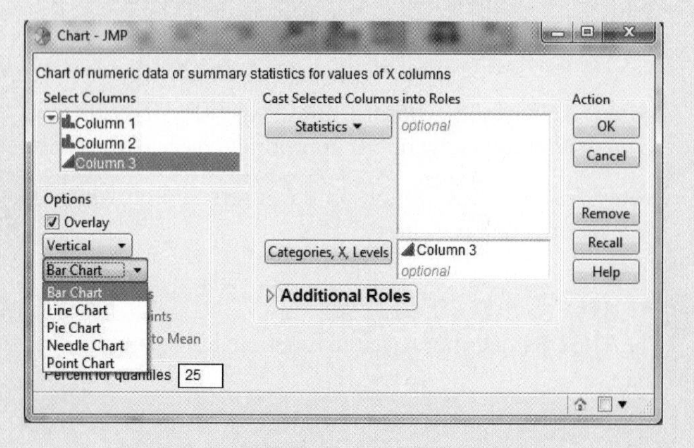

5. Click **OK**

MINITAB

1. Input the raw data into C1
2. Select **Graph** and choose **Dotplot**
3. Highlight Simple under One Y
4. Click **OK**
5. Double click C1 to add it to the **Graph Variables** box
6. Click **OK**

Note: You may add or format titles, axis titles, legends, and so on, by clicking on the **Labels…** button prior to performing Step 6 above.

SPSS

1. Input the raw data into a column
2. Select **Graph** and choose **Chart Builder…**
3. Under **Choose from** highlight Scatter/Dot
4. Click and drag the second option in the second row (Simple Dot Plot) to the Chart preview area
5. Click and drag the variable name from the **Variables** box into the **X-Axis?** box
6. Click **OK**

Excel 2007

Excel 2007 does not have the functionality to create dotplots.

TI-83/84

The TI-83/84 does not have the functionality to create dotplots.

TI-Nspire

1. Enter the data into a data list (to access data lists select the spreadsheet option and press **enter**)

Note: Be sure to title the list by selecting the top row of the column and typing a title.

2. Press the **menu** key then select **3:Data** then select **6:QuickGraph** and press **enter** (a dotplot will appear)

CHAPTER 2

Collecting Data Sensibly

Kwame Zikomo/Purestock/SuperStock

Data and conclusions from data are everywhere—in newspapers, magazines, online resources, and professional publications. But should you believe what you read? For example, should you drink hot chocolate to improve your memory? Will eating beef make you happier? Will thinking positive thoughts add 7 years to your life? These are just three of many claims made in one issue of *Woman's World* (December 23, 2013), a magazine with over 1.5 million readers. The magazine suggests that these claims are based on research studies, but how reliable are these studies? Are the conclusions drawn reasonable, and do they apply to you? These are important questions.

A primary goal of statistical studies is to collect data that can then be used to make informed decisions. It should come as no surprise that the ability to make good decisions depends on the quality of the information available.

Both the type of analysis that is appropriate and the conclusions that can be drawn depend on how the data are collected. In this chapter, we first consider two types of statistical studies and then focus on two widely used methods of data collection: sampling and experimentation.

Chapter 2: Learning Objectives

STUDENTS WILL UNDERSTAND:

- that the types of conclusions that can be drawn from data depend on the way data were collected.
- that bias may be present when data are collected from a sample.
- why random selection is an important component of a sampling plan.
- why random assignment is important when collecting data in an experiment.
- the purposes of a control group and blinding in an experiment.

STUDENTS WILL BE ABLE TO:

- distinguish between an observational study and an experiment.
- distinguish between selection bias, measurement or response bias, and nonresponse bias.
- select a simple random sample from a given population.

- distinguish between simple random sampling, stratified random sampling, cluster sampling, systematic sampling, and convenience sampling.
- describe a procedure for randomly assigning subjects to treatments in an experiment.
- design a completely randomized experiment.
- design a randomized block experiment.

2.1 Statistical Studies: Observation and Experimentation

On September 25, 2009, results from a study of the relationship between spanking and IQ were reported by a number of different news media. Some of the headlines that appeared that day were:

"Spanking lowers a child's IQ" (*Los Angeles Times*)
"Do you spank? Studies indicate it could lower your kid's IQ" (*SciGuy, Houston Chronicle*)
"Spanking can lower IQ" (NBC4i, Columbus, Ohio)
"Smacking hits kids' IQ" (*newscientist.com*)

In the study that these headlines refer to, the investigators followed 806 kids age 2 to 4 and 704 kids age 5 to 9 for 4 years. IQ was measured at the beginning of the study and again 4 years later. The researchers found that at the end of the study, the average IQ of kids in the younger group who were not spanked was 5 points higher than that of kids who were spanked. For the older group, the average IQ of kids who were not spanked was 2.8 points higher.

These headlines all imply that spanking was the cause of the observed difference in IQ. Is this conclusion reasonable? The answer depends in a critical way on the study design. After considering some important aspects of study design, we'll return to these headlines and decide if they are appropriate.

Observation and Experimentation

Data collection is an important step in the data analysis process. When we set out to collect information, it is important to keep in mind the questions we hope to answer on the basis of the resulting data. Sometimes we are interested in answering questions about characteristics of a single existing population or in comparing two or more well-defined populations. To accomplish this, we select a sample from each population under consideration and use the sample information to gain insight into characteristics of those populations.

For example, an ecologist might be interested in estimating the average shell thickness of bald eagle eggs. A social scientist studying a rural community may want to determine whether gender and attitude toward abortion are related. These are examples of studies that are *observational* in nature. In these studies, we want to observe characteristics of members of an existing population or of several populations, and then use the resulting information to draw conclusions. In **observational studies**, it is important to obtain samples that are representative of the corresponding populations.

Sometimes the questions we are trying to answer deal with the effect of certain explanatory variables on some response and cannot be answered using data from an observational study. Such questions are often of the form, "What happens when . . . ?" or, "What is the effect of . . . ?" For example, an educator may wonder what would happen to test scores if the required lab time for a chemistry course were increased from 3 hours to 6 hours per week. To answer such questions, the researcher conducts an experiment to collect relevant data. The value of some response variable (test score in the chemistry example) is recorded under different experimental conditions (3-hour lab and 6-hour lab). In an **experiment**, the researcher manipulates one or more explanatory variables, also sometimes called factors, to create the experimental conditions.

DEFINITION

Observational study: A study in which the investigator observes characteristics of a sample selected from one or more existing populations. The goal of an observational study is usually to draw conclusions about the corresponding population or about differences between two or more populations. In a well-designed observational study, the sample is selected in a way that is designed to produce a sample that is respresentative of the population.

Experiment: A study in which the investigator observes how a response variable behaves when one or more explanatory variables, also called factors, are manipulated. The usual goal of an experiment is to determine the effect of the manipulated explanatory variables (factors) on the response variable. In a well-designed experiment, the composition of the groups that will be exposed to different experimental conditions is determined by random assignment.

The type of conclusion that can be drawn from a statistical study depends on how the study was conducted. Both observational studies and experiments can be used to compare groups, but in an experiment the researcher controls who is in which group, whereas this is not the case in an observational study. This seemingly small difference is critical when it comes to drawing conclusions based on data from the study.

A well-designed experiment can result in data that provide evidence for a cause-and-effect relationship. This is an important difference between an observational study and an experiment. In an observational study, it is impossible to draw clear cause-and-effect conclusions because we cannot rule out the possibility that the observed effect is due to some variable other than the explanatory variable being studied. Such variables are called confounding variables.

DEFINITION

Confounding variable: A variable that is related to both how the experimental groups were formed and the response variable of interest.

Consider the role of confounding variables in the following three studies:

- The article **"Panel Can't Determine the Value of Daily Vitamins" (*San Luis Obispo Tribune*, July 1, 2003)** summarized conclusions from a government advisory panel that investigated the benefits of vitamin use. The panel looked at a large number of studies on vitamin use and concluded that the results were "inadequate or conflicting." A major concern was that many of the studies were observational in nature and the panel worried that people who take vitamins might be healthier just because they tend to take better care of themselves in general. This potential confounding variable prevented the panel from concluding that taking vitamins is the *cause* of observed better health among those who take vitamins.
- Studies have shown that people over age 65 who get a flu shot are less likely to die from a flu-related illness during the following year than those who do not get a flu shot. However, recent research has shown that people over age 65 who get

a flu shot are also less likely to die from *any* cause during the following year than those who don't get a flu shot **(*International Journal of Epidemiology*, December 21, 2005)**. This has led to the speculation that those over age 65 who get flu shots are healthier as a group than those who do not get flu shots. If this is the case, observational studies that compare two groups—those who get flu shots and those who do not—may overestimate the effectiveness of the flu vaccine because general health differs in the two groups. General health is a possible confounding variable in such studies.

- The article **"Heartfelt Thanks to Fido" (*San Luis Obispo Tribune*, July 5, 2003)** summarized a study that appeared in the ***American Journal of Cardiology* (March 15, 2003).** In this study researchers measured heart rate variability (a measure of the heart's ability to handle stress) in patients who had recovered from a heart attack. They found that heart rate variability was higher (which is good and means the heart can handle stress better) for those who owned a dog than for those who did not. Should someone who suffers a heart attack immediately go out and get a dog? Well, maybe not yet. The American Heart Association recommends additional studies to determine if the improved heart rate variability is attributable to dog ownership or due to the fact that dog owners get more exercise. If in fact dog owners do tend to get more exercise than nonowners, level of exercise is a confounding variable that would prevent us from concluding that owning a dog is the *cause* of improved heart rate variability.

Each of the three studies described above illustrates why potential confounding variables make it unreasonable to draw a cause-and-effect conclusion from an observational study.

Let's return to the study on spanking and IQ described at the beginning of this section. Is this study an observational study or an experiment? Two groups were compared (children who were spanked and children who were not spanked), but the researchers did not randomly assign children to the spanking or no-spanking groups. The study is observational, and so cause-and-effect conclusions such as "spanking lowers IQ" are not justified based on the observed data. What we can say is that there is evidence that, as a group, children who are spanked tend to have a lower IQ than children who are not spanked. What we cannot say is that spanking is the *cause* of the lower IQ. It is possible that other variables—such as home or school environment, socio-economic status, or parents' education—are related to both IQ and whether or not a child was spanked. These are examples of possible confounding variables.

Fortunately, not everyone made the same mistake as the writers of the headlines given earlier in this section. Some examples of headlines that got it right are:

"Lower IQ's measured in spanked children" (*world-science.net*)
"Children who get spanked have lower IQs" (*livescience.com*)
"Research suggests an association between spanking and lower IQ in children" (*CBSnews.com*)

Drawing Conclusions from Statistical Studies

In this section, two different types of conclusions have been described. One type involves generalizing from what we have seen in a sample to some larger population, and the other involves reaching a cause-and-effect conclusion about the effect of an explanatory variable on a response. When is it reasonable to draw such conclusions? The answer depends on the way that the data were collected. Table 2.1 summarizes the types of conclusions that can be made with different study designs.

As you can see from Table 2.1, it is important to think carefully about the objectives of a statistical study before planning how the data will be collected. Both observational studies and experiments must be carefully designed if the resulting data are to be useful. The common sampling procedures used in observational studies are considered in Section 2.2. In Sections 2.3 and 2.4, we consider experimentation and explore what constitutes good practice in the design of simple experiments.

TABLE 2.1 Drawing Conclusions from Statistical Studies

Study Description	Reasonable to Generalize Conclusions about Group Characteristics to the Population?	Reasonable to Draw Cause-and-Effect Conclusion?
Observational study with sample selected at random from population of interest	Yes	No
Observational study based on convenience or voluntary response sample (poorly designed sampling plan)	No	No
Experiment with groups formed by random assignment of individuals or objects to experimental conditions		
• Individuals or objects used in study are volunteers or not randomly selected from some population of interest	No	Yes
• Individuals or objects used in study are randomly selected from some population of interest	Yes	Yes
Experiment with groups not formed by random assignment to experimental conditions (poorly designed experiment)	No	No

EXERCISES 2.1 - 2.12

2.1 The article **"How Dangerous Is a Day in the Hospital?" (*Medical Care* [2011]: 1068–1075)** describes a study to determine if the risk of an infection is related to the length of a hospital stay. The researchers looked at a large number of hospitalized patients and compared the proportion who got an infection for two groups of patients—those who were hospitalized overnight and those who were hospitalized for more than one night. Indicate whether the study is an observational study or an experiment. Give a brief explanation for your choice.

2.2 The authors of the paper **"Fudging the Numbers: Distributing Chocolate Influences Student Evaluations of an Undergraduate Course" (*Teaching in Psychology* [2007]: 245–247)** carried out a study to see if events unrelated to an undergraduate course could affect student evaluations. Students enrolled in statistics courses taught by the same instructor participated in the study. All students attended the same lectures and one of six discussion sections that met once a week. At the end of the course, the researchers chose three of the discussion sections to be the "chocolate group." Students in these three sections were offered chocolate prior to having them fill out course evaluations. Students in the other three sections were not offered chocolate.

The researchers concluded that "Overall, students offered chocolate gave more positive evaluations than students not offered chocolate." Indicate whether the study is an observational study or an experiment. Give a brief explanation for your choice.

2.3 The article **"Why We Fall for This" (*AARP Magazine*, May/June 2011)** described a study in which a business professor divided his class into two groups. He showed students a mug and then asked students in one of the groups how much they would pay for the mug. Students in the other group were asked how much they would sell the mug for if it belonged to them. Surprisingly, the average value assigned to the mug was quite different for the two groups! Indicate whether the study is an observational study or an experiment. Give a brief explanation for your choice.

2.4 ▼ The article **"Television's Value to Kids: It's All in How They Use It" (*Seattle Times*, July 6, 2005)** described a study in which researchers analyzed standardized test results and television viewing habits of 1700 children. They found that children who averaged more than 2 hours of television viewing per day when they were younger than 3 tended to score lower on measures of reading ability and short-term memory.

a. Is the study described an observational study or an experiment?

b. Is it reasonable to conclude that watching 2 or more hours of television is the cause of lower reading scores? Explain. (Hint: Look at Table 2.1.)

2.5 The article **"Acupuncture for Bad Backs: Even Sham Therapy Works" (*Time,* May 12, 2009)** summarized a study conducted by researchers at the Group Health Center for Health Studies in Seattle. In this study, 638 adults with back pain were randomly assigned to one of four groups. People in group 1 received the usual care for back pain. People in group 2 received acupuncture at a set of points tailored specifically for each individual. People in group 3 received acupuncture at a standard set of points typically used in the treatment of back pain. Those in group 4 received fake acupuncture—they were poked with a toothpick at the same set of points chosen for the people in group 3!

Two notable conclusions from the study were: (1) patients receiving real or fake acupuncture experienced a greater reduction in pain than those receiving usual care; and (2) there was no significant difference in pain reduction for those who received acupuncture (at individualized or the standard set of points) and those who received fake acupuncture toothpick pokes.

a. Is this study an observational study or an experiment? Explain.

b. Is it reasonable to conclude that receiving either real or fake acupuncture was the cause of the observed reduction in pain in those groups compared to the usual care group? What aspect of this study supports your answer? (Hint: Look at Table 2.1.)

2.6 The article **"Display of Health Risk Behaviors on MySpace by Adolescents" (*Archives of Pediatrics and Adolescent Medicine* [2009]: 27–34)** described a study in which researchers looked at a random sample of 500 publicly accessible MySpace web profiles posted by 18-year-olds. The content of each profile was analyzed. One of the conclusions reported was that displaying sport or hobby involvement was associated with decreased references to risky behavior (sexual references or references to substance abuse or violence).

a. Is the study described an observational study or an experiment?

b. Is it reasonable to generalize the stated conclusion to all 18-year-olds with a publicly accessible MySpace web profile? What aspect of the study supports your answer?

c. Not all MySpace users have a publicly accessible profile. Is it reasonable to generalize the stated conclusion to all 18-year-old MySpace users? Explain.

d. Is it reasonable to generalize the stated conclusion to all MySpace users with a publicly accessible profile? Explain.

2.7 Can choosing the right music make wine taste better? This question was investigated by a researcher at a university in Edinburgh (www.decanter.com/news). Each of 250 volunteers was assigned at random to one of five rooms where they were asked to taste and rate a glass of wine. In one of the rooms, no music was playing and a different style of music was playing in each of the other four rooms. The researchers concluded that cabernet sauvignon is perceived as being richer and more robust when bold music is played than when no music is heard.

a. Is the study described an observational study or an experiment?

b. Can a case be made for the researcher's conclusion that the music played was the cause for the higher rating? Explain.

2.8 **"Fruit Juice May Be Fueling Pudgy Preschoolers, Study Says"** is the title of an article that appeared in the ***San Luis Obispo Tribune* (February 27, 2005).** This article describes a study that found that for 3- and 4-year-olds, drinking something sweet once or twice a day doubled the risk of being seriously overweight one year later. The authors of the study state

> Total energy may be a confounder if consumption of sweet drinks is a marker for other dietary factors associated with overweight **(*Pediatrics*, November 2005).**

Give an example of a dietary factor that might be one of the potentially confounding variables the study authors are worried about.

2.9 The article **"Americans are 'Getting the Wrong Idea' on Alcohol and Health" (Associated Press, April 19, 2005)** reported that observational studies in recent years that have concluded that moderate drinking is associated with a reduction in the risk of heart disease may be misleading. The article refers to a study conducted by the Centers for Disease Control and Prevention that showed that moderate drinkers, as a group, tended to be better educated, wealthier, and more active than nondrinkers.

Explain why the existence of these potentially confounding variables prevents drawing the conclusion that moderate drinking is the cause of reduced risk of heart disease.

2.10 Based on a survey conducted on the eDiets.com web site, investigators concluded that women who regularly watched *Oprah* were only one-seventh as likely to crave fattening foods as those who watched other daytime talk shows (*San Luis Obispo Tribune*, October 14, 2000).

a. Is it reasonable to conclude that watching *Oprah* causes a decrease in cravings for fattening foods? Explain.

b. Is it reasonable to generalize the results of this survey to all women in the United States? To all women who watch daytime talk shows? Explain why or why not.

2.11 ▼ A survey of affluent Americans (those with incomes of $75,000 or more) indicated that 57% would rather have more time than more money (*USA Today*, January 29, 2003).

a. What condition on how the data were collected would make the generalization from the sample to the population of affluent Americans reasonable?

b. Would it be reasonable to generalize from the sample and say that 57% of all Americans would rather have more time than more money? Explain.

2.12 Does living in the South cause high blood pressure? Data from a group of 6278 whites and blacks questioned in the Third National Health and Nutritional Examination Survey between 1988 and 1994 indicates that a greater percentage of Southerners have high blood pressure than do people in any other region of the United States (see CNN.com web site article of January 6, 2000, titled **"High Blood Pressure Greater Risk in U.S. South, Study Says"**). This difference in rate of high blood pressure was found in every ethnic group, gender, and age category studied.

List at least two possible reasons why we cannot conclude that living in the South causes high blood pressure.

Bold exercises answered in back ● Data set available online ▼ Video Solution available

2.2 Sampling

Many studies are conducted in order to generalize the results of the study to the corresponding population. In this case, it is important that the sample be representative of the population. To be reasonably sure of this, we must carefully consider the way in which the sample is selected.

It is sometimes tempting to take the easy way out and gather data in a haphazard way. But if a sample is chosen on the basis of convenience alone, it is not possible to interpret the resulting data with confidence. For example, it might be easy to use the students in your statistics class as a sample of students at your university. However, not all majors include a statistics course in their curriculum, and most students take statistics in their sophomore or junior year. When we attempt to generalize from this convenience sample, the difficulty is that it is not clear how these factors (and others that we might not be aware of) affect any conclusions based on information from such a sample.

> There is no way to tell just by looking at a sample whether it is representative of the population from which it was drawn. Our only assurance comes from the method used to select the sample.

There are many reasons for selecting a sample rather than obtaining information from an entire population (a **census**). Sometimes the process of measuring the characteristics of interest is destructive, as with measuring the lifetime of flashlight batteries or the sugar content of oranges. It would be foolish to study the entire population in situations like these. But the most common reason for selecting a sample is limited resources.

Restrictions on available time or money usually make it impossible to collect data from an entire population.

Bias in Sampling

Bias in sampling is the tendency for samples to differ from the corresponding population in some systematic way. Bias can result from the way in which the sample is selected or from the way in which information is obtained once the sample has been chosen. The most common types of bias encountered in sampling situations are selection bias, measurement or response bias, and nonresponse bias.

Selection bias (sometimes also called undercoverage) is introduced when the way the sample is selected systematically excludes some part of the population of interest. For example, a researcher may wish to generalize from the results of a study to the population consisting of all residents of a particular city, but the method of selecting individuals may tend to exclude the homeless or those without telephones.

If those who are excluded from the sampling process differ in some systematic way from those who are included, the sample is virtually guaranteed to be unrepresentative of the population. If this difference between the included and the excluded occurs on a variable that is important to the study, conclusions based on the sample data may not be valid for the population of interest.

Selection bias also occurs if only volunteers or self-selected individuals are used in a study, because those who choose to participate (for example, in a call-in telephone poll) may differ from those who choose not to participate.

Measurement or response bias occurs when the method of observation tends to produce values that systematically differ from the true value in some way. This might happen if an improperly calibrated scale is used to weigh items or if questions on a survey are worded in a way that tends to influence the response.

For example, a Gallup survey sponsored by the American Paper Institute (***Wall Street Journal*, May 17, 1994**) included the following question:

> "It is estimated that disposable diapers account for less than 2 percent of the trash in today's landfills. In contrast, beverage containers, third-class mail and yard waste are estimated to account for about 21 percent of trash in landfills. Given this, in your opinion, would it be fair to tax or ban disposable diapers?"

It is likely that the wording of this question prompted people to respond in a particular way.

Other things that might contribute to response bias are the appearance or behavior of the person asking the question, the group or organization conducting the study, and the tendency for people not to be completely honest when asked about illegal behavior or unpopular beliefs.

Although the terms *measurement bias* and *response bias* are often used interchangeably, the term *measurement bias* is usually used to describe systematic deviation from the true value as a result of a faulty measurement instrument (as with the improperly calibrated scale). Response bias is typically used to describe systematic deviations from the true value when people provide answers to survey questions.

Nonresponse bias occurs when responses are not obtained from all individuals selected for inclusion in the sample. As with selection bias, nonresponse bias can distort results if those who respond differ in important ways from those who do not respond. Although some level of nonresponse is unavoidable in most surveys, the biasing effect on the resulting sample is lowest when the response rate is high. To minimize nonresponse bias, it is critical that a serious effort be made to follow up with individuals who do not respond to an initial request for information.

The nonresponse rate for surveys or opinion polls varies dramatically, depending on how the data are collected. Surveys are commonly conducted by mail, by phone, and by personal interview. Mail surveys are inexpensive but often have high nonresponse rates. Telephone surveys can also be inexpensive and can be implemented quickly, but they work well only for short surveys and they can also have high nonresponse rates. Personal interviews are generally expensive but tend to have better response rates. Some of the many challenges of conducting surveys are discussed in Section 2.6 (available online).

Types of Bias

Selection Bias
Tendency for samples to differ from the corresponding population as a result of systematic exclusion of some part of the population.

Measurement or Response Bias
Tendency for samples to differ from the corresponding population because the method of observation tends to produce values that differ from the true value.

Nonresponse Bias
Tendency for samples to differ from the corresponding population because data are not obtained from all individuals selected for inclusion in the sample.

It is important to note that bias is introduced by the way in which a sample is selected or by the way in which the data are collected from the sample. Increasing the size of the sample, although possibly desirable for other reasons, does nothing to reduce bias if the method of selecting the sample is flawed or if the nonresponse rate remains high.

Potential sources of bias are illustrated in the following examples.

EXAMPLE 2.1 Are Cell Phone Users Different?

Understand the context ⟩

Many surveys are conducted by telephone and participants are often selected from phone books that include only landline telephones. For many years, it was thought that this was not a serious problem because most cell phone users also had a landline phone and so they still had a chance of being included in the survey. But the number of people with cell phones only is growing, and this trend is a concern for survey organizations.

The article **"Omitting Cell Phone Users May Affect Polls" (*Associated Press*, September 25, 2008)** described a study that examined whether people who only have a cell phone are different from those who have landline phones. One finding from the study was that for people under the age of 30 with only a cell phone, 28% were Republicans compared to 36% of landline users. This suggests that researchers who use telephone surveys need to worry about how selection bias might influence the ability to generalize the results of a survey if only landlines are used. ■

EXAMPLE 2.2 Think Before You Order That Burger!

Understand the context ⟩

The article **"What People Buy from Fast-Food Restaurants: Caloric Content and Menu Item Selection" (*Obesity* [2009]: 1369–1374)** reported that the average number of calories consumed at lunch in New York City fast-food restaurants was 827. The researchers selected 267 fast-food locations at random. The paper states that at each of these locations "adult customers were approached as they entered the restaurant and asked to provide their food receipt when exiting and to complete a brief survey."

Approaching customers as they entered the restaurant and before they ordered may have influenced what they purchased. This introduces the potential for response bias. In addition, some people chose not to participate when approached. If those who chose not to participate differed from those who did participate, the researchers also need to be concerned about nonresponse bias. Both of these potential sources of bias limit the researchers' ability to generalize conclusions based on data from this study. ■

Random Sampling

Most of the methods introduced in this text are based on the idea of random selection. The most straightforward sampling method is called simple random sampling. A **simple random sample** is a sample chosen using a method that ensures that each different possible sample of the desired size has an equal chance of being the one chosen.

For example, suppose that we want a simple random sample of 10 employees chosen from all those who work at a large design firm. For the sample to be a simple random sample, the method used to select the sample must ensure that each of the many different subsets of 10 employees must be equally likely to be selected. A sample taken from only full-time employees would not be a simple random sample of *all* employees, because someone who works part-time has no chance of being selected. Although a simple random sample may, by chance, include only full-time employees, it must be selected in such a way that each possible sample, and therefore *every* employee, has the same chance of inclusion in the sample.

> It is the selection process, not the final sample, which determines whether the sample is a simple random sample.

The letter n is used to denote sample size. It is the number of individuals or objects in the sample. For the design firm scenario just described, $n = 10$ becasue 10 employees were to be selected.

> **DEFINITION**
>
> **Simple random sample of size *n*:** A sample that is selected from a population in a way that ensures that every different possible sample of size n has the same chance of being selected.

The definition of a simple random sample implies that every individual member of the population has an equal chance of being selected. *However, the fact that every individual has an equal chance of selection, by itself, is not enough to guarantee that the sample is a simple random sample.*

For example, suppose that a class is made up of 100 students, 60 of whom are female. A researcher decides to select 6 of the female students by writing all 60 names on slips of paper, mixing the slips, and then picking 6. She then selects 4 male students from the class using a similar procedure. Even though every student in the class has an equal chance of being included in the sample (6 of 60 females are selected and 4 of 40 males are chosen), the resulting sample is *not* a simple random sample because not all different possible samples of 10 students from the class have the same chance of selection. Many possible samples of 10 students—for example, a sample of 7 females and 3 males or a sample of all females—have no chance of being selected. The sample selection method described here is not necessarily a bad choice (in fact, it is an example of stratified sampling, to be discussed in more detail shortly). But it does not produce a simple random sample. When this is the case, it is sometimes necessary to use different methods when generalizing results from the sample to the population. For this reason, the choice of sampling method is an important consideration that must be considered when a method is chosen for analyzing data resulting from such a sampling method.

Selecting a Simple Random Sample

A number of different methods can be used to select a simple random sample. One way is to put the name or number of each member of the population on different but identical slips of paper. The process of thoroughly mixing the slips and then selecting n slips one by one yields a random sample of size n. This method is easy to understand, but it has obvious drawbacks. The mixing must be adequate, and producing the necessary slips of paper can be extremely tedious, even for relatively small populations.

A commonly used method for selecting a random sample is to first create a list, called a **sampling frame**, of the objects or individuals in the population. Each item on the list can then be identified by a number. A table of random digits or a random number generator can then be used to select the sample. A random number generator is a procedure that produces a sequence of numbers that satisfies properties associated with the notion of randomness. Most statistics software packages include a random number generator, as do many calculators. A small table of random digits can be found in Appendix A, Table 1.

For example, suppose a list containing the names of the 427 customers who purchased a new car during 2014 at a large dealership is available. The owner of the dealership wants to interview a sample of these customers to learn about customer satisfaction. She plans to select a simple random sample of 20 customers. Because it would be tedious to write all 427 names on slips of paper, random numbers can be used to select the sample. To do this, we can use three-digit numbers, starting with 001 and ending with 427, to represent the individuals on the list.

The random digits from rows 6 and 7 of Appendix A, Table 1 are shown here:

0 9 3 8 7 6 7 9 9 5 6 2 5 6 5 8 4 2 6 4
4 1 0 1 0 2 2 0 4 7 5 1 1 9 4 7 9 7 5 1

We can use blocks of three digits from this list (underlined in the lists above) to identify the individuals who should be included in the sample. The first block of three digits is 093, so the 93rd person on the list will be included in the sample. The next five blocks of three digits (876, 799, 562, 565, and 842) do not correspond to anyone on the list, so we ignore them. The next block that corresponds to a person on the list is 410, so that person is included in the sample. This process would continue until 20 people have been selected for the sample. We would ignore any three-digit repeats since any particular person should only be selected once for the sample.

Another way to select the sample would be to use computer software or a graphing calculator to generate 20 random numbers. For example, Minitab produced the following numbers when 20 random numbers between 1 and 427 were requested.

289	67	29	26	205	214	422	31	233	98
10	203	346	186	232	410	43	293	25	371

These numbers could be used to determine which 20 customers to include in the sample.

When selecting a random sample, researchers can choose to do the sampling with or without replacement. **Sampling with replacement** means that after each successive item is selected for the sample, the item is "replaced" back into the population and may therefore be selected again at a later stage. In practice, sampling with replacement is rarely used. Instead, the more common method is to not allow the same item to be included in the sample more than once. After being included in the sample, an individual or object would not be considered for further selection. Sampling in this manner is called **sampling without replacement**.

DEFINITION

Sampling without replacement: Once an individual from the population is selected for inclusion in the sample, it may not be selected again in the sampling process. A sample selected without replacement includes *n* distinct individuals from the population.

Sampling with replacement: After an individual from the population is selected for inclusion in the sample and the corresponding data are recorded, the individual is placed back in the population and can be selected again in the sampling process. A sample selected with replacement might include any particular individual from the population more than once.

Although these two forms of sampling are different, when the sample size *n* is small relative to the population size, as is often the case, there is little practical difference between them. In practice, the two methods can be viewed as equivalent if the sample size is less than 10% of the population size.

EXAMPLE 2.3 Selecting a Random Sample of Glass Soda Bottles

Understand the context ⟩

Breaking strength is an important characteristic of glass soda bottles. Suppose that we want to measure the breaking strength of each bottle in a random sample of size $n = 3$ selected from four crates containing a total of 100 bottles (the population). Each crate contains five rows of five bottles each. We can identify each bottle with a number from 1 to 100 by numbering across the rows in each crate, starting with the top row of crate 1, as pictured:

Crate 1

1	2	3	4	5
6	...			

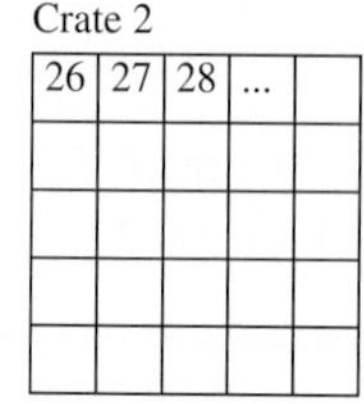

...

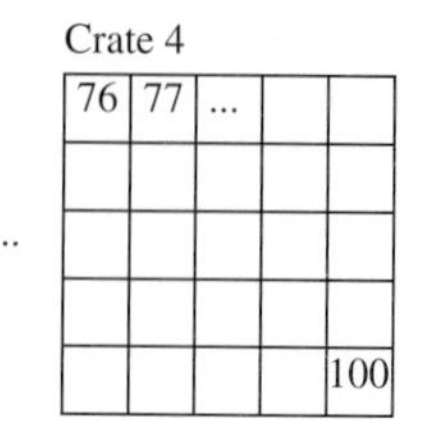

Formulate a plan ⟩

Using a random number generator from a calculator or statistical software package, we could generate three random numbers between 1 and 100 to determine which bottles would be included in the sample. This might result in bottles 15 (row 3 column 5 of crate 1), 89 (row 3 column 4 of crate 4), and 60 (row 2 column 5 of crate 3) being selected. ■

The goal of random sampling is to produce a sample that is likely to be representative of the population. Although random sampling does not *guarantee* that the sample will be representative, it does allow us to assess the risk of an unrepresentative sample. It is the ability to quantify this risk that will enable us to generalize with confidence from a random sample to the corresponding population.

An Important Note Concerning Sample Size

It is a common misconception that if the size of a sample is relatively small compared to the population size, the sample cannot possibly accurately reflect the population. Critics of polls often make statements such as, "There are 14.6 million registered voters in California. How can a sample of 1000 registered voters possibly reflect public opinion when only about 1 in every 14,000 people is included in the sample?" These critics do not understand the power of random selection!

Consider a population consisting of 5000 applicants to a state university, and suppose that we are interested in math SAT scores for this population. A dotplot of the values in this population is shown in Figure 2.1(a). Figure 2.1(b) shows dotplots of the math SAT scores for individuals in five different random samples from the population, ranging in sample size from $n = 50$ to $n = 1000$.

Notice that each of the samples tend to reflect the distribution of scores in the population. If we were interested in using the sample to estimate the population average or to say something about the variability in math SAT scores, even the smallest of the samples ($n = 50$) pictured would provide reliable information.

Although it is possible to obtain a simple random sample that does not do a reasonable job of representing the population, this is likely only when the sample size is very small, and unless the population itself is small, this risk does not depend on what fraction of the population is sampled. The random selection process allows us to be confident that the resulting sample adequately reflects the population, even when the sample consists of only a small fraction of the population.

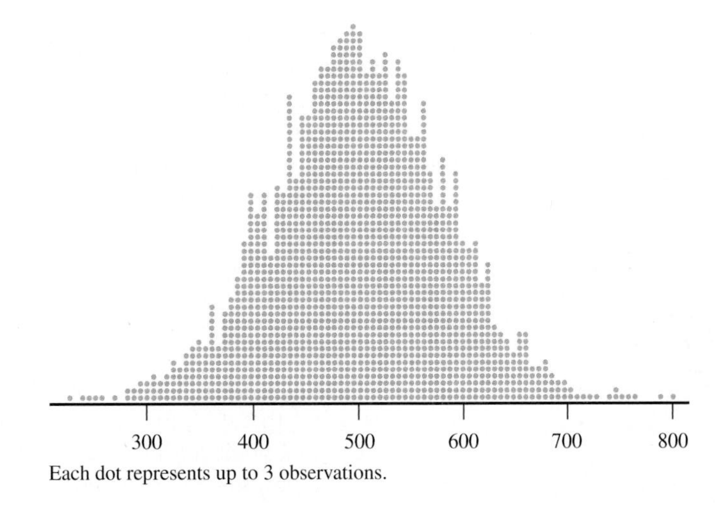

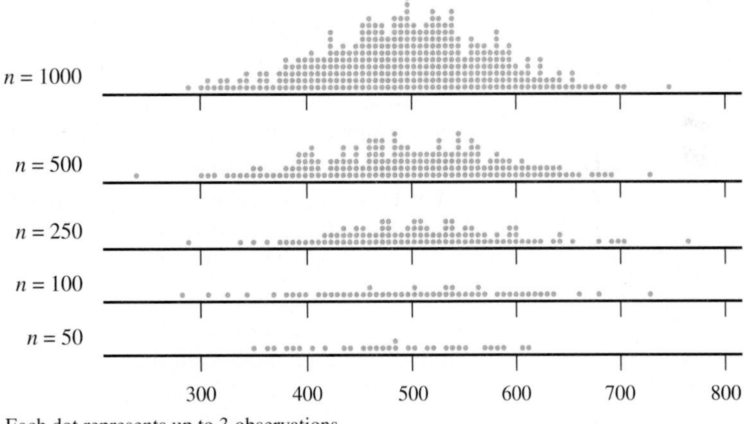

FIGURE 2.1
(a) Dotplot of math SAT scores for the entire population.
(b) Dotplots of math SAT scores for random samples of sizes 50, 100, 250, 500, and 1000.

Other Sampling Methods

Simple random sampling provides researchers with a sampling method that is objective and free of selection bias. In some settings, however, alternative sampling methods may be less costly, easier to implement, and sometimes even more accurate.

Stratified Random Sampling

When the entire population can be divided into a set of nonoverlapping subgroups, a method known as **stratified sampling** often proves easier to implement and more cost-effective than simple random sampling. In stratified random sampling, separate simple random samples are independently selected from each subgroup.

For example, to estimate the average cost of malpractice insurance, a researcher might find it convenient to view the population of all doctors practicing in a particular city as being made up of four subpopulations: (1) surgeons, (2) internists and family practitioners, (3) obstetricians, and (4) a group that includes all other areas of specialization. Rather than taking a random simple sample from the population of all doctors, the researcher could take four separate simple random samples—one from the group of surgeons, another from the internists and family practitioners, and so on. These four samples would provide information about the four subgroups as well as information about the overall population of doctors.

When the population is divided in this way, the subgroups are called **strata** and each individual subgroup is called a stratum (the singular of strata). Stratified sampling entails selecting a separate simple random sample from each stratum. Stratified sampling can be used instead of simple random sampling if it is important to obtain information about characteristics of the individual strata as well as of the entire population, although a stratified sample is not required to do this—subgroup estimates can also be obtained by using an appropriate subset of data from a simple random sample.

The real advantage of stratified sampling is that it often allows us to make more accurate inferences about a population than does simple random sampling. In general, it is much easier to produce relatively accurate estimates of characteristics of a homogeneous group than of a heterogeneous group.

For example, even with a small sample, it is possible to obtain an accurate estimate of the average grade point average (GPA) of students graduating with high honors from a university. The individual GPAs of these students are all quite similar (a homogeneous group), and even a sample of three or four individuals from this subpopulation should be representative. On the other hand, producing a reasonably accurate estimate of the average GPA of *all* seniors at the university, a much more diverse group of GPAs, is a more difficult task. This means that if a varied population can be divided into strata, with each stratum being much more homogeneous than the population with respect to the characteristic of interest, then a stratified random sample can produce more accurate estimates of population characteristics than a simple random sample of the same size.

Cluster Sampling

Sometimes it is easier to select groups of individuals from a population than it is to select individuals themselves. **Cluster sampling** involves dividing the population of interest into nonoverlapping subgroups, called **clusters**. Clusters are then selected at random, and then *all* individuals in the selected clusters are included in the sample.

For example, suppose that a large urban high school has 600 senior students, all of whom are enrolled in a first period homeroom. There are 24 senior homerooms, each with approximately 25 students. If school administrators wanted to select a sample of about 75 seniors to participate in an evaluation of the college and career placement advising available to students, they might find it much easier to select three of the senior homerooms at random and then include all the students in the selected homerooms in the sample. Then a survey could be administered to all students in the selected homerooms at the same time—certainly easier logistically than randomly selecting 75 individual seniors and then administering the survey to these students.

Because whole clusters are selected, the ideal situation for cluster sampling is when each cluster mirrors the characteristics of the population. When this is the case, a small number of clusters results in a sample that is representative of the population. If it is not reasonable to think that the variability present in the population is reflected in each cluster, as is often the case when the cluster sizes are small, then it becomes important to ensure that a large number of clusters are included in the sample.

Be careful not to confuse clustering and stratification. Even though both of these sampling strategies involve dividing the population into subgroups, both the way in which the subgroups are sampled and the optimal strategy for creating the subgroups are different.

In stratified sampling, we sample from every subgroup, whereas in cluster sampling, we include only selected whole clusters in the sample. Because of this difference, to increase the chance of obtaining a sample that is representative of the population, we want to create homogeneous groups for strata and heterogeneous (reflecting the variability in the population) groups for clusters.

Systematic Sampling

Systematic sampling is a procedure that can be used when it is possible to view the population of interest as consisting of a list or some other sequential arrangement. A value k is specified (for example, $k = 50$ or $k = 200$). Then one of the first k individuals is selected at random, after which every kth individual in the sequence is included in the sample. A sample selected in this way is called a **1 in *k* systematic sample**.

For example, a sample of faculty members at a university might be selected from the faculty phone directory. One of the first $k = 20$ faculty members listed could be selected at random, and then every 20th faculty member after that on the list would also be included in the sample. This would result in a 1 in 20 systematic sample.

The value of k for a 1 in k systematic sample is generally chosen to achieve a desired sample size. For example, in the faculty directory scenario just described, if there were 900 faculty members at the university, the 1 in 20 systematic sample described would result in a sample size of 45. If a sample size of 100 was desired, a 1 in 9 systematic sample could be used (because 900/100 = 9).

As long as there are no repeating patterns in the population sequence, systematic sampling works reasonably well. However, if there are such patterns, systematic sampling can result in an unrepresentative sample. For example, suppose that workers at the entry station of a state park have recorded the number of visitors to the park each day for the past 10 years. In a 1 in 70 systematic sample of days from this list, we would pick one of the first 70 days at random and then every 70th day after that. But if the first day selected happened to be a Wednesday, *every* day selected in the entire sample would also be a Wednesday (because there are 7 days a week and 70 is a multiple of 7). It is unlikely that such a sample would be representative of the entire collection of days. The number of visitors is likely to be higher on weekend days, and no Saturdays or Sundays would be included in the sample.

Convenience Sampling: Don't Go There!

It is often tempting to resort to **convenience sampling**—that is, using an easily available or convenient group to form a sample. This is a recipe for disaster! Results from such samples are rarely informative, and it is a mistake to try to generalize from a convenience sample to any larger population.

One common form of convenience sampling is sometimes called **voluntary response sampling**. Such samples rely entirely on individuals who volunteer to be a part of the sample, often by responding to an advertisement, calling a publicized telephone number to register an opinion, or logging on to an Internet site to complete a survey. It is extremely unlikely that individuals participating in such voluntary response surveys are representative of any larger population of interest.

EXERCISES 2.13 - 2.32

2.13 A New York psychologist recommends that if you feel the need to check your e-mail in the middle of a movie or if you sleep with your cell phone next to your bed, it might be time to "power off" (***AARP Bulletin*, September 2010**). Suppose that you want to learn about the proportion of students at your college who would feel the need to check e-mail during the middle of a movie and that you have access to a list of all students enrolled at your college. Describe how you would use this list to select a simple random sample of 100 students.

2.14 As part of a curriculum review, the psychology department would like to select a simple random sample of 20 of last year's 140 graduates to obtain information on how graduates perceived the value of the curriculum. Describe two different methods that might be used to select the sample.

2.15 A petition with 500 signatures is submitted to a university's student council. The council president would like to determine the proportion of those who signed the petition who are actually registered students at the university. There is not enough time to check all 500 names with the registrar, so the council president decides to select a simple random sample of 30 signatures. Describe how this might be done.

2.16 The article **"Bicyclists and Other Cyclists" (*Annals of Emergency Medicine* [2010]: 426)** reported that in 2008, there were 716 bicyclists killed on public roadways in the United States, and that the average age of the cyclists killed was 41 years. These figures were based on an analysis of the records of all traffic-related deaths of bicyclists on U.S. public roadways (this information is kept by the National Highway Traffic Safety Administration).

a. Does the group of 716 bicycle fatalities represent a census or a sample of the 2008 bicycle fatalities?

b. If the population of interest is 2008 bicycle traffic fatalities, is the given average age of 41 years a number that describes a sample or a number that describes the population?

2.17 The article **"Teenage Physical Activity Reduces Risk of Cognitive Impairment in Later Life"** (***Journal of the American Geriatrics Society*** **[2010]**) describes a study of more than 9000 women from Maryland, Minnesota, Oregon, and Pennsylvania. The women were asked about their physical activity as teenagers and at ages 30 and 50. A press release about this study (www.wiley.com) generalized the results of this study to all American women. In the press release, the researcher who conducted the study is quoted as saying

> Our study shows that women who are regularly physically active at any age have lower risk of cognitive impairment than those who are inactive but that being physically active at teenage is most important in preventing cognitive impairment.

Answer the following four questions for this observational study. (Hint: Reviewing Examples 2.1 and 2.2 might be helpful.)

a. What is the population of interest?

b. Was the sample selected in a reasonable way?

c. Is the sample likely to be representative of the population of interest?

d. Are there any obvious sources of bias?

2.18 ▼ For each of the situations described, state whether the sampling procedure is simple random sampling, stratified random sampling, cluster sampling, systematic sampling, or convenience sampling.

a. All first-year students at a university are enrolled in one of 30 sections of a seminar course. To select a sample of freshmen at this university, a researcher selects four sections of the seminar course at random from the 30 sections and all students in the four selected sections are included in the sample.

b. To obtain a sample of students, faculty, and staff at a university, a researcher randomly selects 50 faculty members from a list of faculty, 100 students from a list of students, and 30 staff members from a list of staff.

c. A university researcher obtains a sample of students at his university by using the 85 students enrolled in his Psychology 101 class.

d. To obtain a sample of the seniors at a particular high school, a researcher writes the name of each senior on a slip of paper, places the slips in a box and mixes them, and then selects 10 slips. The students whose names are on the selected slips of paper are included in the sample.

e. To obtain a sample of those attending a basketball game, a researcher selects the 24th person through the door. Then, every 50th person after that is also included in the sample.

2.19 Of the 6500 students enrolled at a community college, 3000 are part time and the other 3500 are full time. The college can provide a list of students that is sorted so that all full-time students are listed first, followed by the part-time students.

a. Describe a procedure for selecting a stratified random sample that uses full-time and part-time students as the two strata and that includes 10 students from each stratum.

b. Does every student at this community college have the same chance of being selected for inclusion in the sample? Explain.

2.20 Briefly explain why it is advisable to avoid the use of convenience samples.

2.21 A sample of pages from this book is to be obtained, and the number of words on each selected page will be determined. For the purposes of this exercise, equations are not counted as words and a number is counted as a word only if it is spelled out—that is, *ten* is counted as a word, but *10* is not.

a. Describe a sampling procedure that would result in a simple random sample of pages from this book.

b. Describe a sampling procedure that would result in a stratified random sample. Explain why you chose the specific strata used in your sampling plan.

c. Describe a sampling procedure that would result in a systematic sample.

d. Describe a sampling procedure that would result in a cluster sample.

e. Using the process you gave in Part (a), select a simple random sample of at least 20 pages, and record the number of words on each of the selected pages. Construct a dotplot of the resulting sample values, and write a sentence or two commenting on what it reveals about the number of words on a page.

f. Using the process you gave in Part (b), select a stratified random sample that includes a total of at least 20 selected pages, and record the number of words on each of the selected pages. Construct a dotplot of the resulting sample values, and write a sentence or two commenting on what it reveals about the number of words on a page.

2.22 In 2000, the chairman of a California ballot initiative campaign to add "none of the above" to the list of ballot options in all candidate races was quite critical of a Field poll that showed his measure trailing by 10 percentage points. The poll was based on a random sample of 1000 registered voters in California. He is quoted by the Associated Press (January 30, 2000) as saying, "Field's sample in that poll equates to one out of 17,505 voters," and he added that this was so dishonest that Field should get out of the polling business! If you worked on the Field poll, how would you respond to this criticism? (Hint: See discussion of sample size on page 40.)

2.23 The authors of the paper "Digital Inequality: Differences in Young Adults' Use of the Internet" (*Communication Research* [2008]: 602–621) were interested in determining if people with higher levels of education use the Internet in different ways than those who do not have as much formal education. To answer this question, they used data from a national telephone survey. Approximately 1300 households were selected for the survey, and 270 of them completed the interview. What type of bias should the researchers be concerned about and why? (Hint: See the box on page 37 that contains the definitions of different types of bias.)

2.24 The authors of the paper "Illicit Use of Psychostimulants among College Students" (*Psychology, Health & Medicine* [2002]: 283–287) surveyed college students about their use of legal and illegal stimulants. The sample of students surveyed consisted of students enrolled in a psychology class at a small, competitive college in the United States.

a. Was this sample a simple random sample, a stratified sample, a systematic sample, or a convenience sample? Explain.

b. Give two reasons why the estimate of the proportion of students who reported using illegal stimulants based on data from this survey should not be generalized to all U.S. college students.

2.25 The paper "Deception and Design: The Impact of Communication Technology on Lying Behavior" (*Computer-Human Interaction* [2009]: 130–136) describes an investigation into whether lying is less common in face-to-face communication than in other forms of communication such as phone conversations or e-mail. Participants in this study were 30 students in an upper-division communications course at Cornell University who received course credit for participation. Participants were asked to record all of their social interactions for a week, making note of any lies told.

Based on data from these records, the authors of the paper concluded that students lie more often in phone conversations than in face-to-face conversations and more often in face-to-face conversations than in e-mail.

Discuss the limitations of this study, commenting on the way the sample was selected and potential sources of bias.

2.26 The authors of the paper "Popular Video Games: Quantifying the Presentation of Violence and its Context" (*Journal of Broadcasting & Electronic Media* [2003]: 58–76) investigated the relationship between video game rating—suitable for everyone (E), suitable for 13 years of age and older (T), and suitable for 17 years of age and older (M)—and the number of violent interactions per minute of play. The sample of games examined consisted of 60 video games—the 20 most popular (by sales) for each of three game systems.

The researchers concluded that video games rated for older children had significantly more violent interactions per minute than did those games rated for more general audiences.

a. Do you think that the sample of 60 games was selected in a way that makes it reasonable to think it is representative of the population of all video games?

b. Is it reasonable to generalize the researchers' conclusion to all video games? Explain why or why not.

2.27 Participants in a study of honesty in online dating profiles were recruited through print and online advertisements in the *Village Voice*, one of New York City's most prominent weekly newspapers, and on Craigslist New York City ("The Truth About Lying in Online Dating Profiles," *Computer-Human Interaction* [2007]: 1–4). The actual height, weight, and age of the participants were compared to what appeared in their online dating profiles. The resulting data was then used to draw conclusions about how common deception was in online dating profiles.

What concerns do you have about generalizing conclusions based on data from this study to the population of all people who have an online dating profile? Be sure to address at least two concerns and give the reason for your concern.

2.28 The report "Undergraduate Students and Credit Cards in 2004: An Analysis of Usage Rates and Trends" (Nellie Mae, May 2005) estimated that 21% of undergraduates with credit cards pay them off each month and that the average outstanding balance on undergraduates' credit cards is $2169. These estimates were based on an online survey that was sent to 1260 students. Responses were received from 132 of these students. Is it reasonable to generalize the reported estimates to the population of all undergraduate students? Address at least two possible sources of bias in your answer.

2.29 The financial aid advisor of a university plans to use a stratified random sample to estimate the average amount of money that students spend on textbooks each term. For each of the following proposed stratification schemes, discuss whether it would be worthwhile to stratify the university students in this manner. (Hint: Remember that it is desirable to create strata that are homogeneous.)

a. Strata corresponding to class standing (freshman, sophomore, junior, senior, graduate student)

b. Strata corresponding to field of study, using the following categories: engineering, architecture, business, other

c. Strata corresponding to the first letter of the last name: A–E, F–K, etc.

2.30 Suppose that you were asked to help design a survey of adult city residents in order to estimate the proportion who would support a sales tax increase. The plan is to use a stratified random sample, and three stratification schemes have been proposed.

Scheme 1: Stratify adult residents into four strata based on the first letter of their last name (A–G, H–N, O–T, U–Z).

Scheme 2: Stratify adult residents into three strata: college students, nonstudents who work full time, nonstudents who do not work full time.

Scheme 3: Stratify adult residents into five strata by randomly assigning residents into one of the five strata.

Which of the three stratification schemes would be best in this situation? Explain.

2.31 The article "High Levels of Mercury Are Found in Californians" (*Los Angeles Times*, February 9, 2006) describes a study in which hair samples were tested for mercury. The hair samples were obtained from more than 6000 people who voluntarily sent hair samples to researchers at Greenpeace and The Sierra Club. The researchers found that nearly one-third of those tested had mercury levels that exceeded the concentration thought to be safe. Is it reasonable to generalize this result to the larger population of U.S. adults? Explain why or why not.

2.32 ▼ Whether or not to continue a Mardi Gras Parade through downtown San Luis Obispo, CA, is a hotly debated topic. The parade is popular with students and many residents, but some celebrations have led to complaints and a call to eliminate the parade. The local newspaper conducted online and telephone surveys of its readers and was surprised by the results. The survey web site received more than 400 responses, with more than 60% favoring continuing the parade, while the telephone response line received more than 120 calls, with more than 90% favoring banning the parade (*San Luis Obispo Tribune*, March 3, 2004). What factors may have contributed to these very different results?

CHAPTER 3

Graphical Methods for Describing Data

©iStockphoto.com/florintt

Most college students (and their parents) are concerned about the cost of a college education. The National Center for Education Statistics (www.nces.ed.gov) reported the average in state tuition and fees for 4-year public institutions in each of the 50 U.S. states for the 2011–2012 academic year. Average tuition and fees (in dollars) are given below for each state.

Several questions could be posed about these data. What is a typical value of average in-state tuition and fees for the 50 states? Are observations concentrated near the typical value, or do average tuition and fees differ quite a bit from state to state? Are there any states whose average tuition and fees are somehow unusual compared to the rest? What proportion of the states have average tuition and fees less than $7500? More than $10,000?

Questions such as these are most easily answered if the data can be organized in a sensible manner. In this chapter, we introduce some techniques for organizing and describing data using tables and graphs.

7502	5957	9021	6367	8907	7167	9069	10,524	4032	6015
7422	5674	11,252	7940	7563	6689	7943	5198	9278	7831
10,104	10,527	9862	5674	7588	6007	6752	4509	13,347	11,596
5293	6192	5701	6414	8800	5538	7975	11,818	9926	10,372
6939	7013	7116	5163	13,078	9373	7701	5241	7851	3501

Chapter 3: **Learning Objectives**

STUDENTS WILL UNDERSTAND:

- that selecting an appropriate graphical display depends on the data type (categorical or numerical) and whether or not the purpose of the display is to compare groups.

continued

- how a graphical display of numerical data is described in terms of center, shape, and spread.
- how a scatterplot is used to investigate the relationship between two numerical variables.
- how time series plots are used to investigate trend over time.

STUDENTS WILL BE ABLE TO:

- construct and interpret graphical displays of categorical data: pie charts and segmented bar charts.
- construct and interpret graphical displays of numerical data: stem-and-leaf displays, histograms, and relative frequency graphs.
- construct and interpret graphical displays designed to compare groups: comparative bar charts and comparative stem-and-leaf displays.
- construct and interpret a scatterplot of bivariate numerical data.
- construct and interpret a time series plot.

3.1 Displaying Categorical Data: Comparative Bar Charts and Pie Charts

Comparative Bar Charts

In Chapter 1 we saw that categorical data could be summarized in a frequency distribution and displayed graphically using a bar chart. Bar charts can also be used to give a visual comparison of two or more groups. This is accomplished by constructing two or more bar charts that use the same set of horizontal and vertical axes, as illustrated in Example 3.1.

EXAMPLE 3.1 How Far Is Far Enough?

Understanding the context ⟩

Step-by-step technology instructions available online

Each year The Princeton Review conducts a survey of high school students who are applying to college and parents of college applicants. The report **"2009 College Hopes & Worries Survey Findings" (www.princetonreview.com/archival/ir/REVU_News_2009_3_25_General.pdf)** included a summary of how 12,715 high school students responded to the question "Ideally how far from home would you like the college you attend to be?" Also included was a summary of how 3007 parents of students applying to college responded to the question "How far from home would you like the college your child attends to be?" The accompanying relative frequency table summarizes the student and parent responses.

Consider the data ⟩

	Frequency		Relative Frequency	
Ideal Distance	Students	Parents	Students	Parents
Less than 250 miles	4450	1594	.35	.53
250 to 500 miles	3942	902	.31	.30
500 to 1000 miles	2416	331	.19	.11
More than 1000 miles	1907	180	.15	.06

Interpret the results ⟩

The comparative bar chart based on the relative frequencies for these data is shown in Figure 3.1. It is easy to see the differences between students and parents. A higher proportion of parents prefer a college close to home, and a higher proportion of students than parents believe that the ideal distance from home would be more than 500 miles.

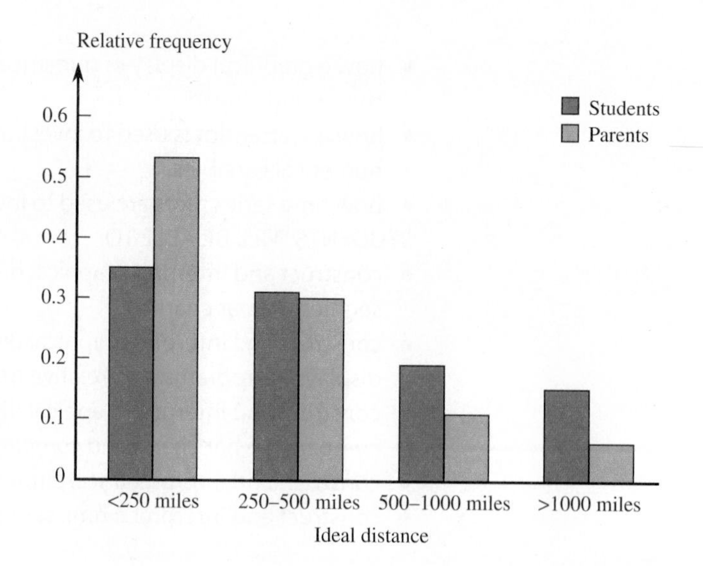

FIGURE 3.1
Comparative bar chart of ideal distance from home.

To see why we use relative frequencies rather than frequencies to compare groups of different sizes, consider the *incorrect* bar chart constructed using the frequencies rather than the relative frequencies (Figure 3.2). The incorrect bar chart conveys a very different and misleading impression of the differences between students and parents.

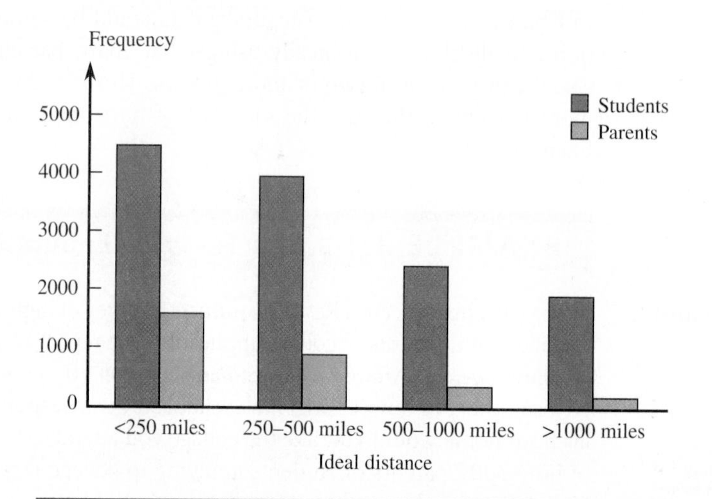

FIGURE 3.2
An **incorrect** comparative bar chart for the data of Example 3.1.

When constructing a comparative bar chart, use relative frequency rather than frequency to construct the scale on the vertical axis so that meaningful comparisons can be made even if the sample sizes are not the same.

Pie Charts

A categorical data set can also be summarized using a pie chart. In a pie chart, a circle is used to represent the whole data set, with "slices" of the pie representing the possible categories. The size of the slice for a particular category is proportional to the corresponding frequency or relative frequency. Pie charts are most effective for summarizing data sets when there are not too many different categories.

EXAMPLE 3.2 Life Insurance for Cartoon Characters??

Understanding the context)

The article **"Fred Flintstone, Check Your Policy" (*The Washington Post,* October 2, 2005)** summarized the results of a survey of 1014 adults conducted by the Life and Health Insurance Foundation for Education. Each person surveyed was asked to select which of five fictional characters, Spider-Man, Batman, Fred Flintstone, Harry Potter, and Marge Simpson, he or she thought had the greatest need for life insurance. The resulting data are summarized in the pie chart of Figure 3.3.

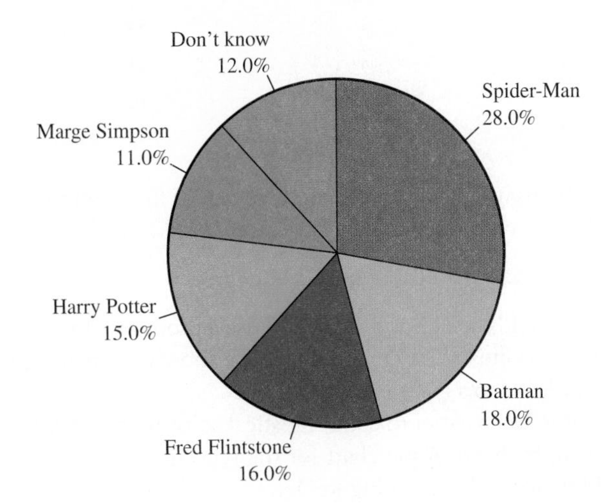

FIGURE 3.3
Pie chart of data on which fictional character most needs life insurance.

Interpret the results)

The survey results were quite different from an insurance expert's assessment. His opinion was that Fred Flintstone, a married father with a young child, was by far the one with the greatest need for life insurance. Spider-Man, unmarried with an elderly aunt, would need life insurance only if his aunt relied on him to supplement her income. Batman, a wealthy bachelor with no dependents, doesn't need life insurance in spite of his dangerous job! ■

Pie Chart for Categorical Data

When to Use Categorical data with a relatively small number of possible categories. Pie charts are most useful for illustrating proportions of the whole data set for various categories.

How to Construct

1. Draw a circle to represent the entire data set.
2. For each category, calculate the "slice" size. Because there are 360 degrees in a circle

   slice size = 360 · (category relative frequency)

3. Draw a slice of appropriate size for each category. This can be tricky, so most pie charts are generated using a graphing calculator or a statistical software package.

What to Look For

- Categories that form large and small proportions of the data set.

Step-by-step technology instructions available online

EXAMPLE 3.3 Watch Those Typos

Understanding the context ⟩

Typos on a résumé do not make a very good impression when applying for a job. Senior executives were asked how many typos in a résumé would make them not consider a job candidate (**"Job Seekers Need a Keen Eye," *USA Today*, August 3, 2009**). The resulting data are summarized in the accompanying relative frequency distribution.

Consider the data ⟩

Number of Typos	Frequency	Relative Frequency
1	60	.40
2	54	.36
3	21	.14
4 or more	10	.07
Don't know	5	.03

Do the work ⟩

To draw a pie chart by hand, we would first compute the slice size for each category. For the one typo category, the slice size would be

$$\text{slice size} = (.40)(360) = 144 \text{ degrees}$$

144 degrees, to represent first attempt category

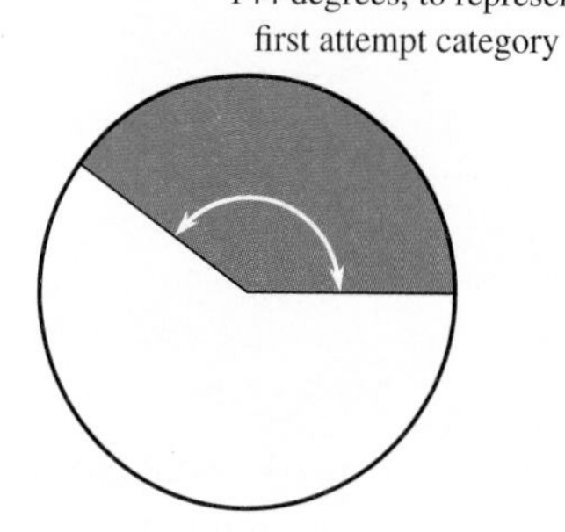

We would then draw a circle and use a protractor to mark off a slice corresponding to about 144°, as illustrated here in the figure shown in the margin. Continuing to add slices in this way leads to a completed pie chart.

It is much easier to use a statistical software package to create pie charts than to construct them by hand. A pie chart for the typo data, created with the statistical software package Minitab, is shown in Figure 3.4.

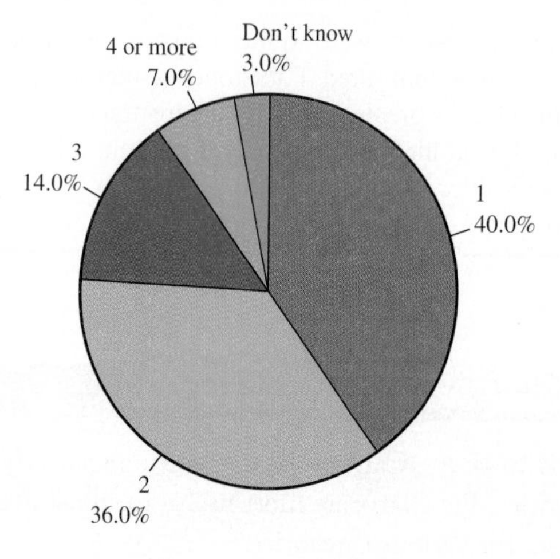

FIGURE 3.4
Pie chart for the typo data of Example 3.3.

Interpret the results ⟩

From the completed pie chart, it is easy to see that even one or two typos would result in many employers not considering a candidate for a job. ■

Pie charts can be used effectively to summarize a single categorical data set if there are not too many different categories. However, pie charts are not usually the best tool if the goal is to compare groups on the basis of a categorical variable. This is illustrated in Example 3.4.

EXAMPLE 3.4 Scientists and Nonscientists Do Not See Eye-to-Eye

Understand the context ⟩

Scientists and nonscientists were asked to indicate if they agreed or disagreed with the following statement: "When something is run by the government, it is usually inefficient and wasteful." The resulting data (from **"Scientists, Public Differ in Outlooks,"** ***USA Today,*** **July 10, 2009**) were used to create the two pie charts in Figure 3.5.

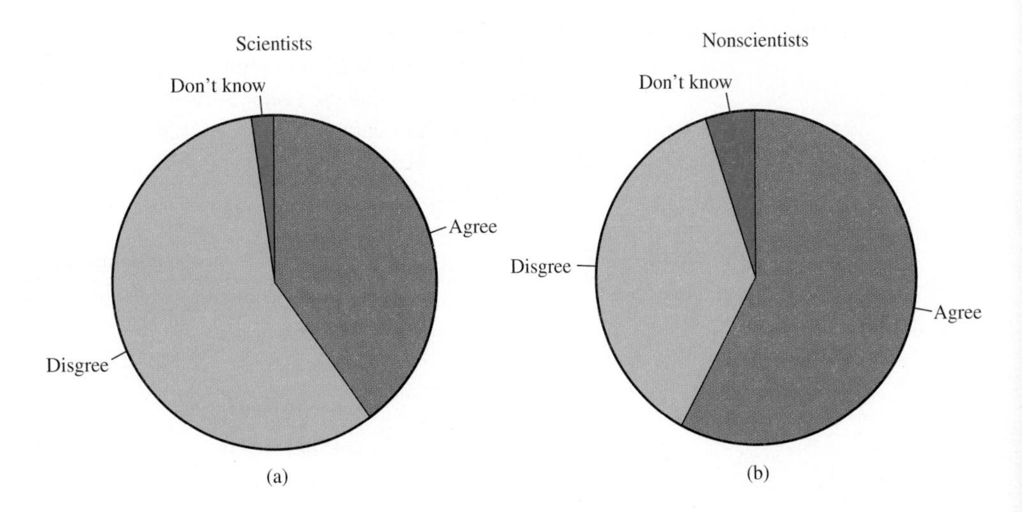

FIGURE 3.5
Pie charts for Example 3.4:
(a) scientist data;
(b) nonscientist data.

Although differences between scientists and nonscientists can be seen by comparing the pie charts of Figure 3.5, it can be difficult to compare category proportions using pie charts. A comparative bar chart (Figure 3.6) makes this type of comparison easier.

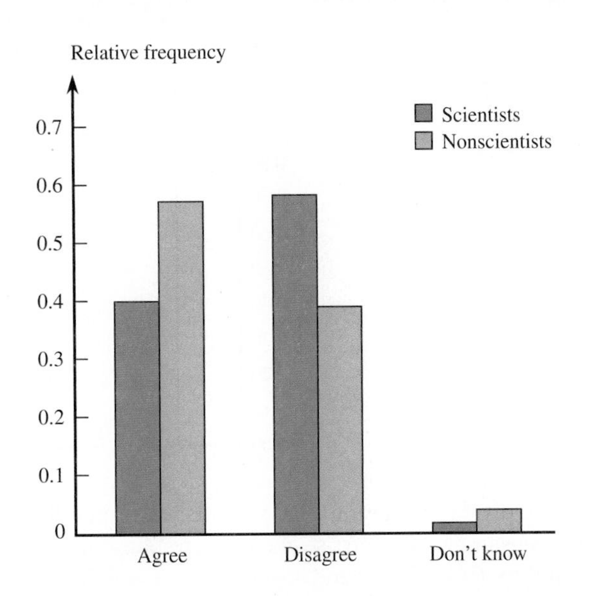

FIGURE 3.6
Comparative bar chart for the scientist and nonscientist data.

Interpret the results)

From the comparative bar chart, it is easy to see that the scientists were much less likely to agree with the statement than people who were not scientists. ■

A Different Type of "Pie" Chart: Segmented Bar Charts

A pie chart can be difficult to construct by hand, and the circular shape sometimes makes it difficult to compare areas for different categories, particularly when the relative frequencies for categories are similar. The **segmented bar chart** (also sometimes called a stacked bar chart) avoids these difficulties by using a rectangular bar rather than a circle to represent the entire data set. The bar is divided into segments, with different segments representing different categories. As with pie charts, the area of the segment for a particular category is proportional to the relative frequency for that category. Example 3.5 illustrates the construction of a segmented bar chart.

EXAMPLE 3.5 How College Seniors Spend Their Time

Understand the context)

Each year, the Higher Education Research Institute conducts a survey of college seniors. In 2008, approximately 23,000 seniors participated in the survey (**"Findings from the 2008 Administration of the College Senior Survey," Higher Education Research Institute, June 2009**). The accompanying relative frequency table summarizes student response to the question: "During the past year, how much time did you spend studying and doing homework in a typical week?"

Consider the data)

Studying/Homework	
Amount of Time	Relative Frequency
2 hours or less	.074
3 to 5 hours	.227
6 to 10 hours	.285
11 to 15 hours	.181
16 to 20 hours	.122
Over 20 hours	.111

Do the work)

To construct a segmented bar chart for these data, first draw a bar of any fixed width and length, and then add a scale that ranges from 0 to 1, as shown.

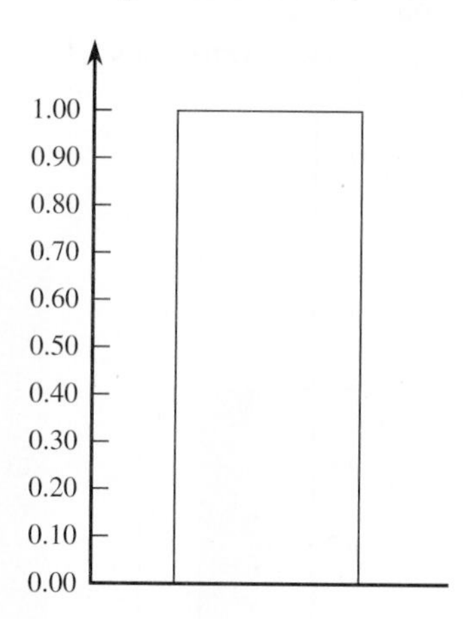

Then divide the bar into six segments, corresponding to the six possible time categories in this example. The first segment, corresponding to the 2 hours or less category, ranges from 0 to .074. The second segment, corresponding to 3 to 5 hours, ranges from .074 to .301 (for a length of .227, the relative frequency for this category), and so on. The segmented bar chart is shown in Figure 3.7.

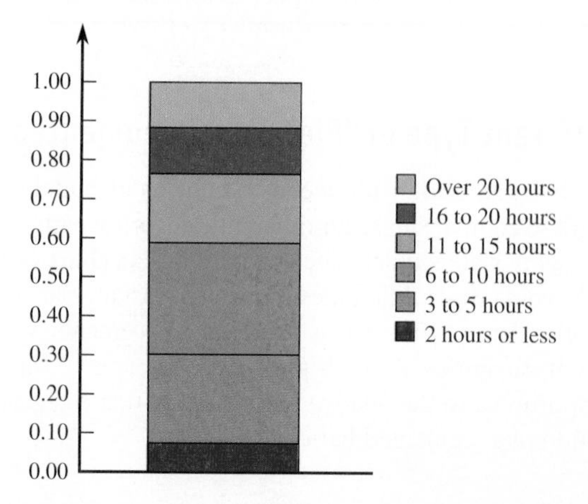

FIGURE 3.7
Segmented bar chart for the study time data of Example 3.5.

The same report also gave data on amount of time spent on exercise or sports in a typical week. Figure 3.8 shows horizontal segmented bar charts (segmented bar charts can be displayed either vertically or horizontally) for both time spent studying and time spent exercising. Viewing these graphs side by side makes it easy to see how students differ with respect to time spent on these two types of activities.

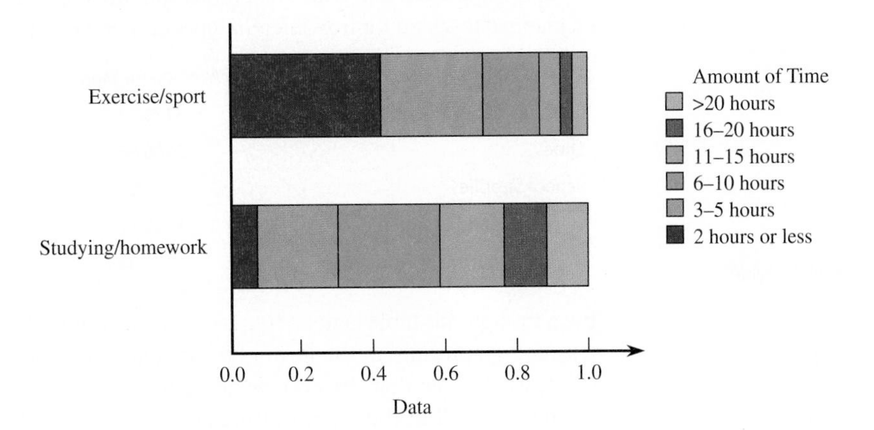

FIGURE 3.8
Segmented bar charts for time spent studying and time spent exercising.

Interpret the results)

The proportion of students who reported spending 5 hours or less per week exercising was much larger than the proportion who reported spending 5 hours or less per week studying. The proportion of students in each of the categories corresponding to 6 or more hours per week was noticeably higher for studying than for exercising. ■

Other Uses of Bar Charts and Pie Charts

As we have seen in previous examples, bar charts and pie charts can be used to summarize categorical data sets. However, they are sometimes also used for other purposes, as illustrated in Examples 3.6 and 3.7.

EXAMPLE 3.6 Grape Production

PhotoLink/Photodisc/Getty Images

● The 2012 **Grape Crush Report for California** gave the following information on grape production for each of four different types of grapes **(California Department of Food and Agriculture, March 8, 2013)**:

Type of Grape	Tons Produced
Red Wine Grapes	2,292,200
White Wine Grapes	1,725,689
Raisin Grapes	270,085
Table Grapes	99,111
Total	**4,387,085**

Raisin
Table
White wine
Red wine

FIGURE 3.9
Pie chart for grape production data.

Although this table is not a frequency distribution, it is common to represent information of this type graphically using a pie chart, as shown in Figure 3.9. The pie represents the total grape production, and the slices show the proportion of the total production for each of the four types of grapes. ■

EXAMPLE 3.7 Back-to-College Spending

The National Retail Federation's **2013 Back to College Survey (www.nrf.com)** asked each person in a sample of college students how much they planned to spend in various categories during the upcoming academic year. The average amounts of money (in dollars) that men and women planned to spend for five different types of purchases are shown in the accompanying table.

● Data set available online

Type of Purchase	Average for Men	Average for Women
Clothing and Accessories	$176.67	$160.68
Shoes	$106.66	$87.70
School Supplies	$87.95	$67.66
Electronics and Computers	$385.11	$401.22
Dorm or Apartment Furnishings	$287.04	$203.66

Even though this table is not a frequency distribution, this type of information is often represented graphically in the form of a bar chart, as illustrated in Figure 3.10. From the bar chart, we can see that the average amount of money that men and women plan to spend is similar for all of the types of purchases except for dorm and apartment furnishings, in which the average for men is noticeably higher than the average for women.

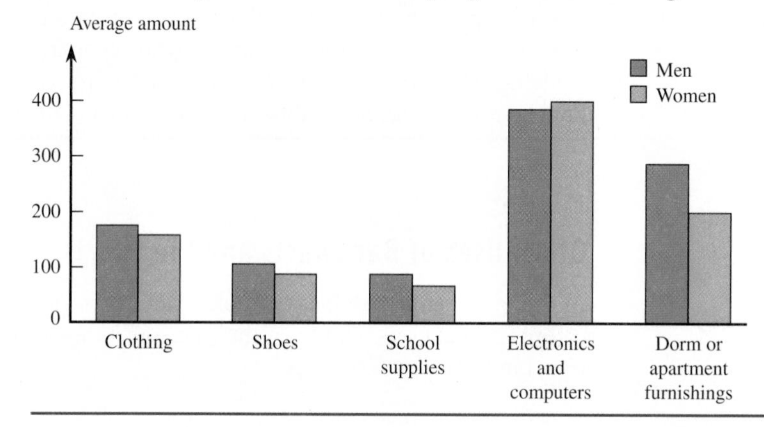

FIGURE 3.10
Comparative bar chart for the back-to-college spending data of men and women.

■

EXERCISES 3.1 - 3.14

3.1 Each person in a nationally representative sample of 1252 young adults age 23 to 28 years old was asked how they viewed their "financial physique" **("2009 Young Adults & Money Survey Findings," Charles Schwab, 2009)**. "Toned and fit" was chosen by 18% of the respondents, while 55% responded "a little bit flabby," and 27% responded "seriously out of shape." Summarize this information in a pie chart. (Hint: See Examples 3.2 and 3.3.)

3.2 The accompanying graphical display is similar to one that appeared in ***USA Today* (October 22, 2009)**. It summarizes survey responses to a question about whether visiting social networking sites is allowed at work. Which of the graph types introduced in this section is used to display the responses? (*USA Today* frequently adds artwork and text to their graphs to try to make them look more interesting.)

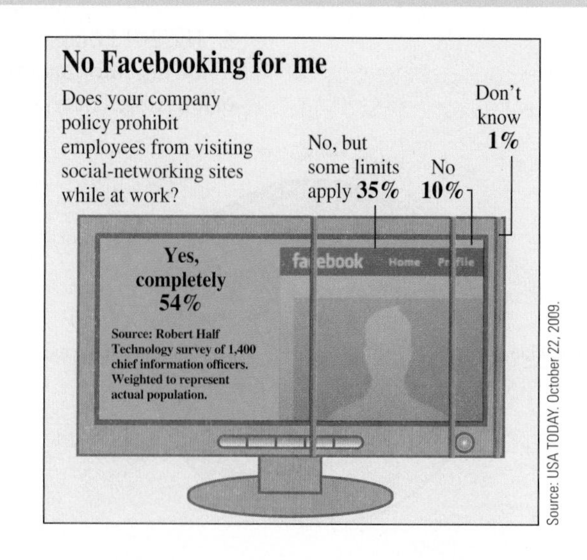

3.3 The survey referenced in the previous exercise was conducted by Robert Half Technology. This company issued a press release **("Whistle—But Don't Tweet—While You Work," www.roberthalftechnology.com, October 6, 2009)** that provided more detail than in the *USA Today* snapshot graph. The actual question asked was "Which of the following most closely describes your company's policy on visiting social networking sites, such as Facebook, MySpace and Twitter, while at work?" The responses are summarized in the following table:

Response Category	Relative Frequency (expressed as percent)
Prohibited completely	54%
Permitted for business purposes only	19%
Permitted for limited personal use	16%
Permitted for any type of personal use	10%
Don't know/no answer	1%

a. Explain how the survey response categories and corresponding relative frequencies were used or modified to produce the graphical display in the previous exercise.

b. Using the original data in the table, construct a segmented bar graph. (Hint: See Example 3.5.)

c. What are two other types of graphical displays that would be appropriate for summarizing these data?

3.4 The National Confectioners Association asked 1006 adults the following question: "Do you set aside a personal stash of Halloween candy?" Fifty-five percent of those surveyed responded no, 41% responded yes, and 4% either did not answer the question or said they did not know **(*USA Today*, October 22, 2009)**. Use the given information to construct a pie chart.

3.5 College student attitudes about e-books were investigated in a survey of 1625 students. Students were asked to indicate their level of agreement with the following statement:

"I would like to be able to get all my textbooks in digital form."

The responses are summarized in the accompanying table. **(*The Chronicle of Higher Education*, August 23, 2013)**

Response	Percentage in This Category
Strongly disagree	21.3
Disagree	29.2
Agree	25.0
Strongly agree	13.2
Don't know	11.3

a. Construct an appropriate graphical display to summarize the information given in the table.

b. Write a headline that would be appropriate for a newspaper article that summarized the results of this survey.

3.6 The Center for Science in the Public Interest evaluated school cafeterias in 20 school districts across the United States. Each district was assigned a numerical score on the basis of rigor of food codes, frequency of food safety inspections, access to inspection information, and the results of cafeteria inspections. Based on the score assigned, each district was also assigned one of four grades.

The scores and grades are summarized in the accompanying table, which appears in the report **"Making the Grade: An Analysis of Food Safety in School Cafeterias" (cspi.us/new/pdf/makingthegrade.pdf, 2007)**.

Top of the Class **Passing** **Barely Passing** **Failing**

Jurisdiction	Overall Score (out of 100)
City of Fort Worth, TX	80
King County, WA	79
City of Houston, TX	78
Maricopa County, AZ	77
City and County of Denver, CO	75
Dekalb County, GA	73
Farmington Valley Health District, CT	72
State of Virginia	72
Fulton County, GA	68
City of Dallas, TX	67
City of Philadelphia, PA	67
City of Chicago, IL	65
City and County of San Francisco, CA	64
Montgomery County, MD	63
Hillsborough County, FL	60
City of Minneapolis, MN	60
Dade County, FL	59
State of Rhode Island	54
District of Columbia	46
City of Hartford, CT	37

a. Two variables are summarized in the figure, grade and overall score. Is overall score a numerical or categorical variable? Is grade (indicated by the different colors in the figure) a numerical or categorical variable?

b. Explain how the figure is equivalent to a segmented bar graph of the grade data.

c. Construct a dotplot of the overall score data. Based on the dotplot, suggest an alternate assignment of grades (top of class, passing, etc.) to the 20 school districts. Explain the reasoning you used to make your assignment. (Hint: Dotplots were covered in Section 1.4.)

3.7 The article **"Housework around the World" (*USA Today*, September 15, 2009)** included the percentage of women who say their spouses never help with household chores for five different countries.

Country	Percentage
Japan	74
France	44
United Kingdom	40
United States	34
Canada	31

a. Display the information in the accompanying table in a bar chart.

b. The article did not state how the author arrived at the given percentages. What are two questions that you would want to ask the author about how the data used to compute the percentages were collected?

c. Assuming that the data that were used to compute these percentages were collected in a reasonable way, write a few sentences describing how the five countries differ in terms of spouses helping their wives with housework.

3.8 The report **"Findings from the 2009 Administration of the College Senior Survey" (Higher Education Research Institute, 2010)** asked a large number of college seniors how they would rate themselves compared to the average person of their age with respect to physical health. The accompanying relative frequency table summarizes the responses for men and women.

	Relative Frequency	
Rating of Physical Health	Men	Women
Highest 10%	.359	.227
Above average	.471	.536
Average	.160	.226
Below average	.009	.007
Lowest 10%	.001	.004

a. Construct a comparative bar graph of the responses that allows you to compare the responses of men and women.

b. There were 8110 men and 15,260 women who responded to the survey. Explain why it is important that the comparative bar graph be constructed using the relative frequencies rather than the actual numbers of people (the frequencies) responding in each category.

c. Write a few sentences commenting on how college seniors perceive themselves with respect to physical health and how men and women differ in their perceptions.

3.9 The article **"Rinse Out Your Mouth" (Associated Press, March 29, 2006)** summarized results from a survey of 1001 adults on the use of profanity. When asked "How many times do you use swear words in conversations?" 46% responded a few or more times per week, 32% responded a few times a month or less, and 21% responded never. Use the given information to construct a segmented bar chart.

3.10 The survey on student attitude toward e-books described in Exercise 3.5 was conducted in 2011. A similar survey was also conducted in 2012 **(*The Chronicle of Higher Education*, August 23, 2013)**. Data from 1588 students who participated in the 2012 survey are summarized in the accompanying table.

Response	Percentage in This Category
Strongly disagree	19.1
Disagree	27.5
Agree	26.3
Strongly agree	16.0
Don't know	11.1

a. Use these data and the data from Exercise 3.5 to construct a comparative bar chart that shows the distribution of responses for the two years. (Hint: See Example 3.1.)

b. Based on your graph from part (a), do you think there was much of a change in attitude toward e-books from 2011 to 2012?

3.11 ▼ Poor fitness in adolescents and adults increases the risk of cardiovascular disease. In a study of 3110 adolescents and 2205 adults **(*Journal of the American Medical Association*, December 21, 2005)**, researchers found 33.6% of adolescents and 13.9% of adults were unfit; the percentage was similar in adolescent males (32.9%) and females (34.4%), but was higher in adult females (16.2%) than in adult males (11.8%).

a. Summarize this information using a comparative bar graph that shows differences between males and females within the two different age groups.

b. Comment on the interesting features of your graphical display.

3.12 A survey of 1001 adults taken by Associated Press–Ipsos asked "How accurate are the weather forecasts in your area?" **(*San Luis Obispo Tribune*, June 15, 2005)**. The responses are summarized in the table below.

Extremely	4%
Very	27%
Somewhat	53%
Not too	11%
Not at all	4%
Not sure	1%

a. Construct a pie chart to summarize these data.

b. Construct a bar chart to summarize these data.

c. Which of these charts—a pie chart or a bar chart—best summarizes the important information? Explain.

3.13 An article about college loans **("New Rules Would Protect Students," *USA Today*, June 16, 2010)** reported the percentage of students who had defaulted on a student loan within 3 years of when they were scheduled to begin repayment. Information was given for public colleges, private non-profit colleges, and for-profit colleges.

	Relative Frequency		
Loan Status	Public Colleges	Private Non-profit Colleges	For-Profit Colleges
Good Standing	.928	.953	.833
In Default	.072	.047	.167

a. Construct a comparative bar chart that would allow you to compare loan status for the three types of colleges.

b. The article states "those who attended for-profit schools were more likely to default than those who attended public or private non-profit schools." What aspect of the comparative bar chart supports this statement?

3.14 The article **"Fraud, Identity Theft Afflict Consumers" (*San Luis Obispo Tribune*, February 2, 2005)** included the accompanying breakdown of identity theft complaints by type.

Type of Complaint	Percent of All Complaints
Credit card fraud	28
Phone or utilities fraud	19
Bank fraud	18
Employment fraud	13
Other	22

Construct a bar chart for these data and write a sentence or two commenting on the most common types of identity theft complaints.

Bold exercises answered in back ● Data set available online ▼ Video Solution available

3.2 Displaying Numerical Data: Stem-and-Leaf Displays

A stem-and-leaf display is an effective and compact way to summarize numerical data. Each number in the data set is broken into two pieces, referred to as a stem and a leaf. The **stem** is the first part of the number and consists of the beginning digit(s). The **leaf** is the last part of the number and consists of the final digit(s). For example, the number 213 might be split into a stem of 2 and a leaf of 13 or a stem of 21 and a leaf of 3. The resulting stems and leaves are then used to construct the display.

EXAMPLE 3.8 Going Wireless

Step-by-step technology instructions available online

Understand the context ❯

● The article **"Going Wireless" (*AARP Bulletin*, June 2009)** reported the estimated percentage of U.S. households with only wireless phone service (no land line) for the 50 states and the District of Columbia. Data for the 19 Eastern states are given here.

Consider the data ❯

State	Wireless %	State	Wireless %	State	Wireless %
CN	5.6	MA	9.3	RI	7.8
DE	5.7	NH	11.6	SC	20.6
DC	20.0	NJ	8.0	VA	10.8
FL	16.8	NY	11.4	VT	5.1
GA	16.5	NC	16.3	WV	11.6
ME	13.4	OH	14.0		
MD	10.8	PA	10.8		

● Data set available online

0	5.6, 5.7, 9.3, 8.0, 7.8, 5.1	
1	6.8, 6.5, 3.4, 0.8, 1.6, 1.4, 6.3, 4.0, 0.8, 0.8, 1.6	Stem: Tens
2	0.0, 0.6	Leaf: Ones

FIGURE 3.11
Stem-and-leaf display of wireless percentage for Eastern states.

Figure 3.11 shows a stem-and-leaf display for the wireless percentage data.

The numbers in the vertical column on the left of the display are the **stems**. Each number to the right of the vertical line is a **leaf** corresponding to one of the observations in the data set. The legend

Stem: Tens
Leaf: Ones

tells us that the observation that had a stem of 0 and a leaf of 5.6 corresponds to a wireless percentage of 05.6 or 5.6 (as opposed to 56 or 0.56). Similarly, the observation with the stem of 1 and leaf of 6.8 corresponds to a wireless percentage of 16.8.

Interpret the results

This display shows that for many Eastern states, the percentage of households with only wireless phone service was in the 10 to 20% range. Two states had percentages of at least 20%. ■

The leaves on each line of the display in Figure 3.11 have not been arranged in order from smallest to largest. Most statistical software packages order the leaves this way, but it is not necessary to do so to get an informative display that still shows many of the important characteristics of the data set, such as shape and spread.

Stem-and-leaf displays can be useful to get a sense of a typical value for the data set, as well as a sense of how spread out the values in the data set are. It is also easy to spot data values that are unusually far from the rest of the values in the data set. Such values are called outliers. The stem-and-leaf display of the wireless percentage data (Figure 3.11) does not show any outliers.

DEFINITION

Outlier: An unusually small or large data value. A precise rule for deciding when an observation is an outlier is given in Chapter 4.

Stem-and-Leaf Displays

When to Use Numerical data sets with a small to moderate number of observations (does not work well for very large data sets)

How to Construct

1. Select one or more leading digits for the stem values. The trailing digits (or sometimes just the first one of the trailing digits) become the leaves.
2. List possible stem values in a vertical column.
3. Record the leaf for every observation beside the corresponding stem value.
4. Indicate the units for stems and leaves someplace in the display.

What to Look For The display conveys information about

- a representative or typical value in the data set
- the extent of spread about a typical value
- the presence of gaps and outliers in the data
- the extent of symmetry in the distribution of values
- the number and location of peaks

EXAMPLE 3.9 Tuition at Public Universities

Understand the context ❯

● The introduction to this chapter gave data on average in-state tuition and fees at public institutions in the year 2012 for the 50 U.S. states. The observations ranged from a low value of $3501 to a high value of $13,347. The data are reproduced here:

Consider the data ❯

7502	5957	9021	6367	8907	7167	9069	10,524	4032	6015
7422	5674	11,252	7940	7563	6689	7943	5198	9278	7831
10,104	10,527	9862	5674	7588	6007	6752	4509	13,347	11,596
5293	6192	5701	6414	8800	5538	7975	11,818	9926	10,372
6939	7013	7116	5163	13,078	9373	7701	5241	7851	3501

Do the work ❯

Notice that some of the data values are only four digits (such as 7502) and others are five digits (such as 10,104). It is easiest to proceed if we first make all of the data values five-digit numbers by adding leading zeros as shown:

07,502	05,957	09,021	06,367	08,907	07,167	09,069	10,524	04,032	06,015
07,422	05,674	11,252	07,940	07,563	06,689	07,943	05,198	09,278	07,831
10,104	10,527	09,862	05,674	07,588	06,007	06,752	04,509	13,347	11,596
05,293	06,192	05,701	06,414	08,800	05,538	07,975	11,818	09,926	10,372
06,939	07,013	07,116	05,163	13,078	09,373	07,701	05,241	07,851	03,501

A natural choice for the stem is the leading two digits. This would result in a display with 11 stems (03, 04, 05, 06, 07, 08, 09, 10, 11, 12, and 13). Using the first three digits of a number of the stem would result in 99 stems (035 to 133). A stem-and-leaf display with 99 stems would not be an effective summary of the data. In general, stem-and-leaf displays that use between 5 and 30 stems tend to work well.

If we choose the first two digits as the stem, the remaining three digits (the hundreds, tens, and ones) would form the leaf. For example, for the first few values in the data set, we would have

07,502 → stem = 07, leaf = 502
05,957 → stem = 05, leaf = 957
09,021 → stem = 09, leaf = 021

The leaves have been entered in the display of Figure 3.12 in the order they are encountered in the data set. Commas are used to separate the leaves only when each leaf has two or more digits. Figure 3.12 shows that most states had average in-state tuition and fees in the $5000 to $8000 range and that the typical average tuition and fees is around $7500. Two states have average in-state tuition and fees at public four-year institutions that are quite a bit higher than most other states (the two states with the highest values were Vermont and New Hampshire).

Interpret the results ❯

03	501
04	032, 509
05	957, 674, 198, 674, 293, 701, 538, 163, 241
06	367, 015, 689, 007, 752, 192, 414, 939
07	502, 167, 422, 940, 563, 943, 831, 588, 975, 013, 116, 701, 851
08	907, 800
09	021, 069, 278, 862, 926, 373
10	524, 104, 527, 372
11	252, 596, 818
12	
13	347, 078

Stem: Thousands
Leaf: Ones

FIGURE 3.12
Stem-and-leaf display of average tuition and fees.

● Data set available online

An alternative display (Figure 3.13) results from dropping all but the first digit of the leaf. This is what most statistical computer packages do when generating a display. Little information about typical value, spread, or shape is lost in this truncation and the display is simpler and more compact.

```
03 | 5
04 | 05
05 | 961627512
06 | 30607149
07 | 5149598590178
08 | 98
09 | 002893
10 | 5153
11 | 258
12 |          Stem: Thousands
13 | 30       Leaf: Hundreds
```

FIGURE 3.13
Stem-and-leaf display of average in-state tuition and fees using truncated leaves.

■

Repeated Stems to Stretch a Display

Sometimes a natural choice of stems gives a display in which too many observations are concentrated on just a few stems. A more informative picture can be obtained by dividing the leaves at any given stem into two groups: those that begin with 0, 1, 2, 3, or 4 (the "low" leaves) and those that begin with 5, 6, 7, 8, or 9 (the "high" leaves). Then each stem value is listed twice when constructing the display, once for the low leaves and once again for the high leaves. It is also possible to repeat a stem more than twice. For example, each stem might be repeated five times, once for each of the leaf groupings {0, 1}, {2, 3}, {4, 5}, {6, 7}, and {8, 9}.

EXAMPLE 3.10 Median Ages in 2030

Understand the context ⟩

● The accompanying data on the Census Bureau's projected median age in 2030 for the 50 U.S. states and Washington, D.C. appeared in the article **"2030 Forecast: Mostly Gray"** (***USA Today*, April 21, 2005**). The median age for a state is the age that divides the state's residents so that half are younger than the median age and half are older than the median age.

Consider the data ⟩

Projected Median Age

41.0	32.9	39.3	29.3	37.4	35.6	41.1	43.6	33.7	45.4	35.6	38.7
39.2	37.8	37.7	42.0	39.1	40.0	38.8	46.9	37.5	40.2	40.2	39.0
41.1	39.6	46.0	38.4	39.4	42.1	40.8	44.8	39.9	36.8	43.2	40.2
37.9	39.1	42.1	40.7	41.3	41.5	38.3	34.6	30.4	43.9	37.8	38.5
46.7	41.6	46.4									

The ages in the data set range from 29.3 to 46.9. Using the first two digits of each data value for the stem results in a large number of stems, while using only the first digit results in a stem-and-leaf display with only three stems.

Do the work ⟩

The stem-and-leaf display using single digit stems and leaves truncated to a single digit is shown in Figure 3.14. A stem-and-leaf display that uses repeated stems is shown in Figure 3.15. Here each stem is listed twice, once for the low leaves (those beginning with 0, 1, 2, 3, 4) and once for the high leaves (those beginning with 5, 6, 7, 8, 9). This display is more informative than the one in Figure 3.14, but is much more compact than a display based on two-digit stems.

```
2 | 9
3 | 02345567777778888899999999
4 | 000000111111222333456666   Stem: Tens
                               Leaf: Ones
```

FIGURE 3.14
Stem-and-leaf display for the projected median age data.

● Data set available online

2H	9
3L	0234
3H	556777777888889999999
4L	00000011111112223334
4H	56666

Stem: Tens
Leaf: Ones

FIGURE 3.15
Stem-and-leaf display for the projected median age data using repeated stems.

Comparative Stem-and-Leaf Displays

Frequently an analyst wishes to see whether two groups of data differ in some fundamental way. A comparative stem-and-leaf display, in which the leaves for one group are listed to the right of the stem values and the leaves for the second group are listed to the left, can provide preliminary visual impressions and insights.

EXAMPLE 3.11 Progress for Children

Understand the context ⟩

● The report **"Progress for Children" (UNICEF, April 2005)** included the accompanying data on the percentage of primary-school-age children who were enrolled in school for 19 countries in Northern Africa and for 23 countries in Central Africa.

Consider the data ⟩

Northern Africa

54.6 34.3 48.9 77.8 59.6 88.5 97.4 92.5 83.9 96.9 88.9
98.8 91.6 97.8 96.1 92.2 94.9 98.6 86.6

Central Africa

58.3 34.6 35.5 45.4 38.6 63.8 53.9 61.9 69.9 43.0 85.0
63.4 58.4 61.9 40.9 73.9 34.8 74.4 97.4 61.0 66.7 79.6
98.9

Formulate a plan ⟩

We will construct a comparative stem-and-leaf display using the first digit of each observation as the stem and the remaining two digits as the leaf. To keep the display simple the leaves will be truncated to one digit. For example, the observation 54.6 would be processed as

54.6 → stem = 5, leaf = 4 (truncated from 4.6)

and the observation 34.3 would be processed as

Do the work ⟩

34.3 → stem = 3, leaf = 4 (truncated from 4.3)

The resulting comparative stem-and-leaf display is shown in Figure 3.16.

Central Africa		Northern Africa
4854	3	4
035	4	8
838	5	49
6113913	6	
943	7	76
5	8	8386
87	9	7268176248

Stem: Tens
Leaf: Ones

FIGURE 3.16
Comparative stem-and-leaf display for percentage of children enrolled in primary school.

Interpret the results ⟩

From the comparative stem-and-leaf display we can see that there is quite a bit of variability in the percentage enrolled in school for both Northern and Central African countries and that the shapes of the two data distributions are quite different. The percentage enrolled in school tends to be higher in Northern African countries than in Central African countries, although the smallest value in each of the two data sets is about the same. For Northern African countries the distribution of values has a single peak in the 90s with the number of observations declining as we move toward the stems corresponding to lower percentages enrolled in school. For Central African countries the distribution is more symmetric, with a typical value in the mid 60s. ■

● Data set available online

EXERCISES 3.15 - 3.21

3.15 ● The **U.S. Department of Health and Human Services** provided the data in the accompanying table in the report **"Births: Preliminary Data for 2007" (*National Vital Statistics Reports,* March 18, 2009)**. Entries in the table are the birth rates (births per 1,000 of population) for the year 2007.

State	*Births	State	*Births
Alabama	14.0	Missouri	13.9
Alaska	16.2	Montana	13.0
Arizona	16.2	Nebraska	15.2
Arkansas	14.6	Nevada	16.1
California	15.5	New Hampshire	10.8
Colorado	14.6	New Jersey	13.4
Connecticut	11.9	New Mexico	15.5
Delaware	14.1	New York	13.1
District of Columbia	15.1	North Carolina	14.5
Florida	13.1	North Dakota	13.8
Georgia	15.9	Ohio	13.2
Hawaii	14.9	Oklahoma	15.2
Idaho	16.7	Oregon	13.2
Illinois	14.1	Pennsylvania	12.1
Indiana	14.2	Rhode Island	11.7
Iowa	13.7	South Carolina	14.3
Kansas	15.1	South Dakota	15.4
Kentucky	14.0	Tennessee	14.1
Louisiana	15.4	Texas	17.1
Maine	10.7	Utah	20.8
Maryland	13.9	Vermont	10.5
Massachusetts	12.1	Virginia	14.1
Michigan	12.4	Washington	13.8
Minnesota	14.2	West Virginia	12.1
Mississippi	15.9	Wisconsin	13.0
		Wyoming	15.1

*Births per 1000 of population

Construct a stem-and-leaf display using stems 10, 11 . . . 20. Comment on the interesting features of the display. (Hint: See Example 3.9.)

3.16 ● ▼ The **National Survey on Drug Use and Health**, conducted in 2006 and 2007 by the Office of Applied Studies, led to the following state estimates of the total number of people ages 12 and older who had used a tobacco product within the last month.

State	Number of People (in thousands)	State	Number of People (in thousands)
Alabama	1,307	Montana	246
Alaska	161	Nebraska	429
Arizona	1,452	Nevada	612
Arkansas	819	New Hampshire	301
California	6,751	New Jersey	1,870
Colorado	1,171	New Mexico	452
Connecticut	766	New York	4,107
Delaware	200	North Carolina	2,263
District of Columbia	141	North Dakota	162
Florida	4,392	Ohio	3,256
Georgia	2,341	Oklahoma	1,057
Hawaii	239	Oregon	857
Idaho	305	Pennsylvania	3,170
Illinois	3,149	Rhode Island	268
Indiana	1,740	South Carolina	1,201
Iowa	755	South Dakota	202
Kansas	726	Tennessee	1,795
Kentucky	1,294	Texas	5,533
Louisiana	1,138	Utah	402
Maine	347	Vermont	158
Maryland	1,206	Virginia	1,771
Massachusetts	1,427	Washington	1,436
Michigan	2,561	West Virginia	582
Minnesota	1,324	Wisconsin	1,504
Mississippi	763	Wyoming	157
Missouri	1,627		

a. Construct a stem-and-leaf display using thousands (of thousands) as the stems and truncating the leaves to the tens (of thousands) digit.

b. Write a few sentences describing the shape of the distribution and any unusual observations.

c. The four largest values were for California, Texas, Florida, and New York. Does this indicate that tobacco use is more of a problem in these states than elsewhere? Explain.

d. If you wanted to compare states on the basis of the extent of tobacco use, would you use the data in the given table? If yes, explain why this would be reasonable. If no, what would you use instead as the basis for the comparison?

3.17 ● The article **"Going Wireless" (*AARP Bulletin*, June 2009)** reported the estimated percentage of households with only wireless phone service (no land line) for the 50 U.S. states and the District of Columbia. In the accompanying data table, each state was also classified into one of three geographical regions—West (W), Middle states (M), and East (E).

Wireless %	Region	State	Wireless %	Region	State
13.9	M	AL	9.2	W	MT
11.7	W	AK	23.2	M	NE
18.9	W	AZ	10.8	W	NV
22.6	M	AR	16.9	M	ND
9.0	W	CA	11.6	E	NH
16.7	W	CO	8.0	E	NJ
5.6	E	CN	21.1	W	NM
5.7	E	DE	11.4	E	NY
20.0	E	DC	16.3	E	NC
16.8	E	FL	14.0	E	OH
16.5	E	GA	23.2	M	OK
8.0	W	HI	17.7	W	OR
22.1	W	ID	10.8	E	PA
16.5	M	IL	7.9	E	RI
13.8	M	IN	20.6	E	SC
22.2	M	IA	6.4	M	SD
16.8	M	KA	20.3	M	TN
21.4	M	KY	20.9	M	TX
15.0	M	LA	25.5	W	UT
13.4	E	ME	10.8	E	VA
10.8	E	MD	5.1	E	VT
9.3	E	MA	16.3	W	WA
16.3	M	MI	11.6	E	WV
17.4	M	MN	15.2	M	WI
19.1	M	MS	11.4	W	WY
9.9	M	MO			

a. Construct a stem-and-leaf display for the wireless percentage using the data from all 50 states and the District of Columbia. What is a typical value for this data set?

b. Construct a back-to-back stem-and-leaf display for the wireless percentage of the states in the West and the states in the East. How do the distributions of wireless percentages compare for states in the East and states in the West? (Hint: See Example 3.11.)

3.18 The article **"Economy Low, Generosity High" (*USA Today*, July 28, 2009)** noted that despite a weak economy in 2008, more Americans volunteered in their communities than in previous years. Based on census data (www.volunteeringinamerica.gov), the top and bottom five states in terms of percentage of the population who volunteered in 2008 were identified. The top five states were Utah (43.5%), Nebraska (38.9%), Minnesota (38.4%), Alaska (38.0%), and Iowa (37.1%). The bottom five states were New York (18.5%), Nevada (18.8%), Florida (19.6%), Louisiana (20.1%), and Mississippi (20.9%).

a. For the data set that includes the percentage who volunteered in 2008 for each of the 50 states, what is the largest value? What is the smallest value?

b. If you were going to construct a stem-and-leaf display for the data set consisting of the percentage who volunteered in 2008 for the 50 states, what stems would you use to construct the display? Explain your choice.

3.19 The U.S. gasoline tax per gallon data for each of the 50 states and the District of Columbia were listed in the article **"Paying at the Pump" (*AARP Bulletin*, June 2010)**.

State	Gasoline tax*	State	Gasoline tax*
Alabama	20.9	Michigan	35.8
Alaska	8.0	Minnesota	27.2
Arizona	19.0	Mississippi	18.8
Arkansas	21.8	Missouri	17.3
California	48.6	Montana	27.8
Colorado	22.0	Nebraska	27.7
Connecticut	42.6	Nevada	33.1
Delaware	23.0	New Hampshire	19.6
District of Columbia	23.5	New Jersey	14.5
Florida	34.4	New Mexico	18.8
Georgia	20.9	New York	44.9
Hawaii	45.1	North Carolina	30.2
Idaho	25.0	North Dakota	23.0
Illinois	40.4	Ohio	28.0
Indiana	34.8	Oklahoma	17.0
Iowa	22.0	Oregon	25.0
Kansas	25.0	Pennsylvania	32.3
Kentucky	22.5	Rhode Island	33.0
Louisiana	20.0	South Carolina	16.8
Maine	31.0	South Dakota	24.0
Maryland	23.5	Tennessee	21.4
Massachusetts	23.5	Texas	20.0

continued

State	Gasoline tax*	State	Gasoline tax*
Utah	24.5	West Virginia	32.2
Vermont	24.7	Wisconsin	32.9
Virginia	19.6	Wyoming	14.0
Washington	37.5		

*Cents per gallon

a. Construct a stem-and-leaf display of these data.

b. Based on the stem-and-leaf display, what do you notice about the center and spread of the data distribution?

c. Do any values in the data set stand out as unusual? If so, which states correspond to the unusual observations, and how do these values differ from the rest?

3.20 ● ▼ A report from **Texas Transportation Institute (Texas A&M University System, 2005)** titled **"Congestion Reduction Strategies"** included the accompanying data on extra travel time for peak travel time in hours per year per traveler for different-sized urban areas.

Very Large Urban Areas	Extra Hours per Year per Traveler
Los Angeles, CA	93
San Francisco, CA	72
Washington DC, VA, MD	69
Atlanta, GA	67
Houston, TX	63
Dallas, Fort Worth, TX	60
Chicago, IL-IN	58
Detroit, MI	57
Miami, FL	51
Boston, MA, NH, RI	51
New York, NY-NJ-CT	49
Phoenix, AZ	49
Philadelphia, PA-NJ-DE-MD	38

Large Urban Areas	Extra Hours per Year per Traveler
Riverside, CA	55
Orlando, FL	55
San Jose, CA	53
San Diego, CA	52
Denver, CO	51
Baltimore, MD	50
Seattle, WA	46
Tampa, FL	46
Minneapolis, St Paul, MN	43
Sacramento, CA	40
Portland, OR, WA	39
Indianapolis, IN	38
St Louis, MO-IL	35
San Antonio, TX	33
Providence, RI, MA	33
Las Vegas, NV	30
Cincinnati, OH-KY-IN	30
Columbus, OH	29
Virginia Beach, VA	26
Milwaukee, WI	23
New Orleans, LA	18
Kansas City, MO-KS	17
Pittsburgh, PA	14
Buffalo, NY	13
Oklahoma City, OK	12
Cleveland, OH	10

continued

a. Construct a comparative stem-and-leaf plot for extra travel time per traveler for each of the two different sizes of urban areas.

b. Is the following statement consistent with the display constructed in Part (a)? Explain.

The larger the urban area, the greater the extra travel time during peak period travel.

3.21 ● The percentage of teens not in school or working in 2010 for the 50 states were given in the **2012 Kids Count Data Book (www.aecf.org)** and are shown in the following table:

State	Rate	State	Rate
Alabama	11%	Kansas	6%
Alaska	11%	Kentucky	11%
Arizona	12%	Louisiana	14%
Arkansas	12%	Maine	7%
California	8%	Maryland	8%
Colorado	7%	Massachusetts	5%
Connecticut	5%	Michigan	9%
Delaware	9%	Minnesota	5%
Florida	10%	Mississippi	13%
Georgia	12%	Missouri	9%
Hawaii	12%	Montana	9%
Idaho	11%	Nebraska	4%
Illinois	8%	Nevada	15%
Indiana	8%	New Hampshire	6%
Iowa	6%	New Jersey	8%

continued

State	Rate	State	Rate
New Mexico	12%	South Dakota	8%
New York	8%	Tennessee	10%
North Carolina	10%	Texas	9%
North Dakota	5%	Utah	9%
Ohio	8%	Vermont	4%
Oklahoma	9%	Virginia	7%
Oregon	10%	Washington	8%
Pennsylvania	7%	West Virginia	14%
Rhode Island	5%	Wisconsin	7%
South Carolina	9%	Wyoming	9%

Note that the percentages range from a low of 4% to a high of 15%. In constructing a stem-and-leaf display for these data, if we regard each percentage as a two-digit number and use the first digit for the stem, then there are only two possible stems, 0 and 1. One solution is to use repeated stems. Consider a scheme that divides the leaf range into five parts: 0 and 1, 2 and 3, 4 and 5, 6 and 7, and 8 and 9. Then, for example, stem 0 could be repeated as

0	with leaves 0 and 1
0t	with leaves 2 and 3
0f	with leaves 4 and 5
0s	with leaves 6 and 7
0*	with leaves 8 and 9

Construct a stem-and-leaf display for this data set that uses stems 0t, 0f, 0s, 0*, and 1, 1t, and 1f. Comment on the important features of the display. (Hint: See Example 3.10.)

Bold exercises answered in back ● Data set available online ▼ Video Solution available

3.3 Displaying Numerical Data: Frequency Distributions and Histograms

A stem-and-leaf display is not always an effective way to summarize data, because it can be unwieldy when the data set contains a large number of observations. Frequency distributions and histograms are displays that work well for large data sets.

Frequency Distributions and Histograms for Discrete Numerical Data

Discrete numerical data almost always result from counting. In such cases, each observation is a whole number. As in the case of categorical data, a frequency distribution for discrete numerical data lists each possible value (either individually or grouped into intervals), the associated frequency, and sometimes the corresponding relative frequency. Recall that relative frequency is calculated by dividing the frequency by the total number of observations in the data set.

EXAMPLE 3.12 Promiscuous Queen Bees

Understand the context ❯

● Queen honey bees mate shortly after they become adults. During a mating flight, the queen usually takes multiple partners, collecting sperm that she will store and use throughout the rest of her life. The authors of the paper **"The Curious Promiscuity of Queen Honey Bees" (*Annals of Zoology* [2001]: 255–265)** studied the behavior of 30 queen honey bees to learn about the length of mating flights and the number of partners a queen takes during a mating flight.

The accompanying data on number of partners were generated to be consistent with summary values and graphs given in the paper.

Consider the data ❯

Number of Partners

12	2	4	6	6	7	8	7	8	11
8	3	5	6	7	10	1	9	7	6
9	7	5	4	7	4	6	7	8	10

● Data set available online

The corresponding relative frequency distribution is given in Table 3.1. The smallest value in the data set is 1 and the largest is 12, so the possible values from 1 to 12 are listed in the table, along with the corresponding frequency and relative frequency.

TABLE 3.1 Relative Frequency Distribution for Number of Partners

Number of Partners	Frequency	Relative Frequency
1	1	.033 ← $\frac{1}{30} = .033$
2	1	.033
3	1	.033
4	3	.100
5	2	.067
6	5	.167
7	7	.233
8	4	.133
9	2	.067
10	2	.067
11	1	.033
12	1	.033
Total	30	.999 ← *Differs from 1 due to rounding*

Interpret the results ⟩

From the relative frequency distribution, we can see that five of the queen bees had six partners during their mating flight. The corresponding relative frequency, $\frac{5}{30} = .167$, tells us that the proportion of queens with six partners is .167, or equivalently 16.7% of the queens had six partners. Adding the relative frequencies for the values 10, 11, and 12 gives

$$.067 + .033 + .033 = .133$$

indicating that 13.3% of the queens had 10 or more partners.

It is possible to create a more compact frequency distribution by grouping some of the possible values into intervals. For example, we might group together 1, 2, and 3 partners to form an interval of 1–3, with a corresponding frequency of 3. The grouping of other values in a similar way results in the relative frequency distribution shown in Table 3.2.

TABLE 3.2 Relative Frequency Distribution of Number of Partners Using Intervals

Number of Partners	Frequency	Relative Frequency
1–3	3	.100
4–6	10	.333
7–9	13	.433
10–12	4	.133

■

A histogram for discrete numerical data is a graph of the frequency or relative frequency distribution, and it is similar to the bar chart for categorical data. Each frequency or relative frequency is represented by a rectangle centered over the corresponding value (or range of values) and the area of the rectangle is proportional to the corresponding frequency or relative frequency.

Histogram for Discrete Numerical Data

When to Use Discrete numerical data. Works well, even for large data sets.

How to Construct

1. Draw a horizontal scale, and mark the possible values of the variable.
2. Draw a vertical scale, and mark it with either frequency or relative frequency.
3. Above each possible value, draw a rectangle centered at that value (so that the rectangle for 1 is centered at 1, the rectangle for 5 is centered at 5, and so on). The height of each rectangle is determined by the corresponding frequency or relative frequency. Often possible values are consecutive whole numbers, in which case the base width for each rectangle is 1.

What to Look For

- Center or typical value
- Extent of spread or variability
- General shape
- Location and number of peaks
- Presence of gaps and outliers

EXAMPLE 3.13 Revisiting Promiscuous Queen Bees

The queen bee data of Example 3.12 were summarized in a frequency distribution. The corresponding histogram is shown in Figure 3.17. Note that each rectangle in the histogram is centered over the corresponding value. When relative frequency instead of frequency is used for the vertical scale, the scale on the vertical axis is different but all essential characteristics of the graph (shape, center, spread) are unchanged.

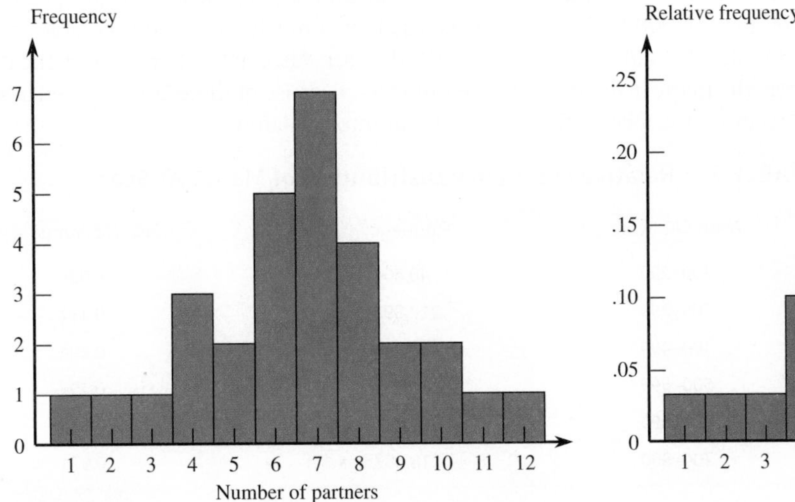

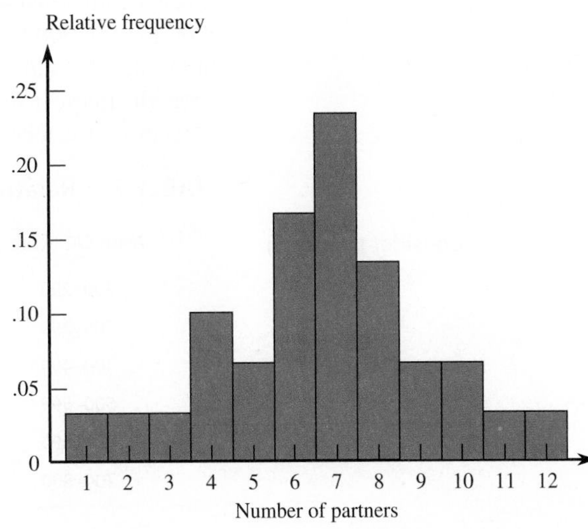

FIGURE 3.17
Histogram and relative frequency histogram of queen bee data.

A histogram based on the grouped frequency distribution of Table 3.2 can be constructed in a similar fashion, and is shown in Figure 3.18. A rectangle represents the frequency or relative frequency for each interval. For the interval 1–3, the rectangle extends from .5 to 3.5 so that there are no gaps between the rectangles of the histogram.

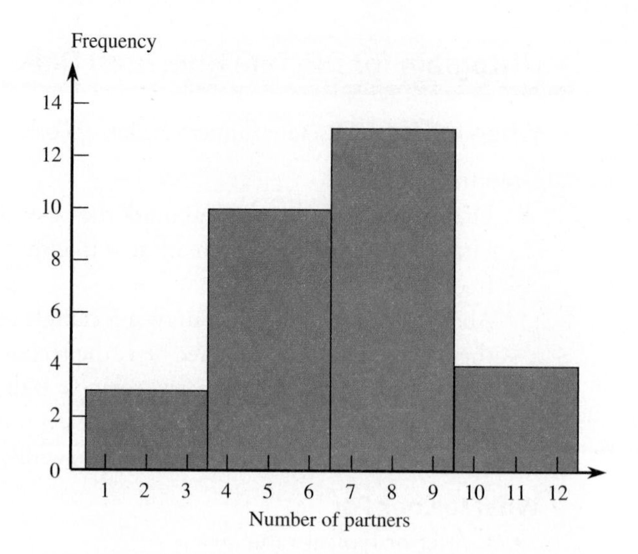

FIGURE 3.18
Histogram of queen bee data using intervals.

Sometimes a discrete numerical data set contains a large number of possible values and perhaps also has a few large or small values that are far away from most of the data. In this case, rather than forming a frequency distribution with a very long list of possible values, it is common to group the observed values into intervals or ranges. This is illustrated in Example 3.14.

EXAMPLE 3.14 Math SAT Score Distribution

Understand the context ❯

Each of the 1,664,479 students who took the math portion of the SAT exam in 2012 received a score between 200 and 800. The score distribution was summarized in a frequency distribution table that appeared in the **College Board** report titled **"2012 College Bound Seniors."** A relative frequency distribution is given in Table 3.3 and the corresponding relative frequency histogram is shown in Figure 3.19. Notice that rather than list each possible individual score value between 200 and 800, the scores are grouped into intervals (200 to 299, 300 to 399, etc.). This results in a much more compact table that still communicates the important features of the data set. Also, notice that because the data set is so large, the frequencies are also large numbers. Because of these large frequencies, it is easier to focus on the relative frequencies in our interpretation.

TABLE 3.3 Relative Frequency Distribution of Math SAT Score

Consider the data ❯

Math SAT Score	Frequency	Relative Frequency
200–299	40,664	0.024
300–399	217,896	0.131
400–499	490,043	0.294
500–599	490,118	0.294
600–699	307,076	0.184
700–800	118,682	0.071

Interpret the results ❯

From the relative frequency distribution and histogram, we can see that while there is a lot of variability in individual math SAT scores, the majority were in the 400 to 600 range and a typical value for math SAT looks to be something in the high 400s.

Before leaving this example, take a second look at the relative frequency histogram of Figure 3.19. Notice that there is one rectangle for each score interval in the relative frequency distribution. For simplicity we have chosen to treat the very last interval, 700 to 800, as if it were 700 to 799 so that all of the score ranges in the frequency distribution

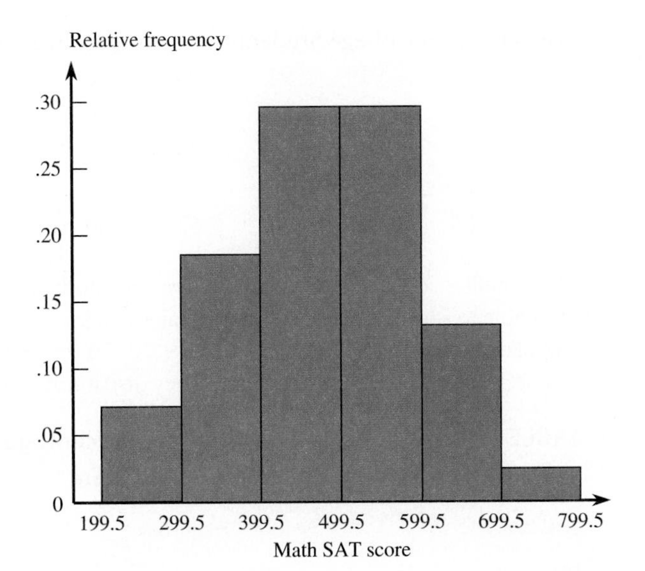

FIGURE 3.19
Relative frequency histogram for the math SAT data.

are the same width. Also note that the rectangle representing the score range 400 to 499 actually extends from 399.5 to 499.5 on the score scale. This is similar to what happens in histograms for discrete numerical data where there is no grouping. For example, in Figure 3.17 the rectangle representing 2 is centered at 2 but extends from 1.5 to 2.5 on the number of partners scale. ■

Frequency Distributions and Histograms for Continuous Numerical Data

The difficulty in constructing tabular or graphical displays with continuous data, such as observations on reaction time (in seconds) or weight of airline passenger carry-on luggage (in pounds), is that there are no natural categories. The way out of this dilemma is to define our own categories. For carry-on luggage weight, we might expect weights up to about 30 pounds. One way to group the weights into 5-pound intervals is shown in Figure 3.20. Then each observed data value could be classified into one of these intervals. The intervals used are sometimes called **class intervals**. The class intervals play the same role that the categories or individual values played in frequency distributions for categorical or discrete numerical data.

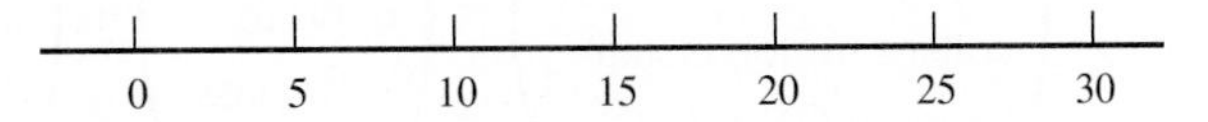

FIGURE 3.20
Suitable class intervals for carry-on luggage weight data.

There is one further difficulty we need to address. Where should we place an observation such as 20, which falls on a boundary between classes? Our convention is to define intervals so that such an observation is placed in the upper rather than the lower class interval. Thus, in a frequency distribution, one class might be 15 to <20, where the symbol < is a substitute for the phrase *less than*. This class interval will contain all observations that are greater than or equal to 15 and less than 20. The observation 20 would then fall in the class interval 20 to <25.

EXAMPLE 3.15 Going Away to College

Understand the context)

● States differ widely in the percentage of college students who attend college in their home state. The percentages of freshmen who attended college in their home state for each of the 50 states are shown here (***The Chronicle of Higher Education*, August 23, 2013**). The data have been ordered from largest to smallest.

● Data set available online

Consider the data)

Percentage of College Students Attending College in Home State

93	92	91	91	90	90	90	90	89	89
89	89	89	89	89	88	87	87	85	85
85	85	84	84	83	81	81	81	80	78
77	77	76	76	76	76	72	72	70	68
67	65	65	64	62	60	58	57	57	50

The smallest observation is 50% (Vermont) and the largest is 93% (Mississippi). It is reasonable to start the first class interval at 40 and let each interval have a width of 10. This gives class intervals of 50 to <60, 60 to <70, 70 to <80, 80 to <90, and 90 to <100.

Table 3.4 displays the resulting frequency distribution, along with the relative frequencies.

TABLE 3.4 Frequency Distribution for Percentage of College Students Attending College in Home State

Class Interval	Frequency	Relative Frequency
50 to <60	4	.08
60 to <70	7	.14
70 to <80	10	0.2
80 to <90	21	.42
90 to <100	8	.16
	50	1.00

Various relative frequencies can be combined to yield other interesting information. For example,

$$\begin{pmatrix}\text{proportion of states with}\\ \text{percent attending college}\\ \text{in home state less than 70}\end{pmatrix} = \begin{pmatrix}\text{proportion in 50}\\ \text{to 60 class}\end{pmatrix} + \begin{pmatrix}\text{proportion in 60}\\ \text{to 70 class}\end{pmatrix}$$

$$= .08 + .14 = .22 \ (22\%)$$

and

$$\begin{pmatrix}\text{proportion of states}\\ \text{with percent attending}\\ \text{college in home state}\\ \text{between 60 and 90}\end{pmatrix} = \begin{pmatrix}\text{proportion}\\ \text{in 60 to}\\ \text{<70 class}\end{pmatrix} + \begin{pmatrix}\text{proportion}\\ \text{in 70 to}\\ \text{<80 class}\end{pmatrix} + \begin{pmatrix}\text{proportion}\\ \text{in 80 to}\\ \text{< 90 class}\end{pmatrix}$$

$$= .14 + .20 + .42 = .76 \ (76\%)$$

■

There are no set rules for selecting either the number of class intervals or the length of the intervals. Using a few relatively wide intervals will bunch the data, whereas using a great many relatively narrow intervals may spread the data over too many intervals, so that no interval contains more than a few observations. Neither type of distribution will give an informative picture of how values are distributed over the range of measurement, and interesting features of the data set may be missed. In general, with a small amount of data, relatively few intervals, perhaps between 5 and 10, should be used. With a large amount of data, a distribution based on 15 to 20 (or even more) intervals is often recommended. The quantity

$$\sqrt{\text{number of observations}}$$

is often used as an estimate of an appropriate number of intervals: 5 intervals for 25 observations, 10 intervals when the number of observations is 100, and so on.

Histograms for Continuous Numerical Data

When the class intervals in a frequency distribution are all of equal width, it is easy to construct a histogram using the information in a frequency distribution.

Histogram for Continuous Numerical Data When the Class Interval Widths Are Equal

When to Use Continuous numerical data. Works well, even for large data sets.

How to Construct

1. Mark the boundaries of the class intervals on a horizontal axis.
2. Use either frequency or relative frequency on the vertical axis.
3. Draw a rectangle for each interval directly above the corresponding interval (so that the edges are at the class interval boundaries). The height of each rectangle is the corresponding frequency or relative frequency.

What to Look For

- Center or typical value
- Extent of spread, variability
- General shape
- Location and number of peaks
- Presence of gaps and outliers

Two people making reasonable and similar choices for the number of intervals, their width, and the starting point of the first interval will usually obtain histograms that are similar in terms of shape, center, and spread.

Step-by-step technology instructions available online

EXAMPLE 3.16 TV Viewing Habits of Children

Understand the data)

The article **"Early Television Exposure and Subsequent Attention Problems in Children"** (***Pediatrics*, April 2004**) investigated the television viewing habits of children in the United States. Table 3.5 gives approximate relative frequencies (read from graphs that appeared in the article) for the number of hours spent watching TV per day for a sample of children at age 1 year and a sample of children at age 3 years. The data summarized in the article were obtained as part of a large-scale national survey.

TABLE 3.5 Relative Frequency Distribution for Number of Hours Spent Watching TV per Day

Consider the data)

TV Hours per Day	Age 1 Year Relative Frequency	Age 3 Years Relative Frequency
0 to <2	.270	.630
2 to <4	.390	.195
4 to <6	.190	.100
6 to <8	.085	.025
8 to <10	.030	.020
10 to <12	.020	.015
12 to <14	.010	.010
14 to <16	.005	.005

Figure 3.21(a) is the relative frequency histogram for the 1-year-old children and Figure 3.21(b) is the relative frequency histogram for 3-year-old children. Because we would like to compare the distributions of TV hours for 1-year-old and 3-year-old children, it is important to use the same scales for the two histograms. Notice that both histograms have a single peak with the majority of children in both age groups concentrated in the

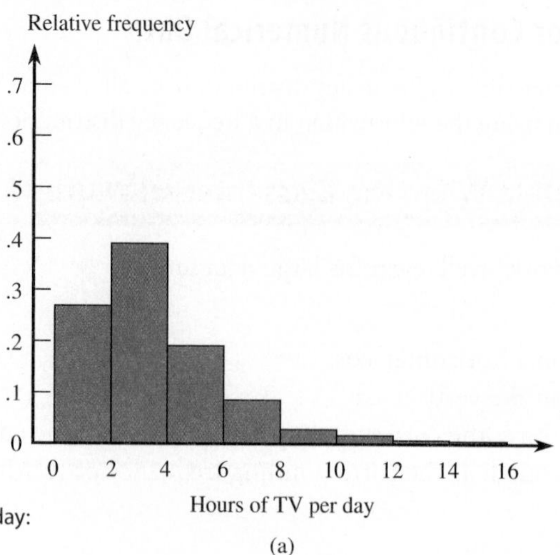

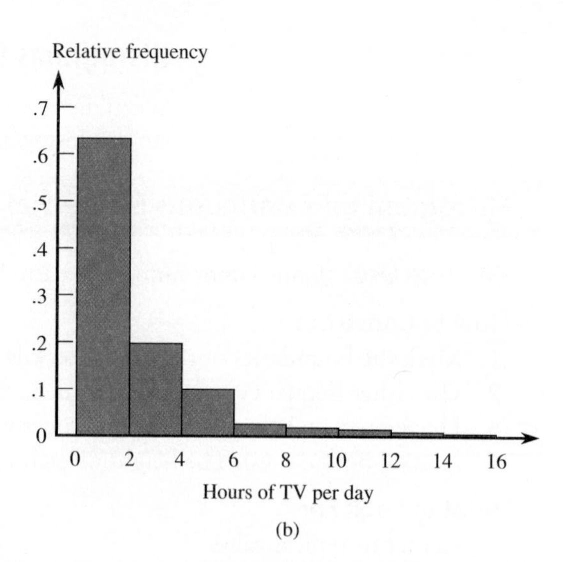

FIGURE 3.21
Histogram of TV hours per day:
(a) 1-year-old children;
(b) 3-year-old children.

smaller TV hours intervals. Both histograms are quite stretched out at the upper end, indicating some young children watch a lot of TV.

Interpret the results)

The big difference between the two histograms is at the low end, with a much greater proportion of 3-year-old children falling in the 0 to 2 TV hours interval than is the case for 1-year-old children. A typical number of TV hours per day for 1-year-old children would be somewhere between 2 and 4 hours, whereas a typical number of TV hours for 3-year-old children is in the 0 to 2 hours interval. ■

Class Intervals of Unequal Widths

Figure 3.22 shows a data set in which a great many observations are concentrated at the center of the data set, with only a few unusual values below and above the main body of data. If a frequency distribution is based on short intervals of equal width, a great many intervals will be required to capture all observations, and many of them will contain no observations, as shown in Figure 3.22(a). On the other hand, only a few wide intervals will capture all values, but then most of the observations will be grouped into a few intervals, as shown in Figure 3.22(b). In such situations, it is best to use a combination of wide class intervals where there are few data points and shorter intervals where there are many data points, as shown in Figure 3.22(c).

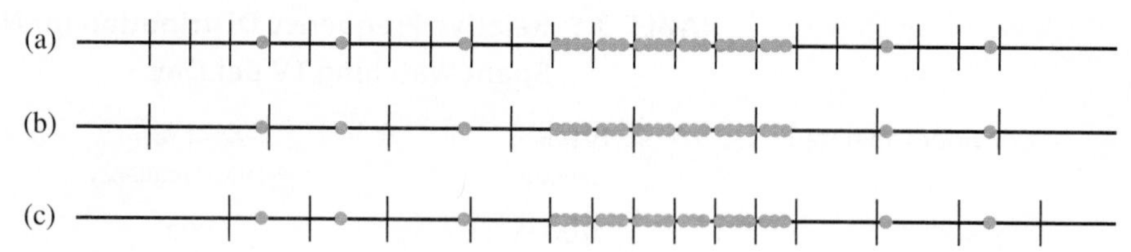

FIGURE 3.22
Three choices of class intervals for a data set with outliers:
(a) many short intervals of equal width;
(b) a few wide intervals of equal width;
(c) intervals of unequal width.

Constructing a Histogram for Continuous Data When Class Interval Widths Are Unequal

When class intervals are not of equal width, frequencies or relative frequencies should not be used on the vertical axis. Instead, the height of each rectangle, called the **density** for the class interval, is given by

$$\text{density} = \text{rectangle height} = \frac{\text{relative frequency of class interval}}{\text{class interval width}}$$

The vertical axis is called the **density scale.**

The use of the density scale to construct the histogram ensures that the area of each rectangle in the histogram will be proportional to the corresponding relative frequency. The formula for density can also be used when class widths are equal. However, when the intervals are of equal width, the extra arithmetic required to obtain the densities is unnecessary.

EXAMPLE 3.17 Misreporting Grade Point Average

Understand the context

When people are asked for the values of characteristics such as age or weight, they sometimes shade the truth in their responses. The article **"Self-Reports of Academic Performance" (*Social Methods and Research* [November 1981]: 165–185)** focused on such characteristics as SAT scores and grade point average (GPA). For each student in a sample, the difference in GPA (reported – actual) was determined. Positive differences resulted from individuals reporting GPAs larger than the correct values. Most differences were close to 0, but there were some rather large errors. Because of this, the frequency distribution based on unequal class widths shown in Table 3.6 gives an informative yet concise summary.

TABLE 3.6 Frequency Distribution for Errors in Reported GPA

Consider the data

Class Interval	Relative Frequency	Width	Density
−2.0 to <−0.4	.023	1.6	0.014
−0.4 to <−0.2	.055	.2	0.275
−0.2 to <−0.1	.097	.1	0.970
−0.1 to <0	.210	.1	2.100
0 to <0.1	.189	.1	1.890
0.1 to <0.2	.139	.1	1.390
0.2 to <0.4	.116	.2	0.580
0.4 to <2.0	.171	1.6	0.107

Figure 3.23 displays two histograms based on this frequency distribution. The histogram in Figure 3.23(a) is correctly drawn, with density used to determine the height of each bar. The histogram in Figure 3.23(b) has height equal to relative frequency and is therefore not correct. In particular, this second histogram considerably exaggerates the incidence of grossly overreported and underreported values—the areas of the two most extreme rectangles are much too large. The eye is naturally drawn to large areas, so it is important that the areas correctly represent the relative frequencies.

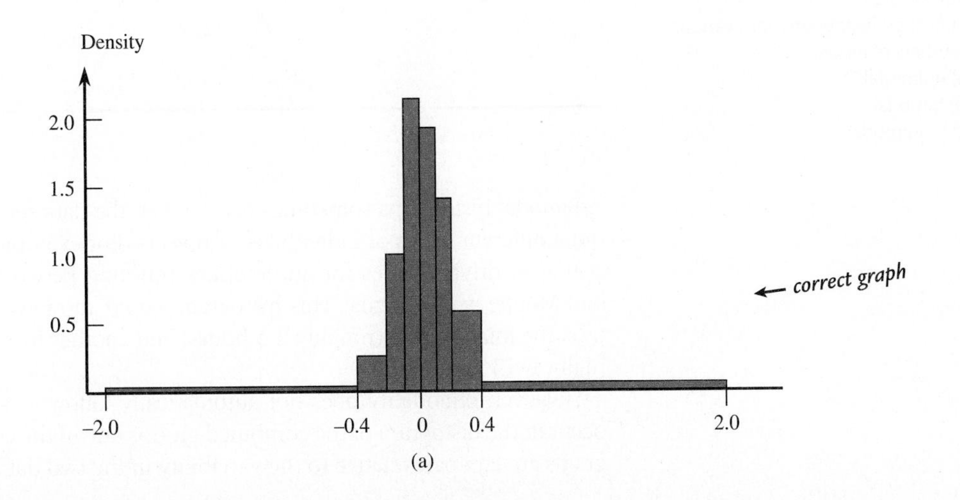

FIGURE 3.23
Histograms for errors in reporting GPA:
(a) a correct histogram (height = density);

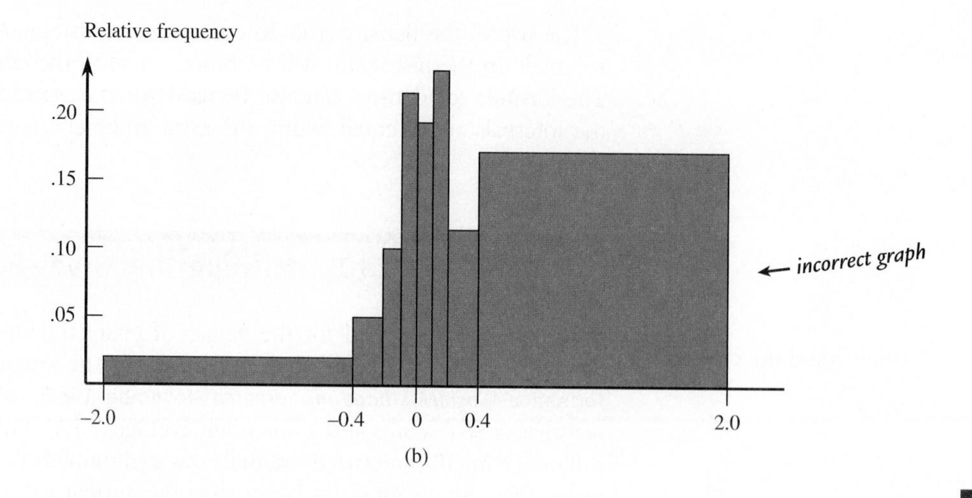

FIGURE 3.23
Histograms for errors in reporting GPA:
(b) an incorrect histogram (height = relative frequency).

■

Histogram Shapes

General shape is an important characteristic of a histogram. In describing various shapes it is convenient to approximate the histogram with a smooth curve (called a *smoothed histogram*). This is illustrated in Figure 3.24.

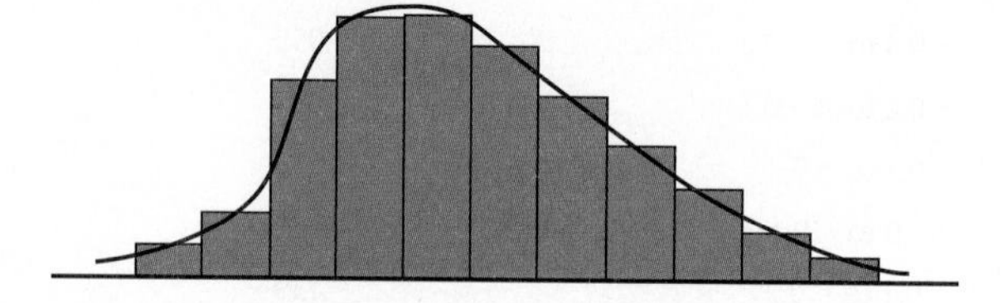

FIGURE 3.24
Approximating a histogram with a smooth curve.

One description of general shape relates to the number of peaks, or **modes**.

DEFINITION

Unimodal: A histogram is **unimodal** if it has a single peak.
Bimodal: A histogram is **bimodal** if it has two peaks.
Multimodal: A histogram is **multimodal** if it has more than two peaks.

These shapes are illustrated in Figure 3.25.

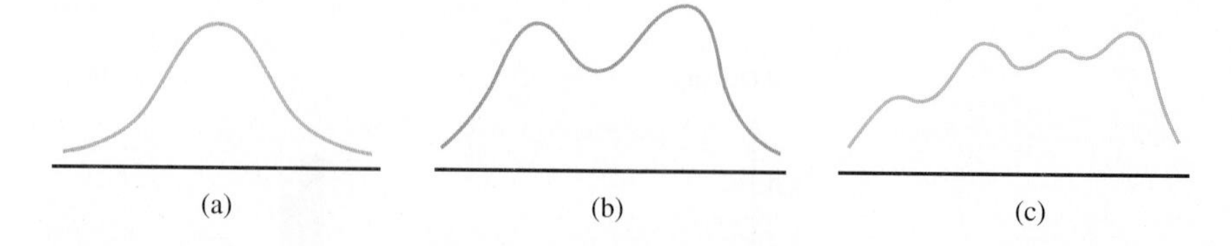

FIGURE 3.25
Smoothed histograms with various numbers of modes:
(a) unimodal;
(b) bimodal;
(c) multimodal.

Bimodal histograms sometimes occur when the data set consists of observations on two quite different kinds of individuals or objects. For example, consider a large data set consisting of driving times for automobiles traveling between San Luis Obispo, California, and Monterey, California. This histogram would show two peaks, one for those cars that took the inland route (roughly 2.5 hours) and another for those cars traveling up the coast highway (3.5–4 hours).

However, bimodality does not automatically follow in such situations. Bimodality will occur in the histogram of the combined groups only if the centers of the two separate histograms are far apart relative to the variability in the two data sets. For example, a large data set consisting of heights of college students would probably not produce a bimodal histogram because the typical height for males (about 69 in.) and the typical height for females (about 66 in.) are not very far apart relative to the variability in the height distributions.

Unimodal histograms come in a variety of shapes. A unimodal histogram is **symmetric** if there is a vertical line of symmetry such that the part of the histogram to the left of the line is a mirror image of the part to the right. (Bimodal and multimodal histograms can also be symmetric in this way.) Several different symmetric smoothed histograms are shown in Figure 3.26.

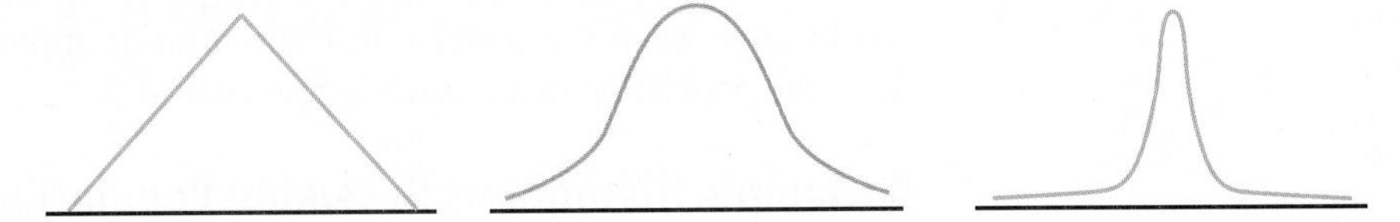

FIGURE 3.26
Several symmetric unimodal smoothed histograms.

Proceeding to the right from the peak of a unimodal histogram, we move into what is called the **upper tail** of the histogram. Going in the opposite direction moves us into the **lower tail**.

DEFINITION

Skewed: A unimodal histogram that is not symmetric is described as **skewed**.

Positively skewed or **right skewed:** A skewed histogram in which the upper tail of the histogram stretches out much farther than the lower tail is described as **positively skewed** or **right skewed**.

Negatively skewed or **left skewed:** A skewed histogram in which the lower tail of the histogram stretches out much farther than the upper tail is described as **negatively skewed** or **left skewed**.

These two types of skewness are illustrated in Figure 3.27. Positive skewness is much more frequently encountered than is negative skewness. An example of positive skewness occurs in the distribution of single-family home prices in Los Angeles County; most homes are moderately priced (at least for California), whereas the relatively few homes in Beverly Hills and Malibu have much higher price tags.

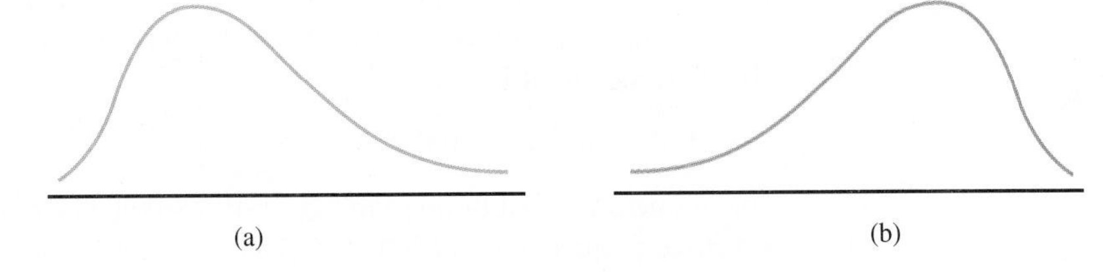

FIGURE 3.27
Two examples of skewed smoothed histograms:
(a) positive skew;
(b) negative skew.

One specific shape, a **normal curve**, arises more frequently than any other in statistical applications. Many histograms can be well approximated by a normal curve (for example, characteristics such as arm span and the weight of an apple). Here we briefly mention several of the most important characteristics of normal curves, postponing a more detailed discussion until Chapter 7. A normal curve is both symmetric and bell-shaped; it looks like the curve in Figure 3.28(a). However, not all bell-shaped curves are normal. In a normal

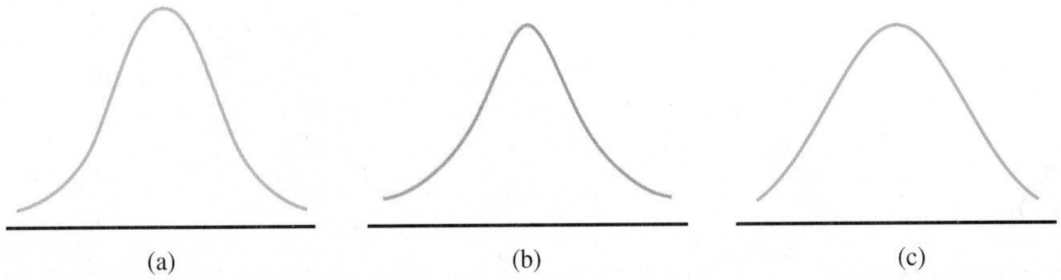

FIGURE 3.28
Three examples of bell-shaped histograms:
(a) normal;
(b) heavy-tailed;
(c) light-tailed.

curve, starting from the top of the bell the height of the curve decreases at a well-defined rate when moving toward either tail. (This rate of decrease is specified by a certain mathematical function.)

A curve with tails that do not decline as rapidly as the tails of a normal curve is called **heavy-tailed** (compared to the normal curve). Similarly, a curve with tails that decrease more rapidly than the normal tails is called **light-tailed**. Figure 3.28(b) and (c) illustrate these possibilities. The reason that we are concerned about the tails in a distribution is that many inferential procedures that work well (i.e., they result in a high proportion of correct conclusions) when the population distribution is approximately normal perform poorly when the population distribution is heavy-tailed.

Do Sample Histograms Resemble Population Histograms?

Sample data are usually collected to make inferences about a population. The resulting conclusions may be in error if the sample is unrepresentative of the population. So how similar might a histogram of sample data be to the histogram of all population values? Will the two histograms be centered at roughly the same place and spread out to about the same extent? Will they have the same number of peaks, and will the peaks occur at approximately the same places?

A related issue concerns the extent to which histograms based on different samples from the same population resemble one another. If two different sample histograms can be expected to differ from one another in obvious ways, then at least one of them might differ substantially from the population histogram. If the sample differs substantially from the population, conclusions about the population based on the sample are likely to be incorrect. **Sampling variability**—the extent to which samples from the same population differ from one another and from the population—is a central idea in statistics. Example 3.18 illustrates sampling variability in histogram shapes.

EXAMPLE 3.18 What You Should Know About Bus Drivers . . .

Understand the context ⟩

● A sample of 708 bus drivers employed by public corporations was selected, and the number of traffic accidents in which each bus driver was involved during a 4-year period was determined (**"Application of Discrete Distribution Theory to the Study of Noncommunicable Events in Medical Epidemiology," in *Random Counts in Biomedical and Social Sciences,* G. P. Patil, ed. [University Park, PA: Pennsylvania State University Press, 1970]**). A listing of the 708 sample observations might look like this:

3 0 6 0 0 2 1 4 1 . . . 6 0 2

The frequency distribution (Table 3.7) shows that 117 of the 708 drivers had no accidents, a relative frequency of 117/708 = .165 (or 16.5%). Similarly, the proportion of sampled drivers who had 1 accident is .222 (or 22.2%). The largest sample observation was 11.

TABLE 3.7 Frequency Distribution for Number of Accidents by Bus Drivers

Number of Accidents	Frequency	Relative Frequency
0	117	.165
1	157	.222
2	158	.223
3	115	.162
4	78	.110
5	44	.062
6	21	.030
7	7	.010
8	6	.008

● Data set available online

continued

Number of Accidents	Frequency	Relative Frequency
9	1	.001
10	3	.004
11	1	.001
	708	.998

Although the 708 observations actually constituted a sample from the population of all bus drivers, we will regard the 708 observations as constituting the entire population. The first histogram in Figure 3.29, then, represents the population histogram. The other four histograms in Figure 3.29 are based on four different samples of 50 observations each

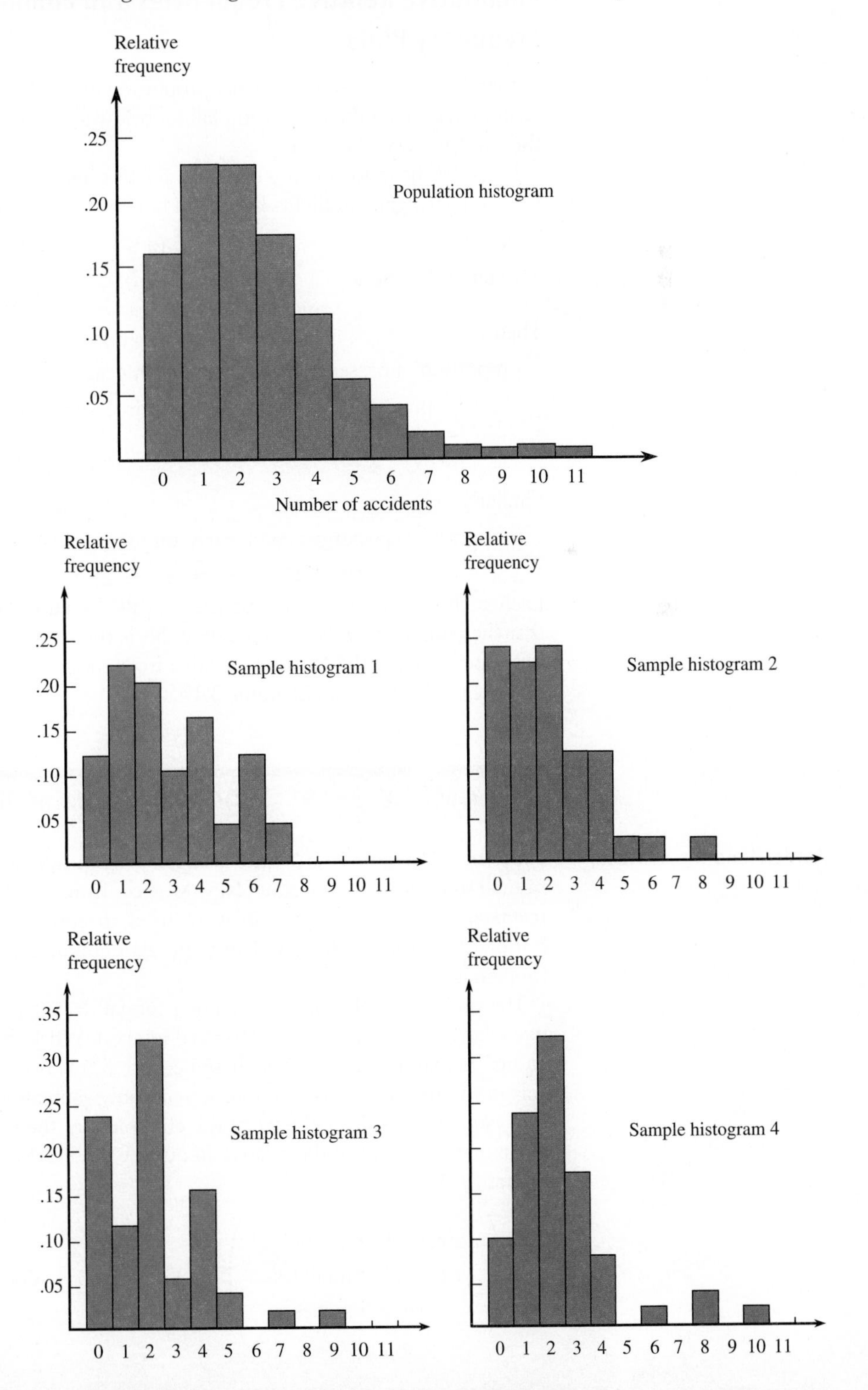

FIGURE 3.29
Comparison of population and sample histograms for number of accidents.

selected at random from this population. The five histograms certainly resemble one another in a general way, but some dissimilarities are also obvious. The population histogram rises to a peak and then declines smoothly, whereas the sample histograms tend to have more peaks, valleys, and gaps. Although the population data set contained an observation of 11, none of the four samples did. In fact, in the first two samples, the largest observations were 7 and 8, respectively. In Chapters 8–15 we will see how sampling variability can be described and taken into account when we use sample data to draw conclusions about a population. ■

Cumulative Relative Frequencies and Cumulative Relative Frequency Plots

Rather than wanting to know what proportion of the data fall in a particular class, we often wish to determine the proportion falling below a specified value. This is easily done when the value is a class boundary.

Consider the following intervals and relative frequencies for carry-on luggage weight (in lbs) for passengers on flights between Phoenix and New York City during October 2009:

Weight	0 to 5	5 to <10	10 to <15	15 to <20	. . .
Relative frequency	.05	.10	.18	.25	. . .

Then

proportion of passengers with carry-on luggage weight less than 15 lbs.

$$= \text{proportion in one of the first three intervals}$$
$$= .05 + .10 + .18$$
$$= .33$$

Similarly,

proportion of passengers with carry-on luggage weight less than 20 lbs.

$$= .05 + .10 + .18 + .25 = .33 + .25 = .58$$

Each such sum of relative frequencies is called a **cumulative relative frequency**. Notice that the cumulative relative frequency .58 is the sum of the previous cumulative relative frequency .33 and the "current" relative frequency .25. The use of cumulative relative frequencies is illustrated in Example 3.19.

EXAMPLE 3.19 Albuquerque Rainfall

Understand the context ⟩

The National Climatic Data Center has been collecting weather data for many years. Annual rainfall totals for Albuquerque, New Mexico, from 1950 to 2008 **(www.ncdc.noaa.gov/oa/climate/research/cag3/city.html)** were used to construct the relative frequency distribution shown in Table 3.8. The table also contains a column of cumulative relative frequencies.

The proportion of years with annual rainfall less than 10 inches is .585, the cumulative relative frequency for the 9 to <10 interval. What about the proportion of years with annual rainfall less than 8.5 inches? Because 8.5 is not the endpoint of one of the intervals in the frequency distribution, we can only estimate this from the information given. The value 8.5 is halfway between the endpoints of the 8 to 9 interval, so it is reasonable to estimate that half of the relative frequency of .172 for this interval belongs in the 8 to 8.5 range. Then

$$\left(\begin{array}{c}\text{estimate of proportion of}\\ \text{years with rainfall less}\\ \text{than 8.5 inches}\end{array}\right) = .052 + .103 + .086 + .103 + \frac{1}{2}(.172) = .430$$

TABLE 3.8 Relative Frequency Distribution for Albuquerque Rainfall Data with Cumulative Relative Frequencies

Annual Rainfall (inches)	Frequency	Relative Frequency	Cumulative Relative Frequency
4 to <5	3	0.052	0.052
5 to <6	6	0.103	0.155 = .052 + .103
6 to <7	5	0.086	0.241 = .052 + .103 + .086 or .155 + .086
7 to <8	6	0.103	0.344
8 to <9	10	0.172	0.516
9 to <10	4	0.069	0.585
10 to <11	12	0.207	0.792
11 to <12	6	0.103	0.895
12 to <13	3	0.052	0.947
13 to <14	3	0.052	0.999

This proportion could also have been computed using the cumulative relative frequencies as

$$\left(\begin{array}{c}\text{estimate of proportion of}\\\text{years with rainfall less}\\\text{than 8.5 inches}\end{array}\right) = .344 + \frac{1}{2}(.172) = .430$$

Similarly, since 11.25 is one-fourth of the way between 11 and 12,

$$\left(\begin{array}{c}\text{estimate of proportion of}\\\text{years with rainfall less}\\\text{than 11.25 inches}\end{array}\right) = .792 + \frac{1}{4}(.103) = .818$$

■

A **cumulative relative frequency** plot is just a graph of the cumulative relative frequencies against the upper endpoint of the corresponding interval. The pairs

(upper endpoint of interval, cumulative relative frequency)

are plotted as points on a rectangular coordinate system, and successive points in the plot are connected by a line segment. For the rainfall data of Example 3.19, the plotted points would be

(5, .052) (6, .155) (7, .241) (8, .344) (9, .516)
(10, .585) (11, .792) (12, .895) (13, .947) (14, .999)

One additional point, the pair (lower endpoint of first interval, 0), is also included in the plot (for the rainfall data, this would be the point (4, 0)), and then points are connected by line segments. Figure 3.30 shows the cumulative relative frequency plot for the rainfall data. The cumulative relative frequency plot can be used to obtain approximate answers to questions such as

What proportion of the observations is smaller than a particular value?

and

What value separates the smallest p percent from the larger values?

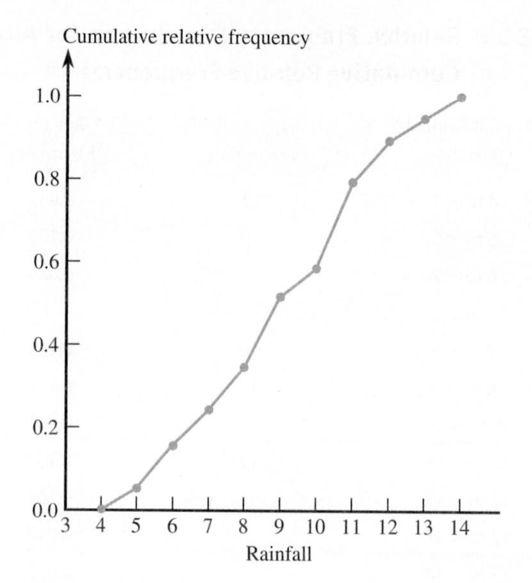

FIGURE 3.30
Cumulative relative frequency plot for the rainfall data of Example 3.19.

For example, to determine the approximate proportion of years with annual rainfall less than 9.5 inches, we would follow a vertical line up from 9.5 on the *x*-axis and then read across to the *y*-axis to obtain the corresponding relative frequency, as illustrated in Figure 3.31(a). Approximately .55, or 55%, of the years had annual rainfall less than 9.5 inches.

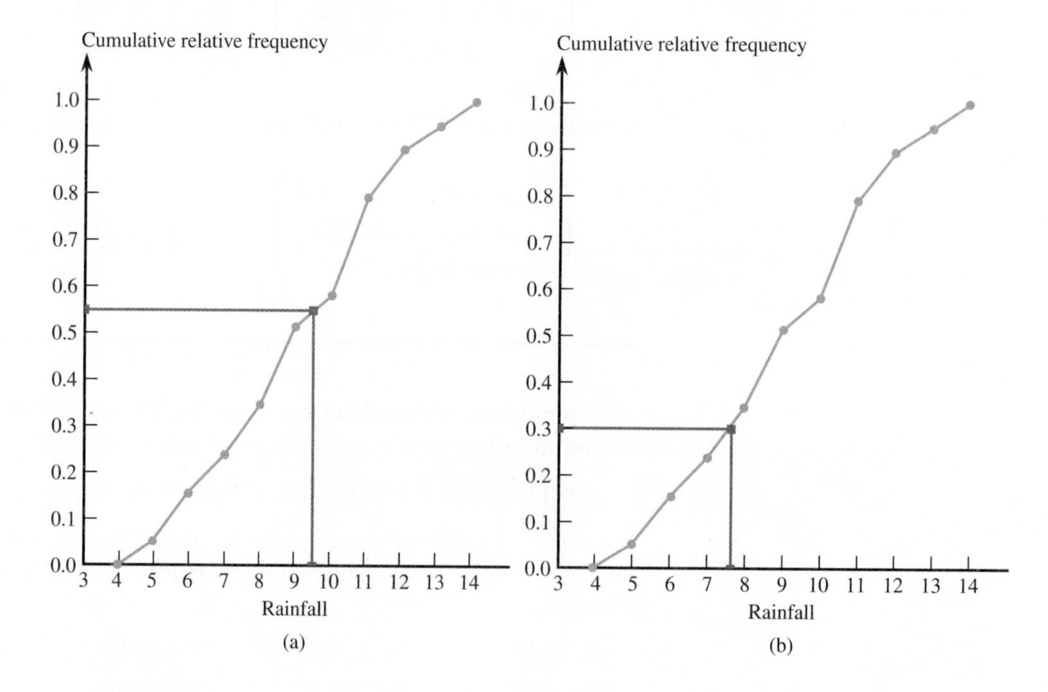

FIGURE 3.31
Using the cumulative relative frequency plot.
(a) Determining the approximate proportion of years with annual rainfall less than 9.5 inches.
(b) Finding the amount of rainfall that separates the 30% of years with the lowest rainfall from the 70% with higher rainfall.

Similarly, to find the amount of rainfall that separates the 30% of years with the smallest annual rainfall from years with higher rainfall, start at .30 on the cumulative relative frequency axis and move across and then down to find the corresponding rainfall amount, as shown in Figure 3.31(b). Approximately 30% of the years had annual rainfall of less than 7.6 inches.

EXERCISES 3.22 - 3.37

3.22 The data in the accompanying table are from ***The Chronicle of Higher Education* (August 23, 2013)**. Entries in the table are the number of people enrolled in college in 2011 per 100,000 people for 39 countries. For example, in the United States, there were 6585 college students for every 100,000 people in the country.

Country	*College Enrollment	Country	*College Enrollment
Argentina	6098	Ireland	4203
Australia	5933	Israel	4900
Austria	4409	Italy	3409
Bangladesh	1284	Japan	3025
Bhutan	969	Mexico	2622
Brazil	3405	New Zealand	6261
Britain	3976	Norway	4805
Cambodia	1511	Pakistan	839
Chile	6285	Panama	4079
China	2342	Poland	5586
Denmark	4361	Singapore	4997
Egypt	2737	Slovenia	5356
Finland	5863	Spain	4172
France	3459	Sweden	5100
Ghana	1153	Switzerland	3373
Greece	5971	Thailand	3743
Hungary	3828	Turkey	4536
Iceland	5843	Uganda	837
India	1768	United States	6585
Iran	5286		

*Per 100,000 people

a. Construct a histogram of these data using the class intervals 0 to < 1000, 1000 to < 2000, ..., 6000 to < 7000. (Hint: See Example 3.16.)

b. Write a few sentences describing the shape, center, and spread of the distribution.

3.23 ● The accompanying data on annual maximum wind speed (in meters per second) in Hong Kong for each year in a 45-year period were given in an article that appeared in the journal ***Renewable Energy* (March, 2007)**.

30.3 39.0 33.9 38.6 44.6 31.4 26.7 51.9 31.9
27.2 52.9 45.8 63.3 36.0 64.0 31.4 42.2 41.1
37.0 34.4 35.5 62.2 30.3 40.0 36.0 39.4 34.4
28.3 39.1 55.0 35.0 28.8 25.7 62.7 32.4 31.9
37.5 31.5 32.0 35.5 37.5 41.0 37.5 48.6 28.1

a. Use the annual maximum wind speed data to construct a histogram.

b. Is the histogram approximately symmetric, positively skewed, or negatively skewed?

c. Would you describe the histogram as unimodal, bimodal, or multimodal?

3.24 ● The accompanying relative frequency table is based on data from the **2012 College Bound Seniors Report for California**.

Score on SAT Critical Reading Exam	Relative Frequency for Males	Relative Frequency for Females
200 to < 300	.037	.031
300 to < 400	.151	.161
400 to < 500	.307	.329
500 to < 600	.298	.295
600 to < 700	.156	.140
700 to < 800	.050	.043

a. Construct a relative frequency histogram for SAT critical reading score for males.

b. Construct a relative frequency histogram for SAT critical reading score for females.

c. Based on the histograms from Parts (a) and (b), write a few sentences commenting on the similarities and differences in the distribution of SAT critical reading scores for males and females.

3.25 ● The data in the accompanying table represents the percentage of workers who are members of a union for each U.S. state and the District of Columbia (***AARP Bulletin*, September 2009**).

State	% of Workers who Belong to a Union	State	% of Workers who Belong to a Union
Alabama	9.8	Hawaii	24.3
Alaska	23.5	Idaho	7.1
Arizona	8.8	Illinois	16.6
Arkansas	5.9	Indiana	12.4
California	18.4	Iowa	10.6
Colorado	8.0	Kansas	7.0
Connecticut	16.9	Kentucky	8.6
Delaware	12.2	Louisiana	4.6
District of Columbia	13.4	Maine	12.3
Florida	6.4	Maryland	15.7
Georgia	3.7	Massachusetts	12.6

continued

State	% of Workers who Belong to a Union	State	% of Workers who Belong to a Union
Michigan	18.8	Oklahoma	6.6
Minnesota	16.1	Oregon	16.6
Mississippi	5.3	Pennsylvania	15.4
Missouri	11.2	Rhode Island	16.5
Montana	12.2	South Carolina	3.9
Nebraska	8.3	South Dakota	5.0
Nevada	16.7	Tennessee	5.5
New Hampshire	3.5	Texas	4.5
New Jersey	6.1	Utah	5.8
New Mexico	10.6	Vermont	4.1
New York	18.3	Virginia	10.4
North Carolina	7.2	Washington	19.8
North Dakota	24.9	West Virginia	13.8
Ohio	14.2	Wisconsin	15.0
		Wyoming	7.7

a. Construct a histogram of these data using class intervals of 0 to <5, 5 to <10, 10 to <15, 15 to <20, and 20 to <25.

b. Construct a dotplot of these data. Comment on the interesting features of the plot. (Hint: Dotplots were covered in Section 1.4.)

c. For this data set, which is a more informative graphical display—the dotplot from Part (b) or the histogram constructed in Part (a)? Explain.

d. Construct a histogram using about twice as many class intervals as the histogram in Part (a). Use 2.5 to <5 as the first class interval. Write a few sentences that explain why this histogram does a better job of displaying this data set than does the histogram in Part (a).

3.26 ● Medicare's new medical plans offer a wide range of variations and choices for seniors when picking a drug plan (***San Luis Obispo Tribune*, November 25, 2005**). The monthly cost for a stand-alone drug plan varies from plan to plan and from state to state. The accompanying table gives the premium for the plan with the lowest cost for each state.

State	Cost per Month (dollars)	State	Cost per Month (dollars)
Alabama	14.08	Colorado	8.62
Alaska	20.05	Connecticut	7.32
Arizona	6.14	Delaware	6.44
Arkansas	10.31	District of Columbia	6.44
California	5.41		

continued

State	Cost per Month (dollars)	State	Cost per Month (dollars)
Florida	10.35	New Jersey	4.43
Georgia	17.91	New Mexico	10.65
Hawaii	17.18	New York	4.10
Idaho	6.33	North Carolina	13.27
Illinois	13.32	North Dakota	1.87
Indiana	12.30	Ohio	14.43
Iowa	1.87	Oklahoma	10.07
Kansas	9.48	Oregon	6.93
Kentucky	12.30	Pennsylvania	10.14
Louisiana	17.06	Rhode Island	7.32
Maine	19.60	South Carolina	16.57
Maryland	6.44	South Dakota	1.87
Massachusetts	7.32	Tennessee	14.08
Michigan	13.75	Texas	10.31
Minnesota	1.87	Utah	6.33
Mississippi	11.60	Vermont	7.32
Missouri	10.29	Virginia	8.81
Montana	1.87	Washington	6.93
Nebraska	1.87	West Virginia	10.14
Nevada	6.42	Wisconsin	11.42
New Hampshire	19.60	Wyoming	1.87

a. Use class intervals of \$0 to <\$3, \$3 to <\$6, \$6 to <\$9, etc., to create a relative frequency distribution for these data.

b. Construct a histogram and comment on its shape.

c. Using the relative frequency distribution or the histogram, estimate the proportion of the states that have a minimum monthly plan of less than \$13.00 a month.

3.27 The following two relative frequency distributions are based on data that appeared in ***The Chronicle of Higher Education* (August 23, 2013)**. The data are from a survey of students at four-year colleges that was conducted in 2012. One relative frequency distribution is for the number of hours spent online at social network sites in a typical week. The second relative frequency distribution is for the number of hours spent playing video and computer games in a typical week.

Number of Hours on Social Networks	Relative Frequency
0 to < 1	.234
1 to < 6	.512
6 to < 21	.211
21 or more	.044

Number of Hours Playing Video and Computer Games	Relative Frequency
0 to< 1	.621
1 to < 6	.252
6 to < 21	.108
21 or more	.020

a. Construct a histogram for the social media data. For purposes of constructing the histogram, assume that none of the students in the sample spent more than 40 hours on social media in a typical week and that the last interval can be regarded as 21 to <40. Be sure to use the density scale when constructing the histogram. (Hint: See Example 3.17.)

b. Construct a histogram for the video and computer game data. Use the same scale that you used for the histogram in Part (a) so that it will be easy to compare the two histograms.

c. Comment on the similarities and differences in the histograms from Parts (a) and (b).

3.28 ● U.S. Census data for San Luis Obispo County, California, were used to construct the following frequency distribution for commute time (in minutes) of working adults (the given frequencies were read from a graph that appeared in the ***San Luis Obispo Tribune*** **[September 1, 2002]** and so are only approximate):

Commute Time	Frequency
0 to <5	5,200
5 to <10	18,200
10 to <15	19,600
15 to <20	15,400
20 to <25	13,800
25 to <30	5,700
30 to <35	10,200
35 to <40	2,000
40 to <45	2,000
45 to <60	4,000
60 to <90	2,100
90 to <120	2,200

a. Notice that not all intervals in the frequency distribution are equal in width. Why do you think that unequal width intervals were used?

b. Construct a table that adds a relative frequency and a density column to the given frequency distribution (Hint: See Example 3.17).

c. Use the densities computed in Part (b) to construct a histogram for this data set. (Note: The newspaper displayed an incorrectly drawn histogram based on frequencies rather than densities!) Write a few sentences commenting on the important features of the histogram.

d. Compute the cumulative relative frequencies, and construct a cumulative relative frequency plot.

e. Use the cumulative relative frequency plot constructed in Part (d) to answer the following questions.

i. Approximately what proportion of commute times were less than 50 minutes?

ii. Approximately what proportion of commute times were greater than 22 minutes?

iii. What is the approximate commute time value that separates the shortest 50% of commute times from the longest 50%?

3.29 The report **"Trends in College Pricing 2012" (www.collegeboard.com)** included the information in the accompanying relative frequency distributions for public and for private not-for-profit four-year college students.

	Public Four-Year College Students	Private Not-for-Profit Four-Year College Students
Tuition and Fees	Proportion of Students (Relative Frequency)	Proportion of Students (Relative Frequency)
0 to < 3,000	.009	.000
3,000 to < 6,000	.107	.066
6,000 to < 9,000	.436	.011
9,000 to< 12,000	.199	.027
12,000 to < 15,000	.124	.032
15,000 to < 18,000	.033	.030
18,000 to < 21,000	.027	.052
21,000 to < 24,000	.019	.074
24,000 to < 27,000	.015	.103
27,000 to < 30,000	.020	.094
30,000 to < 33,000	.005	.103
33,000 to < 36,000	.003	.103
36,000 to < 39,000	.002	.066
39,000 to < 42,000	.002	.080
42,000 to < 45,000	.000	.136
45,000 to < 48,000	.000	.022

a. Construct a relative frequency histogram for tuition and fees for students at public four-year colleges. Write a few sentences describing the distribution of tuition and fees, commenting on center, spread, and shape.

b. Construct a relative frequency histogram for tuition and fees for students at private not-for-profit four-year colleges. Be sure to use the same scale for the vertical and horizontal axes as you used for the histogram in Part (a). Write a few sentences describing the distribution of tuition and fees for students at private not-for-profit four-year colleges.

c. Write a few sentences describing the differences in the distributions.

3.30 An exam is given to students in an introductory statistics course. What is likely to be true of the shape of the histogram of scores if:

a. the exam is quite easy?

b. the exam is quite difficult?

c. half the students in the class have had calculus, the other half have had no prior college math courses, and the exam emphasizes mathematical manipulation?

Explain your reasoning in each case.

3.31 The accompanying frequency distribution summarizes data on the number of times smokers who had successfully quit smoking attempted to quit before their final successful attempt (**"Demographic Variables, Smoking Variables, and Outcome Across Five Studies,"** ***Health Psychology*** **[2007]: 278–287**).

Number of Attempts	Frequency
0	778
1	306
2	274
3–4	221
5 or more	238

Assume that no one had made more than 10 unsuccessful attempts, so that the last entry in the frequency distribution can be regarded as 5–10 attempts. Summarize this data set using a histogram. Be careful—the class intervals are not all the same width, so you will need to use a density scale for the histogram. Also remember that for a discrete variable, the bar for 1 will extend from 0.5 to 1.5. Think about what this will mean for the bars for the 3–4 group and the 5–10 group.

3.32 ● Example 3.19 used annual rainfall data for Albuquerque, New Mexico, to construct a relative frequency distribution and cumulative relative frequency plot. The National Climate Data Center also gave the accompanying annual rainfall (in inches) for Medford, Oregon, from 1950 to 2008.

28.84 20.15 18.88 25.72 16.42 20.18 28.96 20.72 23.58 10.62
20.85 19.86 23.34 19.08 29.23 18.32 21.27 18.93 15.47 20.68
23.43 19.55 20.82 19.04 18.77 19.63 12.39 22.39 15.95 20.46
16.05 22.08 19.44 30.38 18.79 10.89 17.25 14.95 13.86 15.30
13.71 14.68 15.16 16.77 12.33 21.93 31.57 18.13 28.87 16.69
18.81 15.15 18.16 19.99 19.00 23.97 21.99 17.25 14.07

a. Construct a relative frequency distribution for the Medford rainfall data.

b. Use the relative frequency distribution of Part (a) to construct a histogram. Describe the shape of the histogram.

3.33 Use the relative frequency distribution constructed in Exercise 3.32 to answer the following questions.

a. Construct a cumulative relative frequency plot for the Medford rainfall data.

b. Use the cumulative relative frequency plot of Part (a) to answer the following questions:

i. Approximately what proportion of years had annual rainfall less than 15.5 inches?

ii. Approximately what proportion of years had annual rainfall less than 25 inches?

iii. Approximately what proportion of years had annual rainfall between 17.5 and 25 inches?

3.34 The authors of the paper **"Myeloma in Patients Younger than Age 50 Years Presents with More Favorable Features and Shows Better Survival"** (***Blood*** **[2008]: 4039–4047**) studied patients who had been diagnosed with stage 2 multiple myeloma prior to the age of 50. For each patient who received high dose chemotherapy, the number of years that the patient lived after the therapy (survival time) was recorded. The cumulative relative frequencies in the accompanying table were approximated from survival graphs that appeared in the paper.

Years Survived	Cumulative Relative Frequency
0 to <2	.10
2 to <4	.52
4 to <6	.54
6 to <8	.64
8 to <10	.68
10 to <12	.70
12 to <14	.72
14 to <16	1.00

a. Use the given information to construct a cumulative relative frequency plot.

b. Use the cumulative relative frequency plot from Part (a) to answer the following questions:
   i. What is the approximate proportion of patients who lived fewer than 5 years after treatment?
   ii. What is the approximate proportion of patients who lived fewer than 7.5 years after treatment?
   iii. What is the approximate proportion of patients who lived more than 10 years after treatment?

3.35 a. Use the cumulative relative frequencies given in the previous exercise to compute the relative frequencies for each class interval and construct a relative frequency distribution.
b. Summarize the survival time data with a histogram.
c. Based on the histogram, write a few sentences describing survival time of the stage 2 myeloma patients in this study.
d. What additional information would you need in order to decide if it is reasonable to generalize conclusions about survival time from the group of patients in the study to all patients younger than 50 years old who are diagnosed with multiple myeloma and who receive high dose chemotherapy?

3.36 Construct a histogram corresponding to each of the five frequency distributions, I–V, given in the following table, and state whether each histogram is symmetric, bimodal, positively skewed, or negatively skewed:

	Frequency				
Class Interval	I	II	III	IV	V
0 to <10	5	40	30	15	6
10 to <20	10	25	10	25	5
20 to <30	20	10	8	8	6
30 to <40	30	8	7	7	9
40 to <50	20	7	7	20	9
50 to <60	10	5	8	25	23
60 to <70	5	5	30	10	42

3.37 Using the five class intervals 100 to 120, 120 to 140, . . . , 180 to 200, devise a frequency distribution based on 70 observations whose histogram could be described as follows:
a. symmetric
b. bimodal
c. positively skewed
d. negatively skewed

Bold exercises answered in back ● Data set available online ▼ Video Solution available

3.4 Displaying Bivariate Numerical Data

A bivariate data set consists of measurements or observations on two variables, x and y. For example, x might be the distance from a highway and y the lead content of soil at that distance. When both x and y are numerical variables, each observation consists of a pair of numbers, such as (14, 5.2) or (27.63, 18.9). The first number in a pair is the value of x, and the second number is the value of y.

An unorganized list of bivariate data provides little information about the distribution of either the x values or the y values separately and even less information about how the two variables are related to one another. Just as graphical displays can be used to summarize univariate data, they can also help with bivariate data. The most important graph based on bivariate numerical data is a **scatterplot**.

In a scatterplot each observation (pair of numbers) is represented by a point on a rectangular coordinate system, as shown in Figure 3.32(a). The horizontal axis is identified with values of x and is scaled so that any x value can be easily located. Similarly, the vertical or y-axis is marked for easy location of y values. The point corresponding to any particular (x, y) pair is placed where a vertical line from the value on the x-axis meets a horizontal line from the value on the y-axis. Figure 3.32(b) shows the point representing the observation (4.5, 15); it is above 4.5 on the horizontal axis and to the right of 15 on the vertical axis.

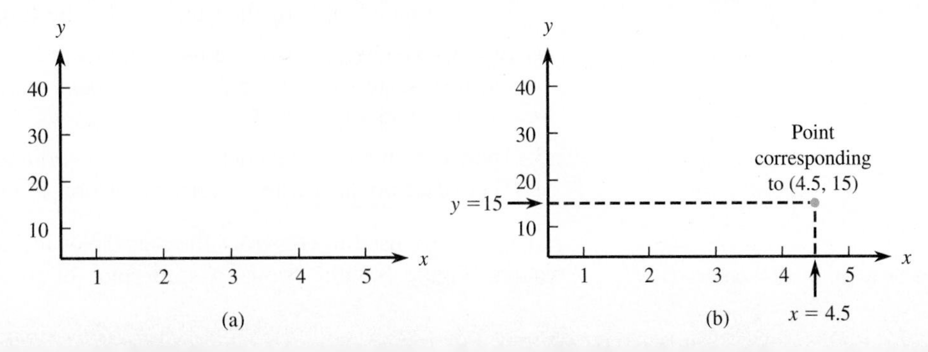

FIGURE 3.32
Constructing a scatterplot:
(a) rectangular coordinate system;
(b) point corresponding to (4.5, 15).

EXAMPLE 3.20 Olympic Figure Skating

Understand the context ⟩

Consider the data ⟩

● Do tall skaters have an advantage when it comes to earning high artistic scores in figure skating competitions? Data on x = height (in cm) and y = artistic score in the free skate for both male and female singles skaters at the 2006 Winter Olympics are shown in the accompanying table. (Data set courtesy of John Walker.)

Name	Gender	Height	Artistic	Name	Gender	Height	Artistic
PLUSHENKO Yevgeny	M	178	41.2100	ARAKAWA Shizuka	F	166	39.3750
BUTTLE Jeffrey	M	173	39.2500	COHEN Sasha	F	157	39.0063
LYSACEK Evan	M	177	37.1700	SLUTSKAYA Irina	F	160	38.6688
LAMBIEL Stephane	M	176	38.1400	SUGURI Fumie	F	157	37.0313
SAVOIE Matt	M	175	35.8600	ROCHETTE Joannie	F	157	35.0813
WEIR Johnny	M	172	37.6800	MEISSNER Kimmie	F	160	33.4625
JOUBERT Brian	M	179	36.7900	HUGHES Emily	F	165	31.8563
VAN DER PERREN Kevin	M	177	33.0100	MEIER Sarah	F	164	32.0313
TAKAHASHI Daisuke	M	165	36.6500	KOSTNER Carolina	F	168	34.9313
KLIMKIN Ilia	M	170	32.6100	SOKOLOVA Yelena	F	162	31.4250
ZHANG Min	M	176	31.8600	YAN Liu	F	164	28.1625
SAWYER Shawn	M	163	34.2500	LEUNG Mira	F	168	26.7000
LI Chengjiang	M	170	28.4700	GEDEVANISHVILI Elene	F	159	31.2250
SANDHU Emanuel	M	183	35.1100	KORPI Kiira	F	166	27.2000
VERNER Tomas	M	180	28.6100	POYKIO Susanna	F	159	31.2125
DAVYDOV Sergei	M	159	30.4700	ANDO Miki	F	162	31.5688
CHIPER Gheorghe	M	176	32.1500	EFREMENKO Galina	F	163	26.5125
DINEV Ivan	M	174	29.2500	LIASHENKO Elena	F	160	28.5750
DAMBIER Frederic	M	163	31.2500	HEGEL Idora	F	166	25.5375
LINDEMANN Stefan	M	163	31.0000	SEBESTYEN Julia	F	164	28.6375
KOVALEVSKI Anton	M	171	28.7500	KARADEMIR Tugba	F	165	23.0000
BERNTSSON Kristoffer	M	175	28.0400	FONTANA Silvia	F	158	26.3938
PFEIFER Viktor	M	180	28.7200	PAVUK Viktoria	F	168	23.6688
TOTH Zoltan	M	185	25.1000	MAXWELL Fleur	F	160	24.5438

Figure 3.33(a) gives a scatterplot of the data. Looking at the data and the scatterplot, we can see that

Interpret the results ⟩

1. Several observations have identical x values but different y values (for example, x = 176 cm for both Stephane Lambiel and Min Zhang, but Lambiel's artistic score was 38.1400 and Zhang's artistic score was 31.8600). Thus, the value of y is *not* determined *solely* by the value of x but by various other factors as well.
2. At any given height there is quite a bit of variability in artistic score. For example, for those skaters with height 160 cm, artistic scores ranged from a low of about 24.5 to a high of about 39.
3. There is no noticeable tendency for artistic score to increase as height increases. There does not appear to be a strong relationship between height and artistic score.

The data set used to construct the scatterplot included data for both male and female skaters. Figure 3.33(b) shows a scatterplot of the (height, artistic score) pairs with

● Data set available online

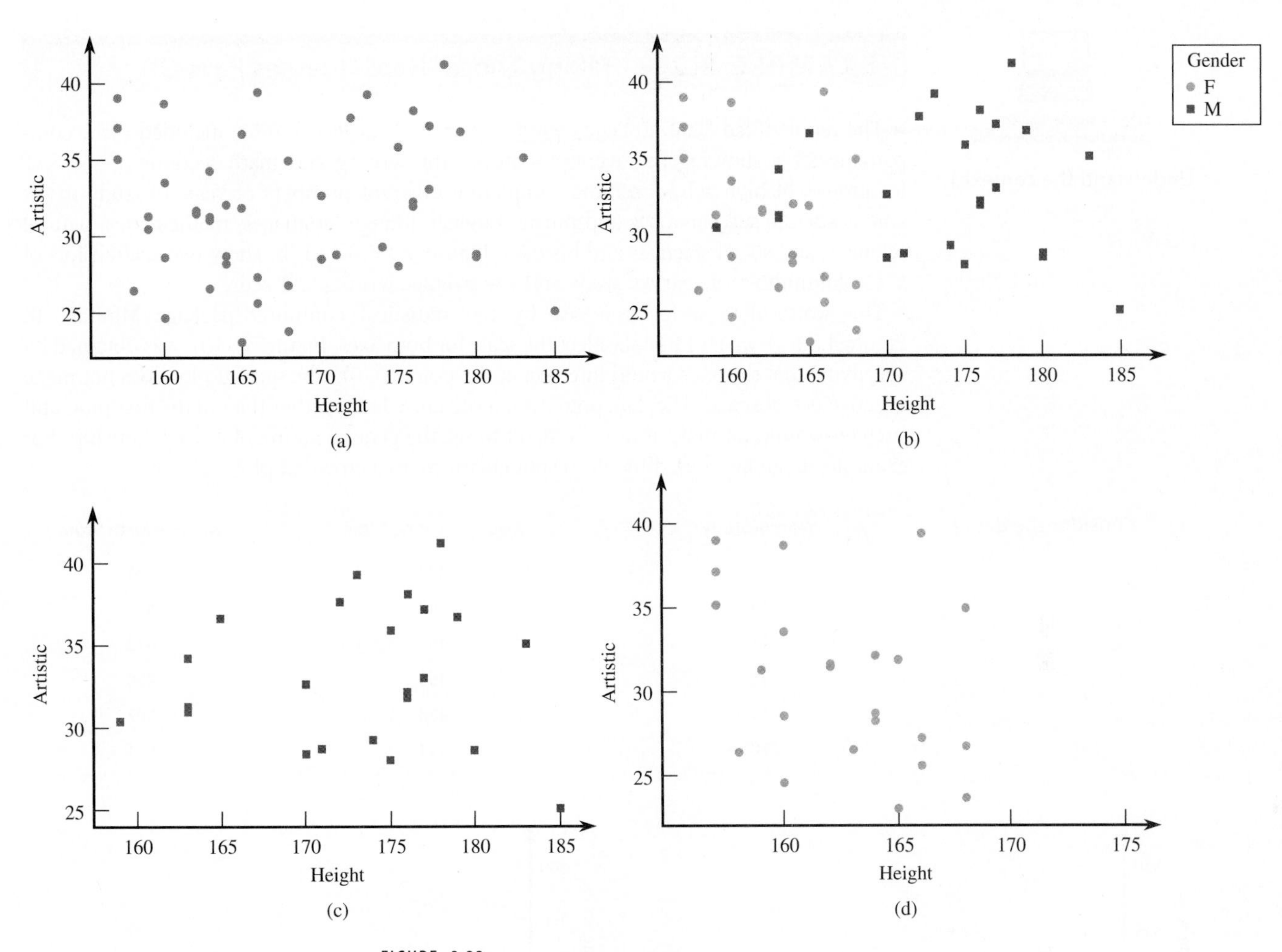

FIGURE 3.33
Scatterplots for the data of Example 3.20:
(a) scatterplot of data;
(b) scatterplot of data with observations for males and females distinguished by color;
(c) scatterplot for male skaters;
(d) scatterplot for female skaters.

observations for male skaters shown in blue and observations for female skaters shown in orange. Not surprisingly, the female skaters tend to be shorter than the male skaters (the observations for females tend to be concentrated toward the left side of the scatterplot). Careful examination of this plot shows that while there was no apparent pattern in the combined (male and female) data set, there may be a relationship between height and artistic score for female skaters.

Figure 3.33(c) and (d) show separate scatterplots for the male and female skaters, respectively. It is interesting to note that it appears that for female skaters, higher artistic scores seem to be associated with smaller height values, but for men there does not appear to be a relationship between height and artistic score. The relationship between height and artistic score for women is not evident in the scatterplot of the combined data. ■

The horizontal and vertical axes in the scatterplots of Figure 3.33 do not intersect at the point (0, 0). In many data sets, the values of x or of y or of both variables differ considerably from 0 relative to the ranges of the values in the data set. For example, a study of how air conditioner efficiency is related to maximum daily outdoor temperature might involve observations at temperatures of 80°, 82°, . . . , 98°, 100°. In such cases, the plot will be more informative if the axes intersect at some point other than (0, 0) and are marked accordingly. This is illustrated in Example 3.21.

Step-by-step technology instructions available online

EXAMPLE 3.21 Taking Those "Hard" Classes Pays Off

Understand the context

● The report titled **"2007 College Bound Seniors" (College Board, 2007)** included the accompanying table showing the average score on the writing and math sections of the SAT for groups of high school seniors completing different numbers of years of study in six core academic subjects (arts and music, English, foreign languages, mathematics, natural sciences, and social sciences and history). Figure 3.34(a) and (b) show two scatterplots of x = total number of years of study and y = average writing SAT score.

The scatterplots were produced by the statistical computer package Minitab. In Figure 3.34(a), we let Minitab select the scale for both axes. Figure 3.34(b) was obtained by specifying that the axes would intersect at the point (0, 0). The second plot does not make effective use of space. The data points are more crowded together than in the first plot, and such crowding can make it more difficult to see the general nature of any relationship. For example, it can be more difficult to spot curvature in a crowded plot.

Consider the data

Years of Study	Average Writing Score	Average Math Score
15	442	461
16	447	466
17	454	473
18	469	490
19	486	507
20	534	551

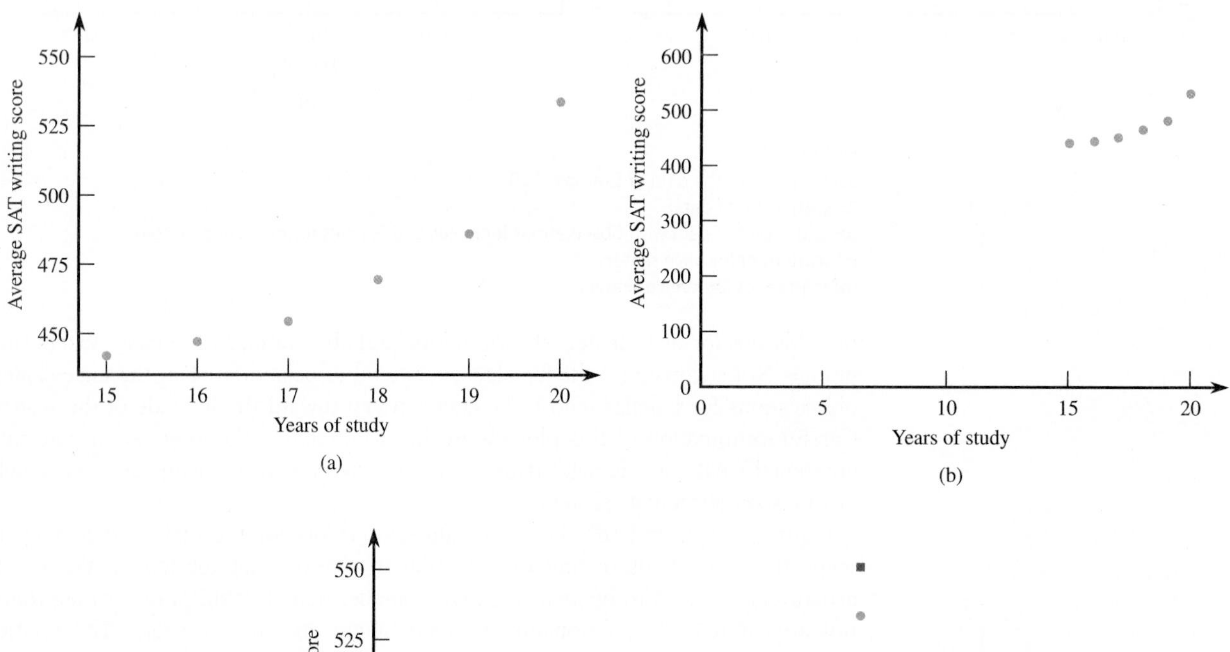

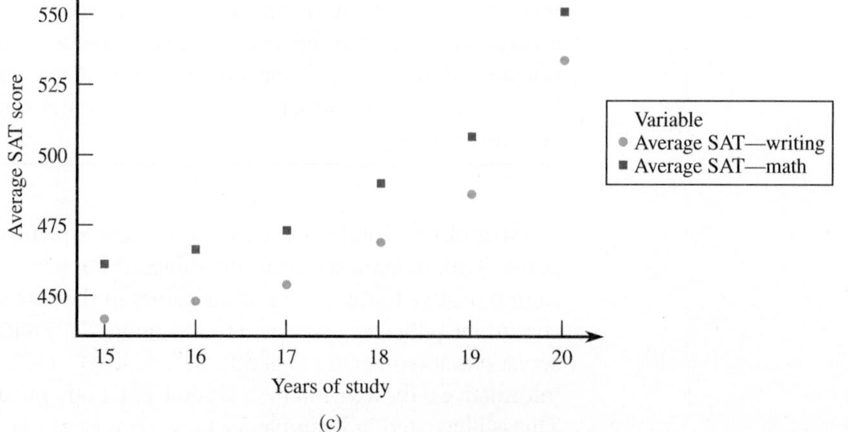

FIGURE 3.34
Minitab scatterplots of data in Example 3.21:
(a) scale for both axes selected by Minitab;
(b) axes intersect at the point (0, 0);
(c) math and writing on same plot.

● Data set available online

Interpret the results)

The scatterplot for average writing SAT score exhibits a fairly strong curved pattern, indicating that there is a strong relationship between average writing SAT score and the total number of years of study in the six core academic subjects. Although the pattern in the plot is curved rather than linear, it is still easy to see that the average writing SAT score increases as the number of years of study increases. Figure 3.34(c) shows a scatterplot with the average writing SAT scores represented by blue squares and the average math SAT scores represented by orange dots. From this plot we can see that while the average math SAT scores tend to be higher than the average writing scores at all of the values of total number of years of study, the general curved form of the relationship is similar. ■

In Chapter 5, methods for summarizing bivariate data when the scatterplot reveals a pattern are introduced. Linear patterns are relatively easy to work with. A curved pattern, such as the one in Example 3.21, is a bit more complicated to analyze, and methods for summarizing such nonlinear relationships are developed in Section 5.4.

Time Series Plots

Data sets often consist of measurements collected over time at regular intervals so that we can learn about change over time. For example, stock prices, sales figures, and other socioeconomic indicators might be recorded on a weekly or monthly basis. A **time series plot** (sometimes also called a time plot) is a simple graph of data collected over time that can be invaluable in identifying trends or patterns that might be of interest.

A time series plot can be constructed by thinking of the data set as a bivariate data set, where y is the variable observed and x is the time at which the observation was made. These (x, y) pairs are plotted as in a scatterplot. Consecutive observations are then connected by a line segment. This helps us to see trends over time.

EXAMPLE 3.22 The Cost of Christmas

Understand the context)

The Christmas Price Index is computed each year by PNC Advisors, and it is a humorous look at the cost of giving all of the gifts described in the popular Christmas song "The Twelve Days of Christmas." The year 2013 was the most costly year since the index began, with the "cost of Christmas" at $27,393.17. Historical data from the PNC web site (www .pncchristmaspriceindex.com) were used to construct the time series plot of Figure 3.35.

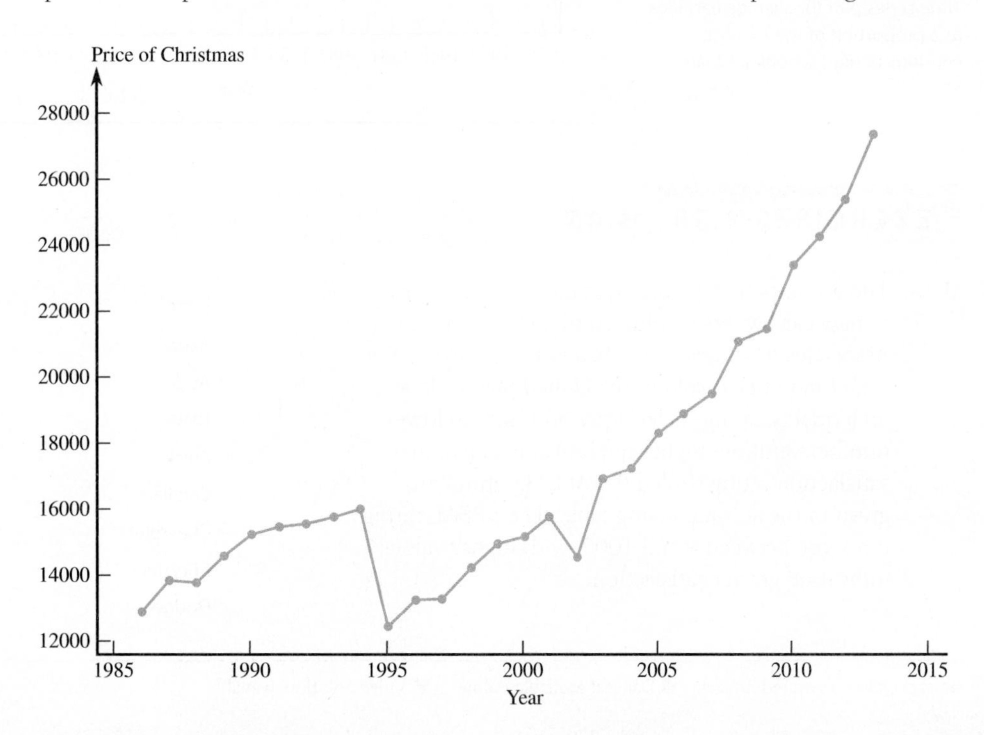

FIGURE 3.35
Time series plot for the Christmas Price Index data of Example 3.22.

Interpret the results ⟩

The plot shows an upward trend in the index from 1984 until 1993. A dramatic drop in the cost occurred between 1993 and 1995, but there has been a clear upward trend in the index since then. You can visit the web site to see individual time series plots for each of the twelve gifts that are used to determine the Christmas Price Index (a partridge in a pear tree, two turtle doves, etc.). See if you can figure out what caused the dramatic decline from 1993 to 1995. ■

EXAMPLE 3.23 Education Level and Income—Stay in School!

Understand the context ⟩

Interpret the results ⟩

The time series plot shown in Figure 3.36 appears on the U.S. Census Bureau web site. It shows the average earnings of workers by educational level as a proportion of the average earnings of a high school graduate over time. For example, we can see from this plot that in 1993 the average earnings for people with bachelor's degrees was about 1.5 times the average for high school graduates. In that same year, the average earnings for those who were not high school graduates was only about 75% (a proportion of .75) of the average for high school graduates. The time series plot also shows that the gap between the average earnings for high school graduates and those with a bachelor's degree or an advanced degree widened during the 1990s.

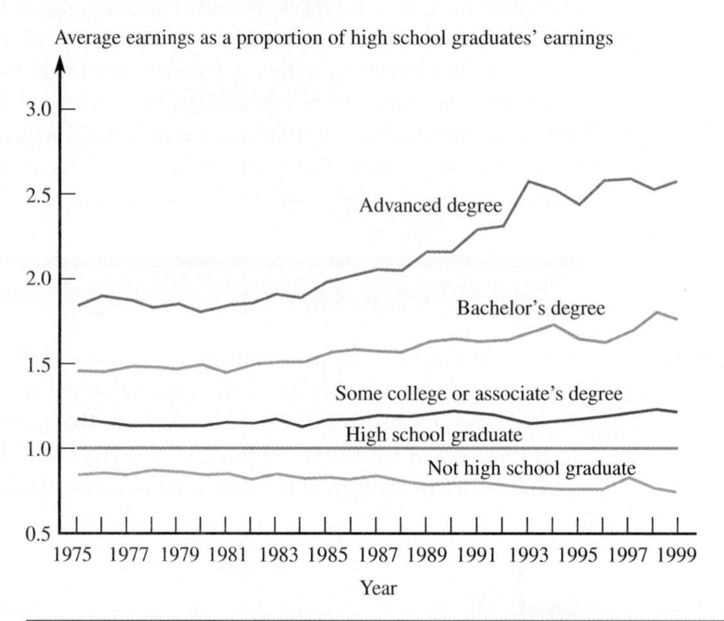

FIGURE 3.36
Time series plot for average earnings as a proportion of the average earnings of high school graduates.

■

EXERCISES 3.38 - 3.45

3.38 The accompanying table gives data from a survey of new car owners conducted by J.D. Power and Associates (***USA Today*, June 16 and July 17, 2010**). For each brand of car sold in the United States, data on a quality rating (defects per 100 cars, so lower numbers indicate higher quality) and a customer satisfaction rating (called the APEAL rating) are given in the accompanying table. The APEAL rating is a score between 0 and 1000, with higher values indicating greater satisfaction.

Brand	Quality Rating	APEAL Rating
Acura	86	822
Audi	111	832
BMW	113	845
Buick	114	802
Cadillac	111	818
Chevrolet	111	789
Chrysler	122	748
Dodge	130	751

continued

Brand	Quality Rating	APEAL Rating
Ford	93	794
GMC	126	792
Honda	95	766
Hyundai	102	760
Infiniti	107	805
Jaguar	130	854
Jeep	129	727
Kia	126	761
Land Rover	170	836
Lexus	88	822
Lincoln	106	820
Mazda	114	774
Mercedes-Benz	87	842
Mercury	113	769
Mini Cooper	133	815
Mitsubishi	146	767
Nissan	111	763
Porsche	83	877
Ram	110	780
Scion	114	764
Subaru	121	755
Suzuki	122	750
Toyota	117	745
Volkswagen	135	797
Volvo	109	795

a. Construct a scatterplot of x = quality rating and y = APEAL rating. (Hint: See Example 3.21.)

b. Does customer satisfaction (as measured by the APEAL rating) appear to be related to car quality? Explain.

3.39 ● ***Consumer Reports Health*** **(www.consumerreports.org)** gave the accompanying data on saturated fat (in grams), sodium (in mg), and calories for 36 fast-food items.

Fat	Sodium	Calories	Fat	Sodium	Calories
2	1042	268	3	450	200
5	921	303	6	800	320
3	250	260	3	1190	420
2	770	660	2	1090	120
1	635	180	5	570	290
6	440	290	3.5	1215	285
4.5	490	290	2.5	1160	390
5	1160	360	0	520	140
3.5	970	300	2.5	1120	330
1	1120	315	1	240	120
2	350	160	3	650	180

continued

Fat	Sodium	Calories	Fat	Sodium	Calories
1	1620	340	1.5	1200	370
4	660	380	2.5	1200	330
3	840	300	3	1250	330
1.5	1050	490	0	1040	220
3	1440	380	0	760	260
9	750	560	2.5	780	220
1	500	230	3	500	230

a. Construct a scatterplot using y = calories and x = fat. Does it look like there is a relationship between fat and calories? Is the relationship what you expected? Explain.

b. Construct a scatterplot using y = calories and x = sodium. Write a few sentences commenting on the difference between the relationship of calories to fat and calories to sodium.

c. Construct a scatterplot using y = sodium and x = fat. Does there appear to be a relationship between fat and sodium?

d. Add a vertical line at $x = 3$ and a horizontal line at $y = 900$ to the scatterplot in Part (c). This divides the scatterplot into four regions, with some of the points in the scatterplot falling into each of the four regions. Which of the four regions corresponds to healthier fast-food choices? Explain.

3.40 The report **"Wireless Substitution: Early Release of Estimates from the National Health Interview Survey" (Center for Disease Control, 2009)** gave the following estimates of the percentage of homes in the United States that had only wireless phone service at 6-month intervals from June 2005 to December 2008.

Date	Percent with Only Wireless Phone Service
June 2005	7.3
December 2005	8.4
June 2006	10.5
December 2006	12.8
June 2007	13.6
December 2007	15.8
June 2008	17.5
December 2008	20.2

a. Construct a time series plot for these data and describe the trend in the percent of homes with only wireless phone service over time. (Hint: See Example 3.22.)

b. Has the percent increased at a fairly steady rate?

3.41 ● The accompanying table gives the cost and an overall quality rating for 15 different brands of bike helmets (www.consumerreports.org).

Cost	Rating
35	65
20	61
30	60
40	55
50	54
23	47
30	47
18	43
40	42
28	41
20	40
25	32
30	63
30	63
40	53

a. Construct a scatterplot using y = quality rating and x = cost.

b. Based on the scatterplot from Part (a), does there appear to be a relationship between cost and quality rating? Does the scatterplot support the statement that the more expensive bike helmets tended to receive higher quality ratings?

3.42 ● The accompanying table gives the cost and an overall quality rating for 10 different brands of men's athletic shoes and nine different brands of women's athletic shoes (www.consumerreports.org).

Cost	Rating	Type
65	71	Men's
45	70	Men's
45	62	Men's
80	59	Men's
110	58	Men's
110	57	Men's
30	56	Men's
80	52	Men's
110	51	Men's
70	51	Men's
65	71	Women's
70	70	Women's
85	66	Women's
80	66	Women's
45	65	Women's
70	62	Women's
55	61	Women's
110	60	Women's
70	59	Women's

a. Using the data for all 19 shoes, construct a scatterplot using y = quality rating and x = cost. Write a sentence describing the relationship between quality rating and cost.

b. Construct a scatterplot of the 19 data points that uses different colors or different symbols to distinguish the points that correspond to men's shoes from those that correspond to women's shoes. How do men's and women's athletic shoes differ with respect to cost and quality rating?

c. Are the relationships between cost and quality rating the same for men and women? If not, how do the relationships differ?

3.43 ● The article **"Cities Trying to Rejuvenate Recycling Efforts" (*USA Today*, October 27, 2006)** states that the amount of waste collected for recycling has grown slowly in recent years. This statement was supported by the data in the accompanying table. Use these data to construct a time series plot. Explain how the plot is or is not consistent with the given statement.

Year	Recycled Waste (in millions of tons)
1990	29.7
1991	32.9
1992	36.0
1993	37.9
1994	43.5
1995	46.1
1996	46.4
1997	47.3
1998	48.0
1999	50.1
2000	52.7
2001	52.8
2002	53.7
2003	55.8
2004	57.2
2005	58.4

3.44 The report **"Trends in Higher Education' (www.collegeboard.com)** gave the accompanying data on smoking rates for people age 25 and older by education level.

	Percent Who Smoke			
Year	Not a High School Graduate	High School Graduate but No College	Some College	Bachelor's Degree or Higher
1960	44	46	48	44
1965	44	45	47	41
1970	42	44	45	37

continued

	Percent Who Smoke			
Year	Not a High School Graduate	High School Graduate but No College	Some College	Bachelor's Degree or Higher
1975	41	42	43	33
1980	38	40	39	28
1985	36	37	36	23
1990	34	34	32	19
1995	32	31	29	16
2000	30	28	26	14
2005	29	27	21	10

a. Construct a time series plot for people who did not graduate from high school and comment on any trend over time.

b. Construct a time series plot that shows trend over time for each of the four education levels. Graph each of the four time series on the same set of axes, using different colors to distinguish the different education levels. Either label the time series in the plot or include a legend to indicate which time series corresponds to which education level.

c. Write a paragraph about your plot from Part (b). Discuss the similarities and differences for the four different education levels.

3.45 The accompanying time series plot of movie box office totals (in millions of dollars) over 18 weeks of summer for both 2001 and 2002 is similar to one that appeared in ***USA Today* (September 3, 2002)**:

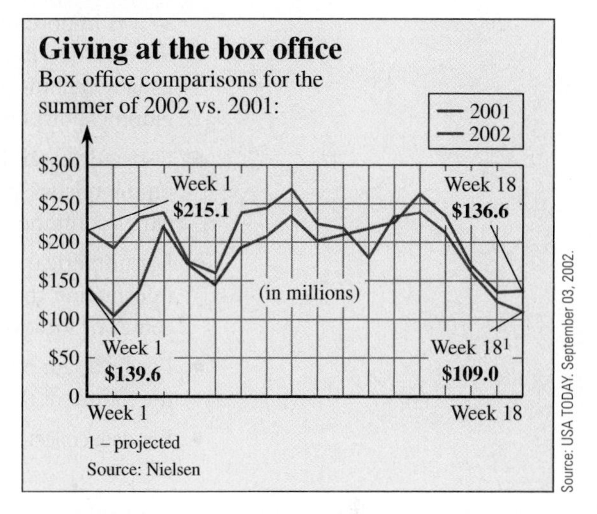

Patterns that tend to repeat on a regular basis over time are called seasonal patterns. Describe any seasonal patterns that you see in the summer box office data. (Hint: Look for patterns that seem to be consistent from year to year.)

Bold exercises answered in back ● Data set available online ▼ Video Solution available

3.5 Interpreting and Communicating the Results of Statistical Analyses

A graphical display, when used appropriately, can be a powerful tool for organizing and summarizing data. By sacrificing some of the detail of a complete listing of a data set, important features of the data distribution are more easily seen and more easily communicated to others.

Communicating the Results of Statistical Analyses

When reporting the results of a data analysis, a good place to start is with a graphical display of the data. A well-constructed graphical display is often the best way to highlight the essential characteristics of the data distribution, such as shape and spread for numerical data sets or the nature of the relationship between the two variables in a bivariate numerical data set.

For effective communication with graphical displays, some things to remember are

- Be sure to select a display that is appropriate for the given type of data.
- Be sure to include scales and labels on the axes of graphical displays.
- In comparative plots, be sure to include labels or a legend so that it is clear which parts of the display correspond to which samples or groups in the data set.
- Although it is sometimes a good idea to have axes that do not cross at (0, 0) in a scatterplot, the vertical axis in a bar chart or a histogram should always start at 0 (see the cautions and limitations later in this section for more about this).

- Keep your graphs simple. A simple graphical display is much more effective than one that has a lot of extra "junk." Most people will not spend a great deal of time studying a graphical display, so its message should be clear and straightforward.
- Keep your graphical displays honest. People tend to look quickly at graphical displays, so it is important that a graph's first impression is an accurate and honest portrayal of the data distribution. In addition to the graphical display itself, data analysis reports usually include a brief discussion of the features of the data distribution based on the graphical display.
- For categorical data, the discussion of the graphical display might be a few sentences on the relative proportion for each category, possibly pointing out categories that were either common or rare compared to other categories.
- For numerical data sets, the discussion of the graphical display usually summarizes the information that the display provides on three characteristics of the data distribution: center or location, spread, and shape.
- For bivariate numerical data, the discussion of the scatterplot would typically focus on the nature of the relationship between the two variables used to construct the plot.
- For data collected over time, any trends or patterns in the time series plot would be described.

Interpreting the Results of Statistical Analyses

When someone uses a web search engine, do they rely on the ranking of the search results returned or do they first scan the results looking for the most relevant? The authors of the paper **"Learning User Interaction Models for Predicting Web Search Result Preferences"** (***Proceedings of the 29th Annual ACM Conference on Research and Development in Information Retrieval,*** **2006**) attempted to answer this question by observing user behavior when they varied the position of the most relevant result in the list of resources returned in response to a web search.

They concluded that people clicked more often on results near the top of the list, even when they were not relevant. They supported this conclusion with the comparative bar graph in Figure 3.37.

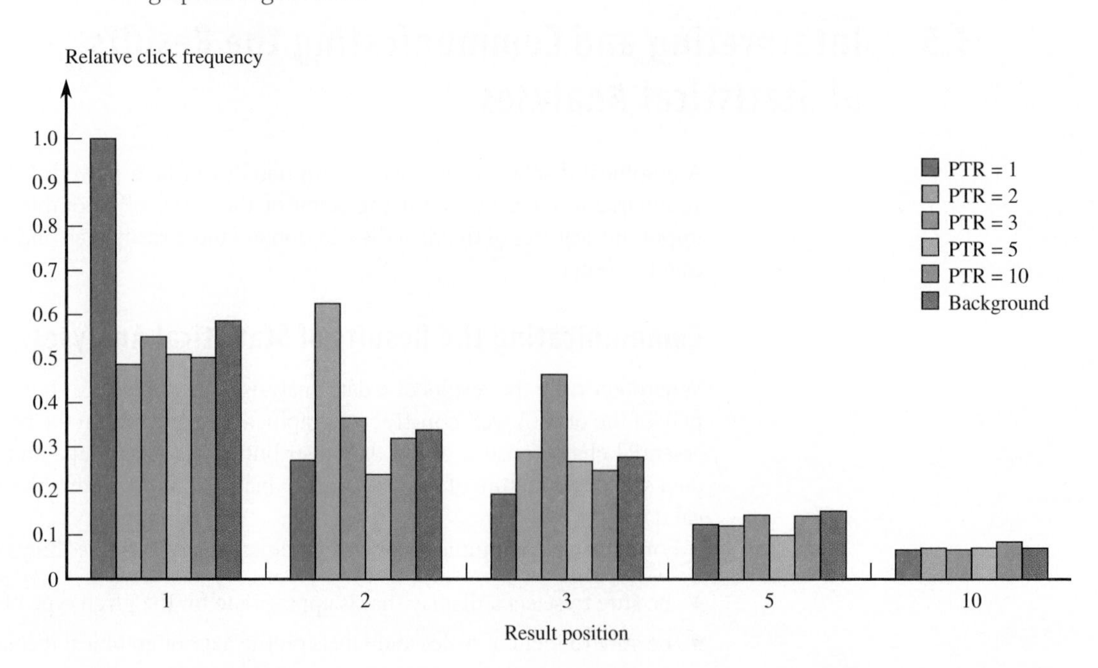

FIGURE 3.37
Comparative bar graph for click frequency data.

Although this comparative bar chart is a bit complicated, we can learn a great deal from this graphical display. Let's start by looking at the first group of bars. The different bars correspond to where in the list of search results the result that was considered to be most relevant was located. For example, in the legend PTR = 1 means that the most relevant result was in position 1 in the list returned. PTR = 2 means that the most relevant result was in the second position in the list returned, and so on. PTR = Background means that the most relevant result was not in the first 10 results returned.

The first group of bars shows the proportion of times users clicked on the first result returned. Notice that all users clicked on the first result when it was the most relevant, but nearly half clicked on the first result when the most relevant result was in the second position and more than half clicked on the first result when the most relevant result was even farther down the list.

The second group of bars represents the proportion of users who clicked on the second result. Notice that the proportion who clicked on the second result was highest when the most relevant result was in that position. Stepping back to look at the entire graphical display, we see that users tended to click on the most relevant result if it was in one of the first three positions, but if it appeared after that, very few selected it. Also, if the most relevant result was in the third or a later position, users were more likely to click on the first result returned, and the likelihood of a click on the most relevant result decreased the farther down the list it appeared. To fully understand why the researchers' conclusions are justified, we need to be able to extract this kind of information from graphical displays.

The use of graphical data displays is quite common in newspapers, magazines, and journals, so it is important to be able to extract information from such displays. For example, data on test scores for a standardized math test given to eighth graders in 37 states, 2 territories (Guam and the Virgin Islands), and the District of Columbia were used to construct the stem-and-leaf display and histogram shown in Figure 3.38. Careful examination of these displays reveals the following:

1. Most of the participating states had average eighth-grade math scores between 240 and 280. We would describe the shape of this display as negatively skewed, because of the longer tail on the low end of the distribution.
2. Three of the average scores differed substantially from the others. These turn out to be 218 (Virgin Islands), 229 (District of Columbia), and 230 (Guam). These three scores could be described as outliers. It is interesting to note that the three unusual values are from the areas that are not states.

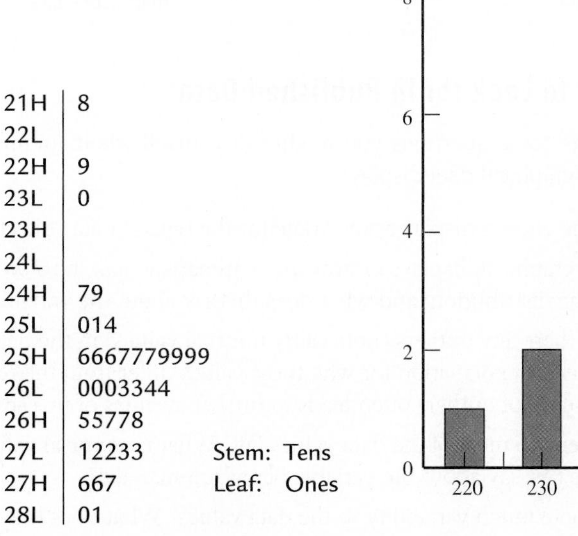

Stem	Leaf
21H	8
22L	
22H	9
23L	0
23H	
24L	
24H	79
25L	014
25H	6667779999
26L	0003344
26H	55778
27L	12233
27H	667
28L	01

Stem: Tens
Leaf: Ones

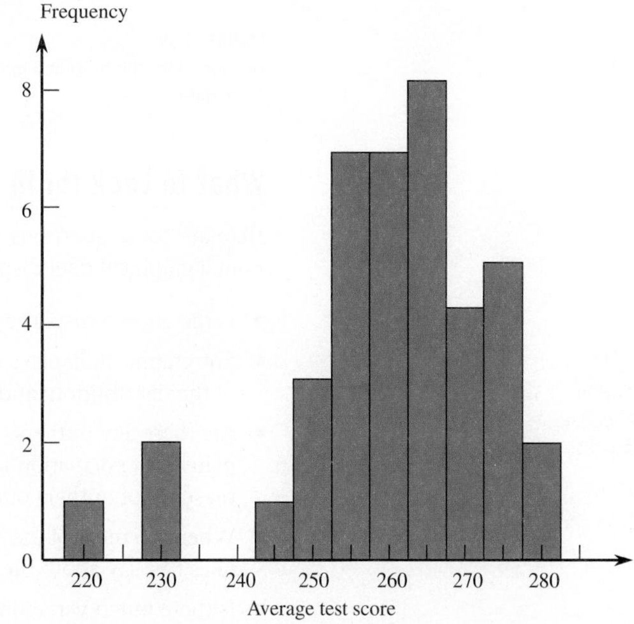

FIGURE 3.38
Stem-and-leaf display and histogram for math test scores.

3. There do not appear to be any outliers on the high side.
4. A "typical" average math score for the 37 states would be somewhere around 260.
5. There is quite a bit of variability in average score from state to state.

How would the displays have been different if the two territories and the District of Columbia had not participated in the testing? The resulting histogram is shown in Figure 3.39. Note that the display is now more symmetric, with no noticeable outliers. The display still reveals quite a bit of state-to-state variability in average score, and 260 still looks reasonable as a "typical" average score.

Now suppose that the two highest values among the 37 states (Montana and North Dakota) had been even higher. The stem-and-leaf display might then look like the one given in Figure 3.40. In this stem-and-leaf display, two values stand out from the main part of the display. This would catch our attention and might cause us to look carefully at these two states to determine what factors may be related to high math scores.

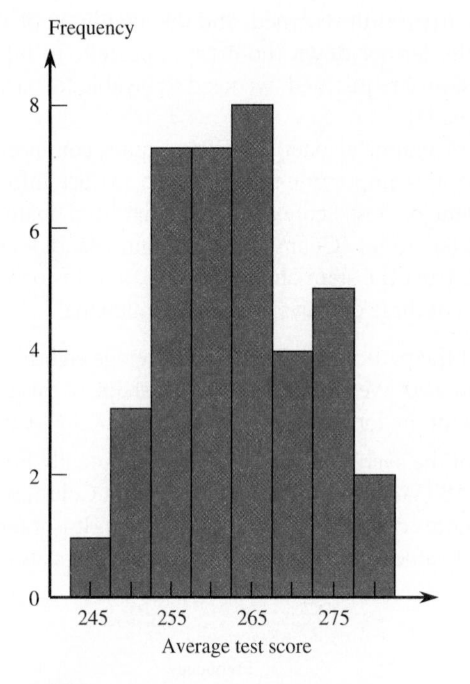

FIGURE 3.39
Histogram for the modified math score data.

Stem	Leaf
24H	79
25L	014
25H	6667779999
26L	0003344
26H	55778
27L	12233
27H	667
28L	
28H	
29L	
29H	68

Stem: Tens
Leaf: Ones

FIGURE 3.40
Stem-and-leaf display for modified math score data.

What to Look for in Published Data

Here are some questions you might ask yourself when attempting to extract information from a graphical data display:

- Is the chosen display appropriate for the type of data collected?
- For graphical displays of univariate numerical data, how would you describe the shape of the distribution, and what does this say about the variable being summarized?
- Are there any outliers (noticeably unusual values) in the data set? Is there any plausible explanation for why these values differ from the rest of the data? (The presence of outliers often leads to further avenues of investigation.)
- Where do most of the data values fall? What is a typical value for the data set? What does this say about the variable being summarized?
- Is there much variability in the data values? What does this say about the variable being summarized?

Of course, you should always think carefully about how the data were collected. If the data were not gathered in a reasonable manner (based on sound sampling methods or experimental design principles), you should be cautious in formulating any conclusions based on the data.

Consider the histogram in Figure 3.41, which is based on data published by the National Center for Health Statistics. The data set summarized by this histogram consisted of infant mortality rates (deaths per 1000 live births) for the 50 states in the United States. A histogram is an appropriate way of summarizing these data (although with only 50 observations, a stem-and-leaf display would also have been reasonable).

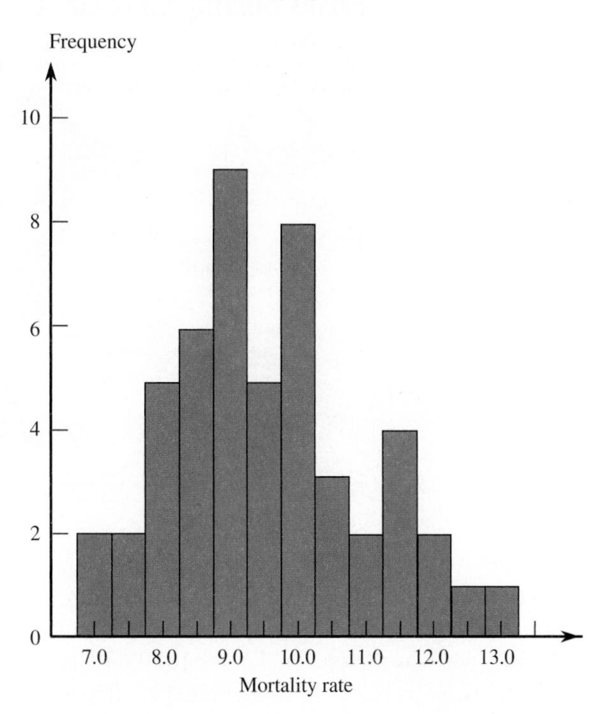

FIGURE 3.41
Histogram of infant mortality rates.

The histogram itself is slightly positively skewed, with most mortality rates between 7.5 and 12. There is quite a bit of variability in infant mortality rate from state to state—perhaps more than we might have expected. This variability might be explained by differences in economic conditions or in access to health care. We may want to look further into these issues.

Although there are no obvious outliers, the upper tail is a little longer than the lower tail. The three largest values in the data set are 12.1 (Alabama), 12.3 (Georgia), and 12.8 (South Carolina)—all Southern states. Again, this may suggest some interesting questions that deserve further investigation.

A typical infant mortality rate would be about 9.5 deaths per 1000 live births. This represents an improvement, because researchers at the National Center for Health Statistics stated that the overall rate for 1988 was 10 deaths per 1000 live births. However, they also point out that the United States still ranked 22 out of 24 industrialized nations surveyed, with only New Zealand and Israel having higher infant mortality rates.

A Word to the Wise: Cautions and Limitations

When constructing and interpreting graphical displays, you need to keep in mind these things:

1. **Areas should be proportional to frequency, relative frequency, or magnitude of the number being represented.** The eye is naturally drawn to large areas in graphical displays, and it is natural for the observer to make informal comparisons based on area. Correctly constructed graphical displays, such as pie charts, bar charts, and histograms, are designed so that the areas of the pie slices or the

bars are proportional to frequency or relative frequency. Sometimes, in an effort to make graphical displays more interesting, designers lose sight of this important principle, and the resulting graphs are misleading. For example, consider the following graph (*USA Today,* **October 3, 2002**):

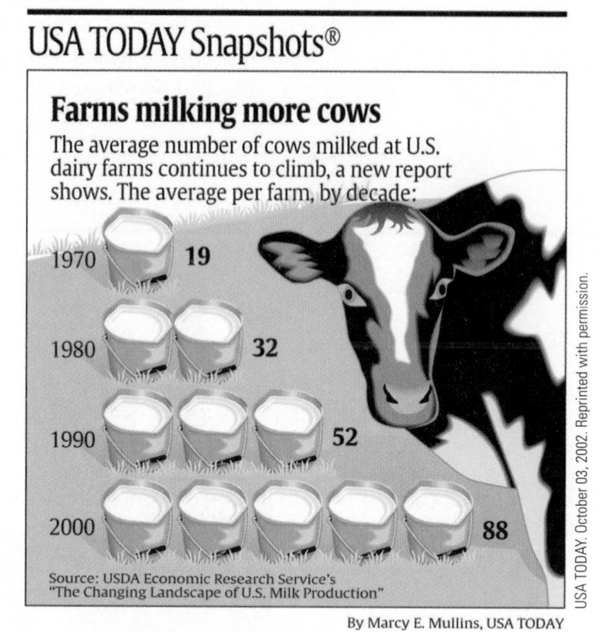

USA TODAY Snapshots®
Farms milking more cows
The average number of cows milked at U.S. dairy farms continues to climb, a new report shows. The average per farm, by decade:
1970
19
1980
32
1990
52
2000
88
Source: USDA Economic Research Service's "The Changing Landscape of U.S. Milk Production"
By Marcy E. Mullins, USA TODAY
USA TODAY. October 03, 2002. Reprinted with permission.

In trying to make the graph more visually interesting by replacing the bars of a bar chart with milk buckets, areas are distorted. For example, the two buckets for 1980 represent 32 cows, whereas the one bucket for 1970 represents 19 cows. This is misleading because 32 is not twice as big as 19. Other areas are distorted as well.

Another common distortion occurs when a third dimension is added to bar charts or pie charts. For example, the pie chart at the bottom left of the page is similar to one that appeared in *USA Today* **(September 17, 2009)**.

Adding the third dimension distorts the areas and makes it much more difficult to interpret correctly. A correctly drawn pie chart is shown below.

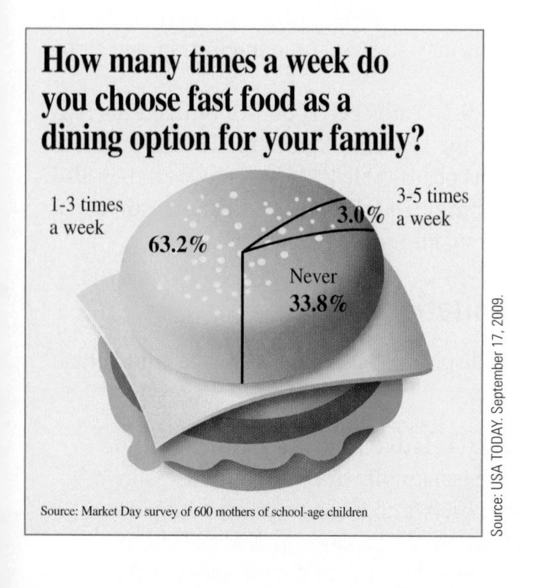

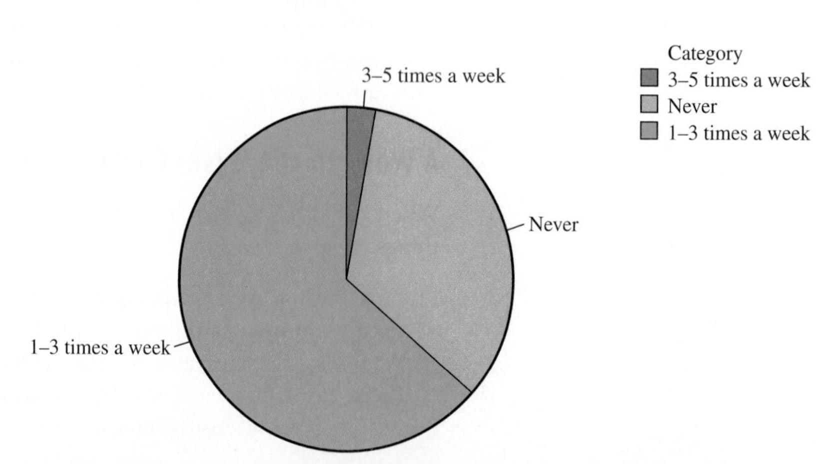

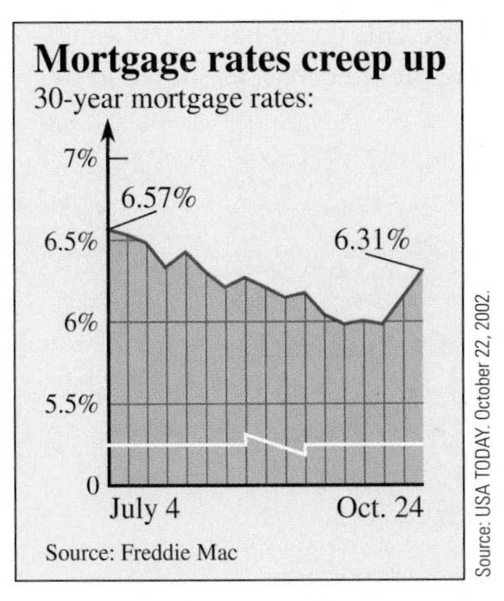

Source: USA TODAY. October 22, 2002.

2. **Be cautious of graphs with broken axes.** Although it is common to see scatterplots with broken axes, be extremely cautious of time series plots, bar charts, or histograms with broken axes. The use of broken axes in a scatterplot does not distort information about the nature of the relationship in the bivariate data set used to construct the display. On the other hand, in time series plots, broken axes can sometimes exaggerate the magnitude of change over time. Although it is not always inadvisable to break the vertical axis in a time series plot, it is something you should watch for, and if you see a time series plot with a broken axis, as in the accompanying time series plot of mortgage rates (similar to a graph appearing in ***USA Today*, October 25, 2002**), you should pay particular attention to the scale on the vertical axis and take extra care in interpreting the graph.

   In bar charts and histograms, the vertical axis (which represents frequency, relative frequency, or density) should *never* be broken. If the vertical axis is broken in this type of graph, the resulting display will violate the "proportional area" principle and the display will be misleading. For example, the accompanying bar chart is similar to one appearing in an advertisement for a software product designed to help teachers raise student test scores. By starting the vertical axis at 50, the gain for students using the software is exaggerated. Areas of the bars are not proportional to the magnitude of the numbers represented—the area for the rectangle representing 68 is more than three times the area of the rectangle representing 55!

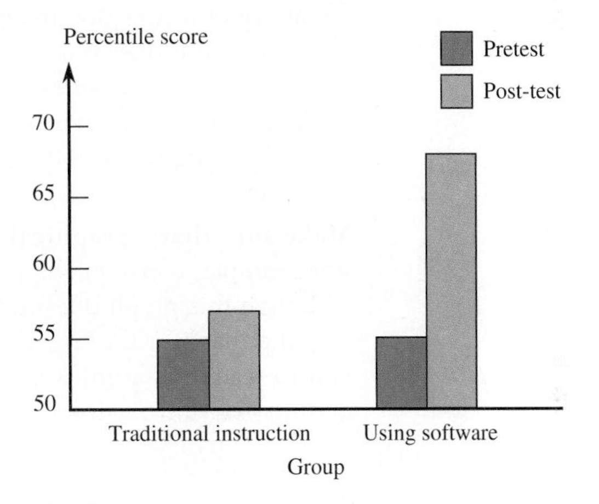

3. **Watch out for unequal time spacing in time series plots.** If observations over time are not made at regular time intervals, special care must be taken in constructing the time series plot. Consider the accompanying time series plot, which is similar to one appearing in the ***San Luis Obispo Tribune* (September 22, 2002)** in an article on online banking:

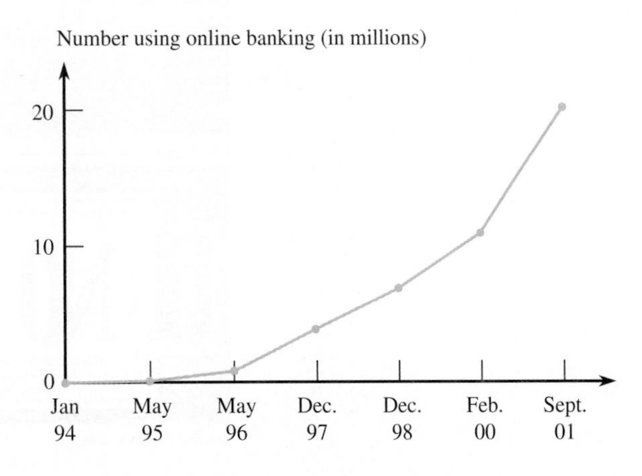

Notice that the intervals between observations are irregular, yet the points in the plot are equally spaced along the time axis. This makes it difficult to make a coherent assessment of the rate of change over time. This could have been remedied by spacing the observations differently along the time axis, as shown in the following plot:

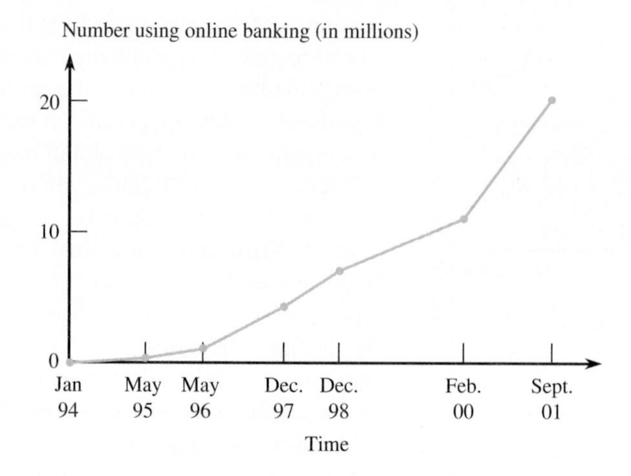

4. **Be careful how you interpret patterns in scatterplots.** A strong pattern in a scatterplot means that the two variables tend to vary together in a predictable way, but it does not mean that there is a cause-and-effect relationship between the two variables. We will consider this point further in Chapter 5, but in the meantime, when describing patterns in scatterplots, be careful not to use wording that implies that changes in one variable *cause* changes in the other.
5. **Make sure that a graphical display creates the right first impression.** For example, consider the graph below from ***USA Today*** **(June 25, 2002)**. Although this graph does not violate the proportional area principle, the way the "bar" for the "none" category is displayed makes this graph difficult to read, and a quick glance at this graph would leave the reader with an incorrect impression.

EXERCISES 3.46 - 3.51

3.46 The accompanying comparative bar chart is from the report **"More and More Teens on Cell Phones" (Pew Research Center, www.pewresearch.org, August 19, 2009)**.

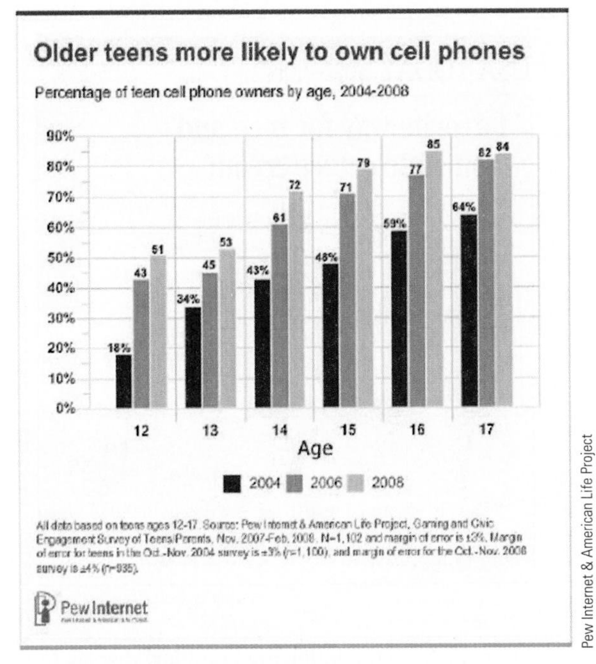

Pew Internet & American Life Project

Suppose that you plan to include this graph in an article that you are writing for your school newspaper. Write a few paragraphs that could accompany the graph. Be sure to address what the graph reveals about how teen cell phone ownership is related to age and how it has changed over time.

3.47 Figure EX-3.47 is from the Fall 2008 Census Enrollment Report at Cal Poly, San Luis Obispo. It uses both a pie chart and a segmented bar graph to summarize data on ethnicity for students enrolled at the university in Fall 2008.

a. Use the information in the graphical display to construct a single segmented bar graph for the ethnicity data.

b. Do you think that the original graphical display or the one you created in Part (a) is more informative? Explain your choice.

c. Why do you think that the original graphical display format (combination of pie chart and segmented bar graph) was chosen over a single pie chart with 7 slices?

3.48 The accompanying graph is similar to one that appeared in ***USA Today* (August 5, 2008)**. This graph is a modified comparative bar graph. Most likely, the modifications (incorporating hands and the earth) were made to try to make a display that readers would find more interesting.

a. Use the information in the *USA Today* graph to construct a traditional comparative bar graph.

b. Explain why the modifications made in the *USA Today* graph may make interpretation more difficult than with the traditional comparative bar graph.

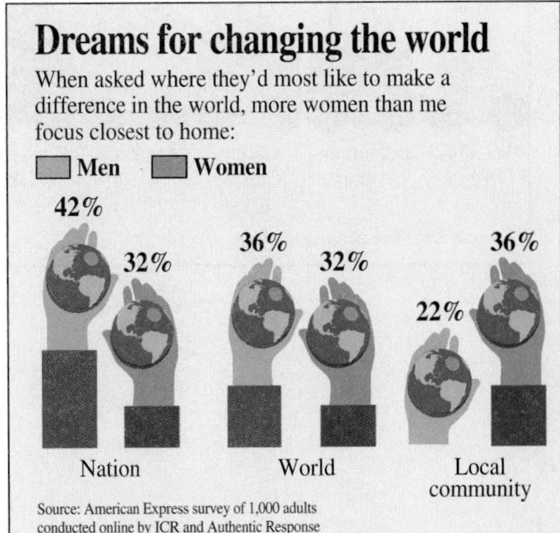

Source: USA TODAY, August 5, 2008.

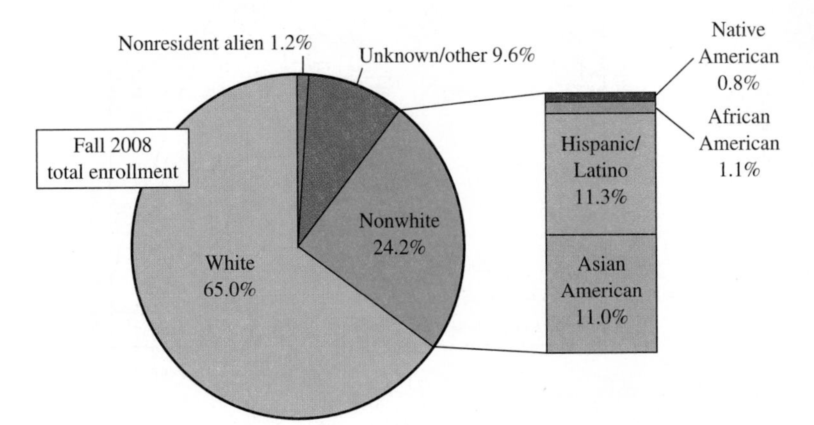

FIGURE EX-3.47

3.49 The two graphical displays below are similar to ones that appeared in ***USA Today* (June 8, 2009 and July 28, 2009)**. One is an appropriate representation and the other is not. For each of the two, explain why it is or is not drawn appropriately.

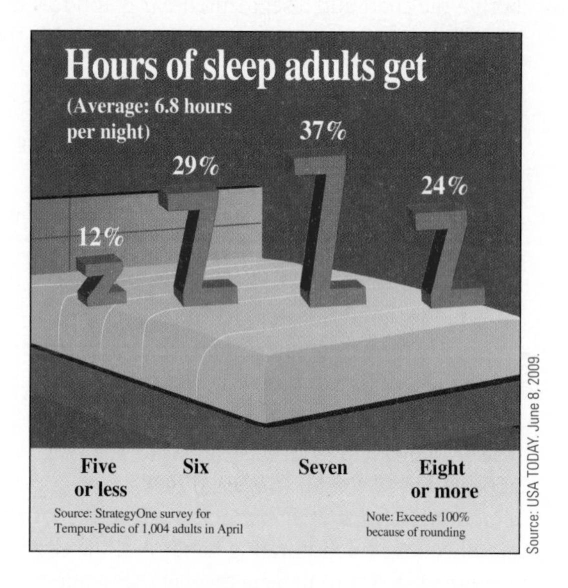

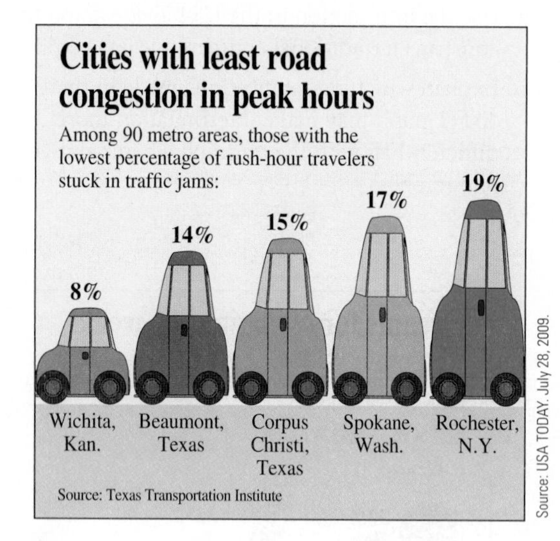

3.50 The following graphical display is meant to be a comparative bar graph (***USA Today*, August 3, 2009**). Do you think that this graphical display is an effective summary of the data? If so, explain why. If not, explain why not and construct a display that makes it easier to compare the ice cream preferences of men and women.

USA TODAY Snapshots®

Favorite way for men and women to eat ice cream

	Men	Women
Cup	50%	41%
Cone	24%	34%
Sundae	17%	18%
Sandwich	2%	2%
Other	8%	5%

Note: Exceeds 100% because of rounding.

Source: Harris Interactive survey of 2,177 adults conducted online June 8-15

By Anne R. Carey and Karl Gelles, USA TODAY

3.51 Explain why the following graphical display (similar to one appearing in ***USA Today*, September 17, 2009**) is misleading.

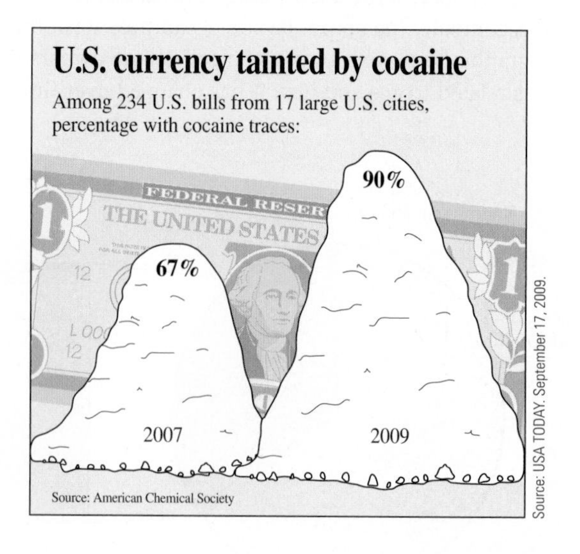

ACTIVITY 3.1 Locating States

Background: A newspaper article bemoaning the state of students' knowledge of geography claimed that more students could identify the island where the 2002 season of the TV show *Survivor* was filmed than could locate Vermont on a map of the United States. In this activity, you will collect data that will allow you to estimate the proportion of students who can correctly locate the states of Vermont and Nebraska.

1. Working as a class, decide how you will select a sample that you think will be representative of the students from your school.
2. Use the sampling method from Step 1 to obtain the subjects for this study. Subjects should be shown the accompanying map of the United States and asked to point out the state of Vermont. After the subject has given his or her answer, ask the subject to point out the state of Nebraska. For each subject, record whether or not Vermont was correctly identified and whether or not Nebraska was correctly identified.

3. When the data collection process is complete, summarize the resulting data in a table like the one shown here:

Response	Frequency
Correctly identified both states	
Correctly identified Vermont but not Nebraska	
Correctly identified Nebraska but not Vermont	
Did not correctly identify either state	

4. Construct a pie chart that summarizes the data in the table from Step 3.
5. What proportion of sampled students were able to correctly identify Vermont on the map?
6. What proportion of sampled students were able to correctly identify Nebraska on the map?
7. Construct a comparative bar chart that shows the proportion correct and the proportion incorrect for each of the two states considered.
8. Which state, Vermont or Nebraska, is closer to the state in which your school is located? Based on the pie chart, do you think that the students at your school were better able to identify the state that was closer than the one that was farther away? Justify your answer.
9. Write a paragraph commenting on the level of knowledge of U.S. geography demonstrated by the students participating in this study.
10. Would you be comfortable generalizing your conclusions in Step 8 to the population of students at your school? Explain why or why not.

ACTIVITY 3.2 Bean Counters!

Materials needed: A large bowl of dried beans (or marbles, plastic beads, or any other small, fairly regular objects) and a coin.

In this activity, you will investigate whether people can hold more in the right hand or in the left hand.

1. Flip a coin to determine which hand you will measure first. If the coin lands heads side up, start with the right hand. If the coin lands tails side up, start with the left hand. With the designated hand, reach into the bowl and grab as many beans as possible. Raise the hand over the bowl and count to 4. If no beans drop during the count to 4, drop the beans onto a piece of paper and record the number of beans grabbed. If any beans drop during the count, restart the count. That is, you must hold the beans for a count of 4 without any beans falling before you can determine the number grabbed. Repeat the process with the other hand, and then record the following information: (1) right-hand number, (2) left-hand number, and (3) dominant hand (left or right, depending on whether you are left- or right-handed).
2. Create a class data set by recording the values of the three variables listed in Step 1 for each student in your class.
3. Using the class data set, construct a comparative stem-and-leaf display with the right-hand counts displayed on the right and the left-hand counts displayed on the left of the stem-and-leaf display. Comment on the interesting features of the display and include a comparison of the right-hand count and left-hand count distributions.

4. Now construct a comparative stem-and-leaf display that allows you to compare dominant-hand count to nondominant-hand count. Does the display support the theory that dominant-hand count tends to be higher than nondominant-hand count?
5. For each observation in the data set, compute the difference

   dominant-hand count − nondominant-hand count

   Construct a stem-and-leaf display of the differences. Comment on the interesting features of this display.
6. Explain why looking at the distribution of the differences (Step 5) provides more information than the comparative stem-and-leaf display (Step 4). What information is lost in the comparative display that is retained in the display of the differences?

SUMMARY Key Concepts and Formulas

TERM OR FORMULA	COMMENT
Frequency distribution	A table that displays frequencies, and sometimes relative and cumulative relative frequencies, for categories (categorical data), possible values (discrete numerical data), or class intervals (continuous data).
Comparative bar chart	Two or more bar charts that use the same set of horizontal and vertical axes.
Pie chart	A graph of a frequency distribution for a categorical data set. Each category is represented by a slice of the pie, and the area of the slice is proportional to the corresponding frequency or relative frequency.
Segmented bar graph	A graph of a frequency distribution for a categorical data set. Each category is represented by a segment of the bar, and the area of the segment is proportional to the corresponding frequency or relative frequency.
Stem-and-leaf display	A method of organizing numerical data in which the stem values (leading digit(s) of the observations) are listed in a column, and the leaf (trailing digit(s)) for each observation is then listed beside the corresponding stem. Sometimes stems are repeated to stretch the display.
Histogram	A graph of the information in a frequency distribution for a numerical data set. A rectangle is drawn above each possible value (discrete data) or class interval. The rectangle's area is proportional to the corresponding frequency or relative frequency.
Histogram shapes	A (smoothed) histogram may be unimodal (a single peak), bimodal (two peaks), or multimodal. A unimodal histogram may be symmetric, positively skewed (a long right or upper tail), or negatively skewed. A frequently occurring shape is one that is approximately normal.
Cumulative relative frequency plot	A graph of a cumulative relative frequency distribution.
Scatterplot	A graph of bivariate numerical data in which each observation (x, y) is represented as a point with respect to a horizontal x-axis and a vertical y-axis.
Time series plot	A graphical display of numerical data collected over time.

CHAPTER REVIEW Exercises 3.52 - 3.71

3.52 The article **"Most Smokers Wish They Could Quit"** (***Gallup Poll Analyses*, November 21, 2002**) noted that smokers and nonsmokers perceive the risks of smoking differently. The accompanying relative frequency table summarizes responses regarding the perceived harm of smoking for each of three groups: a sample of 241 smokers, a sample of 261 former smokers, and a sample of 502 nonsmokers.

a. Construct a comparative bar chart for these data. Do not forget to use relative frequencies in constructing the bar chart because the three sample sizes are different.

b. Comment on how smokers, former smokers, and nonsmokers differ with respect to perceived risk of smoking.

	Frequency		
Perceived Risk of Smoking	Smokers	Former Smokers	Nonsmokers
Very harmful	145	204	432
Somewhat harmful	72	42	50
Not too harmful	17	10	15
Not at all harmful	7	5	5

3.53 Each year the College Board publishes a profile of students taking the SAT. In the report **"2005 College Bound Seniors: Total Group Profile Report,"** the average SAT scores were reported for three groups defined by first language learned. Use the data in the accompanying table to construct a bar chart of the average verbal SAT score for the three groups.

First Language Learned	Average Verbal SAT
English	519
English and another language	486
A language other than English	462

3.54 The report referenced in Exercise 3.53 also gave average math SAT scores for the three language groups, as shown in the following table.

First Language Learned	Average Math SAT
English	521
English and another language	513
A language other than English	521

a. Construct a comparative bar chart for the average verbal and math scores for the three language groups.

b. Write a few sentences describing the differences and similarities between the three language groups as shown in the bar chart.

3.55 ● The Connecticut Agricultural Experiment Station conducted a study of the calorie content of different types of beer. The calorie content (calories per 100 ml) for 26 brands of light beer are (from the web site brewery.org):

29 28 33 31 30 33 30 28 27 41 39 31 29
23 32 31 32 19 40 22 34 31 42 35 29 43

a. Construct a stem-and-leaf display using stems 1, 2, 3, and 4.

b. Write a sentence or two describing the calorie content of light beers.

3.56 The stem-and-leaf display of Exercise 3.55 uses only four stems. Construct a stem-and-leaf display for these data using repeated stems 1H, 2L, 2H, . . . , 4L. For example, the first observation, 29, would have a stem of 2 and a leaf of 9. It would be entered into the display for the stem 2H, because it is a "high" 2—that is, it has a leaf that is on the high end (5, 6, 7, 8, 9).

3.57 ● The article **"A Nation Ablaze with Change" (*USA Today*, July 3, 2001)** gave the accompanying data on percentage increase in population between 1990 and 2000 for the 50 U.S. states. Also provided in the table is a column that indicates for each state whether the state is in the eastern or western part of the United States (the states are listed in order of population size):

State	Percentage Change	East/West
California	13.8	W
Texas	22.8	W
New York	5.5	E
Florida	23.5	E
Illinois	8.6	E
Pennsylvania	3.4	E
Ohio	4.7	E
Michigan	6.9	E
New Jersey	8.9	E
Georgia	26.4	E
North Carolina	21.4	E
Virginia	14.4	E
Massachusetts	5.5	E
Indiana	9.7	E
Washington	21.1	W
Tennessee	16.7	E
Missouri	9.3	E
Wisconsin	9.6	E
Maryland	10.8	E
Arizona	40.0	W
Minnesota	12.4	E
Louisiana	5.9	E
Alabama	10.1	E
Colorado	30.6	W
Kentucky	9.7	E
South Carolina	15.1	E
Oklahoma	9.7	W
Oregon	20.4	W
Connecticut	3.6	E
Iowa	5.4	E
Mississippi	10.5	E
Kansas	8.5	W
Arkansas	13.7	E
Utah	29.6	W
Nevada	66.3	W
New Mexico	20.1	W
West Virginia	0.8	E
Nebraska	8.4	W
Idaho	28.5	W

continued

State	Percentage Change	East/West
Maine	3.9	E
New Hampshire	11.4	E
Hawaii	9.3	W
Rhode Island	4.5	E
Montana	12.9	W
Delaware	17.6	E
South Dakota	8.5	W
North Dakota	0.5	W
Alaska	14.0	W
Vermont	8.2	E
Wyoming	8.9	W

a. Construct a stem-and-leaf display for percentage growth for the data set consisting of all 50 states. Hints: Regard the observations as having two digits to the left of the decimal place. That is, think of an observation such as 8.5 as 08.5. It will also be easier to truncate leaves to a single digit; for example, a leaf of 8.5 could be truncated to 8 for purposes of constructing the display.

b. Comment on any interesting features of the data set. Do any of the observations appear to be outliers?

c. Now construct a comparative stem-and-leaf display for the eastern and western states. Write a few sentences comparing the percentage growth distributions for eastern and western states.

3.58 ● People suffering from Alzheimer's disease often have difficulty performing basic activities of daily living (ADLs). In one study (**"Functional Status and Clinical Findings in Patients with Alzheimer's Disease,"** ***Journal of Gerontology*** **[1992]: 177–182)**, investigators focused on six such activities: dressing, bathing, transferring, toileting, walking, and eating. Here are data on the number of ADL impairments for each of 240 patients:

Number of impairments	0	1	2	3	4	5	6
Frequency	100	43	36	17	24	9	11

a. Determine the relative frequencies that correspond to the given frequencies.

b. What proportion of these patients had at most two impairments?

c. Use the result of Part (b) to determine what proportion of patients had more than two impairments.

d. What proportion of the patients had at least four impairments?

3.59 Does the size of a transplanted organ matter? A study that attempted to answer this question (**"Minimum Graft Size for Successful Living Donor Liver Transplantation,"** ***Transplantation*** **[1999]:1112–1116)** presented a scatterplot much like the following ("graft weight ratio" is the weight of the transplanted liver relative to the ideal size liver for the recipient):

a. Discuss interesting features of this scatterplot.

b. Why do you think the overall relationship is negative?

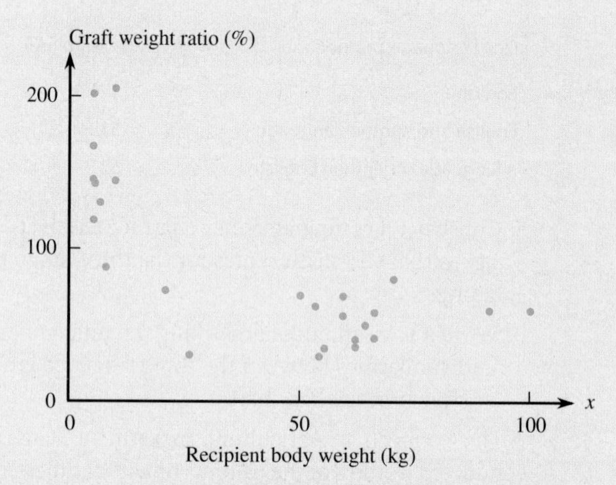

3.60 ● The **National Telecommunications and Information Administration** published a report titled **"Falling Through the Net: Toward Digital Inclusion" (U.S. Department of Commerce, October 2000)** that included the following information on access to computers in the home:

Year	Percentage of Households with a Computer
1985	8.2
1990	15.0
1994	22.8
1995	24.1
1998	36.6
1999	42.1
2000	51.0

a. Construct a time series plot for these data. Be careful—the observations are not equally spaced in time. The points in the plot should not be equally spaced along the x-axis.

b. Comment on any trend over time.

3.61 According to the National Association of Home Builders, the average size of a home in 1950 was 983 ft^2. The average size increased to 1500 ft^2 in 1970, 2080 ft^2 in 1990; and 2330 ft^2 in 2003 (***San Luis Obispo Tribune*, October 16, 2005**).

a. Construct a time series plot that shows how the average size of a home has changed over time.

b. If the trend of the time series plot continued, what would you predict the average home size to have been in 2010?

3.62 The paper **"Community Colleges Start to Ask, Where Are the Men?" (*Chronicle of Higher Education*, June 28, 2002)** gave data on gender for community college students. It was reported that 42% of students enrolled at community colleges nationwide were male and 58% were female. Construct a segmented bar graph for these data.

3.63 ● The article **"Tobacco and Alcohol Use in G-Rated Children's Animated Films" (*Journal of the American Medical Association* [1999]: 1131–1136)** reported exposure to tobacco and alcohol use in all G-rated animated films released between 1937 and 1997 by five major film studios. The researchers found that tobacco use was shown in 56% of the reviewed films. Data on the total tobacco exposure time (in seconds) for films with tobacco use produced by Walt Disney, Inc., were as follows:

223 176 548 37 158 51 299 37 11 165
74 92 6 23 206 9

Data for 11 G-rated animated films showing tobacco use that were produced by MGM/United Artists, Warner Brothers, Universal, and Twentieth Century Fox were also given. The tobacco exposure times (in seconds) for these films was as follows:

205 162 6 1 117 5 91 155 24 55 17

a. Construct a comparative stem-and-leaf display for these data.

b. Comment on the interesting features of this display.

3.64 ● The accompanying data on household expenditures on transportation for the United Kingdom appeared in **"Transport Statistics for Great Britain: 2002 Edition" (in *Family Spending: A Report on the Family Expenditure Survey* [The Stationary Office, 2002])**. Expenditures (in pounds per week) included costs of purchasing and maintaining any vehicles owned by members of the household and any costs associated with public transportation and leisure travel.

Year	Average Transportation	Percentage of Household Expenditures for Transportation
1990	247.20	16.2
1991	259.00	15.3
1992	271.80	15.8
1993	276.70	15.6
1994	283.60	15.1
1995	289.90	14.9
1996	309.10	15.7
1997	328.80	16.7
1998	352.20	17.0
1999	359.40	17.2
2000	385.70	16.7

a. Construct time series plots of the transportation expense data and the percent of household expense data.

b. Do the time series plots of Part (a) support the statement that follows? Explain why or why not. Statement: Although actual expenditures have been increasing, the percentage of the total household expenditures that go toward transportation has remained relatively stable.

3.65 The article **"The Healthy Kids Survey: A Look at the Findings" (*San Luis Obispo Tribune*, October 25, 2002)** gave the accompanying information for a sample of fifth graders in San Luis Obispo County. Responses are to the question:

"After school, are you home alone without adult supervision?"

Response	Percentage
Never	8
Some of the time	15
Most of the time	16
All of the time	61

a. Summarize these data using a pie chart.

b. Construct a segmented bar graph for these data.

c. Which graphing method—the pie chart or the segmented bar graph—do you think does a better job of conveying information about response? Explain.

Bold exercises answered in back ● Data set available online ▼ Video Solution available

3.66 "If you were taking a new job and had your choice of a boss, would you prefer to work for a man or a woman?" That was the question posed to individuals in a sample of 576 employed adults (**Gallup at a Glance, October 16, 2002**). Responses are summarized in the following table:

Response	Frequency
Prefer to work for a man	190
Prefer to work for a woman	92
No difference	282
No opinion	12

a. Construct a pie chart to summarize this data set, and write a sentence or two summarizing how people responded to this question.

b. Summarize the given data using a segmented bar graph.

3.67 ● 2005 was a record year for hurricane devastation in the United States (**San Luis Obispo Tribune, November 30, 2005**). Of the 26 tropical storms and hurricanes in the season, four hurricanes hit the mainland: Dennis, Katrina, Rita, and Wilma. The United States' insured catastrophic losses since 1989 (approximate values read from a graph that appeared in the *San Luis Obispo Tribune*, November 30, 2005) are as follows:

Year	Cost (in billions of dollars)
1989	7.5
1990	2.5
1991	4.0
1992	22.5
1993	5.0
1994	18.0
1995	9.0
1996	8.0
1997	2.6
1998	10.0
1999	9.0
2000	3.0
2001	27.0
2002	5.0
2003	12.0
2004	28.5
2005	56.8

a. Construct a time series plot that shows the insured catastrophic loss over time.

b. What do you think causes the peaks in the graph?

3.68 An article in the **San Luis Obispo Tribune (November 20, 2002)** stated that 39% of those with critical housing needs (those who pay more than half their income for housing) lived in urban areas, 42% lived in suburban areas, and the rest lived in rural areas. Construct a pie chart that shows the distribution of type of residential area (urban, suburban, or rural) for those with critical housing needs.

3.69 ● Living-donor kidney transplants are becoming more common. Often a living donor has chosen to donate a kidney to a relative with kidney disease. The following data appeared in a **USA Today** article on organ transplants (**"Kindness Motivates Newest Kidney Donors," June 19, 2002**):

	Number of Kidney Transplants	
Year	Living-Donor to Relative	Living-Donor to Unrelated Person
1994	2390	202
1995	2906	400
1996	2916	526
1997	3144	607
1998	3324	814
1999	3359	930
2000	3679	1325
2001	3879	1399

a. Construct a time series plot for the number of living-donor kidney transplants where the donor is a relative of the recipient.

b. Describe the trend in this plot.

c. Use the data from 1994 and 2001 to construct a comparative bar chart for the type of donation (relative or unrelated).

d. Write a few sentences commenting on your display.

3.70 ● Many nutritional experts have expressed concern about the high levels of sodium in prepared foods. The following data on sodium content (in milligrams) per frozen meal appeared in the article **"Comparison of 'Light' Frozen Meals" (Boston Globe, April 24, 1991)**:

720 530 800 690 880 1050 340 810 760
300 400 680 780 390 950 520 500 630
480 940 450 990 910 420 850 390 600

Two histograms for these data are shown on the next page.

a. Do the two histograms give different impressions about the distribution of values?

b. Use each histogram to determine approximately the proportion of observations that are less than 800, and compare to the actual proportion.

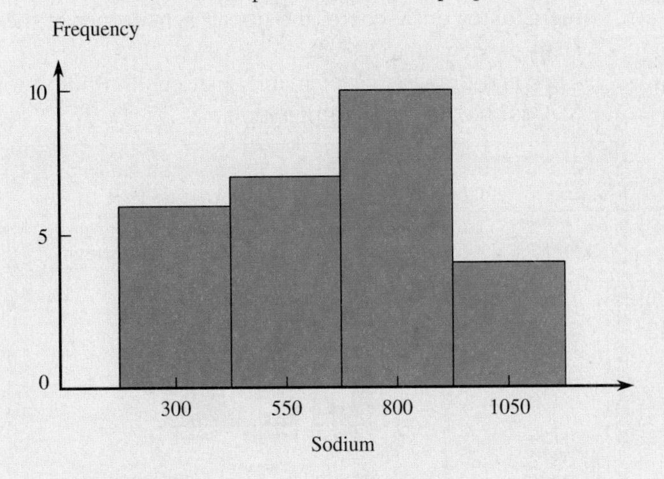

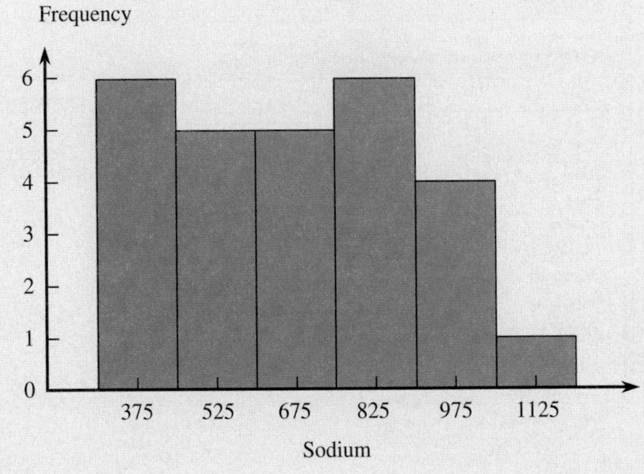

3.71 ● Americium 241 (^{241}Am) is a radioactive material used in the manufacture of smoke detectors. The article **"Retention and Dosimetry of Injected ^{241}Am in Beagles" (*Radiation Research* [1984]: 564–575)** described a study in which 55 beagles were injected with a dose of ^{241}Am (proportional to each animal's weight). Skeletal retention of ^{241}Am (in microcuries per kilogram) was recorded for each beagle, resulting in the following data:

0.196	0.451	0.498	0.411	0.324	0.190	0.489
0.300	0.346	0.448	0.188	0.399	0.305	0.304
0.287	0.243	0.334	0.299	0.292	0.419	0.236
0.315	0.447	0.585	0.291	0.186	0.393	0.419
0.335	0.332	0.292	0.375	0.349	0.324	0.301
0.333	0.408	0.399	0.303	0.318	0.468	0.441
0.306	0.367	0.345	0.428	0.345	0.412	0.337
0.353	0.357	0.320	0.354	0.361	0.329	

a. Construct a frequency distribution for these data, and draw the corresponding histogram.

b. Write a short description of the important features of the shape of the histogram.

Bold exercises answered in back ● Data set available online ▼ Video Solution available

TECHNOLOGY NOTES

Comparative Bar Charts

JMP

1. Enter the raw data into one column
2. In a separate column enter the group information

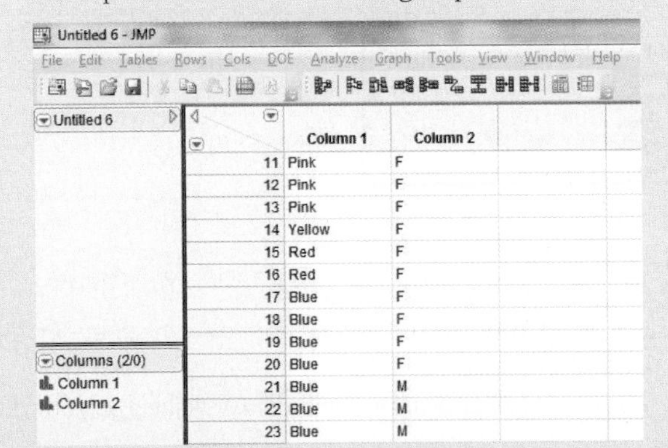

3. Click **Graph** then select **Chart**
4. Click and drag the column containing the raw data from the box under **Select Columns** to the box next to **Categories, X, Levels**
5. Click and drag the column containing the group information from the box under **Select Columns** to the box next to **Categories, X, Labels**
6. Click **OK**

MINITAB

1. Input the group information into C1
2. Input the raw data into C2 (be sure to match the response from the first column with the appropriate group)

Worksheet 1 ***

↓	C1-T	C2-T
1	Male	Blue
2	Female	Pink
3	Male	Red
4	Male	Blue
5	Female	Red
6	Male	Green
7	Female	Blue
8	Male	Red
9	Female	Pink
10	Female	Red

3. Select **Graph** then choose **Bar Chart...**
4. Highlight Cluster
5. Click **OK**
6. Double click C1 to add it to the **Categorical Variables** box
7. Then double click C2 to add it to the **Categorical Variables** box
8. Click **OK**

Note: Be sure to list the grouping variable first in the Categorical Variables box.

Note: You may add or format titles, axis titles, legends, and so on, by clicking on the **Labels...** button prior to performing Step 8 above.

SPSS

1. Enter the raw data into one column
2. Enter the group information into a second column (be sure to match the response from the first column with the appropriate group)

*Untitled1 [DataSet0] - PASW Statistics Student Ve

File Edit View Data Transform Analyze

	VAR00001	VAR000...	var
1	pink	female	
2	blue	male	
3	red	male	
4	green	male	
5	yellow	female	
6	red	female	
7	blue	female	
8	blue	male	
9	pink	female	
10	red	male	
11	green	female	
12	yellow	male	

3. Select **Graph** and choose **Chart Builder...**
4. Under **Choose from** highlight Bar
5. Click and drag the second bar chart in the first column (Clustered Bar) to the Chart preview area
6. Click and drag the group variable (second column) into the **Cluster on X** box in the upper right corner of the chart preview area
7. Click and drag the data variable (first column) into the **X-Axis?** box in the chart preview area

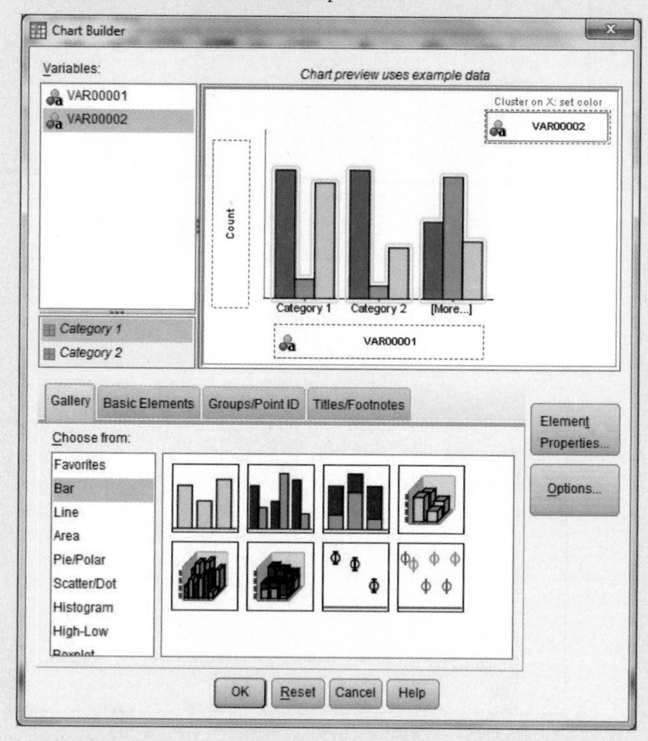

8. Click **Ok**

Excel 2007

1. Enter the category names into column A (you may input the title for the variable in cell A1)
2. Enter the count or percent for different groups into a new column, starting with column B
3. Select all data (including column titles if used)

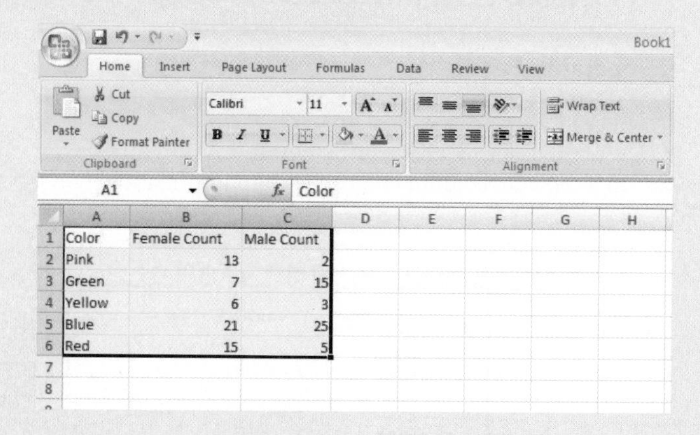

	A	B	C
1	Color	Female Count	Male Count
2	Pink	13	2
3	Green	7	15
4	Yellow	6	3
5	Blue	21	25
6	Red	15	5

4. Click on the **Insert** Ribbon
5. Choose **Column** and select the first chart under 2-D Column (Clustered Column)
6. The chart will appear on the same worksheet as your data

Note: You may add or format titles, axis titles, legends, and so on, by right-clicking on the appropriate piece of the chart.

Note: Using the Bar Option on the Insert Ribbon will produce a horizontal bar chart.

TI-83/84

The TI-83/84 does not have the functionality to produce comparative bar charts.

TI-Nspire

The TI-Nspire does not have the functionality to produce comparative bar charts.

Stem-and-leaf plots

JMP

1. Enter the raw data into a column
2. Click **Analyze** then select **Distribution**
3. Click and drag the column name containing the data from the box under **Select Columns** to the box next to **Y, Columns**
4. Click **OK**
5. Click the red arrow next to the column name

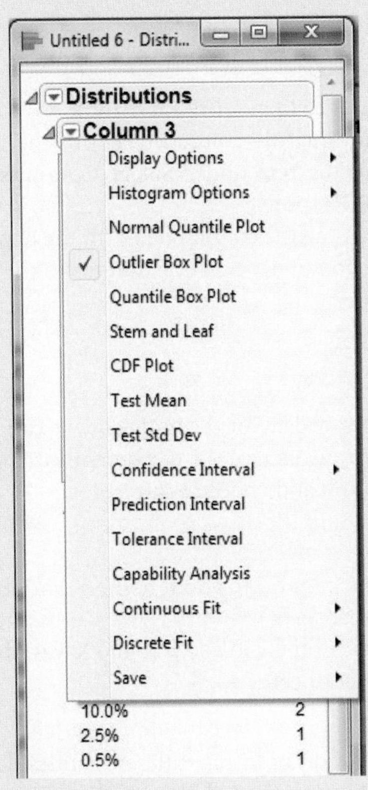

6. Select **Stem and Leaf**

MINITAB

1. Input the raw data into C1
2. Select **Graph** and choose **Stem-and-leaf...**
3. Double click C1 to add it to the **Graph Variables** box
4. Click **OK**

Note: You may add or format titles, axis titles, legends, and so on, by clicking on the **Labels...** button prior to performing Step 4 above.

SPSS

1. Input the raw data into a column
2. Select **Analyze** and choose **Descriptive Statistics** then **Explore**
3. Highlight the variable name from the box on the left
4. Click the arrow to the right of the Dependent List box to add the variable to this box

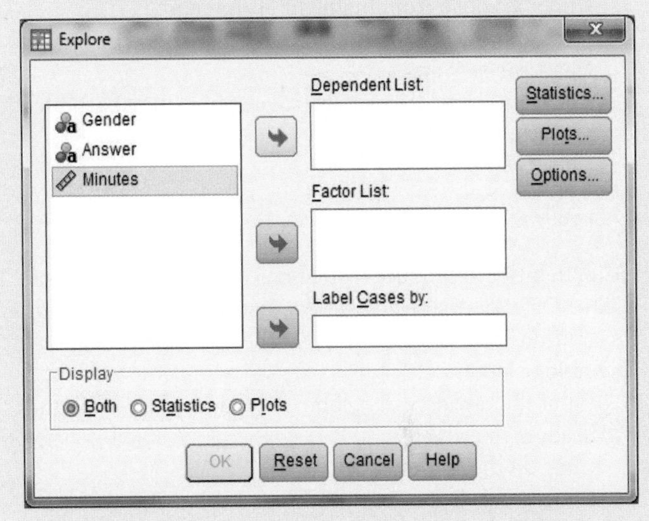

5. Click **OK**
6. **Note:** The stem-and-leaf plot is one of the plots produced along with several other descriptive statistics and plots.

Excel 2007

Excel 2007 does not have the functionality to create stem-and-leaf plots automatically.

TI-83/84

The Ti-83/84 does not have the functionality to create stem-and-leaf plots.

TI-Nspire

The TI-Nspire does not have the functionality to create stem-and-leaf plots.

Histograms

JMP

1. Enter the raw data into a column
2. Click **Analyze** then select **Distribution**
3. Click and drag the column name containing the data from the box under **Select Columns** to the box next to **Y, Columns**
4. Click **OK**
5. Click the red arrow next to the column name
6. Select **Histogram Options** and click **Vertical**

MINITAB

1. Input the raw data into C1
2. Select **Graph** and choose **Histogram...**
3. Highlight Simple
4. Click **OK**
5. Double click C1 to add it to the **Graph Variables** box
6. Click **OK**

Note: You may add or format titles, axis titles, legends, and so on, by clicking on the **Labels...** button prior to performing Step 5 above.

SPSS

1. Input the raw data into a column
2. Select **Graph** and choose **Chart Builder...**
3. Under **Choose from** highlight Histogram
4. Click and drag the first option (Simple Histogram) to the Chart preview area
5. Click and drag the variable name from the **Variables** box into the **X-Axis?** box
6. Click **OK**

Excel 2007

Note: In order to produce histograms in Excel 2007, we will use the Data Analysis Add-On. For instructions on installing this add-on, please see the note at the end of this chapter's Technology Notes.

1. Input the raw data in column A (you may use a column heading in cell A1)
2. Click on the **Data** ribbon
3. Choose the **Data Analysis** option from the **Analysis** group
4. Select **Histogram** from the Data Analysis dialog box and click **OK**
5. Click in the **Input Range:** box and then click and drag to select your data (including the column heading)
   a. If you used a column heading, click the checkbox next to **Labels**
6. Click the checkbox next to **Chart Output**
7. Click **OK**

Note: If you have specified bins manually in another column, click in the box next to **Bin Range:** and select your bin assignments.

TI-83/84

1. Enter the data into **L1** (In order to access lists press the **STAT** key, highlight the option called **Edit...** then press **ENTER**)
2. Press the **2nd** key then the **Y =** key
3. Highlight the option for Plot1 and press **ENTER**
4. Highlight **On** and press **ENTER**
5. For **Type**, select the histogram shape and press **ENTER**

6. Press **GRAPH**

Note: If the graph window does not display appropriately, press the **WINDOW** button and reset the scales appropriately.

TI-Nspire

1. Enter the data into a data list (In order to access data lists select the spreadsheet option and press **enter**)

Note: Be sure to title the list by selecting the top row of the column and typing a title.

2. Press the **menu** key then select **3:Data** then select **6:QuickGraph** and press **enter** (a dotplot will appear)
3. Press the **menu** key and select **1:Plot Type**
4. Select **3:Histogram** and press **enter**

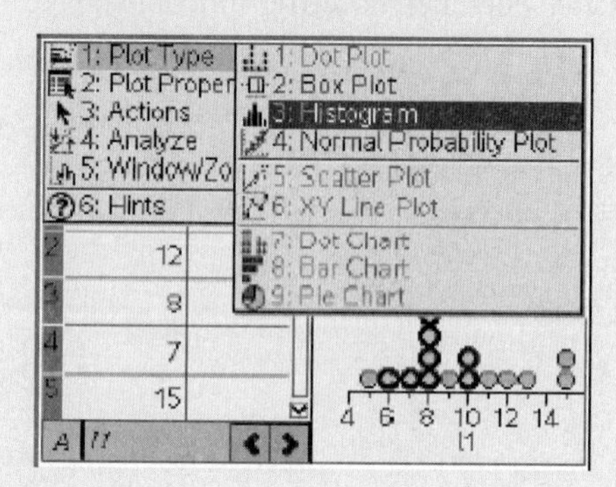

Scatterplots

JMP

1. Input the data for the independent variable into the first column
2. Input the data for the dependent variable into the second column
3. Click **Analyze** then select **Fit Y by X**
4. Click and drag the column name for the independent data from the box under **Select Columns** to the box next to **Y, Response**
5. Click and drag the column name for the dependent data from the box under **Select Columns** to the box next to **X, Factor**
6. Click **OK**

MINITAB

1. Input the raw data for the independent variable into C1
2. Input the raw data for the dependent variable into C2
3. Select **Graph** and choose **Scatterplot**
4. Highlight Simple
5. Click **OK**
6. Double click on C2 to add it to the **Y Variables** column of the spreadsheet
7. Double click on C1 to add it to the **X Variables** column of the spreadsheet
8. Click **OK**

Note: You may add or format titles, axis titles, legends, and so on, by clicking on the **Labels...** button prior to performing Step 7 above.

SPSS

1. Input the raw data into two columns
2. Select **Graph** and choose **Chart Builder...**
3. Under **Choose from** highlight Scatter/Dot
4. Click and drag the first option (Simple Scatter) to the Chart preview area
5. Click and drag the variable name representing the independent variable to the **X-Axis?** box

6. Click and drag the variable name representing the dependent variable to the **Y-Axis?** box
7. Click **OK**

Excel 2007

1. Input the raw data into two columns (you may enter column headings)
2. Select both columns of data
3. Click on the **Insert** ribbon
4. Click **Scatter** and select the first option (Scatter with Markers Only)

Note: You may add or format titles, axis titles, legends, and so on, by right-clicking on the appropriate piece of the chart.

TI-83/84

1. Input the data for the dependent variable into L2 (In order to access lists press the **STAT** key, highlight the option called **Edit…** then press **ENTER**)
2. Input the data for the dependent variable into L1
3. Press the **2nd** key then press the **Y =** key
4. Select the **Plot1** option and press **ENTER**
5. Select **On** and press **ENTER**
6. Select the scatterplot option and press **ENTER**

7. Press the **GRAPH** key

Note: If the graph window does not display appropriately, press the **WINDOW** button and reset the scales appropriately.

TI-Nspire

1. Enter the data for the independent variable into a data list (In order to access data lists select the spreadsheet option and press **enter**)

Note: Be sure to title the list by selecting the top row of the column and typing a title.

2. Enter the data for the dependent variable into a separate data list
3. Highlight both columns of data (by scrolling to the top of the screen to highlight one list then press shift and use the arrow key to highlight the second list)
4. Press the **menu** key and select **3:Data** then select **6:QuickGraph** and press **enter**

Time series plots

JMP

1. Input the raw data into one column
2. Input the time increment data into a second column

Column 4	Column 5
1	1900
2	1901
3	1902
4	1903
5	1904
6	1905
7	1906
8	1907
9	1908
1	1909
2	1910
3	1911
4	1912

3. Click **Graph** then select **Control Chart** then select **Run Chart**
4. Click and drag the name of the column containing the raw data from the box under **Select Columns** to the box next to Process
5. Click and drag the name of the column containing the time increment data from the box under **Select Columns** to the box next to **Sample Label**
6. Click **OK**

MINITAB

1. Input the raw data into C1
2. Select **Graph** and choose **Time Series Plot…**
3. Highlight Simple
4. Click **OK**
5. Double click C1 to add it to the **Series** box
6. Click the Time/Scale button
7. Choose the appropriate time scale under Time Scale (Choose Calendar if you want to use Days, Years, Months, Quarters, etc.)
8. Click to select one set for each variable
9. In the spreadsheet, fill in the start value for the time scale
10. Input the increment value into the box next to **Increment** (i.e., 1 to move by one year or one month, etc.)
11. Click **OK**
12. Click **OK**

Note: You may add or format titles, axis titles, legends, and so on, by clicking on the **Labels…** button prior to performing Step 12 above.

SPSS

1. Input the raw data into two columns: one representing the data and one representing the time increments
2. Select **Analyze** then choose **Forecasting** then **Sequence Charts…**
3. Highlight the column name containing the raw data and add it to the **Variables** box
4. Highlight the column name containing the time data and add it to the **Time Axis Labels** box
5. Click **OK**

Excel 2007

1. Input the data into two columns: one representing the data and one representing the time increments
2. Select an empty cell
3. Click on the **Insert** ribbon
4. Click **Line** and select the first option under 2-D Line (Line)
5. Right-click on the empty chart area that appears and select **Select Data...**
6. Under **Legend Entries (Series)** click **Add**
7. In the **Series name:** box select the column title for the data
8. In the **Series values:** box select the data values
9. Click **OK**
10. Under **Horizontal (Category) Axis Labels** click **Edit**
11. In the **Axis label range:** select the time increment data (do NOT select the column title).
12. Click **OK**
13. Click **OK**

TI-83/84

1. Input the time increment data into **L1**
2. Input the data values for each time into **L2**
3. Press the **2nd** key then the **Y =** key
4. Select **1:Plot1** and press **ENTER**
5. Highlight **On** and press **ENTER**
6. Highlight the second option in the first row and press **ENTER**
7. Press **GRAPH**

Note: If the graph window does not display appropriately, press the **WINDOW** button and reset the scales appropriately.

TI-Nspire

1. Enter the data for the time increments into a data list (In order to access data lists select the spreadsheet option and press **enter**)

Note: Be sure to title the list by selecting the top row of the column and typing a title.

2. Enter the data for each time increment into a separate data list
3. Highlight both columns of data (by scrolling to the top of the screen to highlight one list then press shift and use the arrow key to highlight the second list)
4. Press the **menu** key and select **3:Data** then select **6:QuickGraph** and press **enter**
5. Press the **menu** key and select **1:Plot Type** then select **6:XY Line Plot** and press **enter**

Installing Excel 2007's Data Analysis Add-On

1. Click the **Microsoft Office Button** and then Click **Excel Options**
2. Click **Add-Ins**, then in the **Manage** box, select **Excel Add-ins**.
3. Click **Go**
4. In the **Add-Ins available** box, select the **Analysis ToolPak** checkbox, and then click **OK**

Note: If this option is not listed, click **Browse** to locate it.

Note: If you get prompted that the Analysis ToolPak is not currently installed on your computer, click **Yes** to install it.

5. After you load the Analysis ToolPak, the **Data Analysis** command is available in the **Analysis** group on the **Data** ribbon.

CUMULATIVE REVIEW EXERCISES CR3.1 - CR3.16

CR3.1 Does eating broccoli reduce the risk of prostate cancer? According to an observational study from the Fred Hutchinson Cancer Research Center (see the CNN.com web site article titled **"Broccoli, Not Pizza Sauce, Cuts Cancer Risk, Study Finds," January 5, 2000**), men who ate more cruciferous vegetables (broccoli, cauliflower, brussels sprouts, and cabbage) had a lower risk of prostate cancer. This study made separate comparisons for men who ate different levels of vegetables. According to one of the investigators, "at any given level of total vegetable consumption, as the percent of cruciferous vegetables increased, the prostate cancer risk decreased." Based on this study, is it reasonable to conclude that eating cruciferous vegetables causes a reduction in prostate cancer risk? Explain.

CR3.2 An article that appeared in ***USA Today*** **(August 11, 1998)** described a study on prayer and blood pressure. In this study, 2391 people 65 years or older, were followed for 6 years. The article stated that people who attended a religious service once a week and prayed or studied the Bible at least once a day were less likely to have high blood pressure. The researcher then concluded that "attending religious services lowers blood pressure." The headline for this article was "Prayer Can Lower Blood Pressure." Write a few sentences commenting on the appropriateness of the researcher's conclusion and on the article headline.

CR3.3 Sometimes samples are composed entirely of volunteer responders. Give a brief description of the dangers of using voluntary response samples.

CR3.4 A newspaper headline stated that at a recent budget workshop, nearly three dozen people supported a sales tax increase to help deal with the city's financial deficit (***San Luis Obispo Tribune*****, January 22, 2005**). This conclusion was based on data from a survey acknowledged to be unscientific, in which 34 out of the 43 people who chose to attend the budget workshop recommended raising the sales tax. Briefly discuss why the survey was described as "unscientific" and how this might limit the conclusions that can be drawn from the survey data.

CR3.5 "More than half of California's doctors say they are so frustrated with managed care they will quit, retire early, or leave the state within three years." This conclusion from an article titled **"Doctors Feeling Pessimistic, Study Finds" (*San Luis Obispo Tribune,* July 15, 2001)** was based on a mail survey conducted by the California Medical Association. Surveys were mailed to 19,000 California doctors, and 2000 completed surveys were returned. Describe any concerns you have regarding the conclusion drawn.

CR3.6 Based on observing more than 400 drivers in the Atlanta area, two investigators at Georgia State University concluded that people exiting parking spaces did so more slowly when a driver in another car was waiting for the space than when no one was waiting **("Territorial Defense in Parking Lots: Retaliation Against Waiting Drivers," *Journal of Applied Social Psychology* [1997]: 821-834)**.

a. Describe how you might design an experiment to determine whether this phenomenon is true for your city.
b. What is the response variable?
c. What are some extraneous variables and how does your design control for them?

CR3.7 An article from the Associated Press (May 14, 2002) led with the headline "Academic Success Lowers Pregnancy Risk." The article described an evaluation of a program that involved about 350 students at 18 Seattle schools in high crime areas. Some students took part in a program beginning in elementary school in which teachers showed children how to control their impulses, recognize the feelings of others, and get what they want without aggressive behavior. Others did not participate in the program. The study concluded that the program was effective because by the time young women in the program reached age 21, the pregnancy rate among them was 38%, compared to 56% for the women in the experiment who did not take part in the program. Explain why this conclusion is valid only if the women in the experiment were randomly assigned to one of the two experimental groups.

CR3.8 Researchers at the University of Pennsylvania suggest that a nasal spray derived from pheromones (chemicals emitted by animals when they are trying to attract a mate) may be beneficial in relieving symptoms of premenstrual syndrome (PMS) **(*Los Angeles Times,* January 17, 2003)**.

a. Describe how you might design an experiment using 100 female volunteers who suffer from PMS to determine whether the nasal spray reduces PMS symptoms.
b. Does your design from Part (a) include a placebo treatment? Why or why not?
c. Does your design from Part (a) involve blinding? Is it single-blind or double-blind? Explain.

CR3.9 Students in California are required to pass an exit exam in order to graduate from high school. The pass rate for San Luis Obispo High School has been rising, as have the rates for San Luis Obispo County and the state of California **(*San Luis Obispo Tribune,* August 17, 2004)**. The percentage of students who passed the test was as follows:

Year	District	Pass Rate
2002	San Luis Obispo High School	66%
2003		72%
2004		93%
2002	San Luis Obispo County	62%
2003		57%
2004		85%
2002	State of California	32%
2003		43%
2004		74%

a. Construct a comparative bar chart that allows the change in the pass rate for each group to be compared.
b. Is the change the same for each group? Comment on any difference observed.

CR3.10 A poll conducted by the Associated Press–Ipsos on public attitudes found that most Americans are convinced that political corruption is a major problem **(*San Luis Obispo Tribune,* December 9, 2005)**. In the poll, 1002 adults were surveyed. Two of the questions and the summarized responses to these questions follow:

How widespread do you think corruption is in public service in America?

Hardly anyone	1%
A small number	20%
A moderate number	39%
A lot of people	28%
Almost everyone	10%
Not sure	2%

In general, which elected officials would you say are more ethical?

Democrats	36%
Republicans	33%
Both equally	10%
Neither	15%
Not sure	6%

a. For each question, construct a pie chart summarizing the data.
b. For each question, construct a segmented bar chart displaying the data.
c. Which type of graph (pie chart or segmented bar graph) does a better job of presenting the data? Explain.

CR3.11 ● The article **"Determination of Most Representative Subdivision"** (***Journal of Energy Engineering*** **[1993]: 43–55)** gave data on various characteristics of subdivisions that could be used in deciding whether to provide electrical power using overhead lines or underground lines. Data on the variable x = total length of streets within a subdivision are as follows:

1280	5320	4390	2100	1240	3060	4770	1050
360	3330	3380	340	1000	960	1320	530
3350	540	3870	1250	2400	960	1120	2120
450	2250	2320	2400	3150	5700	5220	500
1850	2460	5850	2700	2730	1670	100	5770
3150	1890	510	240	396	1419	2109	

a. Construct a stem-and-leaf display for these data using the thousands digit as the stem. Comment on the various features of the display.
b. Construct a histogram using class boundaries of 0 to <1000, 1000 to <2000, and so on. How would you describe the shape of the histogram?
c. What proportion of subdivisions has total length less than 2000? between 2000 and 4000?

CR3.12 ● The paper **"Lessons from Pacemaker Implantations"** (***Journal of the American Medical Association*** **[1965]: 231–232)** gave the results of a study that followed 89 heart patients who had received electronic pacemakers. The time (in months) to the first electrical malfunction of the pacemaker was recorded:

24	20	16	32	14	22	2	12	24	6	10	20
8	16	12	24	14	20	18	14	16	18	20	22
24	26	28	18	14	10	12	24	6	12	18	16
34	18	20	22	24	26	18	2	18	12	12	8
24	10	14	16	22	24	22	20	24	28	20	22
26	20	6	14	16	18	24	18	16	6	16	10
14	18	24	22	28	24	30	34	26	24	22	28
30	22	24	22	32							

a. Summarize these data in the form of a frequency distribution, using class intervals of 0 to <6, 6 to <12, and so on.
b. Compute the relative frequencies and cumulative relative frequencies for each class interval of the frequency distribution of Part (a).
c. Show how the relative frequency for the class interval 12 to <18 could be obtained from the cumulative relative frequencies.
d. Use the cumulative relative frequencies to give approximate answers to the following:
   i. What proportion of those who participated in the study had pacemakers that did not malfunction within the first year?
   ii. If the pacemaker must be replaced as soon as the first electrical malfunction occurs, approximately what proportion required replacement between 1 and 2 years after implantation?
e. Construct a cumulative relative frequency plot, and use it to answer the following questions.
   i. What is the approximate time at which about 50% of the pacemakers had failed?
   ii. What is the approximate time at which only about 10% of the pacemakers initially implanted were still functioning?

CR3.13 How does the speed of a runner vary over the course of a marathon (a distance of 42.195 km)? Consider determining both the time (in seconds) to run the first 5 km and the time (in seconds) to run between the 35 km and 40 km points, and then subtracting the 5-km time from the 35–40-km time. A positive value of this difference corresponds to a runner slowing down toward the end of the race. The histogram below is based on times of runners who participated in several different Japanese marathons (**"Factors Affecting Runners' Marathon Performance,"** ***Chance*** **[Fall 1993]: 24–30)**.

a. What are some interesting features of this histogram?
b. What is a typical difference value?
c. Roughly what proportion of the runners ran the late distance more quickly than the early distance?

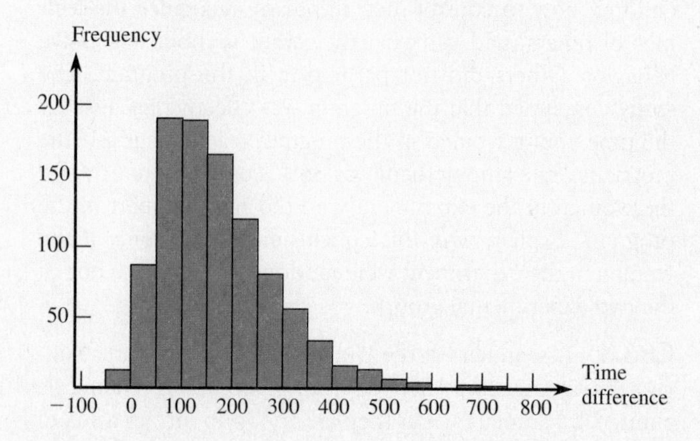

CR3.14 Data on x = poverty rate (%) and y = high school dropout rate (%) for the 50 U.S. states and the District of

Columbia were used to construct the following scatterplot (**Chronicle of Higher Education, August 31, 2001**):

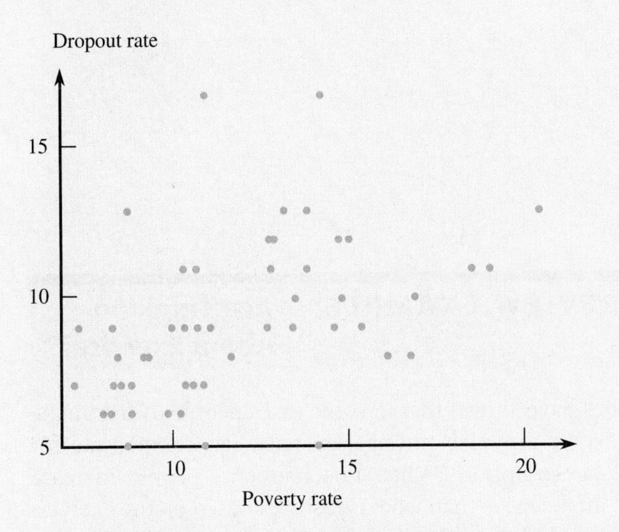

a. Write a few sentences commenting on this scatterplot.
b. Would you describe the relationship between poverty rate and dropout rate as positive (y tends to increase as x increases), negative (y tends to decrease as x increases), or as having no discernible relationship between x and y?

CR3.15 ▼ One factor in the development of tennis elbow, a malady that strikes fear into the hearts of all serious players of that sport, is the impact-induced vibration of the racket-and-arm system at ball contact. It is well known that the likelihood of getting tennis elbow depends on various properties of the racket used. Consider the accompanying scatterplot of x = racket resonance frequency (in hertz) and y = sum of peak-to-peak accelerations (a characteristic of arm vibration, in meters per second per second) for $n = 23$ different rackets (**"Transfer of Tennis Racket Vibrations into the Human Forearm," *Medicine and Science in Sports and Exercise* [1992]: 1134–1140**). Discuss interesting features of the data and of the scatterplot.

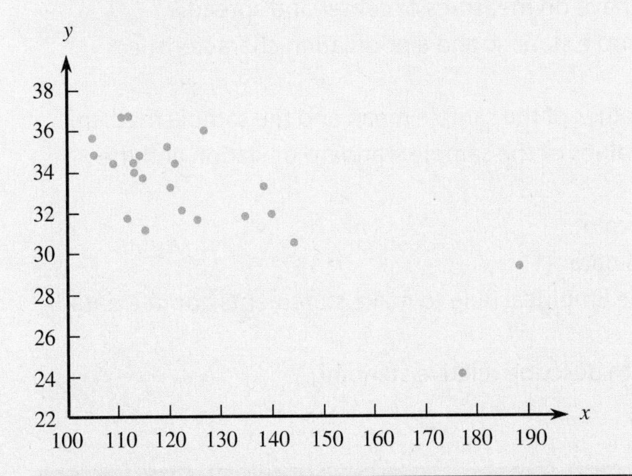

CR3.16 An article that appeared in **USA Today (September 3, 2003)** included a graph similar to the one shown here summarizing responses from polls conducted in 1978, 1991, and 2003 in which a sample of American adults were asked whether or not it was a good time or a bad time to buy a house.

a. Construct a time series plot that shows how the percentage that thought it was a good time to buy a house has changed over time.
b. Add a new line to the plot from Part (a) showing the percentage that thought it was a bad time to buy a house over time. Be sure to label the lines clearly.
c. Which graph, the given bar chart or the time series plot, best shows the trend over time?

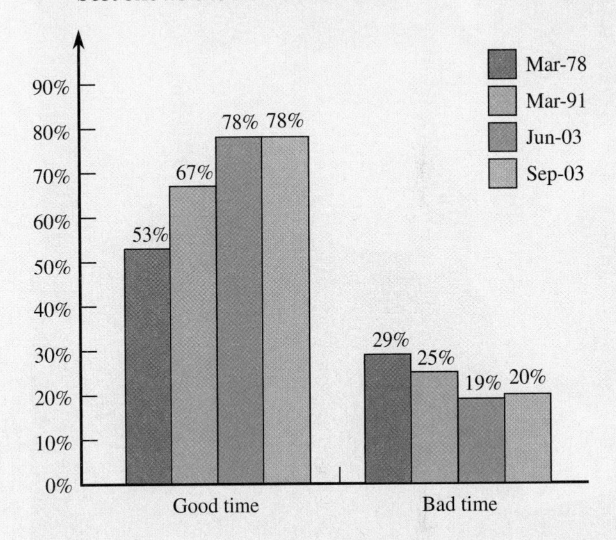

CHAPTER 4

Numerical Methods for Describing Data

©Estudi M6/Shutterstock.com

PREVIEW EXAMPLE: Just Thinking About Exercise??

Studies have shown that in some cases people overestimate the extent to which physical exercise can compensate for food consumption. When this happens, people increase food intake more than what is justified based on the exercise performed. The authors of the paper **"Just Thinking About Exercise Makes Me Serve More Food: Physical Activity and Calorie Compensation" (*Appetite* [2011]: 332–335)** wondered if even *thinking* about exercise would lead to increased food consumption.

They carried out an experiment in which people were offered snacks as a reward for participating in the experiment. People read a short essay and then answered a few questions about the essay. Some participants read an essay that was unrelated to exercise (the control group), some read an essay that described listening to music while taking a

Chapter 4: Learning Objectives

STUDENTS WILL UNDERSTAND:

- how the variance and standard deviation describe variability in a data set.
- the impact that outliers can have on measures of center and spread.
- the difference between a sample statistic and a population characteristic.

STUDENTS WILL BE ABLE TO:

- compute and interpret the values of the sample mean and the sample median.
- compute and interpret the values of the sample standard deviation and the interquartile range.
- construct and interpret a boxplot.
- identify outliers in numerical data.
- use Chebyshev's Rule and the Empirical Rule to make statements about a data distribution.
- use percentiles and z scores to describe relative standing.

30-minute walk (the fun group), and some read an essay that described strenuous exercise (the exercise group).

Participants were then provided with two plastic bags and invited to help themselves to two types of snacks—Chex Mix and M&Ms. After the participants served themselves, the bags were weighed so that the researchers could determine the number of calories in snacks taken.

Data on number of calories consistent with summary values in the paper were used to construct the comparative dotplot shown in Figure 4.1. From the dotplots, it is clear that the number of calories consumed differs from person to person and also tends to be quite a bit higher for those who read about exercise than for those in the control group! If we want to compare the distributions with more precision, the first step is to describe them numerically.

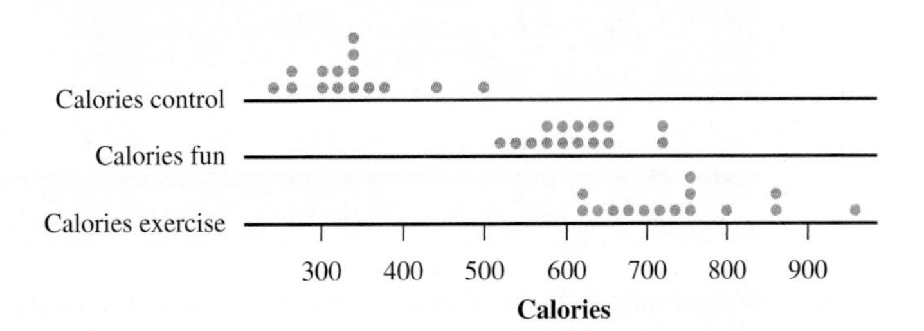

FIGURE 4.1
Comparative dotplot of calories.

How might we summarize this variability numerically? In this chapter, we show how to calculate numerical summary measures that describe both the center and the extent of spread in a data set. In Section 4.1, we introduce the mean and the median, the two most widely used measures of the center of a distribution. The variance and the standard deviation are presented in Section 4.2 as measures of variability. In later sections, we will see some additional ways that measures of center and spread can be used to describe data distributions.

4.1 Describing the Center of a Data Set

When describing numerical data, it is common to report a value that is representative of the observations. Such a number describes roughly where the data are located or "centered" along the number line, and is called a measure of center. The two most widely used measures of center are the mean and the median.

The Mean

The **mean** of a numerical data set is just the familiar arithmetic average: the sum of the observations divided by the number of observations. It is helpful to have concise notation for the variable on which observations are made, for the number of observations in the data set, and for the individual observations:

x = the variable for which we have sample data
n = the number of observations in the data set (the sample size)
x_1 = the first observation in the data set
x_2 = the second observation in the data set
$\vdots$
x_n = the nth (last) observation in the data set

For example, we might have a sample consisting of $n = 4$ observations on x = time it takes to complete an online hotel reservation (in minutes):

$x_1 = 5.9 \quad x_2 = 7.3 \quad x_3 = 6.6 \quad x_4 = 5.7$

Notice that the value of the subscript on x has no relationship to the magnitude of the observation. In this example, x_1 is just the first observation in the data set and not necessarily the smallest observation, and x_n is the last observation but not necessarily the largest.

The sum of $x_1, x_2, \ldots, x_n$ can be denoted by $x_1 + x_2 + \cdots + x_n$, but this is cumbersome. The Greek letter Σ is traditionally used in mathematics to denote summation. In particular, Σx denotes the sum of all the x values in the data set under consideration.*

DEFINITION

Sample mean: The **sample mean** of a sample consisting of numerical observations $x_1, x_2, \ldots, x_n$, denoted by $\bar{x}$, is

$$\bar{x} = \frac{\text{sum of all observations in the sample}}{\text{number of observations in the sample}} = \frac{x_1 + x_2 + \cdots + x_n}{n} = \frac{\Sigma x}{n}$$

EXAMPLE 4.1 Thinking About Exercise Again

Understand the context)

●The example in the chapter introduction described a study that investigated how thinking about exercise might affect food intake. Data on calories in snacks taken for people in each of three groups (control, read about fun walk, and read about strenuous exercise) are given in Table 4.1. Dotplots of these three data sets were given in the chapter introduction and are reproduced here as Figure 4.2.

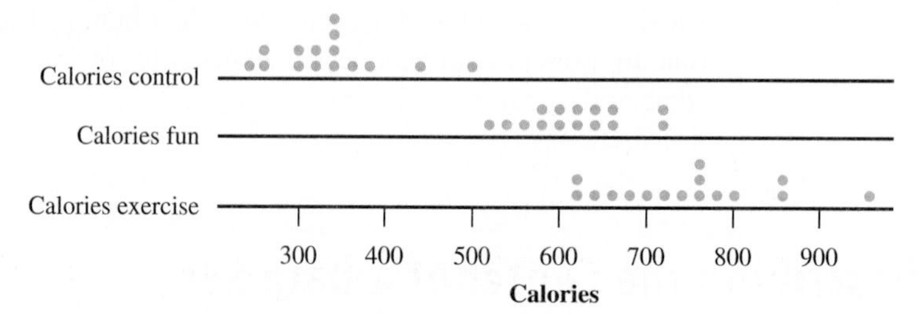

FIGURE 4.2
Dotplot of calories.

TABLE 4.1 Calorie Data

Consider the data)

Control Group	Fun Group	Exercise Group
340	668	626
300	585	802
329	637	768
381	588	738
331	529	751
445	597	715
320	719	670
256	612	630
252	622	862
342	553	761
332	600	648
296	658	956
357	542	671
505	711	854
242	641	703

*It is also common to see Σx written as Σx_i or even as $\sum_{i=1}^{n} x_i$, but for simplicity we will usually omit the summation indices.

● Data set available online

Do the work) For the control group, the sum of the sample data values is

$$\Sigma x = 340 + 300 + \cdots + 242 = 5028$$

and the sample mean number of calories is

$$\bar{x} = \frac{\Sigma x}{n} = \frac{5028}{15} = 335.20$$

Interpret the results) The value of the sample mean describes where the number of calories for the control group is centered along the number line. It can be interpreted as a typical number of calories for people in the control group. For the fun group, the sample mean number of calories is

$$\bar{x} = \frac{\Sigma x}{n} = \frac{9262}{15} = 617.47$$

and for the exercise group the sample mean number of calories is

$$\bar{x} = \frac{\Sigma x}{n} = \frac{11,155}{15} = 743.67$$

Notice that the sample mean number of calories for the control group was much smaller than the means for the other two groups. ■

The data values in Example 4.1 were all integers, yet the means were given as 335.20, 617.47 and 743.67. It is common to use more digits of decimal accuracy for the mean. This allows the value of the mean to fall between possible observable values (for example, the average number of children per family could be 1.8, whereas no single family will have 1.8 children).

The sample mean $\bar{x}$ is computed from sample observations, so it is a characteristic of the particular sample in hand. It is customary to use Roman letters to denote sample characteristics, as we have done with $\bar{x}$. Characteristics of the population are usually denoted by Greek letters. One of the most important of such characteristics is the population mean.

DEFINITION

Population mean: The **population mean**, denoted by μ, is the average of all x values in the entire population.

For example, the average fuel efficiency for *all* 600,000 cars of a certain type under specified conditions might be $\mu = 27.5$ mpg. A sample of $n = 5$ cars might yield efficiencies of 27.3, 26.2, 28.4, 27.9, 26.5, from which we obtain $\bar{x} = 27.26$ for this particular sample (somewhat smaller than μ). However, a second sample might give $\bar{x} = 28.52$, a third $\bar{x} = 26.85$, and so on. The value of $\bar{x}$ varies from sample to sample, whereas there is just one value for μ.

In later chapters, we will see how the value of $\bar{x}$ from a particular sample can be used to draw various conclusions about the value of μ. Example 4.2 illustrates how the value of $\bar{x}$ from a particular sample can differ from the value of μ and how the value of $\bar{x}$ differs from sample to sample.

EXAMPLE 4.2 County Population Sizes

Understand the context) The 50 states plus the District of Columbia contain 3137 counties. Let x denote the number of residents of a county. Then there are 3137 values of the variable x in the population.

The sum of these 3137 values is 293,655,404 (2004 Census Bureau estimate), so the population average value of x is

$$\mu = \frac{293,655,404}{3137} = 93,610.27 \text{ residents per county}$$

We used the Census Bureau web site to select three different random samples from this population of counties, with each sample consisting of five counties. The results appear in Table 4.2, along with the sample mean for each sample. The three $\bar{x}$ values are different from one another because they are based on three different samples and the value of $\bar{x}$ depends on the x values in the sample. Also notice that none of the three values comes close to the value of the population mean, μ. If we did not know the value of μ, we might use one of these $\bar{x}$ values as an *estimate* of μ, but in each case the estimate would be far off the mark.

TABLE 4.2 Three Samples from the Population of All U.S. Counties (x = number of residents)

Sample 1		Sample 2		Sample 3	
County	x Value	County	x Value	County	x Value
Fayette, TX	22,513	Stoddard, MO	29,773	Chattahoochee, GA	13,506
Monroe, IN	121,013	Johnston, OK	10,440	Petroleum, MT	492
Greene, NC	20,219	Sumter, AL	14,141	Armstrong, PA	71,395
Shoshone, ID	12,827	Milwaukee, WI	928,018	Smith, MI	14,306
Jasper, IN	31,624	Albany, WY	31,473	Benton, MO	18,519
	Σx = 208,196 $\bar{x}$ = 41,639.2		Σx = 1,013,845 $\bar{x}$ = 202,769.0		Σx = 118,218 $\bar{x}$ = 23,643.6

Alternatively, we could combine the three samples into a single sample with $n = 15$ observations:

$$x_1 = 22,513, \ldots, x_5 = 31,624, \ldots, x_{15} = 18,519$$

$$\Sigma x = 1,340,259$$

$$\bar{x} = \frac{1,340,259}{15} = 89,350.6$$

This value is closer to the value of μ but is still somewhat unsatisfactory as an estimate. The problem here is that the population of x values exhibits a lot of variability (the largest value is $x = 9,937,739$ for Los Angeles County, California, and the smallest value is $x = 52$ for Loving County, Texas, which evidently few people love). Therefore, it is difficult for a sample of 15 observations, let alone just 5, to be reasonably representative of the population. In Chapter 9, you will see how to take variability into account when deciding on the sample size required to accurately estimate a population mean. ■

One potential drawback to the mean as a measure of center for a data set is that its value can be greatly affected by the presence of even a single *outlier* (an unusually large or small observation) in the data set.

EXAMPLE 4.3 Number of Visits to a Class Web Site

Step-by-step technology instructions available online

Understand the context)

● Forty students were enrolled in a section of a general education course in statistical reasoning during one fall quarter at Cal Poly, San Luis Obispo. The instructor made course materials, grades, and lecture notes available to students on a class web site, and course management software kept track of how often each student accessed any of the web pages on the class site.

● Data set available online

One month after the course began, the instructor requested a report that indicated how many times each student had accessed a web page on the class site. The 40 observations were:

Consider the data

20	37	4	20	0	84	14	36	5	331	19	0
0	22	3	13	14	36	4	0	18	8	0	26
4	0	5	23	19	7	12	8	13	16	21	7
13	12	8	42								

Interpret the results

The sample mean for this data set is $\bar{x} = 23.10$. Figure 4.3 is a Minitab dotplot of the data. Many would argue that 23.10 is not a very representative value for this sample, because 23.10 is larger than most of the observations in the data set. Notice that only 7 of 40 observations, or 17.5%, are larger than 23.10. The two outlying values of 84 and 331 (no, that was *not* a typo!) have a substantial impact on the value of $\bar{x}$.

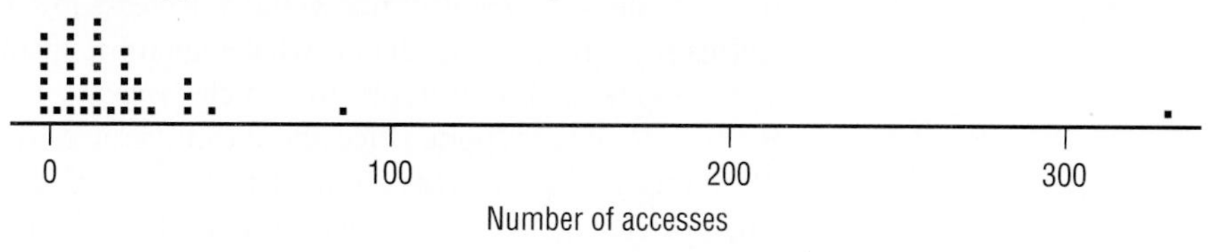

FIGURE 4.3
A Minitab dotplot of the data in Example 4.3.

We now turn our attention to a measure of center that is not as sensitive to outliers—the median.

The Median

The median strip of a highway divides the highway in half, and the median of a numerical data set does the same thing for a data set. Once the data values have been listed in order from smallest to largest, the **median** is the middle value in the list, and it divides the list into two equal parts.

Depending on whether the sample size n is even or odd, the process of determining the median is slightly different. When n is an odd number (say, 5), the sample median is the single middle value. But when n is even (say, 6), there are two middle values in the ordered list, and we average these two middle values to obtain the sample median.

DEFINITION

Sample median: The **sample median** is obtained by first ordering the n observations from smallest to largest (with any repeated values included, so that every sample observation appears in the ordered list). Then

$$\text{sample median} = \begin{cases} \text{the single middle value if } n \text{ is odd} \\ \text{the average of the middle two values if } n \text{ is even} \end{cases}$$

EXAMPLE 4.4 Web Site Data Revised

The sample size for the web site access data of Example 4.3 was $n = 40$, an even number. The median is the average of the 20th and 21st values (the middle two) in the ordered list of the data. Arranging the data in order from smallest to largest produces the following ordered list (with the two middle values highlighted):

0	0	0	0	0	0	3	4	4	4	5	5
7	7	8	8	8	12	12	13	13	13	14	14
16	18	19	19	20	20	21	22	23	26	36	36
37	42	84	331								

The median can now be determined by averaging the two middle values:

Do the work)

$$median = \frac{13 + 13}{2} = 13$$

Interpret the results)

Looking at the dotplot (Figure 4.3), we see that this value appears to be a more typical value for the data set than the sample mean of 23.10. ■

The sample mean can be sensitive to even a single value that lies far above or below the rest of the data. The value of the mean is pulled out toward such an outlying value or values. The median, on the other hand, is quite *in*sensitive to outliers. For example, the largest sample observation (331) in Example 4.4 can be increased by any amount without changing the value of the median. Similarly, an increase in the second or third largest observations does not affect the median, nor would a decrease in several of the smallest observations.

This stability of the median is what sometimes justifies its use as a measure of center in some situations. For example, the article **"Educating Undergraduates on Using Credit Cards" (Nellie Mae, 2005)** reported that the mean credit card debt for undergraduate students in 2001 was $2327, whereas the median credit card debt was only $1770. In this case, the small percentage of students with unusually high credit card debt results in a mean that may not be representative of a typical student's credit card debt.

Comparing the Mean and the Median

Figure 4.4 shows several smoothed histograms that might represent either a distribution of sample values or a population distribution. Pictorially, the median is the value on the measurement axis that separates the smoothed histogram into two parts, with half (50%) of the area under each part of the curve. The mean is a bit harder to visualize. If the histogram were balanced on a triangle (a fulcrum), it would tilt unless the triangle was positioned at the mean. The mean is the balance point for the distribution.

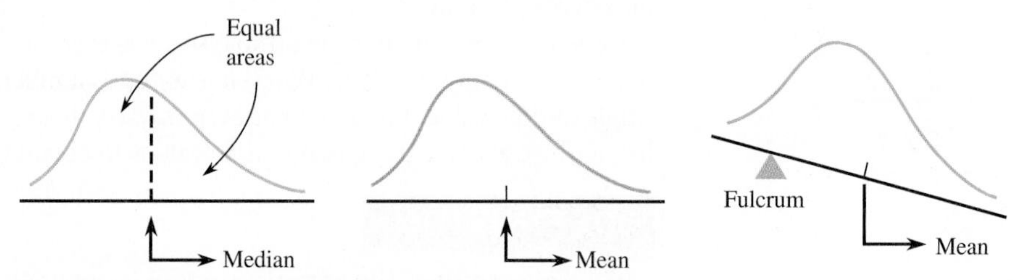

FIGURE 4.4
The mean and the median.

When the histogram is symmetric, the point of symmetry is both the dividing point for equal areas and the balance point, and the mean and the median are equal. However, when the histogram is unimodal (single-peaked) with a longer upper tail (positively skewed), the outlying values in the upper tail pull the mean up, so it generally lies above the median. For example, an unusually high exam score raises the mean but does not affect the median. Similarly, when a unimodal histogram is negatively skewed, the mean is generally smaller than the median (see Figure 4.5).

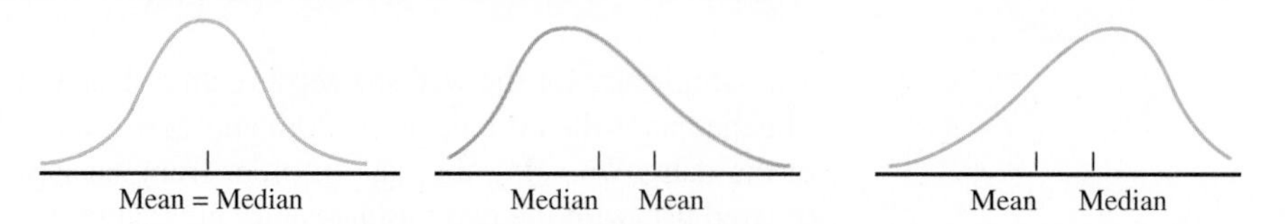

FIGURE 4.5
Relationship between the mean and the median.

EXAMPLE 4.5 NBA Salaries

Understand the context)

● The web site **HoopsHype (hoopshype.com/salaries)** publishes salaries of NBA players. Salaries for the players of the Chicago Bulls in 2009 were

● Data set available online

Consider the data)

Player	2009 Salary
Brad Miller	$12,250,000
Luol Deng	$10,370,425
Kirk Hinrich	$9,500,000
Jerome James	$6,600,000
Tim Thomas	$6,466,600
John Salmons	$5,456,000
Derrick Rose	$5,184,480
Tyrus Thomas	$4,743,598
Joakim Noah	$2,455,680
Jannero Pargo	$2,000,000
James Johnson	$1,594,080
Lindsey Hunter	$1,306,455
Taj Gibson	$1,039,800
Aaron Gray	$1,000,497

A Minitab dotplot of these data is shown in Figure 4.6(a). Because the data distribution is positively skewed and there are outliers, we would expect the mean to be greater than the median.

Do the work)

For this data set, the mean is

$$\bar{x} = \frac{12{,}250{,}000 + \cdots + 1{,}000{,}497}{14} = 4{,}977{,}687$$

The median is the average of the middle two observations in the ordered list of salaries:

$$\text{median} = \frac{5{,}184{,}480 + 4{,}743{,}598}{2} = 4{,}964{,}039$$

Notice that the median is smaller than the mean, and it is probably a better description of a typical salary for this group.

For the L.A., Lakers (see Figure 4.6(b)), the difference between the mean and median is even greater because in 2009, one player on the Lakers earned over $23 million and two players earned well over $10 million (see Figure 4.6(b)).

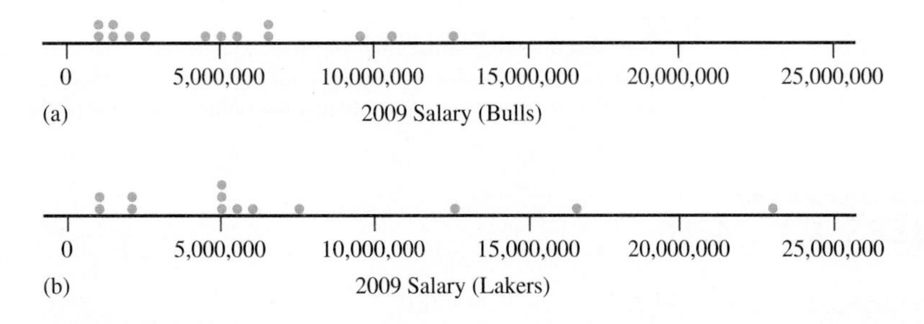

FIGURE 4.6
Minitab dotplots for NBA salary data (a) Bulls (b) Lakers.

Categorical Data

The natural numerical summary quantities for a categorical data set are the relative frequencies for the various categories. Each relative frequency is the proportion (fraction) of responses in the corresponding category. Often there are only two possible responses (a dichotomy)—for example, male or female, does or does not have a driver's license, did or did not vote in the last election. It is convenient in such situations to label one of the two possible responses S (for success) and the other F (for failure). As long as further analysis is consistent with the labeling, it does not matter which category is assigned the S label. When the data set is a sample, the fraction of S's in the sample is called the **sample proportion of successes**.

DEFINITION

Sample proportion of successes: The **sample proportion of successes,** denoted by $\hat{p}$, is

$$\hat{p} = \text{sample proportion of successes} = \frac{\text{number of S's in the sample}}{n}$$

where S is the label used for the response designated as success.

EXAMPLE 4.6 Can You Hear Me Now?

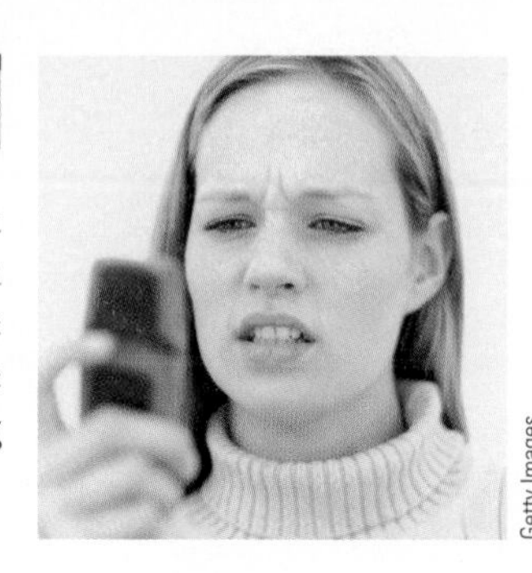
Getty Images

Understand the context ❯ It is not uncommon for a cell phone user to complain about the quality of his or her service provider. Suppose that each person in a sample of $n = 15$ cell phone users is asked if he or she is satisfied with the cell phone service. Each response is classified as S (satisfied) or F (not satisfied). The resulting data are

Consider the data ❯

S F S S S F F S S F
S S S F F

Do the work ❯ This sample contains nine S's, so

$$\hat{p} = \frac{9}{15} = .60$$

Interpret the results ❯ This means that 60% of the sample responses are S's. Of those surveyed, 60% are satisfied with their cell phone service. ■

The letter p is used to denote the **population proportion of S's.*** We will see later how the value of $\hat{p}$ from a particular sample can be used to draw conclusions about the population proportion p.

*Note that this is one situation in which we will not use a Greek letter to denote a population characteristic. Some statistics books use the symbol π for the population proportion and p for the sample proportion. We will not use π in this context so that there is no confusion with the mathematical constant $\pi = 3.14...$.

EXERCISES 4.1 - 4.16

4.1 ● **The Insurance Institute for Highway Safety (www.iihs.org, June 11, 2009)** published data on repair costs for cars involved in different types of accidents. In one study, seven different 2009 models of mini- and micro-cars were driven at 6 mph straight into a fixed barrier. The following table gives the cost of repairing damage to the bumper for each of the seven models.

Model	Repair Cost
Smart Fortwo	$1,480
Chevrolet Aveo	$1,071
Mini Cooper	$2,291
Toyota Yaris	$1,688
Honda Fit	$1,124
Hyundai Accent	$3,476
Kia Rio	$3,701

a. Compute the values of the mean and median.

b. Why are these values so different?

c. Which of the two—mean or median—appears to be better as a description of a typical value for this data set? (Hint: See Example 4.5.)

4.2 ● The article **"Caffeinated Energy Drinks—A Growing Problem"** (***Drug and Alcohol Dependence*** **[2009]: 1–10)** gave the following data on caffeine concentration (mg/ounce) for eight top-selling energy drinks:

Energy Drink	Caffeine Concentration (mg/oz)
Red Bull	9.6
Monster	10.0
Rockstar	10.0
Full Throttle	9.0
No Fear	10.9
Amp	8.9
SoBe Adrenaline Rush	9.5
Tab Energy	9.1

a. What is the value of the mean caffeine concentration for this set of top-selling energy drinks?

b. Coca-Cola has 2.9 mg/ounce of caffeine and Pepsi Cola has 3.2 mg/ounce of caffeine. Write a sentence explaining how the caffeine concentration of top-selling energy drinks compares to that of these colas.

4.3 ● **Consumer Reports Health (www.consumerreports.org/health)** reported the accompanying caffeine concentration (mg/cup) for 12 brands of coffee:

Coffee Brand	Caffeine Concentration (mg/cup)
Eight O'Clock	140
Caribou	195
Kickapoo	155
Starbucks	115
Bucks Country Coffee Co.	195
Archer Farms	180
Gloria Jean's Coffees	110
Chock Full o'Nuts	110
Peet's Coffee	130
Maxwell House	55
Folgers	60
Millstone	60

Use at least one measure of center to compare caffeine concentration for coffee with that of the energy drinks of the previous exercise. (Note: 1 cup = 8 ounces)

4.4 ● **Consumer Reports Health (www.consumerreports.org/health)** reported the sodium content (mg) per 2 tablespoon serving for each of 11 different peanut butters:

120 50 140 120 150 150 150 65
170 250 110

a. Display these data using a dotplot. Comment on any unusual features of the plot.

b. Compute the mean and median sodium content for the peanut butters in this sample.

c. The values of the mean and the median for this data set are similar. What aspect of the distribution of sodium content—as pictured in the dotplot from Part (a)—provides an explanation for why the values of the mean and median are similar? (Hint: See the discussion of Figure 4.4.)

4.5 In August 2009, **Harris Interactive** released the results of the **"Great Schools"** survey. In this survey, 1086 parents of children attending a public or private school were asked approximately how much time they spent volunteering at school per month over the last school year.

For this sample, the mean number of hours per month was 5.6 hours and the median number of hours was 1.0. What does the large difference between the mean and median tell you about this data set?

4.6 ● The accompanying data on number of minutes used for cell phone calls in one month was generated to be consistent with summary statistics published in a report of a marketing study of San Diego residents **(TeleTruth, March 2009)**:

189 0 189 177 106 201 0 212 0 306
0 0 59 224 0 189 142 83 71 165
236 0 142 236 130

Would you recommend the mean or the median as a measure of center for this data set? Give a brief explanation of your choice. (Hint: It may help to look at a graphical display of the data.)

4.7 ● ***USA Today*** **(May 9, 2006)** published the accompanying average weekday circulation for the 6-month period ending March 31, 2006, for the top 20 newspapers in the country:

2,272,815 2,049,786 1,142,464 851,832 724,242
708,477 673,379 579,079 513,387 438,722
427,771 398,329 398,246 397,288 365,011
362,964 350,457 345,861 343,163 323,031

a. Do you think the mean or the median will be larger for this data set? Explain.

b. Compute the values of the mean and the median of this data set.

c. Of the mean and median, which does the best job of describing a typical value for this data set?

d. Explain why it would not be reasonable to generalize from this sample of 20 newspapers to the population of all daily newspapers in the United States.

4.8 Each student in a sample of 20 seniors at a particular university was asked if he or she was registered to vote. With R denoting registered and N denoting not registered, the sample data are:

R R N N N R N R N R
N R N R R R N R R R

a. If being registered to vote is considered a "success," what is the value of the proportion of successes for this sample?

b. When would it be reasonable to generalize from this sample to the population of all seniors at this university?

4.9 ● The U.S. Department of Transportation reported the number of speeding-related crash fatalities for the 20 days of the year that had the highest number of these fatalities between 1994 and 2003 (***Traffic Safety Facts*, July 2005**).

Date	Speeding-Related Fatalities	Date	Speeding-Related Fatalities
Jan 1	521	Aug 17	446
Jul 4	519	Dec 24	436
Aug 12	466	Aug 25	433
Nov 23	461	Sep 2	433
Jul 3	458	Aug 6	431
Dec 26	455	Aug 10	426
Aug 4	455	Sept 21	424
Aug 31	446	Jul 27	422
May 25	446	Sep 14	422
Dec 23	446	May 27	420

a. Compute the mean number of speeding-related fatalities for these 20 days.

b. Compute the median number of speeding-related fatalities for these 20 days.

c. Explain why it is not reasonable to generalize from this sample of 20 days to the other 345 days of the year.

4.10 The ministry of **Health and Long-Term Care in Ontario, Canada,** publishes information on its web site **(www.health.gov.on.ca)** on the time that patients must wait for various medical procedures. For two cardiac procedures completed in fall of 2005, the following information was provided:

	Number of Completed Procedures	Median Wait Time (days)	Mean Wait Time (days)	90% Completed Within (days)
Angioplasty	847	14	18	39
Bypass surgery	539	13	19	42

The median wait time for angioplasty is greater than the median wait time for bypass surgery but the mean wait time is shorter for angioplasty than for bypass surgery. What does this suggest about the distribution of wait times for these two procedures?

4.11 Houses in California are expensive, especially on the Central Coast where the air is clear, the ocean is blue, and the scenery is stunning. The median home price in San Luis Obispo County reached a new high in July 2004, soaring to \$452,272 from \$387,120 in March 2004. (***San Luis Obispo Tribune*, April 28, 2004**).

The article included two quotes from people attempting to explain why the median price had increased. Richard Watkins, chairman of the Central Coast Regional Multiple Listing Services was quoted as saying, "There have been some fairly expensive houses selling, which pulls the median up."

Robert Kleinhenz, deputy chief economist for the California Association of Realtors explained the volatility of house prices by stating: "Fewer sales means a relatively small number of very high or very low home prices can more easily skew medians."

Are either of these statements correct? For each statement that is incorrect, explain why it is incorrect and propose a new wording that would correct any errors in the statement.

4.12 Consider the following statement: More than 65% of the residents of Los Angeles earn less than the average wage for that city. Could this statement be correct? If so, how? If not, why not?

4.13 ▼ A sample consisting of four pieces of luggage was selected from among those checked at an airline counter, yielding the following data on x = weight (in pounds):

$$x_1 = 33.5, x_2 = 27.3, x_3 = 36.7, x_4 = 30.5$$

Suppose that one more piece is selected and denote its weight by x_5. Find a value of x_5 such that $\bar{x}$ = sample median.

4.14 Suppose that 10 patients with meningitis received treatment with large doses of penicillin. Three days later, temperatures were recorded, and the treatment was considered successful if there had been a reduction in a patient's temperature. Denoting success by S and failure by F, the 10 observations are

S S F S S S F F S S

a. What is the value of the sample proportion of successes?

b. Replace each S with a 1 and each F with a 0. Then calculate $\bar{x}$ for this numerically coded sample. How does $\bar{x}$ compare to $\hat{p}$?

c. Suppose that it is decided to include 15 more patients in the study. How many of these would have to be S's to give $\hat{p} = .80$ for the entire sample of 25 patients?

4.15 A study of the lifetime (in hours) for a certain brand of light bulb involved putting 10 light bulbs into operation and observing them for 1000 hours. Eight of the light bulbs failed during that period, and those lifetimes were recorded. The lifetimes of the two light bulbs still functioning after 1000 hours were recorded as 1000+. The resulting sample observations were

480 790 1000+ 350 920 860 570 1000+
170 290

Which of the measures of center discussed in this section can be calculated, and what are the values of those measures?

4.16 An instructor has graded 19 exam papers submitted by students in a class of 20 students, and the average so far is 70. (The maximum possible score is 100.) How high would the score on the last paper have to be to raise the class average by 1 point? By 2 points?

Bold exercises answered in back ● Data set available online ▼ Video Solution available

4.2 Describing Variability in a Data Set

Reporting a measure of center gives only partial information about a data set. It is also important to describe how much the observations differ from one another. The three different samples displayed in Figure 4.7 all have mean = median = 45. There is a lot of variability in the first sample compared to the third sample. The second sample shows less variability than the first and more variability than the third; most of the variability in the second sample is due to the two extreme values being so far from the center.

Sample

1. 20, 40, 50, 30, 60, 70

2. 47, 43, 44, 46, 20, 70

3. 44, 43, 40, 50, 47, 46

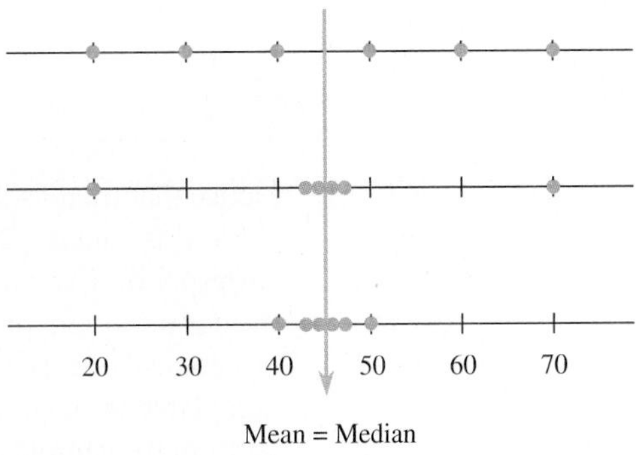

20 30 40 50 60 70

Mean = Median

FIGURE 4.7
Three samples with the same center and different amounts of variability.

The simplest numerical measure of variability is the range.

DEFINITION

Range: The **range** of a data set is defined as

range = largest observation − smallest observation

In general, more variability will result in a larger range. However, variability is a characteristic of the entire data set, and each observation contributes to variability. The first two samples plotted in Figure 4.7 both have a range of 70 − 20 = 50, but there is less variability in the second sample. Because of this, the range is not usually the best measure of variability.

Deviations from the Mean

The most widely used measures of variability describe the extent to which the sample observations deviate from the sample mean $\bar{x}$. Subtracting $\bar{x}$ from each observation gives a set of **deviations from the mean**.

DEFINITION

Deviation from the mean: The ***n*** **deviations from the sample mean** are the differences

$$(x_1 - \bar{x}), (x_2 - \bar{x}), \ldots, (x_n - \bar{x})$$

Notice that a deviation will be positive if the corresponding x value is greater than $\bar{x}$ and negative if the x value is less than $\bar{x}$.

EXAMPLE 4.7 The Big Mac Index

Understand the context ⟩

● McDonald's fast-food restaurants are now found in many countries around the world. But the cost of a Big Mac varies from country to country. Table 4.3 shows data on the cost of a Big Mac (converted to U.S. dollars based on the July 2013 exchange rates) taken from the article **"The Big Mac Index" (*The Economist*, July 11, 2013)**.

TABLE 4.3 Big Mac Prices for 7 Countries

Consider the data ⟩

Country	Big Mac Price in U.S. Dollars
Argentina	3.88
Brazil	5.28
Chile	3.94
Colombia	4.48
Costa Rica	4.31
Peru	3.59
Uruguay	4.98

Notice that there is quite a bit of variability in the Big Mac prices.

For this data set, $\Sigma x = 30.46$ and $\bar{x} = \$4.35$. Table 4.4 displays the data along with the corresponding deviations, formed by subtracting $\bar{x} = 4.35$ from each observation. Three of the deviations are positive because three of the observations are larger than $\bar{x}$. The negative deviations correspond to observations that are smaller than $\bar{x}$. Some of the deviations are quite large in magnitude (0.93 and −0.76, for example), indicating observations that are far from the sample mean.

TABLE 4.4 Deviations from the Mean for the Big Mac Data

Country	Big Mac Price in U.S. Dollars	Deviations from Mean
Argentina	3.88	−0.47
Brazil	5.28	0.93
Chile	3.94	−0.41
Colombia	4.48	0.13
Costa Rica	4.31	−0.04
Peru	3.59	−0.76
Uruguay	4.98	0.63

■

● Data set available online

In general, the greater the amount of variability in the sample data, the larger the magnitudes (ignoring the signs) of the deviations. We now consider how to combine the deviations into a single numerical measure of variability. A first thought might be to calculate the average deviation, by adding the deviations together (this sum can be denoted compactly by $\Sigma(x - \bar{x})$) and then dividing by n. This does not work, though, because negative and positive deviations counteract one another in the summation.

As a result of rounding, the value of the sum of the seven deviations in Example 4.7 is $\Sigma(x - \bar{x}) = 0.01$. If we used even more decimal accuracy in computing $\bar{x}$ the sum would be even closer to zero.

Except for the effects of rounding in computing the deviations, it is always true that

$$\Sigma(x - \bar{x}) = 0$$

Since this sum is zero, the average deviation is always zero and so it cannot be used as a measure of variability.

The Variance and Standard Deviation

One way to prevent negative and positive deviations from counteracting one another is to square them before combining. Then deviations with opposite signs but with the same magnitude, such as +2 and −2, make identical contributions to variability. The squared deviations are $(x_1 - \bar{x})^2, (x_2 - \bar{x})^2, \ldots, (x_n - \bar{x})^2$ and their sum is

$$(x_1 - \bar{x})^2 + (x_2 - \bar{x})^2 + \cdots + (x_n - \bar{x})^2 = \Sigma(x - \bar{x})^2$$

Dividing this sum by the sample size n gives the average squared deviation. Although this seems to be a reasonable measure of variability, we use a divisor slightly smaller than n. (The reason for this will be explained later in this section and in Chapter 9.)

DEFINITION

Sample variance: The **sample variance**, denoted by s^2, is the sum of squared deviations from the mean divided by $n - 1$. That is,

$$s^2 = \frac{\Sigma(x - \bar{x})^2}{n - 1}$$

Sample standard deviation: The **sample standard deviation** is the positive square root of the sample variance and is denoted by **s**.

A large amount of variability in the sample is indicated by a relatively large value of s^2 or s, whereas a value of s^2 or s close to zero indicates a small amount of variability. Notice that the units for the squared deviations and therefore s^2 are the square of whatever unit is used for x. Taking the square root gives a measure expressed in the same units as x. This means that for a sample of heights, the standard deviation might be $s = 3.2$ inches, and for a sample of textbook prices, it might be $s = \$12.43$.

Step-by-step technology instructions available online

EXAMPLE 4.8 Big Mac Revisited

Let's continue using the Big Mac data and the computed deviations from the mean given in Example 4.7 to calculate the sample variance and standard deviation. Table 4.5 shows the observations, deviations from the mean, and squared deviations. Combining the squared deviations to compute the values of s^2 and s gives

$$\Sigma(x - \bar{x})^2 = 2.2469$$

and

$$s^2 = \frac{\Sigma(x - \bar{x})^2}{n - 1} = \frac{2.2469}{7 - 1} = \frac{2.2469}{6} = 0.3745$$

$$s = \sqrt{0.3745} = 0.612$$

TABLE 4.5 Deviations and Squared Deviations for the Big Mac Data

Big Mac Price in U.S. Dollars	Deviations from Mean	Squared Deviations
3.88	−0.47	0.2209
5.28	0.93	0.8649
3.94	−0.41	0.1681
4.48	0.13	0.0169
4.31	−0.04	0.0016
3.59	−0.76	0.5776
4.98	0.63	0.3969
		$\Sigma(x - \bar{x})^2 = 2.2469$

■

The computation of s^2 can be a bit tedious, especially if the sample size is large. Fortunately, many calculators and computer software packages compute the variance and standard deviation upon request. One commonly used statistical computer package is Minitab. The output resulting from using the Minitab Describe command with the Big Mac data follows. Minitab gives a variety of numerical descriptive measures, including the mean, the median, and the standard deviation.

Descriptive Statistics: Big Mac Price in U.S. Dollars

Variable	N	Mean	SE Mean	StDev	Minimum	Q1	Median
Big Mac Price	7	4.351	0.231	0.612	3.590	3.880	4.310

Variable	Q3	Maximum
Big Mac Price	4.980	5.280

The standard deviation can be informally interpreted as the size of a "typical" or "representative" deviation from the mean. In Example 4.8, a typical deviation from $\bar{x}$ is about 0.612; some observations are closer to $\bar{x}$ than 0.612 and others are farther away.

We computed $s = 0.612$ in Example 4.8 without saying whether this value indicated a large or a small amount of variability. At this point, it is better to use s for comparative purposes than for an absolute assessment of variability. If Big Mac prices for a different group of countries resulted in a standard deviation of $s = 1.27$ (this is the standard deviation for all 57 countries for which Big Mac data was available) then we would conclude that our original sample has much less variability than the data set consisting of all 57 countries.

There are measures of variability for the entire population that are analogous to s^2 and s for a sample. These measures are called the **population variance** and the **population standard deviation** and are denoted by σ^2 and σ, respectively. (We again use a lowercase Greek letter for a population characteristic.)

Notation

s^2 sample variance

σ^2 population variance

s sample standard deviation

σ population standard deviation

In many statistical procedures, we would like to use the value of σ, but unfortunately it is not usually known. Therefore, we must estimate σ using a value computed from the sample. The divisor $(n - 1)$ is used in calculating s^2 rather than n because, on average, the resulting value tends to be a bit closer to the value of σ^2. We will say more about this in Chapter 9.

An alternative rationale for using $(n - 1)$ is based on the property $\Sigma(x - \bar{x}) = 0$. Suppose that $(n = 5)$ and that four of the deviations are

$$x_1 - \bar{x} = -4 \quad x_2 - \bar{x} = 6 \quad x_3 - \bar{x} = 1 \quad x_5 - \bar{x} = -8$$

Then, because the sum of these four deviations is -5, the remaining deviation must be $x_4 - \bar{x} = 5$ (so that the sum of all five is zero). More generally, once any $(n - 1)$ of the deviations are known, the value of the remaining deviation is determined. The n deviations actually contain only $(n - 1)$ independent pieces of information about variability. Statisticians express this by saying that s^2 and s are based on $(n - 1)$ *degrees of freedom* (df).

The Interquartile Range

As with $\bar{x}$, the value of s can be greatly affected by the presence of even a single unusually small or large observation. The **interquartile range** is a measure of variability that is resistant to the effects of outliers. It is based on quantities called quartiles. The **lower quartile** separates the bottom 25% of the data set from the upper 75%, and the **upper quartile** separates the top 25% from the bottom 75%. The middle quartile is the median, and it separates the bottom 50% from the top 50%. Figure 4.8 illustrates the locations of these quartiles for a smoothed histogram.

25% 25% 25% 25%

Lower quartile Median Upper quartile

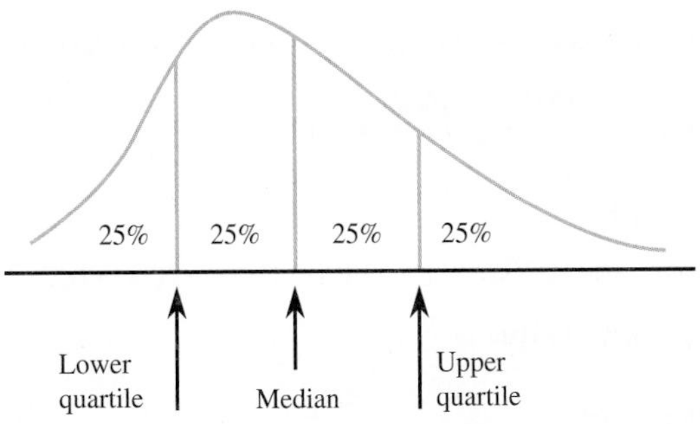

FIGURE 4.8
The quartiles for a smoothed histogram.

The quartiles for sample data are obtained by dividing the n ordered observations into a lower half and an upper half. If n is odd, the median is excluded from both halves. The upper and lower quartiles are then the medians of the two halves. (Note: The median is only temporarily excluded for the purpose of computing quartiles. It is not excluded from the data set.)

Lower quartile: Median of the lower half of the sample

Upper quartile: Median of the upper half of the sample

(If n is odd, the median of the entire sample is excluded from both halves when computing quartiles.)

Interquartile range (iqr): A measure of variability that is not as sensitive to the presence of outliers as the standard deviation. The iqr is calculated as

iqr = upper quartile − lower quartile

*There are several other sensible ways to define quartiles. Some calculators and software packages use an alternative definition.

The resistant nature of the interquartile range follows from the fact that up to 25% of the smallest sample observations and up to 25% of the largest sample observations can be made more extreme without affecting the value of the interquartile range.

EXAMPLE 4.9 Beating That High Score

Understand the context ❯

The authors of the paper **"Striatal Volume Predicts Level of Video Game Skill Acquisition" (*Cerebral Cortex* [2010]: 2522–2530)** studied a number of factors that affect performance on a complex video game. One of the factors investigated was practice strategy. Forty college students who all reported that they played video games less than three hours per week over the two years prior to the study and who had never played the game Space Fortress were assigned at random to one of two practice strategies. In the Space Fortress game, points are awarded for control, velocity, and speed.

Each person completed 20 two-hour practice sessions. Those in one group, the fixed priority group, were told to work on improving their total score at each practice session. Those in the other group, the variable priority group, were told to focus on a particular aspect of the game, such as improving speed score in each practice session, and the focus changed from one practice session to another. The investigators were interested in whether practice strategy makes a difference. They measured the improvement in total score from the first day of the study to the day when the last practice session was completed.

Improvement scores (approximate values read from a graph that appears in the paper) for the 20 people in the variable priority practice strategy group are given here.

Consider the data ❯

1200 1300 2300 3200 3300 3800 4000 4100 4300 4800
5500 5700 5700 5800 6000 6300 6800 6800 6900 7700

The median is the average of the middle two observations, so

$$\text{median} = \frac{4800 + 5500}{2} = 5150$$

The lower half of the data set is

1200 1300 2300 3200 3300 3800 4000 4100 4300 4800

so the lower quartile is

$$\text{lower quartile} = \frac{3300 + 3800}{2} = 3550$$

The upper half of the data set is

5500 5700 5700 5800 6000 6300 6800 6800 6900 7700

so the upper quartile is

$$\text{upper quartile} = \frac{6000 + 6300}{2} = 6150$$

$$\text{lower quartile} = 3550$$
$$\text{upper quartile} = 6150$$
$$\text{iqr} = 6150 - 3550 = 2660$$

The sample mean and standard deviation for this data set are 4775 and 1867, respectively. If we were to change the two largest values from 6900 and 7700 to 10,900 and 11,700 (so that they still remain the two largest values), the median and interquartile range would not be affected, whereas the mean and the standard deviation would change to 7500 and 2390, respectively. The value of the interquartile range is not affected by a few extreme values in the data set. ■

● Data set available online

EXERCISES 4.17 - 4.31

4.17 ● The following data are costs (in cents) per ounce for nine different brands of sliced Swiss cheese (**www.consumerreports.org**):

29 62 37 41 70 82 47 52 49

a. Compute the variance and standard deviation for this data set. (Hint: See Example 4.8.)

b. If a very expensive cheese with a cost per slice of 150 cents was added to the data set, how would the values of the mean and standard deviation change?

4.18 ● Cost per serving (in cents) for six high-fiber cereals rated very good and for nine high-fiber cereals rated good by ***Consumer Reports*** are shown below. Write a few sentences describing how these two data sets differ with respect to center and variability. Use summary statistics to support your statements.

Cereals Rated Very Good
46 49 62 41 19 77

Cereals Rated Good
71 30 53 53 67 43 48 28 54

4.19 ● Combining the cost-per-serving data for high-fiber cereals rated very good and those rated good from the previous exercise gives the following data set:

46 49 62 41 19 77 71 30
53 53 67 43 48 28 54

a. Compute the quartiles and the interquartile range for this combined data set. (Hint: See Example 4.9.)

b. Compute the interquartile range for just the cereals rated good. Is this value greater than, less than, or about equal to the interquartile range computed in Part (a)?

4.20 ● The paper **"Caffeinated Energy Drinks—A Growing Problem" (*Drug and Alcohol Dependence* [2009]: 1–10)** gave the accompanying data on caffeine per ounce for eight top-selling energy drinks and for 11 high-caffeine energy drinks:

Top-Selling Energy Drinks
9.6 10.0 10.0 9.0 10.9 8.9 9.5 9.1

High-Caffeine Energy Drinks
21.0 25.0 15.0 21.5 35.7 15.0
33.3 11.9 16.3 31.3 30.0

The mean caffeine per ounce is clearly higher for the high-caffeine energy drinks, but which of the two groups of energy drinks (top-selling or high-caffeine) is the most variable with respect to caffeine per ounce? Justify your choice.

4.21 The accompanying data are consistent with summary statistics that appeared in the paper **"Shape of Glass and Amount of Alcohol Poured: Comparative Study of Effect of Practice and Concentration" (*British Medical Journal* [2005]: 1512–1514)**. Data represent the actual amount poured (in ml) into a tall, slender glass for bartenders who were asked to pour 44.3 ml (1.5 ounces). Compute and interpret the values of the mean and standard deviation.

44.0 49.6 62.3 28.4 39.1 39.8 60.5 73.0 57.5 56.5 65.0
56.2 57.7 73.5 66.4 32.7 40.4 21.4

4.22 The paper referenced in the previous exercise also gave data on the actual amount poured (in ml) into a short, wide glass for bartenders who were asked to pour 44.3 ml (1.5 ounces).

89.2 68.6 32.7 37.4 39.6 46.8 66.1 79.2 66.3 52.1 47.3
64.4 53.7 63.2 46.4 63.0 92.4 57.8

a. Compute and interpret the values of the mean and standard deviation.

b. What do the values of the mean amount poured in the short, wide glass and the mean computed in the previous exercise suggest about the shape of glasses used?

4.23 ● **The Insurance Institute for Highway Safety (www.iihs.org, June 11, 2009)** published data on repair costs for cars involved in different types of accidents. In one study, seven different models of mini- and micro-cars were driven at 6 mph straight into a fixed barrier. The following table gives the cost of repairing damage to the bumper for each of the seven models:

Model	Repair Cost
Smart Fortwo	$1,480
Chevrolet Aveo	$1,071
Mini Cooper	$2,291
Toyota Yaris	$1,688
Honda Fit	$1,124
Hyundai Accent	$3,476
Kia Rio	$3,701

a. Compute the values of the variance and standard deviation.

b. The standard deviation is fairly large. What does this tell you about the repair costs?

4.24 The **Insurance Institute for Highway Safety** (referenced in the previous exercise) also gave bumper repair costs in a study of six models of minivans **(December 30, 2007)**. Write a few sentences describing how mini- and micro-cars and minivans differ with respect to typical bumper repair cost and bumper repair cost variability.

Model	Repair Cost
Honda Odyssey	$1,538
Dodge Grand Caravan	$1,347
Toyota Sienna	$840
Chevrolet Uplander	$1,631
Kia Sedona	$1,176
Nissan Quest	$1,603

4.25 ● The accompanying data on number of minutes used for cell phone calls in 1 month was generated to be consistent with summary statistics published in a report of a marketing study of San Diego residents **(TeleTruth, March 2009)**:

189 0 189 177 106 201 0 212 0 306
0 0 59 224 0 189 142 83 71 165
236 0 142 236 130

a. Compute the values of the quartiles and the interquartile range for this data set.

b. Explain why the lower quartile is equal to the minimum value for this data set. Will this be the case for every data set? Explain.

4.26 Give two sets of five numbers that have the same mean but different standard deviations, and give two sets of five numbers that have the same standard deviation but different means.

4.27 The article **"Rethink Diversification to Raise Returns, Cut Risk" (*San Luis Obispo Tribune*, January 21, 2006)** included the following paragraph:

> In their research, Mulvey and Reilly compared the results of two hypothetical portfolios and used actual data from 1994 to 2004 to see what returns they would achieve. The first portfolio invested in Treasury bonds, domestic stocks, international stocks, and cash. Its 10-year average annual return was 9.85% and its volatility—measured as the standard deviation of annual returns—was 9.26%. When Mulvey and Reilly shifted some assets in the portfolio to include funds that invest in real estate, commodities, and options, the 10-year return rose to 10.55% while the standard deviation fell to 7.97%. In short, the more diversified portfolio had a slightly better return and much less risk.

Explain why the standard deviation is a reasonable measure of volatility and why it is reasonable to interpret a smaller standard deviation as meaning less risk.

4.28 ● The **U.S. Department of Transportation** reported the accompanying data (see below) on the number of speeding-related crash fatalities during holiday periods for the years from 1994 to 2003 **(*Traffic Safety Facts*, July 20, 2005)**.

a. Compute the standard deviation for the New Year's Day data.

b. Without computing the standard deviation of the Memorial Day data, explain whether the standard deviation for the Memorial Day data would be larger or smaller than the standard deviation of the New Year's Day data.

c. Memorial Day and Labor Day are holidays that always occur on Monday and Thanksgiving always occurs on a Thursday, whereas New Year's Day, July 4th, and Christmas do not always fall on the same day of the week every year. Based on the given data, is there more or less variability in the speeding-related crash fatality numbers from year to year for same day of the week holiday periods than for holidays that can occur on different days of the week? Support your answer with appropriate measures of variability.

Data for Exercise 4.28

	Speeding-Related Fatalities									
Holiday Period	1994	1995	1996	1997	1998	1999	2000	2001	2002	2003
New Year's Day	141	142	178	72	219	138	171	134	210	70
Memorial Day	193	178	185	197	138	183	156	190	188	181
July 4th	178	219	202	179	169	176	219	64	234	184
Labor Day	183	188	166	179	162	171	180	138	202	189
Thanksgiving	212	198	218	210	205	168	187	217	210	202
Christmas	152	129	66	183	134	193	155	210	60	198

4.29 **The Ministry of Health and Long-Term Care in Ontario, Canada,** publishes information on the time that patients must wait for various medical procedures on its web site **(www.health.gov.on.ca)**. For two cardiac procedures completed in fall of 2005, the following information was provided:

Procedure	Number of Completed Procedures	Median Wait Time (days)	Mean Wait Time (days)	90% Completed Within (days)
Angioplasty	847	14	18	39
Bypass surgery	539	13	19	42

a. Which of the following must be true for the lower quartile of the data set consisting of the 847 wait times for angioplasty?

i. The lower quartile is less than 14.

ii. The lower quartile is between 14 and 18.

iii. The lower quartile is between 14 and 39.

iv. The lower quartile is greater than 39.

b. Which of the following must be true for the upper quartile of the data set consisting of the 539 wait times for bypass surgery?

i. The upper quartile is less than 13.

ii. The upper quartile is between 13 and 19.

iii. The upper quartile is between 13 and 42.

iv. The upper quartile is greater than 42.

c. Which of the following must be true for the number of days for which only 5% of the bypass surgery wait times would be longer?

i. It is less than 13.

ii. It is between 13 and 19.

iii. It is between 13 and 42.

iv. It is greater than 42.

4.30 ● In 1997, a woman sued a computer keyboard manufacturer, charging that her repetitive stress injuries were caused by the keyboard **(*Genessey v. Digital Equipment Corporation*)**. The jury awarded about \$3.5 million for pain and suffering, but the court then set aside that award as being unreasonable compensation. In making this determination, the court identified a "normative" group of 27 similar cases and specified a reasonable award as one within 2 standard deviations of the mean of the awards in the 27 cases.

The 27 award amounts were (in thousands of dollars)

37	60	75	115	135	140	149	150
238	290	340	410	600	750	750	750
1050	1100	1139	1150	1200	1200	1250	1576
1700	1825	2000					

What is the maximum possible amount that could be awarded under the "2-standard deviations rule?"

4.31 ● The standard deviation alone does not measure relative variation. For example, a standard deviation of \$1 would be considered large if it is describing the variability from store to store in the price of an ice cube tray. On the other hand, a standard deviation of \$1 would be considered small if it is describing store-to-store variability in the price of a particular brand of freezer.

A quantity designed to give a relative measure of variability is the *coefficient of variation*. Denoted by CV, the coefficient of variation expresses the standard deviation as a percentage of the mean. It is defined by the formula $CV = 100\left(\frac{s}{\bar{x}}\right)$.

Consider two samples. Sample 1 gives the actual weight (in ounces) of the contents of cans of pet food labeled as having a net weight of 8 ounces. Sample 2 gives the actual weight (in pounds) of the contents of bags of dry pet food labeled as having a net weight of 50 pounds. The weights for the two samples are

Sample 1	8.3	7.1	7.6	8.1	7.6
	8.3	8.2	7.7	7.7	7.5
Sample 2	52.3	50.6	52.1	48.4	48.8
	47.0	50.4	50.3	48.7	48.2

a. For each of the given samples, calculate the mean and the standard deviation.

b. Compute the coefficient of variation for each sample. Do the results surprise you? Why or why not?

Bold exercises answered in back ● Data set available online ▼ Video Solution available

4.3 Summarizing a Data Set: Boxplots

In Sections 4.1 and 4.2, we looked at ways of describing the center and variability of a data set using numerical measures. It would be nice to have a method of summarizing data that provides more information than just reporting a measure of center and spread and yet less detail than a stem-and-leaf display or histogram. A **boxplot** is one way to do this. A boxplot is compact, yet it provides information about the center, spread, and symmetry or skewness of the data. We will consider two types of boxplots: the skeletal boxplot and the modified boxplot.

Construction of a Skeletal Boxplot

1. Draw a horizontal (or vertical) measurement scale.
2. Construct a rectangular box with a left (or lower) edge at the lower quartile and a right (or upper) edge at the upper quartile. The box width is then equal to the iqr.
3. Draw a vertical (or horizontal) line segment inside the box at the location of the median.
4. Extend horizontal (or vertical) line segments, called whiskers, from each end of the box to the smallest and largest observations in the data set.

EXAMPLE 4.10 Revisiting Improvement Score Data

Let's reconsider the data from Example 4.9 on improvement score for people in the variable priority practice group. The ordered observations are

Ordered Data

Lower Half:

1200 1300 2300 3200 3300 3800 4000 4100 4300 4800

Median = 5150

Upper Half:

5500 5700 5700 5800 6000 6300 6800 6800 6900 7700

To construct a boxplot of these data, we need the following information: the smallest observation, the lower quartile, the median, the upper quartile, and the largest observation. This collection of summary measures is often referred to as a **five-number summary.** For this data set we have

smallest observation = 1200
lower quartile = median of the lower half = 3550
median = average of the 10th and 11th observations in the ordered list = 5150
upper quartile = median of the upper half = 6150
largest observation = 7700

Figure 4.9 shows the corresponding boxplot. The median line is somewhat closer to the upper edge of the box than to the lower edge, suggesting a concentration of values in the upper part of the middle half. The lower whisker is longer than the upper whisker, suggesting that the distribution of improvement scores for the variable practice strategy group is not symmetric.

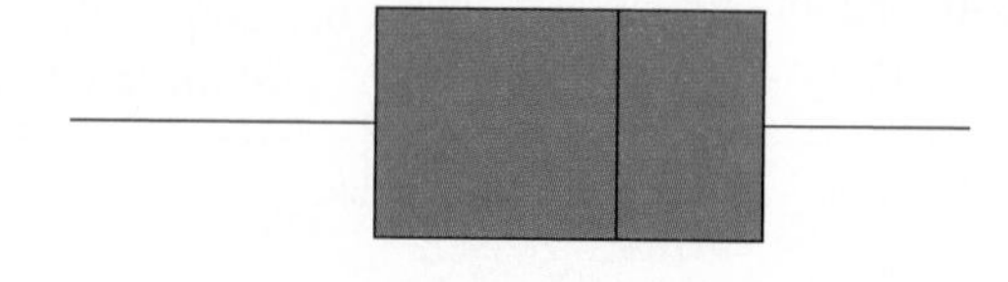

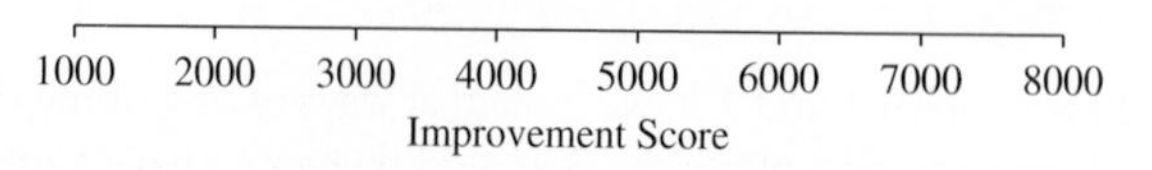

FIGURE 4.9 Skeletal boxplot for the improvement score data of Example 4.10.

■

Boxplots are often used to compare groups. For example, we could compare video game improvement scores for the two different practice strategies described in Example 4.9 using a **comparative boxplot**.

DEFINITION

Comparative boxplot: Two or more boxplots drawn using the same numerical scale.

The video game improvement scores for the two different strategy groups were used to construct the comparative boxplot shown in Figure 4.10.

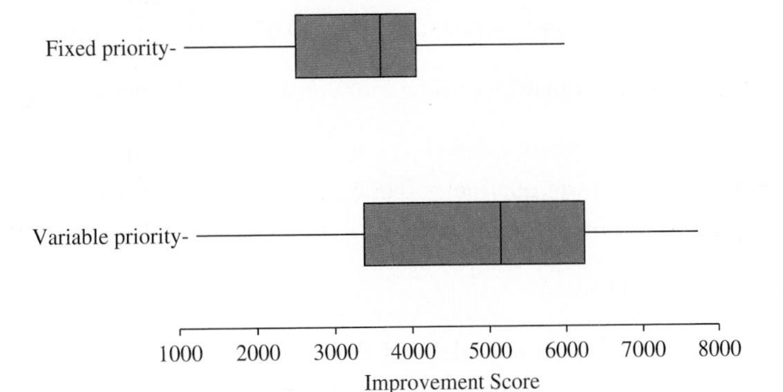

FIGURE 4.10
Comparative boxplots for improvement scores of fixed priority practice group and variable priority practice group.

From the comparative boxplot, we can see that both data distributions are approximately symmetric. The improvement scores tend to be higher for the variable priority practice group than for the fixed priority practice group. However, there is more consistency in the improvement scores for the fixed priority practice group. Improvement scores in the variable priority practice group are more spread out, indicating more variability in improvement scores for this group.

The sequence of steps used to construct a skeletal boxplot is easily modified to give information about outliers.

DEFINITION

Outlier: An observation that is more than 1.5(iqr) away from the nearest quartile (the nearest end of the box).

An outlier is an **extreme outlier** if it is more than 3(iqr) from the nearest quartile and it is a **mild outlier** otherwise.

A **modified boxplot** represents mild outliers by solid circles and extreme outliers by open circles, and the whiskers extend on each end to the most extreme observations that are *not* outliers.

Construction of a Modified Boxplot

1. Draw a horizontal (or vertical) measurement scale.
2. Construct a rectangular box with a left (or lower) edge at the lower quartile and right (or upper) edge at the upper quartile. The box width is then equal to the iqr.
3. Draw a vertical (or horizontal) line segment inside the box at the location of the median.
4. Determine if there are any mild or extreme outliers in the data set.
5. Draw whiskers that extend from each end of the box to the most extreme observation that is *not* an outlier.
6. Draw a solid circle to mark the location of any mild outliers in the data set.
7. Draw an open circle to mark the location of any extreme outliers in the data set.

Step-by-step technology instructions available online

EXAMPLE 4.11 Golden Rectangles

Understand the context)

● The accompanying data came from an anthropological study of rectangular shapes (***Lowie's Selected Papers in Anthropology*, Cora Dubios, ed. [Berkeley, CA: University of California Press, 1960]: 137–142**). Observations were made on the variable x = width/length for a sample of $n = 20$ beaded rectangles used in Shoshoni Indian leather handicrafts:

Consider the data)

0.553	0.570	0.576	0.601	0.606	0.606	0.609	0.611	0.615	0.628
0.654	0.662	0.668	0.670	0.672	0.690	0.693	0.749	0.844	0.933

The quantities needed for constructing the modified boxplot follow:

Do the work)

median = 0.641	iqr = 0.681 − 0.606 = 0.075
lower quartile = 0.606	1.5(iqr) = 0.1125
upper quartile = 0.681	3(iqr) = 0.225

Then,

$$\text{(upper quartile)} + 1.5(\text{iqr}) = 0.681 + 0.1125 = 0.7935$$
$$\text{(lower quartile)} - 1.5(\text{iqr}) = 0.606 - 0.1125 = 0.4935$$

So 0.844 and 0.933 are both outliers on the upper end (because they are larger than 0.7935), and there are no outliers on the lower end (because no observations are smaller than 0.4935). Because

$$\text{(upper quartile)} + 3(\text{iqr}) = 0.681 + 0.225 = 0.906$$

Interpret the results)

0.933 is an extreme outlier and 0.844 is only a mild outlier. The upper whisker extends to the largest observation that is not an outlier, 0.749, and the lower whisker extends to 0.553. The boxplot is shown in Figure 4.11. The median line is not at the center of the box, so there is a slight asymmetry in the middle half of the data. However, the most striking feature is the presence of the two outliers. These two x values considerably exceed the "golden ratio" of 0.618, used since antiquity as an aesthetic standard for rectangles.

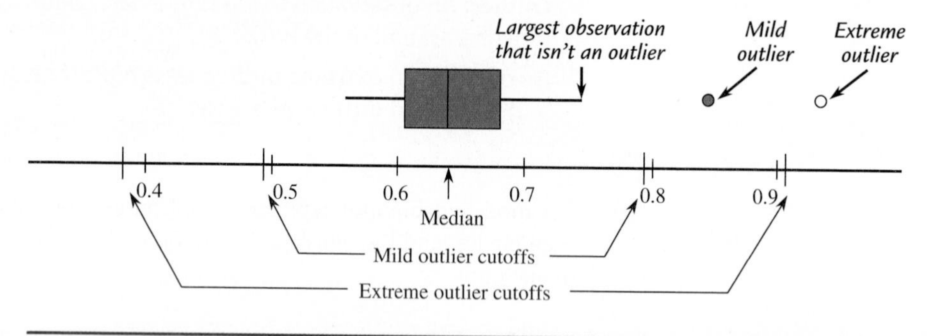

FIGURE 4.11
Boxplot for the rectangle data in Example 4.11.

■

EXAMPLE 4.12 Another Look at Big Mac Prices

Understand the context)

Big Mac prices in U.S. dollars for 57 different countries were given in the article **"The Big Mac Index"** first introduced in Example 4.7. The 57 Big Mac prices were:

Consider the data)

3.88	4.62	5.28	4.02	5.26	3.94	2.61	4.48	4.31	3.49	4.91
2.39	4.66	2.19	3.76	1.50	2.80	4.80	3.20	3.09	3.20	2.30
2.86	4.30	7.51	3.00	3.59	2.65	2.73	2.64	2.67	3.69	1.82
3.43	2.83	6.16	6.72	2.63	2.85	4.34	3.27	2.33	4.56	4.98
7.15	4.36	4.76	3.54	5.27	5.01	4.68	3.34	4.45	4.82	4.44
3.79	4.50									

● Data set available online

Figure 4.12 shows a Minitab boxplot for the Big Mac price data. Note that the upper whisker is longer than the lower whisker and that there is one outlier on the high end (Norway with a Big Mac price of $7.51).

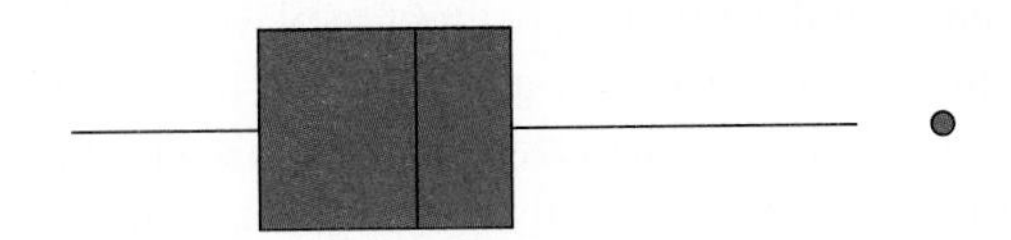

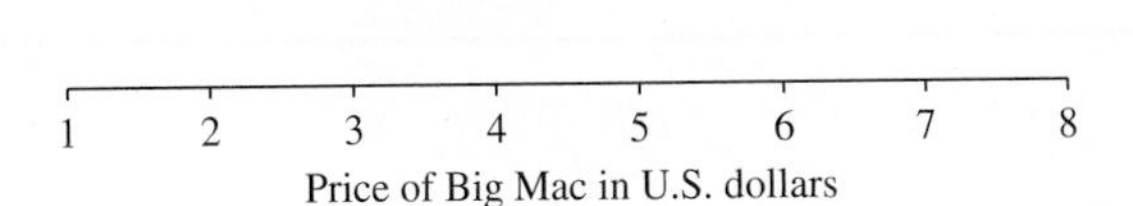

FIGURE 4.12
Minitab boxplot of the Big Mac price data of Example 4.12.

Note that Minitab does not distinguish between mild outliers and extreme outliers in the boxplot. For the Big Mac price data,

Do the work ⟩

lower quartile = 2.84
upper quartile = 4.67
iqr = 4.67 − 2.84 = 1.83

Then

1.5(iqr) = 2.745
3(iqr) = 5.490

We can compute outlier boundaries as follows:

upper quartile + 1.5(iqr) = 4.67 + 2.745 = 7.415
upper quartile + 3(iqr) = 4.67 + 5.49 = 10.16

The observation for Norway (7.51) is a mild outlier because it is greater than 7.415 (the upper quartile + 1.5(iqr)) but less than 10.16 (the upper quartile + 3(iqr)). There are no extreme outliers in this data set. ■

EXAMPLE 4.13 NBA Salaries Revisited

Understand the context ⟩

The 2009–2010 salaries of NBA players published on the web site **hoopshype.com** were used to construct the comparative boxplot of the salary data for five teams shown in Figure 4.13.

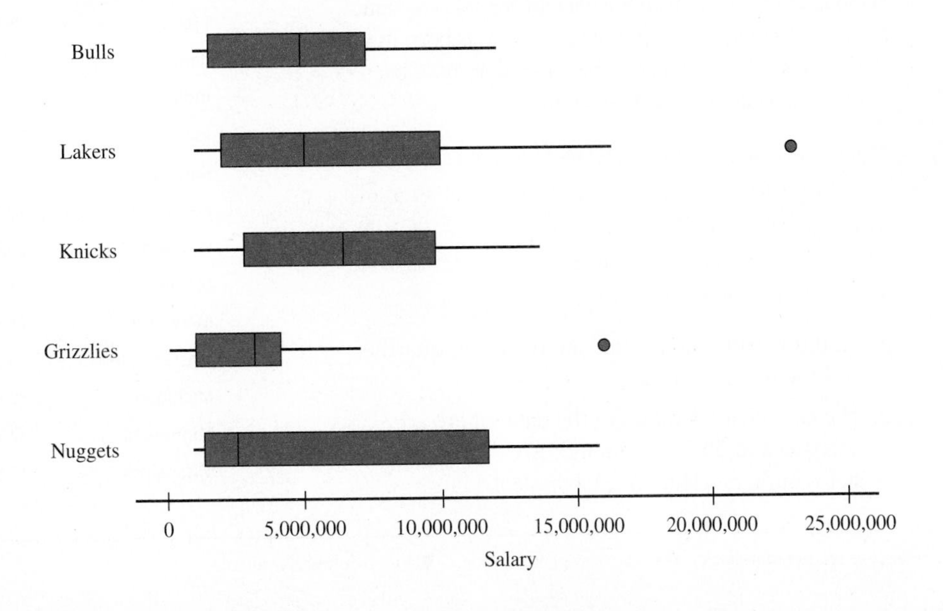

FIGURE 4.13
Comparative boxplot for salaries for five NBA teams.

Interpret the results) The comparative boxplot reveals some interesting similarities and differences in the salary distributions of the five teams. The minimum salary is lower for the Grizzlies, but is about the same for the other four teams. The median salary was lowest for the Nuggets—in fact the median for the Nuggets is about the same as the lower quartile for the Knicks and the Lakers, indicating that half of the players on the Nuggets have salaries less than about \$2.5 million, whereas only about 25% of the Knicks and the Lakers have salaries less than about \$2.5 million. The Lakers had the player with by far the highest salary. The Grizzlies and the Lakers were the only teams that had any salary outliers. With the exception of one highly paid player, salaries for players on the Grizzlies team were noticeably lower than for the other four teams. ■

EXERCISES 4.32 - 4.37

4.32 Based on a large national sample of working adults, the **U.S. Census Bureau** reports the following information on travel time to work for those who do not work at home:

lower quartile = 7 minutes
median = 18 minutes
upper quartile = 31 minutes

Also given was the mean travel time, which was reported as 22.4 minutes.

a. Is the travel time distribution more likely to be approximately symmetric, positively skewed, or negatively skewed? Explain your reasoning based on the given summary quantities.

b. Suppose that the minimum travel time was 1 minute and that the maximum travel time in the sample was 205 minutes. Construct a skeletal boxplot for the travel time data. (Hint: See Example 4.10.)

c. Were there any mild or extreme outliers in the data set? How can you tell? (Hint: See Example 4.11.)

4.33 ● The report **"Who Moves? Who Stays Put? Where's Home?" (*Pew Social and Demographic Trends*, December 17, 2008)** gave the accompanying data for the 50 U.S. states on the percentage of the population that was born in the state and is still living there. The data values have been arranged in order from largest to smallest.

75.8 71.4 69.6 69.0 68.6 67.5 66.7 66.3 66.1 66.0 66.0
65.1 64.4 64.3 63.8 63.7 62.8 62.6 61.9 61.9 61.5 61.1
59.2 59.0 58.7 57.3 57.1 55.6 55.6 55.5 55.3 54.9 54.7
54.5 54.0 54.0 53.9 53.5 52.8 52.5 50.2 50.2 48.9 48.7
48.6 47.1 43.4 40.4 35.7 28.2

a. Find the values of the median, the lower quartile, and the upper quartile.

b. The two smallest values in the data set are 28.2 (Alaska) and 35.7 (Wyoming). Are these two states outliers? (Hint: See Example 4.11.)

c. Construct a boxplot for this data set and comment on the interesting features of the plot.

4.34 Data on the gasoline tax per gallon (in cents) as of April 2010 for the 50 U.S. states and the District of Columbia are shown below **(*AARP Bulletin*, June 2010)**.

State	Gasoline Tax per Gallon	State	Gasoline Tax per Gallon
Alabama	20.9	Missouri	17.3
Alaska	8.0	Montana	27.8
Arizona	19.0	Nebraska	27.7
Arkansas	21.8	Nevada	33.1
California	48.6	New Hampshire	19.6
Colorado	22.0	New Jersey	14.5
Connecticut	42.6	New Mexico	18.8
Delaware	23.0	New York	44.9
District of Columbia	23.5	North Carolina	30.2
Florida	34.4	North Dakota	23.0
Georgia	20.9	Ohio	28.0
Hawaii	45.1	Oklahoma	17.0
Idaho	25.0	Oregon	25.0
Illinois	40.4	Pennsylvania	32.3
Indiana	34.8	Rhode Island	33.0
Iowa	22.0	South Carolina	16.8
Kansas	25.0	South Dakota	24.0
Kentucky	22.5	Tennessee	21.4
Louisiana	20.0	Texas	20.0
Maine	31.0	Utah	24.5
Maryland	23.5	Vermont	24.7
Massachusetts	23.5	Virginia	19.6
Michigan	35.8	Washington	37.5
Minnesota	27.2	West Virginia	32.2
Mississippi	18.8	Wisconsin	32.9
		Wyoming	14.0

a. The smallest value in the data set is 8.0 (Alaska) and the largest value is 48.6 (California). Are these values outliers? Explain. (Hint: See Example 4.11.)

b. Construct a boxplot of the data set and comment on the interesting features of the plot.

4.35 The article **"Going Wireless"** (***AARP Bulletin*, June 2009**) reported the estimated percentage of U.S. households with only wireless phone service (no landline) for the 50 states and the District of Columbia. In the accompanying data table, each state was also classified into one of three geographical regions—West (W), Middle states (M), and East (E).

Wireless %	Region	State	Wireless %	Region	State
13.9	M	AL	9.2	W	MT
11.7	W	AK	23.2	M	NE
18.9	W	AZ	10.8	W	NV
22.6	M	AR	16.9	M	ND
9.0	W	CA	11.6	E	NH
16.7	W	CO	8.0	E	NJ
5.6	E	CT	21.1	W	NM
5.7	E	DE	11.4	E	NY
20.0	E	DC	16.3	E	NC
16.8	E	FL	14.0	E	OH
16.5	E	GA	23.2	M	OK
8.0	W	HI	17.7	W	OR
22.1	W	ID	10.8	E	PA
16.5	M	IL	7.9	E	RI
13.8	M	IN	20.6	E	SC
22.2	M	IA	6.4	M	SD
16.8	M	KA	20.3	M	TN
21.4	M	KY	20.9	M	TX
15.0	M	LA	25.5	W	UT
13.4	E	ME	10.8	E	VA
10.8	E	MD	5.1	E	VT
9.3	E	MA	16.3	W	WA
16.3	M	MI	11.6	E	WV
17.4	M	MN	15.2	M	WI
19.1	M	MS	11.4	W	WY
9.9	M	MO			

a. Display the data in a comparative boxplot that makes it possible to compare wireless percent for the three geographical regions.

b. Does the graphical display in Part (a) reveal any striking differences in wireless percent for the three geographical regions, or are the distributions of wireless percent observations similar for the three regions?

4.36 ● Fiber content (in grams per serving) and sugar content (in grams per serving) for 18 high fiber cereals **(www.consumerreports.com)** are shown below.

Fiber Content

7	10	10	7	8	7	12	12	8
13	10	8	12	7	14	7	8	8

Sugar Content

11	6	14	13	0	18	9	10	19
6	10	17	10	10	0	9	5	11

a. Find the median, quartiles, and interquartile range for the fiber content data set.

b. Find the median, quartiles, and interquartile range for the sugar content data set.

c. Are there any outliers in the sugar content data set?

d. Explain why the minimum value for the fiber content data set and the lower quartile for the fiber content data set are equal.

e. Construct a comparative boxplot and use it to comment on the differences and similarities in the fiber and sugar distributions.

4.37 ● Shown here are the number of auto accidents per year for every 1000 people in each of 40 occupations (***Knight Ridder Tribune*, June 19, 2004**):

Occupation	Accidents per 1000	Occupation	Accidents per 1000
Student	152	Banking, finance	89
Physician	109	Customer service	88
Lawyer	106	Manager	88
Architect	105	Medical support	87
Real estate broker	102	Computer-related	87
Enlisted military	99	Dentist	86
Social worker	98	Pharmacist	85
Manual laborer	96	Proprietor	84
Analyst	95	Teacher, professor	84
Engineer	94	Accountant	84
Consultant	94	Law enforcement	79
Sales	93	Physical therapist	78
Military officer	91	Veterinarian	78
Nurse	90	Clerical, secretary	77
School administrator	90	Clergy	76
Skilled laborer	90	Homemaker	76
Librarian	90	Politician	76
Creative arts	90	Pilot	75
Executive	89	Firefighter	67
Insurance agent	89	Farmer	43

a. Would you recommend using the standard deviation or the iqr as a measure of variability for this data set?

b. Are there outliers in this data set? If so, which observations are mild outliers? Which are extreme outliers?

c. Draw a modified boxplot for this data set.

d. If you were asked by an insurance company to decide which, if any, occupations should be offered a professional discount on auto insurance, which occupations would you recommend? Explain.

Bold exercises answered in back ● Data set available online ▼ Video Solution available

4.4 Interpreting Center and Variability: Chebyshev's Rule, the Empirical Rule, and *z* Scores

The mean and standard deviation can be combined to make informative statements about how the values in a data set are distributed and about the relative position of a particular value in a data set. To do this, it is useful to be able to describe how far away a particular observation is from the mean in terms of the standard deviation. For example, we might say that an observation is 2 standard deviations above the mean or that an observation is 1.3 standard deviations below the mean.

EXAMPLE 4.14 Standardized Test Scores

Consider a data set of scores on a standardized test with a mean and standard deviation of 100 and 15, respectively. We can make the following statements:

1. Because $100 - 15 = 85$, we say that a score of 85 is "1 standard deviation *below* the mean." Similarly, $100 + 15 = 115$ is "1 standard deviation *above* the mean" (see Figure 4.14).

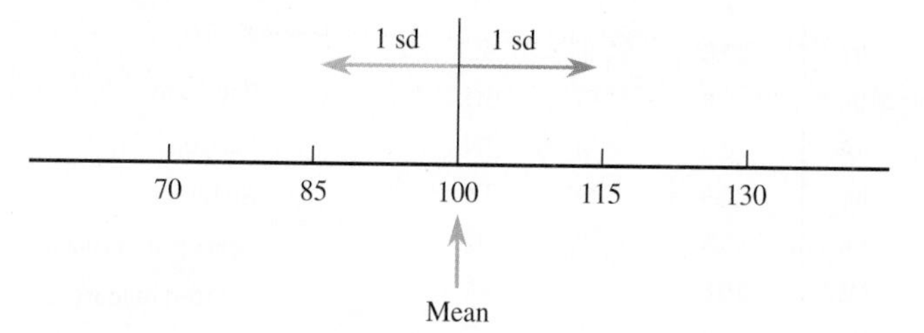

FIGURE 4.14
Values within 1 standard deviation of the mean (Example 4.14).

2. Because 2 times the standard deviation is $2(15) = 30$, and $100 + 30 = 130$ and $100 - 30 = 70$, scores between 70 and 130 are those *within* 2 standard deviations of the mean (see Figure 4.15).

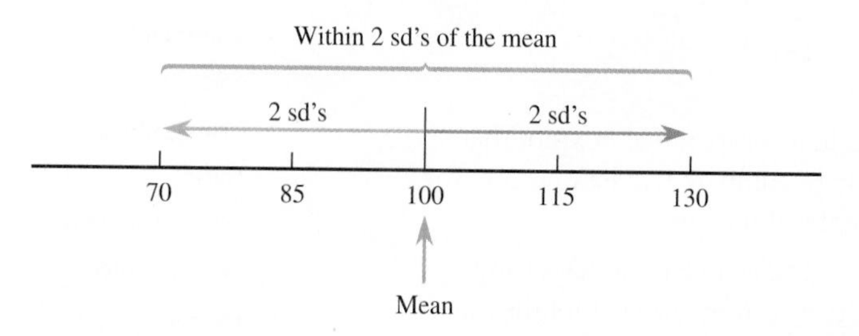

FIGURE 4.15
Values within 2 standard deviations of the mean (Example 4.14).

3. Because $100 + (3)(15) = 145$, scores above 145 are greater than the mean by more than 3 standard deviations. ■

Sometimes in published articles, the mean and standard deviation are reported, but a graphical display of the data is not given. However, using a result called Chebyshev's Rule, it is possible to get a sense of the distribution of data values based on our knowledge of only the mean and standard deviation.

Chebyshev's Rule

Consider any number k, where $k \geq 1$. Then the percentage of observations that are within k standard deviations of the mean is at least $100\left(1 - \frac{1}{k^2}\right)\%$. Substituting selected values of k gives the following results.

Number of Standard Deviations, k	$1 - \frac{1}{k^2}$	Percentage Within k Standard Deviations of the Mean
2	$1 - \frac{1}{4} = .75$	at least 75%
3	$1 - \frac{1}{9} = .89$	at least 89%
4	$1 - \frac{1}{16} = .94$	at least 94%
4.472	$1 - \frac{1}{20} = .95$	at least 95%
5	$1 - \frac{1}{25} = .96$	at least 96%
10	$1 - \frac{1}{100} = .99$	at least 99%

EXAMPLE 4.15 Child Care for Preschool Kids

Understand the context ❭

The article **"Piecing Together Child Care with Multiple Arrangements: Crazy Quilt or Preferred Pattern for Employed Parents of Preschool Children?" (*Journal of Marriage and the Family* [1994]: 669–680)** examined various modes of care for preschool children. For a sample of families with one preschool child, it was reported that the mean and standard deviation of child care time per week were approximately 36 hours and 12 hours, respectively. Figure 4.16 displays values that are 1, 2, and 3 standard deviations from the mean.

Ariel Skelley/Blend Images/Jupiter Images

0	12	24	36	48	60	72
$\bar{x} - 3s$	$\bar{x} - 2s$	$\bar{x} - s$	$\bar{x}$	$\bar{x} + s$	$\bar{x} + 2s$	$\bar{x} + 3s$

FIGURE 4.16
Measurement scale for child care time (Example 4.15).

Chebyshev's Rule allows us to assert the following:

1. At least 75% of the sample observations must be between 12 and 60 hours (within 2 standard deviations of the mean).
2. Because at least 89% of the observations must be between 0 and 72, at most 11% are outside this interval. Time cannot be negative, so we conclude that at most 11% of the observations exceed 72.

3. The values 18 and 54 are 1.5 standard deviations to either side of $\bar{x}$, so using $k = 1.5$ in Chebyshev's Rule implies that at least 55.6% of the observations must be between these two values. This means that at most 44.4% of the observations are less than 18—*not* at most 22.2%, because the distribution of values may not be symmetric. ■

Because Chebyshev's Rule is applicable to any data set (distribution), whether symmetric or skewed, we must be careful when making statements about the proportion above a particular value, below a particular value, or inside or outside an interval that is not centered at the mean. The rule must be used in a conservative fashion. There is another aspect of this conservatism. The rule states that *at least* 75% of the observations are within 2 standard deviations of the mean, but for many data sets substantially more than 75% of the values satisfy this condition. The same sort of understatement is frequently encountered for other values of k (numbers of standard deviations).

EXAMPLE 4.16 IQ Scores

Understand the context ⟩

Figure 4.17 gives a stem-and-leaf display of IQ scores of 112 children in one of the early studies that used the Stanford revision of the Binet–Simon intelligence scale (***The Intelligence of School Children,*** **L. M. Terman [Boston: Houghton-Mifflin, 1919]**).

Summary quantities include

$\bar{x} = 104.5 \qquad s = 16.3 \qquad 2s = 32.6 \qquad 3s = 48.9$

```
 6 | 1
 7 | 25679
 8 | 0000124555668
 9 | 0000112333446666778889
10 | 0001122222333566677778899999
11 | 00001122333344444477899
12 | 01111123445669
13 | 006
14 | 26              Stem: Tens
15 | 2               Leaf: Ones
```

FIGURE 4.17
Stem-and-leaf display of IQ scores used in Example 4.16.

In Figure 4.17, all observations that are within two standard deviations of the mean are shown in blue. Table 4.6 shows how Chebyshev's Rule can sometimes considerably understate actual percentages.

TABLE 4.6 Summarizing the Distribution of IQ Scores

k = Number of sd's	$\bar{x} \pm ks$	Chebyshev	Actual
2	71.9 to 137.1	at least 75%	96% (108)
2.5	63.7 to 145.3	at least 84%	97% (109) ← *the blue leaves in Figure 4.17*
3	55.6 to 153.4	at least 89%	100% (112)

■

Empirical Rule

The fact that statements based on Chebyshev's Rule are frequently conservative suggests that we should look for rules that are less conservative and more precise. One useful rule is the **Empirical Rule**, which can be applied whenever the distribution of data values can be reasonably well described by a normal curve (distributions that are "mound" shaped).

The Empirical Rule

If the histogram of values in a data set can be reasonably well approximated by a normal curve, then

Approximately 68% of the observations are within 1 standard deviation of the mean.
Approximately 95% of the observations are within 2 standard deviations of the mean.
Approximately 99.7% of the observations are within 3 standard deviations of the mean.

The Empirical Rule makes "approximately" instead of "at least" statements, and the percentages for $k = 1, 2$, and 3 standard deviations are much higher than those of Chebyshev's Rule. Figure 4.18 illustrates the percentages given by the Empirical Rule. In contrast to Chebyshev's Rule, dividing the percentages in half is permissible, because a normal curve is symmetric.

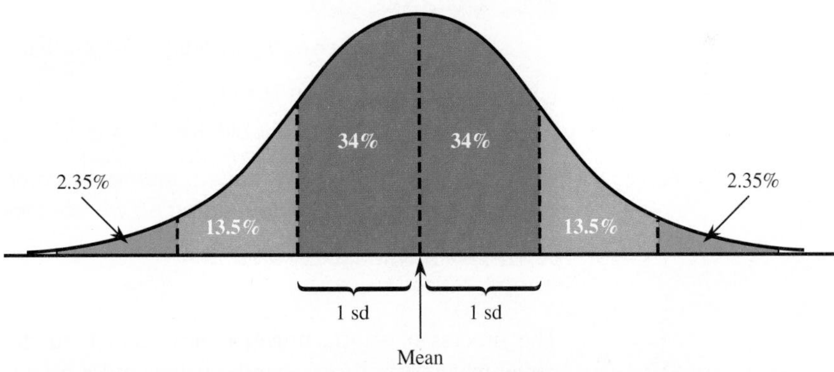

FIGURE 4.18
Approximate percentages implied by the Empirical Rule.

EXAMPLE 4.17 Heights of Mothers and the Empirical Rule

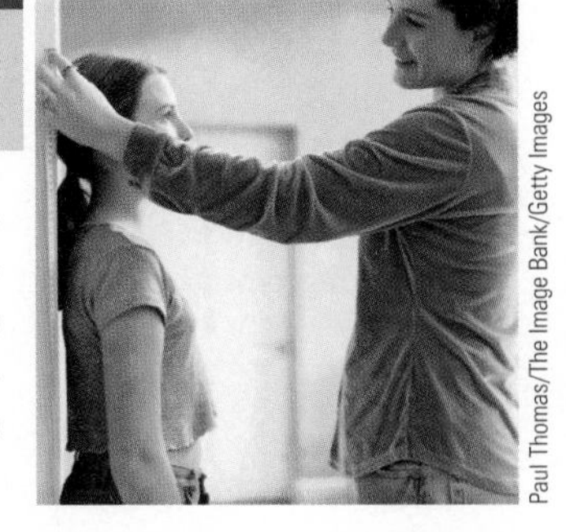

Understand the context)

One of the earliest articles to argue for the wide applicability of the normal distribution was **"On the Laws of Inheritance in Man. I. Inheritance of Physical Characters" (*Biometrika* [1903]: 375–462)**. Among the data sets discussed in the article was one consisting of 1052 measurements of the heights of mothers. The mean and standard deviation were

$\bar{x} = 62.484$ in. $\quad s = 2.390$ in.

The data distribution was described as approximately normal. Table 4.7 contrasts actual percentages with those obtained from Chebyshev's Rule and the Empirical Rule.

TABLE 4.7 Summarizing the Distribution of Mothers' Heights

Number of sd's	Interval	Actual	Empirical Rule	Chebyshev Rule
1	60.094 to 64.874	72.1%	Approximately 68%	At least 0%
2	57.704 to 67.264	96.2%	Approximately 95%	At least 75%
3	55.314 to 69.654	99.2%	Approximately 99.7%	At least 89%

Clearly, the Empirical Rule is much more successful and informative in this case than Chebyshev's Rule. ■

Our detailed study of the normal distribution and areas under normal curves in Chapter 7 will enable us to make statements analogous to those of the Empirical Rule for values other than $k = 1, 2$, or 3 standard deviations. For now, note that it is unusual to see an observation from a normally distributed population that is farther than 2 standard deviations from the mean (only 5%), and it is very surprising to see one that is more than 3 standard deviations away. If you encountered a mother whose height was 72 inches, you might reasonably conclude that she was not part of the population described by the data set in Example 4.17.

Measures of Relative Standing

When you obtain your score after taking a test, you probably want to know how it compares to the scores of others who have taken the test. Is your score above or below the mean, and by how much? Does your score place you among the top 5% of those who took the test or only among the top 25%? Questions of this sort are answered by finding ways to measure the position of a particular value in a data set relative to all values in the set. One measure of relative standing is a ***z* score**.

DEFINITION

***z* score:** The ***z* score** corresponding to a particular value is

$$z \text{ score} = \frac{\text{value} - \text{mean}}{\text{standard deviation}}$$

The *z* score tells us how many standard deviations the value is from the mean. It is positive or negative depending on whether the value is greater than or less than the mean.

The process of subtracting the mean and then dividing by the standard deviation is sometimes referred to as standardizing, and a *z* score is one example of what is called a standardized score.

EXAMPLE 4.18 Relatively Speaking, Which Is the Better Offer?

Understand the context ❯

Suppose that two graduating seniors, one a marketing major and one an accounting major, are comparing job offers. The accounting major has an offer for $45,000 per year, and the marketing student has an offer for $43,000 per year. Summary information about the distribution of offers follows:

Consider the data ❯

Accounting:	mean = 46,000	standard deviation = 1500
Marketing:	mean = 42,500	standard deviation = 1000

Then,

$$\text{accounting } z \text{ score} = \frac{45{,}000 - 46{,}000}{1500} = -.67$$

Do the work ❯

(so $45,000 is .67 standard deviation below the mean), whereas

$$\text{marketing } z \text{ score} = \frac{43{,}000 - 42{,}500}{1000} = .5$$

Interpret the results ❯

Relative to the appropriate data sets, the marketing offer is actually more attractive than the accounting offer (although this may not offer much solace to the marketing major). ■

The z score is particularly useful when the distribution of observations is approximately normal. In this case, from the Empirical Rule, a z score outside the interval from -2 to $+2$ occurs in about 5% of all cases, whereas a z score outside the interval from -3 to $+3$ occurs only about 0.3% of the time.

Percentiles

A particular observation can be located even more precisely by giving the percentage of the data that fall at or below that observation. If, for example, 95% of all test scores are at or below 650, whereas only 5% are above 650, then 650 is called the *95th percentile* of the data set (or of the distribution of scores). Similarly, if 10% of all scores are at or below 400 and 90% are above 400, then the value 400 is the 10th percentile.

DEFINITION

Percentile: For any particular number r between 0 and 100, the **rth percentile** is a value such that r percent of the observations in the data set fall at or below that value.

Figure 4.19 illustrates the 90th percentile. We have already met several percentiles in disguise. The median is the 50th percentile, and the lower and upper quartiles are the 25th and 75th percentiles, respectively.

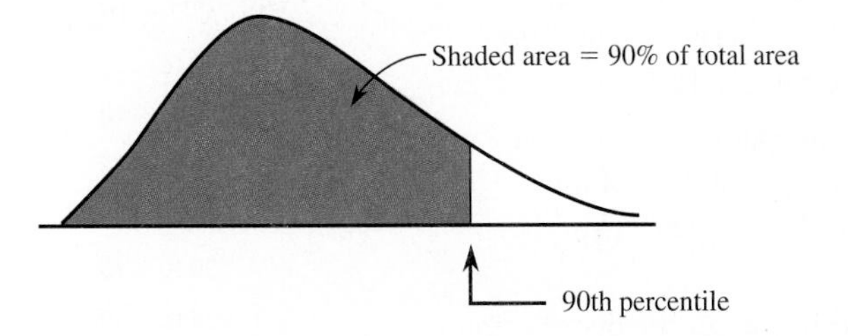

FIGURE 4.19
Ninetieth percentile for a smoothed histogram.

EXAMPLE 4.19 Head Circumference at Birth

Understand the context

In addition to weight and length, head circumference is another measure of health in newborn babies. **The National Center for Health Statistics** reports the following summary values for head circumference (in cm) at birth for boys (approximate values read from graphs on the Center for Disease Control web site):

Consider the data

Percentile	5	10	25	50	75	90	95
Head Circumference (cm)	32.2	33.2	34.5	35.8	37.0	38.2	38.6

Interpret the results

Interpreting these percentiles, we know that half of newborn boys have head circumferences of less than 35.8 cm, because 35.8 is the 50th percentile (the median). The middle 50% of newborn boys have head circumferences between 34.5 cm and 37.0 cm, with about 25% of the head circumferences less than 34.5 cm and about 25% greater than 37.0 cm.

We can tell that the head circumference distribution for newborn boys is not symmetric, because the 5th percentile is 3.6 cm below the median, whereas the 95th percentile is only 2.8 cm above the median. This suggests that the bottom part of the distribution stretches out more than the top part of the distribution. This would be consistent with a distribution that is negatively skewed, as shown in Figure 4.20.

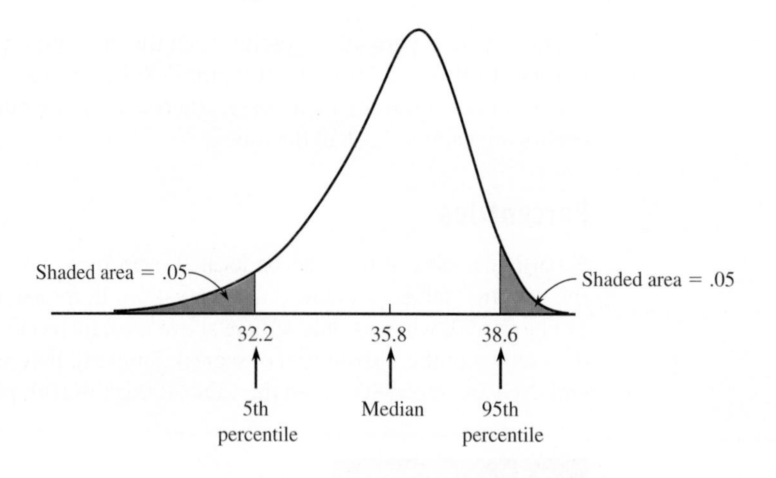

FIGURE 4.20
Negatively skewed distribution.

EXERCISES 4.38 - 4.52

4.38 The average playing time of compact discs in a large collection is 35 minutes, and the standard deviation is 5 minutes.

a. What value is 1 standard deviation above the mean? 1 standard deviation below the mean? What values are 2 standard deviations away from the mean? (Hint: See Example 4.14.)

b. Without assuming anything about the distribution of times, at least what percentage of the times is between 25 and 45 minutes? (Hint: See Example 4.15.)

c. Without assuming anything about the distribution of times, what can be said about the percentage of times that are either less than 20 minutes or greater than 50 minutes?

d. Assuming that the distribution of times is approximately normal, about what percentage of times are between 25 and 45 minutes? less than 20 minutes or greater than 50 minutes? less than 20 minutes?

4.39 ▼ In a study investigating the effect of car speed on accident severity, 5000 reports of fatal automobile accidents were examined, and the vehicle speed at impact was recorded for each one. For these 5000 accidents, the average speed was 42 mph and the standard deviation was 15 mph. A histogram revealed that the vehicle speed at impact distribution was approximately normal.

a. Roughly what proportion of vehicle speeds were between 27 and 57 mph? (Hint: See Example 4.17.)

b. Roughly what proportion of vehicle speeds exceeded 57 mph?

4.40 **The U.S. Census Bureau (2000 census)** reported the following relative frequency distribution for travel time to work for a large sample of adults who did not work at home:

Travel Time (minutes)	Relative Frequency
0 to <5	.04
5 to <10	.13
10 to <15	.16
15 to <20	.17
20 to <25	.14
25 to <30	.05
30 to <35	.12
35 to <40	.03
40 to <45	.03
45 to <60	.06
60 to <90	.05
90 or more	.02

a. Draw the histogram for the travel time distribution. In constructing the histogram, assume that the last interval in the relative frequency distribution (90 or more) ends at 200; so the last interval is 90 to <200. Be sure to use the density scale to determine the heights of the bars in the histogram because not all the intervals have the same width. (Hint: Histograms were covered in Chapter 3.)

b. Describe the interesting features of the histogram from Part (a), including center, shape, and spread.

c. Based on the histogram from Part (a), would it be appropriate to use the Empirical Rule to make statements about the travel time distribution? Explain why or why not.

d. The approximate mean and standard deviation for the travel time distribution are 27 minutes and 24 minutes, respectively. Based on this mean and standard deviation and the fact that travel time cannot be negative, explain why the travel time distribution could not be well approximated by a normal curve.

e. Use the mean and standard deviation given in Part (d) and Chebyshev's Rule to make a statement about

i. the percentage of travel times that were between 0 and 75 minutes

ii. the percentage of travel times that were between 0 and 47 minutes

f. How well do the statements in Part (e) based on Chebyshev's Rule agree with the actual percentages for the travel time distribution? (Hint: You can estimate the actual percentages from the given relative frequency distribution.)

4.41 Mobile homes are tightly constructed for energy conservation. This can lead to a buildup of indoor pollutants. The paper **"A Survey of Nitrogen Dioxide Levels Inside Mobile Homes"** (***Journal of the Air Pollution Control Association*** **[1988]: 647–651)** discussed various aspects of NO_2 concentration in these structures.

a. In one sample of mobile homes in the Los Angeles area, the mean NO_2 concentration in kitchens during the summer was 36.92 ppb, and the standard deviation was 11.34. Making no assumptions about the shape of the NO_2 distribution, what can be said about the percentage of observations between 14.24 and 59.60?

b. Inside what interval is it guaranteed that at least 89% of the concentration observations will lie?

c. In a sample of non–Los Angeles mobile homes, the average kitchen NO_2 concentration during the winter was 24.76 ppb, and the standard deviation was 17.20. Do these values suggest that the histogram of sample observations did not closely resemble a normal curve? (Hint: What is $\bar{x} - 2s$?)

4.42 The article **"Taxable Wealth and Alcoholic Beverage Consumption in the United States"** (***Psychological Reports*** **[1994]: 813–814)** reported that the mean annual adult consumption of wine was 3.15 gallons and that the standard deviation was 6.09 gallons. Would you use the Empirical Rule to approximate the proportion of adults who consume more than 9.24 gallons (i.e., the proportion of adults whose consumption value exceeds the mean by more than 1 standard deviation)? Explain your reasoning.

4.43 A student took two national aptitude tests. The national average and standard deviation were 475 and 100, respectively, for the first test and 30 and 8, respectively, for the second test. The student scored 625 on the first test and 45 on the second test. Use z scores to determine on which exam the student performed better relative to the other test takers. (Hint: See Example 4.18.)

4.44 Suppose that your younger sister is applying for entrance to college and has taken the SAT. She scored at the 83rd percentile on the verbal section of the test and at the 94th percentile on the math section of the test. Because you have been studying statistics, she asks you for an interpretation of these values. What would you tell her? (Hint: See Example 4.19.)

4.45 The report **"Who Borrows Most? Bachelor's Degree Recipients with High Levels of Student Debt" (www.collgeboard.com/trends)** reported the following percentiles for amount of student debt for those graduating with a bachelor's degree in 2010:

10th percentile = \$0 25th percentile = \$0
50th percentile = \$11,000 75th percentile = \$24,600
90th percentile = \$39,300

For each of these percentiles, write a sentence interpreting the value of the percentile.

4.46 The paper **"Modeling and Measurements of Bus Service Reliability"** (***Transportation Research*** **[1978]: 253–256)** studied various aspects of bus service and presented data on travel times (in minutes) from several different routes. The accompanying frequency distribution is for bus travel times from origin to destination on one particular route in Chicago during peak morning traffic periods:

Travel Time	Frequency	Relative Frequency
15 to <16	4	.02
16 to <17	0	.00
17 to <18	26	.13
18 to <19	99	.49
19 to <20	36	.18
20 to <21	8	.04
21 to <22	12	.06
22 to <23	0	.00
23 to <24	0	.00
24 to <25	0	.00
25 to <26	16	.08

a. Construct the corresponding histogram.

b. Approximate the following percentiles:

i. 86th   iv. 95th
ii. 15th   v. 10th
iii. 90th

4.47 An advertisement for the **"30-inch Wonder"** that appeared in the **September 1983** issue of the journal ***Packaging*** claimed that the 30-inch Wonder weighs cases and bags up to 110 pounds and provides accuracy to within 0.25 ounce. Suppose that a 50-ounce weight was repeatedly weighed on this scale and the weight readings recorded. The mean value was 49.5 ounces, and the standard deviation was 0.1. What can be said about the percentage of the time that the scale actually showed a weight that was within 0.25 ounce of the true value of 50 ounces? (Hint: Use Chebyshev's Rule.)

4.48 Suppose that your statistics professor returned your first midterm exam with only a z score written on it. She also told you that a histogram of the scores was approximately normal. How would you interpret each of the following z scores?

a. 2.2
b. 0.4
c. 1.8
d. 1.0
e. 0

4.49 The paper **"Answer Changing on Multiple-Choice Tests" (*Journal of Experimental Education* [1980]: 18–21)** reported that for a group of 162 college students, the average number of responses changed from the correct answer to an incorrect answer on a test containing 80 multiple-choice items was 1.4. The corresponding standard deviation was reported to be 1.5. Based on this mean and standard deviation, what can you tell about the shape of the distribution of the variable *number of answers changed from right to wrong*? What can you say about the number of students who changed at least six answers from correct to incorrect?

4.50 The average reading speed of students completing a speed-reading course is 450 words per minute (wpm). If the standard deviation is 70 wpm, find the z score associated with each of the following reading speeds.

a. 320 wpm
b. 475 wpm
c. 420 wpm
d. 610 wpm

4.51 ● The following data values are 2009 per capita expenditures on public libraries for each of the 50 U.S. states and the District of Columbia **(from www.statemaster.com)**:

16.84	16.17	11.74	11.11	8.65	7.69	7.48
7.03	6.20	6.20	5.95	5.72	5.61	5.47
5.43	5.33	4.84	4.63	4.59	4.58	3.92
3.81	3.75	3.74	3.67	3.40	3.35	3.29
3.18	3.16	2.91	2.78	2.61	2.58	2.45
2.30	2.19	2.06	1.78	1.54	1.31	1.26
1.20	1.19	1.09	0.70	0.66	0.54	0.49
0.30	0.01					

a. Summarize this data set with a frequency distribution. Construct the corresponding histogram.

b. Use the histogram in Part (a) to find approximate values of the following percentiles:

i. 50th
ii. 70th
iii. 10th
iv. 90th
v. 40th

4.52 The accompanying table gives the mean and standard deviation of reaction times (in seconds) for each of two different stimuli:

	Stimulus 1	Stimulus 2
Mean	6.0	3.6
Standard deviation	1.2	0.8

If your reaction time is 4.2 seconds for the first stimulus and 1.8 seconds for the second stimulus, to which stimulus are you reacting (compared to other individuals) relatively more quickly? (Hint: See Example 4.18.)

Bold exercises answered in back ● Data set available online ▼ Video Solution available

4.5 Interpreting and Communicating the Results of Statistical Analyses

As was the case with the graphical displays of Chapter 3, the primary function of the descriptive tools introduced in this chapter is to help us better understand the variables under study. If we have collected data on the amount of money students spend on textbooks at a particular university, most likely we did so because we wanted to learn about the distribution of this variable (amount spent on textbooks) for the population of interest (in this case, students at the university). Numerical measures of center and spread and boxplots help to inform us, and they also allow us to communicate to others what we have learned from the data.

Communicating the Results of Statistical Analyses

When reporting the results of a data analysis, it is common to start with descriptive information about the variables of interest. It is always a good idea to start with a graphical display of the data, and, as we saw in Chapter 3, graphical displays of numerical data are usually described in terms of center, variability, and shape. The numerical measures of this chapter can help you to be more specific in describing the center and spread of a data set.

When describing center and spread, you must first decide which measures to use. Common choices are to use either the sample mean and standard deviation or the sample median and interquartile range (and maybe even a boxplot) to describe center and spread. Because the mean and standard deviation can be sensitive to extreme values in the data set, they are best used when the distribution shape is approximately symmetric and when there are few outliers. If the data set is noticeably skewed or if there are outliers, the median and iqr are generally used to describe center and spread.

Interpreting the Results of Statistical Analyses

It is relatively rare to find raw data in published sources. Typically, only a few numerical summary quantities are reported. We must be able to interpret these values and understand what they tell us about the underlying data set.

For example, a university conducted an investigation of the amount of time required to enter the information contained in an application for admission into the university computer system. One of the individuals who performs this task was asked to note starting time and completion time for 50 randomly selected application forms. The resulting entry times (in minutes) were summarized using the mean, median, and standard deviation:

$$\bar{x} = 7.854$$
$$\text{median} = 7.423$$
$$s = 2.129$$

What do these summary values tell us about entry times? The average time required to enter admissions data was 7.854 minutes, but the relatively large standard deviation suggests that many observations differ substantially from this mean. The median tells us that half of the applications required less than 7.423 minutes to enter. The fact that the mean exceeds the median suggests that some unusually large values in the data set affected the value of the mean. This last conjecture is confirmed by the stem-and-leaf display of the data given in Figure 4.21.

Stem	Leaf
4	8
5	02345679
6	00001234566779
7	223556688
8	23334
9	002
10	011168
11	134
12	2
13	
14	3

Stem: Ones
Leaf: Tenths

FIGURE 4.21
Stem-and-leaf display of data entry times.

The administrators conducting the data-entry study looked at the outlier 14.3 minutes and at the other relatively large values in the data set. They found that the five largest values came from applications that were entered before lunch. After talking with the individual who entered the data, the administrators speculated that morning entry times might differ from afternoon entry times because there tended to be more distractions and interruptions (phone calls, etc.) during the morning hours, when the admissions office generally was busier.

When morning and afternoon entry times were separated, the following summary statistics resulted:

Morning (based on $n = 20$ applications): $\bar{x} = 9.093$ median $= 8.743$ $s = 2.329$
Afternoon (based on $n = 30$ applications): $\bar{x} = 7.027$ median $= 6.737$ $s = 1.529$

Clearly, the average entry time is higher for applications entered in the morning. The individual entry times also differ more from one another in the mornings than in the afternoons (because the standard deviation for morning entry times, 2.329, is about 1.5 times as large as 1.529, the standard deviation for afternoon entry times).

What to Look for in Published Data

Here are a few questions to ask yourself when you interpret numerical summary measures.

- Is the chosen summary measure appropriate for the type of data collected? In particular, watch for inappropriate use of the mean and standard deviation with categorical data that has simply been coded numerically.
- If both the mean and the median are reported, how do the two values compare? What does this suggest about the distribution of values in the data set? If only the mean or the median was used, was the appropriate measure selected?
- Is the standard deviation large or small? Is the value consistent with your expectations regarding variability? What does the value of the standard deviation tell you about the variable being summarized?
- Can anything of interest be said about the values in the data set by applying Chebyshev's Rule or the Empirical Rule?

For example, consider a study that investigated whether people tend to spend more money when they are paying with a credit card than when they are paying with cash. The authors of the paper **"Monopoly Money: The Effect of Payment Coupling and Form on Spending Behavior"** (***Journal of Experimental Psychology: Applied*** **[2008]: 213–225)** randomly assigned each of 114 volunteers to one of two experimental groups. Participants were given a menu for a new restaurant that showed nine menu items. They were then asked to estimate the amount they would be willing to pay for each item. A price index was computed for each participant by averaging the nine prices assigned.

The difference between the two experimental groups was that the menu viewed by one group showed a credit card logo at the bottom of the menu while there was no credit card logo on the menu that those in the other group viewed. The following passage appeared in the results section of the paper:

> On average, participants were willing to pay more when the credit card logo was present (M = \$4.53, SD = 1.15) than when it was absent (M = \$4.11, SD = 1.06). Thus, even though consumers were not explicitly informed which payment mode they would be using, the mere presence of a credit card logo increased the price that they were willing to pay.

The price index data was also described as mound shaped with no outliers for each of the two groups. Because price index (the average of the prices that a participant assigned to the nine menu items) is a numerical variable, the mean and standard deviation are reasonable measures for summarizing center and spread in the data set.

Although the mean for the credit-card-logo group is higher than the mean for the no-logo group, the two standard deviations are similar, indicating similar variability in price index from person to person for the two groups.

Because the distribution of price index values was mound shaped for each of the two groups, we can use the Empirical Rule to tell us a bit more about the distribution. For example, for those in the group who viewed the menu with a credit card logo, approximately 95% of the price index values would have been between

$$4.53 - 2(1.15) = 4.53 - 2.3 = 2.23$$

and

$$4.53 + 2(1.15) = 4.53 + 2.30 = 6.83.$$

A Word to the Wise: Cautions and Limitations

When computing or interpreting numerical descriptive measures, you need to keep in mind the following:

1. Measures of center don't tell all. Although measures of center, such as the mean and the median, do give us a sense of what might be considered a typical value for a variable, this is only one characteristic of a data set. Without additional information about variability and distribution shape, we don't really know much about the behavior of the variable.
2. Data distributions with different shapes can have the same mean and standard deviation. For example, consider the following two histograms:

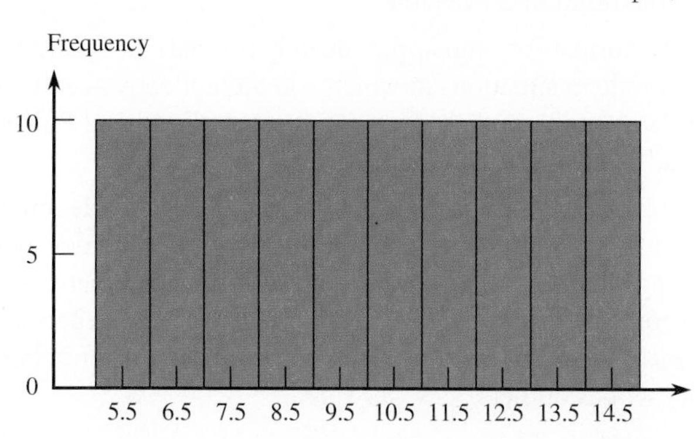

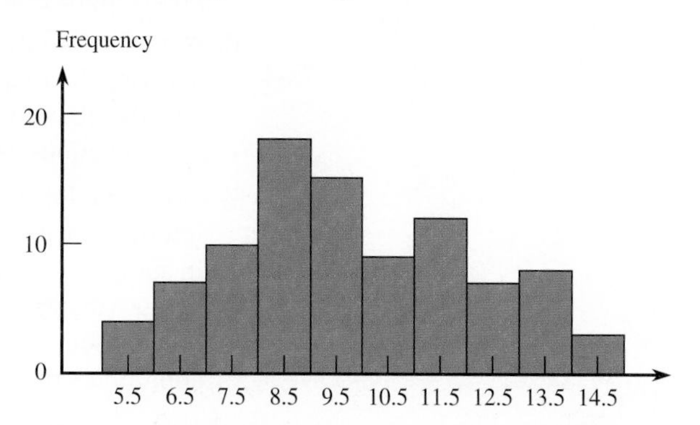

   Both histograms summarize data sets that have a mean of 10 and a standard deviation of 2, yet they have different shapes.
3. Both the mean and the standard deviation are sensitive to extreme values in a data set, especially if the sample size is small. If a data distribution is skewed or if the data set has outliers, the median and the interquartile range may be a better choice for describing center and spread.
4. Measures of center and variability describe the values of the variable studied, not the frequencies in a frequency distribution or the heights of the bars in a histogram. For example, consider the following two frequency distributions and histograms:

Frequency Distribution A		Frequency Distribution B	
Value	Frequency	Value	Frequency
1	10	1	5
2	10	2	10
3	10	3	20
4	10	4	10
5	10	5	5

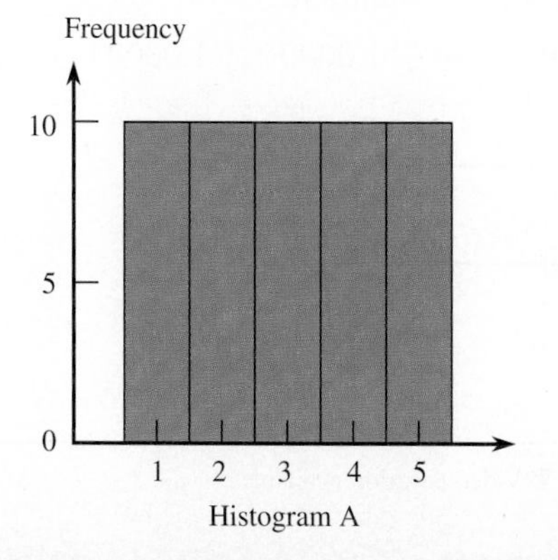

Histogram A

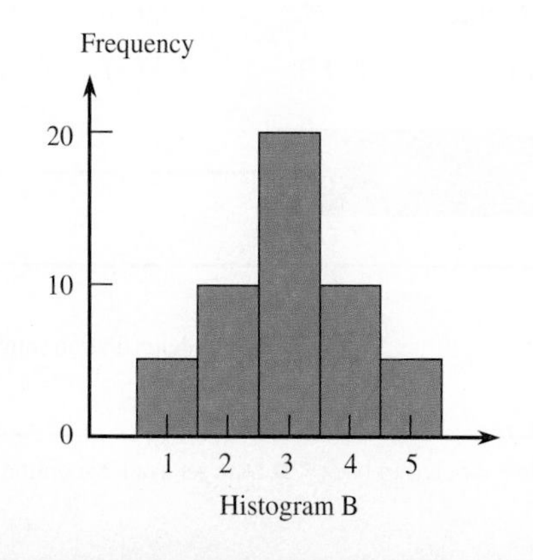

Histogram B

There is more variability in the data summarized by Frequency Distribution and Histogram A than in the data summarized by Frequency Distribution and Histogram B. This is because the values of the variable described by Histogram and Frequency Distribution B are more concentrated near the mean than are the values for the variable described by Histogram and Frequency Distribution A. Don't be misled by the fact that there is no variability in the frequencies in Frequency Distribution A or the heights of the bars in Histogram A.

5. Be careful with boxplots based on small sample sizes. Boxplots convey information about center, variability, and shape, but when the sample size is small, you should be hesitant to overinterpret shape information. It is really not possible to decide whether a data distribution is symmetric or skewed if only a small sample of observations from the distribution is available.
6. Not all distributions are normal (or even approximately normal). Be cautious in applying the Empirical Rule in situations in which you are not convinced that the data distribution is at least approximately normal. Using the Empirical Rule in such situations can lead to incorrect statements.
7. Watch out for outliers! Unusual observations in a data set often provide important information about the variable under study, so it is important to consider outliers in addition to describing what is typical. Outliers can also be problematic—both because the values of some descriptive measures are influenced by outliers and because some of the methods for drawing conclusions from data may not be appropriate if the data set has outliers.

EXERCISES 4.53 - 4.54

4.53 The authors of the paper **"Delayed Time to Defibrillation after In-Hospital Cardiac Arrest"** (***New England Journal of Medicine*** **[2008]: 9–16)** described a study of how survival is related to the length of time it takes from the time of a heart attack to the administration of defibrillation therapy. The following is a statement from the paper:

> We identified 6789 patients from 369 hospitals who had in-hospital cardiac arrest due to ventricular fibrillation (69.7%) or pulseless ventricular trachycardia (30.3%). Overall, the median time to defibrillation was 1 minute (interquartile range [was] 3 minutes).

Data from the paper on time to defibrillation (in minutes) for these 6789 patients was used to produce the following Minitab output and boxplot.

a. Why is there no lower whisker in the given boxplot?

b. How is it possible for the median, the lower quartile, and the minimum value in the data set to all be equal? (Note—this is why you do not see a median line in the box part of the boxplot.)

c. The authors of the paper considered a time to defibrillation of greater than 2 minutes as unacceptable. Based on the given boxplot and summary statistics, is it possible that the percentage of patients having an unacceptable time to defibrillation is greater than 50%? Greater than 25%? Less than 25%? Explain.

d. Is the outlier shown at 7 a mild outlier or an extreme outlier?

Descriptive Statistics: Time to Defibrillation

Variable	N	Mean	StDev	Minimum	Q1	Median	Q3	Maximum
Time	6789	2.3737	2.0713	1.0000	1.0000	1.0000	3.0000	7.0000

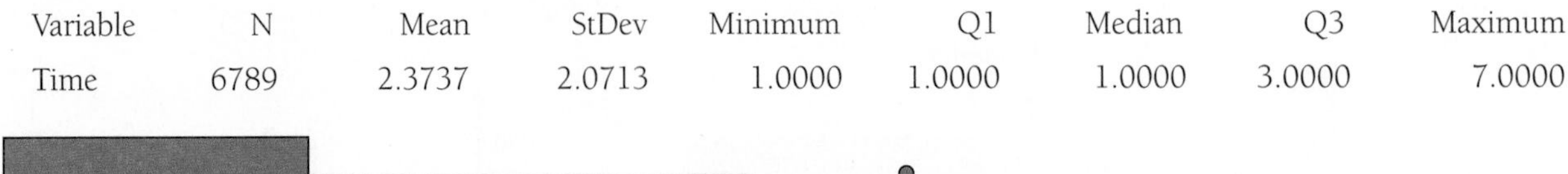

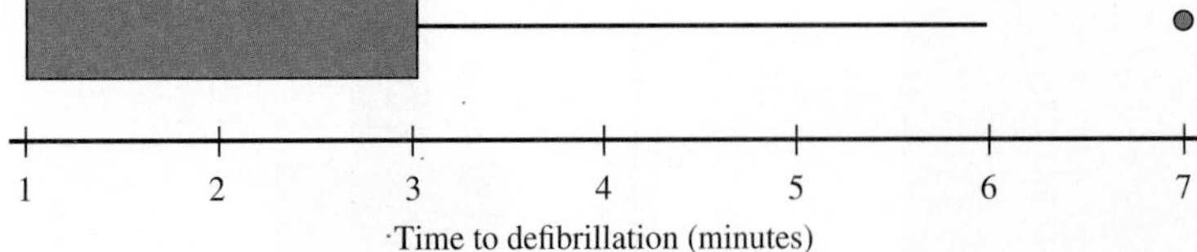

4.54 The paper **"Portable Social Groups: Willingness to Communicate, Interpersonal Communication Gratifications, and Cell Phone Use among Young Adults" (*International Journal of Mobile Communications* [2007]: 139–156)** describes a study of young adult cell phone use patterns.

a. Comment on the following quote from the paper. Do you agree with the authors?

Seven sections of an Introduction to Mass Communication course at a large southern university were surveyed in the spring and fall of 2003. The sample was chosen because it offered an excellent representation of the population under study—young adults.

b. Below is another quote from the paper. In this quote, the author reports the mean number of minutes of cell phone use per week for those who participated in the survey. What additional information would have been provided about cell phone use behavior if the author had also reported the standard deviation?

Based on respondent estimates, users spent an average of 629 minutes (about 10.5 hours) per week using their cell phone on or off line for any reason.

Bold exercises answered in back ● Data set available online ▼ Video Solution available

ACTIVITY 4.1 Collecting and Summarizing Numerical Data

In this activity, you will work in groups to collect data that will provide information about how many hours per week, on average, students at your school spend engaged in a particular activity. You will use the sampling plan designed in Activity 2.1 to collect the data.

1. With your group, pick one of the following activities to be the focus of your study:
   i. Surfing the web
   ii. Studying or doing homework
   iii. Watching TV
   iv. Exercising
   v. Sleeping

   or you may choose a different activity, *subject to the approval of your instructor.*
2. Use the plan developed in Activity 2.1 to collect data on the variable you have chosen for your study.
3. Summarize the resulting data using both numerical and graphical summaries. Be sure to address both center and variability.
4. Write a short article for your school paper summarizing your findings regarding student behavior. Your article should include both numerical and graphical summaries.

ACTIVITY 4.2 Airline Passenger Weights

The article **"Airlines Should Weigh Passengers, Bags, NTSB Says" (*USA Today,* February 27, 2004)** states that the National Transportation Safety Board recommended that airlines weigh passengers and their bags to prevent overloaded planes from attempting to take off. This recommendation was the result of an investigation into the crash of a small commuter plane in 2003, which determined that too much weight contributed to the crash.

Rather than weighing passengers, airlines currently use estimates of average passenger and luggage weights. After the 2003 accident, this estimate was increased by 10 pounds for passengers and 5 pounds for luggage. Although an airplane can fly if it is somewhat overweight if all systems are working properly, if one of the plane's engines fails an overweight plane becomes difficult for the pilot to control.

Assuming that the new estimate of the average passenger weight is accurate, discuss the following questions with a partner and then write a paragraph that answers these questions.

1. What role does variability in passenger weights play in creating a potentially dangerous situation for an airline?
2. Would an airline have a lower risk of a potentially dangerous situation if the variability in passenger weight is large or if it is small?

ACTIVITY 4.3 Boxplot Shapes

In this activity, you will investigate the relationship between boxplot shapes and the corresponding five-number summary. The accompanying figure shows four boxplots, labeled A–D. Also given are 4 five-number summaries, labeled I–IV. Match each five-number summary to the appropriate boxplot. Note that scales are not included on the boxplots, so you will have to think about what the five-number summary implies about characteristics of the boxplot.

Five-Number Summaries

	I	II	III	IV
Minimum	40	4	0.0	10
Lower quartile	45	8	0.1	34
Median	71	16	0.9	44
Upper quartile	88	25	2.2	82
Maximum	106	30	5.1	132

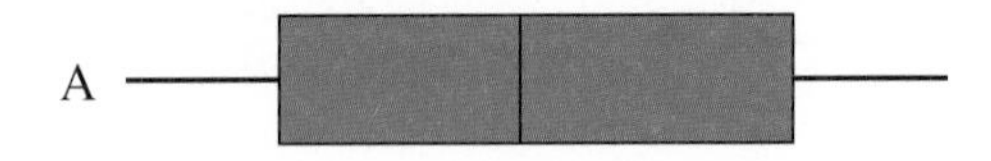

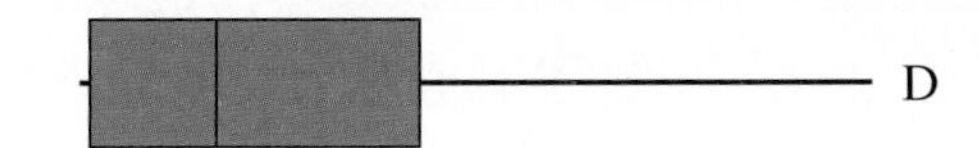

SUMMARY Key Concepts and Formulas

TERM OR FORMULA	COMMENT
$x_1, x_2, \ldots, x_n$	Notation for sample data consisting of observations on a variable *x*, where *n* is the sample size.
Sample mean, $\bar{x}$	The most frequently used measure of center of a sample. It can be very sensitive to the presence of even a single outlier (unusually large or small observation).
Population mean, μ	The average *x* value in the entire population.
Sample median	The middle value in the ordered list of sample observations. (For *n* even, the median is the average of the two middle values.) It is very insensitive to outliers.
Deviations from the mean: $x_1 - \bar{x}, x_2 - \bar{x}, \ldots, x_n - \bar{x}$	Quantities used to assess variability in a sample. Except for rounding effects, $\Sigma(x - \bar{x}) = 0$.
The sample variance $s^2 = \frac{\Sigma(x - \bar{x})^2}{n - 1}$ **and standard deviation** $s = \sqrt{s^2}$	The most frequently used measures of variability for sample data.
The population variance σ^2 and standard deviation σ	Measures of variability for the entire population.
Quartiles and the interquartile range	The lower quartile separates the smallest 25% of the data from the remaining 75%, and the upper quartile separates the largest 25% from the smallest 75%. The interquartile range (iqr), a measure of variability less sensitive to outliers than *s*, is the difference between the upper and lower quartiles.
Chebyshev's Rule	This rule states that for any number $k \geq 1$, *at least* $100\left(1 - \frac{1}{k^2}\right)$% of the observations in *any* data set are within *k* standard deviations of the mean. It is typically conservative in that the actual percentages are often considerably greater than the stated lower bound.
Empirical Rule	This rule gives the approximate percentage of observations within 1 standard deviation (68%), 2 standard deviations (95%), and 3 standard deviations (99.7%) of the mean when the histogram is well approximated by a normal curve.
***z* score**	This quantity gives the distance between an observation and the mean expressed as a certain number of standard deviations. It is positive (negative) if the observation is greater than (less than) the mean.

TERM OR FORMULA	COMMENT	TERM OR FORMULA	COMMENT
rth percentile	The value such that r% of the observations in the data set fall at or below that value.	**Boxplot**	A picture that conveys information about the most important features of a numerical data set: center, spread, extent of skewness, and presence of outliers.
Five-number summary	A summary of a data set that includes the minimum, lower quartile, median, upper quartile, and maximum.		

CHAPTER REVIEW Exercises 4.55 - 4.73

4.55 Research by the Food and Drug Administration (FDA) shows that acrylamide (a possible cancer-causing substance) forms in high-carbohydrate foods cooked at high temperatures and that acrylamide levels can vary widely even within the same brand of food **(Associated Press, December 6, 2002)**. FDA scientists analyzed McDonald's French fries purchased at seven different locations and found the following acrylamide levels:

497 193 328 155 326 245 270

a. Compute the mean acrylamide level and the seven deviations from the mean.

b. Verify that, except for the effect of rounding, the sum of the deviations from the mean is equal to 0 for this data set. (If you rounded the sample mean or the deviations, your sum may not be exactly zero, but it should be close to zero if you have computed the deviations correctly.)

c. Calculate the variance and standard deviation for this data set.

4.56 ● The technical report **"Ozone Season Emissions by State" (U.S. Environmental Protection Agency, 2002)** gave the following nitrous oxide emissions (in thousands of tons) for the 48 states in the continental U.S. states:

76 22 40 7 30 5 6 136 72 33 0
89 136 39 92 40 13 27 1 63 33 60
27 16 63 32 20 2 15 36 19 39 130
40 4 85 38 7 68 151 32 34 0 6
43 89 34 0

a. Use these data to construct a boxplot that shows outliers.

b. Write a few sentences describing the important characteristics of the boxplot.

4.57 The ***San Luis Obispo Telegram-Tribune*** **(October 1, 1994)** reported the following monthly salaries for supervisors from six different counties: \$5354 (Kern), \$5166 (Monterey), \$4443 (Santa Cruz), \$4129 (Santa Barbara), \$2500 (Placer), and \$2220 (Merced). San Luis Obispo County supervisors are supposed to be paid the average of the two counties among these six in the middle of the salary range.

a. Which measure of center determines this salary, and what is its value?

b. Why is the other measure of center featured in this section not as favorable to these supervisors (although it might appeal to taxpayers)?

4.58 ● A sample of 26 offshore oil workers took part in a simulated escape exercise, resulting in the accompanying data on time (in seconds) to complete the escape **("Oxygen Consumption and Ventilation During Escape from an Offshore Platform," *Ergonomics* [1997]: 281–292)**:

389 356 359 363 375 424 325 394 402
373 373 370 364 366 364 325 339 393
392 369 374 359 356 403 334 397

a. Construct a stem-and-leaf display of the data. Will the sample mean or the sample median be larger for this data set?

b. Calculate the values of the sample mean and median.

c. By how much could the largest time be increased without affecting the value of the sample median? By how much could this value be decreased without affecting the sample median?

4.59 Because some homes have selling prices that are much higher than most, the median price is usually used to describe a "typical" home price for a given location. The three accompanying quotes are all from the ***San Luis Obispo Tribune***, but each gives a different interpretation of the median price of a home in San Luis Obispo County. Comment on each of these statements. (Look carefully. At least one of the statements is incorrect.)

a. "So we have gone from 23% to 27% of county residents who can afford the median priced home at $278,380 in SLO County. That means that half of the homes in this county cost less than $278,380 and half cost more." **(October 11, 2001)**

b. "The county's median price rose to $285,170 in the fourth quarter, a 9.6% increase from the same period a year ago, the report said. (The median represents the midpoint of a range.)" **(February 13, 2002)**

c. "'Your median is going to creep up above $300,000 if there is nothing available below $300,000,' Walker said." **(February 26, 2002)**

4.60 ● Although bats are not known for their eyesight, they are able to locate prey (mainly insects) by emitting high-pitched sounds and listening for echoes. A paper appearing in ***Animal Behaviour*** **("The Echolocation of Flying Insects by Bats" [1960]: 141–154)** gave the following distances (in centimeters) at which a bat first detected a nearby insect:

62 23 27 56 52 34 42 40 68 45 83

a. Compute the sample mean distance at which the bat first detects an insect.

b. Compute the sample variance and standard deviation for this data set. Interpret these values.

4.61 For the data in Exercise 4.60, subtract 10 from each sample observation. For the new set of values, compute the mean and the deviations from the mean. How do these deviations compare to the deviations from the mean for the original sample? How does s^2 for the new values compare to s^2 for the old values? In general, what effect does subtracting (or adding) the same number to each observation have on s^2 and s? Explain.

4.62 For the data of Exercise 4.60, multiply each data value by 10. How does s for the new values compare to s for the original values? More generally, what happens to s if each observation is multiplied by the same positive constant c?

4.63 ● The percentage of juice lost after thawing for 19 different strawberry varieties appeared in the article **"Evaluation of Strawberry Cultivars with Different Degrees of Resistance to Red Scale" (*Fruit Varieties Journal* [1991]: 12–17)**:

46 51 44 50 33 46 60 41 55 46 53 53
42 44 50 54 46 41 48

a. Are there any observations that are mild outliers? Extreme outliers?

b. Construct a boxplot, and comment on the important features of the plot.

4.64 ● The risk of developing iron deficiency is especially high during pregnancy. Detecting such a deficiency is complicated by the fact that some methods for determining iron status can be affected by the state of pregnancy itself. Consider the following data on transferrin receptor concentration for a sample of women with laboratory evidence of overt iron-deficiency anemia **("Serum Transferrin Receptor for the Detection of Iron Deficiency in Pregnancy," *American Journal of Clinical Nutrition* [1991]: 1077–1081)**:

15.2 9.3 7.6 11.9 10.4 9.7
20.4 9.4 11.5 16.2 9.4 8.3

a. Compute the values of the sample mean and median. Why are these values different here?

b. Which of the mean and the median do you regard as more representative of the sample, and why?

4.65 ● The paper **"The Pedaling Technique of Elite Endurance Cyclists" (*International Journal of Sport Biomechanics* [1991]: 29–53)** reported the following data on single-leg power at a high workload:

244 191 160 187 180 176 174 205 211
183 211 180 194 200

a. Calculate and interpret the sample mean and median.

b. Suppose that the first observation had been 204, not 244. How would the mean and median change?

4.66 The paper cited in Exercise 4.65 also reported values of single-leg power for a low workload. The sample mean for $n = 13$ observations was $\bar{x} = 119.8$ (actually 119.7692), and the 14th observation, somewhat of an outlier, was 159. What is the value of $\bar{x}$ for the entire sample?

4.67 ● The amount of aluminum contamination (in parts per million) in plastic was determined for a sample of 26 plastic specimens, resulting in the following data **("The Log Normal Distribution for Modeling Quality Data When the Mean Is Near Zero," *Journal of Quality Technology* [1990]: 105–110)**:

30 30 60 63 70 79 87 90 101
102 115 118 119 119 120 125 140 145
172 182 183 191 222 244 291 511

Construct a boxplot that shows outliers, and comment on the interesting features of this plot.

4.68 ● The article **"Can We Really Walk Straight?" (*American Journal of Physical Anthropology* [1992]: 19–27)** reported on an experiment in which each of 20 healthy men was asked to walk as straight as possible to a target 60 m away at normal speed. Consider the following data on cadence (number of strides per second):

0.95 0.85 0.92 0.95 0.93 0.86 1.00 0.92
0.85 0.81 0.78 0.93 0.93 1.05 0.93 1.06
1.06 0.96 0.81 0.96

Use the methods developed in this chapter to summarize the data; include an interpretation or discussion whenever appropriate. (Note: The author of the paper used a rather sophisticated statistical analysis to conclude that people cannot walk in a straight line and suggested several explanations for this.)

4.69 ● The article **"Comparing the Costs of Major Hotel Franchises" (*Real Estate Review* [1992]: 46–51)** gave the following data on franchise cost as a percentage of total room revenue for chains of three different types:

Budget	2.7	2.8	3.8	3.8	4.0	4.1	5.5
	5.9	6.7	7.0	7.2	7.2	7.5	7.5
	7.7	7.9	7.9	8.1	8.2	8.5	
Midrange	1.5	4.0	6.6	6.7	7.0	7.2	7.2
	7.4	7.8	8.0	8.1	8.3	8.6	9.0
First-class	1.8	5.8	6.0	6.6	6.6	6.6	7.1
	7.2	7.5	7.6	7.6	7.8	7.8	8.2
	9.6						

Construct a boxplot for each type of hotel, and comment on interesting features, similarities, and differences.

4.70 ● The accompanying data on milk volume (in grams per day) were taken from the paper **"Smoking During Pregnancy and Lactation and Its Effects on Breast Milk Volume" (*American Journal of Clinical Nutrition* [1991]: 1011–1016)**:

Smoking	621	793	593	545	753	655
mothers	895	767	714	598	693	
Nonsmoking	947	945	1086	1202	973	981
mothers	930	745	903	899	961	

Compare and contrast the two samples

4.71 The ***Los Angeles Times* (July 17, 1995)** reported that in a sample of 364 lawsuits in which punitive damages were awarded, the sample median damage award was \$50,000, and the sample mean was \$775,000. What does this suggest about the distribution of values in the sample?

4.72 ● Age at diagnosis for each of 20 patients under treatment for meningitis was given in the paper **"Penicillin in the Treatment of Meningitis" (*Journal of the American Medical Association* [1984]: 1870–1874)**. The ages (in years) were as follows:

18 18 25 19 23 20 69 18 21 18 20 18
18 20 18 19 28 17 18 18

a. Calculate the values of the sample mean and the standard deviation.

b. Compute the upper quartile, the lower quartile, and the interquartile range.

c. Are there any mild or extreme outliers present in this data set?

d. Construct the boxplot for this data set.

4.73 Suppose that the distribution of scores on an exam is closely described by a normal curve with mean 100. The 16th percentile of this distribution is 80.

a. What is the 84th percentile?

b. What is the approximate value of the standard deviation of exam scores?

c. What z score is associated with an exam score of 90?

d. What percentile corresponds to an exam score of 140?

e. Do you think there were many scores below 40? Explain.

TECHNOLOGY NOTES

Mean

JMP

1. Input the raw data into a column
2. Click **Analyze** and select **Distribution**
3. Click and drag the column name containing the data from the box under **Select Columns** to the box next to **Y, Columns**
4. Click **OK**

Note: These commands also produce the following statistics: standard deviation, minimum, quartile 1, median, quartile 3, and maximum.

Minitab

1. Input the raw data into C1
2. Select **Stat** and choose **Basic Statistics** then choose **Display Descriptive Statistics...**
3. Double-click C1 to add it to the **Variables** list
4. Click **OK**

Note: These commands also produce the following statistics: standard deviation, minimum, quartile 1, median, quartile 3, and maximum.

SPSS

1. Input the raw data into the first column
2. Select **Analyze** and choose **Descriptive Statistics** then choose **Explore...**
3. Highlight the column name for the variable
4. Click the arrow to move the variable to the **Dependent List** box
5. Click **OK**

Note: These commands also produce the following statistics: median, variance, standard deviation, minimum, maximum, interquartile range, and several plots.

Excel 2007

1. Input the raw data into the first column
2. Click on the **Data** ribbon and select **Data Analysis**

   Note: If you do not see **Data Analysis** listed on the Ribbon, see the Technology Notes for Chapter 3 for instructions on installing this add-on.

3. Select **Descriptive Statistics** from the dialog box
4. Click **OK**
5. Click in the box next to **Input Range:** and select the data (if you used and selected column titles, check the box next to **Labels in First Row**)
6. Check the box next to **Summary Statistics**
7. Click **OK**

Note: These commands also produce the following statistics: standard error, median, mode, standard deviation, sample variance, range, minimum, and maximum.

Note: You can also find the mean using the Excel function average.

TI-83/84

1. Input the raw data into **L1** (To access lists press the **STAT** key, highlight the option called **Edit...** then press **ENTER**)
2. Press the **STAT** key
3. Use the arrows to highlight **CALC**
4. Highlight **1-Var Stats** and press **ENTER**
5. Press the **2nd** key and then the **1** key
6. Press **ENTER**

Note: You may need to scroll to view all of the statistics. This procedure also produces the standard deviation, minimum, Q1, median, Q3, maximum.

TI-Nspire

1. Enter the data into a data list (To access data lists select the spreadsheet option and press **enter**)

   Note: Be sure to title the list by selecting the top row of the column and typing a title.

2. Press the **menu** key and select **4:Statistics** then select **1:Stat Calculations** then **1:One-Variable Statistics...**
3. Press **OK**
4. For **X1 List,** select the title for the column containing your data from the drop-down menu
5. Press **OK**

Note: You may need to scroll to view all of the statistics. This procedure also produces the standard deviation, minimum, Q1, median, Q3, maximum.

Median

JMP

1. Input the raw data into a column
2. Click **Analyze** and select **Distribution**
3. Click and drag the column name containing the data from the box under **Select Columns** to the box next to **Y, Columns**
4. Click **OK**

Note: These commands also produce the following statistics: mean, standard deviation, minimum, quartile 1, quartile 3, and maximum.

Minitab

1. Input the raw data into C1
2. Select **Stat** and choose **Basic Statistics** then choose **Display Descriptive Statistics...**
3. Double-click C1 to add it to the **Variables** list
4. Click **OK**

Note: These commands also produce the following statistics: mean, standard deviation, minimum, quartile 1, quartile 3, maximum.

SPSS

1. Input the raw data into the first column
2. Select **Analyze** and choose **Descriptive Statistics** then choose **Explore...**
3. Highlight the column name for the variable
4. Click the arrow to move the variable to the **Dependent List** box
5. Click **OK**

Note: These commands also produce the following statistics: mean, variance, standard deviation, minimum, maximum, interquartile range, and several plots.

Excel 2007

1. Input the raw data into the first column
2. Click on the **Data** ribbon and select **Data Analysis**
3. **Note:** If you do not see **Data Analysis** listed on the Ribbon, see the Technology Notes for Chapter 3 for instructions on installing this add-on.
4. Select **Descriptive Statistics** from the dialog box
5. Click **OK**
6. Click in the box next to **Input Range:** and select the data (if you used and selected column titles, check the box next to **Labels in First Row**)
7. Check the box next to **Summary Statistics**
8. Click **OK**

Note: These commands also produce the following statistics: mean, standard error, mode, standard deviation, sample variance, range, minimum, and maximum.

Note: You can also find the median using the Excel function **median**.

TI-83/84

1. Input the raw data into **L1** (To access lists press the **STAT** key, highlight the option called **Edit...** then press **ENTER**)
2. Press the **STAT** key
3. Use the arrows to highlight **CALC**
4. Highlight **1-Var Stats** and press **ENTER**
5. Press the **2nd** key and then the **1** key
6. Press **ENTER**

Note: You may need to scroll to view all of the statistics. This procedure also produces the mean, standard deviation, minimum, Q1, Q3, maximum.

TI-Nspire

1. Enter the data into a data list (To access data lists select the spreadsheet option and press **enter**)

   Note: Be sure to title the list by selecting the top row of the column and typing a title.

2. Press the **menu** key and select **4:Statistics** then select **1:Stat Calculations** then **1:One-Variable Statistics...**
3. Press **OK**
4. For **X1 List**, select the title for the column containing your data from the drop-down menu
5. Press **OK**

Note: You may need to scroll to view all of the statistics. This procedure also produces the mean, standard deviation, minimum, Q1, Q3, maximum.

Variance

JMP

1. Input the raw data into a column
2. Click **Analyze** and select **Distribution**
3. Click and drag the column name containing the data from the box under **Select Columns** to the box next to **Y, Columns**
4. Click **OK**
5. Click the red arrow next to the column name
6. Click **Display Options** then select **More Moments**

Note: These commands also produce the following statistics: standard deviation, minimum, quartile 1, median, quartile 3, and maximum.

Minitab

1. Input the raw data into C1
2. Select **Stat** and choose **Basic Statistics** then choose **Display Descriptive Statistics...**
3. Double-click C1 to add it to the **Variables** list
4. Click the **Statistics** button
5. Check the box next to **Variance**
6. Click **OK**
7. Click **OK**

SPSS

1. Input the raw data into the first column
2. Select **Analyze** and choose **Descriptive Statistics** then choose **Explore...**
3. Highlight the column name for the variable
4. Click the arrow to move the variable to the **Dependent List** box
5. Click **OK**

Note: These commands also produce the following statistics: mean, median, standard deviation, minimum, maximum, interquartile range, and several plots.

Excel 2007

1. Input the raw data into the first column
2. Click on the **Data** ribbon and select **Data Analysis**
3. **Note:** If you do not see **Data Analysis** listed on the Ribbon, see the Technology Notes for Chapter 3 for instructions on installing this add-on.
4. Select **Descriptive Statistics** from the dialog box
5. Click **OK**
6. Click in the box next to **Input Range:** and select the data (if you used and selected column titles, check the box next to **Labels in First Row**)

7. Check the box next to **Summary Statistics**
8. Click **OK**

Note: These commands also produce the following statistics: mean, standard error, median, mode, standard deviation, range, minimum, and maximum.

Note: You can also find the variance using the Excel function **var**.

TI-83/84

The TI-83/84 does not automatically produce the variance; however, this can be determined by finding the standard deviation and squaring it.

TI-Nspire

The TI-Nspire does not automatically produce the variance; however, this can be determined by finding the standard deviation and squaring it.

Standard Deviation

JMP

1. Input the raw data into a column
2. Click **Analyze** and select **Distribution**
3. Click and drag the column name containing the data from the box under **Select Columns** to the box next to **Y, Columns**
4. Click **OK**

Note: These commands also produce the following statistics: mean, minimum, quartile 1, median, quartile 3, and maximum.

Minitab

1. Input the raw data into C1
2. Select **Stat** and choose **Basic Statistics** then choose **Display Descriptive Statistics...**
3. Double-click C1 to add it to the **Variables** list
4. Click **OK**

Note: These commands also produce the following statistics: standard deviation, minimum, quartile 1, median, quartile 3, maximum.

SPSS

1. Input the raw data into the first column
2. Select **Analyze** and choose **Descriptive Statistics** then choose **Explore...**
3. Highlight the column name for the variable
4. Click the arrow to move the variable to the **Dependent List** box
5. Click **OK**

Note: These commands also produce the following statistics: mean, median, variance, minimum, maximum, interquartile range, and several plots.

Excel 2007

1. Input the raw data into the first column
2. Click on the **Data** ribbon and select **Data Analysis**
3. **Note:** If you do not see **Data Analysis** listed on the Ribbon, see the Technology Notes for Chapter 3 for instructions on installing this add-on.
4. Select **Descriptive Statistics** from the dialog box
5. Click **OK**
6. Click in the box next to **Input Range:** and select the data (if you used and selected column titles, check the box next to **Labels in First Row**)
7. Check the box next to **Summary Statistics**
8. Click **OK**

Note: These commands also produce the following statistics: mean, standard error, median, mode, sample variance, range, minimum, and maximum.

Note: You can also find the standar deviation using the Excel function **sd**.

TI-83/84

1. Input the raw data into **L1** (To access lists press the **STAT** key, highlight the option called **Edit...** then press **ENTER**)
2. Press the **STAT** key
3. Use the arrows to highlight **CALC**
4. Highlight **1-Var Stats** and press **ENTER**
5. Press the **2nd** key and then the **1** key
6. Press **ENTER**

Note: You may need to scroll to view all of the statistics. This procedure also produces the mean, minimum, Q1, median, Q3, maximum.

TI-Nspire

1. Enter the data into a data list (To access data lists select the spreadsheet option and press **enter**)

   Note: Be sure to title the list by selecting the top row of the column and typing a title.

2. Press the **menu** key and select **4:Statistics** then select **1:Stat Calculations** then **1:One-Variable Statistics...**
3. Press **OK**
4. For **X1 List**, select the title for the column containing your data from the drop-down menu
5. Press **OK**

Note: You may need to scroll to view all of the statistics. This procedure also produces the mean, minimum, Q1, median, Q3, maximum.

Quartiles

JMP

1. Input the raw data into a column
2. Click **Analyze** and select **Distribution**
3. Click and drag the column name containing the data from the box under **Select Columns** to the box next to **Y, Columns**
4. Click **OK**

Note: These commands also produce the following statistics: mean, standard deviation, minimum, median, and maximum.

Minitab

1. Input the raw data into C1
2. Select **Stat** and choose **Basic Statistics** then choose **Display Descriptive Statistics...**
3. Double-click C1 to add it to the **Variables** list
4. Click **OK**

Note: These commands also produce the following statistics: standard deviation, minimum, quartile 1, median, quartile 3, maximum.

SPSS

1. Input the raw data into the first column
2. Select **Analyze** and choose **Descriptive Statistics** then choose **Frequencies...**
3. Highlight the column name for the variable
4. Click the arrow to move the variable to the **Dependent List** box
5. Click the **Statistics** button
6. Check the box next to Quartiles
7. Click **Continue**
8. Click **OK**

Excel 2007

1. Input the raw data into the first column
2. Select the cell where you would like to place the first quartile results
3. Click the **Formulas** Ribbon
4. Click **Insert Function**
5. Select the category **Statistical** from the drop-down menu
6. In the **Select a function:** box click **Quartile**
7. Click **OK**
8. Click in the box next to **Array** and select the data
9. Click in the box next to **Quart** and type 1
10. Click **OK**

Note: To find the third quartile, type 3 into the box next to Quart in Step 9.

TI-83/84

1. Input the raw data into **L1** (To access lists press the **STAT** key, highlight the option called **Edit...** then press **ENTER**)
2. Press the **STAT** key
3. Use the arrows to highlight **CALC**
4. Highlight **1-Var Stats** and press **ENTER**
5. Press the **2nd** key and then the **1** key
6. Press **ENTER**

Note: You may need to scroll to view all of the statistics. This procedure also produces the mean, standard deviation, minimum, median, maximum.

TI-Nspire

1. Enter the data into a data list (To access data lists select the spreadsheet option and press **enter**)
   Note: Be sure to title the list by selecting the top row of the column and typing a title.
2. Press the **menu** key and select **4:Statistics** then select **1:Stat Calculations** then **1:One-Variable Statistics...**
3. Press **OK**
4. For **X1 List**, select the title for the column containing your data from the drop-down menu
5. Press **OK**

Note: You may need to scroll to view all of the statistics. This procedure also produces the mean, standard deviation, minimum, median, maximum.

IQR

JMP

JMP does not have the functionality to produce the IQR automatically.

Minitab

1. Input the raw data into C1
2. Select **Stat** and choose **Basic Statistics** then choose **Display Descriptive Statistics...**
3. Double-click C1 to add it to the **Variables** list
4. Click the **Statistics** button
5. Check the box next to **Interquartile range**
6. Click **OK**
7. Click **OK**

SPSS

1. Input the raw data into the first column
2. Select **Analyze** and choose **Descriptive Statistics** then choose **Explore...**
3. Highlight the column name for the variable
4. Click the arrow to move the variable to the **Dependent List** box
5. Click **OK**

Note: These commands also produce the following statistics: mean, median, variance, standard deviation, minimum, maximum, and several plots.

Excel 2007

1. Use the steps under the **Quartiles** section to find both the first and third quartiles
2. Click on an empty cell where you would like the result for IQR to appear
3. Type = into the cell
4. Click on the cell containing the third quartile
5. Type −
6. Click on the cell containing the first quartile
7. Press **Enter**

TI-83/84

The TI-83/84 does not have the functionality to produce the IQR automatically.

TI-Nspire

The TI-Nspire does not have the functionality to produce the IQR automatically.

Boxplot

JMP

1. Input the raw data into a column
2. Click **Analyze** and select **Distribution**
3. Click and drag the column name containing the data from the box under **Select Columns** to the box next to **Y, Columns**
4. Click **OK**

Minitab

1. Input the raw data into C1
2. Select **Graph** then choose **Boxplot...**
3. Highlight Simple under One Y
4. Click **OK**
5. Double-click C1 to add it to the **Graph Variables** box
6. Click **OK**

Note: You may add or format titles, axis titles, legends, and so on by clicking on the **Labels...** button prior to performing Step 6 above.

SPSS

1. Enter the raw data the first column
2. Select **Graph** and choose **Chart Builder...**
3. Under **Choose from** highlight Boxplot
4. Click and drag the first boxplot (Simple Boxplot) to the Chart preview area
5. Click and drag the data variable into the **Y-Axis?** box in the chart preview area
6. Click **OK**

Note: Boxplots are also produced when summary statistics such as mean, median, standard deviation, and so on are produced.

Excel 2007

Excel 2007 does not have the functionality to create boxplots.

TI-83/84

1. Input the raw data into **L1** (To access lists press the **STAT** key, highlight the option called **Edit...** then press **ENTER**)
2. Press the 2nd key and then press the **Y =** key
3. Select **Plot1** and press **ENTER**
4. Highlight **On** and press **ENTER**
5. Highlight the graph option in the second row, second column, and press **ENTER**
6. Press **GRAPH**

Note: If the graph window does not display appropriately, press the **WINDOW** button and reset the scales appropriately.

TI-Nspire

1. Enter the data into a data list (To access data list select the spreadsheet option and press **enter**)

   Note: Be sure to title the list by selecting the top row of the column and typing a title.

2. Press **menu** and select **3:Data** then select **6:QuickGraph**
3. Press **menu** and select **1:Plot Type** then select **2:Box Plot** and press **enter**

Side-by-side Boxplots

JMP

1. Enter the raw data for both groups into a column
2. Enter the group information into a second column

Column 2	Column 3
F	8
F	3
F	6
F	3
F	9
F	2
F	3
F	4
F	8
M	3
M	5
M	4
M	6
M	7
M	2

3. Click **Analyze** then select **Fit Y by X**
4. Click and drag the column name containing the raw data from the box under **Select Columns** to the box next to **Y, Response**
5. Click and drag the column name containing the group information from the box under **Select Columns** to the box next to **X, Factor**
6. Click **OK**
7. Click the red arrow next to **Oneway Analysis of...**
8. Click **Quantiles**

Minitab

1. Input the raw data for the first group into C1
2. Input the raw data for the second group into C2
3. Continue to input data for each group into a separate column
4. Select **Graph** then choose **Boxplot...**
5. Highlight Simple under Multiple Y's
6. Click **OK**
7. Double-click the column names for each column to be graphed to add it to the **Graph Variables** box
8. Click **OK**

SPSS

1. Enter the raw data into the first column
2. Enter the group data into the second column

*Untitled1 [DataSet0] - PASW Statistics

File Edit View Data Transform

	VAR00001	VAR00002
1	3	male
2	6	female
3	5	male
4	7	female
5	5	male
6	8	female
7	2	male
8	3	female
9	6	male
10	5	female

3. Select **Graph** and choose **Chart Builder...**
4. Under **Choose from** highlight Boxplot
5. Click and drag the first boxplot (Simple Boxplot) to the Chart preview area
6. Click and drag the data variable into the **Y-Axis?** box in the chart preview area
7. Click and drag the group variable into the **X-Axis?** box in the chart preview area
8. Click **OK**

Excel 2007

Excel 2007 does not have the functionality to create side-by-side boxplots.

TI-83/84

The TI-83/84 does not have the functionality to create side-by-side boxplots.

TI-Nspire

The TI-Nspire does not have the functionality to create side-by-side boxplots.

CHAPTER 5

Summarizing Bivariate Data

Spencer Platt/Getty Images Sport/Getty Images

Forensic scientists must often estimate the age of an unidentified crime victim. Prior to 2010, this was usually done by analyzing teeth and bones, and the resulting estimates were not very reliable. A groundbreaking study described in the paper **"Estimating Human Age from T-Cell DNA Rearrangements" (*Current Biology* [2010])** examined the relationship between age and a measure based on a blood test. Age and the blood test measure were recorded for 195 people ranging in age from a few weeks to 80 years. By studying this relationship, the investigators hoped to be able to estimate the age of a crime victim from a blood sample.

In this chapter, we will consider methods for describing the relationship between two numerical variables and for assessing the strength of a relationship.

Chapter 5: Learning Objectives

STUDENTS WILL UNDERSTAND:

- that the relationship between two numerical variables can be described in terms of form, strength, and direction.
- the difference between a statistical relationship and a causal relationship (the difference between correlation and causation).
- how a line is used to describe the relationship between two numerical variables.
- the meaning of least-squares in the context of fitting a regression line.
- why it is risky to use the least-squares line to make predictions outside the range of the data.
- why it is important to consider both the standard deviation about the regression line and the value of r^2 when assessing the usefulness of the least-squares regression line.
- the role of a residual plot in assessing whether a line is an appropriate way to describe the relationship between two numerical variables.

STUDENTS WILL BE ABLE TO:

- informally describe the form, direction, and strength of a linear relationship based on a scatterplot.
- compute and interpret the value of the correlation coefficient.
- find the equation of the least-squares regression line and interpret the slope and intercept in context.
- use the least-squares regression line to make predictions.

- compute and interpret the value of the standard deviation about the regression line, s_e.
- compute and interpret the value of r^2, the coefficient of determination.
- construct a residual plot and use it to assess the appropriateness of using a line to describe the relationship between two numerical variables.
- identify outliers and potentially influential observations in a linear regression context.

5.1 Correlation

An investigator is often interested in how two or more variables are related to one another. For example, an environmental researcher might wish to know how the lead content of soil varies with distance from a major highway. Researchers in early childhood education might investigate how vocabulary size is related to age. College admissions officers, who must try to predict whether an applicant will succeed in college, might be interested in the relationship between college grade point average and high school grade point average or score on the ACT or SAT exam.

We can get a visual impression of how strongly two numerical variables are related by looking at a scatterplot. However, to make more precise statements we must go beyond pictures. A **correlation coefficient** (from *co-* and *relation*) is a numerical assessment of the strength of relationship between the x and y values in a bivariate data set consisting of (x, y) pairs. In this section, we introduce the most commonly used correlation coefficient.

Figure 5.1 displays several scatterplots that show different relationships between the x and y values. The plot in Figure 5.1(a) suggests a strong *positive relationship* between x and y. For every pair of points in the plot, the one with the larger x value also has the larger y value. That is, an increase in x is paired with an increase in y.

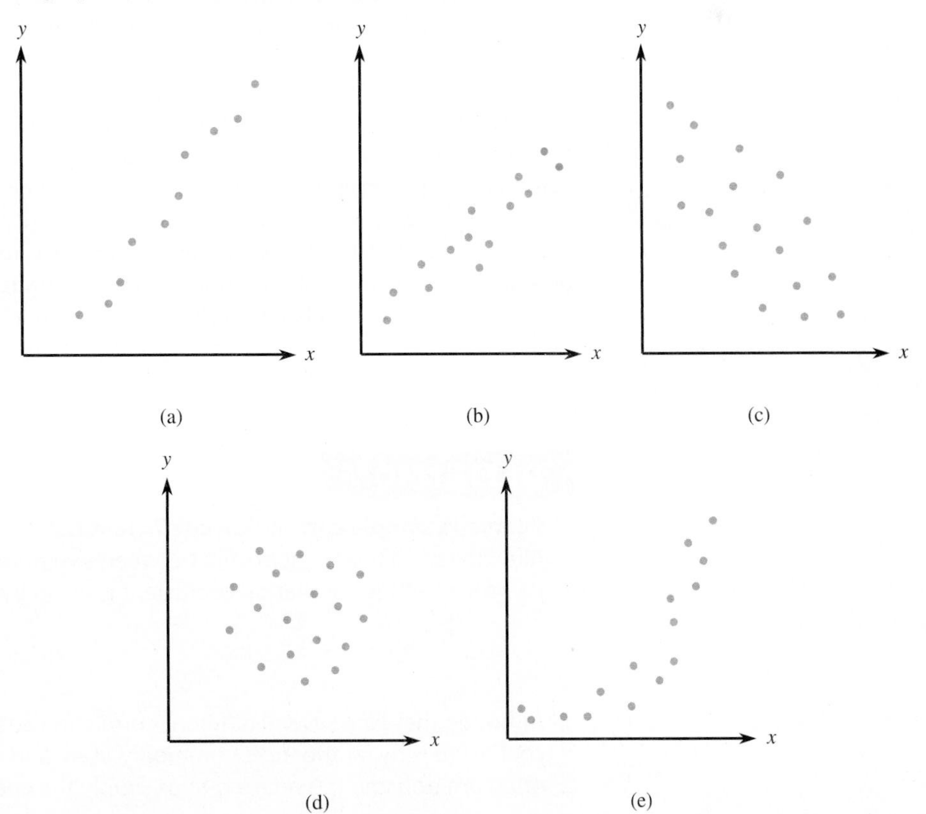

FIGURE 5.1
Scatterplots illustrating various types of relationships:
(a) positive linear relationship;
(b) another positive linear relationship;
(c) negative linear relationship;
(d) no relationship;
(e) curved relationship.

The plot in Figure 5.1(b) shows a strong tendency for y to increase as x does, but there are a few exceptions. For example, the x and y values of the two points with the largest x values (shown in a different color) go in opposite directions (for this pair of points, x increases but y decreases in value). Nevertheless, a plot like this still indicates a fairly strong positive relationship.

Figure 5.1(c) suggests that x and y are *negatively related*. As x increases, y tends to decrease. The negative relationship in this plot is not as strong as the positive relationship in Figure 5.1(b), although both plots show a well-defined linear pattern.

The plot of Figure 5.1(d) indicates that there is not a strong relationship between x and y. There is no tendency for y either to increase or to decrease as x increases. Finally, as illustrated in Figure 5.1(e), a scatterplot can show evidence of a strong relationship that is curved rather than linear.

Pearson's Sample Correlation Coefficient

Pearson's sample correlation coefficient measures the strength of any linear relationship between two numerical variables. It does this by using z scores in a clever way. Consider replacing each x value by the corresponding z score, z_x (by subtracting $\bar{x}$ and then dividing by s_x) and similarly replacing each y value by its z score. Note that x values that are larger than $\bar{x}$ will have positive z scores and those smaller than $\bar{x}$ will have negative z scores. Also y values larger than $\bar{y}$ will have positive z scores and those smaller will have negative z scores. Pearson's sample correlation coefficient is based on the sum of the products of z_x and z_y for each observation in the bivariate data set. This can be written as $\Sigma z_x z_y$.

To see how this works, let's look at some scatterplots. The scatterplot in Figure 5.2(a) indicates a strong positive relationship. A vertical line through $\bar{x}$ and a horizontal line through $\bar{y}$ divide the plot into four regions. In Region I, both x and y are greater than their mean values, so the z score for x and the z score for y are both positive numbers. It follows that $z_x z_y$ is positive. The product of the z scores is also positive for any point in Region III, because both z scores are negative in Region III and multiplying two negative numbers gives a positive number.

In each of the other two regions, one z score is positive and the other is negative, so $z_x z_y$ is negative. But because the points generally fall in Regions I and III, the products of z scores tend to be positive. Thus, the *sum* of the products will be a relatively large positive number.

Similar reasoning for the data displayed in Figure 5.2(b), which exhibits a strong negative relationship, implies that $\Sigma z_x z_y$ will be a relatively large (in magnitude) negative number. When there is no strong relationship, as in Figure 5.2(c), positive and negative products tend to counteract one another, producing a value of $\Sigma z_x z_y$ that is close to zero.

In summary, $\Sigma z_x z_y$ seems to be a reasonable measure of the degree of association between x and y. It can be a large positive number, a large negative number, or a number close to 0, depending on whether there is a strong positive, a strong negative, or no linear relationship.

Pearson's sample correlation coefficient, denoted r, is obtained by dividing $\Sigma z_x z_y$ by $(n - 1)$.

DEFINITION

Pearson's sample correlation coefficient *r*: A measure of the strength and direction of a linear relationship between two numerical variables. Denoted by r, Pearson's sample correlation coefficient is given by

$$r = \frac{\Sigma z_x z_y}{n - 1}$$

Although there are several different correlation coefficients, Pearson's correlation coefficient is by far the most commonly used, and so the name "Pearson's" is often omitted and it is referred to as simply the **correlation coefficient**.

Hand calculation of the correlation coefficient is quite tedious. Fortunately, all statistical software packages and most graphing calculators can compute the value of r once the x and y values have been input.

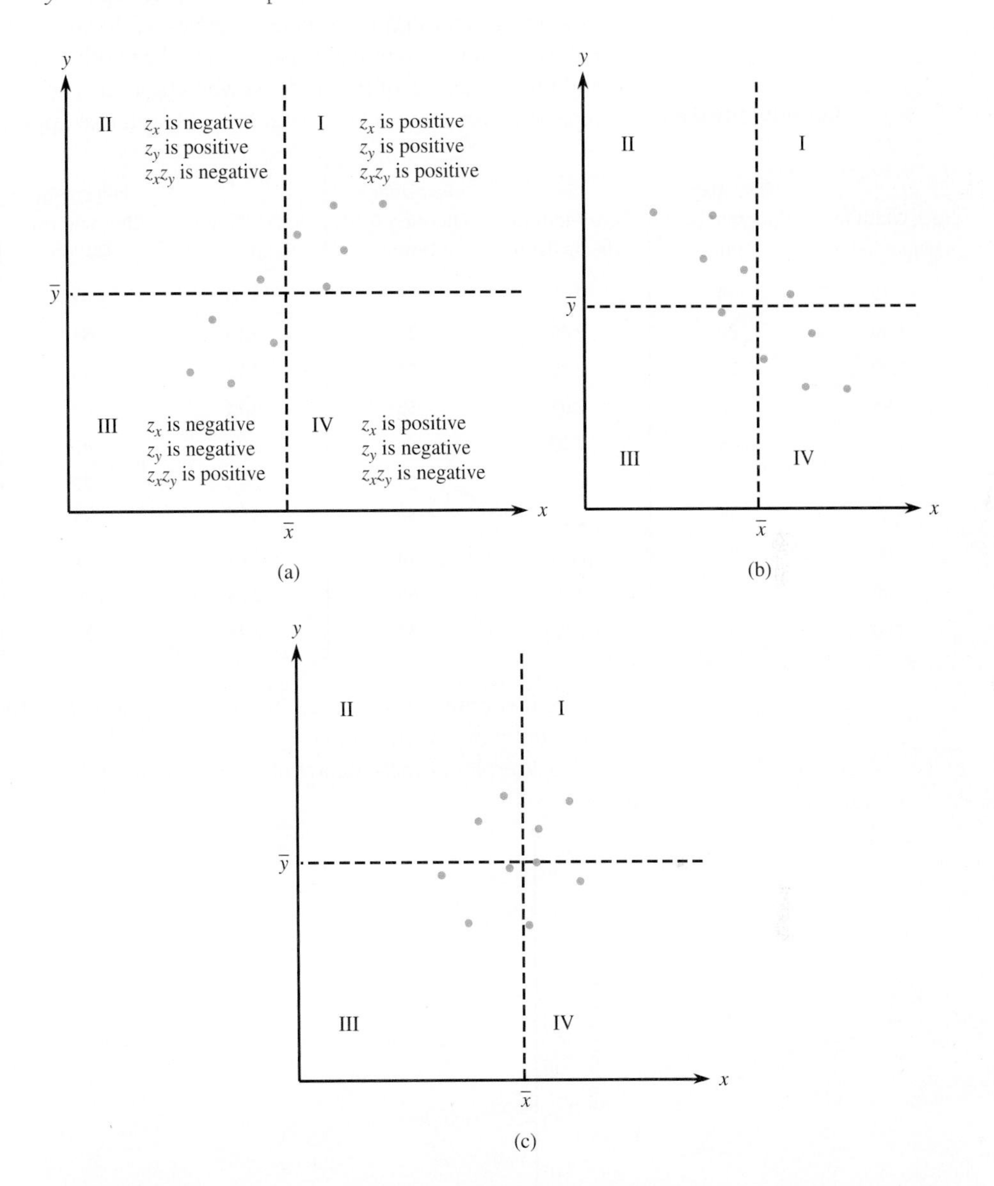

FIGURE 5.2
Viewing a scatterplot according to the signs of z_x and z_y:
(a) a positive relationship;
(b) a negative relationship;
(c) no strong relationship.

EXAMPLE 5.1 A Face You Can Trust

Understand the context ❯

The article **"How to Tell If a Guy Is Trustworthy"** (*LiveScience*, **March 8, 2010**) described an interesting research study published in the journal *Psychological Science* (**"Valid Facial Cues to Cooperation and Trust: Male Facial Width and Trustworthiness,"** ***Psychological Science*** **[2010]: 349–354**). This study investigated whether there is a relationship between facial characteristics and peoples' assessment of trustworthiness. Sixty-two students were each told that they would play a series of two-person games in which they could earn real money. In each game, the student could choose between two options:

1. They could end the game immediately and a total of \$10 would be paid out, with each player receiving \$5.

   or

2. They could "trust" the other player. In this case, \$3 would be added to the money available, but the other player could decide to either split the money fairly with \$6.50 to each player or could decide to keep \$10 and only give \$3 to the first player.

Each student was shown a picture of the person he or she would be playing with. The student then indicated whether they would end the game immediately and take $5 or trust the other player to do the right thing in hopes of increasing the payout to $6.50. This process was repeated for a series of games, each with a photo of a different second player. For each photo, the researchers recorded the width-to-height ratio of the face in the photo and the percentage of the students who chose the trust option when shown that photo. A representative subset of the data (from a graph that appeared in the paper) is given here:

Consider the data ❯

Face Width-to-Height Ratio	Percentage Choosing Trust Option	Face Width-to-Height Ratio	Percentage Choosing Trust Option	Face Width-to-Height Ratio	Percentage Choosing Trust Option	Face Width-to-Height Ratio	Percentage Choosing Trust Option
1.75	58	2.05	8	2.15	42	2.30	43
1.80	80	2.05	35	2.15	31	2.30	8
1.90	30	2.05	50	2.15	15	2.35	41
1.95	50	2.05	58	2.15	65	2.40	53
1.95	65	2.10	64	2.20	42	2.40	35
2.00	10	2.10	25	2.20	22	2.60	55
2.00	28	2.10	35	2.20	14	2.68	20
2.00	62	2.10	50	2.25	9	2.70	11
2.00	70	2.10	66	2.25	20		
2.00	63	2.10	62	2.25	35		

The first observation (1.75, 58) indicates that 58% of the students shown the picture of a person with a width-to-height ration of 1.75 chose the trust option.

A scatterplot of these data is shown in Figure 5.3.

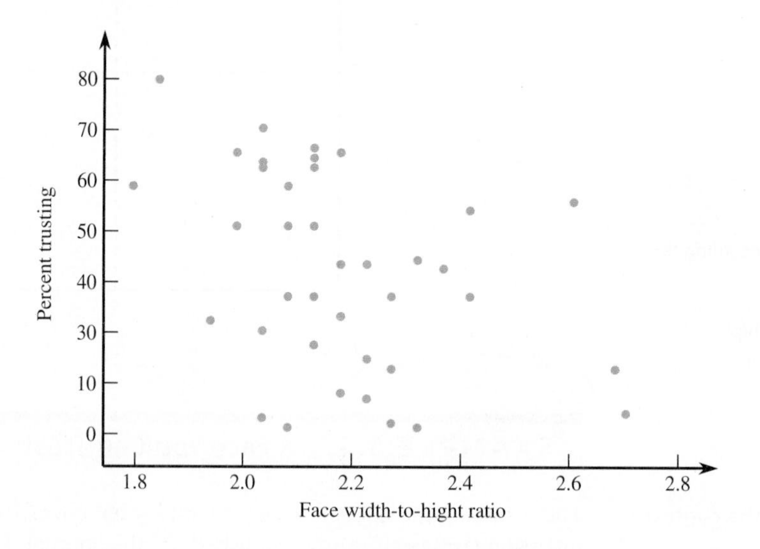

FIGURE 5.3
Scatterplot for the data of Example 5.1.

JMP was used to compute the value of the correlation coefficient, with the following result:

Correlation

Variable	Mean	Std Dev	Correlation	Signif. Prob	Number
Percentage Choosing Trust Option	40.26316	20.58691	-0.39169	0.0150*	38
Face Width-to-Height Ratio	2.153421	0.209598			

The value of the correlation coefficient (–0.39169) is found in the JMP output under the heading "Correlation."

To compute the value of the correlation coefficient by hand, use x to denote the face width-to-height ratio and y to denote the percentage choosing the trust option. It is easy to verify that

$$\bar{x} = 2.153 \quad s_x = 0.210 \quad \bar{y} = 40.26 \quad s_y = 20.59$$

Do the work) Begin by computing z scores for each (x, y) pair in the data set. For example, the first observation is (1.75, 58). The corresponding z scores are

$$z_x = \frac{1.75 - 2.153}{0.210} = -1.92 \qquad z_y = \frac{58 - 40.26}{20.59} = 0.86$$

The following table shows the z scores and the product $z_x z_y$ for each observation:

x	y	z_x	z_y	z_xz_y	x	y	z_x	z_y	z_xz_y	x	y	z_x	z_y	z_xz_y
1.75	58	−1.92	0.86	−1.65	2.05	58	−0.49	0.86	−0.42	2.20	14	0.22	−1.28	−0.28
1.80	80	−1.68	1.93	−3.24	2.10	64	−0.25	1.15	−0.29	2.25	9	0.46	−1.52	−0.7
1.90	30	−1.2	−0.5	0.6	2.10	25	−0.25	−0.74	0.19	2.25	20	0.46	−0.98	−0.45
1.95	50	−0.97	0.47	−0.46	2.10	35	−0.25	−0.26	0.07	2.25	35	0.46	−0.26	−0.12
1.95	65	−0.97	1.2	−1.16	2.10	50	−0.25	0.47	−0.12	2.30	43	0.7	0.13	0.09
2.00	10	−0.73	−1.47	1.07	2.10	66	−0.25	1.25	−0.31	2.30	8	0.7	−1.57	−1.1
2.00	28	−0.73	−0.6	0.44	2.10	62	−0.25	1.06	−0.27	2.35	41	0.94	0.04	0.04
2.00	62	−0.73	1.06	−0.77	2.15	42	−0.01	0.08	0	2.40	53	1.18	0.62	0.73
2.00	70	−0.73	1.44	−1.05	2.15	31	−0.01	−0.45	0	2.40	35	1.18	−0.26	−0.31
2.00	63	−0.73	1.1	−0.8	2.15	15	−0.01	−1.23	0.01	2.60	55	2.13	0.72	1.53
2.05	8	−0.49	−1.57	0.77	2.15	65	−0.01	1.2	−0.01	2.68	20	2.51	−0.98	−2.46
2.05	35	−0.49	−0.26	0.13	2.20	42	0.22	0.08	0.02	2.70	11	2.6	−1.42	−3.69
2.05	50	−0.49	0.47	−0.23	2.20	22	0.22	−0.89	−0.2					

$\Sigma z_x z_y = -14.4$ and $n = 38$, so

$$r = \frac{\Sigma z_x z_y}{n - 1} = \frac{-14.4}{37} = -0.3892$$

The difference between the correlation coefficient reported by JMP and what we obtained is the result of rounding in the z scores when carrying out the calculations by hand.

Interpret the results) Based on the scatterplot and the properties of the correlation coefficient presented in the discussion that follows this example, we conclude that there is a weak negative linear relationship between face width-to-height ratio and the percentage of people who chose to trust the face when playing the game. The relationship is negative, indicating that those with larger face width-to-height ratios (wider faces) tended to be trusted less. Based on this observation, the researchers concluded "It is clear that male facial width ratio is a cue to male trustworthiness and that it predicts trust placed in male faces."

An interesting side note: In another study where subjects were randomly assigned to the player 1 and player 2 roles, the researchers found that those with larger face width-to-height ratios *were* less trustworthy! They found a positive correlation between player 2 face width-to-height ratio and the percentage of the time that player 2 decided to keep \$10 and only give \$3 to player 1. ■

Properties of *r*

1. *The value of r is between* -1 *and* $+1$. A value near the upper limit, $+1$, indicates a strong positive linear relationship, whereas an r close to the lower limit, -1, suggests a strong negative linear relationship. Figure 5.4 shows a useful way to describe the strength of relationship based on r. It may seem surprising that a value of r as extreme as -0.5 or 0.5 should be in the weak category. An explanation for this is given later in the chapter. Even a weak correlation may indicate a meaningful relationship.

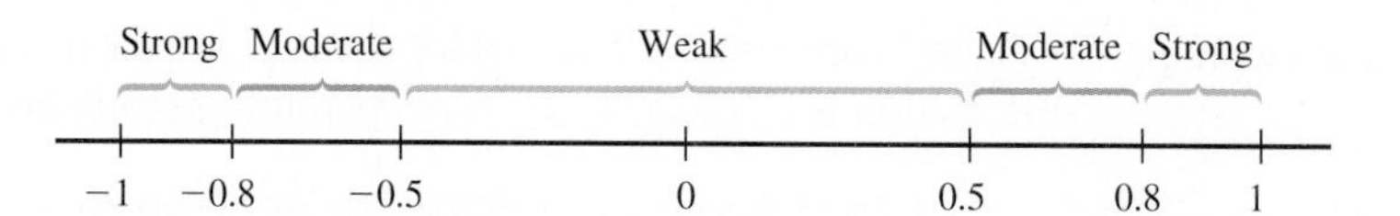

FIGURE 5.4
Describing the strength of a linear relationship.

2. *A correlation coefficient of $r = 1$ occurs only when all the points in a scatterplot of the data lie exactly on a straight line that slopes upward. Similarly, $r = -1$ only when all the points lie exactly on a downward-sloping line.* $r = 1$ or $r = -1$ indicates a perfect linear relationship between x and y in the sample data.
3. *The value of r is a measure of the extent to which x and y are linearly related.* r measures the extent to which the points in the scatterplot fall close to a straight line. A value of r close to 0 does not necessarily mean that there is *no* relationship between x and y. It is possible that there could still be a strong relationship that is not linear.
4. *The value of r does not depend on the unit of measurement for either variable.* For example, if x is height, the corresponding z score is the same whether height is expressed in inches, meters, or miles, and the value of the correlation coefficient is not affected. The correlation coefficient measures the inherent strength of the linear relationship between two numerical variables.
5. *The value of r does not depend on which of the two variables is considered x.* This means that if we had let x = percentage choosing trust option and y = face width-to-height ratio in Example 5.1, the same value, $r = -0.3892$, would have resulted.

EXAMPLE 5.2 Does It Pay to Pay More for a Bike Helmet?

Understand the context

● Are more expensive bike helmets safer than less expensive ones? The accompanying data on x = price and y = quality rating for 12 different brands of bike helmets appeared on the ***Consumer Reports*** web site **(www.consumerreports.org/health)**. Quality rating was a number from 0 (the worst possible rating) to 100, and was determined based on factors that included how well the helmet absorbed the force of an impact, the strength of the helmet, ventilation, and ease of use. Figure 5.5 shows a scatterplot of the data.

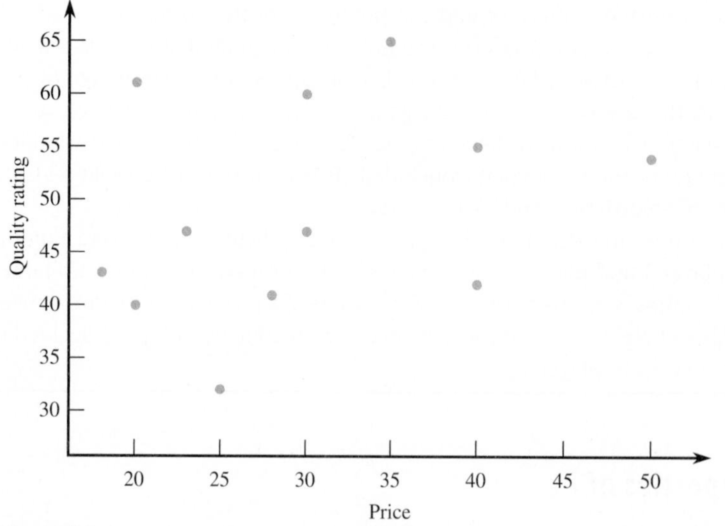

FIGURE 5.5
Minitab scatterplot for the bike helmet data of Example 5.2.

Price	Quality Rating
35	65
20	61
30	60
40	55
50	54
23	47
30	47
18	43
40	42
28	41
20	40
25	32

From the scatterplot, it appears that there is only a weak positive relationship between price and quality rating. The correlation coefficient, obtained using Minitab, is

Correlations: Price, Quality Rating

Pearson correlation of Price and Quality Rating = 0.303

● Data set available online

Interpret the results)

A correlation coefficient of $r = .303$ confirms that there is a tendency for higher quality ratings to be associated with higher priced helmets, but that the relationship is not very strong. In fact, the highest quality rating was for a helmet priced near the middle of the price values. ■

EXAMPLE 5.3 Age and Marathon Times

Understand the context)

● The article **"Master's Performance in the New York City Marathon"** (*British Journal of Sports Medicine* [2004]: 408–412) gave the following data on the average finishing time by age group for female participants in the New York City marathon.

Consider the data)

Age Group	Representative Age	Average Finish Time
10–19	15	302.38
20–29	25	193.63
30–39	35	185.46
40–49	45	198.49
50–59	55	224.30
60–69	65	288.71

The scatterplot of average finish time versus representative age is shown in Figure 5.6.

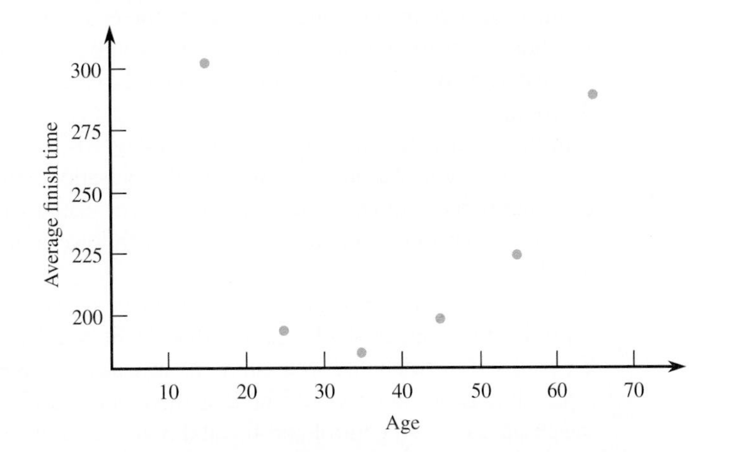

FIGURE 5.6
Scatterplot of y = average finish time and x = age for the data of Example 5.3.

Using Minitab to compute Pearson's correlation coefficient between age and average finish time results in the following:

Correlations: Age, Average Finish Time

Pearson correlation of Age and Average Finish Time = 0.038

Interpret the results)

This example shows the importance of interpreting r as a measure of the strength of a *linear* relationship. Here, r is not large, but there is a strong nonlinear relationship between age and average finish time. This is an important point—we should not conclude that there is no relationship whatsoever simply because the value of r is small in absolute value. Be sure to look at the scatterplot of the data before concluding that there is no relationship between two variables based on a correlation coefficient with a value near 0. ■

The Population Correlation Coefficient

The sample correlation coefficient r measures how strongly the x and y values in a *sample* of pairs are linearly related to one another. There is an analogous measure of how strongly x and y are related in the entire population of pairs from which the sample was obtained. It

● Data set available online

is called the **population correlation coefficient** and is denoted by ρ. (Notice again the use of a Greek letter for a population characteristic and a Roman letter for a sample characteristic.) We will never have to calculate ρ from an entire population of pairs, but it is important to know that ρ satisfies properties paralleling those of r:

1. ρ is a number between -1 and $+1$ that does not depend on the unit of measurement for either x or y, or on which variable is labeled x and which is labeled y.
2. $\rho = +1$ or -1 if and only if all (x, y) pairs in the population lie exactly on a straight line, so ρ measures the extent to which there is a linear relationship in the population.

In Chapter 13, we will see how the sample correlation coefficient r can be used to draw conclusions about the value of the population correlation coefficient ρ.

Correlation and Causation

A value of r close to 1 indicates that the larger values of one variable tend to be associated with the larger values of the other variable. This is different from saying that a large value of one variable *causes* the value of the other variable to be large. Correlation measures the extent of association, but *association does not imply causation*.

It frequently happens that two variables are highly correlated not because one is causally related to the other but because they are both strongly related to a third variable. Among all elementary school children, the relationship between the number of cavities in a child's teeth and the size of his or her vocabulary is strong and positive. Yet no one advocates eating foods that result in more cavities to increase vocabulary size (or working to decrease vocabulary size to protect against cavities). Number of cavities and vocabulary size are both strongly related to age, so older children tend to have higher values of both variables than do younger ones.

In the ABCNews.com series **"Who's Counting?" (February 1, 2001)**, John Paulos reminded readers that correlation does not imply causation and gave the following example: Consumption of hot chocolate is negatively correlated with crime rate (high values of hot chocolate consumption tend to be paired with lower crime rates), but both are responses to cold weather.

Scientific experiments can frequently make a strong case for causality by carefully controlling the values of all variables that might be related to the ones under study. Then, if y is observed to change in a "smooth" way as the experimenter changes the value of x, a plausible explanation would be that there is a causal relationship between x and y. In the absence of such control and the ability to manipulate values of one variable, we must admit the possibility that an unidentified underlying third variable is influencing both variables under investigation. A high correlation in many uncontrolled studies carried out in different settings can also marshal support for causality—as in the case of cigarette smoking and cancer—but proving causality is an elusive task.

EXERCISES 5.1 - 5.13

5.1 For each of the scatterplots shown, answer the following questions:

a. Does there appear to be a relationship between x and y?

b. If so, does the relationship appear to be linear?

c. If so, would you describe the linear relationship as positive or negative?

Scatterplot 1:

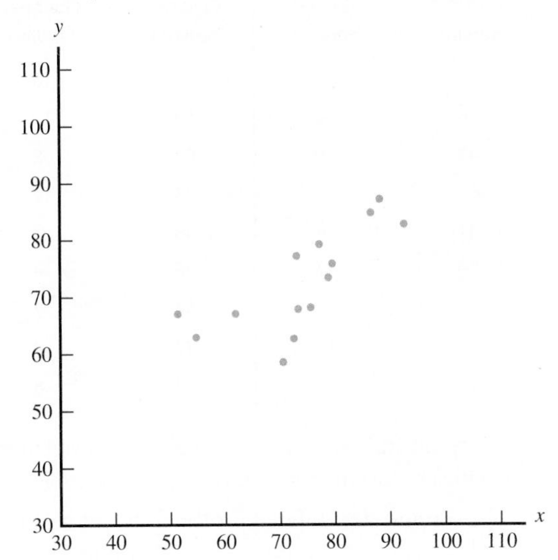

Scatterplot 2:

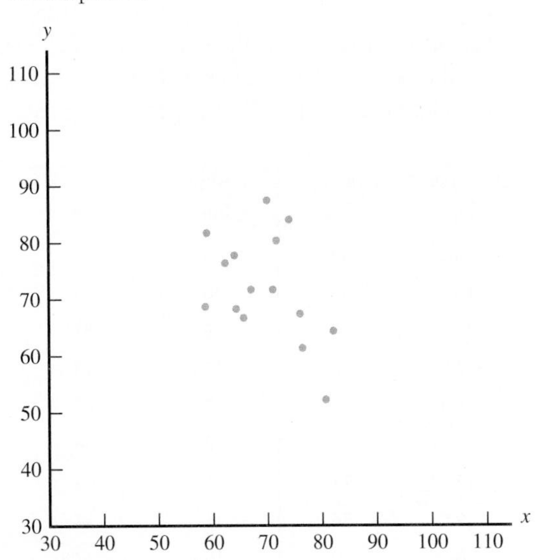

Scatterplot 3:

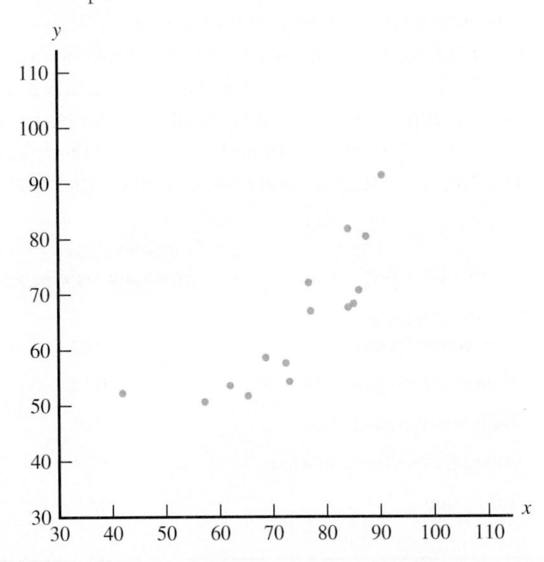

Scatterplot 4:

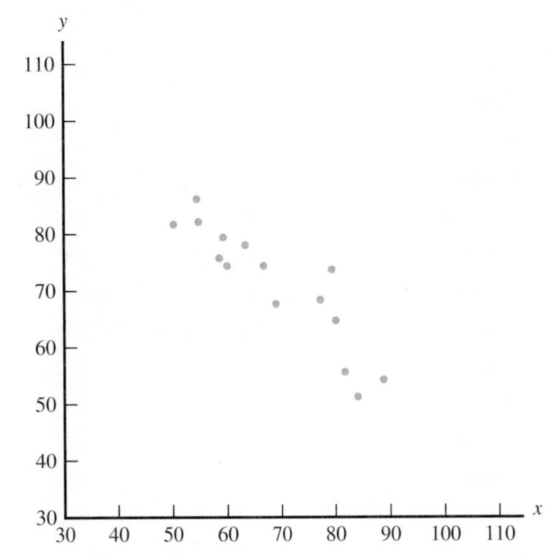

5.2 For each of the following pairs of variables, indicate whether you would expect a positive correlation, a negative correlation, or a correlation close to 0. Explain your choice.

a. Maximum daily temperature and cooling costs

b. Interest rate and number of loan applications

c. Incomes of husbands and wives when both have full-time jobs

d. Height and IQ

e. Height and shoe size

f. Score on the math section of the SAT exam and score on the verbal section of the same test

g. Time spent on homework and time spent watching television during the same day by elementary school children

h. Amount of fertilizer used per acre and crop yield (Hint: As the amount of fertilizer is increased, yield tends to increase for a while but then tends to start decreasing.)

5.3 Is the following statement correct? Explain why or why not. (Hint: See Example 5.3.)

A correlation coefficient of 0 implies that no relationship exists between the two variables under study.

5.4 Draw two scatterplots, one for which $r = 1$ and a second for which $r = -1$.

5.5 Each year J.D. Power and Associates surveys new car owners 90 days after they purchase their cars. This data is used to rate auto brands (such as Toyota and Ford) on quality and customer satisfaction. ***USA Today* (June 16, 2010 and July 17, 2010)** reported a quality rating and a satisfaction score for all 33 brands sold in the United States.

Brand	Quality Rating	Satisfaction Rating
Acura	86	822
Audi	111	832
BMW	113	845
Buick	114	802
Cadillac	111	818
Chevrolet	111	789
Chrysler	122	748
Dodge	130	751
Ford	93	794
GMC	126	792
Honda	95	766
Hyundai	102	760
Infiniti	107	805
Jaguar	130	854
Jeep	129	727
Kia	126	761
Land Rover	170	836
Lexus	88	822
Lincoln	106	820
Mazda	114	774
Mercedes-Benz	87	842
Mercury	113	769
Mini Cooper	133	815
Mitsubishi	146	767
Nissan	111	763
Porsche	83	877
Ram	110	780
Scion	114	764
Subaru	121	755
Suzuki	122	750
Toyota	117	745
Volkswagen	135	797
Volvo	109	795

a. Construct a scatterplot of y = satisfaction rating versus x = quality rating. How would you describe the relationship between x and y?

b. Compute and interpret the value of the correlation coefficient.

5.6 ● The accompanying data are x = cost (cents per serving) and y = fiber content (grams per serving) for 18 high-fiber cereals rated by ***Consumer Reports*** **(www.consumerreports.org/health)**.

Cost per Serving	Fiber per Serving	Cost per Serving	Fiber per Serving
33	7	53	13
46	10	53	10
49	10	67	8
62	7	43	12
41	8	48	7
19	7	28	14
77	12	54	7
71	12	27	8
30	8	58	8

a. Compute and interpret the value of the correlation coefficient for this data set. (Hint: See Example 5.1.)

b. The serving size differed for the different cereals, with serving sizes varying from ½ cup to 1¼ cups. Converting price and fiber content to "per cup" rather than "per serving" results in the accompanying data. Is the correlation coefficient for the per cup data greater than or less than the correlation coefficient for the per serving data?

Cost per Cup	Fiber per Cup	Cost per Cup	Fiber per Cup
9.3	44	13	53
10	46	10	53
10	49	8	67
7	62	12	43
6.4	32.8	7	48
7	19	28	56
12	77	7	54
9.6	56.8	16	54
8	30	10.7	77.3

5.7 The authors of the paper **"Flat-footedness is Not a Disadvantage for Athletic Performance in Children Aged 11 to 15 Years" (*Pediatrics* [2009]: e386–e392)** studied the relationship between y = arch height and scores on a number of different motor ability tests for 218 children. They reported the following correlation coefficients:

Motor Ability Test	Correlation between Test Score and Arch Height
Height of counter movement jump	−0.02
Hopping: average height	−0.10
Hopping: average power	−0.09
Balance, closed eyes, one leg	0.04
Toe flexion	0.05

a. Interpret the value of the correlation coefficient between average hopping height and arch height. What does the fact that the correlation coefficient is negative say about the relationship? Do higher arch heights tend to be paired with higher or lower average hopping heights?

b. The title of the paper suggests that having a small value for arch height (flat-footedness) is not a disadvantage when it comes to motor skills. Do the given correlation coefficients support this conclusion? Explain.

5.8 In a study of 200 Division I athletes, variables related to academic performance were examined. The paper **"Noncognitive Predictors of Student Athletes' Academic Performance"** (***Journal of College Reading and Learning*** **[2000]: e167)** reported that the correlation coefficient for college GPA and a measure of academic self-worth was $r = 0.48$. Also reported were the correlation coefficient for college GPA and high school GPA ($r = 0.46$) and the correlation coefficient for college GPA and a measure of tendency to procrastinate ($r = -0.36$). Higher scores on the measure of self-worth indicate higher self-worth, and higher scores on the measure of procrastination indicate a higher tendency to procrastinate. Write a few sentences summarizing what these correlation coefficients tell you about the academic performance of the 200 athletes in the sample.

5.9 ● Data from the **U.S. Federal Reserve Board (Household Debt Service Burden, 2002)** on the percentage of disposable personal income required to meet consumer loan payments and mortgage payments for selected years are shown in the following table:

Consumer Debt	Household Debt	Consumer Debt	Household Debt
7.88	6.22	6.24	5.73
7.91	6.14	6.09	5.95
7.65	5.95	6.32	6.09
7.61	5.83	6.97	6.28
7.48	5.83	7.38	6.08
7.49	5.85	7.52	5.79
7.37	5.81	7.84	5.81
6.57	5.79		

a. What is the value of the correlation coefficient for this data set?

b. Is it reasonable to conclude in this case that there is no strong relationship between the variables (linear or otherwise)? Use a graphical display to support your answer.

5.10 ● ▼ The accompanying data were read from graphs that appeared in the article **"Bush Timber Proposal Runs Counter to the Record"** (***San Luis Obispo Tribune,*** **September 22, 2002)**. The variables shown are the number of acres burned in forest fires in the western United States and timber sales.

Year	Number of Acres Burned (thousands)	Timber Sales (billions of board feet)
1945	200	2.0
1950	250	3.7
1955	260	4.4
1960	380	6.8
1965	80	9.7
1970	450	11.0
1975	180	11.0
1980	240	10.2
1985	440	10.0
1990	400	11.0
1995	180	3.8

a. Is there a correlation between timber sales and acres burned in forest fires? Compute and interpret the value of the correlation coefficient.

b. The article concludes that "heavier logging led to large forest fires." Do you think this conclusion is justified based on the given data? Explain.

5.11 It may seem odd, but one of the ways biologists can tell how old a lobster is involves measuring the concentration of a pigment called neurolipofuscin in the eyestalk of a lobster. (We are not making this up!) The authors of the paper **"Neurolipofuscin is a Measure of Age in *Panulirus argus,* the Caribbean Spiny Lobster, in Florida"** (***Biological Bulletin*** **[2007]: 55–66)** wondered if it was sufficient to measure the pigment in just one eye stalk, which would be the case if there is a strong relationship between the concentration in the right and left eyestalks.

Pigment concentration (as a percentage of tissue sample) was measured in both eyestalks for 39 lobsters, resulting in the following summary quantities (based on data read from a graph that appeared in the paper):

$$n = 39 \qquad \Sigma x = 88.8 \qquad \Sigma y = 86.1$$

$$\Sigma xy = 281.1 \qquad \Sigma x^2 = 288.0 \qquad \Sigma y^2 = 286.6$$

An alternative formula for computing the correlation coefficient that is based on raw data and is algebraically equivalent to the one given in the text is

$$r = \frac{\Sigma xy - \frac{(\Sigma x)(\Sigma y)}{n}}{\sqrt{\Sigma x^2 - \frac{(\Sigma x)^2}{n}}\sqrt{\Sigma y^2 - \frac{(\Sigma y)^2}{n}}}$$

Use this formula to compute the value of the correlation coefficient, and interpret this value.

5.12 An auction house released a list of 25 recently sold paintings. Eight artists were represented in these sales. The sale price of each painting also appears on the list. Would the correlation coefficient be an appropriate way to summarize the relationship between artist (x) and sale price (y)? Why or why not?

5.13 A sample of automobiles traversing a certain stretch of highway is selected. Each one travels at roughly a constant rate of speed, although speed does vary from auto to auto. Let x = speed and y = time needed to traverse this segment of highway. Would the sample correlation coefficient be closest to .9, .3, −.3, or −.9? Explain.

Bold exercises answered in back ● Data set available online ▼ Video Solution available

5.2 Linear Regression: Fitting a Line to Bivariate Data

The objective of *regression analysis* is to use information about one variable, x, to predict the value of a second variable, y. For example, we might want to predict y = product sales during a given period when the amount spent on advertising is x = $10,000. The two variables in a regression analysis play different roles: y is called the **dependent** or **response variable**, and x is referred to as the **independent**, **predictor**, or **explanatory variable**.

DEFINITION

Dependent variable: In a bivariate data set, the variable whose value we would like to predict. The dependent variable is denoted by y. The dependent variable is also sometimes called the **response variable.**

Independent variable: In a bivariate data set, the variable that will be used to make a prediction of the dependent variable. The independent variable is denoted by x. The independent variable is also sometimes called the **predictor variable** or the **explanatory variable.**

Scatterplots frequently exhibit a linear pattern. When this is the case, it makes sense to summarize the relationship between the variables by finding a line that is as close as possible to the points in the plot. Before seeing how this is done, let's review some elementary facts about lines and linear relationships.

The equation of a line is $y = a + bx$. A particular line is specified by choosing values of a and b. For example, one line is $y = 10 + 2x$; another is $y = 100 - 5x$. If we choose some x values and compute $y = a + bx$ for each value, the points in the plot of the resulting (x, y) pairs will fall exactly on a straight line.

DEFINITION

The equation of a line is

Intercept

$$y = a + bx$$

Slope

Slope: The value of b, called the **slope** of the line, is the amount by which y increases when x increases by 1 unit.

Intercept: The value of a, called the **intercept** (or sometimes the **y-intercept** or **vertical intercept**) of the line, is the height of the line above the value $x = 0$.

The line $y = 10 + 2x$ has slope $b = 2$, so each 1-unit increase in x is paired with an increase of 2 in y. When $x = 0$, $y = 10$, so the height at which the line crosses the vertical axis (where $x = 0$) is 10. This is illustrated in Figure 5.7(a). The slope of the line $y = 100 - 5x$ is -5, so y increases by -5 (or equivalently, decreases by 5) when x increases by 1. The height of the line above $x = 0$ is $a = 100$. The resulting line is pictured in Figure 5.7(b).

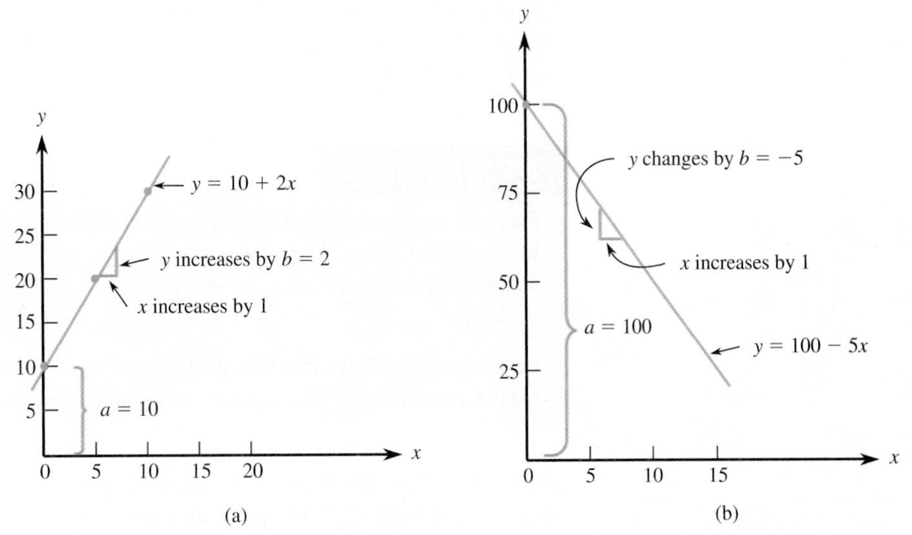

FIGURE 5.7
Graphs of two lines:
(a) slope $b = 2$, intercept $a = 10$;
(b) slope $b = -5$, intercept $a = 100$.

It is easy to draw the line corresponding to any particular linear equation. Choose any two x values and substitute them into the equation to obtain the corresponding y values. Then plot the resulting two (x, y) pairs as two points. The desired line is the one passing through these points. For the equation $y = 10 + 2x$, substituting $x = 5$ yields $y = 20$, whereas using $x = 10$ gives $y = 30$. The resulting two points are then (5, 20) and (10, 30). The line in Figure 5.7(a) passes through these points.

Fitting a Straight Line: The Principle of Least Squares

Figure 5.8 shows a scatterplot with two lines superimposed on the plot. Line II is a better fit to the data than Line I is. In order to measure the extent to which a particular line provides a good fit to data, we focus on the vertical deviations from the line.

Line II in Figure 5.8 has equation $y = 10 + 2x$. The third and fourth points from the left in the scatterplot are (15, 44) and (20, 45). For these two points, the vertical deviations from this line are

$$\begin{aligned}\text{3rd deviation} &= y_3 - \text{height of the line above } x_3 \\ &= 44 - [10 + 2(15)] \\ &= 4\end{aligned}$$

and

$$\text{4th deviation} = 45 - [10 + 2(20)] = -5$$

A positive vertical deviation results from a point that lies above the line, and a negative deviation results from a point that lies below the line. A particular line is said to be a good fit to the data if the deviations from the line are small in magnitude. Line I in Figure 5.8 fits poorly, because all deviations from that line are larger in magnitude (some are much larger) than the corresponding deviations from Line II.

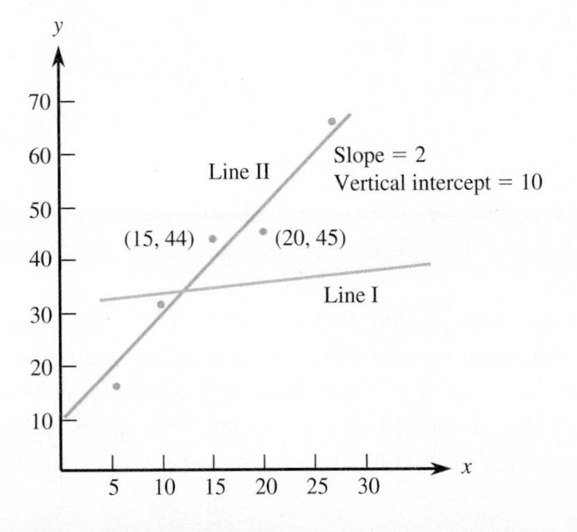

FIGURE 5.8
Line I gives a poor fit and Line II gives a good fit to the data.

To assess the overall fit of a line, we need a way to combine the n deviations into a single measure of fit. The standard approach is to square the deviations (to obtain nonnegative numbers) and then to sum these squared deviations.

DEFINITION

Sum of squared deviations: The most widely used measure of the goodness of fit of a line $y = a + bx$ to bivariate data $(x_1, y_1), \ldots, (x_n, y_n)$ is the **sum of the squared deviations** about the line

$$\Sigma[y - (a + bx)]^2 = [y_1 - (a + bx_1)]^2 + [y_2 - (a + bx_2)]^2 + \cdots + [y_n - (a + bx_n)]^2$$

Least-squares line: The line that minimizes the sum of squared deviations. The least-squares line is also called the **sample regression line.**

Fortunately, the equation of the least-squares line can be obtained without having to calculate deviations from any particular line. The accompanying box gives relatively simple formulas for the slope and intercept of the least-squares line.

The slope of the least-squares line is

$$b = \frac{\Sigma(x - \bar{x})(y - \bar{y})}{\Sigma(x - \bar{x})^2}$$

and the y intercept is

$$a = \bar{y} - b\bar{x}$$

We write the equation of the least-squares line as

$$\hat{y} = a + bx$$

where the ˆ above y indicates that $\hat{y}$ (read as y-hat) is the prediction of y that results from substituting a particular x value into the equation.

Statistical software packages and many calculators can compute the slope and intercept of the least-squares line. If the slope and intercept are to be computed by hand, the following computational formula can be used to reduce the amount of time required to perform the calculations.

Calculating Formula for the Slope of the Least-Squares Line

$$b = \frac{\Sigma xy - \frac{(\Sigma x)(\Sigma y)}{n}}{\Sigma x^2 - \frac{(\Sigma x)^2}{n}}$$

EXAMPLE 5.4 Pomegranate Juice and Tumor Growth

Understand the context

● Pomegranate, a fruit native to Persia, has been used in the folk medicines of many cultures to treat various ailments. Researchers are now studying pomegranate's antioxidant properties to see if it might have any beneficial effects in the treatment of cancer. One such study, described in the paper **"Pomegranate Fruit Juice for Chemoprevention and Chemotherapy of Prostate Cancer"** (***Proceedings of the National Academy of Sciences*** **[October 11,**

● Data set available online

2005]: 14813–14818), investigated whether pomegranate fruit extract (PFE) was effective in slowing the growth of prostate cancer tumors.

In this study, 24 mice were injected with cancer cells. The mice were then randomly assigned to one of three treatment groups. One group of eight mice received normal drinking water, the second group of eight mice received drinking water supplemented with 0.1% PFE, and the third group received drinking water supplemented with 0.2% PFE. The average tumor volume for the mice in each group was recorded at several points in time.

The accompanying data on y = average tumor volume (in mm^3) and x = number of days after injection of cancer cells for the mice that received plain drinking water was approximated from a graph that appeared in the paper:

Consider the data)

x	11	15	19	23	27
y	150	270	450	580	740

A scatterplot of these data (Figure 5.9) shows that the relationship between number of days after injection of cancer cells and average tumor volume could reasonably be summarized by a straight line.

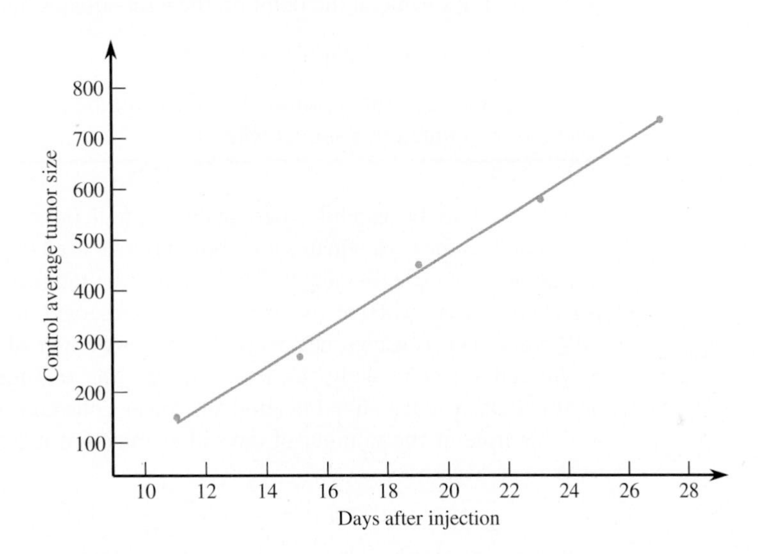

FIGURE 5.9
Scatterplot for the plain drinking water data of Example 5.4.

Calculating the values of the slope and intercept of the least-squares line can be very tedious. This is where a statistics computer package or a graphing calculator can really be helpful. Minitab was used to find the equation of the least-squares line, and the resulting output is shown here:

Regression Analysis: *y* versus *x*

The regression equation is
$y = -270 + 37.3x$

Predictor	Coef	SE Coef	T	P
Constant	−269.75	23.42	−11.52	0.001
x	37.250	1.181	31.53	0.000

From the Minitab output, the equation of the least-squares line is

$$\hat{y} = -270 + 37.3x$$

We could also use the values given in the Coef column for the slope and intercept. These values have more decimal accuracy than those in the given equation.

We could also calculate the values of the slope and intercept of the least-squares line by hand. The summary quantities necessary to compute the equation of the least-squares line are

$$\Sigma x = 95 \qquad \Sigma x^2 = 1965 \qquad \Sigma xy = 47{,}570$$
$$\Sigma y = 2190 \qquad \Sigma y^2 = 1{,}181{,}900$$

From these quantities, we compute

Do the work)

$\bar{x} = 19 \qquad \bar{y} = 438$

$$b = \frac{\Sigma xy - \frac{(\Sigma x)(\Sigma y)}{n}}{\Sigma x^2 - \frac{(\Sigma x)^2}{n}} = \frac{47{,}570 - \frac{(95)(2190)}{5}}{1965 - \frac{(95)^2}{5}} = \frac{5960}{160} = 37.25$$

and

$$a = \bar{y} - b\bar{x} = 438 - (37.25)(19) = -269.75$$

The least-squares line is then

$$\hat{y} = -269.75 + 37.25x$$

This line is also shown on the scatterplot of Figure 5.9.

If we wanted to predict average tumor volume 20 days after injection of cancer cells, we could use the y value of the point on the least-squares line above $x = 20$:

$$\hat{y} = -269.75 + 37.25(20) = 475.25$$

Predicted average tumor volume for other numbers of days after injection of cancer cells could be computed in a similar way. ■

You need to be careful when making predictions. For example, in the context of Example 5.4, the least-squares line should not be used to predict average tumor volume for times much outside the range 11 to 27 days (the range of x values in the data set) because we do not know whether the linear pattern observed in the scatterplot continues outside this range. This is sometimes referred to as the **danger of extrapolation.**

We can see that using the least-squares line to predict average tumor volume for fewer than 10 days after injection of cancer cells can lead to nonsensical predictions. For example, if the number of days after injection is five, the predicted average tumor volume is negative:

$$\hat{y} = -269.75 + 37.25(5) = -83.5$$

Because it is impossible for average tumor volume to be negative, this is a clear indication that the pattern observed for x values in the 11 to 27 range does not continue outside this range. Nonetheless, the least-squares line can be a useful tool for making predictions for x values within the 11- to 27-day range.

USE CAUTION—The Danger of Extrapolation

The least-squares line should not be used to make predictions outside the range of the x values in the data set because we have no evidence that the linear relationship continues outside this range.

EXAMPLE 5.5 More on Pomegranate Juice and Tumor Growth

Figure 5.10 shows a scatterplot for average tumor volume versus number of days after injection of cancer cells for both the group of mice that drank only water and the group that drank water supplemented by 0.2% PFE. Notice that the tumor growth seems to be much slower for the mice that drank water supplemented with PFE. For the 0.2% PFE group, the relationship between average tumor volume and number of days after injection of cancer cells appears to be curved rather than linear. We will see in Section 5.4 how a curve (rather than a straight line) can be used to summarize this relationship.

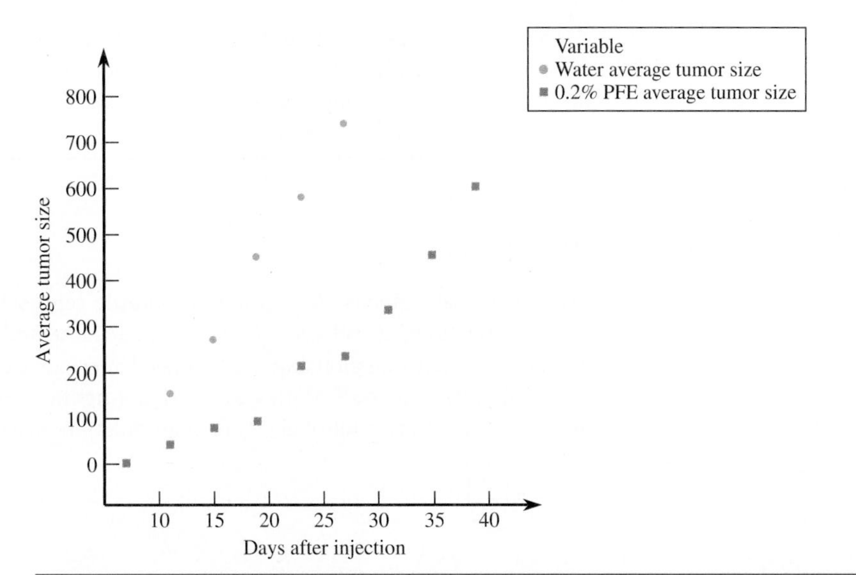

FIGURE 5.10
Scatterplot of average tumor volume versus number of days after injection of cancer cells for the water group and the 0.2% PFE group.

EXAMPLE 5.6 Revisiting a Face You Can Trust

Understand the context ⟩

Data on x = face-to-height ratio and y = percentage choosing the trust option in a game were given in Example 5.1. In that example, we saw that the correlation coefficient was −0.392, indicating a weak negative linear relationship. This linear relationship can be summarized using the least-squares line, as shown in Figure 5.11.

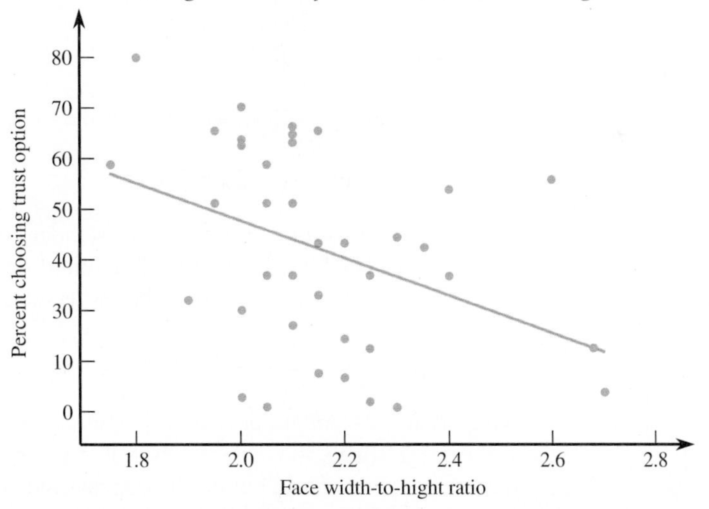

FIGURE 5.11
Scatterplot and least-squares line for the data of Example 5.6.

Minitab was used to fit the least-squares line, and Figure 5.12 shows part of the resulting output. Instead of x and y, the variable labels Face-to-width ratio and Percentage choosing trust option are used. The equation at the top is that of the least-squares line. In the rectangular table just below the equation, the first row gives information about the intercept, a, and the second row gives information concerning the slope, b. In particular, the coefficient column labeled "Coef" contains the values of a and b using more digits than in the rounded values that appear in the equation.

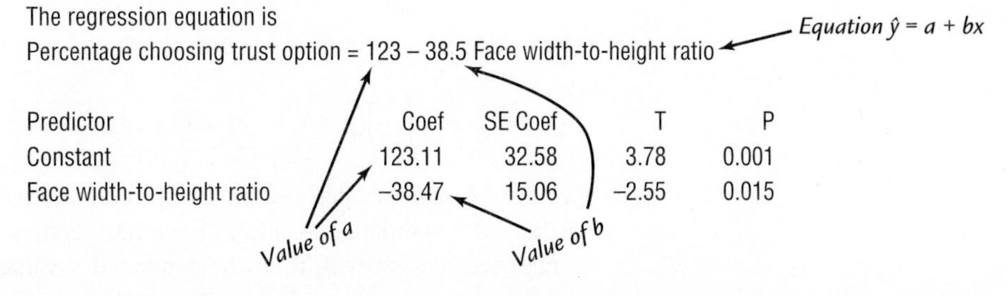
The regression equation is
Percentage choosing trust option = 123 – 38.5 Face width-to-height ratio

Predictor	Coef	SE Coef	T	P
Constant	123.11	32.58	3.78	0.001
Face width-to-height ratio	−38.47	15.06	−2.55	0.015

FIGURE 5.12
Partial Minitab output for Example 5.6.

The least-squares line should not be used to predict the percentage choosing the trust option for faces with width-to-height ratios such as $x = 1.00$ or $x = 4.00$. These x values are well outside the range of the data, and we do not know if the linear relationship continues outside the observed range. ■

Regression

The least-squares line is often called the **sample regression line.** This terminology comes from the relationship between the least-squares line and Pearson's correlation coefficient. To understand this relationship, we first need alternative expressions for the slope b and the equation of the line itself. With s_x and s_y denoting the sample standard deviations of the x's and y's, respectively, a bit of algebraic manipulation gives

$$b = r\left(\frac{s_y}{s_x}\right)$$

$$\hat{y} = \bar{y} + r\left(\frac{s_y}{s_x}\right)(x - \bar{x})$$

You do not need to use these formulas in any computations, but several of their implications are important for appreciating what the least-squares line does.

1. When $x = \bar{x}$ is substituted in the equation of the line, $\hat{y} = \bar{y}$ results. This means that the least-squares line passes through the *point of averages* $(\bar{x}, \bar{y})$.
2. Suppose for the moment that $r = 1$, so that all points lie exactly on the line whose equation is

   $$\hat{y} = \bar{y} + \frac{s_y}{s_x}(x - \bar{x})$$

   Now substitute $x = \bar{x} + s_x$, which is 1 standard deviation above $\bar{x}$:

   $$\hat{y} = \bar{y} + \frac{s_y}{s_x}(\bar{x} + s_x - \bar{x}) = \bar{y} + s_y$$

   We can see that with $r = 1$, when x is 1 standard deviation above its mean, we predict that the associated y value will be 1 standard deviation above its mean. Similarly, if $x = \bar{x} - 2s_x$ (2 standard deviations below its mean), then

   $$\hat{y} = \bar{y} + \frac{s_y}{s_x}(\bar{x} - 2s_x - \bar{x}) = \bar{y} - 2s_y$$

   which is also 2 standard deviations below the mean.

   If $r = -1$, then $x = \bar{x} + s_x$ results in $\hat{y} = \bar{y} - s_y$, so the predicted y is also 1 standard deviation from its mean but on the opposite side of $\bar{y}$ from where x is relative to $\bar{x}$. In general, if x and y are perfectly correlated, the predicted y value associated with a given x value will be the same number of standard deviations (of y) from its mean $\bar{y}$ as x is from its mean $\bar{x}$.
3. Now suppose that x and y are not perfectly correlated. For example, suppose $r = .5$, so the least-squares line has the equation

   $$\hat{y} = \bar{y} + .5\left(\frac{s_y}{s_x}\right)(x - \bar{x})$$

   Then substituting $x = \bar{x} + s_x$ gives

   $$\hat{y} = \bar{y} + .5\left(\frac{s_y}{s_x}\right)(\bar{x} + s_x - \bar{x}) = \bar{y} + .5s_y$$

   For $r = .5$, when x lies 1 standard deviation above its mean, we predict that y will be only 0.5 standard deviation above its mean. Similarly, we can predict y when r is negative. If $r = -.5$, then the predicted y value will be only half the number of standard deviations from $\bar{y}$ that x is from $\bar{x}$ but x and the predicted y will now be on opposite sides of their respective means.

Consider using the least-squares line to predict the value of y associated with an x value some specified number of standard deviations away from $\bar{x}$. Then the predicted y value will be only r times this number of standard deviations from $\bar{y}$. In terms of standard deviations, except when $r = 1$ or -1, the predicted y will always be closer to $\bar{y}$ than x is to $\bar{x}$.

Using the least-squares line for prediction results in a predicted y that is pulled back in, or regressed, toward the mean of y compared to where x is relative to the mean of x. This regression effect was first noticed by Sir Francis Galton (1822–1911), a famous biologist, when he was studying the relationship between the heights of fathers and their sons. He found that predicted heights of sons whose fathers were above average in height were also above average (because r is positive here) but not by as much as the father's height. He found a similar relationship for fathers whose heights were below average. This regression effect has led to the term **regression analysis** for the collection of methods involving the fitting of lines, curves, and more complicated functions to bivariate and multivariate data.

The alternative form of the regression (least-squares) line emphasizes that predicting y from knowledge of x is not the same problem as predicting x from knowledge of y. The slope of the least-squares line for predicting x is $r(s_x/s_y)$ rather than $r(s_y/s_x)$ and the intercepts of the lines are almost always different. For purposes of prediction, it makes a difference whether y is regressed on x, as we have done, or x is regressed on y.

The regression line of y on x should not be used to predict x, because it is not the line that minimizes the sum of squared deviations in the x direction.

EXERCISES 5.14 - 5.28

5.14 Two scatterplots are shown below. Explain why it makes sense to use the least-squares regression line to summarize the relationship between x and y for one of these data sets but not the other. (Hint: See Example 5.5.)

Scatterplot 1:

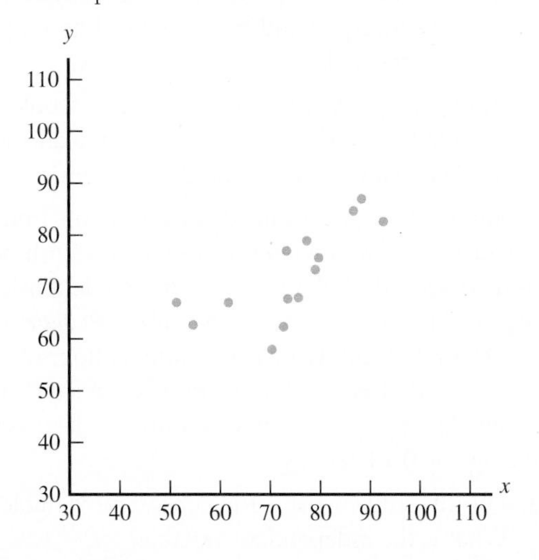

Scatterplot 2:

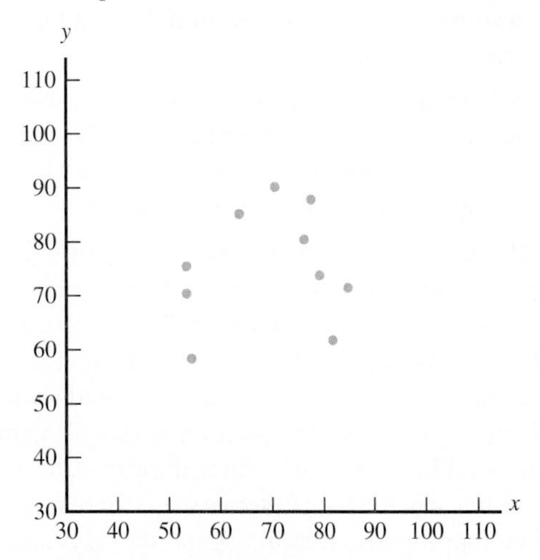

5.15 The authors of the paper **"Statistical Methods for Assessing Agreement Between Two Methods of Clinical Measurement"** (***International Journal of Nursing Studies* [2010]: 931–936**) compared two different instruments for measuring a person's ability to breathe out air.

(This measurement is helpful in diagnosing various lung disorders.) The two instruments considered were a Wright peak flow meter and a mini-Wright peak flow meter. Seventeen people participated in the study, and for each person air flow was measured once using the Wright meter and once using the mini-Wright meter.

Subject	Mini-Wright Meter	Wright Meter	Subject	Mini-Wright Meter	Wright Meter
1	512	494	10	445	433
2	430	395	11	432	417
3	520	516	12	626	656
4	428	434	13	260	267
5	500	476	14	477	478
6	600	557	15	259	178
7	364	413	16	350	423
8	380	442	17	451	427
9	658	650			

a. Suppose that the Wright meter is considered to provide a better measure of air flow, but the mini-Wright meter is easier to transport and to use. If the two types of meters produce different readings but there is a strong relationship between the readings, it would be possible to use a reading from the mini-Wright meter to predict the reading that the larger Wright meter would have given. Use the given data to find an equation to predict Wright meter reading using a reading from the mini-Wright meter. (Hint: See Example 5.4.)

b. What would you predict for the Wright meter reading of a person whose mini-Wright meter reading was 500?

c. What would you predict for the Wright meter reading of a person whose mini-Wright meter reading was 300? (Hint: See discussion of extrapolation that follows Example 5.4.)

5.16 ● The authors of the paper **"Evaluating Existing Movement Hypotheses in Linear Systems Using Larval Stream Salamanders" (*Canadian Journal of Zoology* [2009]: 292–298)** investigated whether water temperature was related to how far a salamander would swim and whether it would swim upstream or downstream. Data for 14 streams with different mean water temperatures where salamander larvae were released are given (approximated from a graph that appeared in the paper).

The two variables of interest are x = mean water temperature (°C) and y = net directionality, which was defined as the difference in the relative frequency of the released salamander larvae moving upstream and the relative frequency of released salamander larvae moving downstream. A positive value of net directionality means a higher proportion were moving upstream than downstream. A negative value of net directionality means a higher proportion were moving downstream than upstream.

Mean Temperature (x)	Net Directionality (y)
6.17	−0.08
8.06	0.25
8.62	−0.14
10.56	0.00
12.45	0.08
11.99	0.03
12.50	−0.07
17.98	0.29
18.29	0.23
19.89	0.24
20.25	0.19
19.07	0.14
17.73	0.05
19.62	0.07

a. Construct a scatterplot of the data. How would you describe the relationship between x and y?

b. Find the equation of the least-squares line describing the relationship between y = net directionality and x = mean water temperature.

c. What value of net directionality would you predict for a stream that had mean water temperature of 15 °C?

d. The authors state that "when temperatures were warmer, more larvae were captured moving upstream, but when temperatures were cooler, more larvae were captured moving downstream." Do the scatterplot and least-squares line support this statement? Explain.

e. Approximately what mean temperature would result in a prediction of the same number of salamander larvae moving upstream and downstream?

5.17 A sample of 548 ethnically diverse students from Massachusetts were followed over a 19-month period from 1995 and 1997 in a study of the relationship between TV viewing and eating habits **(*Pediatrics* [2003]: 1321–1326)**. For each additional hour of television viewed per day, the number of fruit and vegetable servings per day was found to decrease on average by 0.14 serving.

a. For this study, what is the dependent variable? What is the independent variable?

b. Would the least-squares line for predicting number of servings of fruits and vegetables using number of hours spent watching TV as a predictor have a positive or negative slope? Explain.

5.18 The relationship between hospital patient-to-nurse ratio and various characteristics of job satisfaction and patient care has been the focus of a number of research studies. Suppose x = patient-to-nurse ratio is the independent variable. For each of the following potential dependent variables, indicate whether you expect the slope of the least-squares line to be positive or negative and give a brief explanation for your choice.

a. y = a measure of nurse's job satisfaction (higher values indicate higher satisfaction)

b. y = a measure of patient satisfaction with hospital care (higher values indicate higher satisfaction)

c. y = a measure of patient quality of care

5.19 The article **"No-Can-Do Attitudes Irk Fliers" (*USA Today*, September 15, 2010)** included the accompanying data on airline quality score and a ranking based on the number of passenger complaints per 100,000 passengers boarded. In the complaint ranking, a rank of 1 is best, corresponding to fewest complaints. Similarly, for the J.D. Powers quality score, lower scores correspond to higher quality.

Airline	Passenger Complaint Rank	J.D. Powers Quality Score
Airtran	6	3
Alaska	2	4
American	8	7
Continental	7	6
Delta	12	8
Frontier	5	5
Hawaiian	3	Not rated
JetBlue	4	1
Northwest	9	Not rated
Southwest	1	2
United	11	9
US Airways	10	10

Two of the airlines were not scored by J.D. Powers. Use the ranks for the other 10 airlines to fit a least-squares regression line.

5.20 Use the least-squares line from Exercise 5.19 to predict the J.D. Powers Quality Score for Hawaiian Airlines and Northwest Airlines.

5.21 Studies have shown that people who suffer sudden cardiac arrest have a better chance of survival if a defibrillator shock is administered very soon after cardiac arrest. How is survival rate related to the time between when cardiac arrest occurs and when the defibrillator shock is delivered? This question is addressed in the paper **"Improving Survival from Sudden Cardiac Arrest: The Role of Home Defibrillators" (by J. K. Stross, University of Michigan, February 2002; available at www.heartstarthome.com)**.

The accompanying data give y = survival rate (percent) and x = mean call-to-shock time (minutes) for a cardiac rehabilitation center (in which cardiac arrests occurred while victims were hospitalized and so the call-to-shock time tended to be short) and for four communities of different sizes:

Mean call-to-shock time, x	2	6	7	9	12
Survival rate, y	90	45	30	5	2

a. Construct a scatterplot for these data. How would you describe the relationship between mean call-to-shock time and survival rate?

b. Find the equation of the least-squares line.

c. Use the least-squares line to predict survival rate for a community with a mean call-to-shock time of 10 minutes.

5.22 The data given in the previous exercise on x = call-to-shock time (in minutes) and y = survival rate (percent) were used to compute the equation of the least-squares line, which was

$$\hat{y} = 101.33 - 9.30x$$

The newspaper article **"FDA OKs Use of Home Defibrillators" (*San Luis Obispo Tribune*, November 13, 2002)** reported that "every minute spent waiting for paramedics to arrive with a defibrillator lowers the chance of survival by 10 percent." Is this statement consistent with the given least-squares line? Explain.

5.23 An article on the cost of housing in California that appeared in the ***San Luis Obispo Tribune* (March 30, 2001)** included the following statement: "In Northern California, people from the San Francisco Bay area pushed into the Central Valley, benefiting from home prices that dropped on average \$4000 for every mile traveled east of the Bay area." If this statement is correct, what is the slope of the least-squares regression line, $\hat{y} = a + bx$, where y = house price (in dollars) and x = distance east of the Bay (in miles)? Explain.

5.24 ● The following data on sale price, size, and land-to-building ratio for 10 large industrial properties appeared in the paper **"Using Multiple Regression**

Analysis in Real Estate Appraisal" (*Appraisal Journal* [2002]: 424–430):

Property	Sale Price (millions of dollars)	Size (thousands of sq. ft.)	Land-to-Building Ratio
1	10.6	2166	2.0
2	2.6	751	3.5
3	30.5	2422	3.6
4	1.8	224	4.7
5	20.0	3917	1.7
6	8.0	2866	2.3
7	10.0	1698	3.1
8	6.7	1046	4.8
9	5.8	1108	7.6
10	4.5	405	17.2

a. Calculate and interpret the value of the correlation coefficient between sale price and size.

b. Calculate and interpret the value of the correlation coefficient between sale price and land-to-building ratio.

c. If you wanted to predict sale price and you could use either size or land-to-building ratio as the basis for making predictions, which would you use? Explain.

d. Based on your choice in Part (c), find the equation of the least-squares regression line you would use for predicting y = sale price.

5.25 Explain why it can be dangerous to use the least-squares line to obtain predictions for x values that are substantially larger or smaller than those contained in the sample.

5.26 The sales manager of a large company selected a random sample of $n = 10$ salespeople and determined for each one the values of x = years of sales experience and y = annual sales (in thousands of dollars). A scatterplot of the resulting (x, y) pairs showed a linear pattern.

a. Suppose that the sample correlation coefficient is $r = .75$ and that the average annual sales is $\bar{y} = 100$. If a particular salesperson is 2 standard deviations above the mean in terms of experience, what would you predict for that person's annual sales?

b. If a particular person whose sales experience is 1.5 standard deviations below the average experience is predicted to have an annual sales value that is 1 standard deviation below the average annual sales, what is the value of r?

5.27 Explain why the slope b of the least-squares line always has the same sign (positive or negative) as the sample correlation coefficient r.

5.28 ● The accompanying data resulted from an experiment in which weld diameter x and shear strength y (in pounds) were determined for five different spot welds on steel. A scatterplot shows a strong linear pattern. With $\Sigma(x - \bar{x})^2 = 1000$ and $\Sigma(x - \bar{x})(y - \bar{y}) = 8577$, the least-squares line is $\hat{y} = -936.22 + 8.577x$.

x	200.1	210.1	220.1	230.1	240.0
y	813.7	785.3	960.4	1118.0	1076.2

a. Because 1 lb = 0.4536 kg, strength observations can be re-expressed in kilograms through multiplication by this conversion factor: new y = 0.4536(old y). What is the equation of the least-squares line when y is expressed in kilograms?

b. More generally, suppose that each y value in a data set consisting of n (x, y) pairs is multiplied by a conversion factor c (which changes the units of measurement for y). What effect does this have on the slope b (i.e., how does the new value of b compare to the value before conversion), on the intercept a, and on the equation of the least-squares line? Verify your conjectures by using the given formulas for b and a. (Hint: Replace y with cy, and see what happens—and remember, this conversion will affect $\bar{y}$.)

Bold exercises answered in back ● Data set available online ▼ Video Solution available

5.3 Assessing the Fit of a Line

Once we have found the equation of the least-squares regression line, the next step is to consider how effectively the line summarizes the relationship between x and y. Important questions to consider are

1. Is a line an appropriate way to summarize the relationship between the two variables?
2. Are there any unusual aspects of the data set that we need to consider before proceeding to use the regression line to make predictions?
3. If we decide that it is reasonable to use the regression line as a basis for prediction, how accurate can we expect predictions based on the regression line to be?

In this section, we look at graphical and numerical methods that will allow us to answer these questions.

Most of these methods are based on the vertical deviations of the data points from the regression line. These vertical deviations are called residuals, and each represents the difference between an actual y value and the corresponding predicted value, $\hat{y}$, that would result from using the regression line to make a prediction.

Predicted Values and Residuals

The predicted value corresponding to the first observation in a data set is obtained by substituting that value, x_1, into the regression equation to obtain $\hat{y}_1$, where

$$\hat{y}_1 = a + bx_1$$

The difference between the actual y value for the first observation, y_1, and the corresponding predicted value is

$$y_1 - \hat{y}_1$$

This difference is the **residual**, which is the vertical deviation of a point in the scatterplot from the regression line.

An observation falling above the line results in a positive residual, whereas a point falling below the line results in a negative residual. This is shown in Figure 5.13.

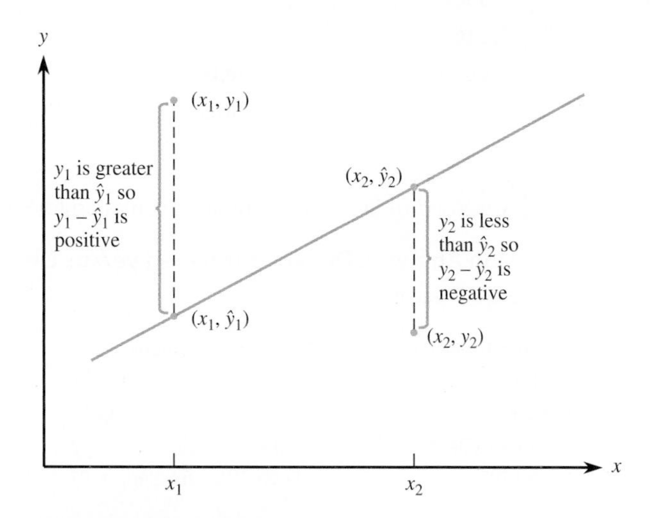

FIGURE 5.13
Positive and negative deviations from the least-squares line (residuals).

DEFINITION

Predicted values: The **predicted values** result from substituting each sample x value into the equation for the least-squares line. This gives

$\hat{y}_1 =$ *first predicted value* $= a + bx_1$

$\hat{y}_2 =$ *second predicted value* $= a + bx_2$

$\vdots$

$\hat{y}_n =$ *nth predicted value* $= a + bx_n$

Residuals: The **residuals** from the least-squares line are the n quantities

$y_1 - \hat{y}_1 =$ *first residual*

$y_2 - \hat{y}_2 =$ *second residual*

$\vdots$

$y_n - \hat{y}_n =$ *nth residual*

Each residual is the difference between an observed y value and the corresponding predicted y value.

EXAMPLE 5.7 It May Be a Pile of Debris to You, but It Is Home to a Mouse

Understand the context)

● The accompanying data is a subset of data read from a scatterplot that appeared in the paper **"Small Mammal Responses to Fine Woody Debris and Forest Fuel Reduction in Southwest Oregon"** (***Journal of Wildlife Management*** **[2005]: 625–632**). The authors of the paper were interested in how the distance a deer mouse will travel for food is related to the distance from the food to the nearest pile of fine woody debris. Distances were measured in meters. The data are given in Table 5.1.

TABLE 5.1 Predicted Values and Residuals for the Data of Example 5.7

Consider the data)

Distance from Debris (x)	Distance Traveled (y)	Predicted Distance Traveled ($\hat{y}$)	Residual ($y - \hat{y}$)
6.94	0.00	14.76	−14.76
5.23	6.13	9.23	−3.10
5.21	11.29	9.16	2.13
7.10	14.35	15.28	−0.93
8.16	12.03	18.70	−6.67
5.50	22.72	10.10	12.62
9.19	20.11	22.04	−1.93
9.05	26.16	21.58	4.58
9.36	30.65	22.59	8.06

Minitab was used to fit the least-squares regression line. Partial computer output follows:

Regression Analysis: Distance Traveled versus Distance to Debris

The regression equation is

Distance Traveled = −7.7 + 3.23 Distance to Debris

Predictor	Coef	SE Coef	T	P
Constant	−7.69	13.33	−0.58	0.582
Distance to Debris	3.234	1.782	1.82	0.112
S = 8.67071	R-Sq = 32.0%	R-Sq(adj) = 22.3%		

The resulting least-squares line is $\hat{y} = -7.69 + 3.234x$.

A plot of the data that also includes the regression line is shown in Figure 5.14. The residuals for this data set are the signed vertical distances from the points to the line.

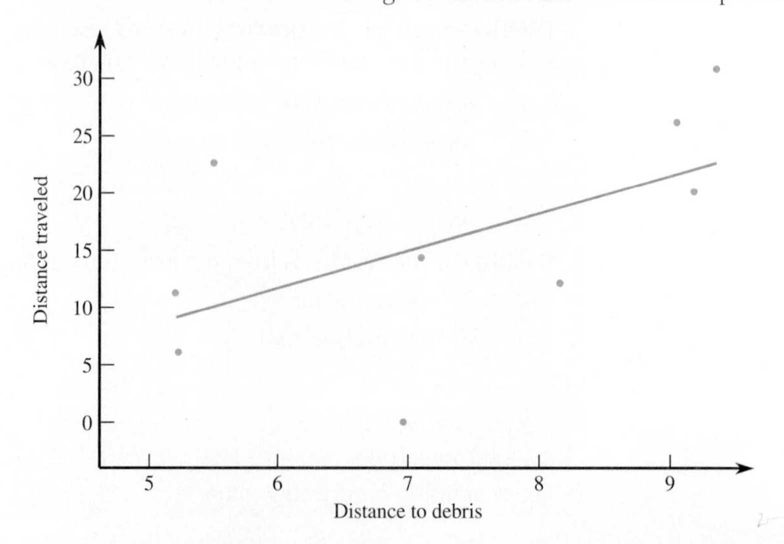

FIGURE 5.14
Scatterplot for the data of Example 5.7.

● Data set available online

Do the work ⟩ For the mouse with the smallest x value (the third observation with $x_3 = 5.21$ and $y_3 = 11.29$), the corresponding predicted value and residual are

$$\text{predicted value} = \hat{y}_3 = -7.69 + 3.234(x_3) = -7.69 + 3.234(5.21) = 9.16$$

$$\text{residual} = y_3 - \hat{y}_3 = 11.29 - 9.16 = 2.13$$

The other predicted values and residuals are computed in a similar manner and are included in Table 5.1.

Computing the predicted values and residuals by hand can be tedious, but Minitab and other statistical software packages, as well as many graphing calculators, include them as part of the output, as shown in Figure 5.15. The predicted values and residuals can be found in the table at the bottom of the Minitab output in the columns labeled "Fit" and "Residual," respectively.

The regression equation is
Distance Traveled = – 7.7 + 3.23 Distance to Debris

Predictor	Coef	SE Coef	T	P
Constant	–7.69	13.33	–0.58	0.582
Distance to Debris	3.234	1.782	1.82	0.112

S = 8.67071 R-Sq = 32.0% R-Sq(adj) = 22.3%

Analysis of Variance

Source	DF	SS	MS	F	P
Regression	1	247.68	247.68	3.29	0.112
Residual Error	7	526.27	75.18		
Total	8	773.95			

Obs	Distance to Debris	Distance Traveled	Fit	SE Fit	Residual	St Resid
1	6.94	0.00	14.76	2.96	–14.76	–1.81
2	5.23	6.13	9.23	4.69	–3.10	–0.42
3	5.21	11.29	9.16	4.72	2.13	0.29
4	7.10	14.35	15.28	2.91	–0.93	–0.11
5	8.16	12.03	18.70	3.27	–6.67	–0.83
6	5.50	22.72	10.10	4.32	12.62	1.68
7	9.19	20.11	22.04	4.43	–1.93	–0.26
8	9.05	26.16	21.58	4.25	4.58	0.61
9	9.36	30.65	22.59	4.67	8.06	1.10

FIGURE 5.15 Minitab output for the data of Example 5.7.

■

Plotting the Residuals

A careful look at residuals can reveal many potential problems. A **residual plot** is a good place to start when assessing the appropriateness of the regression line.

DEFINITION

Residual plot: A scatterplot of the (x, residual) pairs.

Isolated points or a pattern of points in the residual plot indicate potential problems.

A desirable residual plot is one that exhibits no particular pattern, such as curvature. Curvature in the residual plot is an indication that the relationship between x and y is not linear and that a curve would be a better choice than a line for describing the relationship between x and y. This is sometimes easier to see in a residual plot than in a scatterplot of y versus x, as illustrated in Example 5.8.

EXAMPLE 5.8 Heights and Weights of American Women

Understand the context

● Consider the accompanying data on x = height (in inches) and y = average weight (in pounds) for American females, age 30–39 (from ***The World Almanac and Book of Facts***). The scatterplot displayed in Figure 5.16(a) appears rather straight. However, when the residuals from the least-squares line ($\hat{y} = 98.23 + 3.59x$) are plotted (Figure 5.16(b)), substantial curvature is apparent (even though $r \approx .99$). It is not accurate to say that weight increases in direct proportion to height (linearly with height). Instead, average weight increases somewhat more rapidly for relatively large heights than it does for relatively small heights.

x	58	59	60	61	62	63	64	65
y	113	115	118	121	124	128	131	134

x	66	67	68	69	70	71	72
y	137	141	145	150	153	159	164

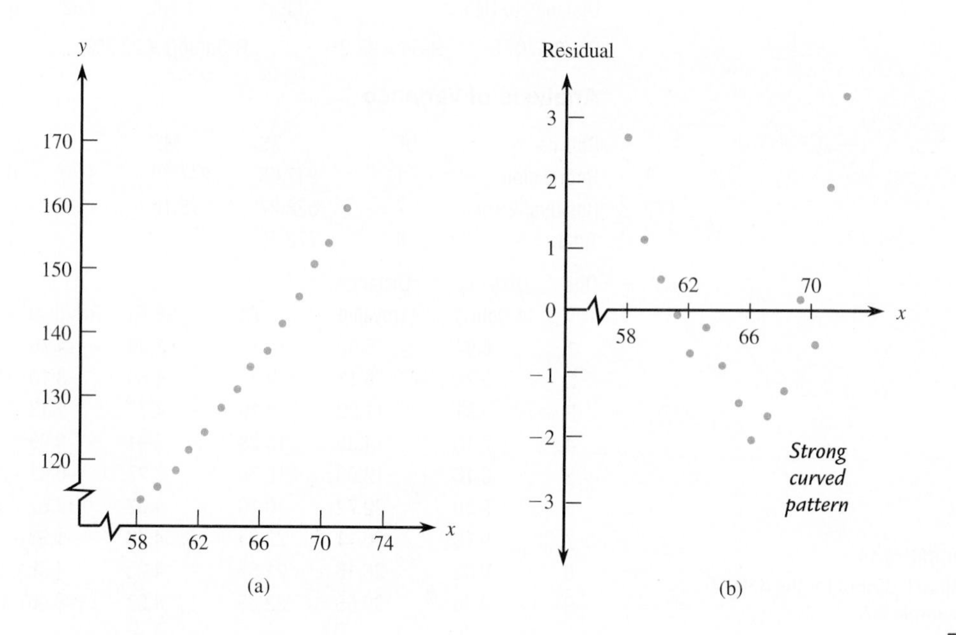

FIGURE 5.16
Plots for the data of Example 5.8:
(a) scatterplot;
(b) residual plot.

■

There is another common type of residual plot—one that plots the residuals versus the corresponding $\hat{y}$ values rather than versus the x values. Because $\hat{y} = a + bx$ is simply a linear function of x, the only real difference between the two types of residual plots is the scale on the horizontal axis. The pattern of points in the residual plots will be the same, and it is this pattern of points that is important, not the scale. The two plots give equivalent information, as can be seen in Figure 5.17, which gives both plots for the data of Example 5.7.

It is also important to look for unusual values in the scatterplot or in the residual plot. A point falling far above or below the horizontal line at height 0 corresponds to a large residual, which may indicate some unusual circumstance such as a recording error, a non-standard experimental condition, or an atypical experimental subject.

A point whose x value differs greatly from others in the data set may have exerted excessive influence in determining the fitted line. One method for assessing the impact of such an isolated point is to delete it from the data set, recompute the least-squares line, and evaluate the extent to which the equation of the line has changed.

● Data set available online

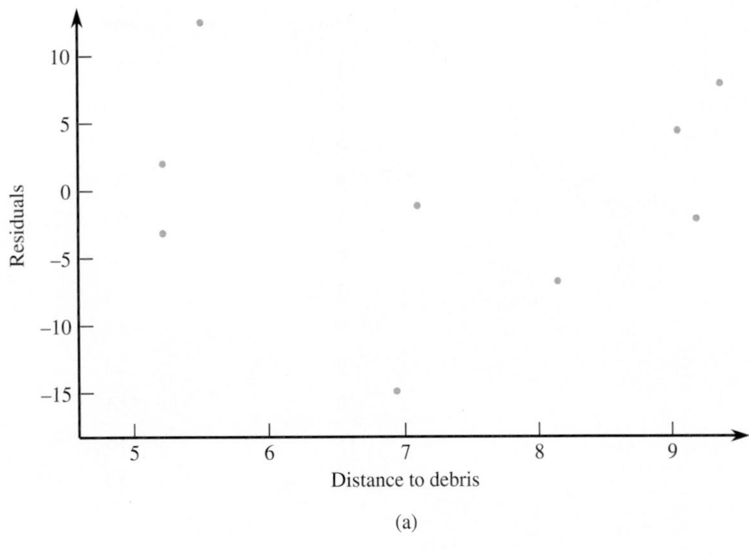

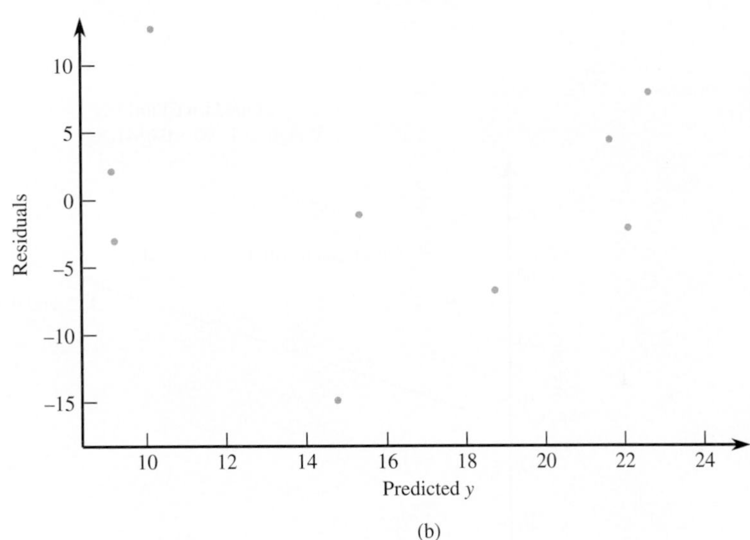

FIGURE 5.17
Plots for the data of Example 5.7.
(a) Plot of residuals versus x;
(b) plot of residuals versus $\hat{y}$.

EXAMPLE 5.9 Older Than Your Average Bear

Understand the context ❭

● The accompanying data on x = age (in years) and y = weight (in kg) for 12 black bears appeared in the paper **"Habitat Selection by Black Bears in an Intensively Logged Boreal Forest"** (***Canadian Journal of Zoology*** **[2008]: 1307–1316**).

A scatterplot and residual plot are shown in Figures 5.18(a) and 5.18(b), respectively. One bear in the sample was much older than the other bears (bear 3 with an age of x = 28.5 years and a weight of y = 62.00 kg). This results in a point in the scatterplot that is far to the right of the other points in the scatterplot.

Because the least-squares line minimizes the sum of squared residuals, the line is pulled toward this observation. This single observation plays a big role in determining the slope of the least-squares line, and it is therefore called an **influential observation**. Notice that this influential observation is not necessarily one with a large residual, because the least-squares line actually passes near this point. Figure 5.19 shows what happens when the influential observation is removed from the data set. Both the slope and intercept of the least-squares line are quite different from the slope and intercept of the line with this influential observation included.

● Data set available online

Bear	Age	Weight
1	10.5	54
2	6.5	40
3	28.5	62
4	10.5	51
5	6.5	55
6	7.5	56
7	6.5	62
8	5.5	42
9	7.5	40
10	11.5	59
11	9.5	51
12	5.5	50

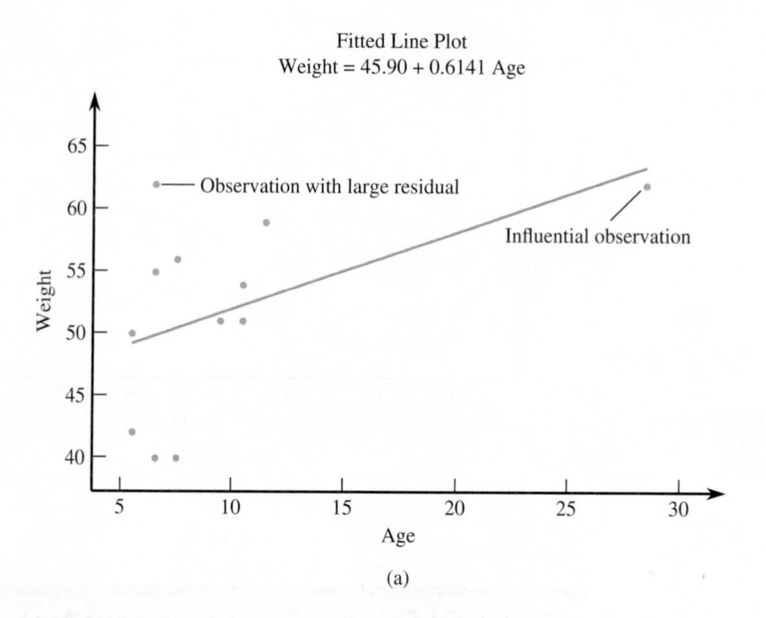

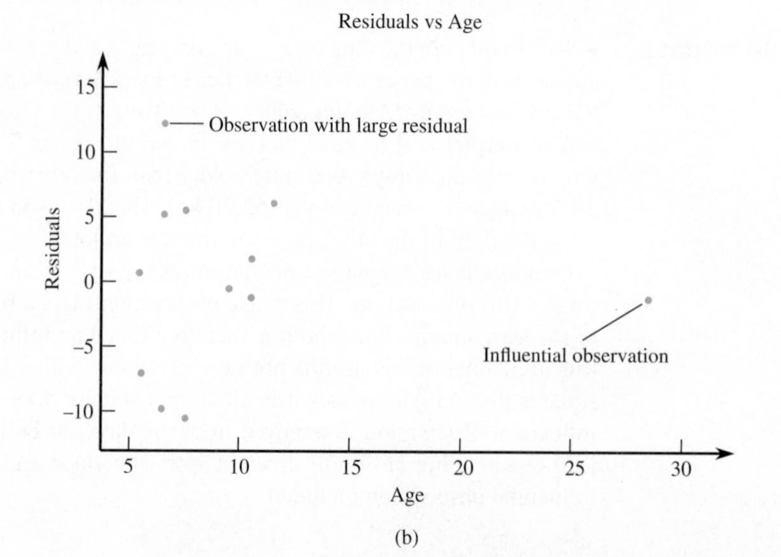

FIGURE 5.18
Plots for the bear data of Example 5.9:
(a) scatterplot;
(b) residual plot.

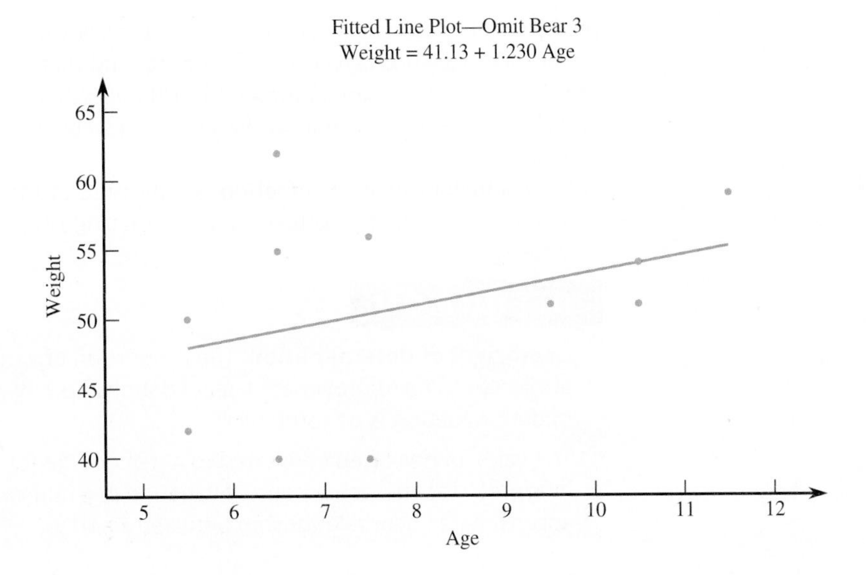

FIGURE 5.19
Scatterplot and least-squares line with bear 3 removed from data set.

Some points in a scatterplot may fall far from the least-squares line in the y direction, resulting in a large residual. These points are sometimes referred to as **outliers**. In Example 5.9, the observation with the largest residual is bear 7 with an age of $x = 6.5$ years and a weight of $y = 62.00$ kg. This observation is labeled in Figure 5.18. Even though this observation has a large residual, this observation is not very influential. The equation of the least-squares line for the data set consisting of all 12 observations is $\hat{y} = 45.90 + 0.6141x$, which is not much different from the equation that results from deleting bear 7 from the data set ($\hat{y} = 43.81 + 0.7131x$).

DEFINITION

Unusual points in a bivariate data set are those that fall away from most of the other points in the scatterplot in either the x direction or the y direction.

Influential observation: An observation is potentially influential if it has an x value that is far away from the rest of the data (separated from the rest of the data in the x direction). To determine if the observation is in fact influential, we assess whether removal of this observation has a large impact on the value of the slope or intercept of the least-squares line.

Outlier: An observation that has a large residual. Outlier observations fall far away from the least-squares line in the y direction.

Careful examination of a scatterplot and a residual plot can help us determine the appropriateness of a line for summarizing a relationship. If we decide that a line is appropriate, the next step is to think about assessing the accuracy of predictions based on the least-squares line and whether these predictions (based on the value of x) are better in general than those made without knowledge of the value of x. Two numerical measures that are helpful in this assessment are the coefficient of determination and the standard deviation about the regression line.

Coefficient of Determination, r^2

Suppose that we would like to predict the price of homes in a particular city. A random sample of 20 homes that are for sale is selected, and y = price and x = size (in square feet) are recorded for each house in the sample. There will be variability in house price (the houses will differ with respect to price), and it is this variability that makes accurate prediction of price a challenge. How much of the variability in house price can be explained by the fact that price is related to house size and that houses differ in size? If differences in size account for a large proportion of the variability in price, a price prediction that takes house size into account will be a big improvement over a prediction that is not based on size.

The **coefficient of determination** is a measure of the proportion of variability in the *y* variable that can be "explained" by a linear relationship between *x* and *y*.

DEFINITION

Coefficient of determination: The proportion of variation in *y* that can be attributed to an approximate linear relationship between *x* and *y*. The coefficient of determination is denoted by r^2.

The value of r^2 is often converted to a percentage (by multiplying by 100) and interpreted as the percentage of variation in *y* that can be explained by an approximate linear relationship between *x* and *y*.

To understand how r^2 is computed, we first consider variation in the *y* values. Variation in *y* can effectively be explained by an approximate straight-line relationship when the points in the scatterplot fall close to the least-squares line—that is, when the residuals are small in magnitude. A natural measure of variation about the least-squares line is the sum of the squared residuals. (Squaring before combining prevents negative and positive residuals from counteracting one another.) A second sum of squares assesses the total amount of variation in observed *y* values by considering how spread out the *y* values are from the mean *y* value.

DEFINITION

Total sum of squares: The **total sum of squares**, denoted by **SSTo**, is defined as

$$\text{SSTo} = (y_1 - \bar{y})^2 + (y_2 - \bar{y})^2 + \cdots + (y_n - \bar{y})^2 = \Sigma(y - \bar{y})^2$$

Residual sum of squares: The **residual sum of squares** (sometimes referred to as the error sum of squares), denoted by SSResid, is defined as

$$\text{SSResid} = (y_1 - \hat{y}_1)^2 + (y_2 - \hat{y}_2)^2 + \cdots + (y_n - \hat{y}_n)^2 = \Sigma(y - \hat{y})^2$$

These sums of squares can be found as part of the regression output from most standard statistical packages or can be obtained using the following computational formulas:

$$\text{SSTo} = \Sigma y^2 - \frac{(\Sigma y)^2}{n}$$

$$\text{SSResid} = \Sigma y^2 - a\Sigma y - b\Sigma xy$$

EXAMPLE 5.10 Revisiting the Deer Mice Data

Figure 5.20 displays part of the Minitab output that results from fitting the least-squares line to the data on *y* = distance traveled for food and *x* = distance to nearest woody debris pile from Example 5.7. From the output,

$$\text{SSTo} = 773.95 \text{ and } \text{SSResid} = 526.27$$

Notice that SSResid is fairly large relative to SSTo.

Regression Analysis: Distance Traveled versus Distance to Debris

The regression equation is
Distance Traveled = – 7.7 + 3.23 Distance to Debris

Predictor	Coef	SE Coef	T	P
Constant	−7.69	13.33	−0.58	0.582
Distance to Debris	3.234	1.782	1.82	0.112

S = 8.67071   R-Sq = 32.0%   R-Sq(adj) = 22.3%

Analysis of Variance

Source	DF	SS	MS	F	P
Regression	1	247.68	247.68	3.29	0.112
Residual Error	7	526.27	75.18		
Total	8	773.95			

SSTo   SSResid

FIGURE 5.20
Minitab output for the data of Example 5.10.

■

The residual sum of squares is the sum of squared vertical deviations from the least-squares line. As Figure 5.21 illustrates, SSTo is also a sum of squared vertical deviations from a line—the horizontal line at height $\bar{y}$. The least-squares line is, by definition, the one having the smallest sum of squared deviations. It follows that SSResid ≤ SSTo. The two sums of squares are equal only when the least-squares line *is* the horizontal line.

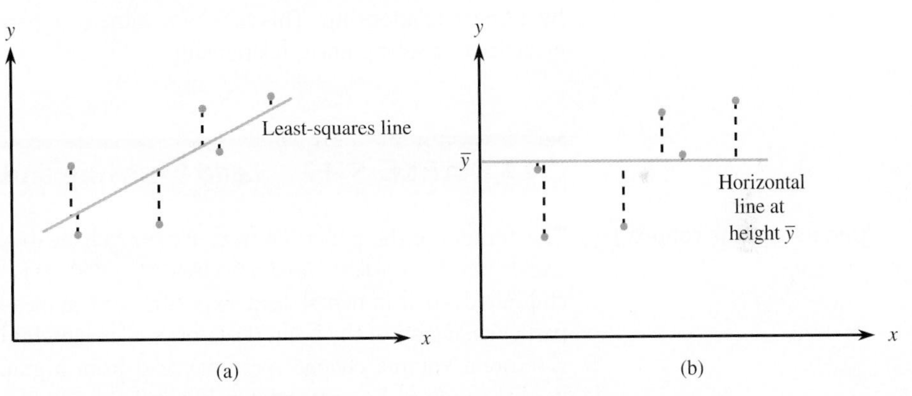

FIGURE 5.21
Interpreting sums of squares: (a) SSResid = sum of squared vertical deviations from the least-squares line; (b) SSTo = sum of squared vertical deviations from the horizontal line at $\bar{y}$.

SSResid is often referred to as a measure of unexplained variation—the amount of variation in y that cannot be attributed to the linear relationship between x and y. The more the points in the scatterplot deviate from the least-squares line, the larger the value of SSResid and the greater the amount of y variation that cannot be explained by the approximate linear relationship. Similarly, SSTo is interpreted as a measure of total variation. The larger the value of SSTo, the greater the amount of variability in $y_1, y_2, \ldots, y_n$.

The ratio SSResid/SSTo is the fraction or proportion of total variation that is unexplained by a straight-line relation. Subtracting this ratio from 1 gives the proportion of total variation that *is* explained:

DEFINITION

Coefficient of determination, r^2: The coefficient of determination is defined as

$$r^2 = 1 - \frac{\text{SSResid}}{\text{SSTo}}$$

Multiplying r^2 by 100 gives the percentage of y variation attributable to the approximate linear relationship. The closer this percentage is to 100%, the more successful the linear relationship is in explaining variation in y.

EXAMPLE 5.11 r^2 for the Deer Mice Data

Do the work ❯ For the data on distance traveled for food and distance to nearest debris pile from Example 5.10, we found SSTo = 773.95 and SSResid = 526.27. It follows that

$$r^2 = 1 - \frac{\text{SSResid}}{\text{SSTo}} = 1 - \frac{526.27}{773.95} = .32$$

Interpret the results ❯ This means that only 32% of the observed variability in distance traveled for food can be explained by an approximate linear relationship between distance traveled for food and distance to nearest debris pile. Note that the r^2 value can be found in the Minitab output of Figure 5.20, labeled "R-Sq." ■

The symbol r was used in Section 5.1 to denote Pearson's sample correlation coefficient. It is not a coincidence that r^2 is used to represent the coefficient of determination. The notation suggests how these two quantities are related:

(correlation coefficient)^2 = coefficient of determination

This means that if $r = .8$ or $r = -.8$, then $r^2 = .64$, so 64% of the observed variation in the dependent variable can be explained by the linear relationship. Because the value of r does not depend on which variable is labeled x, the same is true of r^2. The coefficient of determination is one of the few quantities computed in a regression analysis whose value remains the same when the roles of the dependent and independent variables are interchanged. When $r = .5$, we get $r^2 = .25$, so only 25% of the observed variation is explained by a linear relationship. This is why a value of r between $-.5$ and $.5$ is not considered evidence of a strong linear relationship.

EXAMPLE 5.12 Lead Exposure and Brain Volume

Understand the context ❯ The authors of the paper **"Decreased Brain Volume in Adults with Childhood Lead Exposure"** (***Public Library of Science Medicine*** **[May 27, 2008]: e112)** studied the relationship between childhood environmental lead exposure and a measure of brain volume change in a particular region of the brain. Data on x = mean childhood blood lead level (μg/dL) and y = brain volume change (percent) read from a graph that appeared in the paper was used to produce the scatterplot in Figure 5.22. The least-squares line is also shown on the scatterplot.

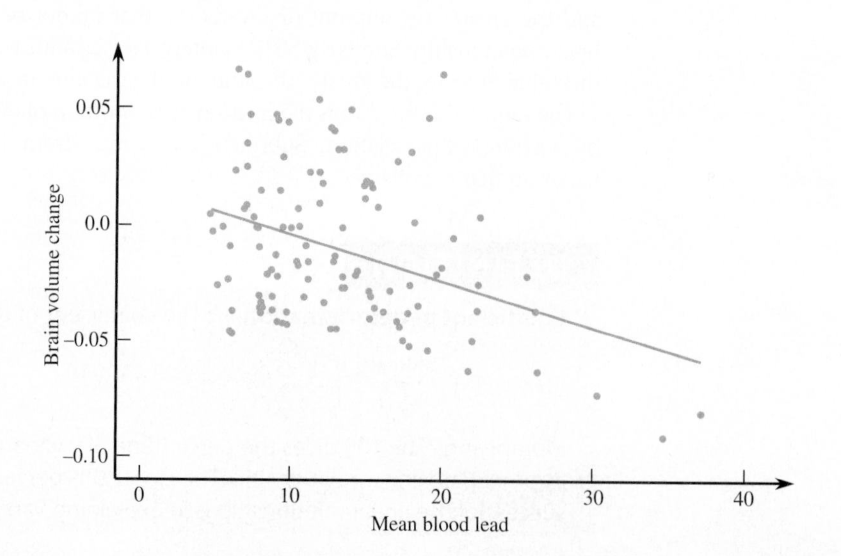

FIGURE 5.22
Scatterplot and least-squares line for the data of Example 5.12.

Regression Analysis: Brain Volume Change versus Mean Blood Lead

The regression equation is
Brain Volume Change = 0.01559 – 0.001993 Mean Blood Lead

S = 0.0310931 R-Sq = 13.6% R-Sq(adj) = 12.9%

Analysis of Variance

Source	DF	SS	MS	F	P
Regression	1	0.016941	0.0169410	17.52	0.000
Error	111	0.107313	0.0009668		
Total	112	0.124254			

FIGURE 5.23
Minitab output for the data of Example 5.12.

Interpret the results)

Figure 5.23 displays part of the Minitab output that results from fitting the least-squares line to the data. Notice that although there is a slight tendency for smaller y values (corresponding to a brain volume decrease) to be paired with higher values of mean blood lead levels, the relationship is weak. The points in the plot are widely scattered around the least-squares line.

From the computer output, we see that $100r^2 = 13.6\%$, so $r^2 = .136$. This means that differences in childhood mean blood lead level explain only 13.6% of the variability in adult brain volume change. Because the coefficient of determination is the square of the correlation coefficient, we can compute the value of the correlation coefficient by taking the square root of r^2. In this case, we know that the correlation coefficient will be negative (because there is a negative relationship between x and y), so we want the negative square root:

$$r = -\sqrt{.136} = -.369$$

Based on the values of the correlation coefficient and the coefficient of determination, we would conclude that there is a weak negative linear relationship and that childhood mean blood lead level explains only about 13.6% of adult change in brain volume. ■

Standard Deviation About the Least-Squares Line, s_e

The coefficient of determination measures the extent of variation about the least-squares line *relative* to overall variation in y. A large value of r^2 does not by itself promise that the deviations from the line are small in an absolute sense. A typical observation could deviate from the line by quite a bit, but these deviations might still be small relative to overall y variation.

Recall that in Chapter 4 the sample standard deviation

$$s = \sqrt{\frac{\Sigma (x - \bar{x})^2}{n - 1}}$$

was used as a measure of variability in a single sample. Roughly speaking, s is the typical amount by which a sample observation deviates from the sample mean. There is an analogous measure of variability when a least-squares line is fit.

DEFINITION

Standard deviation about the least-squares line: The standard deviation about the least-squares line is defined as

$$s_e = \sqrt{\frac{\text{SSResid}}{n - 2}}$$

Roughly speaking, s_e is the typical amount by which an observation deviates from the least-squares line.

The standard deviation about the least-squares line is given as part of the usual regression output when statistical software or a graphing calculator is used. For the brain volume data of Example 5.12, the value of s_e can be found in the Minitab output of Figure 5.23 labeled as "S." For the brain volume data, $s_e \approx 0.031$.

EXAMPLE 5.13 Predicting Graduation Rates

Understand the context)

● Consider the accompanying data on six-year graduation rate (%), student-related expenditure per full-time student, and median SAT score for the 38 primarily undergraduate public universities and colleges in the United States with enrollments between 10,000 and 20,000 (**Source: College Results Online, The Education Trust**).

Consider the data)

Graduation Rate	Expenditure	Median SAT	Graduation Rate	Expenditure	Median SAT
81.2	7462	1160	46.5	6744	1010
66.8	7310	1115	45.3	8842	1223
66.4	6959	1070	45.2	7743	990
66.1	8810	1205	43.7	5587	1010
64.9	7657	1135	43.5	7166	1010
63.7	8063	1060	42.9	5749	950
62.6	8352	1130	42.1	6268	955
62.5	7789	1200	42.0	8477	985
61.2	8106	1015	38.9	7076	990
59.8	7776	1100	38.8	8153	990
56.6	8515	990	38.3	7342	910
54.8	7037	1085	35.9	8444	1075
52.7	8715	1040	32.8	7245	885
52.4	7780	1040	32.6	6408	1060
52.4	7198	1105	32.3	4981	990
50.5	7429	975	31.8	7333	970
49.9	7551	1030	31.3	7984	905
48.9	8112	1030	31.0	5811	1010
48.1	8149	950	26.0	7410	1005

Figure 5.24 displays scatterplots of graduation rate versus student-related expenditure and graduation rate versus median SAT score. The least-squares lines and the values of r^2 and s_e are also shown.

Interpret the results)

Notice that while there is a positive linear relationship between student-related expenditure and graduation rate, the relationship is weak. The value of r^2 is only .119 (11.9%), indicating that only about 11.9% of the variability in graduation rate from university to university can be explained by student-related expenditures. The standard deviation about the regression line is $s_e = 12.2846$, which is larger than s_e for the predictor median SAT, a reflection of the fact that the points in the scatterplot of graduation rate versus student-related expenditure tend to fall farther from the regression line than is the case for the line that describes graduation rate versus median SAT.

The value of r^2 for graduation rate versus median SAT is .421 (42.1%) and $s_e = 9.95214$, indicating that the predictor median SAT does a better job of explaining variability in graduation rates. The corresponding least-squares line would be expected to produce more accurate estimates of graduation rates than would be the case for the predictor student-related expenditure.

Based on the values of r^2 and s_e, median SAT would be a better choice for predicting graduation rates than student-related expenditures. It is also possible to develop a prediction equation that would incorporate both potential predictors—techniques for doing this are introduced in Chapter 14.

● Data set available online

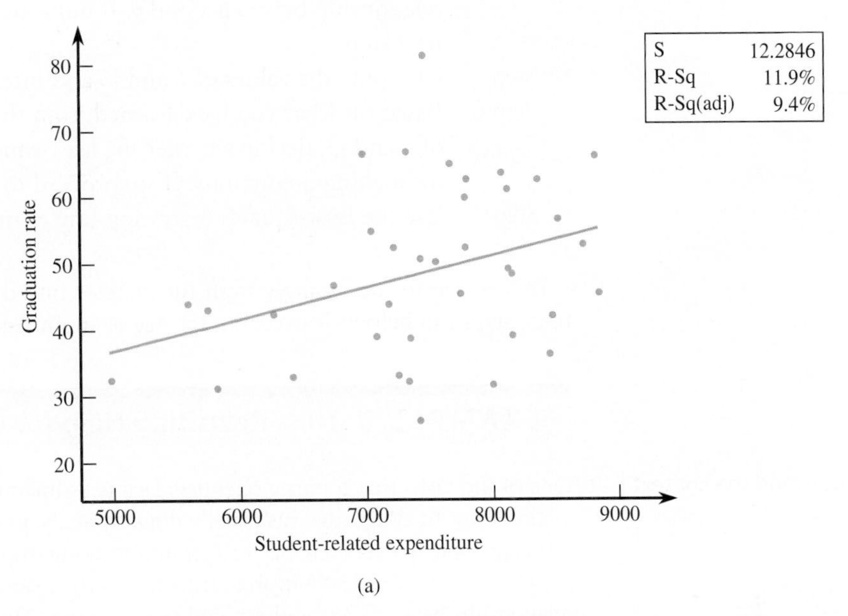

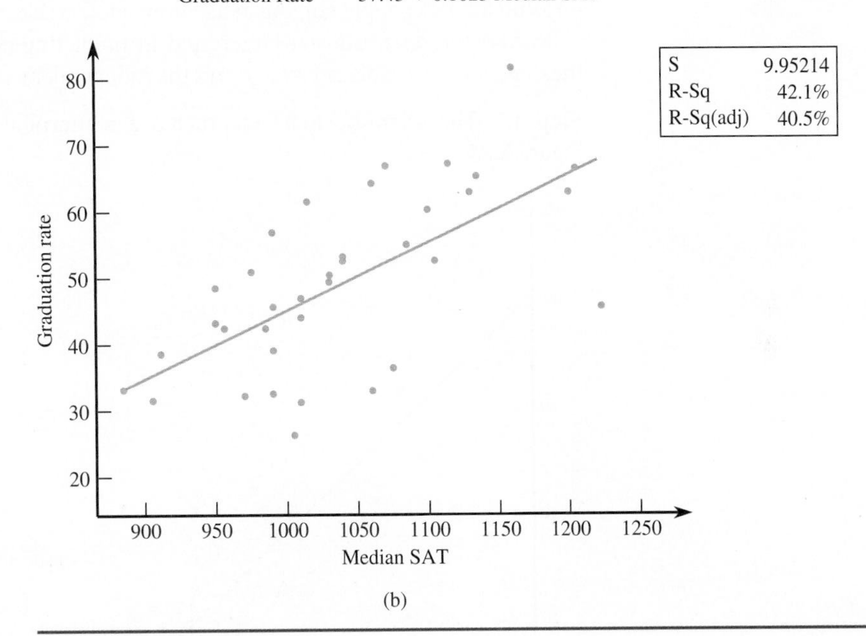

FIGURE 5.24
Scatterplots for the data of Example 5.13:
(a) graduation rate versus student-related expenditure;
(b) graduation rate versus median SAT.

■

Now that we have considered all of the parts of a linear regression analysis, let's put all the parts together. The steps in a linear regression analysis are summarized in the accompanying box.

Steps in a Linear Regression Analysis

Given a bivariate numerical data set consisting of observations on a dependent variable y and an independent variable x:

Step 1. Summarize the data graphically by constructing a scatterplot.
Step 2. Based on the scatterplot, decide if it looks like the relationship between x and y is approximately linear. If so, proceed to the next step.
Step 3. Find the equation of the least-squares regression line.
Step 4. Construct a residual plot and look for any patterns or unusual features that may indicate that a line is not the best way to summarize the

relationship between x and y. If none are found, proceed to the next step.

Step 5. Compute the values of s_e and r^2 and interpret them in context.

Step 6. Based on what you have learned from the residual plot and the values of s_e and r^2, decide whether the least-squares regression line is useful for making predictions. If so, proceed to the last step.

Step 7. Use the least-squares regression line to make predictions.

Let's return to the example from the chapter introduction and look at how following these steps can help us learn about the age of an unidentified crime victim.

EXAMPLE 5.14 Revisiting Help for Crime Scene Investigators

Understand the context

One of the tasks that forensic scientists face is estimating the age of an unidentified crime victim. Prior to 2010, this was usually done by analyzing teeth and bones, and the resulting estimates were not very reliable. In a groundbreaking study described in the paper **"Estimating Human Age from T-Cell DNA Rearrangements" (*Current Biology* [2010])**, scientists examined the relationship between age and a blood test measure. They recorded age and the blood test measure for 195 people ranging in age from a few weeks to 80 years.

Because the scientists were interested in predicting age using the blood test measure, the dependent variable is y = age and the independent variable is x = blood test measure.

Step 1: The scientists first constructed a scatterplot of the data, which is shown in Figure 5.25.

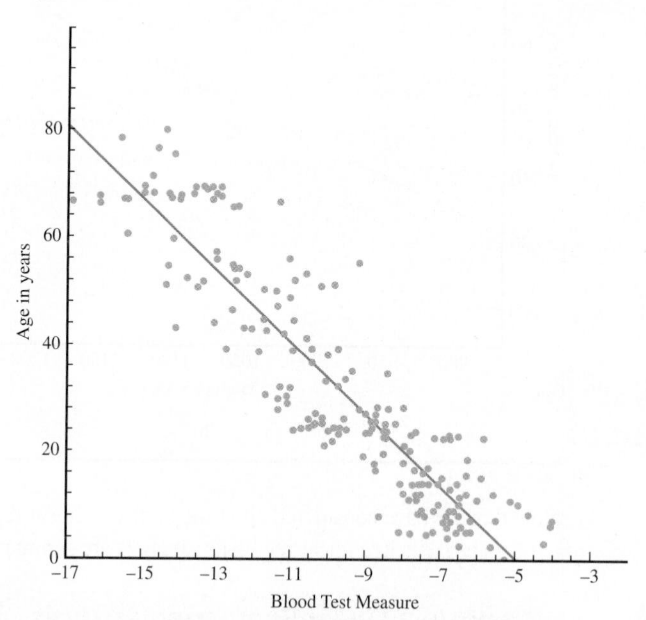

FIGURE 5.25
Scatterplot of age versus blood test measure.

Step 2: Based on the scatterplot, it does appear that there is a reasonably strong negative linear relationship between age and the blood test measure. The scientists also reported that the correlation coefficient for this data set was $r = -0.92$, which is consistent with the strong negative linear pattern in the scatterplot.

Step 3: The scientists calculated the equation of the least-squares regression line to be

$$\hat{y} = -33.65 - 6.74x$$

Step 4: A residual plot constructed from these data showed a few observations with large residuals, but these observations were not far removed from the rest of the data in the x direction. These observations were not judged to be influential. Also, there were no unusual patterns in the residual plot that would suggest a nonlinear relationship between age and the blood test measure.

Step 5: The values of s_e and r^2 were

$$s_e = 8.9 \text{ and } r^2 = 0.835$$

This means that approximately 83.5% of the variability in age can be explained by the linear relationship between age and blood test measure. The value of s_e tells us that if the least-squares regression line is used to predict age from the blood test measure, a typical difference between the predicted age and the actual age would be about 9 years.

Step 6: Based on the residual plot, the large value of r^2, and the relatively small value of s_e, the scientists proposed using the blood test measure and the least-squares regression line as a way to estimate ages of crime victims.

Step 7: To illustrate predicting age, suppose that a blood sample is take from an unidentified crime victim and that the value of the blood test measure is determined to be -10. The predicted age of the victim would be

$$\begin{aligned}\hat{y} &= -33.65 - 6.74(-10)\\ &= -33.65 - (-67.4)\\ &= 33.75 \text{ years}\end{aligned}$$

This is just an estimate of the actual age, and a typical prediction error is about 9 years. ■

EXERCISES 5.29 - 5.43

5.29 Does it pay to stay in school? The report ***Trends in Higher Education* (The College Board, 2010)** looked at the median hourly wage gain per additional year of schooling in 2007. The report states that workers with a high school diploma had a median hourly wage that was 10% higher than those who had only completed 11 years of school. Workers who had completed 1 year of college (13 years of education) had a median hourly wage that was 11% higher than that of the workers who had completed only 12 years of school. The added gain in median hourly wage for each additional year of school is shown in the accompanying table. The entry for 15 years of schooling has been intentionally omitted from the table.

Years of Schooling	2007 Median Hourly Wage Gain for the Additional Year (percent)
12	10
13	11
14	13
16	16
17	18
18	19

a. Use the given data to predict the median hourly wage gain for the 15th year of schooling.

b. The actual wage gain for 15th year of schooling was 14%. How close was the actual value to the predicted wage gain percent from Part (a)?

5.30 ● The data in the accompanying table is from the paper **"Six-Minute Walk Test in Children and Adolescents" (*The Journal of Pediatrics* [2007]: 395–399)**. Two hundred and eighty boys completed a test that measures the distance that the subject can walk on a flat, hard surface in 6 minutes. For each age group shown in the table, the median distance walked by the boys in that age group is also given.

Age Group	Representative Age (Midpoint of Age Group)	Median Six-minute Walk Distance (meters)
3–5	4	544.3
6–8	7	584.0
9–11	10	667.3
12–15	13.5	701.1
16–18	17	727.6

a. With x = representative age and y = median distance walked in 6 minutes, construct a scatterplot. Does the pattern in the scatterplot look linear?

b. Find the equation of the least-squares regression line that describes the relationship between median distance walked in 6 minutes and representative age.

c. Compute the five residuals and construct a residual plot. (Hint: See Examples 5.7 and 5.8)

d. Are there any unusual features in the plot?

5.31 ● The paper referenced in the previous exercise also gave the 6-minute walk distances for 248 girls age 3 to 18 years. The median 6-minute walk times for girls for the five age groups were

492.4 578.3 655.8 657.6 660.9

a. With x = representative age and y = median distance walked in 6 minutes, construct a scatterplot.

b. How does the pattern in the scatterplot for girls differ from the pattern in the scatterplot for boys from Exercise 5.30?

c. Find the equation of the least-squares regression line that describes the relationship between median distance walked in 6 minutes and representative age for girls.

d. Compute the five residuals and construct a residual plot.

e. The authors of the paper decided to use a curve rather than a straight line to describe the relationship between median distance walked in 6 minutes and age for girls. What aspect of the residual plot supports this decision? (Hint: See Example 5.8.)

5.32 ● Northern flying squirrels eat lichen and fungi, which makes for a relatively low quality diet. The authors of the paper **"Nutritional Value and Diet Preference of Arboreal Lichens and Hypogeous Fungi for Small Mammals in the Rocky Mountains" (*Canadian Journal of Zoology* [2008]: 851–862)** measured nitrogen intake and nitrogen retention in six flying squirrels that were fed the fungus *Rhizopogon*.

Data read from a graph that appeared in the paper are given in the table below. (The negative value for nitrogen retention for the first squirrel represents a net loss in nitrogen.)

Nitrogen Intake, x (grams)	Nitrogen Retention, y (grams)
0.03	−0.04
0.10	0.00
0.07	0.01
0.06	0.01
0.07	0.04
0.25	0.11

a. Construct a scatterplot of these data.

b. Find the equation of the least-squares regression line.

c. Based on this line, what would you predict nitrogen retention to be for a flying squirrel whose nitrogen intake is 0.06 gram?

d. What is the residual associated with the observation (0.06, 0.01)?

e. Look again at the scatterplot from Part (a). Which observation is potentially influential? Explain the reason for your choice. (Hint: See Example 5.9.)

f. When the potentially influential observation is deleted from the data set, the equation of the least-squares regression line fit to the remaining five observations is $\hat{y} = -0.037 + 0.627x$. Use this equation to predict nitrogen retention for a flying squirrel whose nitrogen intake is 0.06. Is this prediction much different from the prediction made in Part (c)?

5.33 Some types of algae have the potential to cause damage to river ecosystems. The accompanying data on algae colony density (y) and rock surface area (x) for nine rivers is a subset of data that appeared in a scatterplot in a paper in the journal *Aquatic Ecology* (2010: 33–40).

x	50	55	50	79	44	37	70	45	49
y	152	48	22	35	38	171	13	185	25

a. Compute the equation of the least-squares regression line.

b. What is the value of r^2 for this data set? Write a sentence interpreting this value in context. (Hint: See Example 5.12.)

c. What is the value of s_e for this data set? Write a sentence interpreting this value in context. (Hint: See Example 5.13.)

d. Is the linear relationship between rock surface area and algae colony density positive or negative? Is it weak, moderate, or strong? Justify your answer.

5.34 ● The relationship between x = total number of salmon in a creek and y = percentage of salmon killed by bears that were transported away from the stream prior to the bear eating the salmon was examined in the paper **"Transportation of Pacific Salmon Carcasses from Streams to Riparian Forests by Bears" (*Canadian Journal of Zoology* [2009]: 195–203).**

Data for the 10 years from 1999 to 2008 is given in the accompanying table.

Total Number	Percentage Transported
19,504	77.8
3,460	28.7
1,976	28.9
8,439	27.9
11,142	55.3
3,467	20.4
3,928	46.8
20,440	76.3
7,850	40.3
4,134	24.1

a. Construct a scatterplot of the data.

b. Does there appear to be a relationship between the total number of salmon in the stream and the percentage of salmon killed by bears that are transported away from the stream?

c. Find the equation of the least-squares regression line. Draw the regression line on the scatterplot from Part (a).

d. The residuals from the least-squares line are shown in the accompanying table. The observation (3928, 46.8) has a large residual. Is this data point also an influential observation?

Total Number	Percent Transported	Residual
19,504	77.8	3.43
3,460	28.7	0.30
1,976	28.9	4.76
8,439	27.9	−14.76
11,142	55.3	4.89
3,467	20.4	−8.02
3,928	46.8	17.06
20,440	76.3	−0.75
7,850	40.3	−0.68
4,134	24.1	−6.23

e. The two points with unusually large x values (19,504 and 20,440) were not thought to be influential observations even though they are far removed in the x direction from the rest of the points in the scatterplot. Explain why these two points are not influential.

f. Partial Minitab output resulting from fitting the least-squares line is shown here. What is the value of s_e? Write a sentence interpreting this value.

Regression Analysis: Percent Transported versus Total Number

The regression equation is
Percent Transported = 18.5 + 0.00287 Total Number

Predictor	Coef	SE Coef	T	P
Constant	18.483	4.813	3.84	0.005
Total Number	0.0028655	0.0004557	6.29	0.000

S = 9.16217 R-Sq = 83.2% R-Sq(adj) = 81.1%

g. What is the value of r^2 for this data set (see Minitab output in Part (f))? Is the value of r^2 large or small? Write a sentence interpreting the value of r^2.

5.35 ● The paper **"Effects of Age and Gender on Physical Performance" (*Age* [2007]: 77–85)** describes a study of the relationship between age and 1-hour swimming performance. Data on age and swim distance for over 10,000 men participating in a national long-distance 1-hour swimming competition are summarized in the accompanying table.

Age Group	Representative Age (Midpoint of Age Group)	Average Swim Distance (meters)
20–29	25	3913.5
30–39	35	3728.8
40–49	45	3579.4
50–59	55	3361.9
60–69	65	3000.1
70–79	75	2649.0
80–89	85	2118.4

a. Find the equation of the least-squares line with x = representative age and y = average swim distance.

b. Compute the seven residuals and use them to construct a residual plot.

c. What does the residual plot suggest about the appropriateness of using a line to describe the relationship between representative age and swim distance?

d. Would it be reasonable to use the least-squares line from Part (a) to predict the average swim distance for women age 40 to 49 by substituting the representative age of 45 into the equation of the least-squares line? Explain.

5.36 ● Cost-to-charge ratio (the percentage of the amount billed that represents the actual cost) for inpatient and outpatient services at 11 Oregon hospitals is shown in the following table **(Oregon Department of Health Services, 2002)**. A scatterplot of the data is also shown.

Hospital	Cost-to-Charge Ratio: Outpatient Care	Cost-to-Charge Ratio: Inpatient Care
1	62	80
2	66	76
3	63	75
4	51	62
5	75	100
6	65	88
7	56	64
8	45	50
9	48	54
10	71	83
11	54	100

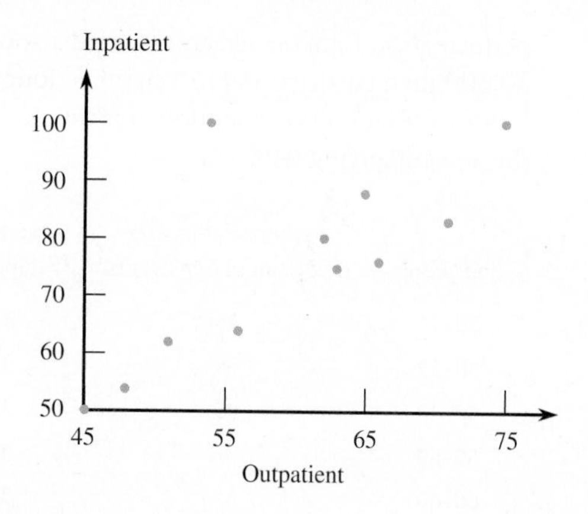

The least-squares regression line with y = inpatient cost-to-charge ratio and x = outpatient cost-to-charge ratio is $\hat{y} = -1.1 + 1.29x$.

a. Is the observation for Hospital 11 an influential observation? Justify your answer.

b. Is the observation for Hospital 11 an outlier? Explain.

c. Is the observation for Hospital 5 an influential observation? Justify your answer.

d. Is the observation for Hospital 5 an outlier? Explain.

5.37 The article **"Examined Life: What Stanley H. Kaplan Taught Us About the SAT" (*The New Yorker* [December 17, 2001]: 86–92)** included a summary of findings regarding the use of SAT I scores, SAT II scores, and high school grade point average (GPA) to predict first-year college GPA. The article states that "among these, SAT II scores are the best predictor, explaining 16 percent of the variance in first-year college grades. GPA was second at 15.4 percent, and SAT I was last at 13.3 percent."

a. If the data from this study were used to fit a least-squares line with y = first-year college GPA and x = high school GPA, what would be the value of r^2?

b. The article stated that SAT II was the best predictor of first-year college grades. Do you think that predictions based on a least-squares line with y = first-year college GPA and x = SAT II score would be very accurate? Explain why or why not.

5.38 ● The paper **"Accelerated Telomere Shortening in Response to Life Stress" (*Proceedings of the National Academy of Sciences* [2004]: 17312–17315)** described a study that examined whether stress accelerates aging at a cellular level. The accompanying data on a measure of perceived stress (x) and telomere length (y) were read from a scatterplot that appeared in the paper. Telomere length is a measure of cell longevity.

Perceived Stress	Telomere Length	Perceived Stress	Telomere Length
5	1.25	20	1.22
6	1.32	20	1.30
6	1.5	20	1.32
7	1.35	21	1.24
10	1.3	21	1.26
11	1	21	1.30
12	1.18	22	1.18
13	1.1	22	1.22
14	1.08	22	1.24
14	1.3	23	1.18
15	0.92	24	1.12
15	1.22	24	1.50
15	1.24	25	0.94
17	1.12	26	0.84
17	1.32	27	1.02
17	1.4	27	1.12
18	1.12	28	1.22
18	1.46	29	1.30
19	0.84	33	0.94

a. Compute the equation of the least-squares line.

b. What is the value of r^2?

c. Does the linear relationship between perceived stress and telomere length account for a large or small proportion of the variability in telomere length? Justify your answer.

5.39 ● The article **"California State Parks Closure List Due Soon" (*The Sacramento Bee,* August 30, 2009)** gave the following data on y = number of employees in fiscal year 2007–2008 and x = total size of parks (in acres) for the 20 state park districts in California:

Number of Employees, y	Total Park Size, x
95	39,334
95	324
102	17,315
69	8,244
67	620,231
77	43,501
81	8,625
116	31,572
51	14,276
36	21,094
96	103,289
71	130,023

continued

Number of Employees, y	Total Park Size, x
76	16,068
112	3,286
43	24,089
87	6,309
131	14,502
138	62,595
80	23,666
52	35,833

a. Construct a scatterplot of the data.

b. Find the equation of the least-squares line.

c. Do you think the least-squares line gives accurate predictions? Explain.

d. Delete the observation with the largest x value from the data set and recalculate the equation of the least-squares line. Does this observation greatly affect the equation of the line?

5.40 ● The article referenced in the previous exercise also gave data on the percentage of operating costs covered by park revenues for the 2007–2008 fiscal year.

Number of Employees, x	Percent of Operating Cost Covered by Park Revenues, y
95	37
95	19
102	32
69	80
67	17
77	34
81	36
116	32
51	38
36	40
96	53
71	31
76	35
112	108
43	34
87	97
131	62
138	36
80	36
52	34

a. Find the equation of the least-squares line relating y = percent of operating costs covered by park revenues and x = number of employees.

b. Based on the values of r^2 and s_e, do you think that the least-squares regression line does a good job of describing the relationship between y = percent of operating costs covered by park revenues and x = number of employees? Explain.

c. The graph in Figure EX-5.40 is a scatterplot of y = percent of operating costs covered by park revenues and x = number of employees. The least-squares line is also shown. Which observations are outliers? Do the observations with the largest residuals correspond to the park districts with the largest number of employees?

5.41 A study was carried out to investigate the relationship between the hardness of molded plastic (y, in Brinell units) and the amount of time elapsed since the plastic was molded (x, in hours). Summary quantities include $n = 15$, SSResid = 1235.470,

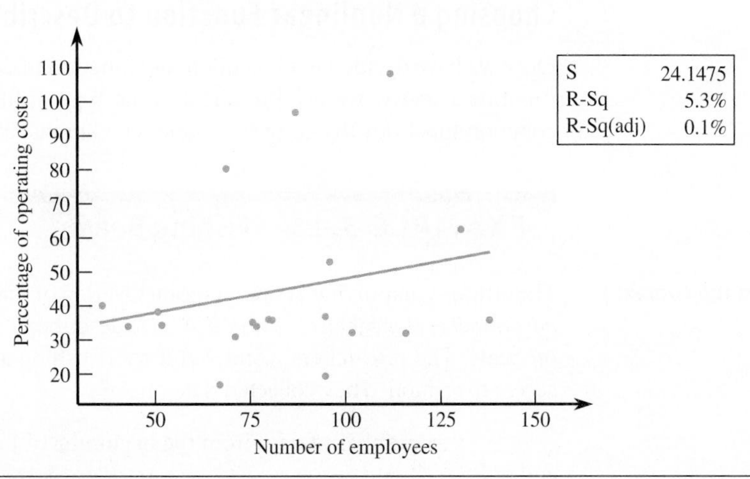

FIGURE EX-5.40

and SSTo = 25,321.368. Calculate and interpret the coefficient of determination.

5.42 Both r^2 and s_e are used to assess the fit of a line.

a. Is it possible that both r^2 and s_e could be large for a bivariate data set? Explain. (A picture might be helpful.)

b. Is it possible that a bivariate data set could yield values of r^2 and s_e that are both small? Explain. (Again, a picture might be helpful.)

c. Explain why it is desirable to have r^2 large and s_e small if the relationship between two variables x and y is to be described using a straight line.

5.43 With a bit of algebra, we can show that

$$\text{SSResid} = (1 - r^2)\Sigma(y - \bar{y})^2$$

from which it follows that

$$s_e = \sqrt{\frac{n-1}{n-2}}\sqrt{1 - r^2}\, s_y$$

Unless n is quite small, $(n - 1)/(n - 2) \approx 1$, so

$$s_e \approx \sqrt{1 - r^2}\, s_y$$

a. For what value of r is s_e as large as s_y? What is the least-squares line in this case?

b. For what values of r will s_e be much smaller than s_y?

c. A study by the Berkeley Institute of Human Development (see the book *Statistics* by Freedman et al., listed in the back of the book) reported the following summary data for a sample of $n = 66$ California boys:

$r \approx .80$

At age 6, average height $\approx$ 46 inches, standard deviation $\approx$ 1.7 inches.

At age 18, average height $\approx$ 70 inches, standard deviation $\approx$ 2.5 inches.

What would s_e be for the least-squares line used to predict 18-year-old height from 6-year-old height?

d. Referring to Part (c), suppose that you wanted to predict the past value of 6-year-old height from knowledge of 18-year-old height. Find the equation for the appropriate least-squares line. What is the corresponding value of s_e?

Bold exercises answered in back ● Data set available online ▼ Video Solution available

5.4 Nonlinear Relationships and Transformations

When the points in a scatterplot exhibit a linear pattern and the residual plot does not reveal any problems with the linear fit, the least-squares line is a sensible way to summarize the relationship between two numerical variables. But what should we do if a scatterplot or residual plot exhibits a curved pattern, indicating a more complicated relationship between x and y? When this happens, we consider using a nonlinear function to summarize the relationship between x and y. The process we will use is similar to what we have done for linear relationships, but now we will use a curve instead of a line. After choosing an appropriate curve, we will evaluate the fit of the curve and use a residual plot to assess whether the curve is an appropriate way to describe the relationship. If we decide the curve is a useful summary of the relationship between x and y, we can then use the curve to make predictions.

Choosing a Nonlinear Function to Describe a Relationship

Once we have decided to use a nonlinear function to describe the relationship between two variables x and y, we need to decide what type of function should be considered. Some common functions that are used to model relationships are shown in Figure 5.26.

EXAMPLE 5.15 Fishing Bears

Understand the context ❯

The article "Quantifying Spatiotemporal Overlap of Alaskan Brown Bears and People" (***Journal of Wildlife Management* [2005]: 810–817**) describes a study of the effect of human activity on bears. The researchers wondered if sport fishing and boating might be limiting bears' access to salmon. They collected data on

x = number of days from the beginning of June 2003 (June 1st = 1)

and

y = total fishing time (in bear-hours) for the day at a particular location in Alaska

	Function	Equation	Looks Like
(a)	Quadratic	$\hat{y} = a + b_1x + x^2$	
(b)	Square Root	$\hat{y} = a + b\sqrt{x}$	
(c)	Reciprocal	$\hat{y} = a + b\left(\frac{1}{x}\right)$	

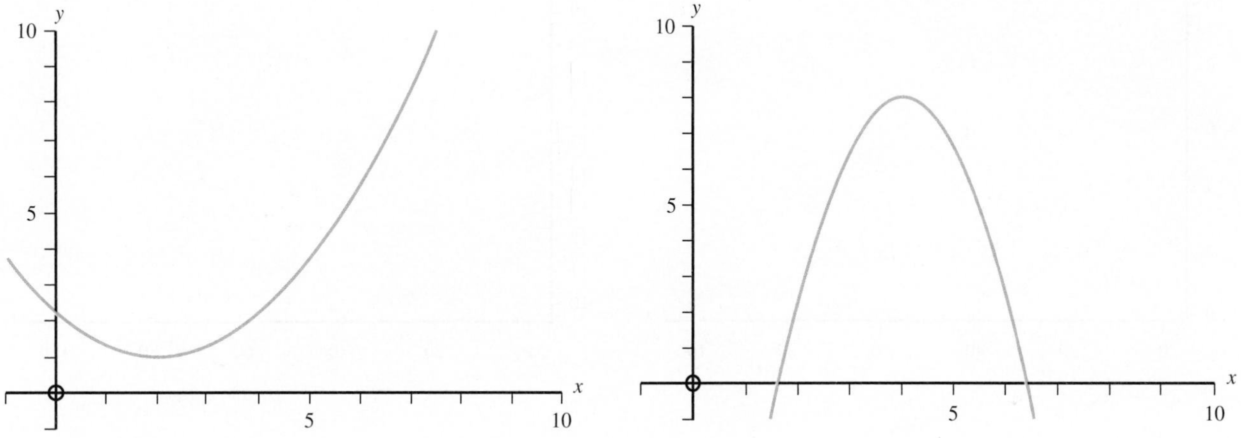

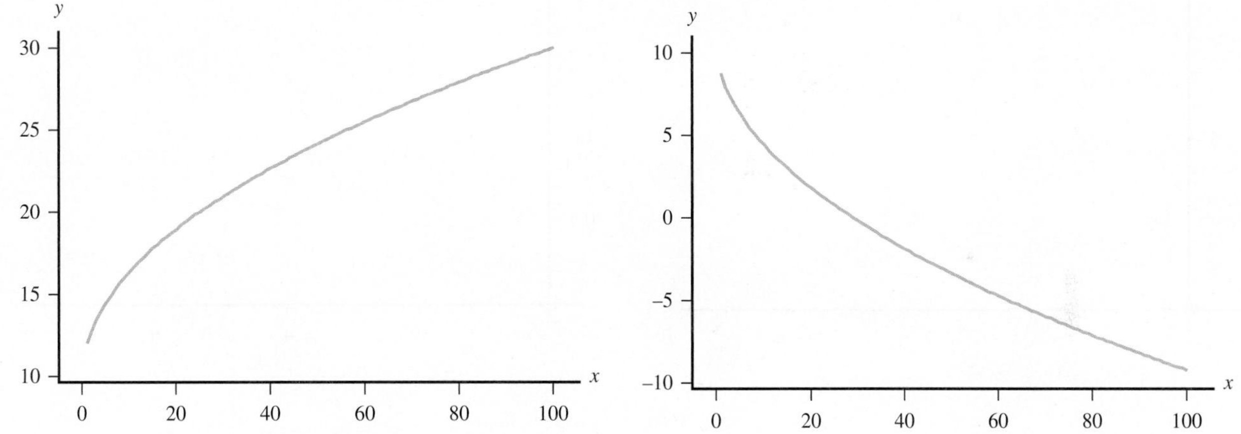

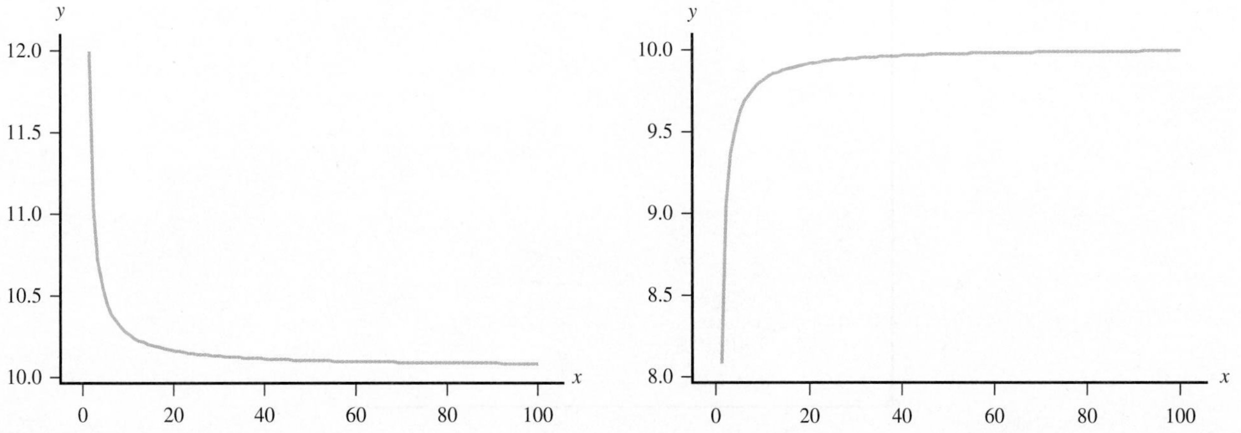

FIGURE 5.26
Common nonlinear functions

continued

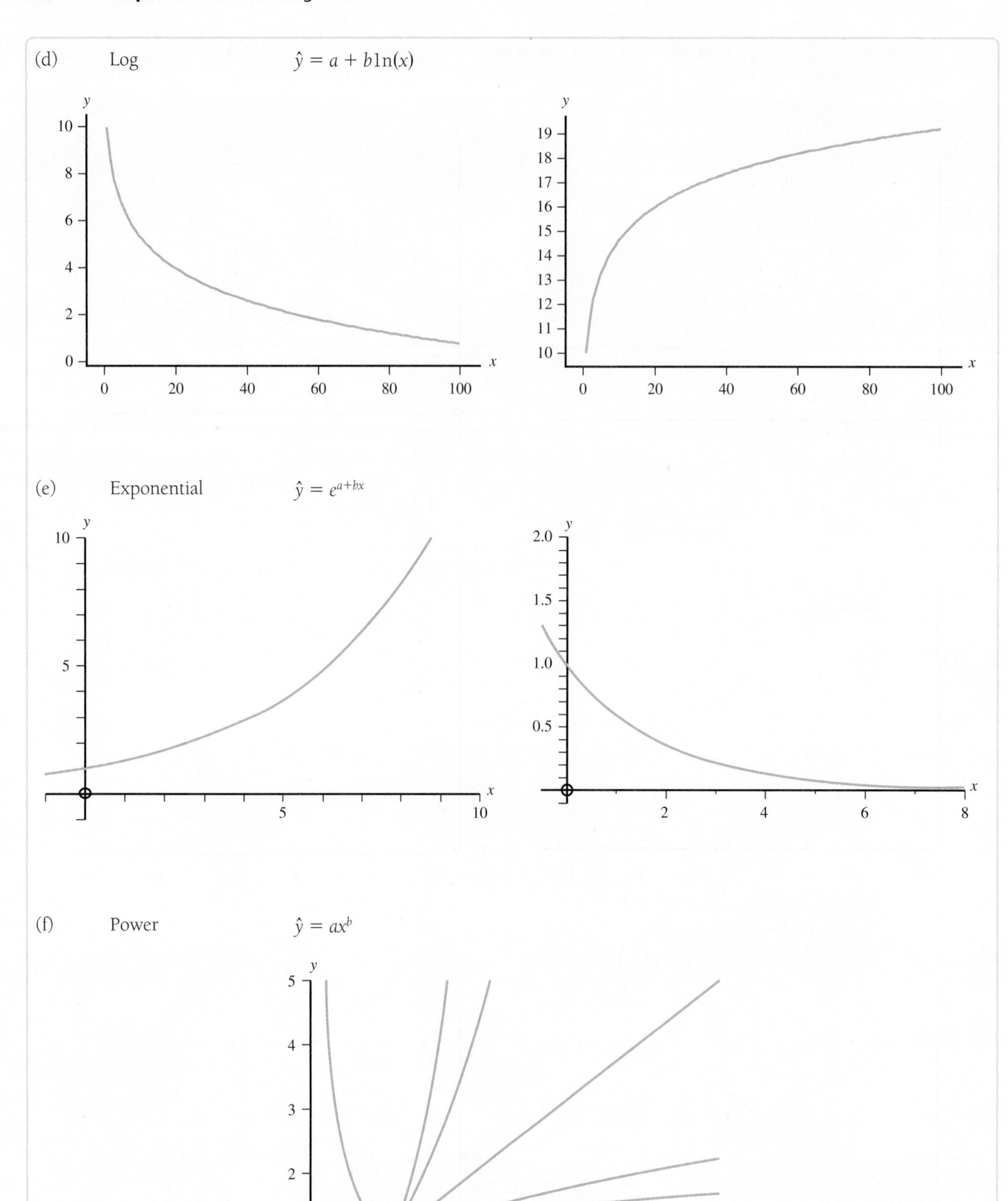

FIGURE 5.26
Common nonlinear functions

For example, the data pair (33, 45.7) corresponds to the day that is 32 days after June 1 (which is July 2). On this day, the total fishing time for bears was 45.7 bear-hours. This is the sum of the number of hours spent fishing for all bears fishing at this location on July 2, 2003.

Consider the data)

Date (x) (June 1 = 1)	Fishing Time (bear-hours)	Date (x) (June 1 = 1)	Fishing Time (bear-hours)	Date (x) (June 1 = 1)	Fishing Time (bear-hours)
11	11.3	24	11.3	40	22.0
12	15.1	25	18.4	41	26.0
13	6.6	27	16.2	44	10.5
14	12.9	28	19.5	45	18.6
15	12.1	31	35.8	46	21.1
17	18.1	32	37.1	49	11.9
18	20.9	33	45.7	50	13.7
19	17.6	36	34.8	51	13.7
20	11.0	37	25.6	54	6.3
21	24.6	38	26.7	55	1.8

Figure 5.27 is a scatterplot of these data.

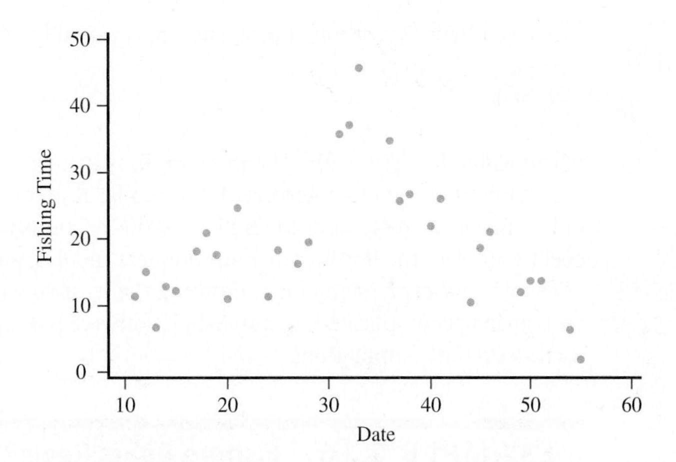

FIGURE 5.27
Scatterplot of fishing time versus date

Formulate a plan)

The pattern in the scatterplot is clearly nonlinear. Because the shape of the pattern looks like a quadratic curve (see Figure 5.26(a)), a quadratic function could be used to describe this relationship. The model to consider is

$$\hat{y} = a + b_1x + b_2x^2$$

■

Once we have decided on a potential nonlinear model, such as quadratic or exponential, the next step is to estimate the coefficients in the model. For example, for the quadratic model $\hat{y} = a + b_1x + b_2x^2$, we would need to use the sample data to determine appropriate values for a, b_1, and b_2. The process used to estimate the coefficients in a nonlinear model is a bit different for quadratic models than for the other nonlinear models listed in Figure 5.26. Let's begin by considering quadratic models.

Quadratic Regression Models

The general form of the quadratic regression model is $\hat{y} = a + b_1x + b_2x^2$. What are the best choices for the values of a, b_1, and b_2? In fitting a line to data, the principle of least-squares was used to determine the values of the slope and intercept for the "best fit" line. Least-squares can also be used to fit a quadratic function. The deviations, $y - \hat{y}$, are still represented by vertical distances in the scatterplot, but now they are vertical distances from the points to a parabola (the graph of a quadratic function) rather than to a line, as shown in Figure 5.28. Values for the coefficients in the quadratic function are then chosen to make the sum of the squared deviations as small as possible.

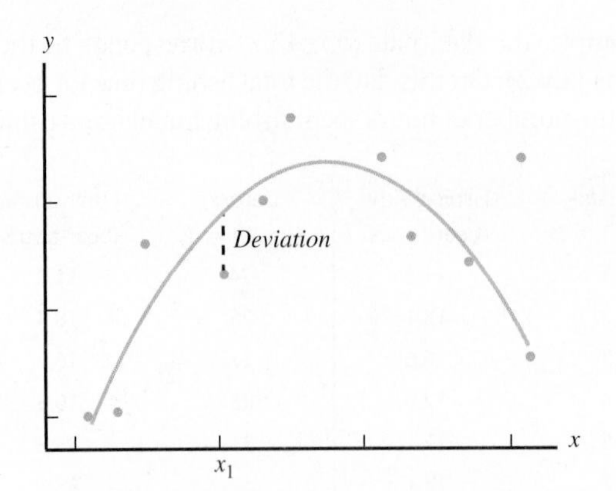

FIGURE 5.28
Deviation for a quadratic function

For a quadratic regression, the least-squares estimates of a, b_1, and b_2 are those values that minimize the sum of squared deviations, $\Sigma(y - \hat{y})^2$, where $\hat{y} = a + b_1x + b_2x^2$.

For quadratic regression, a measure that is useful for assessing fit is

$$R^2 = 1 - \frac{SSResid}{SSTo}$$

where $SSResid = \Sigma(y - \hat{y})^2$. The measure R^2 is defined in a way similar to r^2 for the linear regression model and is interpreted in a similar fashion. The lowercase notation r^2 is used only with linear regression to emphasize the relationship between r^2 and the correlation coefficient, r, in the linear case. For nonlinear models, an uppercase R^2 is used.

The general expressions for computing the least-squares quadratic regression estimates are somewhat complicated, so a statistical software package or a graphing calculator will be used to do the computations.

EXAMPLE 5.16 Fishing Bears Revisited—Fitting a Quadratic Model

For the bear fishing data of Example 5.15, the scatterplot showed a curved pattern. If a least-squares line is fit to these data, it is not surprising that the line does not do a good job of describing the relationship ($r^2 = 0.000$, and $s_e = 10.1$). Both the scatterplot and the residual plot show a distinct curved pattern (see Figure 5.29).

FIGURE 5.29
Plots for the bear fishing data of Examples 5.15 and 5.16:
(a) least-squares regression line;
(b) residual plot for linear model.

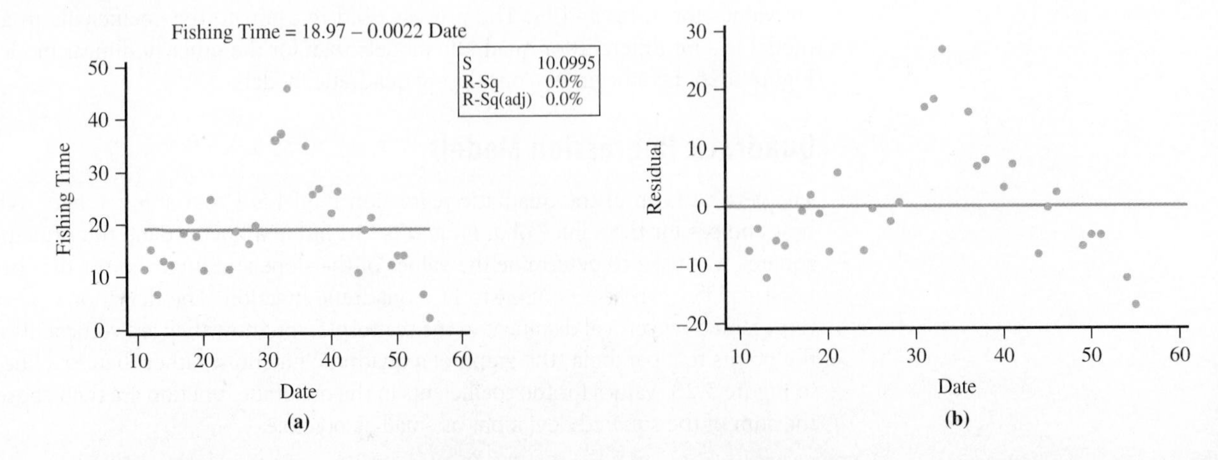

Do the work ❯ JMP output resulting from fitting the least-squares quadratic regression to these data is shown in Figure 5.30.

Polynomial Fit Degree=2

Usage = −20.96713 + 2.9957567 * Date − 0.0463151 * Date^2

Summary of Fit

RSquare	0.555541
RSquare Adj	0.522618
Root Mean Square Error	6.856681
Mean of Response	18.89667
Observations (or Sum Wgts)	30

Analysis of Variance

Parameter Estimates

Term	Estimate	Std Error	t Ratio	Prob>[t]
Intercept	−20.96713	7.567052	−2.77	0.0100*
Date	2.9957567	0.524223	5.71	<.0001*
Date^2	−0.046315	0.007973	−5.81	<.0001*

FIGURE 5.30
JMP output for the quadratic regression of Example 5.16.

From the JMP output the least-squares quadratic is

$$\hat{y} = -20.967 + 2.996x - 0.046x^2$$

Interpret the results ❯ The curve and the corresponding residual plot for the quadratic regression are shown in Figure 5.31. Notice that there is no strong pattern in this residual plot, unlike the linear case. For the quadratic regression, $R^2 = 0.556$ (as opposed to essentially zero for the linear model). This means that 55.6% of the variability in the bear fishing time can be explained by an approximate quadratic relationship between fishing time and date.

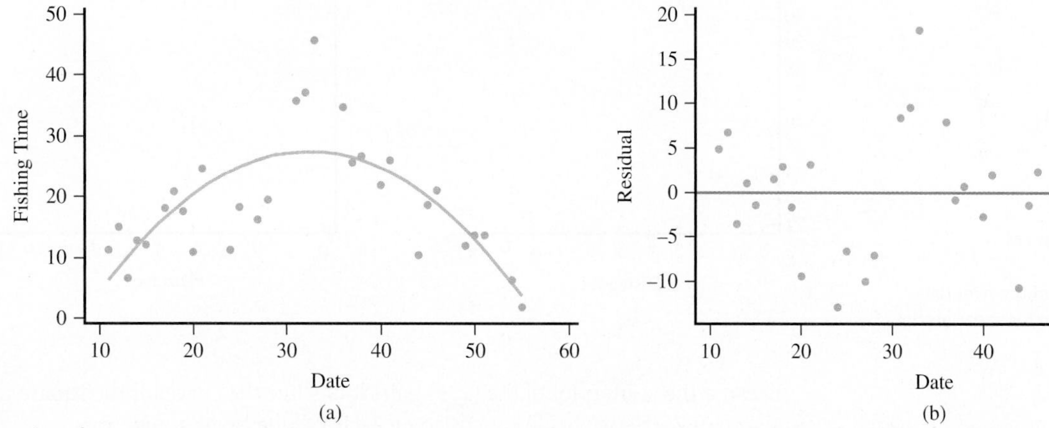

FIGURE 5.31
Quadratic regression of Example 5.16:
(a) scatterplot with curve;
(b) residual plot for quadratic model.

To use the quadratic regression model to make a prediction, we substitute a value of x into the regression equation. For example, the predicted fishing time for June 18 ($x = 18$) is

$$\begin{aligned}\hat{y} &= -20.967 + 2.996x - 0.046x^2 \\ &= -20.967 + 2.996(18) - 0.046(18)^2 \\ &= -20.967 + 53.928 - 14.904 \\ &= 18.057 \text{ bear-hours}\end{aligned}$$

■

Other Nonlinear Regression Models: Using Transformations

Now that we have seen how to use a quadratic regression model to describe relationships that look like the curves in Figure 5.26(a), let's consider some other nonlinear models. The square root function of Figure 5.26(b) is

$$\hat{y} = a + b\sqrt{x}$$

Notice that if you define a new variable x' where

$$x' = \sqrt{x}$$

the square root model $\hat{y} = a + b\sqrt{x}$ can be written as $\hat{y} = a + bx'$, so y is a linear function of x'. This suggests that if a scatterplot of (x, y) pairs looks like the curve in Figure 5.26(b), a scatterplot of (x', y) pairs should look linear. You can then use a line to describe the relationship between y and x'—something you already know how to do. This is illustrated in the following example.

EXAMPLE 5.17 River Water Velocity and Distance from Shore

Understand the context)

As fans of white-water rafting know, a river flows more slowly close to its banks (because of friction between the riverbank and the water). To study the nature of the relationship between water velocity and the distance from the shore, data were gathered on velocity (in centimeters per second) of a river at different distances (in meters) from the bank. Suppose that the resulting data were as follows:

Consider the data)

Distance	0.5	1.5	2.5	3.5	4.5	5.5	6.5	7.5	8.5	9.5
Velocity	22.00	23.18	25.48	25.25	27.15	27.83	28.49	28.18	28.50	28.63

A scatterplot of the data exhibits a curved pattern, as seen in both the scatterplot (Figure 5.32(a)) and the residual plot from a linear fit (Figure 5.32(b)).

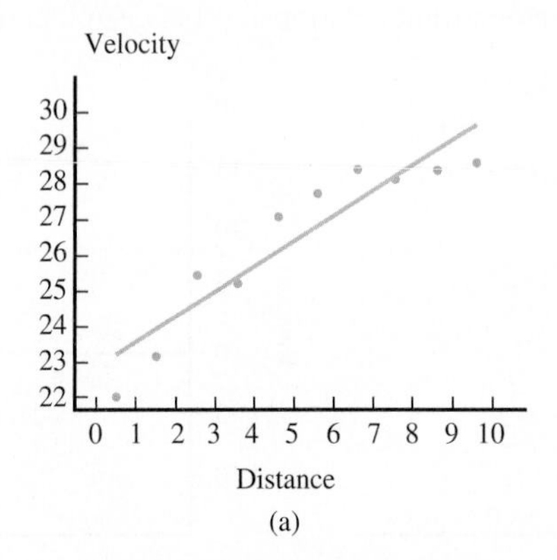

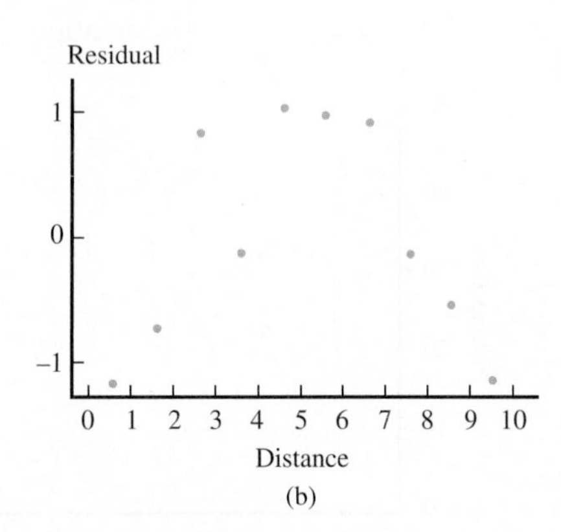

FIGURE 5.32
Plots for the data of Example 5.17:
(a) scatterplot of the river data;
(b) residual plot for linear model.

Formulate a plan)

Because the scatterplot of the (x, y) pairs looks like the curve for the square root function in Figure 5.26(b), we begin by replacing each x value by its square root:

$$x' = \sqrt{x}$$

This is called transforming the x values. The original and transformed data are shown in Table 5.2.

Do the work)

TABLE 5.2 Original and Transformed Data of Example 5.17

Original Data		Transformed Data		Original Data		Transformed Data	
x	y	x'	y	x	y	x'	y
0.5	22.00	0.7071	22.00	5.5	27.83	2.3452	27.83
1.5	23.18	1.2247	23.18	6.5	28.49	2.5495	28.49
2.5	25.48	1.5811	25.48	7.5	28.18	2.7386	28.18
3.5	25.25	1.8708	25.25	8.5	28.50	2.9155	28.50
4.5	27.15	2.1213	27.15	9.5	28.63	3.0822	28.63

Figure 5.33(a) shows a scatterplot of y versus x' (or equivalently y versus $\sqrt{x}$). The pattern of points in this plot looks linear, and so you can fit the least-squares regression line using the transformed data. The Minitab output from this regression is as follows:

Regression Analysis

The regression equation is

velocity = 20.1 + 3.01 sqrt distance

Predictor	Coef	SE Coef	T	P
Constant	20.1103	0.6097	32.99	0.000
sqrt distance	3.0085	0.2726	11.03	0.000

S = 0.629242 R-Sq = 93.8% R-Sq(adj) = 93.1%

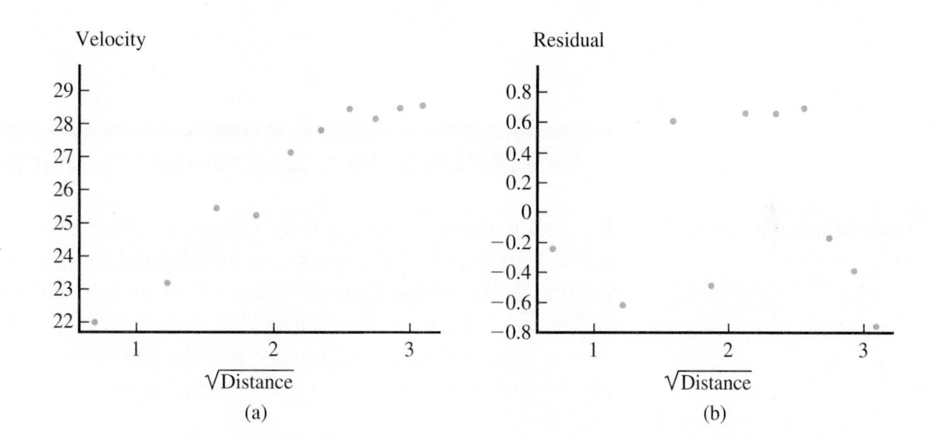

FIGURE 5.33 Plots for the transformed data of Example 5.17: (a) scatterplot of y versus x'; (b) residual plot resulting from a linear fit to the transformed data.

The resulting regression equation is

$$\hat{y} = 20.1 + 3.01x'$$

or, equivalently,

$$\hat{y} = 20.1 + 3.01\sqrt{x}$$

Interpret the results ⟩

This model is evaluated using R^2, s_e, and a residual plot. The residual plot (shown in Figure 5.33(b)) does not have any patterns or unusual features. The values of R^2 and s_e (see the Minitab output) indicate that a line is a reasonable way to describe the relationship between y and x'.

To predict velocity of the river at a distance of 9 meters from shore, we first compute $x' = \sqrt{x} = \sqrt{9} = 3$ and then use the least-squares regression line to obtain a predicted y value:

$$\hat{y} = 20.1 + 3.01x' = 20.1 + (3.01)(3) = 29.13 \text{ cm per second}$$

■

A general strategy for fitting a nonlinear model is to find a way to transform the x and/or y values so that a scatterplot of the transformed data has a linear appearance. A **transformation** (sometimes called a re-expression) involves using a simple function of a variable in place of the variable itself (like $x' = \sqrt{x}$ in Example 5.17). Common transformations involve taking square roots, logarithms, or reciprocals.

The nonlinear models (other than the quadratic model) given in Figure 5.26 (square root, reciprocal, log, exponential, and power) can all be fit by transforming variables and then fitting a line to the transformed data. It is convenient to separate these models into two types:

1. Models that involve transforming *only* the x variable (square root, reciprocal, and log).
2. Models that involve transforming the y variable (exponential and power).

Models That Involve Transforming Only *x*

The square root, reciprocal, and log models all have the form

$$\hat{y} = a + b(\text{some function of } x)$$

where the function of x is square root, reciprocal, or log. Notice that these models all describe a linear relationship between y and a function of x. This suggests that if the pattern in the scatterplot of (x, y) pairs looks like one of the curves in Figure 5.26(b)–(d), an appropriate transformation of the x values should result in transformed data that shows a linear pattern.

Model	Transformation
Square root	$x' = \sqrt{x}$
Reciprocal	$x' = \frac{1}{x}$
Log	$x' = \ln(x)$

EXAMPLE 5.18 Electromagnetic Radiation

Understand the context ⟩

Is electromagnetic radiation from phone antennae associated with declining bird populations? This is one of the questions investigated by the authors of the paper "The Urban Decline of the House Sparrow (***Passer domesticus***): A Possible Link with Electromagnetic Radiation" (***Electromagnetic Biology and Medicine*** **[2007]: 141–151**). The accompanying data on x = electromagnetic field strength (Volts per meter) and y = sparrow density (sparrows per hectare) were read from a graph in the paper.

Consider the data ⟩

Field Strength	Sparrow Density	Field Strength	Sparrow Density
0.11	41.71	0.90	16.29
0.20	33.60	1.20	16.97
0.29	24.74	1.30	12.83
0.40	19.50	1.41	13.17
0.50	19.42	1.50	4.64
0.61	18.74	1.80	2.11
1.01	24.23	1.90	0.00
1.10	22.04	3.01	0.00
0.70	16.29	3.10	14.69
0.80	14.69	3.41	0.00

A scatterplot of these data is shown in Figure 5.34.

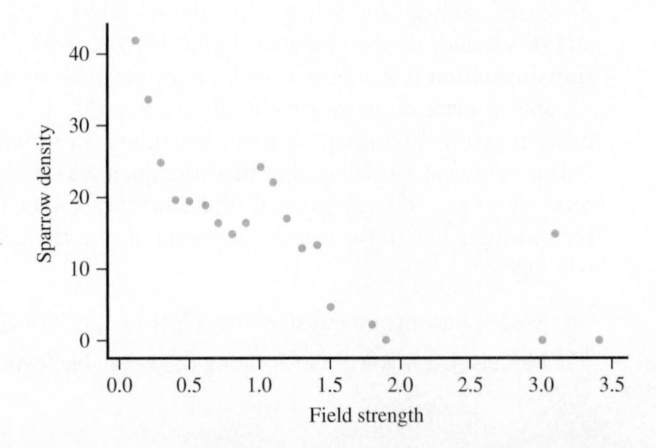

FIGURE 5.34
Scatterplot of the sparrow data.

Formulate a plan ⟩

Notice that there is a curved pattern in the scatterplot that has a shape that is similar to the first graph in Figure 5.26(d). This suggests that a reasonable choice for describing the relationship between sparrow density and field strength is a log model

$\hat{y} = a + b\ln(x)$

To fit this model, begin by transforming the x values using

$x' = \ln(x)$

Do the work ⟩

In this text, natural logs (denoted by ln or $\log_e$) are used, but in practice, either the natural log or log base 10 (denoted by log or $\log_{10}$) can be used. The transformed data is given here.

ln(Field Strength)	Sparrow Density	ln(Field Strength)	Sparrow Density
−2.207	41.71	−0.105	16.29
−1.609	33.60	0.182	16.97
−1.238	24.74	0.262	12.83
−0.916	19.50	0.344	13.17
−0.693	19.42	0.405	4.64
−0.494	18.74	0.588	2.11
0.001	24.23	0.642	0.00
0.095	22.04	1.102	0.00
−0.357	16.29	1.131	14.69
−0.223	14.69	1.227	0.00

A scatterplot of the (x', y) pairs is shown in Figure 5.35. The pattern in this plot looks linear, so you can find the least-squares regression line using the transformed data. From the resulting Minitab output (Figure 5.36), the equation of the least-squares regression line is

$$\hat{y} = 14.8 - 10.5x'$$

or, in terms of x

$$\hat{y} = 14.8 - 10.5\ln(x)$$

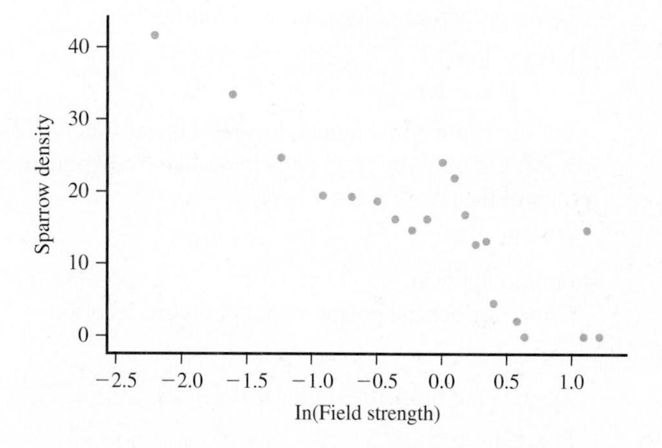

FIGURE 5.35
Scatterplot of the transformed data of Example 5.18.

Regression Analysis: Sparrow Density Versus ln (Field Strength)

The regression equation is
Sparrow Density = 14.8 – 10.5 ln (Field Strength)

Predictor	Coef	SE Coef	T	P
Constant	14.805	1.238	11.96	0.000
ln (Field Strength)	−10.546	1.389	−7.59	0.000

S = 5.50641 R–Sq = 76.2% R–Sq(adj) = 74.9%

FIGURE 5.36
Minitab output for the log model of Example 5.18.

Interpret the results

The value of R^2 for this model is 0.762 and $s_e = 5.5$. A residual plot from the least-squares regression line fit to the transformed data is shown in Figure 5.37. There are no apparent patterns or unusual features in the residual plot, so it appears that the log model is a reasonable choice for describing the relationship between sparrow density and field strength.

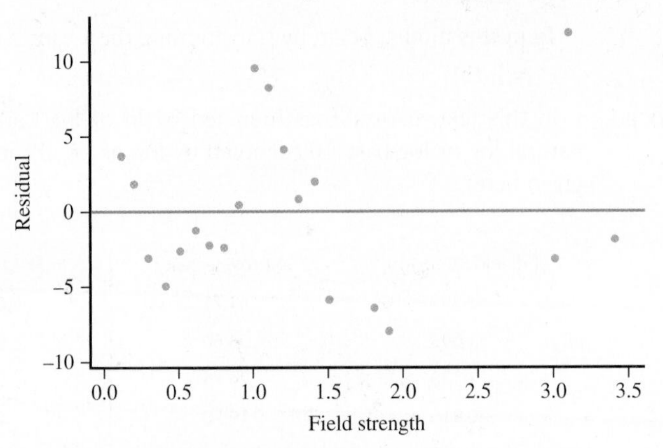

FIGURE 5.37
Residual plot for the log model fit to the sparrow data of Example 5.18.

This model can now be used to predict sparrow density from field strength. For example, if the field strength is 1.6 Volts per meter, we would predict that sparrow density would be

$$\begin{aligned}\hat{y} &= 14.8 - 10.5x' \\ &= 14.8 - 10.5\ln(x) \\ &= 14.8 - 10.5\ln(1.6) \\ &= 14.8 - 10.5\,(0.470) \\ &= 9.685 \text{ sparrows per hectare}\end{aligned}$$

■

Models That Involve Transforming *y*

There are two nonlinear models in Figure 5.26 that have not yet been considered—the exponential and power models. Let's begin by considering the exponential model

$$y = e^{a+bx}$$

Using properties of logarithms, it follows that

$$\begin{aligned}\ln(y) &= \ln e^{a+bx} \\ &= a + bx\end{aligned}$$

For the exponential model, $\ln(y)$ is a linear function of x. This means that if the pattern in a scatterplot of the (x, y) pairs looks like the exponential curves of Figure 5.26(e), a scatterplot of the (x, y') pairs, where

$$y' = \ln y$$

should look linear.

Now consider the power model of Figure 5.26(f),

$$y = ax^b$$

Again using properties of logarithms,

$$\begin{aligned}\ln y &= \ln(ax^b) \\ &= \ln a + \ln x^b \\ &= \ln a + b\ln x\end{aligned}$$

If a power model is an appropriate way to describe the relationship between x and y, a linear model would describe the relationship between $x' = \ln x$ and $y' = \ln y$.

Model	Transformation
Exponential	$y' = \ln y$
Power	$x' = \ln x$ $y' = \ln y$

EXAMPLE 5.19 Loons in Acidic Lakes

Understand the context ›

A study of factors that affect the survival of loon chicks is described in the paper "Does Prey Biomass or Mercury Exposure Affect Loon Chick Survival in Wisconsin?" (***The Journal of Wildlife Management*** **[2005]: 57–67**). In this study, a relationship between the pH of lake water and blood mercury level in loon chicks was observed.

The researchers thought that it is possible that the pH of the lake water could be related to the type of fish that the loons ate. The accompanying data (Table 5.3) on x = lake pH and y = blood mercury level (mg/g) for 37 loon chicks from different lakes in Wisconsin were read from a graph in the paper. A scatterplot is shown in Figure 5.38(a).

Consider the data ›

TABLE 5.3 Data and Transformed Data from Example 5.19

Lake pH (x)	Blood Mercury Level (y)	y′ = ln(y)	Lake pH (x)	Blood Mercury Level (y)	y′ = ln(y)	Lake pH (x)	Blood Mercury Level (y)	y′ = ln(y)
5.28	1.10	0.0953	6.17	0.55	−0.5978	7.03	0.12	−2.1203
5.69	0.76	−0.2744	6.22	0.43	−0.8440	7.20	0.15	−1.8971
5.56	0.74	−0.3011	6.15	0.40	−0.9163	7.89	0.11	−2.2073
5.51	0.60	−0.5108	6.05	0.33	−1.1087	7.93	0.11	−2.2073
4.90	0.48	−0.7340	6.04	0.26	−1.3471	7.99	0.09	−2.4079
5.02	0.43	−0.8440	6.24	0.18	−1.7148	7.99	0.06	−2.8134
5.02	0.29	−1.2379	6.30	0.16	−1.8326	8.30	0.09	−2.4079
5.04	0.09	−2.4079	6.80	0.45	−0.7985	8.42	0.09	−2.4079
5.30	0.10	−2.3026	6.58	0.30	−1.2040	8.42	0.04	−3.2189
5.33	0.20	−1.6094	6.65	0.28	−1.2730	8.95	0.12	−2.1203
5.64	0.28	−1.2730	7.06	0.22	−1.5141	9.49	0.14	−1.9661
5.83	0.17	−1.7720	6.99	0.21	−1.5606			
5.83	0.18	−1.7148	6.97	0.13	−2.0402			

Formulate a plan ›

The pattern in this scatterplot is typical of exponential decay and looks like the second curve in Figure 5.26(e). The change in y as x increases is much smaller for large x values than for small x values. We can see that a change of 1 in pH is associated with a much larger change in blood mercury level in the part of the plot where the x values are small than in the part of the plot where the x values are large.

FIGURE 5.38
Scatterplots for the loon data of Example 5.19:
(a) scatterplot oforiginal data;
(b) scatterplot of transformed data with $y' = \ln y$.

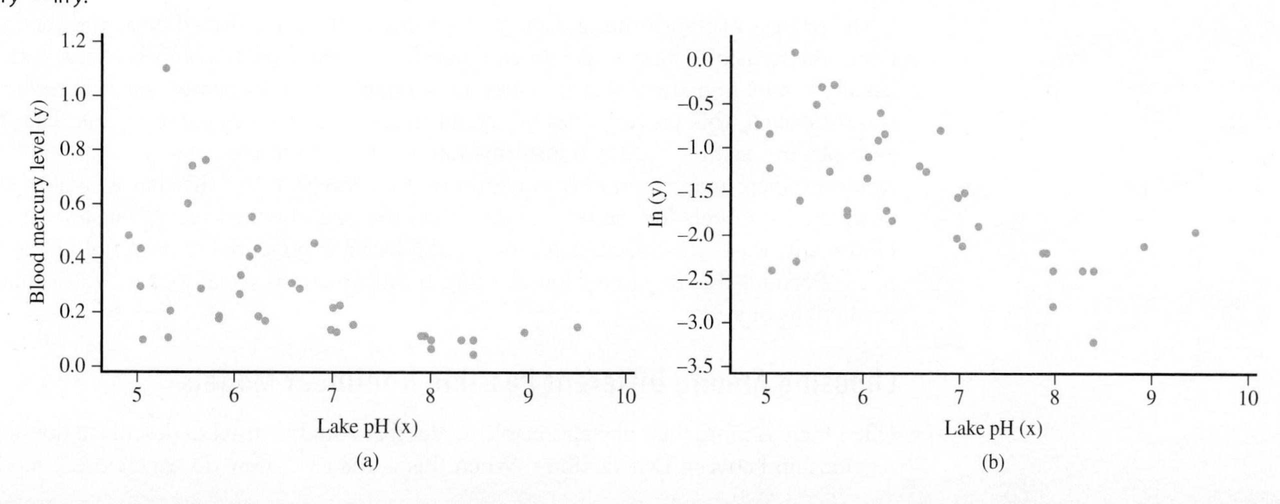

To fit an exponential model, begin by transforming the y values using $y' = \ln y$. In addition to the original data, Table 5.3 also displays the transformed data. A scatterplot of the (x, y') pairs is shown in Figure 5.38(b). The pattern in the scatterplot of the transformed data looks linear, so it is reasonable to use the least-squares regression line to describe the relationship between y' and x.

The following Minitab output shows the result of fitting the least-squares line to the transformed data:

Do the work

The regression equation is

ln(y) = 1.06 − 0.396 Lake pH (x)

Predictor	Coef	SE Coef	T	P
Constant	1.0550	0.5535	1.91	0.065
Lake pH (x)	−0.39564	0.08264	−4.79	0.000

S = 0.605645 R-Sq = 39.6% R-Sq(adj) = 37.8%

The resulting least-squares regression line is

$$\hat{y}' = 1.06 - 0.396x$$

or equivalently

$$\widehat{\ln y} = 1.06 - 0.396x$$

■

Notice that the linear model in Example 5.19 could be used to predict values of ln(y). To translate this linear model back to the form of the exponential model, you can take antilogarithms of each side of the equation. This is illustrated in Example 5.20.

EXAMPLE 5.20 Revisiting the Loon Data

For the loon data of Example 5.19, $y' = \ln y$ and the least-squares regression line describing the relationship between y' and x was $\hat{y}' = 1.06 - 0.396x$ or $\widehat{\ln y} = 1.06 - 0.396x$. To "undo" the transformation, we can take the antilog of both sides of this equation:

$$e^{\ln y} = e^{1.06 - 0.396x}$$

Using properties of logs and exponents, $e^{\ln y} = e^{1.06 - 0.396x} = (e^{1.06})(e^{-0.396x})$. Then

$$\hat{y} = e^{1.06}\, e^{-0.396x} = 2.886e^{-0.396x}$$

The equation $\hat{y} = 2.886e^{-0.396x}$ can be used to predict the y value (blood mercury level) for a given x (lake pH). For example, the predicted blood mercury level when lake pH is 6 is

$$\hat{y} = 2.886e^{-0.396x} = 2.886e^{0.396(6)} = 2.886e^{-2.376} = 2.886(0.093) = 0.268$$

■

The process of transforming data, fitting a line to the transformed data, and then undoing the transformation to get an equation for a curved relationship between x and y usually results in a curve that provides a reasonable fit to the sample data. But when y is transformed, this process does not result in the least-squares curve for the data. For example, in Example 5.20, a transformation was used to fit the curve $\hat{y} = 2.886e^{-0.396x}$. However, there may be another equation of the form $\hat{y} = ae^{bx}$ that has a smaller sum of squared residuals for the *original* data than the one obtained using transformations. Finding the least-squares estimates for a and b for an exponential or power model is difficult. Fortunately, the curves found using transformations usually provide reasonable predictions of y.

Choosing Among Different Possible Nonlinear Models

Often there is more than one reasonable model that could be used to describe a nonlinear relationship between two variables. When this is the case, how do we choose a model?

The choice often involves considering how well the model fits the data and what scientific theory suggests the form of the relationship should be. In the absence of any scientific theory, we want to choose a model that has small residuals (small s_e) and accounts for a large proportion of the variability in y (large R^2). On the other hand, if scientific theory suggests that the relationship should have a particular form, we would choose a model that is consistent with the suggested form. The following two examples illustrate these two different situations.

EXAMPLE 5.21 Look for Salamanders Under Those Rocks

Understand the context ❭

The paper "The Relationship Between Rock Density and Salamander Density in a Mountain Stream" (***Herpetologica* [1987]: 357–361**) describes a study of factors that influence population density of salamanders. In this study, researchers created a range of salamander habitats at different locations in a small stream in the Appalachian Mountains by using different sized rocks and pebbles. Three months later, they returned to measure salamander density. This resulted in data on

$$x = \text{rock density (rocks per 1.4 square meters)}$$

and

$$y = \text{salamander density (salamanders per 1.4 square meters)}$$

Consider the data ❭

A scatterplot of these data is shown in Figure 5.39.

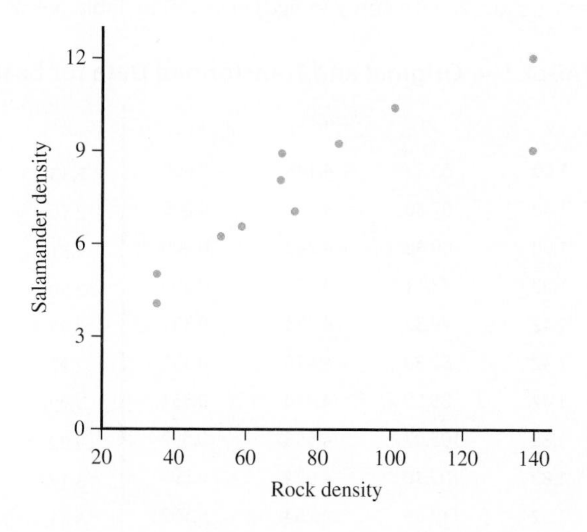

FIGURE 5.39
Scatterplot of salamander density versus rock density.

Formulate a plan ❭

Looking at the curves in Figure 5.26, we might think about using a square root model or a reciprocal model. The researchers chose the reciprocal regression model, $y = a + b\left(\frac{1}{x}\right)$, for two reasons: (1) it provided a good fit to the data, and (2) the reciprocal model made sense scientifically. The investigators felt that there would be an upper limit to the population density, since the streambed is a nonrenewable resource and the stream therefore had a limit in the number of salamanders that could be sustained. This limit is known as the "carrying capacity" of an environment and is estimated by the value of the intercept, a, in the reciprocal model.

Do the work ❭

Using the transformation $x' = \frac{1}{x}$, a least-squares regression line was fit using the transformed data. The resulting model was

$$y = 12.37 - 292.6x' = 12.37 - 292.6\left(\frac{1}{x}\right)$$

The value of R^2 for this model is $R^2 = 0.82$. Figure 5.40(a) shows a scatterplot of the transformed data and the least-squares regression line. Figure 5.40(b) is a scatterplot of the original data that also shows the curve corresponding to the reciprocal model.

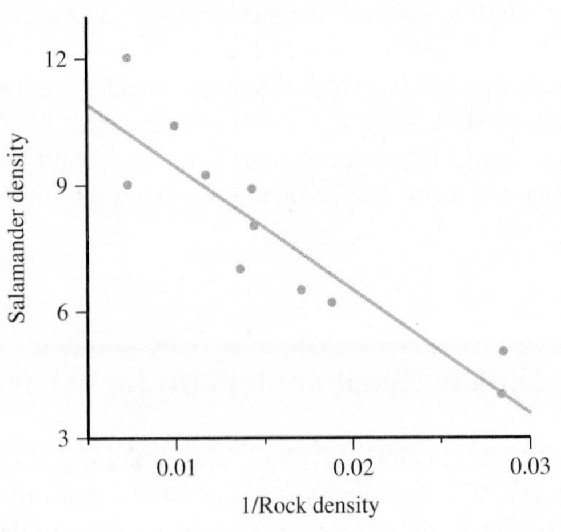

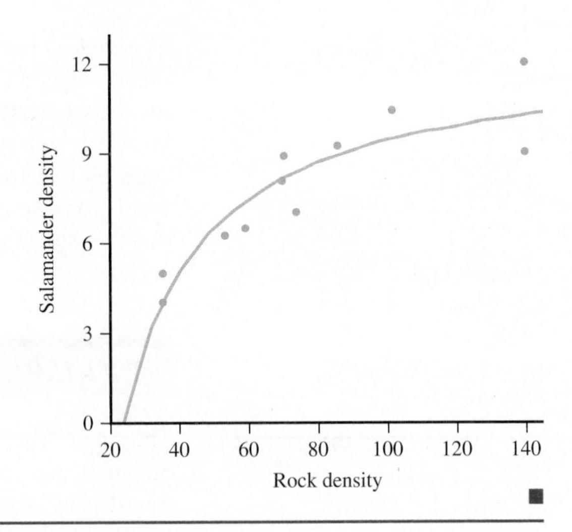

FIGURE 5.40
Scatterplots for the salamander data of Example 5.21:
(a) transformed data;
(b) original data.

EXAMPLE 5.22 How Old Is That Lobster?

Understand the context

Can you tell how old a lobster is by its size? This question was investigated by the authors of a paper that appeared in the ***Biological Bulletin*** **(August, 2007)**. Researchers measured carapace (the exterior shell) length (in mm) of 27 laboratory-raised lobsters of known age. The data on x = carapace length and y = age (in years) in Table 5.4 were read from a graph in the paper.

Consider the data

TABLE 5.4 Original and Transformed Data for Example 5.22

y	x	ln(x)	ln(y)	y	x	ln(x)	ln(y)
1.00	63.32	4.148	0.000	2.33	138.47	4.931	0.846
1.00	67.50	4.212	0.000	2.50	133.95	4.897	0.916
1.00	69.58	4.242	0.000	2.51	125.25	4.830	0.920
1.00	73.41	4.296	0.000	2.50	123.51	4.816	0.916
1.42	79.32	4.373	0.351	2.93	146.82	4.989	1.075
1.42	82.80	4.416	0.351	2.92	139.17	4.936	1.072
1.42	85.59	4.450	0.351	2.92	136.73	4.918	1.072
1.82	105.07	4.655	0.599	2.92	122.81	4.811	1.072
1.82	107.16	4.674	0.599	3.17	142.30	4.958	1.154
1.82	117.25	4.764	0.599	3.41	152.73	5.029	1.227
2.18	109.24	4.694	0.779	3.42	145.78	4.982	1.230
2.18	110.64	4.706	0.779	3.75	148.21	4.999	1.322
2.17	118.99	4.779	0.775	4.08	152.04	5.024	1.406
2.17	122.81	4.811	0.775				

Formulate a plan

The scatterplot of the data in Figure 5.41 shows a clear curved pattern. Considering the curves in Figure 5.26, there are several models that could be considered, including the exponential model and the power model.

In this situation, there is no scientific theory to suggest what the form of the relationship might be. A reasonable way to proceed is to fit both models and then choose the one that provides the best fit. Let's start with the exponential model. Using the transformation

$$y' = \ln y$$

results in the transformed y values given in Table 5.4. A scatterplot of the (x, y') pairs is shown in Figure 5.42(a). Notice that the relationship between y' and x looks linear.

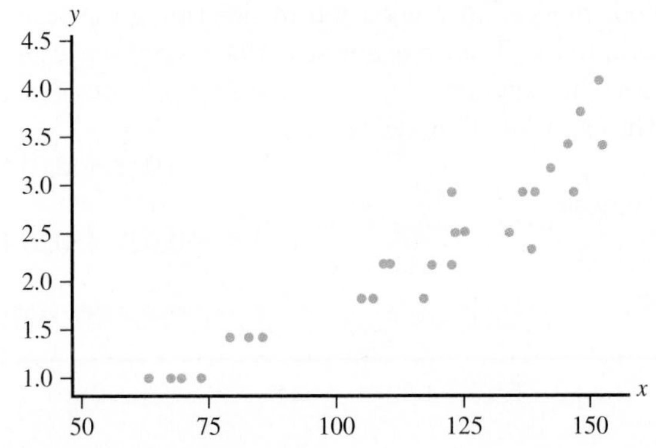

FIGURE 5.41
Scatterplot of age versus carapace length.

Minitab was used to fit the least-squares regression line to the (x, y') data, resulting in the following output:

Do the work)

Regression Analysis: ln(*y*) versus *x*

The regression equation is
ln(y) = −0.927 + 0.0145 x

Predictor	Coef	SE Coef	T	P
Constant	−0.92704	0.08692	−10.67	0.000
x	0.0144903	0.0007311	19.82	0.000

S = 0.105917 R-Sq = 94.0% R-Sq(adj) = 93.8%

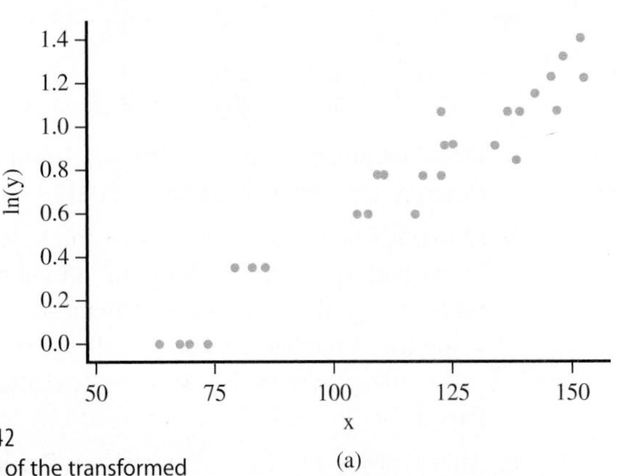

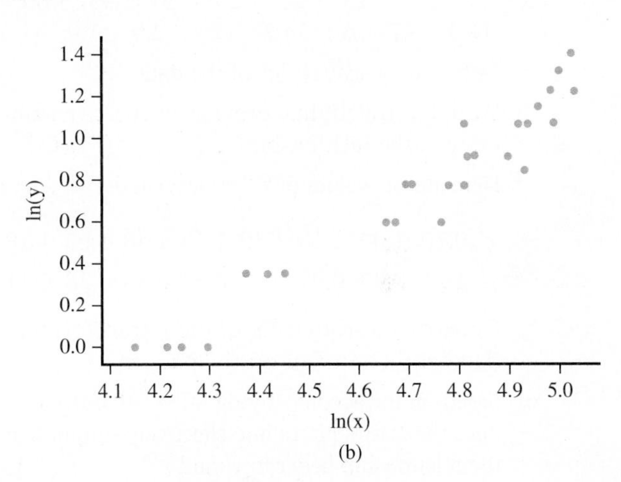

FIGURE 5.42
Scatterplot of the transformed lobster data:
(a) x = carapace length, y' = ln(age);
(b) x' = ln(carapace length), y' = ln(age).

Figure 5.42(b) shows a scatterplot of the transformed data that would be used to fit a power model ($x' = \ln$, $y' = \ln y$). The pattern in this scatterplot also looks linear. Minitab output from fitting a least-squares regression line using the (x', y') data is shown here:

Regression Analysis: ln(*y*) versus ln(*x*)

The regression equation is
ln(y) = −6.32 + 1.50 ln(x)

Predictor	Coef	SE Coef	T	P
Constant	−6.3205	0.3687	−17.14	0.000
ln(x)	1.49866	0.07805	19.20	0.000

S = 0.109120 R-Sq = 93.6% R-Sq(adj) = 93.4%

Notice that both the exponential model and the power model have large R^2 values. Comparing the fit of the two models results in

Model	R^2	S_e
Exponential	0.940	0.106
Power	0.936	0.109

Interpret the results) Both models do a good job in describing the relationship between age and carapace length. In the absence of any scientific theory suggesting the form of the model, we would choose the exponential model because it provides a (slightly) better fit.

The exponential model is

$$\hat{y}' = -0.927 + 0.0145x$$

or equivalently

$$\ln \hat{y} = -0.927 + 0.0145x$$

or

$$\hat{y} = e^{-0.927}e^{0.0145x} = 0.396e^{0.0145x}$$

■

EXERCISES 5.44 - 5.50

5.44 The following data on x = frying time (in seconds) and y = moisture content (%) appeared in the paper "Thermal and Physical Properties of Tortilla Chips as a Function of Frying Time" (***Journal of Food Processing and Preservation* [1995]: 175–189**):

x	5	10	15	20	25	30	45	60
y	16.3	9.7	8.1	4.2	3.4	2.9	1.9	1.3

a. Construct a scatterplot of the data.

b. Would a straight line provide an effective summary of the relationship?

c. Here are the values of $x' = \log(x)$ and $y' = \log(y)$:

x'	0.70	1.00	1.18	1.30	1.40	1.48	1.65	1.78
y'	1.21	0.99	0.91	0.62	0.53	0.46	0.28	0.11

Construct a scatterplot of these transformed data, and comment on the pattern.

d. Based on the accompanying MINITAB output, does the least-squares line effectively summarize the relationship between y' and x'?

The regression equation is
log(moisture) = 2.02 − 1.05 log(time)

Predictor	Coef	SE Coef	T	P
Constant	2.01780	0.09584	21.05	0.000
log(time)	−1.05171	0.07091	−14.83	0.000

S 5 0.0657067 R-Sq 5 97.3% R-Sq(adj) 5 96.9%

Analysis of Variance

Source	DF	SS	MS	F	P
Regression	1	0.94978	0.94978	219.99	0.000
Residual Error	6	0.02590	0.00432		
Total	7	0.97569			

e. Use the MINITAB output to predict moisture content when frying time is 35 sec.

5.45 The paper "Aspects of Food Finding by Wintering Bald Eagles" (***The Auk* [1983]: 477–484**) examined the relationship between the time that eagles spend aloft searching for food (indicated by the percentage of eagles soaring) and relative food availability. The accompanying data were taken from a scatterplot that appeared in this paper. Salmon availability is denoted by x and the percentage of eagles in the air is denoted by y.

x	0	0	0.2	0.5	0.5	1.0
y	28.2	69.0	27.0	38.5	48.4	31.1

x	1.2	1.9	2.6	3.3	4.7	6.5
y	26.9	8.2	4.6	7.4	7.0	6.8

a. Draw a scatterplot for this data set. Would you describe the plot as linear or curved?

b. One possible transformation that might lead to a linear pattern involves taking the square root of both the x and y values. Construct a scatterplot using the variables $\sqrt{x}$ and $\sqrt{y}$. Is the pattern in this scatterplot more linear than the scatterplot in Part (a)?

c. After considering your scatterplot in Part (a), suggest another transformation that might be used to straighten the original plot. Use your suggestion to construct a scatterplot for your x' and/or y'. Which transformation (the one from this part or the square root transformation from Part (b)) do you prefer? Explain.

5.46 Food intake of grazing animals is limited by the rate grass can be chewed and swallowed, as well as the rate at which food can be digested. The authors of the paper "What Constrains Daily Intake in Thomson's Gazelles?" (***Ecology* [1999]: 2338–2347**) observed the grazing activity of captive Thomson's gazelles. They recorded grazing rate (amount of grass eaten, in grams per minute) and biomass of the grazing area (food density, in grams per square meter). Scatterplots of these data with four possible functions that might be used to describe the relationship between grazing rate and biomass are shown in the

Figure EX-5.46 the bottom of the page. Which of these functions would you recommend? Explain your reasoning in a few sentences. (Hint: See Example 5.21.)

5.47 A study, described in the paper "Prediction of Defibrillation Success from a Single Defibrillation Threshold Measurement" **(*Circulation* [1988]: 1144–1149)** investigated the relationship between defibrillation success and the energy of the defibrillation shock (expressed as a multiple of the defibrillation threshold). The accompanying data are from this study.

Energy of Shock	Success (%)
0.5	33.3
1.0	58.3
1.5	81.8
2.0	96.7
2.5	100.0

a. Construct a scatterplot of y = success and x = energy of shock. Does the relationship appear to be linear or nonlinear?

b. Fit a least-squares line to the given data, and construct a residual plot. Does the residual plot support your conclusion in Part (a)? Explain.

c. Consider transforming the data by leaving y unchanged and using either $x' = \sqrt{x}$ or $x'' = \ln(x)$. Which of these transformations would you recommend? Justify your choice by appealing to appropriate graphical displays.

d. Using the transformation you recommended in Part (c), find the equation of the least-squares line that describes the relationship between y and the transformed x.

e. What would you predict success to be when the energy of shock is 1.75 times the threshold level? When it is 0.8 times the threshold level?

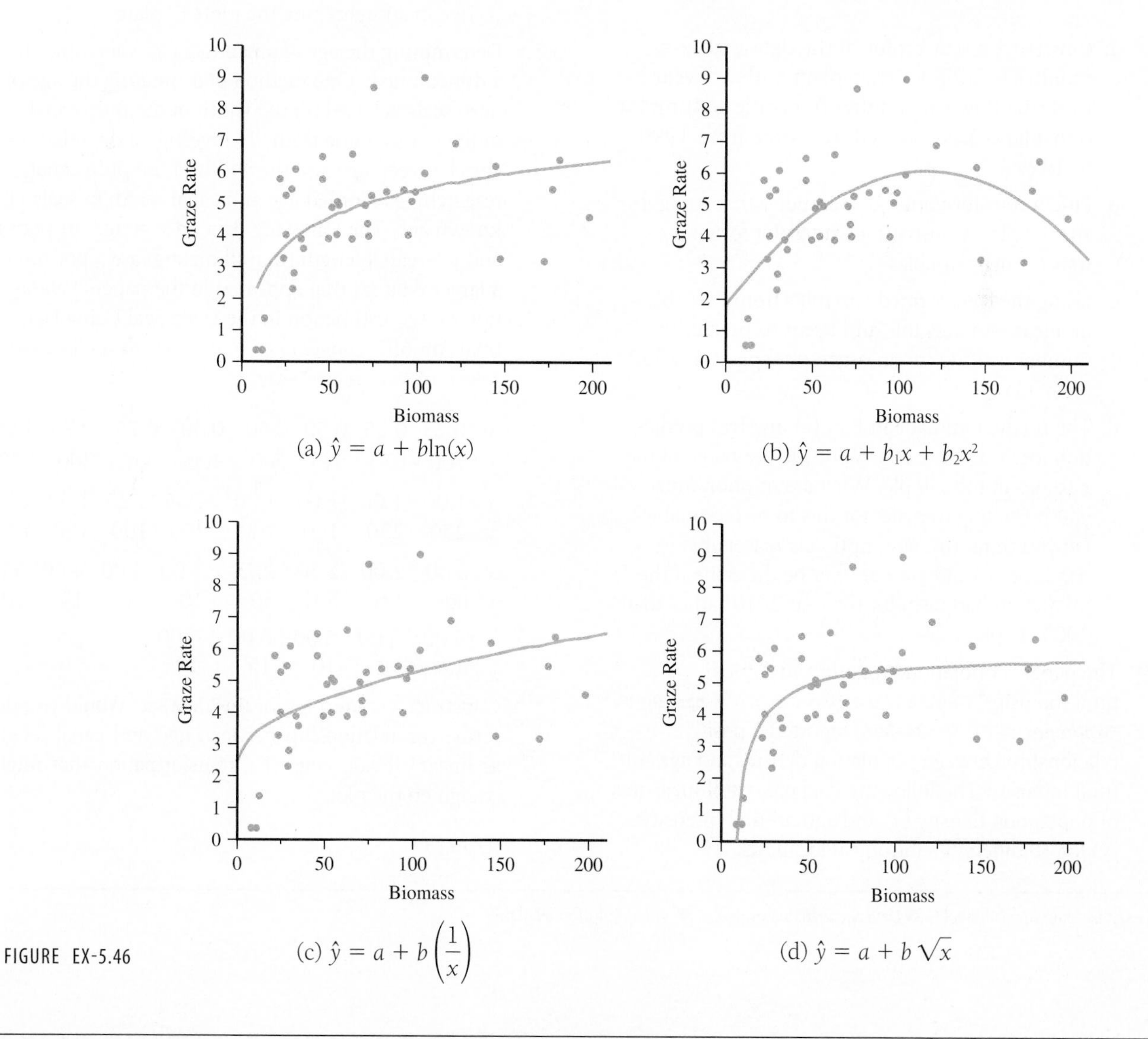

FIGURE EX-5.46

5.48 The article **"Organ Transplant Demand Rises Five Times as Fast as Existing Supply" (*San Luis Obispo Tribune*, February 23, 2001)** included a graph that showed the number of people waiting for organ transplants each year from 1990 to 1999. The following data are approximate values and were read from the graph in the article:

Year	Number Waiting for Transplant (in thousands)
1 (1990)	22
2	25
3	29
4	33
5	38
6	44
7	50
8	57
9	64
10 (1999)	72

a. Construct a scatterplot of the data with y = number waiting for transplant and x = year. Describe how the number of people waiting for transplants has changed over time from 1990 to 1999.

b. Find a transformation of x and/or y that straightens the plot. Construct a scatterplot for your transformed variables.

c. Using the transformed variables from Part (b), fit a least-squares line and use it to predict the number waiting for an organ transplant in 2000 (Year 11).

d. The prediction made in Part (c) involves prediction for an x value that is outside the range of the x values in the sample. What assumption must you be willing to make for this to be reasonable? Do you think this assumption is reasonable in this case? Would your answer be the same if the prediction had been for the year 2010 rather than 2000? Explain.

5.49 The paper "Population Pressure and Agricultural Intensity" **(*Annals of the Association of American Geographers* [1977]: 384–396)** reported a positive relationship between population density and agricultural intensity. The following data consist of measures of population density (x) and agricultural intensity (y) for 18 different subtropical locations:

x	1.0	26.0	1.1	101.0	14.9	134.7
y	9	7	6	50	5	100

x	3.0	5.7	7.6	25.0	143.0	27.5
y	7	14	14	10	50	14

x	103.0	180.0	49.6	140.6	140.0	233.0
y	50	150	10	67	100	100

a. Construct a scatterplot of y versus x. Is the scatterplot compatible with the statement of positive relationship made in the paper?

b. Try a scatterplot that uses y and x^2. Does this transformation straighten the plot?

c. Draw a scatterplot that uses log(y) and x. The log(y) values, given in order corresponding to the y values, are 0.95, 0.85, 0.78, 1.70, 0.70, 2.00, 0.85, 1.15, 1.15, 1.00, 1.70, 1.15, 1.70, 2.18, 1.00, 1.83, 2.00, and 2.00. How does this scatterplot compare with that of Part (b)?

d. Now consider a scatterplot that uses transformations on both x and y: log(y) and x^2. Is this effective in straightening the plot? Explain.

5.50 Determining the age of an animal can sometimes be a difficult task. One method of estimating the age of harp seals is based on the width of the pulp canal in the seal's canine teeth. To investigate the relationship between age and the width of the pulp canal, researchers recorded age and canal width in seals of known age. The following data on x = age (in years) and y = canal length (in millimeters) are a portion of a larger data set that appeared in the paper "Validation of Age Estimation in the Harp Seal Using Dentinal Annuli" **(*Canadian Journal of Fisheries and Aquatic Science* [1983]: 1430–1441)**:

x	0.25	0.25	0.50	0.50	0.50	0.75	0.75	1.00
y	700	675	525	500	400	350	300	300

x	1.00	1.00	1.00	1.00	1.25	1.25	1.50	1.50
y	250	230	150	100	200	100	100	125

x	2.00	2.00	2.50	2.75	3.00	4.00	4.00	5.00
y	60	140	60	50	10	10	10	10

x	5.00	5.00	5.00	6.00	6.00
y	15	10	10	15	10

Construct a scatterplot for this data set. Would you describe the relationship between age and canal length as linear? If not, suggest a transformation that might straighten the plot.

CHAPTER 6

Probability

Doug Menuez/Stockbyte/Getty Images

We make decisions based on uncertainty every day. Should you buy an extended warranty for your new iPod? It depends on the likelihood that it will fail during the warranty period. Should you allow 45 minutes to get to your 8 A.M. class, or is 35 minutes enough? From experience, you may know that most mornings you can drive to school and park in 25 minutes or less. Most of the time, the walk from your parking space to class is 5 minutes or less. But how often will the drive to school or the walk to class take longer than you expect? When it takes longer than usual to drive to campus, is it more likely that it will also take longer to walk to class? less likely? Or are the driving and walking times unrelated?

Some questions involving uncertainty are more serious: If an artificial heart has four key parts, how likely is each one to fail? How likely is it that at least one will fail? We can answer questions like these using the ideas and methods of probability, the systematic study of uncertainty.

Chapter 6: Learning Objectives

STUDENTS WILL UNDERSTAND:

- basic properties of probabilities.
- what it means for two events to be mutually exclusive.
- what it means for two events to be independent.
- the difference between an unconditional and a conditional probability.

STUDENTS WILL BE ABLE TO:

- interpret a probability in context as a long-run relative frequency of occurrence.
- identify the sample space for a given chance experiment.
- calculate the probability of events, including events that are the union of, the intersection of, or the complement of other events.
- calculate conditional probabilities.
- estimate probabilities empirically and using simulation.

6.1 Chance Experiments and Events

The basic ideas and terminology of probability are most easily introduced in situations that are both familiar and reasonably simple. Because of this, some of our initial examples involve such elementary activities as tossing a coin, selecting cards from a deck, and rolling a die. However, after considering a few of these simplistic examples, we will move on to more interesting and realistic situations.

Chance Experiments

When a single coin is tossed, there are two possible outcomes. The coin can land with its heads side up or its tails side up. The selection of a single card from a well-mixed standard deck may result in the ace of spades, the five of diamonds, or any one of the other 50 possibilities. Consider rolling both a red die and a green die. One possible outcome is the red die lands with four dots facing up and the green die shows one dot on its upturned face. Another outcome is the red die lands with three dots facing up and the green die also shows three dots on its upturned face. There are 36 possible outcomes in all, and we do not know in advance what the result of a particular roll will be. Situations such as these are referred to as **chance experiments**.

DEFINITION

Chance experiment: Any activity or situation in which there is uncertainty about which of two or more possible outcomes will result.

When the term chance experiment is used in a probability setting, we mean something different from what was meant by the term experiment in Chapter 2 (where experiments investigated the effect of two or more treatments on a response). For example, in an opinion poll or survey, there is uncertainty about whether an individual selected at random from the population of interest supports a school bond, and when a die is rolled, there is uncertainty about which face will land upturned. Both of these situations fit the definition of a *chance* experiment.

Consider a chance experiment to investigate whether men or women are more likely to choose a hybrid car over a traditional internal combustion engine car when purchasing a Honda Civic at a particular car dealership. The Honda Civic is available with either a hybrid or a traditional engine. In this chance experiment, a customer will be selected at random from those who purchased a Honda Civic. The type of vehicle purchased (hybrid or traditional) will be determined and the customer's gender will be recorded. Before the customer is selected the outcome of this chance experiment is unknown to us. We do know, however, what the possible outcomes are. This set of possible outcomes is called the **sample space**.

DEFINITION

Sample space: The collection of all possible outcomes of a chance experiment.

The sample space of a chance experiment can be represented in many ways. One representation is a simple list of all the possible outcomes. For the car-purchase chance experiment, the possible outcomes are:

1. A male buying hybrid
2. A female buying hybrid
3. A male buying traditional
4. A female buying traditional

We can also use set notation and ordered pairs. A male purchasing a hybrid is represented as (male, hybrid). The sample space is then

$$\text{sample space} = \begin{Bmatrix} \text{(male, hybrid), (female, hybrid),} \\ \text{(male, traditional), (female, traditional)} \end{Bmatrix}$$

Another useful representation of the sample space is a tree diagram. A tree diagram (shown in Figure 6.1) for the outcomes of the car-purchase chance experiment has two sets of branches corresponding to the two pieces of information that were gathered. To identify any particular outcome in the sample space, traverse the tree by first selecting a branch corresponding to gender and then a branch identified with a type of car.

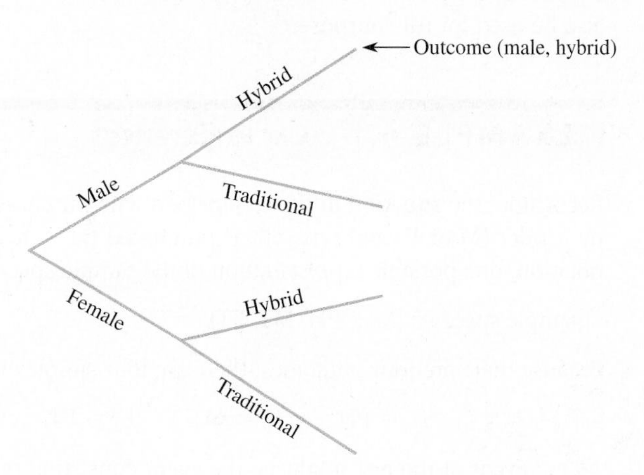

FIGURE 6.1
Tree diagram for the car purchase example.

In the tree diagram of Figure 6.1, there is no particular reason for having the gender as the first generation branch and the type of car as the second generation branch. Some chance experiments involve making observations in a particular order, in which case the order of the branches in the tree does matter. In this example, however, it would be acceptable to represent the sample space with the tree diagram shown in Figure 6.2.

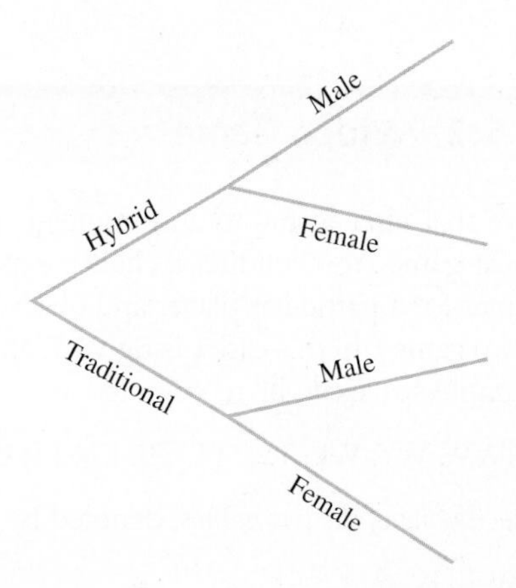

FIGURE 6.2
Another tree diagram for the car purchase example.

As we have seen, the sample space for a chance experiment can be represented in several ways, but the representations all have one thing in common: *Every* possible outcome of the chance experiment is included in the representation.

Events

In the car-purchase chance experiment, we might be interested in which particular outcome will result. Or we might focus on a group of outcomes that involve the purchase of a hybrid—the group consisting of (male, hyrid) and (female, hybrid). When we form a collection of one or more individual outcomes, we are creating what is known as an *event*.

DEFINITION

Event: Any collection of outcomes from the sample space of a chance experiment.

Simple event: An event consisting of exactly one outcome.

We usually represent an event by an uppercase letter, such as A, B, C, and so on. On some occasions, the same letter with different numerical subscripts, such as E_1, E_2, E_3, ... may also be used for this purpose.

EXAMPLE 6.1 Car Preferences

Reconsider the situation in which a person who purchased a Honda Civic was categorized by gender (M or F) and type of car purchased (H = hybrid, T = traditional). Using this notation, one possible representation of the sample space is

sample space = {MH, FH, MT, FT}

Because there are four outcomes, there are four simple events:

E_1 = MH E_2 = FH E_3 = MT E_4 = FT

One event of interest might be the event consisting of all outcomes for which a hybrid was purchased. A symbolic description of the event *hybrid* is

hybrid = {MH, FH}

Another event is the event that the purchaser is male,

male = {MH, MT} ■

EXAMPLE 6.2 Video Game

Suppose you believe that after losing to an opponent in a video game, a player is more likely to lose the next game. You conduct a chance experiment that consists of watching two consecutive games for a particular player and observing whether the player won, tied, or lost each of the two games. In this case (using W, T, and L to represent win, tie, and loss, respectively), the sample space can be represented as

sample space = {WW, WT, WL, TW, TT, TL, LW, LT, LL}

The event *lose exactly one of the two games*, denoted by L_1, could then be defined as

L_1 = {WL, TL, LW, LT} ■

Only one outcome (one simple event) occurs when a chance experiment is performed. We say that a given event occurs whenever one of the outcomes making up the event occurs. For example, if the outcome in the car purchase example is MH, then the simple event *male purchasing a hybrid* has occurred, and so has the nonsimple event *hybrid purchased.*

Forming New Events

Once some events have been specified, they can be manipulated in several useful ways to create new events.

DEFINITION

Let *A* and *B* denote two events.

***Not A*:** The event that consists of all experimental outcomes that are not in event *A*. *Not A* is sometimes called the *complement* of *A* and is usually denoted by A^C, A', or $\bar{A}$.

***A or B*:** The event that consists of all experimental outcomes that are in at least one of the two events, that is, in *A* or in *B* or in both of these. *A or B* is called the *union* of the two events and is denoted by $A \cup B$.

***A and B*:** The event that consists of all experimental outcomes that are in both of the events *A* and *B*. *A and B* is called the *intersection* of the two events and is denoted by $A \cap B$.

EXAMPLE 6.3 Turning Directions

Grant Faint/The Image Bank/Getty Images

A traffic engineer has been asked to consider whether a stop sign at the bottom of a freeway off-ramp should be replaced by a traffic light. To help in this decision, she plans to observe traffic patterns for this off-ramp. Suppose she were to record the turning direction (L = left or R = right) of each of three successive vehicles. This is a chance experiment and the sample space contains eight outcomes:

all 3 cars turn left   *the 1st car turns left and the next 2 turn right*

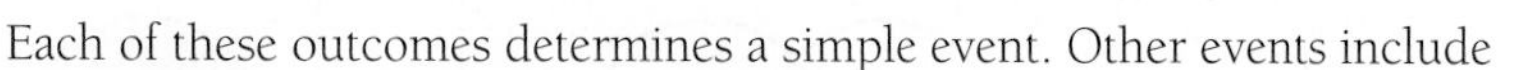

{LLL, RLL, LRL, LLR, RRL, RLR, LRR, RRR}

Each of these outcomes determines a simple event. Other events include

A = event that exactly one of the cars turns right = {RLL, LRL, LLR}
B = event that at most one of the cars turns right = {LLL, RLL, LRL, LLR}
C = event that all cars turn in the same direction = {LLL, RRR}

Some other events that can be formed from those just defined are

not C = C^C = event that not all cars turn in the same direction
= {RLL, LRL, LLR, RRL, RLR, LRR}

A or C = $A \cup C$ = event that exactly one of the cars turns right or all cars turn in the same direction
= {RLL, LRL, LLR, LLL, RRR}

B and C = $B \cap C$ = event that at most one car turns right and all cars turn in the same direction
= {LLL} ■

EXAMPLE 6.4 More on Video Games

In Example 6.2, in addition to the event that a video game player loses exactly one of the two games, L_1 = {WL, TL, LW, LT}, we could also define events corresponding to L_0 = neither game is lost and L_2 = both games are lost. Then L_0 = {WW, WT, TW, TT} and L_2 = {LL}.

Some other events that can be formed from those just defined include

L_2^C = event that at most one game was lost
= {WW, WT, WL, TW, TT, TL, LW, LT}

$L_1 \cup L_2$ = event that at least one game was lost
= {WL, TL, LW, LT, LL}

$L_1 \cap L_2$ = event that exactly one game was lost and two games were lost
= the empty set ■

It frequently happens that two events have no common outcomes, as was the case for events L_1 and L_2 in the video game example. Such situations are described using special terminology.

DEFINITION

Mutually exclusive: Two events are **mutually exclusive** if they have no outcomes in common. The term **disjoint** is also sometimes used to describe events that have no outcomes in common.

Saying that two events are mutually exclusive means that they can't both occur when the chance experiment is performed once.

It is sometimes useful to draw an informal picture of events to visualize relationships. In a **Venn diagram**, the collection of all possible outcomes is typically shown as the interior of a rectangle. Other events are then identified by specified regions inside this rectangle. Figure 6.3 illustrates several Venn diagrams.

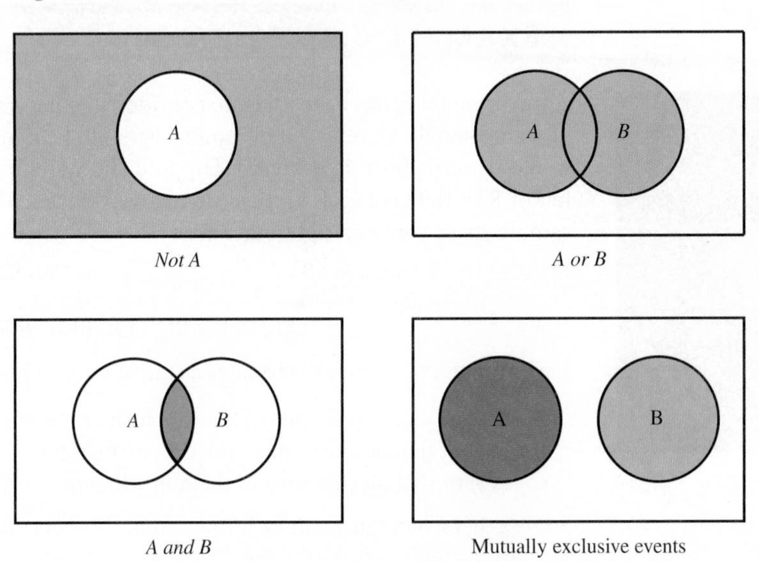

FIGURE 6.3
Venn diagrams.

The use of the *or* and *and* operations can be extended to form new events from more than two initially specified events (as illustrated in Figure 6.4).

Let $A_1, A_2, \ldots, A_k$ denote *k* events.

1. The event A_1 *or* A_2 *or* … *or* A_k consists of all outcomes in at least one of the individual events $A_1, A_2, \ldots, A_k$.
2. The event A_1 *and* A_2 *and* … *and* A_k consists of all outcomes that are simultaneously in every one of the individual events $A_1, A_2, \ldots, A_k$.

These *k* events are mutually exclusive if no two of them have any common outcomes.

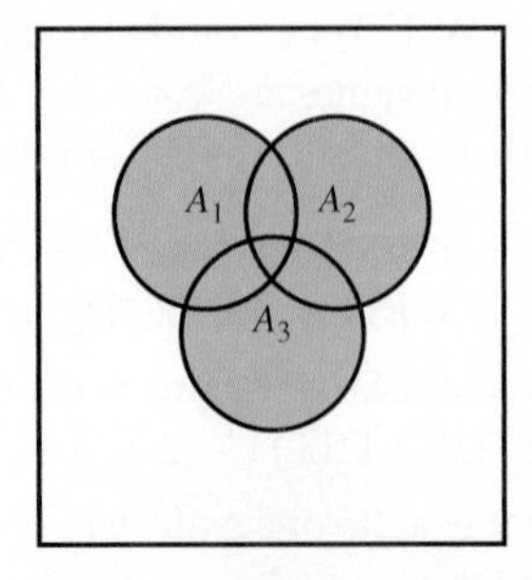

A or B or C

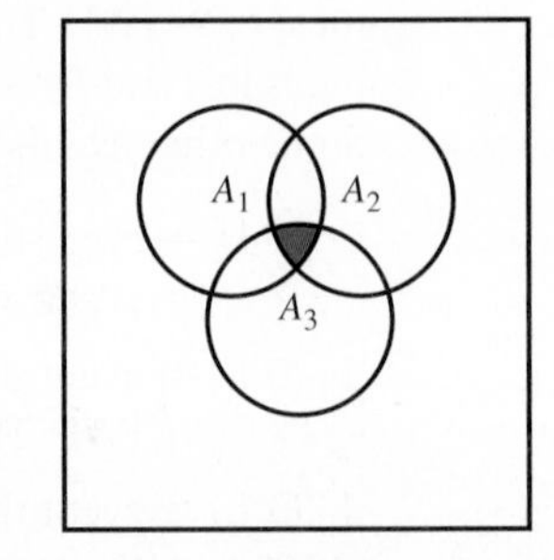

A and B and C

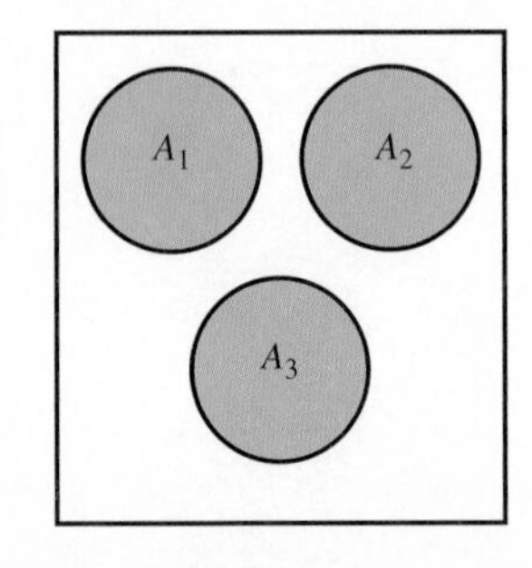

Three mutually exclusive events

FIGURE 6.4
Venn diagrams with three events.

EXAMPLE 6.5 Asking Questions

The instructor in a seminar class consisting of four students has an unusual way of asking questions. Four slips of paper numbered 1, 2, 3, and 4 are placed in a box. The instructor determines the student to whom any particular question is to be addressed by selecting one of these four slips. Suppose that one question is to be posed during each of the next two class meetings. One possible outcome could be represented as (3, 1)—the first question is addressed to Student 3 and the second question to Student 1. There are 15 other possibilities. Consider the following events:

the event that the same student is asked both questions → $A = \{(1,1), (2,2), (3,3), (4,4)\}$

the event that Student 1 is asked at least one of the two questions → $B = \{(1,1), (1,2), (1,3), (1,4), (2,1), (3,1), (4,1)\}$

$C = \{(3,1), (2,2), (1,3)\}$

$D = \{(3,3), (3,4), (4,3)\}$

$E = \{(1,1), (1,3), (2,2), (3,1), (4,2), (3,3), (2,4), (4,4)\}$

$F = \{(1,1), (1,2), (2,1)\}$

Then

$$A \text{ or } C \text{ or } D = \{(1,1), (2,2), (3,3), (4,4), (3,1), (1,3), (3,4), (4,3)\}$$

The outcome (3,1) is contained in each of the events B, C, and E, as is the outcome (1,3). These are the only two common outcomes in all three events, so

$$B \text{ and } C \text{ and } E = \{(3,1), (1,3)\}$$

The events C, D, and F are mutually exclusive because no outcome in any one of these events is contained in either of the other two events. ■

EXERCISES 6.1 - 6.12

6.1 Define the term *chance experiment*, and give an example of a chance experiment with four possible outcomes.

6.2 Define the term *sample space*, and then give the sample space for the chance experiment you described in Exercise 6.1.

6.3 Consider the chance experiment in which the type of transmission—automatic (A) or manual (M)—is recorded for each of the next two cars purchased from a certain dealer.

a. What is the set of all possible outcomes (the sample space)? (Hint: See Example 6.2.)

b. Display the possible outcomes in a tree diagram.

c. List the outcomes in each of the following events. Which of these events are simple events? (Hint: See Example 6.3.)

i. B the event that at least one car has an automatic transmission

ii. C the event that exactly one car has an automatic transmission

iii. D the event that neither car has an automatic transmission

d. What outcomes are in the event B *and* C? In the event B *or* C?

6.4 A tennis shop sells five different brands of rackets, each of which comes in either a midsize version or an oversize version. Consider the chance experiment in which brand and size are noted for the next racket purchased. One possible outcome is Head midsize, and another is Prince oversize. Possible outcomes correspond to cells in the following table:

	Head	Prince	Slazenger	Wimbledon	Wilson
Midsize					
Oversize					

a. Let A denote the event that an oversize racket is purchased. List the outcomes in A.

b. Let B denote the event that the name of the brand purchased begins with a W. List the outcomes in B.

c. List the outcomes in the event *not B*.

d. Head, Prince, and Wilson are U.S. companies and the others are not. Let C denote the event that the racket purchased is made by a U.S. company. List the outcomes in the event *B or C*. (Hint: See Example 6.4.)

e. List outcomes in *B and C*.

f. Display the possible outcomes on a tree diagram, with a first-generation branch for each brand.

6.5 A new model of laptop computer can be ordered with one of three screen sizes (10 inches, 12 inches, 15 inches) and one of four hard drive sizes (50 GB, 100 GB, 150 GB, and 200 GB). Consider the chance experiment in which a laptop order is selected and the screen size and hard drive size are recorded.

a. Display possible outcomes using a tree diagram.

b. Let A be the event that the order is for a laptop with a screen size of 12 inches or smaller. Let B be the event that the order is for a laptop with a hard drive size of at most 100 GB. What outcomes are in **i.** A^C? **ii.** $A \cup B$? **iii.** $A \cap B$?

c. Let C denote the event that the order is for a laptop with a 200 GB hard drive. Are A and C mutually exclusive events? Are B and C mutually exclusive?

6.6 A college library has four copies of a certain book; the copies are numbered 1, 2, 3, and 4. Two of these are selected at random. The first selected book is placed on 2-hour reserve, and the second book can be checked out overnight.

a. Construct a tree diagram to display the 12 outcomes in the sample space.

b. Let A denote the event that at least one of the books selected is an even-numbered copy. What outcomes are in A?

c. Suppose that copies 1 and 2 are hardcover books, whereas copies 3 and 4 are softcover books. Let B denote the event that exactly one of the copies selected is a hardcover book. What outcomes are contained in B?

6.7 A library has five copies of a certain textbook on reserve of which two copies (1 and 2) are hardcover books and the other three (3, 4, and 5) are softcover books. A student examines these books in random order, stopping only when a softcover book has been selected.

a. Display the possible outcomes in a tree diagram.

b. What outcomes are contained in the event A, that exactly one book is examined before the chance experiment terminates?

c. What outcomes are contained in the event C, that the chance experiment terminates with the examination of book 5?

6.8 Suppose that, starting at a certain time, batteries coming off an assembly line are examined one by one to see whether they are defective (let D = defective and N = not defective). The chance experiment terminates as soon as a nondefective battery is obtained.

a. Give five possible outcomes for this chance experiment.

b. What can be said about the number of outcomes in the sample space?

c. What outcomes are in the event E, that the number of batteries examined is an even number?

6.9 Refer to the previous exercise and now suppose that the chance experiment terminates only when two nondefective batteries have been obtained.

a. Let A denote the event that at most three batteries must be examined before the chance experiment terminates. What outcomes are contained in A?

b. Let B be the event that exactly four batteries must be examined before the chance experiment terminates. What outcomes are in B?

c. What can be said about the number of possible outcomes for this chance experiment?

6.10 A family consisting of three people—P_1, P_2, and P_3—belongs to a medical clinic that always has a physician at each of stations 1, 2, and 3. During a certain week, each member of the family visits the clinic exactly once and is randomly assigned to a station. One experimental outcome is (1, 2, 1), which means that P_1 is assigned to station 1, P_2 to station 2, and P_3 to station 1.

a. List the 27 possible outcomes. (Hint: First list the nine outcomes in which P_1 goes to station 1, then the nine in which P_1 goes to station 2, and finally the nine in which P_1 goes to station 3; a tree diagram might help.)

b. List all outcomes in the event A, that all three people go to the same station.

c. List all outcomes in the event B, that all three people go to different stations.

d. List all outcomes in the event C, that no one goes to station 2.

e. Identify outcomes in each of the following events:
i. B^C ii. C^C iii. $A \cup B$ iv. $A \cap B$ v. $A \cap C$

6.11 An engineering construction firm is currently working on power plants at three different sites. Define events E_1, E_2, and E_3 as follows:

E_1 = the plant at Site 1 is completed by the contract date
E_2 = the plant at Site 2 is completed by the contract date
E_3 = the plant at Site 3 is completed by the contract date

The following Venn diagram pictures the relationships among these events:

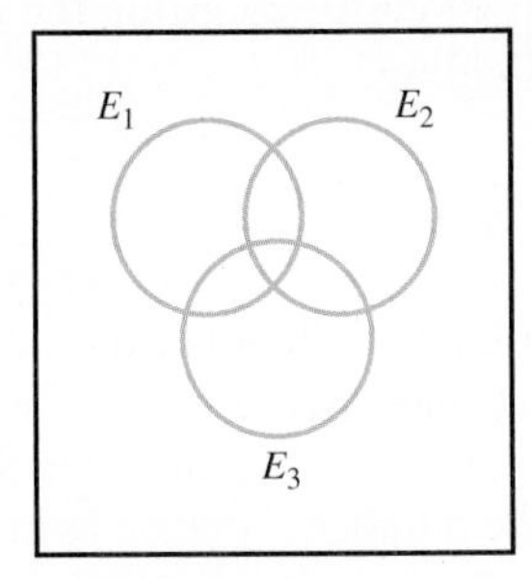

Shade the region in the Venn diagram corresponding to each of the following events. Redraw the Venn diagram for each part of the problem. (Hint: See the discussion of Venn diagrams)

a. At least one plant is completed by the contract date.
b. All plants are completed by the contract date.
c. None of the plants are completed by the contract date.
d. Only the plant at Site 1 is completed by the contract date.
e. Exactly one of the three plants is completed by the contract date.
f. Either the plant at Site 1 or both of the other two plants are completed by the contract date.

6.12 Consider a Venn diagram picturing two events A and B that are not mutually exclusive.

a. Shade the event $(A \cup B)^C$. On a separate Venn diagram shade the event $A^C \cap B^C$. How are these two events related?
b. Shade the event $(A \cap B)^C$. On a separate Venn diagram shade the event $A^C \cup B^C$. How are these two events related? (Note: These two relationships together are called DeMorgan's laws.)

Bold exercises answered in back ● Data set available online ▼ Video Solution available

6.2 Definition of Probability

Reasoning about games of chance and probabilistic thinking share a long history. Archeologists have found evidence that Egyptians used a small bone in mammals, the astragalus, as a sort of four-sided die as early as 3500 B.C. Games of chance were common in Greek and Roman times and during the Renaissance of Western Europe.

The formal study of probability began with the correspondence between Blaise Pascal (1623–62) and Pierre de Fermat (1601–65) in 1654. Their letters discussed some problems related to gambling that were posed by the French nobleman Chevalier de Mère. The methods and solutions that resulted from this exchange greatly enhanced the study of probability.

From this early work, new interpretations of probability have evolved, including an empirical approach preferred by many statisticians today. We begin our discussion of probability with a look at some of the different approaches to probability: the classical, relative frequency, and subjective approaches.

Classical Approach to Probability

Early mathematicians' development of the theory of probability was primarily in the context of gambling and reflected the peculiarities of games of chance. A characteristic common to most games of chance is a physical device used to generate different outcomes. In children's games, dice or a "spinner" might be used to create chance outcomes.

Dice are constructed so that the physical characteristics of each face are alike (except, of course, for the number of dots), ensuring that the different outcomes for an individual die are very close to equally likely. Therefore, it seems quite natural to assume, for example, that the probability of getting a five when a single six-sided die is rolled is one-sixth (1 chance in 6). In general, if there are N *equally likely* outcomes in a game of chance, the probability of each of the outcomes is $1/N$.

Classical Approach to Probability for Equally Likely Outcomes

When the outcomes in the sample space of a chance experiment are equally likely, the **probability of an event *E*, denoted by *P*(*E*),** is the ratio of the number of outcomes favorable to E to the total number of outcomes in the sample space:

$$P(E) = \frac{\text{number of outcomes favorable to } E}{\text{number of outcomes in the sample space}}$$

According to this definition, the calculation of a probability consists of counting the number of outcomes that make up an event, counting the number of outcomes in the sample space, and then dividing.

Chance experiments that involve tossing fair coins, rolling fair dice, or selecting cards from a well-mixed deck have equally likely outcomes. For example, if a fair die is rolled once, each outcome (simple event) has probability 1/6. With E denoting the event that the number rolled is even, $P(E) = 3/6$. This is just the number of outcomes in E divided by the total number of possible outcomes.

EXAMPLE 6.6 Calling the Toss

On some football teams, the honor of calling the toss at the beginning of a football game is determined by random selection. Suppose that this week a member of the offensive team will call the toss. There are 5 interior linemen on the 11-player offensive team. If we define the event L as the event that an interior lineman is selected to call the toss, 5 of the 11 possible outcomes are included in L. The probability that an interior lineman will be selected is then

$$P(L) = \frac{5}{11}$$

■

EXAMPLE 6.7 Math Contest

Four students (Adam, Betina, Carlos, and Debra) submitted correct solutions to a math contest with two prizes. The contest rules specify that if more than two correct responses are submitted, the winners will be selected at random from those submitting correct responses. In this case, the set of possible outcomes for the chance experiment that consists of selecting the two winners from the four correct responses is

{(A, B), (A, C), (A, D), (B, C), (B, D), (C, D)}

Because the winners are selected at random, the six outcomes are equally likely and the probability of each individual outcome is 1/6.

Let E be the event that both selected winners are the same gender. Then

E = {(A, C), (B, D)}

Because E contains two outcomes, $P(E) = 2/6 = .333$.

If F denotes the event that at least one of the selected winners is female, then F consists of all outcomes except (A, C) and $P(F) = 5/6 = .833$. ■

Limitations of the Classical Approach to Probability

The classical approach to probability works well with games of chance or other situations with a finite set of outcomes that can be regarded as equally likely. However, some situations that are clearly probabilistic in nature do not fit the classical model. Consider the accident rates of young adults driving cars. Information of this kind is used by insurance companies to set car insurance rates. Suppose that we have two individuals, one 18 years old and one 28 years old. A person purchasing a standard accident insurance policy at age 28 should have a lower annual premium than a person purchasing the same policy at age 18, because the 28-year-old has a much lower chance of an accident, based on prior experience.

However, the classical probabilist, playing by the classical rules, would have to consider that there are two outcomes for a policy year—having one or more accidents or not having an accident—*and would consider those outcomes equally likely for both the 28-year-old and the 18-year-old.* This certainly seems to defy reason and experience.

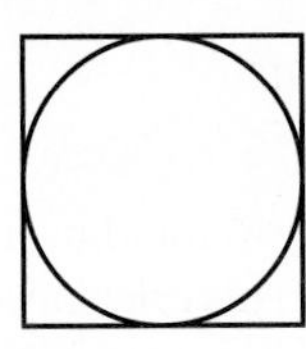

Another example of a situation that cannot be handled using the classical approach has a geometric flavor. Suppose that you have a square that is 8 inches on each side and that a circle with a diameter of 8 inches is drawn inside the square, touching the square on all four sides, as shown.

This drawing will be used as a target in a dart game. A dart will be thrown at the target and any dart that completely misses the square is a "do-over" and will be thrown again. It seems reasonable to think about the probability of a dart landing in the circle as a ratio of the area of the circle to the area of the square. That ratio is

$$\frac{\text{area of circle}}{\text{area of square}} = \frac{\pi r^2}{s^2} = \frac{\pi(4)^2}{8^2} = \frac{\pi}{4} = .785$$

A practiced dart player who is aiming for the center of the target might have an even greater probability of hitting the inside of the circle. In any case, it is not reasonable to think that the two outcomes *inside the circle* and *not inside the circle* are equally likely, so the classical approach to probability will not help here.

From the standpoint of statistics, a major limitation of the classical approach to probability is that there seems to be no way to allow past experience to inform our expectation of the future. To return to the car insurance example, suppose that in the experience of the insurance companies, 0.5% of 18-year-olds have accidents during their 19th year. It seems reasonable that, all things being equal, we could expect this percentage to be stable enough to set the prices for an insurance policy. Although the classical approach's assumption of equal probabilities is based on and consistent with the experience with dice, it cannot accommodate using an observed proportion in real life to estimate a probability. For a statistician, this is a problem. Without an understanding of probability that allows such a generalization, statistical inference would be impossible.

Relative Frequency Approach to Probability

Early scientific investigators were aware that chance experiments do not always give the same results when repeated. However, even though it is not possible to predict the outcome of a chance experiment in any particular instance, there is a dependable and stable regularity when a chance experiment is repeated many times. This became the basis for fair wagering in games of chance.

For example, suppose that two friends, Chris and Jay, meet to play a game. A fair coin is flipped. If it lands heads up, Chris pays Jay $1; otherwise, Jay pays the same amount to Chris. After many repetitions, the proportion of the time that Chris wins will be close to one-half, and the two friends will have simply enjoyed the pleasure of each other's company for a few hours. This happy circumstance occurs because *in the long run* (i.e., over many repetitions) the proportion of heads is .5. In the long run, half the time Chris wins, and half the time Jay wins.

When any given chance experiment is performed, some events are relatively likely to occur, whereas others are not as likely to occur. For a specified event E, we want to assign a probability that gives a precise indication of how likely it is that E will occur. In the relative frequency approach to probability, the probability describes how frequently E occurs when the chance experiment is performed many times.

EXAMPLE 6.8 Tossing a Coin

Frequently, we hear a coin described as fair, or we are told that there is a 50% chance of a coin landing heads up. What is the meaning of the expression *fair*? It cannot refer to the result of a single toss, because a single toss cannot result in both a head and a tail. Might "fairness" and "50%" refer to 10 successive tosses yielding exactly five heads and five tails? Not really, because it is easy to imagine a fair coin landing heads up on only 3 or 4 of the 10 tosses.

Consider the chance experiment of tossing a coin just once. Define the simple events

H = event that the coin lands with its heads side facing up
T = event that the coin lands with its tails side facing up

Now suppose that we take a fair coin and begin to toss it over and over. After each toss, we compute the relative frequency of heads observed so far. This calculation gives the value of the ratio

$$\frac{\text{number of times event } H \text{ occurs}}{\text{number of tosses}}$$

The results of the first 10 tosses might be as follows:

Toss	1	2	3	4	5	6	7	8	9	10
Cumulative number of *H*'s	0	1	2	3	3	3	4	5	5	5
Relative frequency of *H*'s	0	.5	.667	.75	.6	.5	.571	.625	.556	.5

Figure 6.5 illustrates how the relative frequency of heads fluctuates during a sample sequence of 50 tosses. *As the number of tosses increases, the relative frequency of heads does not continue to fluctuate wildly but instead stabilizes and approaches some fixed number (the limiting value).* This stabilization is illustrated for a sequence of 1000 tosses in Figure 6.6.

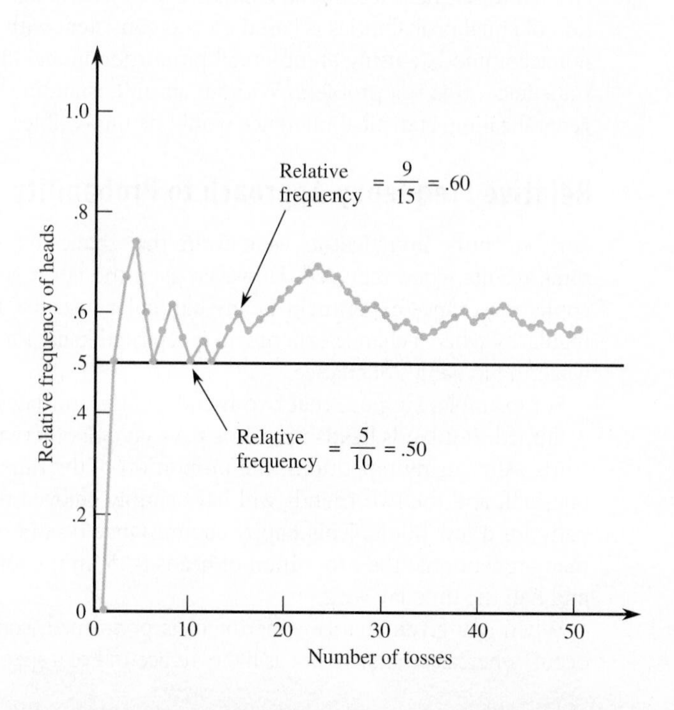

FIGURE 6.5
Relative frequency of heads in the first 50 of a long series of coin tosses.

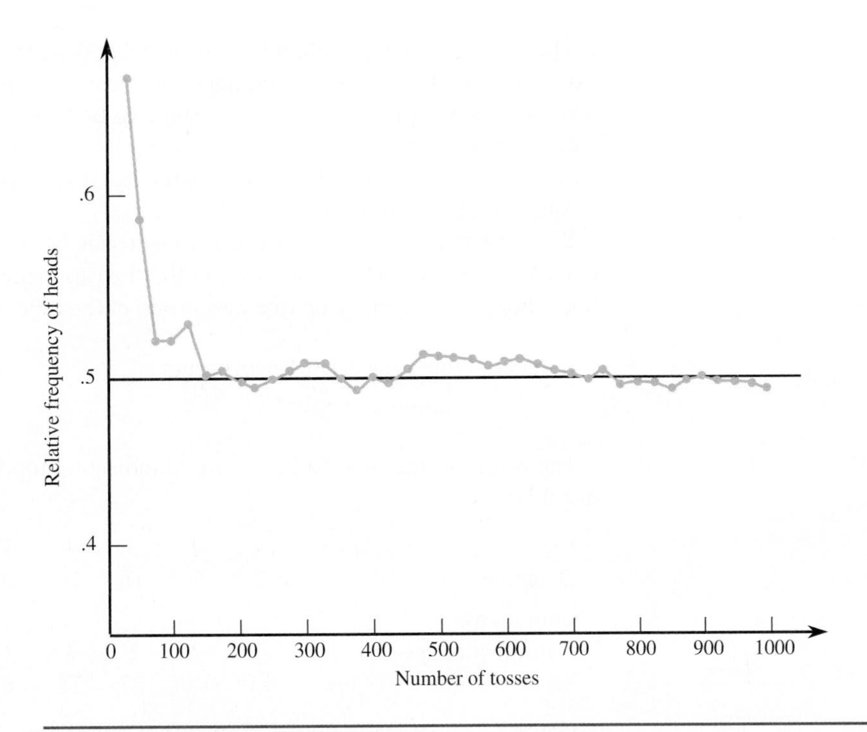

FIGURE 6.6
Stabilization of the relative frequency of heads in coin tossing.

■

It may seem natural to you that the proportion of *H*'s would get closer and closer to the "real" probability of .5. That an empirically observed proportion could behave like this in real life also seemed reasonable to James Bernoulli in the early 18th century. He focused his considerable mathematical power on this topic and was able to prove mathematically what we now know as the law of large numbers.*

Law of Large Numbers

As the number of repetitions of a chance experiment increases, the chance that the relative frequency of occurrence for an event will differ from the true probability of the event by more than any small number approaches 0.

The law of large numbers tells us that, as the number of repetitions of a chance experiment increases, the proportion of the time an event E occurs gets close and stays close to the real probability of E occurring in a single chance experiment *even if the value of this probability is not known*. This means that we can observe the outcomes of repetitions of a chance experiment and then use the observed outcomes to estimate probabilities.

Relative Frequency Approach To Probability

The **probability of an event *E***, denoted by ***P*(*E*)**, is defined to be the value approached by the relative frequency of occurrence of *E* when a chance experiment is performed many times. If the number of times the chance experiment is performed is quite large,

$$P(E) \approx \frac{\text{number of times } E \text{ occurs}}{\text{number of times the experiment is performed}}$$

* Technically, this should be referred to more specifically as the weak law of large numbers for Bernoulli trials. After Bernoulli's proof, mathematical statisticians proved more general laws of large numbers.

The relative frequency definition of probability depends on being able to repeat a chance experiment under identical conditions. Suppose that we perform a chance experiment that consists of flipping a cap from a 20-ounce bottle of soda and noting whether the cap lands with the open side up or down. The crucial difference from the previous coin-tossing chance experiment is that there is no particular reason to believe the cap is equally likely to land top up or top down.

If we assume that the chance experiment can be repeated under similar conditions (which seems reasonable), then we can flip the cap many times and compute the relative frequency of the event *top up* (the proportion of the time the event has occurred so far):

$$\frac{\text{number of times the event } \textit{top up} \text{ occurs}}{\text{number of flips}}$$

The results of the first 10 flips, with U indicating top up and D indicating top down, might be:

Flip	1	2	3	4	5	6	7	8	9	10
Outcome	U	U	D	U	D	D	U	D	U	U
Cumulative number of *ups*	1	2	2	3	3	3	4	4	5	6
Relative frequency of *ups*	1.0	1.0	.67	.75	.6	.5	.57	.5	.56	.6

Figure 6.7 illustrates how the relative frequency of the event *top up* fluctuates during a sequence of 100 flips. Based on these results and faith in the law of large numbers, it is reasonable to think that the probability of the cap landing top up is about .7.

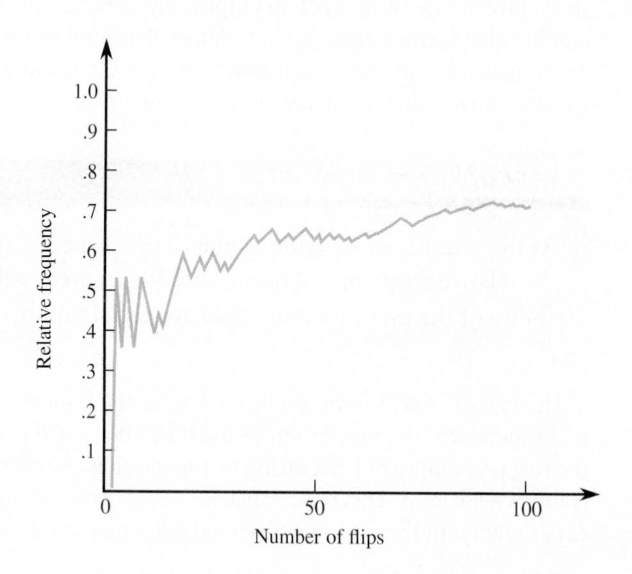

FIGURE 6.7
Stabilization of the relative frequency of a bottle cap landing top up.

The relative frequency approach to probability is based on observation. From repeated observation, we can obtain stable relative frequencies that will, in the long run, provide good estimates of the probabilities of different events. The relative frequency interpretation of probability is intuitive, widely used, and most relevant for the inferential procedures introduced in later chapters.

Subjective Approach to Probability

A third distinct approach to probability is based on subjective judgments. In this view, a probability is interpreted as a personal measure of the strength of belief that a particular outcome will occur. A probability of 1 represents a belief that the outcome will certainly occur. A probability of 0 represents a belief that the outcome will certainly not occur—that it is impossible. Other probabilities are placed somewhere between 0 and 1, based on the strength of one's beliefs.

For example, an airline passenger might report her subjective assessment of the probability of being denied a seat as a result of overbooking on a particular flight as .01. Because this probability is close to 0, she believes it is very unlikely that she will be denied a seat.

The subjective interpretation of probability presents some difficulties. For one thing, different people may assign different probabilities to the same outcome, because each could have a different subjective belief. Although subjective probabilities are useful in studying and analyzing decision making, they are of limited use because they are personal and not generally replicable by others.

How Are Probabilities Determined?

Probability statements are frequently encountered in newspapers and magazines (and, of course, in this textbook). What is the basis for these probability statements? In most cases, research allows reasonable estimates based on observation and analysis. In general, if a probability is stated, it is based on one of the following approaches:

1. *The classical approach:* This approach is appropriate only for modeling chance experiments with equally likely outcomes.
2. *The subjective approach:* In this case, the probability represents an individual's judgment based on facts combined with personal evaluation of other information.
3. *The relative frequency approach:* An estimate is based on observing the outcomes of a chance experiment.

When you see probabilities in this book and in other published sources, you should consider how the probabilities were produced. Be particularly cautious in the case of subjective probabilities.

6.3 Basic Properties of Probability

The view of a probability as a long-run relative frequency that can be approximated empirically has some limitations when it comes to calculating probabilities in real life. The most obvious problem is time. If computing probabilities even for simple chance experiments were to require hundreds or thousands of observations, even the most ardent statistics students would avoid this task! An alternative approach that simplifies things in some situations is to study fundamental properties of probability that can be used to find probabilities of complex events. These fundamental properties are given in the accompanying box. A discussion of each property follows.

Fundamental Properties of Probability

1. For any event E, $0 \leq P(E) \leq 1$.
2. If S is the sample space for an experiment, $P(S) = 1$.
3. If two events E and F are mutually exclusive, then $P(E \text{ or } F) = P(E) + P(F)$.
4. For any event E, $P(E) + P(\text{not } E) = 1$. Therefore

$$P(\text{not } E) = 1 - P(E)$$

and

$$P(E) = 1 - P(\text{not } E).$$

Property 1: For any event E, $0 \leq P(E) \leq 1$

To understand the first property of probability, recall the previous bottle cap chance experiment. You may remember that we were keeping track of the number of flips landing with the top of the bottle cap facing up. Suppose that, after flipping the bottle cap N times, we have observed x occurrences of top up. What are the possible values of x?

The fewest that could have been counted is 0, and the most that could have been counted is N. Therefore, the relative frequency of top up falls between two numbers:

$$\frac{0}{N} \le \text{relative frequency} \le \frac{N}{N}$$

This means that the relative frequency must always be greater than or equal to 0 and less than or equal to 1. As N increases, the long-run value of the relative frequency, which defines the probability, must also lie between 0 and 1.

Property 2: If *S* is the sample space for a chance experiment, *P*(*S*) = 1

The probability of any event is the proportion of time an outcome in the event will occur in the long run. Because the sample space consists of all possible outcomes for a chance experiment, in the long run an outcome that is in S must occur 100% of the time. This mean that $P(S) = 1$.

Property 3: If two events *E* and *F* are mutually exclusive, then *P*(*E* or *F*) = *P*(*E*) + *P*(*F*)

This is one of the most important properties of probability because it provides a method for computing probabilities if the number of possible outcomes (simple events) in the sample space is finite. Any nonsimple event is just a collection of outcomes from the sample space—some outcomes are in the event and others are not. Because simple events are mutually exclusive, any event can be viewed as a union of mutually exclusive events.

For example, suppose that a chance experiment has a sample space that consists of six outcomes, $O_1, O_2, \ldots, O_6$. Then $P(O_1)$ can be interpreted as the long run proportion of times that the outcome O_1 will occur, and the probabilities of the other outcomes can be interpreted in a similar way. Suppose that an event E is made up of outcomes O_1, O_2, and O_4. Then E will occur whenever O_1, O_2, or O_4 occurs. Because O_1, O_2, and O_4 are mutually exclusive, the long-run proportion of time that E will occur is just the sum of the proportion of time that each of the outcomes O_1, O_2, and O_4 will occur.

Because it is always possible to express any event as a collection of mutually exclusive simple events in this way, finding the probability of an event can be reduced to finding the probabilities of the simple events (outcomes) that make up the event and then adding. If the probabilities of all the simple events are known, it is then easy to compute the probability of more complex events.

Property 4: For any event *E*, *P*(*E*) + *P*(*not E*) = 1. Therefore *P*(*not E*) = 1 − *P*(*E*) and *P*(*E*) = 1 − *P*(*not E*)

Property 4 follows from Properties 2 and 3. Property 2 tells us that $P(E)$ is the sum of the probabilities in E and that $P(\textit{not } E)$ is the sum of the probabilities for simple events corresponding to outcomes in *not* E. Every outcome in the sample space is in either E or *not* E. We know from Property 3 that the sum of the probabilities of all the simple events is 1. It follows that $P(E) + P(\textit{not } E)$ must equal 1.

The implication of Property 4, namely, that $P(E) = 1 - P(\textit{not } E)$ is surprisingly useful. There are many situations in which calculation of $P(\textit{not } E)$ is much easier than calculating $P(E)$ directly.

EXAMPLE 6.9 Medical Errors

Understand the context ❯

Diagnostic errors in medical settings are of concern because they often lead to serious consequences and sometimes even death. The paper **"Diagnostic Error in Internal Medicine"** (***Archives of Internal Medicine* [2005]: 1493–1499**) describes possible outcomes when a doctor specializing in internal medicine reaches a diagnosis for a patient. The possible diagnostic outcomes are

O_1 Correct diagnosis
O_2 No-fault diagnostic error (a misdiagnosis due to unusual symptoms or due to an uncooperative or deceptive patient)

O_3 System-related diagnostic error (a misdiagnosis due to equipment failure, faulty lab results, etc.)

O_4 Cognitive diagnostic error (a misdiagnosis due to faulty or inadequate doctor knowledge or failure to synthesize available information correctly)

O_5 Misdiagnosis resulting from a combination of system-related error and cognitive error

Based on a study of a large number of patients, the paper suggests that

85% of patients are correctly diagnosed
1% are misdiagnosed because of a no-fault error
3% are misdiagnosed because of a system-related error
4% are misdiagnosed because of a cognitive error
7% are misdiagnosed because of a combination of system and cognitive errors.

The following table displays the probabilities of the simple events for the chance experiment in which the diagnostic outcome is observed for a randomly selected patient:

Simple Event (Outcome)	O_1	O_2	O_3	O_4	O_5
Probability	.85	.01	.03	.04	.07

Let's create event E, the event that *a randomly selected patient is misdiagnosed.* This event consists of the outcomes O_2, O_3, O_4, and O_5. It follows that

$$\begin{aligned} P(E) &= P(O_2 \cup O_3 \cup O_4 \cup O_5) \\ &= P(O_2) + P(O_3) + P(O_4) + P(O_5) \\ &= .01 + .03 + .04 + .07 \\ &= .15 \end{aligned}$$

That is, in the long run, 15% of all patients seen by internal medicine specialists are misdiagnosed. Another way to compute this probability is to notice that the event *not E* consists only of the outcome O_1. Another way to calculate $P(E)$ is

$$P(E) = 1 - P(not\ E) = 1 - P(O_1) = 1 - .85 = .15$$

If we were interested in the probability of a misdiagnosis that involved a cognitive error in some way, we could define the event F, where F is the event that *a randomly selected patient is misdiagnosed in a way that involved a cognitive error.* Then

$$F = O_4 \cup O_5$$

and

$$P(F) = P(O_4 \cup O_5) = P(O_4) + P(O_5) = .04 + .07 = .11$$

This means that, in the long run, cognitive errors play a role in the misdiagnosis of about 11% of patients. ■

Having established the fundamental properties of probability, we now present some practical probability rules that can be used to evaluate probabilities in some situations. An important aspect of each rule is the set of conditions necessary in order to use the rule. A common error for beginning statistics students is to believe that the rules can be used in *any* probability calculation. This is *not* the case. You must be careful to use the rules only after verifying that any necessary conditions are met.

Equally Likely Outcomes

The first probability rule that we consider applies only when events are equally likely. It is *not* always true for all events in all situations. Chance experiments involving tossing fair coins, rolling fair dice, or selecting cards from a well-mixed deck are examples of chance experiments that have equally likely outcomes. If a fair die is rolled once, each possible outcome (1, 2, 3, 4, 5, and 6) has probability 1/6. With E denoting the event that the outcome is an even number,

$$P(E) = P(2 \text{ or } 4 \text{ or } 6) = P(2) + P(4) + P(6) = \frac{1}{6} + \frac{1}{6} + \frac{1}{6} = \frac{3}{6}$$

This is just the ratio of the number of outcomes in E to the total number of possible outcomes. The following box presents the generalization of this result.

Calculating Probabilities When Outcomes Are Equally Likely

Consider an experiment that can result in any one of N possible outcomes. Denote the corresponding simple events by $O_1, O_2, \dots, O_N$. If these simple events are equally likely to occur, then

1. $P(O_1) = \frac{1}{N}, P(O_2) = \frac{1}{N}, \dots, P(O_N) = \frac{1}{N}$

2. For *any* event E,

$$P(E) = \frac{\text{number of outcomes in } E}{N}$$

Addition Rule for Mutually Exclusive Events

We have seen previously that the probability of an event can be calculated by adding together probabilities of the simple events that correspond to the outcomes making up the event. This addition process is also legitimate when calculating the probability of the union of two events that are mutually exclusive.

The Addition Rule for Mutually Exclusive Events

Let E and F be two mutually exclusive events. One of the basic properties (axioms) of probability is

$$P(E \text{ or } F) = P(E \cup F) = P(E) + P(F)$$

This property of probability is known as the addition rule for mutually exclusive events. More generally, if events $E_1, E_2, \dots, E_k$ are all mutually exclusive, then

$$P(E_1 \text{ or } E_2 \text{ or} \dots \text{or } E_k) = P(E_1 \cup E_2 \cup \cdots \cup E_k) = P(E_1) + P(E_2) + \cdots + P(E_k)$$

In words, the probability that any of these k mutually exclusive events occurs is the sum of the probabilities of the individual events.

Consider the chance experiment that consists of rolling a pair of fair dice. There are 36 possible outcomes for this chance experiment, such as (1, 1), (1, 2), and so on. Because these outcomes are equally likely, we can compute the probability of observing a total of 5 on the two dice by counting the number of outcomes that result in a total of 5. There are four such outcomes—(1, 4), (2, 3), (3, 2), and (4, 1)—giving

$$P(\text{total of } 5) = \frac{4}{36}$$

The probabilities for the other totals can be computed in a similar fashion, and they are shown in the following table:

Total	Probability	Total	Probability
2	1/36	8	5/36
3	2/36	9	4/36
4	3/36	10	3/36
5	4/36	11	2/36
6	5/36	12	1/36
7	6/36		

What is the probability of getting a total of 3 or 5? Consider the two events

E = *total is 3*

and F = *total is 5*

Clearly E and F are mutually exclusive because the sum cannot simultaneously be both 3 and 5. Notice also that neither of these events is simple, because neither consists of only a single outcome from the sample space. We can apply the addition rule for mutually exclusive events as follows:

$$P(E \text{ or } F) = P(E \cup F) = P(E) + P(F) = \frac{2}{36} + \frac{4}{36} = \frac{6}{36}$$

EXAMPLE 6.10 Car Choices

Understand the context)

A large auto center sells cars made by a number of different manufacturers. Three of these manufacturers are Japanese: Honda, Nissan, and Toyota. Consider a chance experiment that consists of observing the make and model of the next car purchased at this auto center. The outcomes in the sample space would then be simple events such as Nissan Altima and Toyota Prius. Let's define events E_1, E_2, and E_3 by

E_1 = Honda

E_2 = Nissan

and E_3 = Toyota

Notice that E_1 is not a simple event because there is more than one model Honda (for example, Civic, Accord, and Insight).

Based on several years of sales data, the probabilities of these three events have been estimated empirically as $P(E_1) = .25$, $P(E_2) = .18$, and $P(E_3) = .14$. Because E_1, E_2, and E_3 are mutually exclusive, the addition rule gives

$$\begin{aligned} P(\text{Honda } or \text{ Nissan } or \text{ Toyota}) &= P(E_1 \cup E_2 \cup E_3) \\ &= P(E_1) + P(E_2) + P(E_3) \\ &= .25 + .18 + .14 = .57 \end{aligned}$$

The probability that the next car purchased is *not* made by one of these three manufacturers is

$$P(not\,(E_1 \text{ or } E_2 \text{ or } E_3)) = P((E_1 \cup E_2 \cup E_3)^C) = 1 - .57 = .43$$

■

In Section 6.6, we will show how $P(E \cup F)$ can be calculated when the two events E and F are not mutually exclusive.

EXERCISES 6.13 - 6.29

6.13 A large department store offers online ordering. When a purchase is made online, the customer can select one of four different delivery options: expedited overnight delivery, expedited second-business-day delivery, standard delivery, or delivery to the nearest store for customer pick-up. Consider the chance experiment that consists of observing the selected delivery option for a randomly selected online purchase.

a. What are the four simple events that make up the sample space for this experiment?

b. Suppose that the probability of an overnight delivery selection is .1, the probability of a second-day delivery selection is .3, and the probability of a standard-delivery selection is .4. Find the following probabilities. (Hint: See Example 6.9.)

i. the probability that a randomly selected online purchase selects delivery to the nearest store for customer pick-up.

ii. the probability that the customer selects a form of expedited delivery.

iii. the probability that either standard delivery or delivery to the nearest store is selected.

6.14 The manager of a music store has kept records of the number of CDs bought in a single transaction by customers who make a purchase at the store. The accompanying table gives six possible outcomes and the estimated probability associated with each of these outcomes for the chance experiment that consists of observing the number of CDs purchased by a randomly selected customer at the store.

Number of CDs purchased	1	2	3	4	5	6 or more
Estimated probability	.45	.25	.10	.10	.07	.03

a. What is the estimated probability that the next customer purchases three or fewer CDs?

b. What is the estimated probability that the next customer purchases at most three CDs? How does this compare to the probability computed in Part (a)?

c. What is the estimated probability that the next customer purchases five or more CDs?

d. What is the estimated probability that the next customer purchases one or two CDs?

e. What is the estimated probability that the next customer purchases more than two CDs? Show two different ways to compute this probability that use the probability rules of this section.

6.15 A bookstore sells two types of books (fiction and nonfiction) in several formats (hardcover, paperback, digital, and audio). For the chance experiment that consists of observing the type and format of a single-book purchase, two of the eight possible outcomes are a hardcover fiction book and an audio nonfiction book.

a. There are eight outcomes in the sample space for this experiment. List these possible outcomes.

b. Do you think it is reasonable to think that the outcomes for this experiment would be equally likely? Explain.

c. For customers who purchase a single book, the estimated probabilities for the different possible outcomes are given in the cells of the accompanying table. What is the probability that a randomly selected single-book purchase will be for a book in print format (hardcover or paperback)?

	Hardcover	Paperback	Digital	Audio
Fiction	.15	.45	.10	.10
Nonfiction	.08	.04	.02	.06

d. Show two different ways to compute the probability that a randomly selected single-book purchase will be for a book that is not in a print format. (Hint: See Example 6.9.)

e. Find the probability that a randomly selected single-book purchase will be for a work of fiction.

6.16 ▼ Medical insurance status—covered (C) or not covered (N)—is determined for each individual arriving for treatment at a hospital's emergency room. Consider the chance experiment in which this determination is made for two randomly selected patients.

The simple events are O_1 = (C, C), meaning that the first patient selected was covered and the second patient selected was also covered, O_2 = (C, N), O_3 = (N, C), and O_4 = (N, N). Suppose that probabilities are $P(O_1) = .81$, $P(O_2) = .09$, $P(O_3) = .09$, and $P(O_4) = .01$.

a. What outcomes are contained in A, the event that at most one patient is covered, and what is $P(A)$?

b. What outcomes are contained in B, the event that the two patients have the same status with respect to coverage, and what is $P(B)$?

6.17 Roulette is a game of chance that involves spinning a wheel that is divided into 38 equal segments, as shown in the accompanying picture.

©Anna Baburkina/Shutterstock.com

A metal ball is tossed into the wheel as it is spinning, and the ball eventually lands in one of the 38 segments. Each segment has an associated color. Two segments are green. Half of the other 36 segments are red and the others are black. When a balanced roulette wheel is spun, the ball is equally likely to land in any one of the 38 segments.

a. When a balanced roulette wheel is spun, what is the probability that the ball lands in a red segment?

b. In the roulette wheel shown, black and red segments alternate. Suppose instead that the red segments were side-by-side and that the black segments were together. Does this increase the probability that the ball will land in a red segment? Explain.

c. Suppose that you watch 1000 spins of a roulette wheel and note the color that results from each spin. What would be an indication that the wheel was not balanced?

6.18 Phoenix is a hub for a large airline. Suppose that on a particular day, 8000 passengers arrived in Phoenix on this airline. Phoenix was the final destination for 1800 of these passengers. The others were all connecting to flights to other cities. On this particular day, several inbound flights were late, and 480 connecting passengers missed their connecting flight and were delayed in Phoenix. Of the 480 who were delayed, 75 were delayed overnight and had to spend the night in Phoenix. Consider the chance experiment of choosing a passenger at random from these 8000 passengers. Compute the following probabilities:

a. the probability that the selected passenger had Phoenix as a final destination.

b. the probability that the selected passenger did not have Phoenix as a final destination.

c. the probability that the selected passenger was connecting and missed the connecting flight.

d. the probability that the selected passenger was a connecting passenger and did not miss the connecting flight.

e. the probability that the selected passenger either had Phoenix as a final destination or was delayed overnight in Phoenix.

f. An independent customer satisfaction survey is planned. The company carrying out the survey plans to contact 50 passengers selected at random from the 8000 passengers who arrived in Phoenix on the day described above. The airline knows that the survey results will not be favorable if too many people who were delayed overnight are included in the survey. Should the airline be worried? Write a few sentences explaining whether or not you think the airline should be worried, using relevant probabilities to support your answer.

6.19 A professor assigns five problems to be completed as homework. At the next class meeting, two of the five problems will be selected at random and collected for grading. You only completed the first three problems.

a. What is the probability that you will be able to turn in both of the problems selected? (Hint: You can think of the problems as being labeled A, B, C, D, and E. Then one possible selection of two problems is A and B. If these are the two problems selected and you did problems A, B, and C, you will be able to turn in both problems. There are nine other possible selections to consider.)

b. Does the probability that you will be able to turn in both problems change if you had completed the last three problems instead of the first three problems? Explain.

c. What happens to the probability that you will be able to turn in both problems selected if you had completed four of the problems rather than just three?

6.20 Refer to the following information on full-term births in the United States over a given period of time:

Type of Birth	Number of Births
Single birth	41,500,000
Twins	500,000
Triplets	5,000
Quadruplets	100

Use this information to estimate the probability that a randomly selected pregnant woman who reaches full term

a. Delivers twins

b. Delivers quadruplets

c. Gives birth to more than a single child

6.21 Suppose you want to estimate the probability that a customer at a particular grocery store will pay by credit card. Over the past three months, 80,500 purchases were made, and 37,100 of them were paid for by credit card. What is the estimated probability that a customer will pay by credit card?

6.22 A Nielsen survey of teens between the ages of 13 and 17 found that 83% use text messaging and 56% use picture messaging (**"How Teens Use Media," Nielsen, June 2009**). Use these percentages to explain why the two events

T = event that a randomly selected teen uses text messaging

and

P = event that a randomly selected teen uses picture messaging

cannot be mutually exclusive.

6.23 According to ***The Chronicle for Higher Education* (Aug. 26, 2011)**, there were 787,325 associate degrees awarded by U.S. community colleges in the 2008–2009 academic year. A total of 488,142 of these degrees were awarded to women.

a. If a person who received a degree in 2008–2009 was selected at random, what is the probability that the selected student will be female?

b. What is the probability that the selected student will be male?

6.24 The same issue of ***The Chronicle for Higher Education*** referenced in the previous exercise also reported the following information for degrees awarded by U.S. colleges to Hispanic students in the 2008–2009 academic year:

- A total of 274,515 degrees were awarded to Hispanic students.
- 97,921 of these degrees were associate degrees.
- 129,526 of these degrees were bachelor's degrees.
- The remaining degrees were either graduate or professional degrees.

What is the probability that a randomly selected Hispanic student who received a degree in 2008–2009

a. received an associate degree?

b. received a graduate or professional degree?

c. did not receive a bachelor's degree?

6.25 A deck of 52 cards is mixed well, and 5 cards are dealt.

a. It can be shown that (disregarding the order in which the cards are dealt) there are 2,598,960 possible five-card hands, of which only 1287 are hands consisting entirely of spades. What is the probability that a hand will consist entirely of spades? What is the probability that a hand will consist entirely of a single suit?

b. It can be shown that exactly 63,206 hands contain only spades and clubs, with both suits represented. What is the probability that a hand consists entirely of spades and clubs with both suits represented?

c. Using the result of Part (b), what is the probability that a hand contains cards from exactly two suits?

6.26 After all students have left the classroom, a statistics professor notices that four copies of the text were left under desks. At the beginning of the next lecture, the professor distributes the four books at random to the four students (1, 2, 3, and 4) who claim to have left books. One possible outcome is that 1 receives 2's book, 2 receives 4's book, 3 receives his or her own book, and 4 receives 1's book. This outcome can be abbreviated (2, 4, 3, 1).

a. List the 23 other possible outcomes.

b. Which outcomes are contained in the event that exactly two of the books are returned to their correct owners? Assuming equally likely outcomes, what is the probability of this event?

c. What is the probability that exactly one of the four students receives his or her own book?

d. What is the probability that exactly three receive their own books?

e. What is the probability that at least two of the four students receive their own books?

6.27 The student council for a school of science and math has one representative from each of the five academic departments: biology (B), chemistry (C), mathematics (M), physics (P), and statistics (S). Two of these students are to be randomly selected for inclusion on a university-wide student committee (by placing five slips of paper in a bowl, mixing, and drawing out two of them).

a. What are the 10 possible outcomes (simple events)?

b. From the description of the selection process, all outcomes are equally likely. What is the probability of each simple event?

c. What is the probability that one of the committee members is the statistics department representative?

d. What is the probability that both committee members come from laboratory science departments?

6.28 A student placement center has requests from five students for interviews regarding employment with a particular consulting firm. Three of these students are math majors, and the other two students are statistics majors. Unfortunately, the interviewer has time to talk to only two of the students. These two will be randomly selected from among the five.

a. What is the probability that both selected students are statistics majors?

b. What is the probability that both students are math majors?

c. What is the probability that at least one of the students selected is a statistics major?

d. What is the probability that the selected students have different majors?

6.29 Suppose that a six-sided die is "loaded" so that any particular even-numbered face is twice as likely to land face up as any particular odd-numbered face. Consider the chance experiment that consists of rolling this die.

a. What are the probabilities of the six simple events? (Hint: Denote these events by $O_1, \ldots, O_6$. Then $P(O_1) = p$, $P(O_2) = 2p$, $P(O_3) = p, \ldots$, $P(O_6) = 2p$. Now use a condition on the sum of these probabilities to determine p.)

b. What is the probability that the number showing is an odd number? at most three?

c. Now suppose that the die is loaded so that the probability of any particular simple event is proportional to the number showing on the corresponding upturned face; that is, $P(O_1) = c$, $P(O_2) = 2c, \ldots, P(O_6) = 6c$. What are the probabilities of the six simple events? Calculate the probabilities of Part (b) for this die.

Bold exercises answered in back ● Data set available online ▼ Video Solution available

6.4 Conditional Probability

Sometimes the knowledge that one event has occurred changes our assessment of the likelihood that another event occurs. For example, consider a population in which 0.1% of all individuals have a certain disease. The presence of the disease cannot be discerned from outward appearances, but there is a diagnostic test available. Unfortunately, the test is not always correct. Of those with positive test results, 80% actually have the disease; and the other 20% who show positive test results are false-positives.

To put this in probability terms, consider the chance experiment in which an individual is randomly selected from the population. Define the following events:

E = event that the individual has the disease
F = event that the individual's diagnostic test is positive

We will use $P(E|F)$ to denote the probability of the event E *given that* the event F is known to have occurred. A new symbol has been used to indicate that a probability calculation has been made conditional on the occurrence of another event. The standard symbol for this is a vertical line, and it is read "given." For example, we would say, "the probability that an individual has the disease *given* that the diagnostic test is positive," and represent this symbolically as P(has disease|positive test) or $P(E|F)$. This probability is called a **conditional probability**.

The information provided then implies that

$$P(E) = .001$$
$$P(E|F) = .8$$

This means that before we have diagnostic test information, the occurrence of E is unlikely. However, once it is known that the test result is positive, the likelihood of the disease increases dramatically. (If this were not so, the diagnostic test would not be very useful!)

EXAMPLE 6.11 College Housing Options

Understand the context ⟩

At a small liberal arts college, students have three housing options. They can live in on-campus housing, off-campus housing, or at home with family. The following table gives the number of students in each housing option by year in school.

Consider the data)

	Freshman	Sophomore	Junior	Senior	Total
On-campus housing	150	160	140	150	**600**
Off-campus housing	100	90	120	125	**435**
Home with family	125	140	150	150	**565**
Total	**375**	**390**	**410**	**425**	**1600**

Consider each of the following statements, and make sure that you see how each follows from the information in the table:

1. There are 160 sophomores who live in on-campus housing.
2. The number of seniors who live in off-campus housing is 125.
3. There are 435 students who live in off-campus housing.
4. There are 410 juniors at the college.
5. The total number of students at the college is 1600.

Each week, the college president selects a students at random and invites him or her to have lunch with her to discuss various issues that might be of concern to them. She feels that random selection will give her the greatest chance of hearing from a diverse group of students. What is the probability that a randomly selected student is a senior who lives on campus? Assuming that each student is equally likely to be selected, we can calculate this probability as follows:

Do the work)

$$P(\text{senior who lives on campus}) = \frac{\text{number of seniors who live on campus}}{\text{total number of students}} = \frac{150}{1600} = .09375$$

Now suppose that the president's assistant records not only the student's name but also the student's year in school. The assistant has indicated that the selected student is a senior. Does this information change our assessment of the likelihood that the selected student lives on campus? Because 150 of the 425 seniors live on campus, this suggests that

$$P(\text{live on campus}|\text{senior}) = \frac{\text{number of seniors who live on campus}}{\text{total number of seniors}} = \frac{150}{425} = .3529$$

Interpret the results)

The probability is calculated in this way because we know that the selected student is one of 425 seniors, each of whom is equally likely to have been the one selected. The interpretation of this conditional probability is that if we were to repeat the chance experiment of selecting a student at random, about 35.29% of the selections that resulted in a senior being selected would also result in the selection of someone who lives on campus. ■

EXAMPLE 6.12 GFI Switches

Understand the context)

A GFI (ground fault interrupt) switch turns off power to a system in the event of an electrical malfunction. A spa manufacturer currently has 25 spas in stock, each equipped with a single GFI switch. Two different companies supply the switches, and some of the switches are defective, as summarized in the following table:

Consider the data)

	Nondefective	Defective	Total
Company 1	10	5	**15**
Company 2	8	2	**10**
Total	**18**	**7**	

A spa is randomly selected for testing. Let

E = event that GFI switch in selected spa is from Company 1
F = event that GFI switch in selected spa is defective

Do the work) Using the information in the table, we can calculate the following probabilities:

$$P(E) = \frac{15}{25} = .60 \quad P(F) = \frac{7}{25} = .28 \quad P(E \text{ and } F) = P(E \cap F) = \frac{5}{25} = .20$$

Now suppose that testing reveals a defective switch. (This means that the chosen spa is one of the seven in the "defective" column.) How likely is it that the switch came from the first company? Because five of the seven defective switches are from Company 1,

$$P(E|F) = P(\text{company } 1|\text{defective}) = \frac{5}{7} = .714$$

This is larger than the unconditional probability $P(E)$. This is because Company 1 has a much higher defective rate than Company 2.

An alternative expression for the conditional probability is

$$P(E|F) = \frac{5}{7} = \frac{5/25}{7/25} = \frac{P(E \text{ and } F)}{P(F)} = \frac{P(E \cap F)}{P(F)}$$

Notice that $P(E|F)$ is a ratio of two previously specified probabilities. It is the probability that both events occur divided by the probability of the "conditioning event" F. Additional insight comes from the Venn diagram of Figure 6.8. Once it is known that the outcome lies in F, the chance that E also occurs is the "size" of (E *and* F) relative to the size of F.

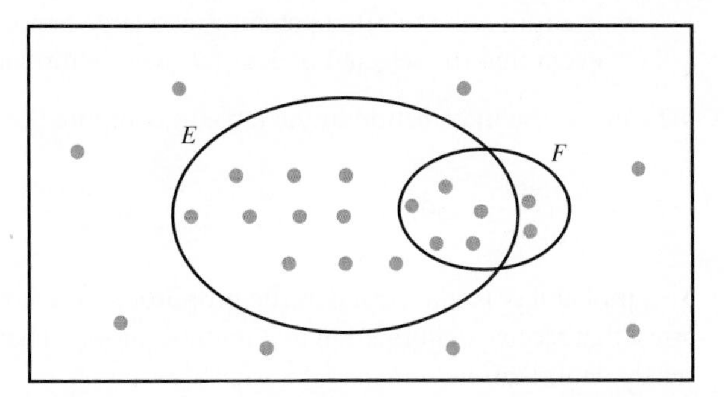

FIGURE 6.8
Venn diagram for Example 6.12 (each dot represents one GFI switch).

■

The results of the previous example lead us to a general definition of conditional probability.

DEFINITION

Conditional probability: Suppose that E and F are *two events with* $P(F) > 0$. **The conditional probability of the event E given that the event F has occurred,** denoted by $\boldsymbol{P(E|F)}$, is

$$P(E|F) = \frac{P(E \cap F)}{P(F)}$$

Notice the requirement that $P(F) > 0$. In addition to the standard warning about division by 0, there is another reason for requiring $P(F)$ to be positive. If the probability of F were 0, the event F would never occur. Therefore, it would not make sense to calculate the probability of another event conditional on F having occurred.

EXAMPLE 6.13 Surviving a Heart Attack

Understand the context

Medical guidelines recommend that a hospitalized patient who suffers cardiac arrest should receive defibrillation (an electric shock to the heart) within 2 minutes. The paper **"Delayed Time to Defibrillation After In-Hospital Cardiac Arrest" (*The New England Journal of Medicine* [2008]: 9–17)** describes a study of the time to defibrillation for hospitalized patients in hospitals of various sizes.

The authors examined medical records of 6716 patients who suffered cardiac arrest while hospitalized, recording the size of the hospital and whether or not defibrillation occurred in 2 minutes or less. Data from this study are summarized in the accompanying table.

Consider the data

	Time to Defibrillation		
Hospital Size	2 minutes or less	More than 2 minutes	Total
Small (Less than 250 beds)	1124	576	**1700**
Medium (250–499 beds)	2178	886	**3064**
Large (500 or more beds)	1387	565	**1952**
Total	**4689**	**2027**	**6716**

We will assume that these data are representative of the larger group of all hospitalized patients who suffer a cardiac arrest. Suppose that a hospitalized patient who suffered a cardiac arrest is selected at random. The following events are of interest:

S = event that the selected patient is at a small hospital
M = event that the selected patient is at a medium-sized hospital
L = event that the selected patient is at a large hospital
D = event that the selected patient receives defibrillation in 2 minutes or less

Do the work

We can use the information in the table to compute

$$P(D) = \frac{4689}{6716} = .698$$

This probability is interpreted as the proportion of hospitalized patients who suffer cardiac arrest that receive defibrillation in 2 minutes or less. That is, 69.8% of these patients receive timely defibrillation.

Now suppose it is known that the selected patient was at a small hospital. How likely is it that this patient received defibrillation in 2 minutes or less? To answer this question, we need to compute $P(D|S)$, the probability of defibrillation in 2 minutes or less *given* that the patient is at a small hospital. Using the formula for conditional probability, we know

$$P(D|S) = \frac{P(D \cap S)}{P(S)}$$

From the information in the table, we can compute

$$P(D \cap S) = \frac{1124}{6716} = .167$$

$$P(S) = \frac{1700}{6716} = .253$$

and so

$$P(D|S) = \frac{P(D \cap S)}{P(S)} = \frac{.167}{.253} = .660$$

Notice that this is smaller than the unconditional probability, $P(D) = .698$. This tells us that there is a smaller probability of timely defibrillation at a small hospital.

Two other conditional probabilities of interest are

$$P(D|M) = \frac{P(D \cap M)}{P(M)} = \frac{2178/6716}{3064/6716} = .711$$

and

$$P(D|L) = \frac{P(D \cap L)}{P(L)} = \frac{1387/6716}{1952/6716} = .711$$

From this, we see that the probability of timely defibrillation is the same for patients at medium-sized and large hospitals, and that this probability is higher than that for patients at small hospitals.

It is also possible to compute $P(L|D)$, the probability that a patient is at a large hospital given that the patient received timely defibrillation:

$$P(L|D) = \frac{P(L \cap D)}{P(D)} = \frac{1387/6716}{4689/6716} = .296$$

Interpret the results ⟩

Let's look carefully at the interpretation of some of these probabilities:

1. $P(D) = .698$ is interpreted as the proportion *of all hospitalized patients* suffering cardiac arrest who would receive timely defibrillation. Approximately 69.8% of these patients would receive timely defibrillation.
2. $P(D \cap L) = \frac{1387}{6716} = .207$ gives the proportion *of all hospitalized patients* who suffer cardiac arrest who are at a large hospital *and* who would receive timely defibrillation.
3. $P(D|L) = .711$ is the proportion *of patients at large hospitals* suffering cardiac arrest who would receive timely defibrillation.
4. $P(L|D) = .296$ is the proportion *of patients who receive timely defibrillation* who were at large hospitals.

Notice the difference between the unconditional probabilities in Interpretations 1 and 2 and the conditional probabilities in Interpretations 3 and 4. The reference point for the unconditional probabilities is the entire group of interest (all hospitalized patients suffering cardiac arrest), whereas the conditional probabilities are interpreted in a more restricted context defined by the "given" event. ■

Example 6.14 demonstrates the calculation of conditional probabilities and also makes the point that we must be careful when translating real-world problems—especially probability problems—into mathematical form. Not only are probability problems sometimes difficult to formulate precisely, but the answers are sometimes surprising.

EXAMPLE 6.14 Two-Kid Families

Understand the context ⟩

Consider the population of all families with two children. Representing the gender of each child using G for girl and B for boy results in four possibilities: BB, BG, GB, GG. The gender information is sequential, with the first letter indicating the gender of the older sibling. Thus, a family having a girl first and then a boy is denoted GB. If we assume that a child is equally likely to be male or female, each of the four possibilities in the sample space for the chance experiment that selects at random from families with two children is equally likely. Consider the following two questions:

1. What is the probability that the selected family has two girls, given that the family has at least one girl?

2. What is the probability that the selected family has two girls, given that the older child is a girl?

To many people, these questions *appear* to be identical. However, by computing the appropriate probabilities, we can see that they are actually different.

For question 1 we obtain

$$P(\text{family with two girls} \mid \text{family has at least one girl})$$
$$= \frac{P(\text{family with two girls } \textit{and} \text{ family with at least one girl})}{P(\text{family with at least one girl})}$$

Do the work ⟩

$$= \frac{P(GG)}{P(GG \text{ or } BG \text{ or } GB)}$$
$$= \frac{1/4}{3/4} = \frac{.25}{.75} = .3333$$

For question 2 we calculate

$$P(\text{family with two girls} \mid \text{family with older child a girl})$$
$$= \frac{P(\text{family with two girls } \textit{and} \text{ family with older child a girl})}{P(\text{family with older child a girl})}$$
$$= \frac{P(GG)}{P(GB \text{ or } GG)}$$
$$= \frac{1/4}{1/2} = \frac{.25}{.50} = .50$$

The moral of this story is: Correct solutions to probability problems, especially conditional probability problems, are much more the result of careful consideration of the sample spaces and the probabilities than they are a product of what your intuition tells you! ■

As we mentioned in the opening paragraphs of this section, one of the most important practical uses of conditional probability is in making diagnoses. Your mechanic diagnoses your car by hooking it up to a machine and reading the pressures and speeds of the various components. A meteorologist diagnoses the weather by looking at temperatures, isobars, and wind speeds.

Doctors observe characteristics of their patients in an attempt to determine whether or not their patients have a certain disease. Many diseases are not actually observable—or at least not easily so—and often the doctor must make a probabilistic judgment. As we will see, conditional probability plays a large role in evaluating diagnostic techniques.

EXAMPLE 6.15 Diagnosing Tuberculosis

Understand the context ⟩

To illustrate the calculations involved in evaluating a diagnostic test, we consider the case of tuberculosis (TB), an infectious disease that typically attacks lung tissue. Before 1998, culturing was the standard method for diagnosing TB. This method always resulted in a correct diagnosis, but it took 10 to 15 days to produce a test result. In 1998, investigators evaluated a DNA technique that turned out to be much faster (**"LCx: A Diagnostic Alternative for the Early Detection of *Mycobacterium tuberculosis* Complex," *Diagnostic Microbiology and Infectious Diseases* [1998]: 259–264**).

The DNA technique for detecting tuberculosis was evaluated by comparing results from the test to the existing gold standard, with the following results for 207 patients exhibiting symptoms:

Consider the data)

	Standard Test Shows Has Tuberculosis	Standard Test Shows Does Not Have Tuberculosis
DNA Positive Indication	14	0
DNA Negative Indication	12	181

Converting these data to proportions and adding the column and row totals into the table, we get the following information:

Do the work)

	Has TB	Does Not Have TB	Total
DNA+	.0676	.0000	**.0676**
DNA−	.0580	.8744	**.9324**
Total	**.1256**	**.8744**	**1.0000**

A quick look at the table indicates that the DNA technique seems to be working in a manner consistent with what we expect from a diagnostic test. Samples that tested positive with the technique agreed with the standard test in every case. Samples that tested negative were generally in agreement with the standard test, but the table also indicates some false-negative results.

Now think about a randomly selected individual who is tested for TB. Define the following events:

T = event that the individual has tuberculosis
N = event that the DNA test is negative

Then $P(T|N)$ denotes the probability of the event T given that the event N has occurred. We calculate this probability as follows:

$$P(T|N) = P(\text{tuberculosis}|\text{negative DNA test})$$
$$= \frac{P(\text{tuberculosis} \cap \text{negative DNA test})}{P(\text{negative DNA test})}$$
$$= \frac{.0580}{.9324}$$
$$= .0622$$

Interpret the results)

Notice that .1256 is the proportion of those tested who had tuberculosis. The added information provided by the diagnostic test has altered the probability—and provided some measure of relief for the patients who test negative. Once it is known that the test result is negative, the estimated likelihood of the disease is cut in half. If the diagnostic test did not significantly alter the probability, the test would not be very useful to the doctor or patient. ■

EXERCISES 6.30 - 6.42

6.30 Two different airlines have a flight from Los Angeles to New York that departs each weekday morning at a certain time. Let E denote the event that the first airline's flight is fully booked on a particular day, and let F denote the event that the second airline's flight is fully booked on that same day. Suppose that $P(E) = .7$, $P(F) = .6$, and $P(E \cap F) = .54$.

a. Calculate $P(E|F)$, the probability that the first airline's flight is fully booked given that the second airline's flight is fully booked. (Hint: See Example 6.12.)

b. Calculate $P(F|E)$.

6.31 The article **"Chances Are You Know Someone with a Tattoo, and He's Not a Sailor" (*Associated Press*, June 11, 2006)** included results from a survey of adults aged 18 to 50. The accompanying data are consistent with summary values given in the article.

	At Least One Tattoo	No Tattoo
Age 18–29	18	32
Age 30–50	6	44

Assuming these data are representative of adult Americans and that an adult American is selected at random, use the given information to estimate the following probabilities. (Hint: See Example 6.13.)

a. $P(tattoo)$

b. $P(tattoo|age\ 18–29)$

c. $P(tattoo|age\ 30–50)$

d. $P(age\ 18–29|tattoo)$

6.32 The accompanying data are from the article **"Characteristics of Buyers of Hybrid Honda Civic IMA: Preferences, Decision Process, Vehicle Ownership, and Willingness-to-Pay" (Institute for Environmental Decisions, November 2006)**. Each of 311 people who purchased a Honda Civic was classified according to gender and whether the car purchased had a hybrid engine or not.

	Hybrid	Not Hybrid
Male	77	117
Female	34	83

Suppose one of these 311 individuals is to be selected at random.

a. Find the following probabilities:

i. $P(male)$

ii. $P(hybrid)$

iii. $P(hybrid|male)$

iv. $P(hybrid|female)$

v. $P(female|hybrid)$

b. For each of the probabilities calculated in Part (a), write a sentence interpreting the probability. (Hint: See Example 6.13.)

c. Are the probabilities $P(hybrid|male)$ and $P(male|hybrid)$ equal? If not, write a sentence or two explaining the difference between these two probabilities.

6.33 The following graphical display is similar to one that appeared in ***USA Today* (January 8, 2010)**.

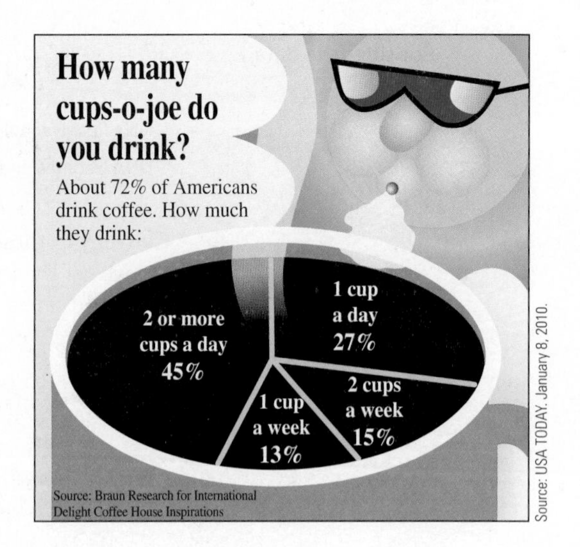

Use the information in this display to answer the following questions. Assume that the percentages in the graph are representative of adult Americans.

a. What is the probability that a randomly selected adult American drinks coffee?

b. The display associates 45% with the category "two or more cups a day." For the chance experiment that consists of selecting an adult American at random, is .45 the probability that the selected individual drinks two or more cups of coffee a day, or is it the conditional probability that the selected individual drinks two or more cups a day given that he or she drinks coffee? Explain.

6.34 Delayed diagnosis of cancer is a problem because it can delay the start of treatment. The paper **"Causes of Physician Delay in the Diagnosis of Breast Cancer" (*Archives of Internal Medicine* [2002]: 1343–1348)** examined possible causes for delayed diagnosis for women with breast cancer. The accompanying table summarizes data on the initial written mammogram report (benign or suspicious) and whether or not diagnosis was delayed for 433 women with breast cancer.

	Diagnosis Delayed	Diagnosis Not Delayed
Mammogram Report Benign	32	89
Mammogram Report Suspicious	8	304

Consider the following events:

B = the event that the mammogram report says benign

S = event that the mammogram report says suspicious

D = event that diagnosis is delayed

a. Assume that these data are representative of the larger group of all women with breast cancer. Use the data in the table to estimate and interpret the following probabilities. (Hint: See Example 6.15.)
   i. $P(B)$
   ii. $P(S)$
   iii. $P(D|B)$
   iv. $P(D|S)$

b. Remember that all of the 433 women in this study actually had breast cancer, so benign mammogram reports were, by definition, in error. Write a few sentences explaining whether this type of error in the reading of mammograms is related to delayed diagnosis of breast cancer.

6.35 The events E and T_i are defined as E = the event that someone who is out of work and actively looking for work will find a job within the next month and T_i = the event that someone who is currently out of work has been out of work for i months. For example, T_2 is the event that someone who is out of work has been out of work for 2 months.

The following conditional probabilities are approximate and were read from a graph in the paper **"The Probability of Finding a Job" (*American Economic Review: Papers & Proceedings* [2008]: 268–273)**:

$P(E|T_1) = .30$ $P(E|T_2) = .24$
$P(E|T_3) = .22$ $P(E|T_4) = .21$
$P(E|T_5) = .20$ $P(E|T_6) = .19$
$P(E|T_7) = .19$ $P(E|T_8) = .18$
$P(E|T_9) = .18$ $P(E|T_{10}) = .18$
$P(E|T_{11}) = .18$ $P(E|T_{12}) = .18$

a. Interpret the following two probabilities:
   i. $P(E|T_1) = .30$
   ii. $P(E|T_6) = .19$

b. Construct a graph of $P(E|T_i)$ versus i. That is, plot $P(E|T_i)$ on the y-axis and $i = 1, 2, \dots, 12$ on the x-axis.

c. Write a few sentences about how the probability of finding a job in the next month changes as a function of length of unemployment.

6.36 The newspaper article **"Folic Acid Might Reduce Risk of Down Syndrome" (*USA Today,* September 29, 1999)** makes the following statement: "Older women are at a greater risk of giving birth to a baby with Down Syndrome than are younger women. But younger women are more fertile, so most children with Down Syndrome are born to mothers under 30."

Let D = event that a randomly selected baby is born with Down Syndrome and Y = event that a randomly selected baby is born to a young mother (under age 30). For each of the following probability statements, indicate whether the statement is consistent with the quote from the article, and if not, explain why not.

a. $P(D|Y) = .001,\ P(D|Y^C) = .004,\ P(Y) = .7$
b. $P(D|Y) = .001,\ P(D|Y^C) = .001,\ P(Y) = .7$
c. $P(D|Y) = .004,\ P(D|Y^C) = .004,\ P(Y) = .7$
d. $P(D|Y) = .001,\ P(D|Y^C) = .004,\ P(Y) = .4$
e. $P(D|Y) = .001,\ P(D|Y^C) = .001,\ P(Y) = .4$
f. $P(D|Y) = .004,\ P(D|Y^C) = .004,\ P(Y) = .4$

6.37 Suppose that an individual is randomly selected from the population of all adult males living in the United States. Let A be the event that the selected individual is over 6 feet in height, and let B be the event that the selected individual is a professional basketball player. Which do you think is larger, $P(A|B)$ or $P(B|A)$? Why?

6.38 ▼ Is ultrasound a reliable method for determining the gender of an unborn baby? The accompanying data on 1000 births are consistent with summary values that appeared in the ***Journal of Statistics Education* ("New Approaches to Learning Probability in the First Statistics Course," 2001)**.

	Ultrasound Predicted Female	Ultrasound Predicted Male
Actual Gender Is Female	432	48
Actual Gender Is Male	130	390

a. Use the given information to estimate the probability that a newborn baby is female, given that the ultrasound predicted the baby would be female.

b. Use the given information to estimate the probability that a newborn baby is male, given that the ultrasound predicted the baby would be male.

c. Based on your answers to Parts (a) and (b), do you think that a prediction that a baby is male and a prediction that a baby is female are equally reliable? Explain.

6.39 The table at the top of the next page summarizes data on smoking status and perceived risk of smoking and is consistent with summary quantities obtained in a Gallup Poll conducted in November 2002. Assume that it is reasonable to consider these data as representative of the adult American population.

	Perceived Risk			
Smoking Status	Very Harmful	Somewhat Harmful	Not Too Harmful	Not at All Harmful
Current Smoker	60	30	5	1
Former Smoker	78	16	3	2
Never Smoked	86	10	2	1

a. What is the probability that a randomly selected adult American is a former smoker?

b. What is the probability that a randomly selected adult American views smoking as very harmful?

c. What is the probability that a randomly selected adult American views smoking as very harmful given that the selected individual is a current smoker?

d. What is the probability that a randomly selected adult American views smoking as very harmful given that the selected individual is a former smoker?

e. What is the probability that a randomly selected adult American views smoking as very harmful given that the selected individual never smoked?

f. How do the probabilities computed in Parts (c), (d), and (e) compare? Does this surprise you? Explain.

6.40 ***USA Today* (June 6, 2000)** gave information on seat belt usage by gender. The proportions in the following table are based on a survey of a large number of adult men and women in the United States.

	Male	Female
Uses Seat Belts Regularly	.10	.175
Does Not Use Seat Belts Regularly	.40	.325

Assume that these proportions are representative of adult Americans and that an adult American is selected at random.

a. What is the probability that the selected adult regularly uses a seat belt?

b. What is the probability that the selected adult regularly uses a seat belt given that the individual selected is male?

c. What is the probability that the selected adult does not use a seat belt regularly given that the selected individual is female?

d. What is the probability that the selected individual is female given that the selected individual does not use a seat belt regularly?

e. Are the probabilities from Parts (c) and (d) equal? Write a couple of sentences explaining why this is so.

6.41 The ***USA Today*** article referenced in the previous exercise also gave information on seat belt usage by age, which is summarized in the following table:

Age	Does Not Use Seat Belt Regularly	Uses Seat Belt Regularly
18–24	59	41
25–34	73	27
35–44	74	26
45–54	70	30
55–64	70	30
65 and older	82	18

Consider the following events: S = event that a randomly selected individual uses a seat belt regularly, A_1 = event that a randomly selected individual is in age group 18–24, and A_6 = event that a randomly selected individual is in age group 65 and older.

a. Convert the counts to proportions and then use them to compute the following probabilities:

i. $P(A_1)$ **ii.** $P(A_1 \cap S)$ **iii.** $P(A_1|S)$

iv. $P(not\ A_1)$ **v.** $P(S|A_1)$ **vi.** $P(S|A_6)$

b. Using the probabilities $P(S|A_1)$ and $P(S|A_6)$ computed in Part (a), comment on how 18–24-year-olds and seniors differ with respect to seat belt usage.

6.42 The paper **"Good for Women, Good for Men, Bad for People: Simpson's Paradox and the Importance of Sex-Specific Analysis in Observational Studies" (*Journal of Women's Health and Gender-Based Medicine* [2001]: 867–872)** described the results of a medical study in which one treatment was shown to be better for men and better for women than a competing treatment. However, if the data for men and women are combined, it appears as though the competing treatment is better.

To see how this can happen, consider the accompanying data tables constructed from information in the paper. Subjects in the study were given either Treatment A or Treatment B, and survival was noted. Let S be the event that a patient selected at random survives, A be the event that a patient selected at random received Treatment A, and B be the event that a patient selected at random received Treatment B.

a. The following table summarizes data for men and women combined:

	Survived	Died	Total
Treatment A	215	85	**300**
Treatment B	241	59	**300**
Total	**456**	**144**	

i. Find $P(S)$.
ii. Find $P(S|A)$.
iii. Find $P(S|B)$.
iv. Which treatment appears to be better?

b. Now consider the summary data for the men who participated in the study:

	Survived	Died	Total
Treatment A	120	80	**200**
Treatment B	20	20	**40**
Total	**140**	**100**	

i. Find $P(S)$.
ii. Find $P(S|A)$.
iii. Find $P(S|B)$.
iv. Which treatment appears to be better?

c. Now consider the summary data for the women who participated in the study:

	Survived	Died	Total
Treatment A	95	5	**100**
Treatment B	221	39	**260**
Total	**316**	**144**	

i. Find $P(S)$.
ii. Find $P(S|A)$.
iii. Find $P(S|B)$.
iv. Which treatment appears to be better?

d. You should have noticed from Parts (b) and (c) that for both men and women, Treatment A appears to be better. But in Part (a), when the data for men and women are combined, it looks like Treatment B is better. This is an example of what is called Simpson's paradox. Write a brief explanation of why this apparent inconsistency occurs for this data set. (Hint: Do men and women respond similarly to the two treatments?)

Bold exercises answered in back ● Data set available online ▼ Video Solution available

6.5 Independence

In Section 6.4, we saw that knowledge of the occurrence of one event can alter our assessment of the likelihood that some other event has occurred. For example, we saw how information about conditional probabilities could be used in medical diagnosis to revise assessments of patients in light of the outcome of a diagnostic procedure. However, it is also possible that knowledge that one event has occurred will not change our assessment of the probability of occurrence of a second event.

EXAMPLE 6.16 MORTGAGE CHOICES

Understand the context ❯

A large lending institution issues both adjustable-rate and fixed-rate mortgage loans on residential property, which it classifies into three categories: single-family houses, condominiums, and multifamily dwellings. The following table, sometimes called a *joint probability table*, displays probabilities based on the bank's long-run lending behavior:

Consider the data ❯

	Single-Family	Condo	Multifamily	Total
Adjustable	.40	.21	.09	**.70**
Fixed	.10	.09	.11	**.30**
Total	**.50**	**.30**	**.20**	

From the table we see that 70% of all mortgages are adjustable rate, 50% of all mortgages are for single-family properties, 40% of all mortgages are adjustable rate for single-family properties (adjustable-rate *and* single-family), and so on.

Define the events E and F by

E = event that a mortgage is adjustable rate
F = event that a mortgage is for a single-family house

Do the work ⟩ Then $P(E) = .70$ and

$$P(E|F) = \frac{P(E \text{ and } F)}{P(F)} = \frac{.40}{.50} = .80$$

Interpret the results ⟩ That is, 80% of loans made for single-family houses are adjustable-rate loans. Notice that $P(E|F)$ is larger than the original (unconditional) probability $P(E) = .70$. Also,

$$P(F|E) = \frac{P(E \text{ and } F)}{P(E)} = \frac{.40}{.70} = .571$$

which is larger than the unconditional probability $P(F) = .5$. Knowing that E has occurred has changed our assessment of how likely it is that F has also occurred.

Now consider another event C defined as

C = event that a mortgage is for a condominium

then

$$P(E|C) = \frac{P(E \text{ and } C)}{P(C)} = \frac{.21}{.30} = .70$$

Notice that $P(E|C) = P(E)$. In this case, knowing that a mortgage is for a condominium doesn't change our assessment of the probability that the mortgage has an adjustable interest rate. ■

When two events E and F are such that $P(E|F) = P(E)$, the probability that event E has occurred is the same after we learn that F has occurred as it was before we knew that F had occurred. In this case, we say that E and F are independent of one another.

DEFINITION

Independent events: Two events E and F are said to be **independent** if

$P(E|F) = P(E)$

Dependent events: If two events E and F are not independent, they are said to be **dependent** events.

If $P(E|F) = P(E)$, it is also true that $P(F|E) = P(F)$, and vice versa.

Independence of events E and F also implies the following additional three relationships:

$P(\text{not } E|F) = P(\text{not } E)$
$P(E|\text{not } F) = P(E)$
$P(\text{not } E|\text{not } F) = P(\text{not } E)$

This means that if E and F are independent, nothing we learn about F will change the likelihood of E or of *not* E.

Recall that the formula for conditional probability is

$$P(E|F) = \frac{P(E \cap F)}{P(F)}$$

which can be rearranged to give

$$P(E \cap F) = P(E|F)P(F)$$

When E and F are independent, $P(E|F) = P(E)$, so it follows that if E and F are independent,

$$P(E \cap F) = P(E|F)P(F) = P(E)P(F)$$

This result is called the multiplication rule for two independent events.

Multiplication Rule for Two Independent Events

The events E and F are independent if and only if

$$P(E \cap F) = P(E)P(F)$$

EXAMPLE 6.17 Hitchhiker's Thumb

Understand the context ⟩

In humans, there is a gene that controls a characteristic known as hitchhiker's thumb. Hitchhiker's thumb is the ability to bend the last joint of the thumb back at an angle of 60° or more. Whether a child has hitchhiker's thumb is determined by two random events: which of two alleles is contributed by the father and which of two alleles is contributed by the mother. You can think of these alleles as a parental vote of yes or no on the hitchhiker's thumb gene. If the votes by the two parents disagree, the dominant allele wins and the child does not have hitchhiker's thumb. These two random events, the results of cell division in two different biological parents, are independent of each other.

Do the work ⟩

Suppose that there is a .10 probability that a parent contributes a positive hitchhiker's thumb allele. Because the events are independent, the probability that each parent contributes a positive hitchhiker's thumb allele, H+, to the offspring is

$$\begin{aligned} &P(\textit{mother contributes } \text{H+} \cap \textit{father contributes } \text{H+}) \\ &\quad = P(\textit{mother contributes } \text{H+})P(\textit{father contributes } \text{H+}) \\ &\quad = (.10)(.10) \\ &\quad = .01 \end{aligned}$$

Interpret the results ⟩

This means that only about 1% of all children in the population would have hitchhiker's thumb. ■

EXAMPLE 6.18 Curious Guppies

Maximilian Weinzier/Alamy

Let's look at another example, this time from the field of animal behavior, to illustrate how an investigator could judge whether two events are independent. In a number of fish species, including guppies, a phenomenon known as predator inspection has been reported. It is thought that predator inspection allows a guppy to assess the risk posed by a potential predator. In a typical inspection a guppy moves toward a predator, presumably to acquire information and then (hopefully) depart to inspect again another day.

Understand the context ⟩

Investigators have observed that guppies sometimes approach and inspect a predator in pairs. Suppose that it is not known whether these predator inspections are independent or whether the guppies are operating as a team. Denote the probability that an individual guppy will inspect a predator by p. Let event E_1 be the event that guppy 1 will approach and inspect a predator and E_2 be the event that guppy 2 will approach and inspect a predator. Then the probability of guppies 1 and 2, approaching the predator at the same time by chance if they are acting *independently* is

$$P(E_1 \cap E_2) = P(E_1)P(E_2) = p \cdot p = p^2$$

Interpret the results ⟩

Based on our analysis, if the inspections are in fact independent, we would expect the proportion of times that two guppies happen to simultaneously inspect a predator to be equal to the square of the proportion of times that a single fish does so. For example, if the probability a single guppy inspects a predator is .3, we would expect the proportion of the time that two guppies would inspect a predator simultaneously to be $(.3)^2 = .09$. Based on observations of the inspection behavior of a large number of guppies, scientists found that inspections by two guppies occurred much more often than 9% of the time. As a result, they concluded that the inspection behavior of guppies does not appear to be independent. ■

The concept of independence extends to more than two events. Consider three events, E_1, E_2, and E_3. Then independence means not only that

$$P(E_1|E_2) = P(E_1)$$
$$P(E_3|E_2) = P(E_3)$$

and so on but also that

$$P(E_1|E_2 \text{ and } E_3) = P(E_1)$$
$$P(E_1 \text{ and } E_3|E_2) = P(E_1 \text{ and } E_3)$$

and so on. There is also a multiplication rule for more than two independent events.

The independence of more than two events is an important concept in studying complex systems with many components. If these components are critical to the operation of a machine, an examination of the probability of the machine's failure is undertaken by analyzing the failure probabilities of the components.

Multiplication Rule for *k* Independent Events

Events $E_1, E_2, \ldots, E_k$ are **independent** if knowledge that any of the events have occurred does not change the probabilities that any particular one or more of the other events has occurred.

Independence implies that

$$P(E_1 \cap E_2 \cap \cdots \cap E_k) = P(E_1)P(E_2) \cdots P(E_k)$$

This means that when events are independent, the probability that all occur together is the product of the individual probabilities. This relationship also holds if one or more of the events is replaced by its complement.

In Example 6.19 we take a rather simplified view of a desktop computer to illustrate the use of the multiplication rule.

EXAMPLE 6.19 Computer Configurations

Understand the context)

Suppose that a desktop computer system consists of a monitor, a mouse, a keyboard, the computer processor itself, and storage devices such as a disk drive. Most computer system problems due to manufacturing defects occur soon in the system's lifetime. Purchasers of new computer systems are advised to turn their computers on as soon as they are purchased and then to let them run for a few hours to see if any problems crop up.

Let

E_1 = event that a newly purchased monitor is not defective
E_2 = event that a newly purchased mouse is not defective
E_3 = event that a newly purchased disk drive is not defective
E_4 = event that a newly purchased computer processor is not defective

Suppose these four events are independent, with

$$P(E_1) = P(E_2) = .98 \qquad P(E_3) = .95 \qquad P(E_4) = .99$$

Do the work ⟩ The probability that all these components are not defective and that the system will operate properly is then

$$\begin{aligned} P(E_1 \cap E_2 \cap E_3 \cap E_4) &= P(E_1)P(E_2)P(E_3)P(E_4) \\ &= (.98)(.98)(.94)(.99) \\ &= .89 \end{aligned}$$

Interpret the result ⟩ We interpret this probability as follows: In the long run, 89% of such systems will run properly when tested shortly after purchase. (In reality, the reliability of these components is much higher than the numbers used in this example!) The probability that all components except the monitor will run properly is

$$\begin{aligned} P(E_1^C \cap E_2 \cap E_3 \cap E_4) &= P(E_1^C)P(E_2)P(E_3)P(E_4) \\ &= (1 - P(E_1))P(E_2)P(E_3)P(E_4) \\ &= (.02)(.98)(.94)(.99) \\ &= .018 \end{aligned}$$

■

Sampling With and Without Replacement

One area of statistics where the rules of probability are important is sampling. As we saw in Chapter 2, a well-designed sampling plan allows investigators to make inferences about a population based on information from a sample. Sampling methods can be classified into two categories: **sampling with replacement** and **sampling without replacement**.

Most inferential methods presented in an introductory statistics course are based on the assumption of sampling *with* replacement, but when sampling from real populations, we almost always sample *without* replacement. This seemingly contradictory practice can be a source of confusion. Fortunately, under certain conditions, the distinction between sampling with and without replacement is not important.

DEFINITION

Sampling with replacement: Once selected, an individual or object is put back into the population before the next selection.

Sampling without replacement: Once selected, an individual or object is not returned to the population prior to subsequent selections.

EXAMPLE 6.20 Sampling With and Without Replacement

Consider the process of selecting three cards from a standard deck of cards. This selection can be made in two ways. One method is to shuffle the cards and then deal three cards off the top of the deck. This would be sampling without replacement. A second method, rarely seen in real games, is to select a card at random, note which card is observed, replace it in the deck, and shuffle before selecting the next card. This method is sampling with replacement.

From the standpoint of probability, sampling with and without replacement are analyzed differently. To see this, consider these events:

H_1 = event that the first card is a heart
H_2 = event that the second card is a heart
H_3 = event that the third card is a heart

For sampling with replacement, the probability of H_3 is .25, regardless of whether either H_1 or H_2 occurs, because replacing selected cards gives the same deck for the third selection as for the first two selections. Whether either of the first two cards is a heart has no bearing on the third card selected, so the three events, H_1, H_2, and H_3, are independent.

When sampling is without replacement, the chance of getting a heart on the third draw does depend on the results of the first two draws. If both H_1 and H_2 occur, only 11 of the

50 remaining cards are hearts. Because any one of these 50 has the same chance of being selected, the probability of H_3 in this case is

$$P(H_3|H_1 \text{ and } H_2) = \frac{11}{50} = .22$$

Alternatively, if neither of the first 2 cards is a heart, then all 13 hearts remain in the deck for the third draw, so

$$P(H_3|\text{ not } H_1 \text{ and not } H_2) = \frac{13}{50} = .26$$

Information about the occurrence of H_1 and H_2 affects the chance that H_3 has occurred. For sampling without replacement, the three events described here are not independent. ■

In opinion polls and other types of surveys, sampling is virtually always done without replacement. For this method of sampling, the results of successive selections are not independent of one another. However, Example 6.21 suggests that, *under certain circumstances*, the fact that selections in sampling without replacement are not independent is not a cause for concern.

EXAMPLE 6.21 Independence and Sampling Without Replacement

A lot of 10,000 computer chips used in graphing calculators consists of 2500 manufactured by one firm and 7500 manufactured by a second firm, all mixed together. Three of the chips will be selected at random *without* replacement. Let

E_1 = event that first chip selected was manufactured by Firm 1
E_2 = event that second chip selected was manufactured by Firm 1
E_3 = event that third chip selected was manufactured by Firm 1

Following the reasoning we used in Example 6.20, we get

$$P(E_3|E_1 \text{ and } E_2) = \frac{2498}{9998} = .24985$$

$$P(E_3|\text{not } E_1 \text{ and not } E_2) = \frac{2500}{9998} = .25005$$

Although these two probabilities differ slightly, when rounded to three decimal places they are both .250. We conclude that the occurrence or nonoccurrence of E_1 or E_2 has virtually no effect on the chance that E_3 will occur. *For practical purposes*, the three events can be considered independent. ■

The essential difference between the situations of Example 6.20 and Example 6.21 is the size of the sample relative to the size of the population. In Example 6.20 a relatively large proportion of the population was sampled (3 out of 52), whereas in Example 6.21 the proportion of the population sampled was quite small (only 3 out of 10,000).

In most settings, the sample size is small compared to the size of the population. The theory of sampling with replacement required for many inferential methods can coexist with the practice of sampling without replacement because of the following principle:

If a random sample of size *n* is taken from a population of size *N*, the theoretical probabilities of successive selections calculated on the basis of sampling with replacement and on the basis of sampling without replacement differ by insignificant amounts when *n* is small compared to *N*.

In practice, independence can be assumed for the purpose of calculating probabilities as long as *n* is not larger than 5% of *N*.

This principle justifies the assumption of independence in many statistical problems. The phrase *assumption of independence* does not signify that the investigators are in some sense fooling themselves. They are recognizing that, for all practical purposes, the results will not differ from the "right" answers.

In some sampling situations, the sample size might be more than 5% of the population. For example, a newspaper editor at a small school might easily sample more than 5% of the students at the school to assess student opinion on a particular issue. In that instance, the editors would be wise to consult a statistician before proceeding. Does this mean that an investigator should never sample more than 5% of a population? Certainly not! It is almost always the case that a larger sample results in better inferences about the population. The only disadvantage of sampling more than 5% of the population (other than an increase in the time and resources required) is that the analysis of the resulting data is slightly more complicated.

EXERCISES 6.43 - 6.58

6.43 Many fire stations handle emergency calls for medical assistance as well as calls requesting firefighting equipment. A particular station says that the probability that an incoming call is for medical assistance is .85. This can be expressed as *P(call is for medical assistance)* = .85.

a. Give a relative frequency interpretation of the given probability.

b. What is the probability that a call is not for medical assistance?

c. Assuming that successive calls are independent of one another, calculate the probability that two successive calls will both be for medical assistance. (Hint: See Example 6.18.)

d. Still assuming independence, calculate the probability that for two successive calls, the first is for medical assistance and the second is not for medical assistance.

e. Still assuming independence, calculate the probability that exactly one of the next two calls will be for medical assistance. (Hint: There are two different possibilities. The one call for medical assistance might be the first call, or it might be the second call.)

f. Do you think that it is reasonable to assume that the requests made in successive calls are independent? Explain.

6.44 The paper **"Predictors of Complementary Therapy Use Among Asthma Patients: Results of a Primary Care Survey" (*Health and Social Care in the Community* [2008]: 155–164)** included the accompanying table. The table summarizes the responses given by 1077 asthma patients to two questions:

Question 1: Do conventional asthma medications usually help your asthma symptoms?

Question 2: Do you use complementary therapies (such as herbs, acupuncture, aroma therapy) in the treatment of your asthma?

	Doesn't Use Complementary Therapies	Does Use Complementary Therapies
Conventional Medications Usually Help	816	131
Conventional Medications Usually Do Not Help	103	27

Consider a chance experiment that consists of randomly selecting one of the 1077 survey participants.

a. Construct a joint probability table by dividing the count in each cell of the table by the sample size $n = 1077$. (Hint: See Example 6.16.)

b. The joint probability in the upper left cell of the table from Part (a) is $\frac{816}{1077} = .758$. This represents the probability of conventional medications usually help *and* does not use complementary therapies. Interpret the other three probabilities in the joint probability table of Part (a).

c. Are the events

CH = event that the selected participant reports that conventional medications usually help

and

CT = event that the selected participant reports using complementary therapies

independent events?

Use a probability argument to justify your choice. (Hint: See Example 6.18.)

6.45 The report **"TV Drama/Comedy Viewers and Health Information" (www.cdc.gov/Healthmarketing)** describes the results of a large survey involving approximately 3500 people that was conducted for the Center for Disease Control. The sample was selected in a way that the Center for Disease Control believed would result in a sample that was representative of adult Americans.

One question on the survey asked respondents if they had learned something new about a health issue or disease from a TV show in the previous 6 months. Data from the survey was used to estimate the following probabilities, where

L = event that a randomly selected adult American reports learning something new about a health issue or disease from a TV show in the previous 6 months

and

F = event that a randomly selected adult American is female

$P(L) = .58 \qquad P(L \cap F) = .31$

Assume that $P(F) = .5$. Are the events L and F independent events? Use probabilities to justify your answer.

6.46 The article **"SUVs Score Low in New Federal Rollover Ratings" (*San Luis Obispo Tribune*, January 6, 2001)** gave information on death rates for various kinds of accidents by vehicle type for accidents reported to the police. Suppose that we randomly select an accident reported to the police and consider the following events:

R = event that the selected accident is a single-vehicle rollover
F = event that the selected accident is a frontal collision
D = event that the selected accident results in a death

Information in the article indicates that the following probability estimates are reasonable:

$P(R) = .06$, $P(F) = .60$, $P(R|D) = .30$, $P(F|D) = .54$.

a. Interpret the value of $P(R|D)$.
b. Interpret the value of $P(F|D)$.
c. Are the events R and D independent? Justify your answer.
d. Is $P(F \cap D) = P(F)P(D)$? Explain why or why not.
e. Is $F = R^C$? Explain how you can tell.

6.47 A Gallup survey of 2002 adults found that 46% of women and 37% of men experience pain daily **(*San Luis Obispo Tribune*, April 6, 2000)**. Suppose that this information is representative of adult Americans. If an adult American is selected at random, are the events *selected adult is male* and *selected adult experiences pain daily* independent or dependent? Explain.

6.48 In a small city, approximately 15% of those eligible are called for jury duty in any one calendar year. People are selected for jury duty at random from those eligible, and the same individual cannot be called more than once in the same year.

a. What is the probability that a particular eligible person in this city is selected in both of the next 2 years?
b. What is the probability that a particular eligible person in this city is selected in all three of the next 3 years?

6.49 Jeanie is a bit forgetful, and if she doesn't make a "to do" list, the probability that she forgets something she is supposed to do is .1. Tomorrow she intends to run three errands, and she fails to write them on her list.

a. What is the probability that Jeanie forgets all three errands? What assumptions did you make to calculate this probability?
b. What is the probability that Jeanie remembers at least one of the three errands?
c. What is the probability that Jeanie remembers the first errand but not the second or third?

6.50 ▼ Approximately 30% of the calls to an airline reservation phone line result in a reservation being made.

a. Suppose that an operator handles 10 calls. What is the probability that none of the 10 calls result in a reservation?
b. What assumption did you make to calculate the probability in Part (a)?
c. What is the probability that at least one call results in a reservation being made? (Hint: Use your answer to Part (a).)

6.51 Consider a system consisting of four components, as pictured in the following diagram:

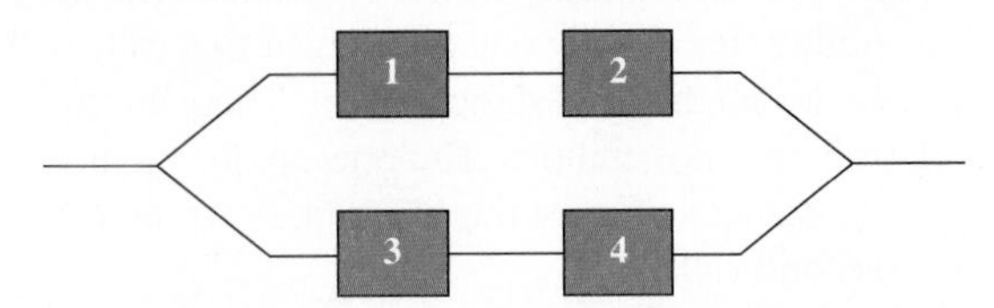

Components 1 and 2 form a series subsystem, as do Components 3 and 4. The two subsystems are connected in parallel. Suppose that

$P(1 \text{ works}) = .9$
$P(2 \text{ works}) = .9$
$P(3 \text{ works}) = .9$
$P(4 \text{ works}) = .9$

and that the four components work independently of one another.

a. The 1–2 subsystem works only if both components work. What is the probability of this happening?

b. What is the probability that the 1–2 subsystem doesn't work? that the 3–4 subsystem doesn't work?

c. The system won't work if the 1–2 subsystem doesn't work and if the 3–4 subsystem also doesn't work. What is the probability that the system won't work? that it will work?

d. How would the probability of the system working change if a 5–6 subsystem were added in parallel with the other two subsystems?

e. How would the probability that the system works change if there were three components in series in each of the two subsystems?

6.52 Information from a poll of registered voters in Cedar Rapids, Iowa, to assess voter support for a new school tax was the basis for the following statements (***Cedar Rapids Gazette,*** **August 28, 1999**):

> The poll showed 51% of the respondents in the Cedar Rapids school district are in favor of the tax. The approval rating rises to 56% for those with children in public schools. It falls to 45% for those with no children in public schools. The older the respondent, the less favorable the view of the proposed tax: 36% of those over age 56 said they would vote for the tax compared with 72% of 18- to 25-year-olds.

Suppose that a registered voter from Cedar Rapids is selected at random, and define the following events:

F = event that the selected individual favors the school tax
C = event that the selected individual has children in the public schools
O = event that the selected individual is over 56 years old
Y = event that the selected individual is 18–25 years old

a. Use the given information to estimate the values of the following probabilities:

i. $P(F)$
ii. $P(F|C)$
iii. $P(F|C^C)$
iv. $P(F|O)$
v. $P(F|Y)$

b. Are F and C independent? Justify your answer.

c. Are F and O independent? Justify your answer.

6.53 The following case study was reported in the article **"Parking Tickets and Missing Women,"** which appeared in an early edition of the book ***Statistics: A Guide to the Unknown***. In a Swedish trial on a charge of overtime parking, a police officer testified that he had noted the position of the two air valves on the tires of a parked car: To the closest hour, one was at the one o'clock position and the other was at the six o'clock position. After the allowable time for parking in that zone had passed, the policeman returned, noted that the valves were in the same position, and ticketed the car. The owner of the car claimed that he had left the parking place in time and had returned later. The valves just happened by chance to be in the same positions.

An "expert" witness computed the probability of this occurring as (1/12)(1/12) = 1/144.

a. What reasoning did the expert use to arrive at the probability of 1/144?

b. Can you spot the error in the reasoning that leads to the stated probability of 1/144?

c. What effect does this error have on the probability of occurrence?

d. Do you think that 1/144 is larger or smaller than the correct probability of occurrence?

6.54 Three friends (A, B, and C) will participate in a round-robin tournament in which each one plays both of the others. Suppose that

P(A beats B) = .7
P(A beats C) = .8
P(B beats C) = .6

and that the outcomes of the three matches are independent of one another.

a. What is the probability that A wins both her matches and that B beats C?

b. What is the probability that A wins both her matches?

c. What is the probability that A loses both her matches?

d. What is the probability that each person wins one match? (Hint: There are two different ways for this to happen.)

6.55 A shipment of 5000 printed circuit boards contains 40 that are defective. Two boards will be chosen at random, without replacement. Consider the two events E_1 = event that the first board selected is defective and E_2 = event that the second board selected is defective.

a. Are E_1 and E_2 dependent events? Explain in words.

b. Let *not* E_1 be the event that the first board selected is not defective (the event E_1^C). What is $P(not\ E_1)$?

c. How do the two probabilities $P(E_2|E_1)$ and $P(E_2|not\ E_1)$ compare?

d. Based on your answer to Part (c), would it be reasonable to view E_1 and E_2 as approximately independent?

6.56 A store sells two different brands of dishwasher soap, and each brand comes in three different sizes: small (S), medium (M), and large (L). The proportions of the two brands and of the three sizes purchased are displayed as marginal totals in the following table.

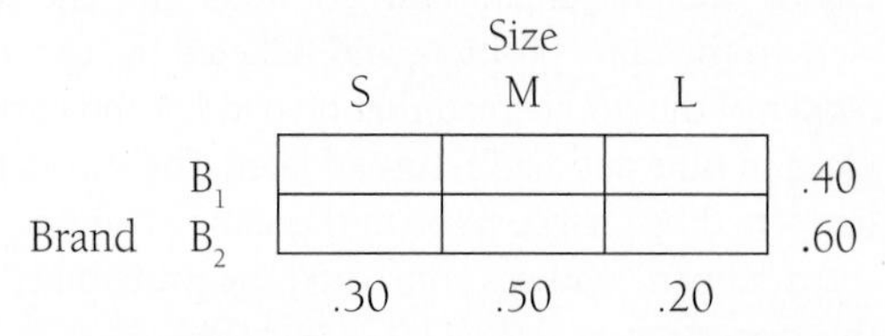

		Size			
		S	M	L	
Brand	B_1				.40
	B_2				.60
		.30	.50	.20	

Suppose that any event involving brand is independent of any event involving size. What is the probability of the event that a randomly selected purchaser buys the small size of Brand B_1 (the event $B_1 \cap S$)? What are the probabilities of the other brand–size combinations?

6.57 The **National Public Radio** show ***Car Talk*** has a feature called "The Puzzler." Listeners are asked to send in answers to some puzzling questions—usually about cars but sometimes about probability (which, of course, must account for the incredible popularity of the program!).

Suppose that for a car question, 800 answers are submitted, of which 50 are correct.

a. Suppose that the hosts randomly select two answers from those submitted *with replacement*. Calculate the probability that both selected answers are correct. (For purposes of this problem, keep at least five digits to the right of the decimal.)

b. Suppose now that the hosts select the answers at random but *without replacement*. Use conditional probability to evaluate the probability that both answers selected are correct. How does this probability compare to the one computed in Part (a)?

6.58 Refer to the previous exercise. Suppose now that for a probability question, 100 answers are submitted, of which 50 are correct. Calculate the probabilities in Parts (a) and (b) of the previous exercise for the probability question.

Bold exercises answered in back ● Data set available online ▼ Video Solution available

6.6 Some General Probability Rules

In previous sections, we saw how the probability of $P(E \cup F)$ could be easily computed when E and F are mutually exclusive and how $P(E \cap F)$ could be computed when E and F are independent. In this section, we develop more general rules: an addition rule that can be used even when events are not mutually exclusive and a multiplication rule that can be used even when events are not independent.

General Addition Rule

Calculating $P(E \cup F)$ when the two events are not mutually exclusive is a bit more complicated than in the case of mutually exclusive events. Consider Figure 6.9, in which E and F overlap. The area of the colored region ($E \cup F$) is not the sum of the area of E and the area of F, because when the two individual areas are added, the area of the intersection ($E \cap F$) is counted twice. Similarly, $P(E) + P(F)$ includes $P(E \cap F)$ twice, so this intersection probability must then be subtracted from the sum to obtain $P(E \cup F)$. This reasoning leads to the general addition rule.

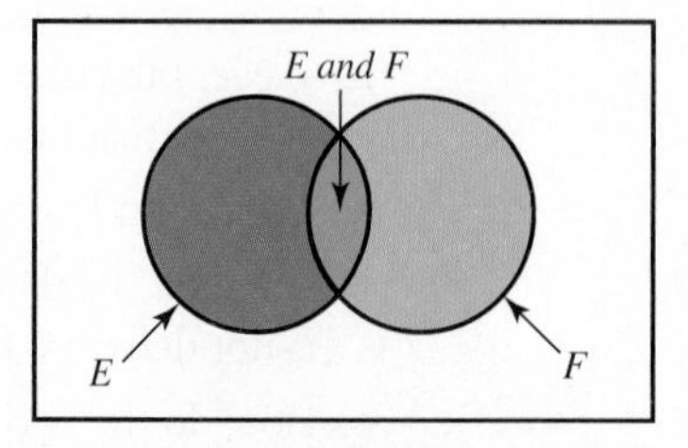

FIGURE 6.9
The colored region is $P(E \cup F)$, and $P(E \cup F) \neq P(E) + P(F)$.

General Addition Rule for Two Events

For any two events E and F,

$$P(E \cup F) = P(E) + P(F) - P(E \cap F)$$

When E and F are mutually exclusive, the general addition rule simplifies to the previous rule for mutually exclusive events. This is because when E and F are mutually exclusive, $E \cap F$ contains no outcomes and $P(E \cap F) = 0$. The general addition rule can be used to determine any one of the four probabilities $P(E)$, $P(F)$, $P(E \cap F)$, or $P(E \cup F)$ provided that the other three probabilities are known.

EXAMPLE 6.22 Cable Services

Understand the context ❯

Suppose that 60% of all customers of a large cable company subscribe to Internet service, 40% subscribe to phone service, and 25% have both types of services. If a customer is selected at random, what is the probability that he or she has at least one of these two types of service? We can define the following events:

E = event that a selected customer has Internet service
F = event that a selected customer has phone service

Do the work ❯

The given information implies that

$$P(E) = .60 \quad P(F) = .40 \quad P(E \cap F) = .25$$

from which we obtain

$$\begin{aligned} &P(\text{customer has at least one of the two types of services}) \\ &\quad = P(E \cup F) \\ &\quad = P(E) + P(F) - P(E \cap F) \\ &\quad = .60 + .40 - .25 \\ &\quad = .75 \end{aligned}$$

The event that the customer has neither type of service is $(E \cup F)^C$, so

$$P(\text{customer has neither type of service}) = 1 - P(E \cup F) = .25$$

Now let's determine the probability that the selected customer has exactly one type of service. Referring to the Venn diagram in Figure 6.10, we see that the event *at least one* can be thought of as consisting of two mutually exclusive parts: *exactly one* and *both*. This means that

$$\begin{aligned} P(E \cup F) &= P(\text{at least one}) \\ &= P(\text{exactly one} \cup \text{both}) \\ &= P(\text{exactly one}) + P(\text{both}) \\ &= P(\text{exactly one}) + P(E \cap F) \end{aligned}$$

It follows that

$$\begin{aligned} P(\text{exactly one}) &= P(E \cup F) - P(E \cap F) \\ &= .75 - .25 \\ &= .50 \end{aligned}$$

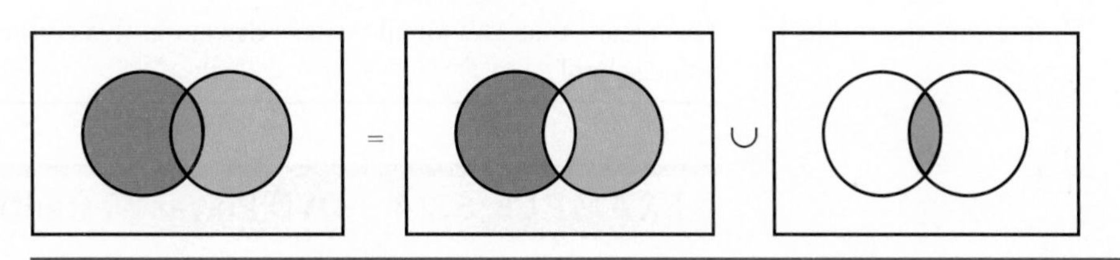

FIGURE 6.10
Representing $P(E \cup F)$ as the union of two mutually exclusive events.

■

The general addition rule for more than two events is rather complicated. For example, in the case of three events,

$$\begin{aligned} P(E \cup F \cup G) = P(E) + P(F) + P(G) - P(E \cap F) - P(E \cap G) \\ - P(F \cap G) + P(E \cap F \cap G) \end{aligned}$$

For more than three events, you should consult a book with more extensive coverage of probability.

General Multiplication Rule

In Section 6.4, we used the formula

$$P(E|F) = \frac{P(E \cap F)}{P(F)}$$

to compute conditional probabilities when $P(E \cap F)$ is known. Sometimes, however, conditional probabilities are known or can be estimated. When this is the case, they can be used to calculate the probability of the intersection of two events.

Multiplying both sides of the conditional probability formula by $P(F)$ gives a useful expression for the probability that both events E and F will occur.

General Multiplication Rule for Two Events

For any two events E and F,

$$P(E \cap F) = P(E|F)P(F)$$

EXAMPLE 6.23 Traffic School

Understand the context ⟩ Suppose that 20% of all teenage drivers in a certain county received a citation for a moving violation in 2014. Also suppose that 80% of those receiving such a citation attended traffic school so that the citation would not appear on their driving record. If a teenage driver from this county is randomly selected, what is the probability that he or she received a citation and attended traffic school?

Formulate a plan ⟩ Let's define two events E and F as follows:

E = selected driver attended traffic school
F = selected driver received such a citation

The question posed can then be answered by calculating $P(E \cap F)$. The percentages given in the problem imply that

$$P(F) = .20$$
and $$P(E|F) = .80.$$

Do the work ⟩ Notice the difference between $P(E)$, which is the proportion in the entire population who attended traffic school (not given), and $P(E|F)$, which is the proportion of those receiving a citation that attended traffic school. Using the multiplication rule, we calculate

$$\begin{aligned} P(E \text{ and } F) &= P(E|F)P(F) \\ &= (.80)(.20) \\ &= .16 \end{aligned}$$

Interpret the results ⟩ This means that 16% of all teenage drivers in this county received a citation *and* attended traffic school. ■

EXAMPLE 6.24 DVD Player Warranties

Understand the context ⟩ The following table gives information on DVD players sold by a large electronics store:

	Percentage of Customers Purchasing	Of Those Who Purchase, Percentage Who Purchase Extended Warranty
Brand 1	70	20
Brand 2	30	40

Suppose a purchaser is randomly selected from among all those who bought a DVD player from this store. What is the probability that the selected customer purchased a Brand 1 model and an extended warranty?

Formulate a plan) To answer this question, we first define the following events:

B_1 = event that Brand 1 is purchased
B_2 = event that Brand 2 is purchased
E = event that an extended warranty is purchased

The information in the table implies that

$$P(\text{Brand 1 purchased}) = P(B_1) = .70$$
$$P(\text{extended warranty}|\text{Brand 1 purchased}) = P(E|B_1) = .20$$

Do the work) Notice that the 20% is identified with a *conditional* probability. *Among purchasers of Brand 1*, this is the percentage opting for an extended warranty. Substituting these numbers into the general multiplication rule yields

$$\begin{aligned} P(B_1 \textit{ and } E) &= P(E|B_1)P(B_1) \\ &= (.20)(.70) \\ &= .14 \end{aligned}$$

The tree diagram of Figure 6.11 gives a nice visual display of how the general multiplication rule is used here. The two first-generation branches are labeled with events B_1 and B_2 along with their probabilities. Two second-generation branches extend from each first-generation branch. These correspond to the two events E and *not* E. The *conditional* probabilities $P(E|B_1)$, $P(\textit{not } E|B_1)$, $P(E|B_2)$, and $P(\textit{not } E|B_2)$, appear on these branches. Application of the multiplication rule then consists of multiplying probabilities along the branches of the tree diagram. For example,

$$\begin{aligned} P(\text{Brand 2 and warranty purchased}) &= P(B_2 \textit{ and } E) \\ &= P(B_2 \cap E) \\ &= P(E|B_2)P(B_2) \\ &= (.4)(.3) \\ &= .12 \end{aligned}$$

and this probability is displayed to the right of the E branch that comes from the B_2 branch.

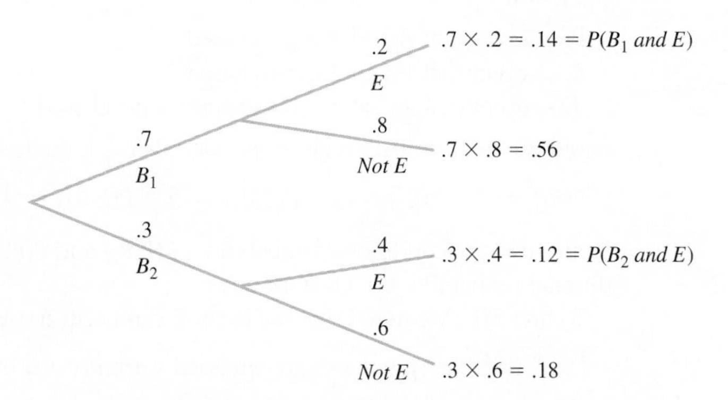

FIGURE 6.11
A tree diagram for the probability calculations of Example 6.24.

We can now easily calculate $P(E)$, the probability that an extended warranty is purchased. The event E can occur in two different ways: Buy Brand 1 *and* warranty, or buy Brand 2 *and* warranty. Symbolically, these events are $B_1 \cap E$ and $B_2 \cap E$. Furthermore, if each customer purchased a single DVD player, he or she could not have simultaneously purchased both Brand 1 and Brand 2, so the two events $B_1 \cap E$ and $B_2 \cap E$ are mutually exclusive.

It follows that

$$\begin{aligned} P(E) &= P(B_1 \cap E) + P(B_2 \cap E) \\ &= P(E|B_1)P(B_1) + P(E|B_2)P(B_2) \\ &= (.2)(.7) + (.4)(.3) \\ &= .14 + .12 \\ &= .26 \end{aligned}$$

Interpret the results) This probability is the sum of two of the probabilities shown on the right-hand side of the tree diagram. This means that 26% of all DVD player purchasers selected an extended warranty. ■

The general multiplication rule can be extended to give an expression for the probability that several events occur together. In the case of three events E, F, and G, we have

$$P(E \cap F \cap G) = P(E|F \cap G)P(F|G)P(G)$$

When the events are all independent, $P(E|F \cap G) = P(E)$ and $P(F|G) = P(F)$ so the right-hand side of the equation for $P(E \cap F \cap G)$ simplifies to the product of the three unconditional probabilities.

EXAMPLE 6.25 Lost Luggage

Understand the context) Twenty percent of all passengers who fly from Los Angeles (LA) to New York (NY) do so on Airline G. This airline misplaces luggage for 10% of its passengers, and 90% of this lost luggage is subsequently recovered. If a passenger who has flown from LA to NY is randomly selected, what is the probability that the selected individual flew on Airline G (event G), had luggage misplaced (event F), and subsequently recovered the misplaced luggage (event E)?

Do the work) The given information implies that

$$P(G) = .20 \quad P(F|G) = .10 \quad P(E|F \cap G) = .90$$

Then

$$P(E \cap F \cap G) = P(E|F \cap G)P(F|G)P(G) = (.90)(.10)(.20) = .018$$

Interpret the results) This means that about 1.8% of passengers flying from LA to NY fly on Airline G, have their luggage misplaced, and subsequently recover the lost luggage. ■

Law of Total Probability

Let's reconsider the information on DVD player sales from Example 6.24. In this example, the following events were defined:

B_1 = event that Brand 1 is purchased
B_2 = event that Brand 2 is purchased
E = event that an extended warranty is purchased

Based on the information given in Example 6.24, the following probabilities are known:

$$P(B_1) = .7 \quad P(B_2) = .3 \quad P(E|B_1) = .2 \quad P(E|B_2) = .4$$

Notice that the conditional probabilities $P(E|B_1)$ and $P(E|B_2)$ are known but that the unconditional probability $P(E)$ is not known.

To find $P(E)$ we noted that the event E can occur in two ways:

(1) A customer purchases an extended warranty *and* buys Brand 1 $(E \cap B_1)$; or

(2) a customer purchases an extended warranty *and* buys Brand 2 $(E \cap B_2)$.

Because these are the only ways in which E can occur, we can write the event E as

$$E = (E \cap B_1) \cup (E \cap B_2)$$

The two events $(E \cap B_1)$ and $(E \cap B_2)$ are mutually exclusive (since B_1 and B_2 are mutually exclusive), so using the addition rule for mutually exclusive events gives

$$\begin{aligned} P(E) &= P((E \cap B_1) \cup (E \cap B_2)) \\ &= P(E \cap B_1) + P(E \cap B_2) \end{aligned}$$

Finally, using the general multiplication rule to evaluate $P(E \cap B_1)$ and $P(E \cap B_2)$ results in

$$\begin{aligned} P(E) &= P(E \cap B_1) + P(E \cap B_2) \\ &= P(E|B_1)P(B_1) + P(E|B_2)P(B_2) \end{aligned}$$

Substituting in the known probabilities gave

$$\begin{aligned} P(E) &= P(E|B_1)P(B_1) + P(E|B_2)P(B_2) \\ &= (.2)(.7) + (.4)(.3) \\ &= .26 \end{aligned}$$

We concluded that 26% of the DVD player customers purchased an extended warranty.

As we have just illustrated, when conditional probabilities are known, they can sometimes be used to compute unconditional probabilities. The **law of total probability** formalizes this use of conditional probabilities.

The Law of Total Probability

If B_1 and B_2 are mutually exclusive events with $P(B_1) + P(B_2) = 1$, then for any event E

$$\begin{aligned} P(E) &= P(E \cap B_1) + P(E \cap B_2) \\ &= P(E|B_1)P(B_1) + P(E|B_2)P(B_2) \end{aligned}$$

More generally, if $B_1, B_2, \ldots, B_k$ are mutually exclusive events with $P(B_1) + P(B_2) + \cdots + P(B_k) = 1$, then for any event E

$$\begin{aligned} P(E) &= P(E \cap B_1) + P(E \cap B_2) + \cdots + P(E \cap B_k) \\ &= P(E|B_1)P(B_1) + P(E|B_2)P(B_2) + \cdots + P(E|B_k)P(B_k) \end{aligned}$$

EXAMPLE 6.26 Which Way to Jump?

Understand the context ⟩

The paper **"Action Bias among Elite Soccer Goalkeepers: The Case of Penalty Kicks"** (***Journal of Economic Psychology*** **[2007]: 606–621**) presents an interesting analysis of 286 penalty kicks in televised championship soccer games from around the world. In a penalty kick, the only players involved are the kicker and the goalkeeper from the opposing team. The kicker tries to kick a ball into the goal from a point located 11 meters away. The goalkeeper tries to block the ball from reaching the goal.

For each penalty kick analyzed, the researchers recorded the direction that the goalkeeper moved (jumped to the left, stayed in the center, or jumped to the right) and whether or not the penalty kick was successfully blocked.

Consider the following events:

L = the event that the goalkeeper jumps to the left
C = the event that the goalkeeper stays in the center
R = the event that the goalkeeper jumps to the right
B = the event that the penalty kick is blocked

Consider the data ⟩

Based on an analysis of the penalty kicks, the authors of the paper gave the following probability estimates:

$P(B|L) = .142$ $\quad$ $P(B|C) = .333$ $\quad$ $P(B|R) = .126$
$P(L) = .493$ $\quad$ $P(C) = .063$ $\quad$ $P(R) = .444$

Formulate a plan ⟩

What proportion of penalty kicks were blocked? We can use the law of total probability to answer this question. Here, the three events L, C, and R play the role of B_1, B_2, and B_3 and B plays the role of E in the formula for the law of total probability.

Substituting into the formula, we get

Do the work ⟩

$$\begin{aligned} P(B) &= P(B \cap L) + P(B \cap C) + P(B \cap R) \\ &= P(B|L)P(L) + P(B|C)P(C) + P(B|R)P(R) \\ &= (.142)(.493) + (.333)(.063) + (.126)(.444) \\ &= .070 + .021 + .056 \\ &= .147 \end{aligned}$$

Interpret the results ⟩

This means that only 14.7% of penalty kicks were successfully blocked. Two other interesting findings of this study were

1. The direction that the goalkeeper moves appears to be independent of whether the kicker kicked the ball to the left, center, or right of the goal. This was attributed to the fact that goalkeepers have to choose their action before they can clearly observe the direction of the kick.
2. Based on the three conditional probabilities—$P(B|L) = .142$, $P(B|C) = .333$, and $P(B|R) = .126$—the optimal strategy for a goalkeeper appears to be to stay in the center of the goal. However, staying in the center was only chosen 6.3% of the time—much less often that jumping left or right. The authors believe that this is because a goalkeeper does not feel as bad about not successfully blocking a kick if some action (jumping left or right) is taken compared to if no action (staying in the center) is taken. This is the "action bias" referred to in the title of the paper. ■

Bayes' Rule

We conclude our discussion of probability rules by considering a formula discovered by the Reverend Thomas Bayes (1702–1761), an English Presbyterian minister. He discovered what is now known as Bayes' rule (or Bayes' theorem). Bayes' rule is a solution to what Bayes called the converse problem. To see what he meant by this, we return to the field of medical diagnosis.

EXAMPLE 6.27 Lyme Disease

Understand the context ⟩

Lyme disease is the leading tick-borne disease in the United States and Europe. Diagnosis of the disease is difficult and is aided by a test that detects particular antibodies in the blood. The article **"Laboratory Considerations in the Diagnosis and Management of Lyme Borreliosis"** (***American Journal of Clinical Pathology*** **[1993]: 168–174)** used the following notation:

$+$ represents a positive result on the blood test
$-$ represents a negative result on the blood test
L represents the event that the patient actually has Lyme disease
L^C represents the event that the patient actually does not have Lyme disease

The following probabilities were reported in the article:

Consider the data ⟩

Probability	Interpretation
$P(L) = .00207$	The prevalence of Lyme disease in the population. About .207% of the population actually has Lyme disease.
$P(L^C) = .99793$	99.793% of the population does not have Lyme disease.
$P(+\|L) = .937$	93.7% of those with Lyme disease test positive.
$P(-\|L) = .063$	6.3% of those with Lyme disease test negative.
$P(+\|L^C) = .03$	3% of those who do not have Lyme disease test positive.
$P(-\|L^C) = .97$	97% of those who do not have Lyme disease test negative.

Notice the form of the known conditional probabilities. For example, $P(+|L)$ is the probability of a positive test *given* that a person selected at random from the population actually has Lyme disease. Bayes' converse problem poses a question of a different form: Given that a person tests positive for the disease, what is the probability that he or she actually has Lyme disease? This converse problem is the one that is of primary interest in medical diagnosis problems.

Bayes reasoned as follows to obtain the answer to the converse problem of finding $P(L|+)$. We know from the definition of conditional probability that

$$P(L|+) = \frac{P(L \cap +)}{P(+)}$$

Because $P(L \cap +) = P(+ \cap L)$, we can use the general multiplication rule to get

$$P(L \cap +) = P(+ \cap L) = P(+|L)P(L)$$

This helps, because both $P(+|L)$ and $P(L)$are known. We now have

$$P(L|+) = \frac{P(+|L)P(L)}{P(+)}$$

The denominator $P(+)$ can be evaluated using the law of total probability, because L and L^C are mutually exclusive with $P(L) + P(L^C) = 1$. Applying the law of total probability to the denominator, we obtain

$$\begin{aligned} P(+) &= P(+ \cap L) + P(+ \cap L^C) \\ &= P(+|L)P(L) + P(+|L^C)P(L^C) \end{aligned}$$

We now have all we need to answer the converse problem:

Do the work ⟩

$$\begin{aligned} P(L|+) &= \frac{P(+|L)P(L)}{P(+|L)P(L) + P(+|L^C)P(L^C)} \\ &= \frac{(.937)(.00207)}{(.937)(.00207) + (.03)(.99793)} \\ &= \frac{.0019}{.0319} \\ &= .0596 \end{aligned}$$

Interpret the results ⟩

The probability $P(L|+)$ is a conditional probability. $P(L|+) = .0596$ means that, in the long run, only 5.96% of those who test positive actually have the disease. Notice the difference between $P(L|+)$ and the previously reported conditional probability $P(+|L) = .937$, which means that 93.7% of those with Lyme disease test positive. ■

The accompanying box formalizes this reasoning in the statement of Bayes' rule.

Bayes' Rule

If B_1 and B_2 are mutually exclusive events with $P(B_1) + P(B_2) = 1$, then for any event E

$$P(B_1|E) = \frac{P(E|B_1)P(B_1)}{P(E|B_1)P(B_1) + P(E|B_2)P(B_2)}$$

More generally, if $B_1, B_2, \ldots, B_k$ are mutually exclusive events with $P(B_1) + P(B_2) + \cdots + P(B_k) = 1$ then for any event E,

$$P(B_i|E) = \frac{P(E|B_i)P(B_i)}{P(E|B_1)P(B_1) + P(E|B_2)P(B_2) + \cdots + P(E|B_k)P(B_k)}$$

EXAMPLE 6.28 Internet Addiction

Understand the context ⟩

Internet addiction has been defined by researchers as a disorder characterized by excessive time and effort spent on the Internet, impaired judgment and decision-making ability, social withdrawal, and depression. The paper **"The Association between Aggressive Behaviors and Internet Addiction and Online Activities in Adolescents"** (***Journal of Adolescent Health*** **[2009]: 598–605)** reported on a study of more than 9400 adolescents.

Consider the data ⟩ Each participant in the study was assessed using the Chen Internet Addiction Scale to determine if he or she suffered from Internet addiction. The following statements are based on the survey results:

1. 51.8% of the study participants were female and 48.2% were male.
2. 13.1% of the females suffered from Internet addiction.
3. 24.8% of the males suffered from Internet addiction.

Consider the chance experiment that consists of selecting a study participant at random. Let's define the following events:

F = the event that the selected participant is female
M = the event that the selected participant is male
I = the event that the selected participant suffers from Internet addiction

The three statements from the paper define the following probabilities:

$P(F) = .518 \quad P(M) = .482$
$P(I|F) = .131 \quad P(I|M) = .248$

Formulate a plan ⟩ Suppose that we want to know the proportion of those who suffer from Internet addiction who are female. This is equivalent to $P(F|I)$.

Notice that we know $P(I|F)$ but not $P(F|I)$. We can use Bayes' rule to evaluate $P(F|I)$ as follows (with F and M playing the role of B_1 and B_2 and I playing the role of E in the formula for Bayes' rule):

Do the work ⟩

$$P(F|I) = \frac{P(I|F)P(F)}{P(I|F)P(F) + P(I|M)P(M)}$$

$$= \frac{(.131)(.518)}{(.131)(.518) + (.482)(.248)}$$

$$= \frac{.068}{.068 + .120}$$

$$= \frac{068}{.188}$$

$$= .362$$

Interpret the results ⟩ This tells us that 36.2% of those who suffered from Internet addiction were female. ■

EXERCISES 6.59 - 6.77

6.59 ▼ A certain university has 10 vehicles available for use by faculty and staff. Six of these are vans and four are cars. On a particular day, only two requests for vehicles have been made. Suppose that the two vehicles to be assigned are chosen in a completely random fashion from among the 10.

a. Let E denote the event that the first vehicle assigned is a van. What is $P(E)$?

b. Let F denote the event that the second vehicle assigned is a van. What is $P(F|E)$?

c. Use the results of Parts (a) and (b) to calculate $P(E \textit{ and } F)$ (Hint: See Example 6.23.)

6.60 A construction firm bids on two different contracts. Let E_1 be the event that the bid on the first contract is successful, and define E_2 analogously for the second contract. Suppose that $P(E_1) = .4$ and $P(E_2) = .3$ and that E_1 and E_2 are independent events.

a. Calculate the probability that both bids are successful (the probability of the event E_1 *and* E_2).

b. Calculate the probability that neither bid is successful (the probability of the event (*not* E_1) *and* (*not* E_2)).

c. What is the probability that the firm is successful in at least one of the two bids? (Hint: See Example 6.22.)

6.61 There are two traffic lights on the route used by a certain individual to go from home to work. Let E denote the event that the individual must stop at the first light, and define the event F in a similar manner for the second light. Suppose that $P(E) = .4$, $P(F) = .3$, and $P(E \cap F) = .15$.

a. What is the probability that the individual must stop at at least one light; that is, what is the probability of the event $E \cup F$?

b. What is the probability that the individual doesn't have to stop at either light?

c. What is the probability that the individual must stop at exactly one of the two lights?

d. What is the probability that the individual must stop just at the first light? (Hint: How is the probability of this event related to $P(E)$ and $P(E \cap F)$? A Venn diagram might help.)

6.62 Let F denote the event that a randomly selected registered voter in a certain city has signed a petition to recall the mayor. Also, let E denote the event that the randomly selected registered voter actually votes in the recall election. Describe the event $E \cap F$ in words. If $P(F) = .10$ and $P(E|F) = .80$, determine $P(E \cap F)$.

6.63 According to a July 31, 2013, posting on cnn.com, a 2010 study in the journal ***Pediatrics*** found that 8% of children younger than age 18 in the United States have at least one food allergy. Among those with food allergies, about 39% had a history of severe reaction.

a. If a child younger than 18 is randomly selected, what is the probability that he or she has at least one food allergy and has a history of severe reaction?

b. It was also reported that 30% of those with an allergy in fact are allergic to multiple foods. If a child younger than 18 is randomly selected, what is the probability that he or she is allergic to multiple foods?

6.64 Blue Cab operates 15% of the taxis in a certain city, and Green Cab operates the other 85%. After a nighttime hit-and-run accident involving a taxi, an eyewitness said the vehicle was blue. Suppose, though, that under night vision conditions, only 80% of individuals can correctly distinguish between a blue and a green vehicle. What is the probability that the taxi at fault was blue? (Hint: A tree diagram might help.)

6.65 A large cable company reports the following:

- 80% of its customers subscribe to cable TV service
- 42% of its customers subscribe to Internet service
- 32% of its customers subscribe to telephone service
- 25% of its customers subscribe to both cable TV and Internet service
- 21% of its customers subscribe to both cable TV and phone service
- 23% of its customers subscribe to both Internet and phone service
- 15% of its customers subscribe to all three services

Consider the chance experiment that consists of selecting one of the cable company customers at random. Find and interpret the following probabilities:

a. P(cable TV only)

b. P(Internet | cable TV)

c. P(exactly two services)

d. P(Internet and cable TV only)

6.66 Refer to the information given in the previous exercise about customers of a large cable company.

a. Suppose two customers are to be selected at random. Would it be reasonable to consider the events C_1 = event that the first customer selected subscribes to cable TV and C_2 = event that the second customer selected subscribes to cable TV as independent events? Explain.

b. With C_1 and C_2 as defined in Part (a), find $P(C_1 \cap C_2)$.

6.67 The authors of the paper **"Do Physicians Know when Their Diagnoses Are Correct?"** (***Journal of General Internal Medicine*** **[2005]: 334–339)** presented detailed case studies to medical students and to faculty at medical schools. Each participant was asked to provide a diagnosis in the case and also to indicate whether his or her confidence in the correctness of the diagnosis was high or low. Define the events C, I, and H as follows:

C = event that diagnosis is correct
I = event that diagnosis is incorrect
H = event that confidence in the correctness of the diagnosis is high

a. Data appearing in the paper were used to estimate the following probabilities for medical students:

$P(C) = .261$ $\quad P(I) = .739$
$P(H|C) = .375$ $\quad P(H|I) = .073$

Use Bayes' rule to compute the probability of a correct diagnosis given that the student's confidence level in the correctness of the diagnosis is high.

b. Data from the paper were also used to estimate the following probabilities for medical school faculty:

$P(C) = .495$ $\quad P(I) = .505$
$P(H|C) = .537$ $\quad P(H|I) = .252$

Compute $P(C|H)$ for medical school faculty. How does the value of this probability compare to the value of $P(C|H)$ for students computed in Part (a)?

6.68 A study of how people are using online services for medical consulting is described in the paper **"Internet Based Consultation to Transfer Knowledge for Patients Requiring Specialized Care" (*British Medical Journal* [2003]: 696–699)**. Patients using a particular online site could request any combination of three services: specialist opinion, assessment of imaging studies (such as X-ray and MRI), and assessment of pathology results.

The accompanying table shows the combinations of services that were requested by 79 patients in their online consultations.

Combination of Services Provided	Number of Patients
Specialist opinion only	37
Assessment of pathology results only	1
Specialist opinion and assessment of pathology results only	11
Specialist opinion and assessment of imaging studies only	14
Specialist opinion, assessment of imaging studies, and assessment of pathology results	16

For a randomly selected patient from this study, define the events O, I, and A as follows:

O = event that the online consultation involves a specialist opinion
I = event that the online consultation involves the assessment of imaging studies
A = event that the online consultation involves the assessment of pathology results

Use the given information to find the following probabilities:

a. $P(O)$
b. $P(not\ A)$
c. $P(O \cap I)$
d. $P(I|O)$
e. $P(O|I)$

6.69 The report **"Twitter in Higher Education: Usage Habits and Trends of Today's College Faculty" (*Magna Publications,* September 2009)** describes results of a survey of nearly 2000 college faculty. The report indicates the following:

- 30.7% reported that they use Twitter and 69.3% said that they did not use Twitter.
- Of those who use Twitter, 39.9% said they sometimes use Twitter to communicate with students.
- Of those who use Twitter, 27.5% said that they sometimes use Twitter as a learning tool in the classroom.

Consider the chance experiment that selects one of the study participants at random and define the following events:

T = event that selected faculty member uses Twitter
C = event that selected faculty member sometimes uses Twitter to communicate with students
L = event that selected faculty member sometimes uses Twitter as a learning tool in the classroom

a. Use the given information to determine the following probabilities:
   i. $P(T)$
   ii. $P(T^C)$
   iii. $P(C|T)$
   iv. $P(L|T)$
   v. $P(C \cap T)$

b. Interpret each of the probabilities computed in Part (a).

c. What proportion of the faculty surveyed sometimes use Twitter to communicate with students? [Hint: Use the law of total probability to find $P(C)$.]

d. What proportion of faculty surveyed sometimes use Twitter as a learning tool in the classroom?

6.70 The accompanying table summarizes data from a medical expenditures survey carried out by the National Center for Health Statistics **("Assessing the Effects of Race and Ethnicity on Use of Complementary and Alternative Therapies in the USA," *Ethnicity and Health* [2005]: 19–32)**.

Use of Alternative Therapies by Education Level

Education Level	Percent Using Alternative Therapies
High school or less	4.3%
College—1 to 4 years	8.2%
College—5 years or more	11.0%

These percentages were based on data from 7320 people whose education level was high school or less, 4793 people with 1 to 4 years of college, and 1095 people with 5 or more years of college.

a. Use the information given to determine the *number* of respondents falling into each of the six cells of the table below.

	Uses Alternative Therapies	Does Not Use Alternative Therapies	Total
HS/less			**7320**
College: 1–4 yrs			**4793**
College: ≥5 yrs			**1095**

b. Construct a table of estimated probabilities by dividing the count in each of the six table cells by the total sample size, $n = 13{,}208$.

c. The authors of the study indicated that the sample was selected in a way that makes it reasonable to regard the estimated probabilities in the table from Part (b) as representative of the adult population in the United States. Use the information in that table to estimate the following probabilities for adults in the United States.

   i. The probability that a randomly selected individual has 5 or more years of college.

   ii. The probability that a randomly selected individual uses alternative therapies.

   iii. The probability that a randomly selected individual uses alternative therapies given that he or she has 5 or more years of college.

   iv. The probability that a randomly selected individual uses alternative therapies given that he or she has an education level of high school or less.

   v. The probability that a randomly selected individual who uses alternative therapies has an education level of high school or less.

   vi. The probability that a randomly selected individual with some college uses alternative therapies.

d. Are the events H = event that a randomly selected individual has an education level of high school or less and A = event that a randomly selected individual uses alternative therapies independent events? Explain.

6.71 Suppose that we define the following events:

C = event that a randomly selected driver is observed to be using a cell phone
A = event that a randomly selected driver is observed driving a passenger automobile
V = event that a randomly selected driver is observed driving a van or SUV
T = event that a randomly selected driver is observed driving a pickup truck

Based on the article **"Three Percent of Drivers on Hand-Held Cell Phones at Any Given Time" (*San Luis Obispo Tribune*, July 24, 2001)**, the following probability estimates are reasonable:

$P(C) = .03$
$P(C|A) = .026$
$P(C|V) = .048$
$P(C|T) = .019$

Explain why $P(C)$ is not just the average of the three given conditional probabilities.

6.72 The article **"Checks Halt over 200,000 Gun Sales" (*San Luis Obispo Tribune*, June 5, 2000)** reported that required background checks blocked 204,000 gun sales in 1999. The article also indicated that state and local police reject a higher percentage of would-be gun buyers than does the FBI, stating,

> "The FBI performed 4.5 million of the 8.6 million checks, compared with 4.1 million by state and local agencies. The rejection rate among state and local agencies was 3%, compared with 1.8% for the FBI."

Define the following events:

F = event that a randomly selected gun purchase background check is performed by the FBI
S = event that a randomly selected gun purchase background check is performed by a state or local agency
R = event that a randomly selected gun purchase background check results in a blocked sale

a. Use the given information to estimate the following probabilities:

   i. $P(F)$

   ii. $P(S)$

   iii. $P(R|F)$

   iv. $P(R|S)$

b. Use the probabilities from Part (a) to evaluate $P(S|R)$, and write a sentence interpreting this value in the context of this problem.

6.73 Radiologists are often asked to predict the gender of a baby from ultrasound images made during pregnancy. The authors of the paper **"The Use of Three-Dimensional Ultrasound for Fetal Gender Determination in the First Trimester" (*The British Journal of Radiology*, [2003]: 448–451)** followed up on 159 predictions made by a particular radiologist (Radiologist 1) to determine whether or not they were correct. Data from the paper is summarized in the accompanying table.

	Radiologist 1	
	Predicted Male	Predicted Female
Baby Is Male	74	12
Baby Is Female	14	59

a. Assuming that these data are representative of gender predictions made by Radiologist 1, estimate the probability that a gender prediction is correct, given that the baby is male.

b. Assuming that these data are representative of gender predictions made by Radiologist 1, estimate the probability that a gender prediction is correct, given that the baby is female.

c. For Radiologist 1, is a gender prediction more likely to be correct if the baby is male? Explain.

d. Estimate the probability that a gender prediction made by Radiologist 1 is correct.

6.74 The paper referenced in the previous exercise also included data for a second radiologist, Radiologist 2. Based on the data from the previous exercise for Radiologist 1 and the data in the accompanying table for Radiologist 2, write a paragraph comparing the accuracy of gender predictions made by these two radiologists.

	Radiologist 2	
	Predicted Male	Predicted Female
Baby Is Male	81	8
Baby Is Female	7	58

6.75 In an article that appears on the web site of the American Statistical Association (www.amstat.org), Carlton Gunn, a public defender in Seattle, Washington, wrote about how he uses statistics in his work as an attorney. He states:

> I personally have used statistics in trying to challenge the reliability of drug testing results. Suppose the chance of a mistake in the taking and processing of a urine sample for a drug test is just 1 in 100. And your client has a "dirty" (i.e., positive) test result. Only a 1 in 100 chance that it could be wrong? Not necessarily. If the vast majority of all tests given—say 99 in 100—are truly clean, then you get one false dirty and one true dirty in every 100 tests, so that half of the dirty tests are false.

Define the following events as:

TD = event that the test result is dirty
TC = event that the test result is clean
D = event that the person tested is actually dirty
C = event that the person tested is actually clean

a. Using the information in the quote, what are the values of

i. $P(TD|D)$
ii. $P(TD|C)$
iii. $P(C)$
iv. $P(D)$

b. Use the law of total probability to find $P(TD)$.

c. Use Bayes' rule to evaluate $P(C|TD)$. Is this value consistent with the argument given in the quote? Explain.

6.76 According to a study released by the federal Substance Abuse and Mental Health Services Administration (**Knight Ridder Tribune, September 9, 1999**), approximately 8% of all adult full-time workers are drug users and approximately 70% of adult drug users are employed full-time.

a. Is it possible for both of the reported percentages to be correct? Explain.

b. Define the events D and E as D = event that a randomly selected adult is a drug user and E = event that a randomly selected adult is employed full-time. What are the estimated values of $P(D|E)$ and $P(E|D)$?

c. Is it possible to determine P(D), the probability that a randomly selected adult is a drug user, from the information given? If not, what additional information would be needed?

6.77 Only 0.1% of the individuals in a certain population have a particular disease (an incidence rate of .001). Of those who have the disease, 95% test positive when a certain diagnostic test is applied. Of those who do not have the disease, 90% test negative when the test is applied. Suppose that an individual from this population is randomly selected and given the test.

a. Construct a tree diagram having two first-generation branches, for *has disease* and *doesn't have disease*, and two second-generation branches leading out from each of these, for *positive test* and *negative test*. Then enter appropriate probabilities on the four branches.

b. Use the general multiplication rule to calculate P(has disease *and* positive test).

c. Calculate P(positive test).

d. Calculate P(has disease | positive test). Does the result surprise you? Give an intuitive explanation for why this probability is small.

6.7 Estimating Probabilities Empirically Using Simulation

In the examples presented so far, reaching conclusions required knowledge of the probabilities of various outcomes. In some cases, this is reasonable, and we know the actual long-run proportion of the time that each outcome will occur. In other situations, these probabilities are not known. Sometimes probabilities can be determined analytically, by using mathematical rules and probability properties, including the basic ones introduced in this chapter. However, when an analytical approach is impossible, impractical, or just beyond the limited probability tools of the introductory course, we can *estimate* probabilities empirically through observation or by simulation.

Estimating Probabilities Empirically

It is fairly common practice to use observed long-run proportions to estimate probabilities. The process of estimating probabilities empirically is simple:

1. Observe a very large number of chance outcomes under controlled circumstances.
2. Estimate the probability of an event to be the observed proportion of occurrence by appealing to the interpretation of probability as a long-run relative frequency and to the law of large numbers.

This process is illustrated in Examples 6.29 and 6.30.

EXAMPLE 6.29 Fair Hiring Practices

Understand the context ❯

The Biology Department at a university plans to recruit a new faculty member and intends to advertise for someone with a Ph.D. in biology and at least 10 years of college-level teaching experience. A member of the department expresses the belief that the experience requirement will exclude many potential applicants and will exclude far more female applicants than male applicants. The Biology Department would like to determine the probability that an applicant with a Ph.D. in biology would be eliminated from consideration because of the experience requirement.

A similar university just completed a search in which there was no requirement for prior teaching experience, but the information about prior teaching experience was recorded. The 410 applications yielded the following data:

Consider the data ❯

	Number of Applicants		
	Less Than 10 Years of Experience	10 Years of Experience or More	Total
Male	178	112	**290**
Female	99	21	**120**
Total	**277**	**133**	**410**

Let's assume that the populations of applicants for the two positions can be regarded as the same. We can use the available information to approximate the probability that an applicant will fall into each of the four gender–experience combinations.

Do the work ❯

The estimated probabilities (obtained by dividing the number of applicants for each gender–experience combination by 410) are given in Table 6.1. From Table 6.1, the estimate of P(candidate excluded because of the experience requirement) $= .4341 + .2415 = .6756$.

TABLE 6.1 Estimated Probabilities for Example 6.29

	Less Than 10 Years of Experience	10 Years of Experience or More
Male	.4341	.2732
Female	.2415	.0512

Interpret the results ❯

We can also assess the impact of the experience requirement separately for male and for female applicants. From the given information, the proportion of male applicants who have less than 10 years of experience is $178/290 = .6138$, whereas the corresponding proportion for females is $99/120 = .8250$. Therefore, approximately 61% of the male applicants would be eliminated by the experience requirement, and about 83% of the female applicants would be eliminated.

These subgroup proportions—.6138 for males and .8250 for females—are estimates of conditional probabilities, which show how the original probability changes in light of new information. In this example, the probability that a potential candidate has less than 10 years of experience is .6756, but this probability changes to .8250 if we know that a candidate is female. These probabilities can be expressed as

P(less than 10 years of experience) = .6756 (an unconditional probability)

and

P(less than 10 years of experience|female) = .8250 (a conditional probability) ■

EXAMPLE 6.30 Who Has the Upper Hand?

Understand the context ❯

Men and women frequently express intimacy through the simple act of holding hands. Some researchers have suggested that hand-holding is not only an expression of intimacy but also communicates status differences. For two people to hold hands, one must assume an overhand grip and one an underhand grip. Research in this area has shown that it is predominantly the male who assumes the overhand grip.

In the view of some investigators, the overhand grip is seen to imply status or superiority. The authors of the paper **"Men and Women Holding Hands: Whose Hand Is Uppermost?" (*Perceptual and Motor Skills* [1999]: 537–549)** investigated an alternative explanation—perhaps the positioning of hands is a function of the heights of the individuals. Because men, on average, tend to be taller than women, maybe comfort, not status, dictates the positioning.

Investigators at two separate universities observed hand-holding male–female pairs and recorded the data in the accompanying table.

Consider the data ❯

Number of Hand-Holding Couples

	Sex of Person with Uppermost Hand		
	Male	Female	Total
Man Taller	2149	299	**2448**
Equal Height	780	246	**1026**
Woman Taller	241	205	**446**
Total	**3170**	**750**	**3920**

Do the work ❯

Assuming that these hand-holding couples are representative of hand-holding couples in general, we can use the available information to estimate various probabilities. For example, if a hand-holding couple is selected at random, then

$$\text{estimate of } P(\text{man's hand uppermost}) = \frac{3170}{3920} = 0.809$$

For a randomly selected hand-holding couple, if the man is taller, then the probability that the male has the uppermost hand is

$$2149/2448 = 0.878.$$

On the other hand—so to speak—if the woman is taller, the probability that the female has the uppermost hand is

$$205/446 = 0.460.$$

Interpret the results ❯

Notice that these are estimates of the conditional probabilities P(male uppermost | male taller) and P(female uppermost | female taller), respectively. Also, because P(male uppermost | male taller) is not equal to P(male uppermost), the events *male uppermost* and *male taller* are not independent events. Even when the female is taller, the male is still more likely to have the upper hand! ■

Estimating Probabilities Using Simulation

Simulation provides a way to estimate probabilities when we are unable to determine probabilities analytically and when it is impractical to estimate them empirically by observation. Simulation is a method that generates "observations" by performing a chance experiment that is as similar as possible in structure to the real situation of interest.

To illustrate the idea of simulation, consider the situation in which a professor wishes to estimate the probabilities of different possible scores on a 20-question true–false quiz when students are just guessing at the answers. Because each question is a true–false question, a person who is guessing should be equally likely to answer correctly or incorrectly on any given question.

Rather than asking a guessing student to select true or false and then comparing the choice to the correct answer, an equivalent process would be to pick a ball at random from a box that contains half red balls and half blue balls, with a blue ball representing a correct answer. Making 20 selections from the box (with replacement) and then counting the number of correct choices (the number of times a blue ball is selected) is a physical substitute for an observation from a student who has guessed at the answers to 20 true–false questions. Any particular number of blue balls in 20 selections should have the same probability as the same number of correct responses to the quiz when a student is guessing.

For example, 20 selections of balls might produce the following results:

Selection	1	2	3	4	5	6	7	8	9	10
	R	R	B	R	B	B	R	R	R	B

Selection	11	12	13	14	15	16	17	18	19	20
	R	R	B	R	R	B	B	R	R	B

Because 8 of the 20 selections resulted in a blue ball, this would correspond to a quiz with eight correct responses, and it would provide us with one observation for estimating the probabilities of interest. This process could then be repeated a large number of times to generate additional observations. For example, we might find the following:

Repetition	Number of "Correct" Responses
1	8
2	11
3	10
4	12
⋮	⋮
1000	11

The 1000 simulated quiz scores could then be used to construct a table of estimated probabilities.

Taking this many balls out of a box and writing down the results would be cumbersome and tedious. The process can be simplified by using random digits to substitute for drawing balls from the box. For example, a single digit could be selected at random from the 10 digits 0, 1, 2, 3, 4, 5, 6, 7, 8, 9. When using random digits, each of the 10 possibilities is equally likely to occur, so we can use the even digits (including 0) to indicate a correct response and the odd digits to indicate an incorrect response. This would maintain the important property that a correct response and an incorrect response are equally likely, because correct and incorrect are each represented by 5 of the 10 digits.

To aid in carrying out such a simulation, tables of random digits (such as Appendix A Table 1) or computer-generated random digits can be used. The numbers in Appendix A Table 1 were generated using a computer's random number generator.

To see how a table of random numbers can be used to carry out a simulation, let's reconsider the quiz example. We use a random digit to represent the guess on a single question, with an even digit representing a correct response. A series of 20 digits represents the answers to the 20 quiz questions. We pick an arbitrary starting point in Appendix A Table 1. Suppose that we start at row 10 and take the 20 digits in a row to represent one quiz. The first five "quizzes" and the corresponding number correct (number of even digits) are:

Quiz	Random Digits	Number Correct
1	9 4 6 0 6 9 7 8 8 2 5 2 9 6 0 1 4 6 0 5	13
2	6 6 9 5 7 4 4 6 3 2 0 6 0 8 9 1 3 6 1 8	12
3	0 7 1 7 7 7 2 9 7 8 7 5 8 8 6 9 8 4 1 0	9
4	6 1 3 0 9 7 3 3 6 6 0 4 1 8 3 2 6 7 6 8	11
5	2 2 3 6 2 1 3 0 2 2 6 6 9 7 0 2 1 2 5 8	13

This process would be repeated to generate a large number of observations, which would then be used to construct a table of estimated probabilities.

The method for generating observations must preserve the important characteristics of the actual process being considered if simulation is to be successful. For example, it would be easy to adapt the simulation procedure for the true–false quiz to one for a multiple-choice quiz. Suppose that each of the 20 questions on the quiz has five possible responses, only one of which is correct. For any particular question, we would expect a student to be able to guess the correct answer only one-fifth of the time in the long run.

To simulate this situation, we could select at random from a box that contained four red balls and only one blue ball (or, more generally, four times as many red balls as blue balls). If we are using random digits for the simulation, we could use 0 and 1 to represent a correct response and 2, 3, . . . , 9 to represent an incorrect response.

Using Simulation to Approximate a Probability

1. Design a method that uses a random mechanism (such as a random number generator or table, the selection of a ball from a box, the toss of a coin, etc.) to represent an observation. Be sure that the important characteristics of the actual process are preserved.
2. Generate an observation using the method from Step 1, and determine whether the outcome of interest has occurred.
3. Repeat Step 2 a large number of times.
4. Calculate the estimated probability by dividing the number of observations for which the outcome of interest occurred by the total number of observations generated.

The simulation process is illustrated in Examples 6.31–6.33.

EXAMPLE 6.31 Building Permits

Understand the context ❭

Many California cities limit the number of building permits that are issued each year. Because of limited water resources, one such city plans to issue permits for only 10 dwelling units in the upcoming year. The city will decide who is to receive permits by holding a lottery.

Suppose that you are one of 39 individuals who apply for permits. Thirty of these individuals are requesting permits for a single-family home, eight are requesting permits for a duplex (which counts as two dwelling units), and one person is requesting a permit for a small apartment building with eight units (which counts as eight dwelling units). Each request will be entered into the lottery.

Requests will be selected at random one at a time, and if there are enough permits remaining, the request will be granted. This process will continue until all 10 permits have been issued. If your request is for a single-family home, what is the probability that you receive a permit? We can use simulation to estimate this probability. (It is not easy to determine analytically.)

To carry out the simulation, we can view the requests as being numbered from 1 to 39 as follows:

01–30	Requests for single-family homes
31–38	Requests for duplexes
39	Request for 8-unit apartment

For ease of discussion, let's assume that your request is number 01.

One method for simulating the permit lottery consists of these three steps:

Formulate a plan

1. Choose a random number between 01 and 39 to indicate which permit request is selected first, and grant this request.
2. Select another random number between 01 and 39 to indicate which permit request is considered next. Determine the number of dwelling units for the selected request. Grant the request only if there are enough permits remaining to satisfy the request.
3. Repeat Step 2 until permits for 10 dwelling units have been granted.

Do the work

We used Minitab to generate random numbers between 01 and 39 to imitate the lottery drawing. (The random number table in Appendix A Table 1 could also be used by selecting two digits and ignoring 00 and any value over 39.) The first sequence generated by Minitab was

Random Number	Type of Request	Total Number of Units So Far
25	Single-family home	1
07	Single-family home	2
38	Duplex	4
31	Duplex	6
26	Single-family home	7
12	Single-family home	8
33	Duplex	10

We would stop at this point, because permits for 10 units would have been issued. In this simulated lottery, Request 01 was not selected, so you would not have received a permit.

The next simulated lottery (using Minitab to generate the selections) was as follows:

Random Number	Type of Request	Total Number of Units So Far
38	Duplex	2
16	Single-family home	3
30	Single-family home	4
39	Apartment—not granted, since there are not 8 permits remaining	4
14	Single-family home	5
26	Single-family home	6
36	Duplex	8
13	Single-family home	9
15	Single-family home	10

Again, Request 01 was not selected, so you would not have received a permit in this simulated lottery.

Now that a strategy for simulating a lottery has been devised, the tedious part of the simulation begins. We would now simulate a large number of lottery drawings, determining for each one whether Request 01 was granted. We simulated 500 such drawings and found that Request 01 was selected in 85 of the lotteries. This results in

$$\text{estimated probability of receiving a building permit} = \frac{85}{500} = .17$$ ■

EXAMPLE 6.32 One-Boy Family Planning

Understand the context ❯

Suppose that couples who wanted children were to continue having children until a boy is born. Assuming that each newborn child is equally likely to be a boy or a girl, would this behavior change the proportion of boys in the population? This question was posed in an article that appeared in ***The American Statistician*** **("What Some Puzzling Problems Teach About the Theory of Simulation and the Use of Resampling" [1994]: 290–293)**, and many people answered the question incorrectly.

We will use simulation to estimate the long-run proportion of boys in the population if families were to continue to have children until they have a boy. This proportion is an estimate of the probability that a randomly selected child from this population is a boy. Notice that in this population, every sibling group would have exactly one boy.

Formulate a plan ❯

We use a single-digit random number to represent a child. The odd digits (1, 3, 5, 7, 9) will represent a male birth, and the even digits will represent a female birth. An observation is constructed by selecting a sequence of random digits. If the first random number obtained is odd (a boy), the observation is complete. If the first selected number is even (a girl), another digit is chosen. We continue in this way until an odd digit is obtained. For example, reading across row 15 of the random number table (Appendix A Table 1), the first 10 digits are

0 7 1 7 4 2 0 0 0 1

Do the work ❯

Using these numbers to simulate sibling groups, we get

Sibling group 1	0 7	girl, boy
Sibling group 2	1	boy
Sibling group 3	7	boy
Sibling group 4	4 2 0 0 0 1	girl, girl, girl, girl, girl, boy

Continuing along row 15 of the random number table,

Sibling group 5	3	boy
Sibling group 6	1	boy
Sibling group 7	2 0 4 7	girl, girl, girl, boy
Sibling group 8	8 4 1	girl, girl, boy

Interpret the results ❯

After simulating eight sibling groups, we have 8 boys among 19 children. The proportion of boys is 8/19, which is close to .5. Continuing the simulation to obtain a large number of observations suggests that the long-run proportion of boys in the population would still be .5, which is indeed the case. ■

EXAMPLE 6.33 ESP?

Understand the context ❯

Can a close friend read your mind? Try the following chance experiment. Write the word *blue* on one piece of paper and the word *red* on another, and place the two slips of paper in a box. Select one slip of paper from the box, look at the word written on it, and then try to convey the word by sending a mental message to a friend who is seated in the same room.

Ask your friend to select either red or blue, and record whether the response is correct. Repeat this 10 times and determine the number of correct responses. How did your friend do? Is your friend receiving your mental messages or just guessing?

Formulate a plan) Let's investigate by using simulation to get the approximate probabilities of the various possible numbers of correct responses for someone who is guessing. Someone who is guessing should have an equal chance of responding correctly or incorrectly. We can use a random digit to represent a response, with an even digit representing a correct response (C) and an odd digit representing an incorrect response (X). A sequence of 10 digits can be used to simulate performing the chance experiment one time.

Do the work) For example, using the last 10 digits in row 25 of the random number table (Appendix A Table 1) gives

5	2	8	3	4	3	0	7	3	5
X	C	C	X	C	X	C	X	X	X

which is a simulated chance experiment resulting in four correct responses. We used Minitab to generate 150 sequences of 10 random digits and obtained the following results:

Sequence Number	Digits	Number Correct
1	3996285890	**5**
2	1690555784	**3**
3	9133190550	**2**
⋮	⋮	⋮
149	3083994450	**5**
150	9202078546	**7**

Table 6.2 summarizes the results of our simulation.

Interpret the results) The estimated probabilities in Table 6.2 are based on the assumption that a correct and an incorrect response are equally likely (guessing). Evaluate your friend's performance in light of the information in Table 6.2. Is it likely that someone who is guessing would have been able to get as many correct as your friend did? Do you think your friend was receiving your mental messages? How are the estimated probabilities in Table 6.2 used to support your answer?

TABLE 6.2 Estimated Probabilities for Example 6.33

Number Correct	Number of Sequences	Estimated Probability
0	0	.0000
1	1	.0067
2	8	.0533
3	16	.1067
4	30	.2000
5	36	.2400
6	35	.2333
7	17	.1133
8	7	.0467
9	0	.0000
10	0	.0000
Total	**150**	**1.0000**

■

EXERCISES 6.78 - 6.86

6.78 The ***Los Angeles Times* (June 14, 1995)** reported that the U.S. Postal Service is getting speedier, with higher overnight on-time delivery rates than in the past. The Price Waterhouse accounting firm conducted an independent audit by seeding the mail with letters and recording on-time delivery rates for these letters.

Suppose that the results were as follows (these numbers are fictitious but are compatible with summary values given in the article):

	Number of Letters Mailed	Number of Letters Arriving on Time
Los Angeles	500	425
New York	500	415
Washington, D.C.	500	405
Nationwide	6000	5220

Use the given information to estimate the following probabilities. (Hint: See Example 6.30.)

a. The probability of an on-time delivery in Los Angeles

b. The probability of late delivery in Washington, D.C.

c. The probability that two letters mailed in New York are both delivered on time

d. The probability of on-time delivery nationwide

6.79 Five hundred first-year students at a state university were classified according to both high school GPA and whether they were on academic probation at the end of their first semester. The data are summarized in the accompanying table.

	High School GPA			
Probation	2.5 to <3.0	3.0 to <3.5	3.5 and Above	Total
Yes	50	55	30	**135**
No	45	135	185	**365**
Total	**95**	**190**	**215**	**500**

a. Construct a table of the estimated probabilities for each GPA–probation combination. (Hint: See Example 6.25.)

b. Use the table constructed in Part (a) to approximate the probability that a randomly selected first-year student at this university will be on academic probation at the end of the first semester.

c. What is the estimated probability that a randomly selected first-year student at this university had a high school GPA of 3.5 or above?

d. Are the two events *selected student has a high school GPA of 3.5 or above and selected student is on academic probation at the end of the first semester* independent events? How can you tell?

e. Estimate the proportion of first-year students with high school GPAs between 2.5 and 3.0 who are on academic probation at the end of the first semester.

f. Estimate the proportion of those first-year students with high school GPAs 3.5 and above who are on academic probation at the end of the first semester.

6.80 ▼ The table for Exercise 6.80 at the bottom of the page describes (approximately) the distribution of students by gender and college at a mid-sized public university in the West. Suppose that we will randomly select one student from this university:

a. What is the probability that the selected student is a male?

b. What is the probability that the selected student is in the College of Agriculture?

c. What is the probability that the selected student is a male in the College of Agriculture?

d. What is the probability that the selected student is a male who is not from the College of Agriculture?

6.81 On April 1, 2010, the Bureau of the Census in the United States attempted to count every U.S. resident. Suppose that the counts in the table for Exercise 6.81 at the top of the next page are obtained for four counties in one region.

Table for Exercise 6.80

	College						
Gender	Education	Engineering	Liberal Arts	Science and Math	Agriculture	Business	Architecture
Male	200	3200	2500	1500	2100	1500	200
Female	300	800	1500	1500	900	1500	300

Table for Exercise 6.81

	Race/Ethnicity				
County	Caucasian	Hispanic	Black	Asian	American Indian
Monterey	163,000	139,000	24,000	39,000	4,000
San Luis Obispo	180,000	37,000	7,000	9,000	3,000
Santa Barbara	230,000	121,000	12,000	24,000	5,000
Ventura	430,000	231,000	18,000	50,000	7,000

a. If one person is selected at random from this region, what is the estimated probability that the selected person is from Ventura County?

b. If one person is selected at random from Ventura County, what is the estimated probability that the selected person is Hispanic?

c. If one Hispanic person is selected at random from this region, what is the estimated probability that the selected individual is from Ventura County?

d. If one person is selected at random from this region, what is the estimated probability that the selected person is an Asian from San Luis Obispo County?

e. If one person is selected at random from this region, what is the estimated probability that the person is either Asian or from San Luis Obispo County?

f. If one person is selected at random from this region, what is the estimated probability that the person is Asian or from San Luis Obispo County but not both?

g. If two people are selected at random from this region, what is the estimated probability that both are Caucasians?

h. If two people are selected at random from this region, what is the estimated probability that neither is Caucasian?

i. If two people are selected at random from this region, what is the estimated probability that exactly one is a Caucasian?

j. If two people are selected at random from this region, what is the estimated probability that both are residents of the same county?

k. If two people are selected at random from this region, what is the estimated probability that both are from different racial/ethnic groups?

6.82 A medical research team wishes to evaluate two different treatments for a disease. Subjects are selected two at a time, and then one of the pair is assigned to each of the two treatments. The treatments are applied, and each is either a success (S) or a failure (F).

The researchers keep track of the total number of successes for each treatment. They plan to continue the chance experiment until the number of successes for one treatment exceeds the number of successes for the other treatment by 2. For example, they might observe the results in the table for Exercise 6.82 given below. The chance experiment would stop after the sixth pair, because Treatment 1 has two more successes than Treatment 2. The researchers would conclude that Treatment 1 is preferable to Treatment 2.

Suppose that Treatment 1 has a success rate of .7 (that is, $P(\text{success}) = .7$ for Treatment 1) and that

Table for Exercise 6.82

Pair	Treatment 1	Treatment 2	Total Number of Successes for Treatment 1	Total Number of Successes for Treatment 2
1	S	F	1	0
2	S	S	2	1
3	F	F	2	1
4	S	S	3	2
5	F	F	3	2
6	S	F	4	2

Treatment 2 has a success rate of .4. Use simulation to estimate the probabilities in Parts (a) and (b) by using the following procedure:

1. Use a pair of random digits to simulate one pair of subjects. Let the first digit represent Treatment 1 and use 1–7 as an indication of a success and 8, 9, and 0 to indicate a failure. Let the second digit represent Treatment 2, with 1–4 representing a success. For example, if the two digits selected to represent a pair were 8 and 3, you would record failure for Treatment 1 and success for Treatment 2.
2. Continue to select pairs, keeping track of the total number of successes for each treatment. Stop the trial as soon as the number of successes for one treatment exceeds that for the other by 2. This would complete one trial.
3. Repeat this whole process until you have results for at least 20 trials (more is better).
4. Use the simulation results to estimate the desired probabilities.

a. Estimate the probability that more than five pairs must be treated before a conclusion can be reached. (Hint: P(more than 5) $= 1 - P$(5 or fewer).)

b. Estimate the probability that the researchers will incorrectly conclude that Treatment 2 is the better treatment.

6.83 Many cities regulate the number of taxi licenses, and there is a great deal of competition for both new and existing licenses. Suppose that a city has decided to sell 10 new licenses for $25,000 each. A lottery will be held to determine who gets the licenses, and no one may request more than three licenses.

Twenty individuals and taxi companies have entered the lottery. Six of the 20 entries are requests for 3 licenses, 9 are requests for 2 licenses, and the rest are requests for a single license. The city will select requests at random, filling as many of the requests as possible.

For example, the city might fill requests for 2, 3, 1, and 3 licenses and then select a request for 3. Because there is only one license left, the last request selected would receive a license, but only one.

a. An individual has put in a request for a single license. Use simulation to approximate the probability that the request will be granted. Perform *at least* 20 simulated lotteries (more is better!).

b. Do you think that this is a fair way of distributing licenses? Can you propose an alternative procedure for distribution?

6.84 Four students must work together on a group project. They decide that each will take responsibility for a particular part of the project, as follows:

Person	Maria	Alex	Juan	Jacob
Task	Survey design	Data collection	Analysis	Report writing

Because of the way the tasks have been divided, one student must finish before the next student can begin work.

To ensure that the project is completed on time, a schedule is established, with a deadline for each team member. If any one of the team members is late, the timely completion of the project is jeopardized. Assume the following probabilities:

1. The probability that Maria completes her part on time is .8.
2. If Maria completes her part on time, the probability that Alex completes on time is .9, but if Maria is late, the probability that Alex completes on time is only .6.
3. If Alex completes his part on time, the probability that Juan completes on time is .8, but if Alex is late, the probability that Juan completes on time is only .5.
4. If Juan completes his part on time, the probability that Jacob completes on time is .9, but if Juan is late, the probability that Jacob completes on time is only .7.

Use simulation (with at least 20 trials) to estimate the probability that the project is completed on time. Think carefully about this one. For example, you might use a random digit to represent each part of the project (four in all). For the first digit (Maria's part), 1–8 could represent *on time* and 9 and 0 could represent *late*. Depending on what happened with Maria (late or on time), you would then look at the digit representing Alex's part. If Maria was on time, 1–9 would represent *on time* for Alex, but if Maria was late, only 1–6 would represent *on time*. The parts for Juan and Jacob could be handled similarly.

6.85 In Exercise 6.84, the probability that Maria completes her part on time was .8. Suppose that this probability is really only .6. Use simulation (with at least 20 trials) to estimate the probability that the project is completed on time.

6.86 Refer to Exercises 6.84 and 6.85. Suppose that the probabilities of timely completion are as in Exercise 6.84 for Maria, Alex, and Juan, but that Jacob has a probability of completing on time of .7 if Juan is on time and .5 if Juan is late.

a. Use simulation (with at least 20 trials) to estimate the probability that the project is completed on time.

b. Compare the probability from Part (a) to the one computed in Exercise 6.85. Which decrease in the probability of on-time completion (Maria's or Jacob's) made the bigger change in the probability that the project is completed on time?

ACTIVITY 6.1 Kisses

Background: The paper **"What Is the Probability of a Kiss? (It's Not What You Think)"** (***Journal of Statistics Education*** **(online) [2002]**) posed the following question: What is the probability that a Hershey's Kiss will land on its base (as opposed to its side) if it is flipped onto a table? Unlike flipping a coin, there is no reason to believe that this probability would be .5.

Working as a class, develop a plan that would enable you to estimate this probability empirically.

Once you have an acceptable plan, carry it out and use the resulting data to produce an estimate of the desired probability. Do you think that a kiss is equally likely to land on its base or on its side? Explain.

ACTIVITY 6.2 A Crisis for European Sports Fans?

Background: The ***New Scientist*** **(January 4, 2002)** reported on a controversy surrounding the Euro coins that have been introduced as a common currency across Europe. Each country mints its own coins, but these coins are accepted in any of the countries that have adopted the Euro as their currency.

A group in Poland claims that the Belgium-minted Euro does not have an equal chance of landing heads or tails. This claim was based on 250 tosses of the Belgium-minted Euro, of which 140 (56%) came up heads. Should this be cause for alarm for European sports fans, who know that "important" decisions are made by the flip of a coin?

In this activity, we will investigate whether this should be cause for alarm by examining whether observing 140 heads out of 250 tosses is an unusual outcome if the coin is fair.

1. For this first step, you can either (a) flip a U.S. penny 250 times, keeping a tally of the number of heads and tails observed (this won't take as long as you think), or (b) simulate 250 coin tosses by using your calculator or a statistics software package to generate random numbers (if you choose this option, give a brief description of how you carried out the simulation).
2. For your sequence of 250 tosses, calculate the proportion of heads observed.
3. Form a data set that consists of the values for proportion of heads observed in 250 tosses of a fair coin for the entire class. Summarize this data set by constructing a graphical display.
4. Working with a partner, write a paragraph explaining why European sports fans should or should not be worried by the results of the Polish experiment. Your explanation should be based on the observed proportion of heads from the Polish experiment and the graphical display constructed in Step 3.

ACTIVITY 6.3 The "Hot Hand" in Basketball

Background: Consider a mediocre basketball player who has consistently made only 50% of his free throws over several seasons. If we were to examine his free throw record over the last 50 free throw attempts, is it likely that we would see a streak of 5 in a row where he is successful in making the free throw? In this activity, we will investigate this question. We will assume that the outcomes of successive free throw attempts are independent and that the probability that the player is successful on any particular attempt is .5.

1. Begin by simulating a sequence of 50 free throws for this player. Because this player has probability of success of .5 for each attempt and the attempts are independent, we can model a free throw by tossing a coin. Using heads to represent a successful free throw and tails to represent a missed free throw, simulate 50 free throws by tossing a coin 50 times, recording the outcome of each toss.
2. For your sequence of 50 tosses, identify the longest streak by looking for the longest string of heads in your sequence. Determine the length of this longest streak.
3. Combine your longest streak value with those from the rest of the class and construct a histogram or dotplot of these longest streak values.
4. Based on the graph from Step 3, does it appear likely that a player of this skill level would have a streak of 5 or more successes sometime during a sequence of 50 free throw attempts? Justify your answer based on the graph from Step 3.
5. Use the combined class data to estimate the probability that a player of this skill level has a streak of at least 5 somewhere in a sequence of 50 free throw attempts.

6. Using basic probability rules, we can calculate the probability that a player of this skill level is successful on the *next* 5 free throw attempts:

$$P(SSSSS) = \left(\frac{1}{2}\right)\left(\frac{1}{2}\right)\left(\frac{1}{2}\right)\left(\frac{1}{2}\right)\left(\frac{1}{2}\right) = \left(\frac{1}{2}\right)^5 = .031$$

which is relatively small. At first this might seem inconsistent with your answer in Step 5, but the estimated probability from Step 5 and the computed probability of .031 are really considering different situations. Explain why it is plausible that both probabilities could be correct.

7. Do you think that the assumption that the outcomes of successive free throws are independent is reasonable? Explain. (This is a hotly debated topic among both sports fans and statisticians!)

SUMMARY Key Concepts and Formulas

TERM OR FORMULA	COMMENT
Chance experiment	Any experiment for which there is uncertainty concerning the resulting outcome.
Sample space	The collection of all possible outcomes from a chance experiment.
Event	Any collection of possible outcomes from a chance experiment.
Simple event	An event that consists of a single outcome.
Events	
1. *not A*, A^C	1. The event consisting of all outcomes not in *A*.
2. *A or B*, $A \cup B$	2. The event consisting of all outcomes in at least one of the two events.
3. *A and B*, $A \cap B$	3. The event consisting of outcomes common to both events.
Mutually exclusive events	Events that have no outcomes in common.
Fundamental properties of probability	Fundamental properties of probability 1. The probability of any event must be a number between 0 and 1. 2. If *S* is the sample space for a chance experiment, $P(S) = 1$ 3. If *E* and *F* are mutually exclusive events, $P(E \cup F) = P(E) + P(F)$ 4. $P(E) + P(E^C) = 1$
$P(E) = \frac{\text{number of outcomes in } E}{N}$	*P*(*E*) when the outcomes are *equally likely* and where *N* is the number of outcomes in the sample space.
$P(E \cup F) = P(E) + P(F)$ $P(E_1 \cup \cdots \cup E_k) = P(E_1) + \cdots + P(E_k)$	Addition rules when events are *mutually exclusive*.
$P(E\|F) = \frac{P(E \cap F)}{P(F)}$	The conditional probability of the event *E* given that the event *F* has occurred.
Independence of events *E* and *F* $P(E\|F) = P(E)$	Events *E* and *F* are independent if the probability that *E* has occurred given *F* is the same as the probability that *E* will occur with no knowledge of *F*.
$P(E \cap F) = P(E)P(F)$ $P(E_1 \cap \cdots \cap E_k) = P(E_1)P(E_2)\cdots P(E_k)$	Multiplication rules for *independent* events.
$P(E \cup F) = P(E) + P(F) - P(E \cap F)$	The general addition rule for two events.
$P(E \cap F) = P(E\|F)P(F)$	The general multiplication rule for two events.
$P(E) = P(E\|B_1)P(B_1) + P(E\|B_2)P(B_2) + \cdots + P(E\|B_k)P(B_k)$	The law of total probability, where $B_1, B_2, \ldots, B_k$ are mutually exclusive events with $P(B_1) + P(B_2) + \cdots + P(B_k) = 1$
$P(B_i\|E) = \frac{P(E\|B_i)P(B_i)}{P(E\|B_1)P(B_1) + P(E\|B_2)P(B_2) + \cdots + P(E\|B_k)P(B_k)}$	Bayes' rule, where $B_1, B_2, \ldots, B_k$ are mutually exclusive events with $P(B_1) + P(B_2) + \cdots + P(B_k) = 1$

CHAPTER REVIEW Exercises 6.87 - 6.103

6.87 A company uses three different assembly lines—A_1, A_2, and A_3—to manufacture a particular component. Of those manufactured by A_1, 5% need rework to remedy a defect, whereas 8% of A_2's components and 10% of A_3's components need rework. Suppose that 50% of all components are produced by A_1, 30% are produced by A_2 and 20% are produced by A_3.

a. Construct a tree diagram with first-generation branches corresponding to the three lines. Leading from each branch, draw one branch for rework (R) and another for no rework (N). Then enter appropriate probabilities on the branches.

b. What is the probability that a randomly selected component came from A_1 and needed rework?

c. What is the probability that a randomly selected component needed rework?

6.88 Consider the following information about passengers on a cruise ship on vacation: 40% check work e-mail, 30% use a cell phone to stay connected to work, 25% bring a laptop with them on vacation, 23% both check work e-mail and use a cell phone to stay connected, and 51% neither check work e-mail nor use a cell phone to stay connected nor bring a laptop. In addition, 88% of those who bring a laptop also check work e-mail and 70% of those who use a cell phone to stay connected also bring a laptop. With

E = event that a traveler on vacation checks work e-mail

C = event that a traveler on vacation uses a cell phone to stay connected

L = event that a traveler on vacation brought a laptop

use the given information to determine the following probabilities. A Venn diagram may help.

a. $P(E)$

b. $P(C)$

c. $P(L)$

d. $P(E \text{ and } C)$

e. $P(E^C \text{ and } C^C \text{ and } L^C)$

f. $P(E \text{ and } C \text{ or } L)$

g. $P(E|L)$

h. $P(L|C)$

i. $P(E \text{ and } C \text{ and } L)$

j. $P(E \text{ and } L)$

k. $P(C \text{ and } L)$

l. $P(C|E \text{ and } L)$

6.89 Online chat rooms allow people from all over the world to exchange opinions on various topics of interest. A side effect of such conversations is "flaming," which is negative criticism of others' contributions to the conversation. The paper **"Criticism on the Internet: An Analysis of Participant Reactions" (*Communication Research Reports* [1998]: 180–187)** investigated the effect that personal criticism has on an individual. Data from a survey of 193 chat room users from this study are summarized here:

	Have Been Personally Criticized	Have Not Been Personally Criticized	Total
Have Criticized Others	19	8	**27**
Have Not Criticized Others	23	143	**166**
Total	**42**	**151**	**193**

Assume that this data is representative of the larger group of all chat room users and that the frequencies in the table are indicative of the long-run behavior of chat room users. Suppose that a chat room user is selected at random, and define events

C = event that the selected individual has criticized others

O = event that the selected individual has been personally criticized by others

Find and interpret each of the following probabilities:

a. $P(C)$

b. $P(O)$

c. $P(C \cap O)$

d. $P(C|O)$

e. $P(O|C)$

6.90 The Associated Press **(*San Luis Obispo Telegram-Tribune*, August 23, 1995)** reported on the results of mass screening of schoolchildren for tuberculosis (TB). For Santa Clara County, California, the proportion of all tested kindergartners who were found to have TB was .0006. The corresponding proportion for recent immigrants (thought to be a high-risk group) was .0075. Suppose that a Santa Clara County kindergartner is selected at random. Are the events *selected student is a recent immigrant* and *selected* *student has TB* independent or dependent events? Justify your answer using the given information.

6.91 The Australian newspaper ***The Mercury* (May 30, 1995)** reported that, based on a survey of 600 reformed and current smokers, 11.3% of those who had attempted to quit smoking in the previous 2 years had used a nicotine aid (such as a nicotine patch). It also reported that 62% of those who quit smoking without a nicotine aid began smoking again within 2 weeks and 60% of those who used a nicotine aid began smoking again within 2 weeks.

If a smoker who is trying to quit smoking is selected at random, are the events *selected smoker who is trying to quit uses a nicotine aid* and *selected smoker who has attempted to quit begins smoking again within 2 weeks* independent or dependent events? Justify your answer using the given information.

6.92 A certain company sends 40% of its overnight mail parcels by means of express mail service A_1. Of these parcels, 2% arrive after the guaranteed delivery time. What is the probability that a randomly selected overnight parcel was shipped by mail service A_1 and was late?

6.93 Return to the context of the previous exercise and suppose that 50% of the overnight parcels are sent by means of express mail service A_2 and the remaining 10% are sent by means of A_3. Of those sent by means of A_2, only 1% arrived late, whereas 5% of the parcels handled by A_3 arrived late.

a. What is the probability that a randomly selected parcel arrived late? (Hint: A tree diagram should help.)

b. Suppose that a randomly selected overnight parcel arrived late. What is the probability that the parcel was shipped using mail service A_1? That is, what is the probability of A_1 given L, denoted $P(A_1|L)$?

c. What is $P(A_2|L)$?

d. What is $P(A_3|L)$?

6.94 Two individuals, A and B, are finalists for a chess championship. They will play a sequence of games, each of which can result in a win for A, a win for B, or a draw. Suppose that the outcomes of successive games are independent, with $P(\text{A wins game}) = .3$, $P(\text{B wins game}) = .2$, and $P(\text{draw}) = .5$. Each time a player wins a game, he earns 1 point and his opponent earns no points. The first player to win 5 points wins the championship.

a. What is the probability that A wins the championship in just five games?

b. What is the probability that it takes just five games to obtain a champion?

c. If a draw earns a half-point for each player, describe how you would perform a simulation to estimate $P(\text{A wins the championship})$. Assume that the championship will end in a draw if both players obtain 5 points at the same time.

6.95 A single-elimination tournament with four players is to be held. In Game 1, the players seeded (rated) first and fourth play. In Game 2, the players seeded second and third play. In Game 3, the winners of Games 1 and 2 play, with the winner of Game 3 declared the tournament winner. Suppose that the following probabilities are given:

$P(\text{seed 1 defeats seed 4}) = .8$
$P(\text{seed 1 defeats seed 2}) = .6$
$P(\text{seed 1 defeats seed 3}) = .7$
$P(\text{seed 2 defeats seed 3}) = .6$
$P(\text{seed 2 defeats seed 4}) = .7$
$P(\text{seed 3 defeats seed 4}) = .6$

a. Describe how you would use random digits to simulate Game 1 of this tournament.

b. Describe how you would use random digits to simulate Game 2 of this tournament.

c. How would you use random digits to simulate Game 3 in the tournament? (This will depend on the outcomes of Games 1 and 2.)

d. Simulate one complete tournament, giving an explanation for each step in the process.

e. Simulate 10 tournaments, and use the resulting information to estimate the probability that the first seed wins the tournament.

f. Ask four classmates for their simulation results. Along with your own results, this should give you information for 50 simulated tournaments. Use this information to estimate the probability that the first seed wins the tournament.

g. Why do the estimated probabilities from Parts (e) and (f) differ? Which do you think is a better estimate of the true probability? Explain.

6.96 In a school machine shop, 60% of all machine breakdowns occur on lathes and 15% occur on drill presses. Let E denote the event that the next machine breakdown is on a lathe, and let F denote the event that a drill press is the next machine to break down. With $P(E) = .60$ and $P(F) = .15$, calculate:

a. $P(E^C)$

b. $P(E \cup F)$

c. $P(E^C \cap F^C)$

6.97 There are five faculty members in a certain academic department. These individuals have 3, 6, 7, 10, and 14 years of teaching experience. Two of these individuals are randomly selected to serve on a personnel review committee. What is the probability that the chosen representatives have a total of at least 15 years of teaching experience? (Hint: Consider all possible committees.)

6.98 The general addition rule for three events states that

$$P(A \text{ or } B \text{ or } C) = P(A) + P(B) + P(C) - P(A \text{ and } B) - P(A \text{ and } C) - P(B \text{ and } C) + P(A \text{ and } B \text{ and } C)$$

A new magazine publishes columns entitled "Art" (A), "Books" (B), and "Cinema" (C). Suppose that

14% of all subscribers read A
23% read B
37% read C
8% read A and B
9% read A and C
13% read B and C
5% read all three columns

What is the probability that a randomly selected subscriber reads at least one of these three columns?

6.99 A theater complex is currently showing four R-rated movies, three PG-13 movies, two PG movies, and one G movie. The following table gives the number of people at the first showing of each movie on a certain Saturday:

Theater	Rating	Number of Viewers
1	R	600
2	PG-13	420
3	PG-13	323
4	R	196
5	G	254
6	PG	179
7	PG-13	114
8	R	205
9	R	139
10	PG	87

Suppose that one of these people is randomly selected.

a. What is the probability that the selected individual saw a PG movie?

b. What is the probability that the selected individual saw a PG or a PG-13 movie?

c. What is the probability that the selected individual did not see an R movie?

6.100 Refer to Exercise 6.99, but now suppose that two viewers are randomly selected (without replacement). Let R_1 and R_2 denote the events that the first and second individuals, respectively, watched an R-rated movie. Are R_1 and R_2 independent events? Explain. From a practical point of view, can these events be regarded as independent? Explain.

6.101 Suppose that a box contains 25 light bulbs, of which 20 are good and the other 5 are defective. Consider randomly selecting three bulbs without replacement. Let E denote the event that the first bulb selected is good, F be the event that the second bulb is good, and G represent the event that the third bulb selected is good.

a. What is $P(E)$?

b. What is $P(F|E)$?

c. What is $P(G|E \cap F)$?

d. What is the probability that all three selected bulbs are good?

6.102 Return to Exercise 6.101, and suppose that 4 bulbs are randomly selected from the 25.

a. What is the probability that all 4 are good?

b. What is the probability that at least 1 selected bulb is bad?

6.103 A transmitter is sending a message using a binary code (a sequence of 0's and 1's). Each transmitted bit (0 or 1) must pass through three relays to reach the receiver. At each relay, the probability is .20 that the bit sent on is different from the bit received (a reversal). Assume that the relays operate independently of one another:

transmitter → relay 1 → relay 2 → relay 3 → receiver

a. If a 1 is sent from the transmitter, what is the probability that a 1 is sent on by all three relays?

b. If a 1 is sent from the transmitter, what is the probability that a 1 is received by the receiver? (Hint: The eight experimental outcomes can be displayed on a tree diagram with three generations of branches, one generation for each relay.)

CHAPTER 7

Random Variables and Probability Distributions

This chapter is the first of two chapters that together link the basic ideas of probability explored in Chapter 6 with the methods of statistical inference. Chapter 6 used probability to describe the long-run relative frequency of occurrence of various types of outcomes. In this chapter we introduce probability models that can be used to describe the distribution of values of a variable.

In Chapter 8, we will see how these same probability models can be used to describe the behavior of sample statistics. Such models play an important role in drawing conclusions based on sample data.

In a chance experiment, we often focus on some numerical aspect of the outcome. For example, an environmental scientist who obtains an air sample from a specified location might be especially concerned with the concentration of ozone (a major constituent of smog). A quality control inspector who must decide whether to accept a large shipment of components may base the decision on the number of defective components in a group of 20 components randomly selected from the shipment.

Before selection of the air sample, the value of the ozone concentration is uncertain. Similarly, the number of defective components among the 20 selected might be any whole number between 0 and 20. Because the value of a variable quantity such as ozone concentration or number of defective components is subject to uncertainty, such variables are called *random variables.*

In this chapter we begin by distinguishing between discrete and continuous numerical random variables. We show how the behavior of both discrete and continuous numerical random variables can be described by a probability distribution. This distribution can then be used to make probability statements about values of the random variable. Special emphasis is given to three commonly encountered probability distributions: the binomial, geometric, and normal distributions.

Chapter 7: Learning Objectives

STUDENTS WILL UNDERSTAND:

- that a probability distribution describes the long-run behavior of a random variable.
- that areas under a density curve for a continuous random variable are interpreted as probabilities.

STUDENTS WILL BE ABLE TO:

- distinguish between discrete and continuous random variables.
- construct the probability distribution of a discrete random variable.
- compute and interpret the mean and standard deviation of a discrete random variable.
- distinguish between binomial and geometric random variables.
- compute and interpret binomial probabilities.
- compute probabilities involving continuous random variables whose density curves have a simple form.
- find an area under a normal curve and interpret this area as a probability.
- construct and interpret a normal probability plot.

7.1 Random Variables

In most chance experiments, an investigator focuses attention on one or more variable quantities. For example, consider a management consultant who is studying the operation of a supermarket. A chance experiment might involve randomly selecting a customer leaving the store. One interesting numerical variable might be the number of items purchased by the customer. We can denote this variable using a letter, such as x. Possible values of this variable are 0, 1, 2, 3, and so on. The possible values of x are isolated points on the number line. Until a customer is selected and the number of items counted, the value of x is uncertain.

Another variable of potential interest might be the time y (in minutes) spent in a checkout line. One possible value of y is 3.0 minutes and another is 4.0 minutes, but *any* other number between 3.0 and 4.0 is also a possibility. The possible y values form an entire interval (a continuum) on the number line.

DEFINITION

Random variable: A numerical variable whose value depends on the outcome of a chance experiment. A random variable associates a numerical value with each outcome of a chance experiment.

Discrete random variable: A random variable is discrete if its set of possible values is a collection of isolated points along the number line.

Continuous random variable: A random variable is continuous if its set of possible values includes an entire interval on the number line.

We use lowercase letters, such as x and y, to represent random variables.* Figure 7.1 shows a set of possible values for each type of random variable. In practice, a discrete

*In some books, uppercase letters are used to name random variables, with lowercase letters representing a particular value that the variable might assume. We have chosen to use a simpler and less formal notation.

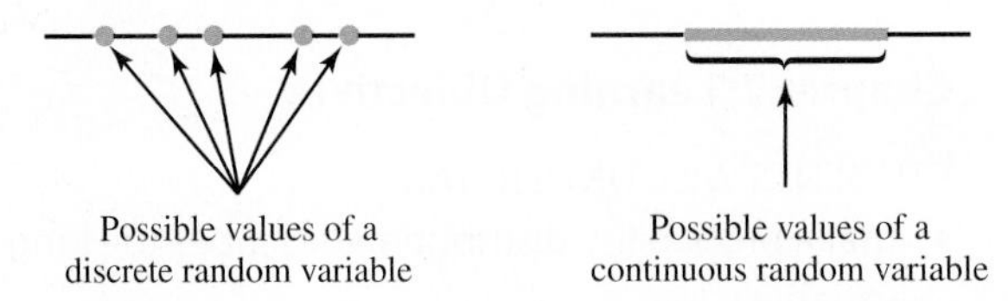

FIGURE 7.1
Two different types of random variables.

random variable almost always arises in connection with counting (for example, the number of items purchased, the number of gas pumps in use, or the number of broken eggs in a carton). A continuous random variable is one whose value is typically obtained by measurement (temperature in a freezer compartment, weight of a pineapple, amount of time spent in a store, etc.).

Because there is a limit to the accuracy of any measuring instrument, such as a watch or a scale, it may seem that any variable should be regarded as discrete. For example, when weight is measured to the nearest pound, the observed values appear to be isolated points along the number line, such as 2 pounds, 3 pounds, etc. But this is just a function of the accuracy with which weight is recorded and not because a weight between 2 and 3 pounds is impossible. In this case, the variable is continuous.

EXAMPLE 7.1 Book Sales

Consider an experiment in which the type of book, print (P) or digital (D), chosen by each of three successive customers making a purchase from on an online bookstore is noted. Define a random variable x by

x = number of customers purchasing a book in digital format

The experimental outcome in which the first and third customers purchase a digital book and the second customer purchases a print book can be abbreviated DPD. The associated x value is 2, because two of the three customers selected a digital book. Similarly, the x value for the outcome DDD (all three purchase a digital book) is 3.

The eight possible experimental outcomes and the corresponding value of x are displayed in the following table:

Outcome	PPP	DPP	PDP	PPD	DDP	DPD	PDD	DDD
x value	0	1	1	1	2	2	2	3

There are only four possible x values—0, 1, 2, and 3—and these are isolated points on the number line. This means that x is a discrete random variable. ■

In some situations, the random variable of interest is discrete, but the number of possible values is not finite. This is illustrated in Example 7.2.

EXAMPLE 7.2 This Could Be a Long Game . . .

Two friends agree to play a game that consists of a sequence of trials. The game continues until one player wins two trials in a row. One random variable of interest might be

x = number of trials required to complete the game

Let A denote a win for Player 1 and B denote a win for Player 2. The simplest possible experimental outcomes are AA (the case in which Player 1 wins the first two trials and the game ends) and BB (the case in which Player 2 wins the first two trials). With either of these two outcomes, $x = 2$. There are also two outcomes for which $x = 3$: ABB and BAA. Some other possible outcomes and associated x values are

Outcomes	x value
AA, BB	2
BAA, ABB	3
ABAA, BABB	4
ABABB, BABAA	5
⋮	⋮
ABABABABAA, BABABABABB	10

and so on.

Any positive integer that is 2 or greater is a possible value. Because the values 2, 3, 4, ... are isolated points on the number line, x is a discrete random variable even though there is no upper limit to the number of possible values. ■

EXAMPLE 7.3 Stress

In an engineering stress test, pressure is applied to a thin 1-foot-long bar until the bar snaps. The precise location where the bar will snap is uncertain. Let x be the distance from the left end of the bar to the break. Then $x = 0.25$ is one possibility, $x = 0.9$ is another, and in fact any number between 0 and 1 is a possible value of x. (Figure 7.2 shows the case of the outcome $x = 0.6$.) The set of possible values is an entire interval on the number line, so x is a continuous random variable.

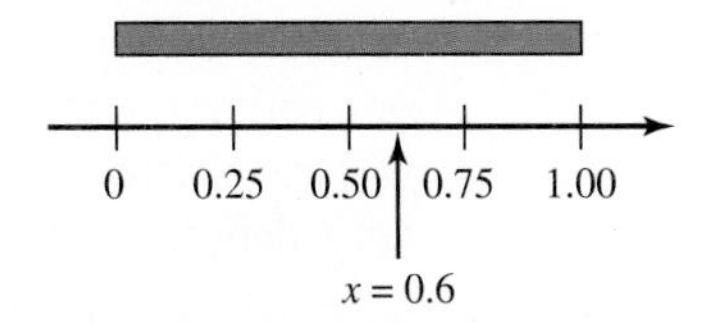

FIGURE 7.2
The bar for Example 7.3 and the outcome $x = 0.6$.

Even though in practice we may only be able to measure the distance to the nearest tenth of an inch or hundredth of an inch, the *actual* distance could be any number between 0 and 1. So, even though the recorded values might be rounded because of the accuracy of the measuring instrument, the variable is still continuous. ■

In data analysis, random variables often arise in the context of summarizing sample data when a sample is selected from some population. This is illustrated in Example 7.4.

EXAMPLE 7.4 College Plans

Suppose that a counselor plans to select a random sample of 50 seniors at a large high school and to ask each student in the sample whether he or she plans to attend college after graduation. The process of sampling is a chance experiment. The sample space for this experiment consists of all the different possible random samples of size 50 that might result (there is a very large number of these), and for simple random sampling, each of these outcomes is equally likely.

Suppose that the counselor is interested in the number of students in the sample who plan to attend college. We can use the letter x to represent this number:

x = number of students in the sample who plan to attend college

Then x is a random variable, because it associates a numerical value with each of the possible outcomes (random samples) that might occur. Possible values of x are 0, 1, 2, . . . , 50, and x is a discrete random variable. ■

EXERCISES 7.1 - 7.7

7.1 State whether each of the following random variables is discrete or continuous:

a. The number of defective tires on a car

b. The body temperature of a hospital patient

c. The number of pages in a book

d. The number of draws (with replacement) from a deck of cards until a heart is selected

e. The lifetime of a lightbulb

7.2 Classify each of the following random variables as either discrete or continuous:

a. The fuel efficiency (mpg) of an automobile

b. The amount of rainfall at a particular location during the next year

c. The distance that a person throws a baseball

d. The number of questions asked during a 1-hour lecture

e. The tension (in pounds per square inch) at which a tennis racket is strung

f. The amount of water used by a household during a given month

g. The number of traffic citations issued by the highway patrol in a particular county on a given day

7.3 Starting at a particular time, each car entering an intersection is observed to see whether it turns left (L) or right (R) or goes straight ahead (S). The experiment terminates as soon as a car is observed to go straight. Let y denote the number of cars observed.

a. What are possible y values?

b. List five different outcomes and their associated y values. (Hint: See Example 7.2.)

7.4 A point is randomly selected from the interior of the square pictured.

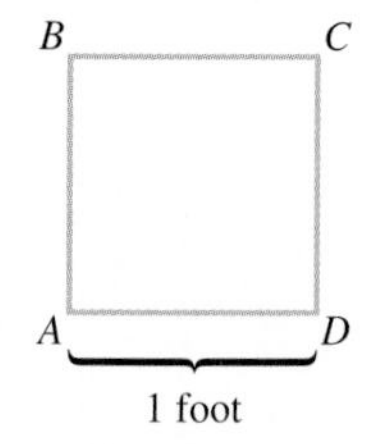

Let x denote the distance from the lower left-hand corner A of the square to the selected point.

a. What are possible values of x?

b. Is x a discrete or a continuous variable? (Hint: See Example 7.3.)

7.5 A point is randomly selected on the surface of a lake that has a maximum depth of 100 feet. Let y be the depth of the lake at the randomly chosen point.

a. What are possible values of y?

b. Is y discrete or continuous?

7.6 A person stands at the corner marked A of the square pictured in Exercise 7.4 and tosses a coin. If it lands heads up, the person moves one corner clockwise, to B. If the coin lands tails up, the person moves one corner counterclockwise, to D. This process is then repeated until the person arrives back at A. Let y denote the number of coin tosses.

a. What are possible values of y?

b. Is y discrete or continuous?

7.7 A box contains four slips of paper marked 1, 2, 3, and 4. Two slips are selected without replacement. List the possible values for each of the following random variables:

a. x = sum of the two numbers

b. y = difference between the first and second numbers

c. z = number of slips selected that show an even number

d. w = number of slips selected that show a 4

Bold exercises answered in back ● Data set available online ▼ Video Solution available

7.2 Probability Distributions for Discrete Random Variables

The probability distribution for a random variable is a model that describes the long-run behavior of the variable. For example, suppose that the Department of Animal Regulation in a particular county is interested in studying the variable

x = number of licensed dogs or cats for a household

County regulations prohibit more than five dogs or cats per household. If we consider the chance experiment of randomly selecting a household in this county, then x is a discrete random variable because it associates a numerical value (0, 1, 2, 3, 4, or 5) with each of the possible outcomes (households) in the sample space.

Although we know what the possible values for x are, it would also be useful to know how this variable behaves in repeated observation. What would be the most common value? What proportion of the time would $x = 5$ be observed? $x = 3$? A probability distribution provides this type of information about the long-run behavior of a random variable.

The **probability distribution of a discrete random variable *x*** gives the probability associated with each possible x value. Each probability is the long-run relative frequency of occurrence of the corresponding x value when the chance experiment is performed a very large number of times.

Common ways to display a probability distribution for a discrete random variable are a table, a probability histogram, or a formula.

Notation: If one possible value of x is 2, we often write $p(2)$ in place of $P(x = 2)$. Similarly, $p(5)$ denotes the probability that $x = 5$, and so on.

EXAMPLE 7.5 Energy Efficient Refrigerators

Understand the context

Suppose that each of four randomly selected customers purchasing a refrigerator at a large appliance store chooses either an energy efficient model (E) or one from a less expensive group of models (G) that do not have an energy efficient rating. Assume that these customers make their choices independently of one another and that 40% of all customers select an energy efficient model. This implies that for any particular one of the four customers, $P(E) = .4$ and $P(G) = .6$.

One possible experimental outcome is EGGE, where the first and fourth customers select energy efficient models and the other two choose less expensive models. Because the customers make their choices independently, the multiplication rule for independent events implies that

$$\begin{aligned} P(\text{EGGE}) &= P(\text{1st chooses E } \textit{and} \text{ 2nd chooses G } \textit{and} \text{ 3rd chooses G } \textit{and} \text{ 4th chooses E}) \\ &= P(E)P(G)P(G)P(E) \\ &= (.4)(.6)(.6)(.4) \\ &= .0576 \end{aligned}$$

Similarly,

$$\begin{aligned} P(\text{EGEG}) &= P(E)P(G)P(E)P(G) \\ &= (.4)(.6)(.4)(.6) \\ &= .0576 \quad (\text{identical to } P(\text{EGGE})) \end{aligned}$$

and

$$P(\text{GGGE}) = (.6)(.6)(.6)(.4) = .0864$$

The number among the four customers who purchase an energy efficient model is a random variable. We can denote this variable by x:

x = the number of energy efficient refrigerators purchased by the four customers

Table 7.1 displays the 16 possible experimental outcomes, the probability of each outcome, and the value of the random variable x that is associated with each outcome.

Do the work

The probability distribution of x is easily obtained from this information. Consider the smallest possible x value, 0. The only outcome for which $x = 0$ is GGGG, so

$$p(0) = P(x = 0) = P(\text{GGGG}) = .1296$$

TABLE 7.1 Outcomes and Probabilities for Example 7.5

Outcome	Probability	x Value	Outcome	Probability	x Value
GGGG	.1296	0	GEEG	.0576	2
EGGG	.0864	1	GEGE	.0576	2
GEGG	.0864	1	GGEE	.0576	2
GGEG	.0864	1	GEEE	.0384	3
GGGE	.0864	1	EGEE	.0384	3
EEGG	.0576	2	EEGE	.0384	3
EGEG	.0576	2	EEEG	.0384	3
EGGE	.0576	2	EEEE	.0256	4

There are four different outcomes for which $x = 1$, so $p(1)$ results from adding the four corresponding probabilities:

$$\begin{aligned} p(1) &= P(x = 1) = P(\text{EGGG } or \text{ GEGG } or \text{ GGEG } or \text{ GGGE}) \\ &= P(\text{EGGG}) + P(\text{GEGG}) + P(\text{GGEG}) + P(\text{GGGE}) \\ &= .0864 + .0864 + .0864 + .0864 \\ &= 4(.0864) \\ &= .3456 \end{aligned}$$

Similarly,

$$\begin{aligned} p(2) &= P(\text{EEGG}) + \cdots + P(\text{GGEE}) = 6(.0576) = .3456 \\ p(3) &= 4(.0384) = .1536 \\ p(4) &= .0256 \end{aligned}$$

The probability distribution of x is summarized in the following table:

x Value	0	1	2	3	4
$p(x)$ = Probability of Value	.1296	.3456	.3456	.1536	.0256

Interpret the results)

To interpret $p(3) = .1536$, think of performing the chance experiment repeatedly, each time with a new group of four customers. In the long run, 15.36% of these groups will have exactly three customers purchasing an energy efficient refrigerator.

The probability distribution can be used to determine probabilities of various events involving x. For example, the probability that at least two of the four customers choose energy efficient models is

$$\begin{aligned} P(x \geq 2) &= P(x = 2 \text{ or } x = 3 \text{ or } x = 4) \\ &= p(2) + p(3) + p(4) \\ &= .5248 \end{aligned}$$

This means that, in the long run, a group of four refrigerator purchasers will include at least two who select energy efficient models 52.48% of the time. ■

A probability distribution table for a discrete variable shows the possible x values and also $p(x)$ for each possible x value. Because $p(x)$ is a probability, it must be a number between 0 and 1, and because the probability distribution lists all possible x values, the sum of all the $p(x)$ values must equal 1. These properties of discrete probability distributions are summarized in the following box.

Properties of Discrete Probability Distributions

Properties of Discrete Probability Distributions

1. For every possible x value, $0 \leq p(x) \leq 1$.
2. $\sum_{\text{all } x \text{ values}} p(x) = 1$

A graphical representation of a discrete probability distribution is called a *probability histogram*. The picture has a rectangle centered above each possible value of x, and the area of each rectangle is proportional to the probability of the corresponding value. Figure 7.3 displays the probability histogram for the probability distribution of Example 7.5.

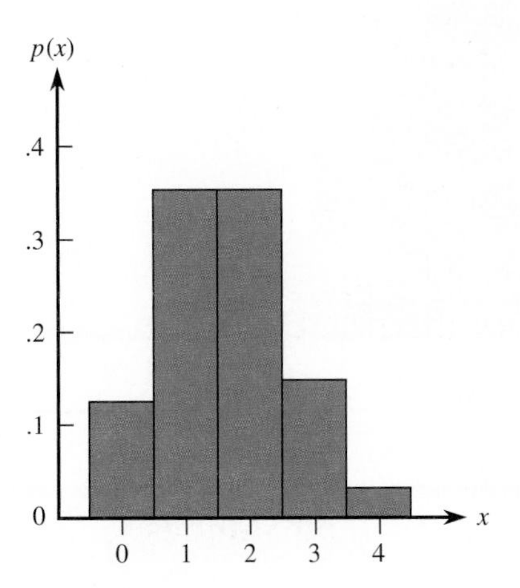

FIGURE 7.3
Probability histogram for the distribution of Example 7.5.

In Example 7.5, the probability distribution was derived by starting with a chance experiment and using probability rules. When a derivation from fundamental probabilities is not possible because of the complexity of the experimental situation, an investigator often proposes an approximate probability distribution consistent with empirical evidence and prior knowledge. Probability distributions based on empirical evidence must still be consistent with rules of probability. Specifically,

1. $p(x) \geq 0$ for every x value.
2. $\sum_{\text{all } x \text{ values}} p(x) = 1$

This is illustrated in Examples 7.6 and 7.7

EXAMPLE 7.6 Paint Flaws

Understand the context ❯

In automobile manufacturing, one of the last steps in the process of assembling a new car is painting. Some minor blemishes in the paint surface are considered acceptable, but if there are too many, it becomes noticeable to a potential customer and the car must be repainted. Cars coming off the assembly line are carefully inspected and the number of minor blemishes in the paint surface is determined. Let x denote the number of minor blemishes on a randomly selected car from a particular manufacturing plant. A large number of automobiles were evaluated, and a probability distribution consistent with these observations is

x	0	1	2	3	4	5	6	7	8	9	10
$p(x)$	.041	.010	.209	.223	.178	.114	.061	.028	.011	.004	.001

Interpret the results ❯

The corresponding probability histogram appears in Figure 7.4. The probabilities in this distribution reflect the car manufacturer's experience. For example, $p(3) = .223$ indicates that 22.3% of new automobiles had 3 minor paint blemishes. The probability that the number of minor paint blemishes is between 2 and 5 inclusive is

$$P(2 \leq x \leq 5) = p(2) + p(3) + p(4) + p(5) = .724$$

If car after car of this type were examined, in the long run, 72.4% would have 2, 3, 4, or 5 minor paint blemishes.

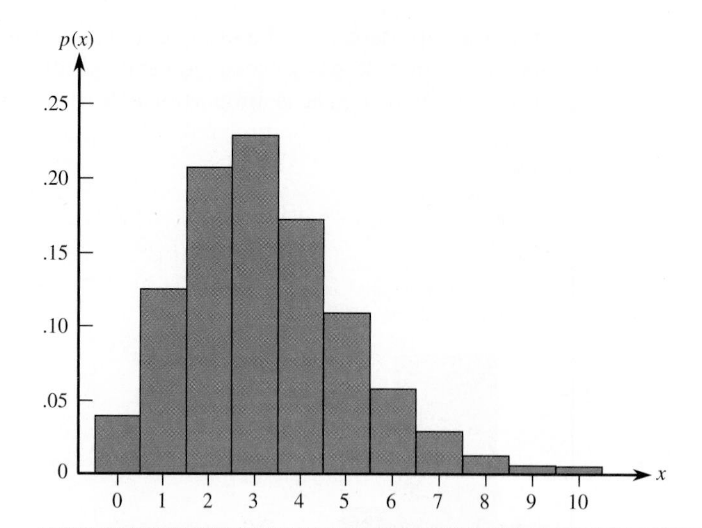

FIGURE 7.4
Probability histogram for the distribution of the number of minor paint blemishes on a randomly selected car.

EXAMPLE 7.7 iPod Shuffles

Understand the context

The paper **"Does Your iPod *Really* Play Favorites?" (*American Statistician* [2009]: 263–268)** investigated the shuffle feature of the iPod. The shuffle feature takes a group of songs, called a playlist, and plays them in a random order. Some have questioned the "randomness" of the shuffle, citing examples of situations where several songs from the same artist were played in close proximity to each other. One such example appeared in a *Newsweek* article in 2005, where the author states that Steely Dan songs always seem to pop up in the first hour of play. (For those young readers unfamiliar with Steely Dan, see http://www.steelydan.com.) Is this consistent with a "random" shuffle?

To investigate the alleged non-randomness, suppose we create a playlist of 3000 songs that includes 50 songs by Steely Dan (this is the situation considered by the authors of the *American Statistician* paper). We could then carry out a simulation by creating a shuffle of 20 songs (about one hour of playing time) from the playlist and noting the number of songs by Steely Dan that were among the 20.

This could be done by thinking of the songs in the playlist as being numbered from 1 to 3000, with numbers 1 to 50 representing the Steely Dan songs. A random number generator could then be used to select 20 random numbers between 1 and 3000. We would then count how many times a number between 1 and 50 was included in this list. This would correspond to the number of Steely Dan songs in this particular shuffle of 20 songs. Repeating this process a large number of times would enable us to estimate the probabilities needed for the probability distribution of

x = number of Steely Dan songs in a random shuffle consisting of 20 songs

The probabilities in the accompanying probability distribution are from the *American Statistician* paper. Even though possible values of x are 0, 1, 2, . . . , 20, the probability of x taking on a value of 4 or greater is very small, and so 4, 5, . . . , 20 have been grouped into a single entry in the probability distribution table.

Probability Distribution of x = number of Steely Dan songs in a 20-song shuffle

x	$p(x)$
0	.7138
1	.2435
2	.0387
3	.0038
4 or more	.0002

Interpret the results) Notice that $P(x \geq 1) = 1 - .7138 = .2862$. This means that about 28.6% of the time, at least one Steely Dan song would occur in a random shuffle of 20 songs. Given that there are only 50 Steely Dan songs in the playlist of 3000 songs, this result surprises many people! ■

We have seen examples in which the probability distribution of a discrete random variable has been given as a table or as a probability histogram. It is also possible to give a formula that allows calculation of the probability for each possible value of the random variable. Examples of this approach are given in Section 7.5.

EXERCISES 7.8 - 7.19

7.8 Let x be the number of courses for which a randomly selected student at a certain university is registered. The probability distribution of x appears in the following table:

x	1	2	3	4	5	6	7
$p(x)$	.02	.03	.09	.25	.40	.16	.05

a. What is $P(x = 4)$?

b. What is $P(x \leq 4)$?

c. What is the probability that the selected student is taking at most five courses?

d. What is the probability that the selected student is taking at least five courses? more than five courses?

e. Calculate $P(3 \leq x \leq 6)$ and $P(3 < x < 6)$. Explain in words why these two probabilities are different.

7.9 ▼ Let y denote the number of broken eggs in a randomly selected carton of one dozen eggs. Suppose that the probability distribution of y is as follows:

y	0	1	2	3	4
$p(y)$	.65	.20	.10	.04	?

a. Only y values of 0, 1, 2, 3, and 4 have positive probabilities. What is $p(4)$? (Hint: Consider the properties of a discrete probability distribution.)

b. How would you interpret $p(1) = .20$?

c. Calculate $P(y \leq 2)$, the probability that the carton contains at most two broken eggs, and interpret this probability.

d. Calculate $P(y < 2)$, the probability that the carton contains *fewer than* two broken eggs. Why is this smaller than the probability in Part (c)?

e. What is the probability that the carton contains exactly 10 unbroken eggs?

f. What is the probability that at least 10 eggs are unbroken?

7.10 Suppose that fund-raisers at a university call recent graduates to request donations for campus outreach programs. They report the following information for last year's graduates:

Size of donation	\$0	\$10	\$25	\$50
Proportion of calls	0.45	0.30	0.20	0.05

Three attempts were made to contact each graduate. A donation of \$0 was recorded both for those who were contacted but who declined to make a donation and for those who were not reached in three attempts. Consider the variable x = amount of donation for the population of last year's graduates of this university.

a. Write a few sentences describing what you think you might see if the value of x was observed for each of 1000 graduates.

b. What is the most common value of x in this population?

c. What is $P(x \geq 25)$?

d. What is $P(x > 0)$?

7.11 Airlines sometimes overbook flights. Suppose that for a plane with 100 seats, an airline takes 110 reservations. Define the variable x as the number of people who actually show up for a sold-out flight. From past experience, the probability distribution of x is given in the following table:

x	95	96	97	98	99	100	101	102
$p(x)$	.05	.10	.12	.14	.24	.17	.06	.04

x	103	104	105	106	107	108	109	110
$p(x)$	.03	.02	.01	.005	.005	.005	.0037	.0013

a. What is the probability that the airline can accommodate everyone who shows up for the flight?

b. What is the probability that not all passengers can be accommodated?

c. If you are trying to get a seat on such a flight and you are number 1 on the standby list, what is the probability that you will be able to take the flight? What if you are number 3?

7.12 Suppose that a computer manufacturer receives computer boards in lots of five. Two boards are selected from each lot for inspection. We can represent possible outcomes of the selection process by pairs. For example, the pair (1,2) represents the selection of Boards 1 and 2 for inspection.

a. List the 10 different possible outcomes.

b. Suppose that Boards 1 and 2 are the only defective boards in a lot of five. Two boards are to be chosen at random. Define x to be the number of defective boards observed among those inspected. Find the probability distribution of x.

7.13 Simulate the chance experiment described in the previous exercise using five slips of paper, with two marked *defective* and three marked *nondefective*. Place the slips in a box, mix them well, and draw out two. Record the number of defective boards. Replace the slips and repeat until you have 50 observations on the variable x. Construct a relative frequency distribution for the 50 observations, and compare this with the probability distribution obtained in the previous exercise.

7.14 Of all airline flight requests received by a certain discount ticket broker, 70% are for domestic travel (D) and 30% are for international flights (I). Let x be the number of requests among the next three requests received that are for domestic flights. Assuming independence of successive requests, determine the probability distribution of x. (Hint: See Example 7.5.)

7.15 Suppose that 20% of all homeowners in an earthquake-prone area of California are insured against earthquake damage. Four homeowners are selected at random; let x denote the number among the four who have earthquake insurance.

a. Find the probability distribution of x. (Hint: Let S denote a homeowner who has insurance and F one who does not. Then one possible outcome is SFSS, with probability (.2)(.8)(.2)(.2) and associated x value of 3. There are 15 other outcomes.)

b. What is the most likely value of x?

c. What is the probability that at least two of the four selected homeowners have earthquake insurance?

7.16 A box contains five slips of paper, marked \$1, \$1, \$1, \$10, and \$25. The winner of a contest selects two slips of paper at random and then gets the larger of the dollar amounts on the two slips. Define a random variable w by w = amount awarded. Determine the probability distribution of w. (Hint: Think of the slips as numbered 1, 2, 3, 4, and 5, so that an outcome of the experiment consists of two of these numbers.)

7.17 Components coming off an assembly line are either free of defects (S, for success) or defective (F, for failure). Suppose that 70% of all such components are defect-free. Components are independently selected and tested one by one. Let y denote the number of components that must be tested until a defect-free component is obtained.

a. What is the smallest possible y value, and what experimental outcome gives this y value? What is the second smallest y value, and what outcome gives rise to it?

b. What is the set of all possible y values?

c. Determine the probability of each of the five smallest y values. You should see a pattern that leads to a simple formula for $p(y)$, the probability distribution of y.

7.18 A contractor is required by a county planning department to submit anywhere from one to five forms (depending on the nature of the project) in applying for a building permit. Let y be the number of forms required of the next applicant. The probability that y forms are required is known to be proportional to y; that is, $p(y) = ky$ for $y = 1, \ldots, 5$.

a. What is the value of k? (Hint: $\Sigma p(y) = 1$.)

b. What is the probability that at most three forms are required?

c. What is the probability that between two and four forms (inclusive) are required?

d. Could $p(y) = y^2/50$ for $y = 1, 2, 3, 4, 5$ be the probability distribution of y? Explain.

7.19 A library subscribes to two different weekly news magazines, each of which is supposed to arrive in Wednesday's mail. In actuality, each one could arrive on Wednesday (W), Thursday (T), Friday (F), or Saturday (S). Suppose that the two magazines arrive independently of one another and that for each magazine

$P(W) = .4,\ P(T) = .3,\ P(F) = .2$, and $P(S) = .1$

Define a random variable y by

y = the number of days beyond Wednesday that it takes for both magazines to arrive.

For example, if the first magazine arrives on Friday and the second magazine arrives on Wednesday, then $y = 2$, whereas $y = 1$ if both magazines arrive on Thursday. Determine the probability distribution of y. (Hint: Draw a tree diagram with two generations of branches, the first labeled with arrival days for Magazine 1 and the second for Magazine 2.)

7.3 Probability Distributions for Continuous Random Variables

A continuous random variable is one that has possible values that form an entire interval on the number line. An example is the weight x (in pounds) of a full-term newborn child. Suppose for the moment that weight is recorded only to the nearest pound. Then a reported weight of 7 pounds would be used for any weight greater than or equal to 6.5 pounds and less than 7.5 pounds.

The probability distribution can be pictured as a probability histogram in which rectangles are centered at 4, 5, and so on. The area of each rectangle is the probability of the corresponding weight value. The total area of all the rectangles is 1, and the probability that a weight (to the nearest pound) is between two values, such as 6 and 8, is the sum of the corresponding rectangular areas. Figure 7.5(a) illustrates this.

Now suppose that weight is measured to the nearest tenth of a pound. There are many more possible reported weight values than before, such as 5.0, 5.1, 5.7, 7.3, and 8.9. As shown in Figure 7.5(b), the rectangles in the probability histogram are much narrower, and this histogram has a much smoother appearance than the first one. Again, this histogram can be drawn so that the area of each rectangle equals the corresponding probability associated with the interval, and the total area of all the rectangles is 1.

Figure 7.5(c) shows what happens as weight is measured to a greater and greater degree of accuracy. The sequence of probability histograms approaches a smooth curve. The curve cannot go below the horizontal measurement scale, and the total area under the curve is 1 (because this is true of each probability histogram). The probability that x falls in an interval such as $6 \leq x \leq 8$ is represented by the area under the curve and above that interval.

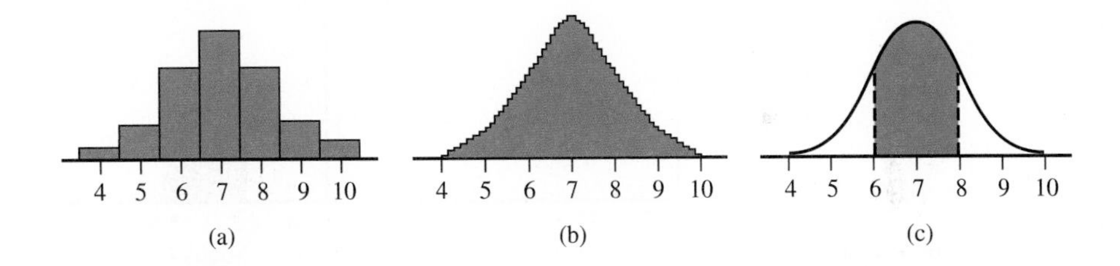

FIGURE 7.5
Probability distribution for birth weight:
(a) weight measured to the nearest pound;
(b) weight measured to the nearest tenth of a pound;
(c) limiting curve as measurement accuracy increases; shaded area = $P(6 \leq \text{weight} \leq 8)$.

A **probability distribution for a continuous random variable *x*** is specified by a curve called a **density curve.** The function that defines this curve is denoted by $f(x)$ and it is called the **density function.**

The following are properties of all continuous probability distributions:

1. $f(x) \geq 0$ (the curve cannot dip below the horizontal axis).
2. The total area under the density curve is equal to 1.

The probability that *x* falls in any particular interval is equal to the area under the density curve and above the interval.

Many probability calculations for continuous random variables involve the following three types of events:

1. $a < x < b$, the event that the random variable x assumes a value between two given numbers, a and b
2. $x < a$, the event that the random variable x assumes a value less than a given number a
3. $x > b$, the event that the random variable x assumes a value greater than a given number b (this can also be written as $b < x$)

Figure 7.6 illustrates how the probabilities of these events are represented by areas under a density curve.

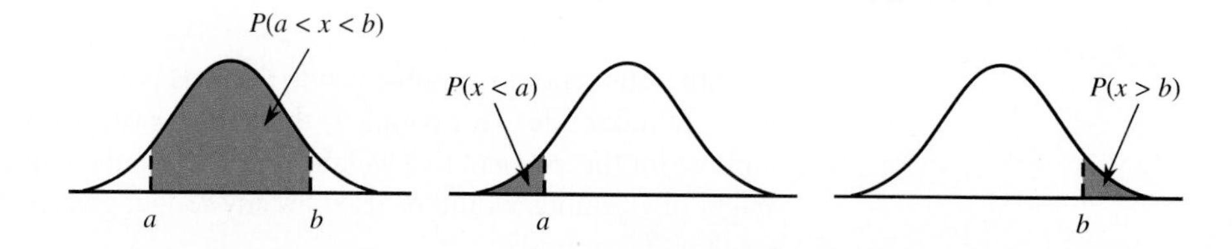

FIGURE 7.6
Probabilities as areas under a probability density curve.

EXAMPLE 7.8 Application Processing Times

Understand the context ❯

Suppose x represents the continuous random variable

x = amount of time (in minutes) taken by a clerk to process a certain type of application form

Suppose that x has a probability distribution with density function

$$f(x) = \begin{cases} .5 & 4 < x < 6 \\ 0 & \text{otherwise} \end{cases}$$

The graph of $f(x)$, the density curve, is shown in Figure 7.7(a). It is especially easy to use this density curve to calculate probabilities, because it just requires finding the area of rectangles using the formula

$$\text{area} = (\text{base})(\text{height})$$

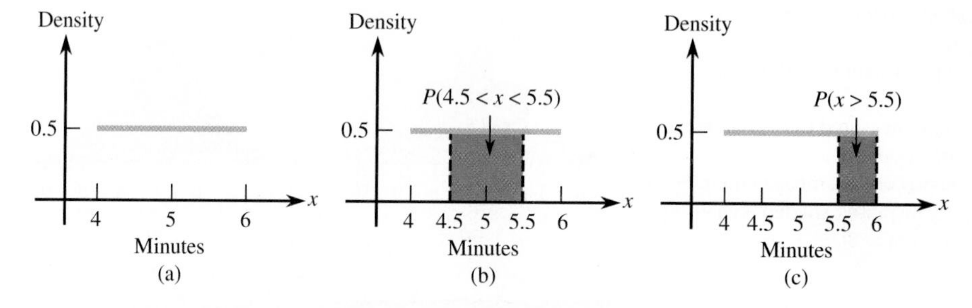

FIGURE 7.7
The probability distribution for Example 7.8.

Do the work ❯

The curve has positive height, 0.5, only between $x = 4$ and $x = 6$. The total area under the curve is just the area of the rectangle with base extending from 4 to 6 and with height 0.5. This gives

$$\text{area} = (6 - 4)(0.5) = 1$$

as required.

When the density is constant over an interval (resulting in a horizontal density curve), the probability distribution is called a **uniform distribution**.

As illustrated in Figure 7.7(b), the probability that x is between 4.5 and 5.5 is

$$\begin{aligned} P(4.5 < x < 5.5) &= \text{area of shaded rectangle} \\ &= (\text{base width})(\text{height}) \\ &= (5.5 - 4.5)(.5) \\ &= .5 \end{aligned}$$

Similarly (see Figure 7.7(c)), because in this context $x > 5.5$ is equivalent to $5.5 < x \leq 6$, we have

$$P(x > 5.5) = (6 - 5.5)(.5) = .25$$

Inerpret the results ❯

According to this model, in the long run, 25% of all forms that are processed will have processing times that exceed 5.5 min. ■

The probability that a *discrete* random variable x lies in the interval between two limits a and b can depend on whether either limit is included in the interval. For example, suppose that x is the number of major defects on a new automobile. Then

$$P(3 \leq x \leq 7) = p(3) + p(4) + p(5) + p(6) + p(7)$$

whereas

$$P(3 < x < 7) = p(4) + p(5) + p(6)$$

However, if x is a *continuous* random variable, such as task completion time, then

$$P(3 \leq x \leq 7) = P(3 < x < 7)$$

because the area under a density curve and above a single value such as 3 or 7 is 0. Geometrically, we can think of finding the area above a single point as finding the area of a rectangle with width = 0. For a continuous random variable, the area above an interval of values, therefore, does not depend on whether either endpoint is included.

If x is a continuous random variable, then for any two numbers a and b with $a < b$,

$$P(a \leq x \leq b) = P(a < x \leq b) = P(a \leq x < b) = P(a < x < b)$$

Probabilities for continuous random variables are often calculated using cumulative areas. A cumulative area is all of the area under the density curve to the left of a particular value. Figure 7.8 illustrates the cumulative area to the left of .5, which represents $P(x < .5)$. The probability that x is in any particular interval, $P(a < x < b)$, can be expressed as the difference between two cumulative areas.

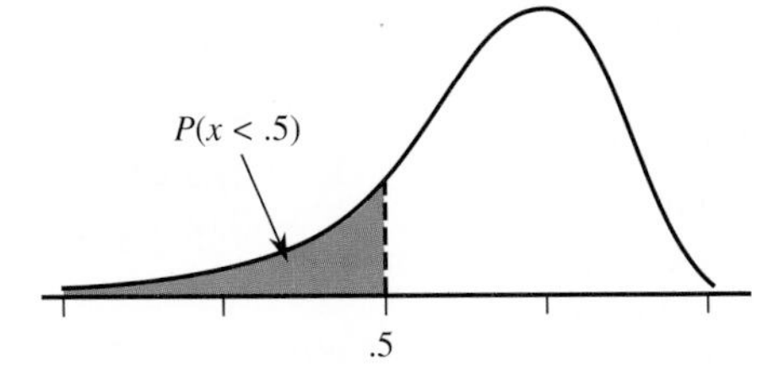

FIGURE 7.8
A cumulative area under a density curve.

The probability that a continuous random variable x lies between a lower limit a and an upper limit b is

$$P(a < x < b) = (\text{cumulative area to the left of } b) - (\text{cumulative area to the left of } a)$$
$$= P(x < b) - P(x < a)$$

This property is illustrated in Figure 7.9 for the case of $a = .25$ and $b = .75$. We will use this result often in Section 7.6 when we calculate probabilities for random variables that have a normal distribution.

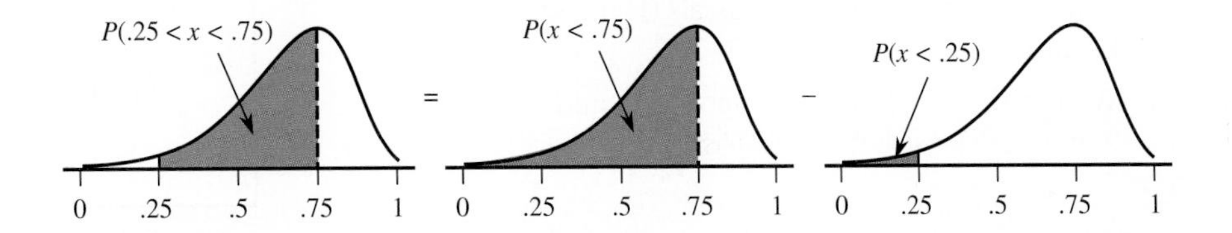

FIGURE 7.9
Calculation of $P(a < x < b)$ using cumulative areas.

For some continuous distributions, cumulative areas can be calculated using methods from the branch of mathematics called integral calculus. However, because we are not assuming knowledge of calculus, we will rely on technology or tables that have been constructed for the commonly encountered continuous probability distributions. Most graphing calculators and statistical software packages will compute areas for the most widely used continuous probability distributions.

EXERCISES 7.20 - 7.26

7.20 Let x denote the lifetime (in thousands of hours) of a certain type of fan used in diesel engines. The density curve of x is as pictured:

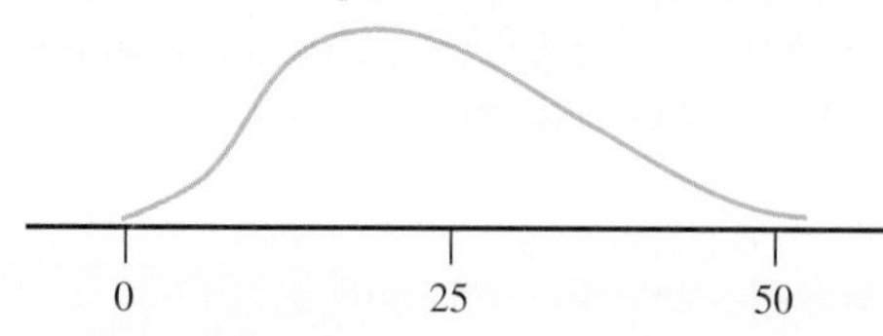

Shade the area under the curve corresponding to each of the following probabilities (draw a new curve for each part). (Hint: See Example 7.8.)

a. $P(10 < x < 25)$

b. $P(10 \leq x \leq 25)$

c. $P(x < 30)$

d. The probability that the lifetime is at least 25,000 hours

e. The probability that the lifetime exceeds 25,000 hours

7.21 A particular professor never dismisses class early. Let x denote the amount of time past the hour (minutes) that elapses before the professor dismisses class. Suppose that x has a uniform distribution on the interval from 0 to 10 minutes. The density curve is shown in the following figure:

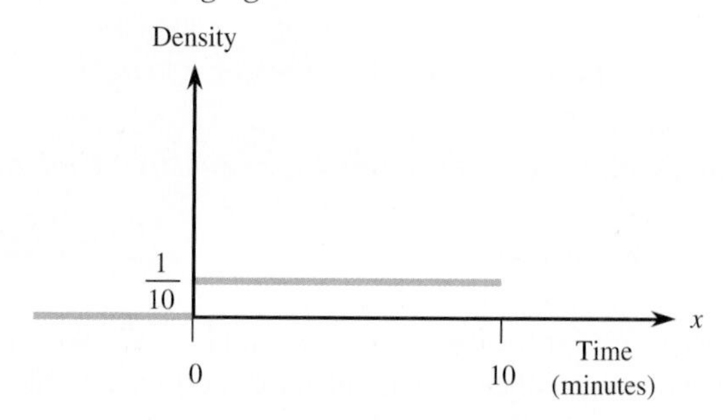

a. What is the probability that at most 5 minutes elapse before dismissal? (Hint: See Example 7.8.)

b. What is the probability that between 3 and 5 minutes elapse before dismissal?

7.22 Refer to the probability distribution given in the previous exercise. Put the following probabilities in order, from smallest to largest:

$P(2 < x < 3)$, $P(2 \leq x \leq 3)$, $P(x < 2)$, $P(x > 7)$

Explain your reasoning.

7.23 The article **"Modeling Sediment and Water Column Interactions for Hydrophobic Pollutants" (*Water Research* [1984]: 1169–1174)** suggests the uniform distribution on the interval from 7.5 to 20 as a model for x = depth (in centimeters) of the bioturbation layer in sediment for a certain region.

a. Draw the density curve for x. (Hint: See Example 7.8)

b. What is the height of the density curve?

c. What is the probability that x is at most 12?

d. What is the probability that x is between 10 and 15?

e. What is the probability that x is between 12 and 17?

f. Why are the two probabilities computed in Parts (d) and (e) equal?

7.24 Let x denote the amount of gravel sold (in tons) during a randomly selected week at a particular sales facility. Suppose that the density curve has height $f(x)$ above the value x, where

$$f(x) = \begin{cases} 2(1 - x) & 0 \leq x \leq 1 \\ 0 & \text{otherwise} \end{cases}$$

The density curve (the graph of $f(x)$) is shown in the following figure:

Density
2
0
1
x
Tons

Use the fact that the area of a triangle $= \frac{1}{2}$ (base)(height) to calculate each of the following probabilities. (Hint: Drawing a picture and shading the appropriate area will help.)

a. $P\left(x < \frac{1}{2}\right)$

b. $P\left(x \leq \frac{1}{2}\right)$

c. $P\left(x < \frac{1}{4}\right)$

d. $P\left(\frac{1}{4} < x < \frac{1}{2}\right)$ (Hint: Use the results of Parts (a)–(c).)

e. The probability that gravel sold exceeds $\frac{1}{2}$ ton

f. The probability that gravel sold is at least $\frac{1}{4}$ ton

7.25 Let x be the amount of time (in minutes) that a particular San Francisco commuter must wait for a BART train. Suppose that the density curve is as pictured (a uniform distribution):

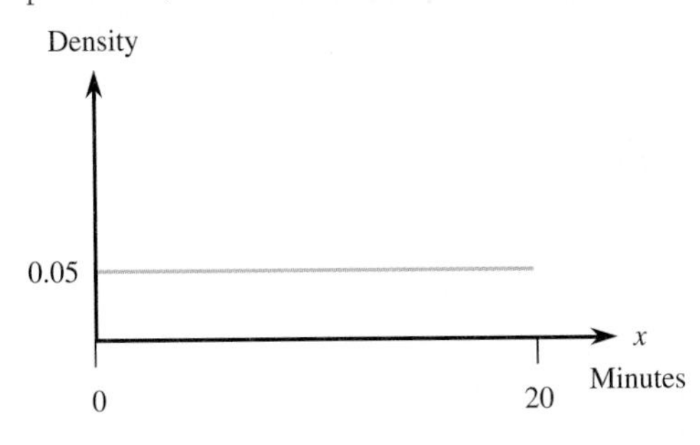

a. What is the probability that x is less than 10 minutes? more than 15 minutes?

b. What is the probability that x is between 7 and 12 minutes?

c. Find the value c for which $P(x < c) = .9$.

7.26 Referring to the previous exercise, let x and y be waiting times on two independently selected days. Define a new random variable w by

$$w = x + y$$

the sum of the two waiting times. The set of possible values for w is the interval from 0 to 40 (because both x and y can range from 0 to 20). It can be shown that the density curve of w is as pictured (this curve is called a triangular distribution, for obvious reasons!):

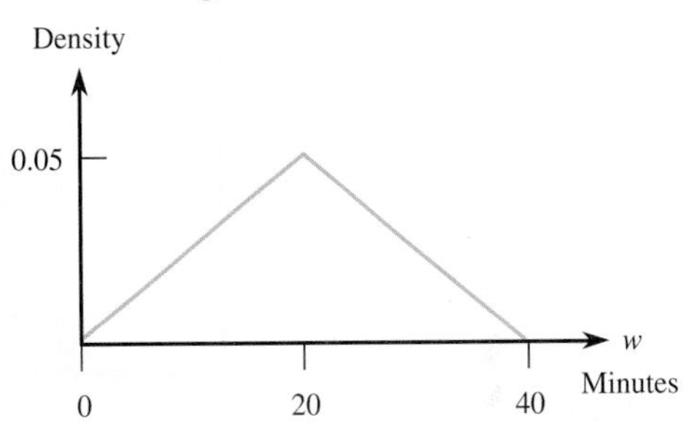

a. Verify that the total area under the density curve is equal to 1. (Hint: The area of a triangle is $= \frac{1}{2}$ (base)(height).)

b. What is the probability that w is less than 20?

c. What is the probability that w is less than 10?

d. What is the probability that w is greater than 30?

e. What is the probability that w is between 10 and 30? (Hint: It might be easier first to find the probability that w is not between 10 and 30.)

Bold exercises answered in back ● Data set available online ▼ Video Solution available

7.4 Mean and Standard Deviation of a Random Variable

We study a random variable x, such as the number of insurance claims made by a homeowner (a discrete variable) or the birth weight of a baby (a continuous variable), to learn something about how its values are distributed along the measurement scale. The sample mean $\bar{x}$ and sample standard deviation s summarize center and spread for the values in a sample. Similarly, the mean value and standard deviation of a random variable describe where the variable's probability distribution is centered and the extent to which it spreads out about the center.

The **mean value of a random variable x**, denoted by μ_x, describes where the probability distribution of x is centered.

The **standard deviation of a random variable x**, denoted by σ_x, describes variability in the probability distribution. When the value of σ_x is small, observed values of x will tend to be close to the mean value (little variability). When the value of σ_x is large, there will be more variability in observed x values.

Consider the probability distributions in Figure 7.10.

1. Figure 7.10(a) shows two discrete probability distributions with the same standard deviation (spread) but different means (center). One distribution has a mean of $\mu_x = 6$ and the other has $\mu_x = 10$. Which has the mean of 6 and which has the mean of 10?
2. Figure 7.10(b) shows two continuous probability distributions that have the same mean but different standard deviations. Which distribution—(i) or (ii)—has the larger standard deviation?
3. Figure 7.10(c) shows three continuous distributions with different means and standard deviations. Which of the three distributions has the largest mean? Which has a mean of about 5? Which distribution has the smallest standard deviation?

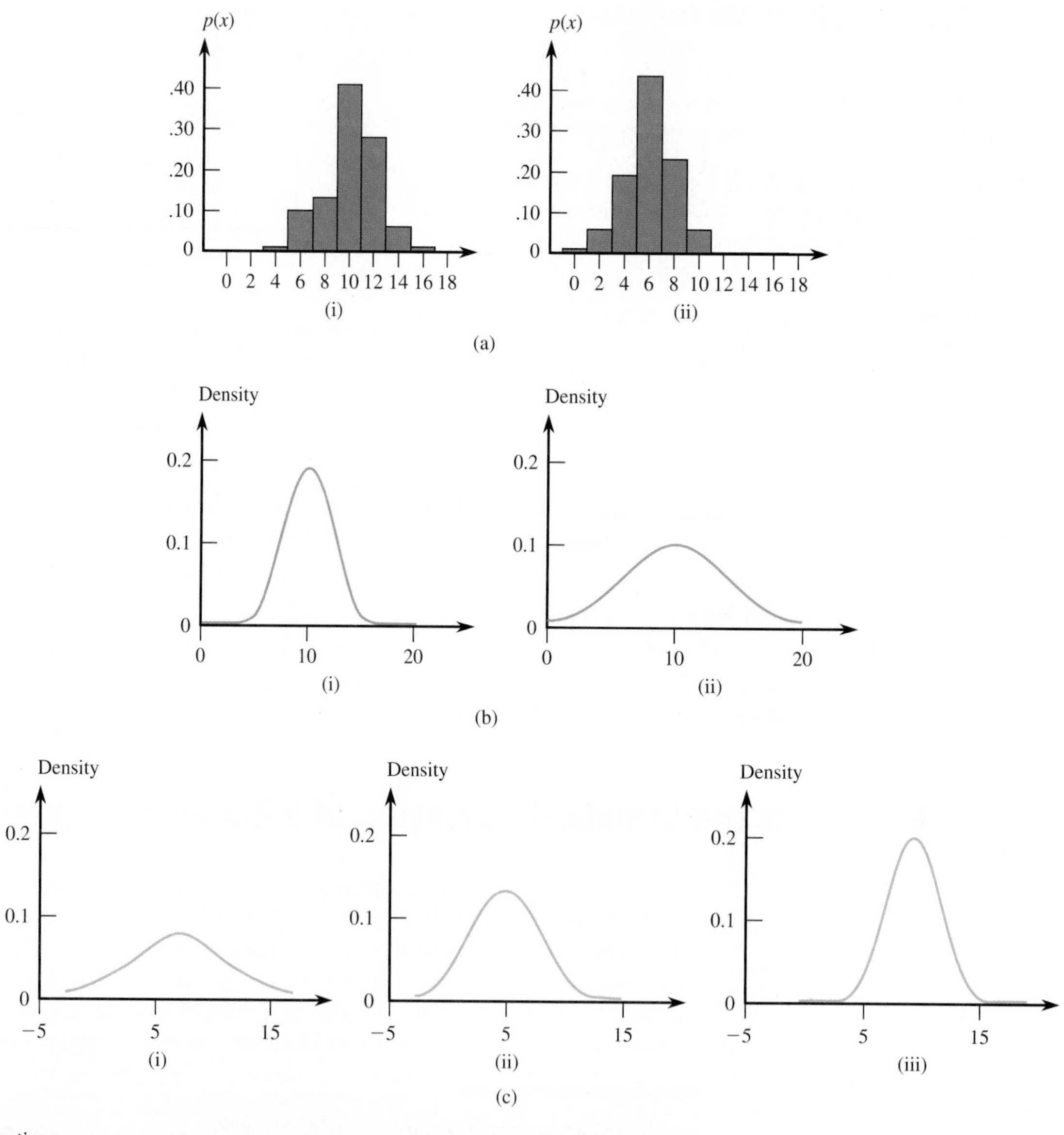

FIGURE 7.10
Some probability distributions:
(a) different values of μ_x with the same value of σ_x;
(b) different values of σ_x with the same value of μ_x;
(c) different values of μ_x and σ_x.

The answers to these questions are the following: Figure 7.10(a)(ii) has a mean of 6, and Figure 7.10(a)(i) has a mean of 10; Figure 7.10(b)(ii) has the larger standard deviation; Figure 7.10(c)(iii) has the largest mean, Figure 7.10(c)(ii) has a mean of about 5, and Figure 7.10(c)(iii) has the smallest standard deviation.

It is customary to use the terms *mean of the random variable* x and *mean of the probability distribution of* x interchangeably. Similarly, the standard deviation of the random variable x and the standard deviation of the probability distribution of x refer to the same thing. Although the mean and standard deviation are computed differently for discrete and continuous random variables, the interpretation is the same in both cases.

Mean Value of a Discrete Random Variable

Consider a chance experiment consisting of randomly selecting an automobile licensed in a particular state. Consider the discrete random variable x defined as

x = the number of low-beam headlights on the selected car that need adjustment

Possible x values are 0, 1, and 2, and the probability distribution of x might be as follows:

x value	0	1	2
Probability	.5	.3	.2

The corresponding probability histogram appears in Figure 7.11. In a sample of 100 cars, the sample relative frequencies might differ somewhat from the given probabilities (which are the limiting relative frequencies). For example, we might see:

x value	0	1	2
Frequency	46	33	21

The sample average value of x for these 100 observations is then the sum of 46 zeros, 33 ones, and 21 twos, all divided by 100:

$$\begin{aligned}\bar{x} &= \frac{(46)(0) + (33)(1) + (21)(2)}{100} \\ &= \left(\frac{46}{100}\right)(0) + \left(\frac{33}{100}\right)(1) + \left(\frac{21}{100}\right)(2) \\ &= (\text{rel. freq. of } 0)(0) + (\text{rel. freq. of } 1)(1) + (\text{rel. freq. of } 2)(2) \\ &= .75\end{aligned}$$

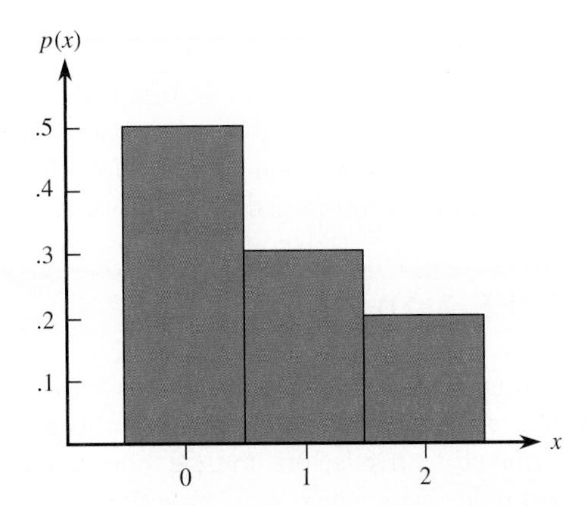

FIGURE 7.11
Probability histogram for the distribution of the number of headlights needing adjustment.

As the sample size increases, each relative frequency approaches the corresponding probability. In a very long sequence of experiments, the value of $\bar{x}$ approaches

$$\begin{aligned}\text{mena value of } x &= P(x = 0)(0) + P(x = 1)(1) + P(x = 2)(2) \\ &= (.5)(0) + (.3)(1) + (.2)(2) \\ &= .70\end{aligned}$$

Notice that the expression for $\bar{x}$ is a weighted average of possible x values. The weight of each value is the observed relative frequency. Similarly, the mean value of the random variable x is a weighted average, but now the weights are the probabilities from the probability distribution.

DEFINITION

Mean value of a discrete random variable: The mean value of a discrete random variable x, denoted by μ_x, is computed by first multiplying each possible x value by the probability of observing that value and then adding the resulting quantities. Symbolically,

$$\mu_x = \sum_{\substack{\text{all possible} \\ x \text{ values}}} x \cdot p(x)$$

The term **expected value** is sometimes used in place of mean value, and $E(x)$ is alternative notation for μ_x.

EXAMPLE 7.9 Exam Attempts

Understand the context ⟩

Individuals applying for a certain license are allowed up to four attempts to pass the licensing exam. Consider the random variable x defined as

x = the number of attempts made by a randomly selected applicant

Suppose the probability distribution of x is as follows:

x	1	2	3	4
$p(x)$	.10	.20	.30	.40

Do the work ⟩

Then x has mean value

$$\begin{aligned}\mu_x &= \sum_{x=1,2,3,4} x \cdot p(x) \\ &= (1)p(1) + (2)p(2) + (3)p(3) + (4)p(4) \\ &= (1)(.10) + (2)(.20) + (3)(.30) + (4)(.40) \\ &= .10 + .40 + .90 + 1.60 \\ &= 3.00\end{aligned}$$

Interpret the results ⟩

This means that the mean number of exam attempts for individuals applying for the license is 3. ■

It is no accident that the symbol μ_x for the mean value is the same symbol used previously for a population mean. When the probability distribution describes how x values are distributed among the members of a population (and therefore the probabilities are population relative frequencies), the mean value of x is the average value of x in the population.

EXAMPLE 7.10 Apgar Scores

Understand the context ⟩

At 1 minute after birth and again at 5 minutes, each newborn child is given a numerical rating called an Apgar score. Possible values of this score are 0, 1, 2, . . . , 9, 10. A child's score is determined by five factors: muscle tone, skin color, respiratory effort, strength of heartbeat, and reflex, with a high score indicating a healthy infant.

Consider the random variable x defined as

x = the Apgar score (at 1 minute) of a randomly selected newborn infant at a particular hospital

Suppose that x has the following probability distribution:

x	0	1	2	3	4	5	6	7	8	9	10
$p(x)$	.002	.001	.002	.005	.02	.04	.17	.38	.25	.12	.01

Do the work ⟩

The mean value of x is

$$\begin{aligned}\mu_x &= (0)p(0) + (1)p(1) + \cdots + (9)p(9) + (10)p(10) \\ &= (0)(.002) + (1)(.001) + \cdots + (9)(.12) + (10)(.01) \\ &= 7.16\end{aligned}$$

Interpret the results

The average Apgar score for a *sample* of newborn children born at this hospital may be $\bar{x} = 7.05$, $\bar{x} = 8.30$, or any one of a number of other possible values between 0 and 10. However, as child after child is born and rated, the average score will approach the value 7.16. This value can be interpreted as the mean Apgar score for the population of all babies born at this hospital. ■

Standard Deviation of a Discrete Random Variable

The mean value μ_x provides only a partial summary of a probability distribution. Two different distributions can have the same value of μ_x, but a long sequence of observations from one distribution might exhibit considerably more variability than a long sequence of observations from the other distribution.

EXAMPLE 7.11 Glass Panels

Understand the context

Flat screen TVs require high quality glass with very few flaws. A television manufacturer receives glass panels from two different suppliers. Let x and y denote the number of flaws in a randomly selected glass panel from the first and second suppliers, respectively. Suppose that the probability distributions for x and y are as follows:

x	0	1	2	3	4
$p(x)$	.4	.3	.2	.1	0

y	0	1	2	3	4
$p(y)$	.2	.6	.2	0	0

Probability histograms for x and y are given in Figure 7.12.

It is easy to verify that the mean values of both x and y are 1, so for either supplier the long-run average number of flaws per panel is 1. However, the two probability histograms show that the probability distribution for the second supplier is concentrated closer to the mean value than is the first supplier's distribution.

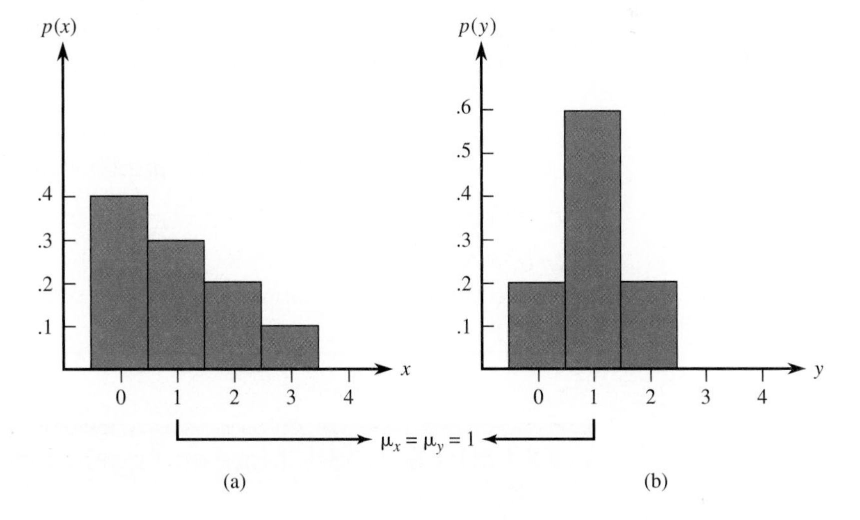

FIGURE 7.12 Probability distribution for the number of flaws in a glass panel in Example 7.11: (a) Supplier 1; (b) Supplier 2.

Interpret the results

The greater spread of the first distribution implies that there will be more variability in a long sequence of observed x values than in an observed sequence of y values. For example, the y sequence will contain no 3's, whereas in the long run, 10% of the observed x values will be 3. ■

As with s^2 and s, the variance and standard deviation of a random variable x involve squared deviations from the mean. A value far from the mean results in a large squared deviation. However, such a value does not contribute substantially to variability in x if the probability associated with that value is small. For example, if $\mu_x = 1$ and $x = 25$ is a possible value, then the squared deviation is $(25 - 1)^2 = 576$. However, if $P(x = 25) = .000001$, the value

25 will hardly ever be observed, so it won't contribute much to variability in a long sequence of observations. This is why each squared deviation is multiplied by the probability associated with the value to obtain a measure of variability.

DEFINITION

Variance of a discrete random variable: The variance of a discrete random variable x, denoted by σ_x^2, is computed by

1. subtracting the mean from each possible x value to obtain the deviations
2. squaring each deviation
3. multiplying each squared deviation by the probability of the corresponding x value
4. adding these quantities

Symbolically,

$$\sigma_x^2 = \sum_{\substack{\text{all possible} \\ x \text{ values}}} (x-\mu_x)^2 p(x)$$

Standard deviation of a discrete random variable: The standard deviation of a discrete random variable x, denoted by σ_x, is the square root of the variance.

When the probability distribution describes how x values are distributed among members of a population (so that the probabilities are population relative frequencies), σ_x^2 and σ_x are the population variance and standard deviation of x, respectively.

EXAMPLE 7.12 Glass Panels Revisited

For x = number of flaws in a glass panel from the first supplier in Example 7.11,

$$\begin{aligned}\sigma_x^2 &= (0-1)^2p(0) + (1-1)^2p(1) + (2-1)^2p(2) + (3-1)^2p(3) \\ &= (1)(.4) + (0)(.3) + (1)(.2) + (4)(.1) \\ &= 1.0\end{aligned}$$

Therefore $\sigma_x = 1.0$.

For y = the number of flaws in a glass panel from the second supplier,

$$\sigma_y^2 = (0-1)^2(.2) + (1-1)^2(.6) + (2-1)^2(.2) = .4$$

Then $\sigma_y = \sqrt{.4} = .632$.

The fact that $\sigma_x > \sigma_y$ confirms the impression conveyed by Figure 7.12 concerning the variability of x and y. ■

EXAMPLE 7.13 More on Apgar Scores

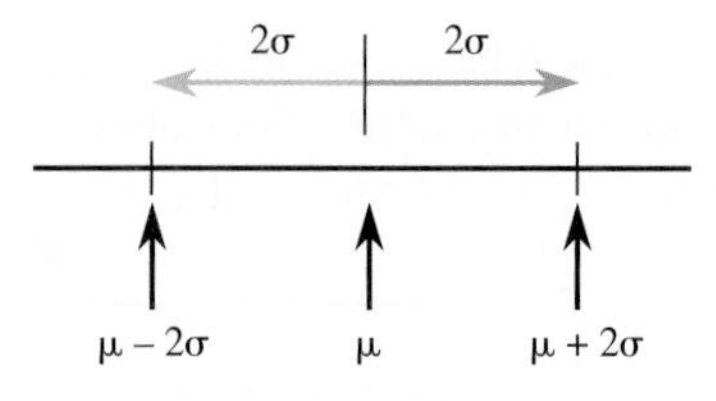

FIGURE 7.13
Values within 2 standard deviations of the mean.

Reconsider the distribution of Apgar scores for children born at a certain hospital, introduced in Example 7.10. What is the probability that a randomly selected child's score will be within 2 standard deviations of the mean score? As Figure 7.13 shows, values of x within 2 standard deviations of the mean are those for which

$$\mu - 2\sigma < x < \mu + 2\sigma$$

From Example 7.10 we already have $\mu_x = 7.16$. The variance is

$$\begin{aligned}\sigma^2 &= \Sigma(x-\mu)^2p(x) = \Sigma(x-7.16)^2p(x) \\ &= (0-7.16)^2(.002) + (1-7.16)^2(.001) + \cdots + (10-7.16)^2(.01) \\ &= 1.5684\end{aligned}$$

and the standard deviation is

$$\sigma = \sqrt{1.5684} = 1.25$$

This gives (using the probabilities given in Example 7.10)

$$\begin{aligned} P(\mu - 2\sigma < x < \mu + 2\sigma) &= P(7.16 - 2.50 < x < 7.16 + 2.50) \\ &= P(4.66 < x < 9.66) \\ &= p(5) + \cdots + p(9) \\ &= .96 \end{aligned}$$

■

Mean and Standard Deviation When *x* Is Continuous

For continuous probability distributions, μ_x and σ_x can be defined and computed using methods from calculus. The details need not concern us. What is important is knowing that μ_x and σ_x play exactly the same role here as they do in the discrete case. The mean value μ_x locates the center of the continuous distribution and gives the approximate long-run average of many observed x values. The standard deviation σ_x measures the extent that the continuous distribution (density curve) spreads out about μ_x and gives information about the amount of variability that can be expected in a long sequence of observed x values.

EXAMPLE 7.14 A "Concrete" Example

Understand the context ⟩

A company receives concrete of a certain type from two different suppliers. Consider the random variables x and y defined as follows:

x = compressive strength of a randomly selected batch from Supplier 1
y = compressive strength of a randomly selected batch from Supplier 2

Suppose that

$\mu_x = 4650$ pounds/inch^2 $\sigma_x = 200$ pounds/inch^2
$\mu_y = 4500$ pounds/inch^2 $\sigma_y = 275$ pounds/inch^2

Interpret the results ⟩

The long-run average strength per batch for many, many batches from Supplier 1 will be roughly 4650 pounds/inch^2. This is 150 pounds/inch^2 greater than the long-run average for batches from Supplier 2. In addition, a long sequence of batches from Supplier 1 will exhibit substantially less variability in compressive strength values than will a long sequence from Supplier 2. The first supplier is preferred to the second both in terms of mean value and variability. Figure 7.14 displays density curves that are consistent with this information.

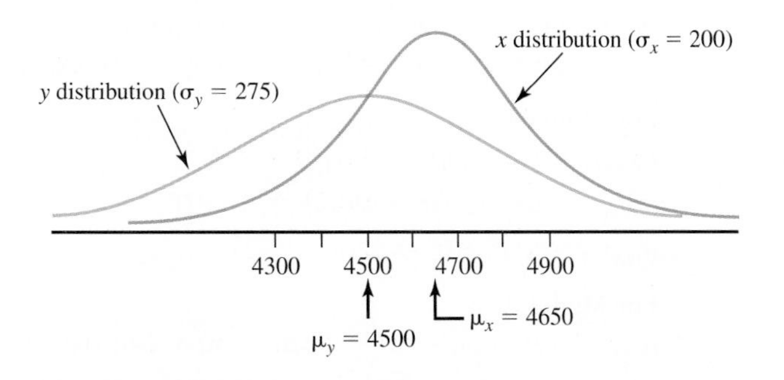

FIGURE 7.14
Density curves for Example 7.14.

■

Mean and Variance of Linear Functions and Linear Combinations

We have seen how the mean and standard deviation of one or more random variables provide useful information about the variables' long-run behavior, but we might also be interested in the behavior of some function of these variables.

For example, consider the chance experiment in which a customer of a propane gas company is randomly selected. Suppose that the mean and standard deviation of the random variable

x = number of gallons required to fill a customer's propane tank

are known to be 318 gallons and 42 gallons, respectively. The company is considering two different pricing models:

Model 1: \$3 per gallon
Model 2: service charge of \$50 + \$2.80 per gallon

The company is interested in the variable

y = amount billed

For each of the two models, y can be expressed as a function of the random variable x:

Model 1: $y_{model\ 1} = 3x$
Model 2: $y_{model\ 2} = 50 + 2.8x$

Both of these equations are examples of a linear function of x. The mean and standard deviation of a linear function of x can be computed from the mean and standard deviation of x, as described in the following box.

The Mean, Variance, and Standard Deviation of a Linear Function

If x is a random variable with mean μ_x and variance σ_x and a and b are numerical constants, the random variable y defined by

$$y = a + bx$$

is called a **linear function of the random variable *x*.**
The mean of $y = a + bx$ is

$$\mu_y = \mu_{a+bx} = a + b\mu_x$$

The variance of y is

$$\sigma_y^2 = \sigma_{a+bx}^2 = b^2\sigma_x^2$$

from which it follows that the standard deviation of y is

$$\sigma_y = \sigma_{a+bx} = |b|\sigma_x$$

We can use the results in the preceding box to compute the mean and standard deviation of the billing amount variable for the propane gas example, as follows:

For Model 1:

$\mu_{model\ 1} = \mu_{3x} = 3\mu_x = 3(318) = 954$
$\sigma_{model\ 1}^2 = \sigma_{3x}^2 = 3^2\sigma_x^2 = 9(42)^2 = 15{,}876$
$\sigma_{model\ 1} = \sqrt{15{,}876} = 126 = 3(42)$

For Model 2:

$\mu_{model\ 2} = \mu_{50+2.8x} = 50 + 2.8\mu_x = 50 + 2.8(318) = 940.40$
$\sigma_{model\ 2}^2 = \sigma_{50+2.8x}^2 = 2.8^2\sigma_x^2 = (2.8)^2(42)^2 = 13{,}829.76$
$\sigma_{model\ 2} = \sqrt{13{,}829.76} = 117.60$, which is equal to 2.8(42)

The mean billing amount for Model 1 is a bit higher than for Model 2, as is the variability in billing amounts. Model 2 results in slightly more consistency from bill to bill in the amount charged.

Linear Combinations

Now let's consider a different type of problem. Suppose that you have three tasks that you plan to complete on the way home from school: stop at the library to return an overdue book for which you must pay a fine, deposit your most recent paycheck at the bank, and stop by the office supply store to purchase paper for your computer printer. Define the following variables:

x_1 = time required to return book and pay fine
x_2 = time required to deposit paycheck
x_3 = time required to buy printer paper

We can then define a new variable, y, to represent the total amount of time to complete these tasks:

$$y = x_1 + x_2 + x_3$$

Defined in this way, y is an example of a linear combination of random variables.

If $x_1, x_2, \ldots, x_n$ are random variables and $a_1, a_2, \ldots, a_n$ are numerical constants, the random variable y defined as

$$y = a_1x_1 + a_2x_2 + \cdots + a_nx_n$$

is a **linear combination of the x_i's.**

For example, $y = 10x_1 - 5x_2 + 8x_3$ is a linear combination of x_1, x_2, and x_3 with $a_1 = 10$, $a_2 = -5$ and $a_3 = 8$.

It is easy to compute the mean of a linear combination of x_i's if the individual means $\mu_1, \mu_2, \ldots, \mu_n$ are known. The variance and standard deviation of a linear combination of the x_i's are also easily computed *if the x_i's are independent*. Two random variables x_i and x_j are independent if any event defined solely by x_i is independent of any event defined solely by x_j. When the x_i's are not independent, computation of the variance and standard deviation of a linear combination of the x_i's is more complicated, and this case is not considered here.

Mean, Variance, and Standard Deviation for Linear Combinations

If $x_1, x_2, \ldots, x_n$ are random variables with means $\mu_1, \mu_2, \ldots, \mu_n$ and variances $\sigma_1^2, \sigma_2^2, \ldots, \sigma_n^2$, respectively, and

$$y = a_1 x_1 + a_2 x_2 + \cdots + a_n x_n$$

then

1. $\mu_y = \mu_{a_1x_1+a_2x_2+\cdots+a_nx_n} = a_1\mu_1 + a_2\mu_2 + \cdots + a_n\mu_n$

   This result is true regardless of whether the x_i's are independent.

2. When $x_1, x_2, \ldots, x_n$ are *independent* random variables,

   $$\sigma_y^2 = \sigma^2_{a_1x_1+a_2x_2+\cdots+a_nx_n} = a_1^2\sigma_1^2 + a_2^2\sigma_2^2 + \cdots + a_n^2\sigma_n^2$$

   $$\sigma_y = \sigma_{a_1x_1+a_2x_2+\cdots+a_nx_n} = \sqrt{a_1^2\sigma_1^2 + a_2^2\sigma_2^2 + \cdots + a_n^2\sigma_n^2}$$

   These formulas are appropriate only when the x_i's are independent.

Examples 7.15–7.17 illustrate the use of these rules.

EXAMPLE 7.15 Freeway Traffic

Understand the context)

Three different roads feed into a particular freeway entrance. Suppose that during a fixed time period, the number of cars coming from each road onto the freeway is a random variable with mean values as follows:

Road	1	2	3
Mean	800	1000	600

With x_i representing the number of cars entering from road i, we can define the total number of cars entering the freeway y as

$$y = x_1 + x_2 + x_3.$$

Do the work)

The mean value of y is

$$\begin{aligned}\mu_y &= \mu_{x_1+x_2+x_3}\\ &= \mu_{x_1} + \mu_{x_2} + \mu_{x_3}\\ &= 800 + 1000 + 600\\ &= 2400\end{aligned}$$

■

EXAMPLE 7.16 Combining Exam Subscores

Understand the context)

A nationwide standardized exam consists of a multiple-choice section and a free response section. For each section, the mean and standard deviation are reported to be as shown in the following table:

	Mean	Standard Deviation
Multiple Choice	38	6
Free Response	30	7

Let's define x_1 and x_2 as the multiple-choice score and the free-response score, respectively, of a student selected at random from those taking this exam. We are also interested in the variable y = overall exam score.

Suppose that the free-response score is given twice the weight of the multiple-choice score in determining the overall exam score. Then the overall score is computed as

$$y = x_1 + 2x_2$$

Do the work)

What are the mean and standard deviation of y?

Because $y = x_1 + 2x_2$ is a linear combination of x_1 and x_2, the mean of y is

$$\begin{aligned}\mu_y &= \mu_{x_1+2x_2}\\ &= \mu_{x_1} + 2\mu_{x_2}\\ &= 38 + 2(30)\\ &= 98\end{aligned}$$

Interpret the results)

This means that the average overall score for students taking this exam is 98.

What about the variance and standard deviation of y? To use the formulas in the preceding box, x_1 and x_2 must be independent. It is unlikely that the value of x_1 (a student's multiple-choice score) would be unrelated to the value of x_2 (the same student's free-response score), because it seems probable that students who score well on one section of the exam will also tend to score well on the other section. Therefore, it would not be appropriate to use the formulas mentioned above to calculate the variance and standard deviation. ■

EXAMPLE 7.17 Baggage Weights

Understand the context ❯

A commuter airline flies small planes between San Luis Obispo and San Francisco. For small planes, the baggage weight is a concern, especially on foggy mornings, because the weight of the plane has an effect on how quickly the plane can ascend. Suppose that it is known that the variable x = weight (in pounds) of baggage checked by a randomly selected passenger has a mean of 42 and standard deviation of 16.

Consider a flight on which 10 passengers, all traveling alone, are flying. If we use x_i to denote the baggage weight for passenger i (for i ranging from 1 to 10), the total weight of checked baggage, y, is then

$$y = x_1 + x_2 + \cdots + x_{10}$$

Do the work ❯

Notice that y is a linear combination of the x_i's. The mean value of y is

$$\begin{aligned}\mu_y &= \mu_{x_1} + \mu_{x_2} + \cdots + \mu_{x_{10}} \\ &= 42 + 42 + \cdots + 42 \\ &= 420\end{aligned}$$

Since the 10 passengers are all traveling alone, it is reasonable to think that the 10 baggage weights are unrelated and that the x_i's are independent. (This might not be a reasonable assumption if the 10 passengers were not traveling alone.) Then the variance of y is

$$\begin{aligned}\sigma_y^2 &= \sigma_{x_1}^2 + \sigma_{x_2}^2 + \cdots + \sigma_{x_{10}}^2 \\ &= 16^2 + 16^2 + \cdots + 16^2 \\ &= 2650\end{aligned}$$

and the standard deviation of y is

$$\sigma_y = \sqrt{2650} = 50.596$$

Interpret the results ❯

This means that the average total weight of checked baggage is 420 pounds. The large standard deviation of 50.596 pounds indicates that there will be a lot of variability in total checked baggage weight from flight to flight. ■

One Last Note on Linear Functions and Linear Combinations

In Example 7.17, the random variable of interest was x = weight of baggage checked by a randomly selected airline passenger. However, when we considered a flight with multiple passengers, we added subscripts to create variables like

$$x_1 = \text{baggage weight for passenger 1}$$

and

$$x_2 = \text{baggage weight for passenger 2}$$

It is important to note that the sum of the baggage weights for two different passengers does not result in the same value as doubling the baggage weight of a single passenger. *That is, the linear combination $x_1 + x_2$ is different from the linear function $2x$.*

Although the mean of $x_1 + x_2$ and the mean of $2x$ are the same, the variances and standard deviations are different. For example, for the baggage weight distribution described in Example 7.17 (mean of 42 and standard deviation of 16), the mean of $2x$ is

$$\mu_{2x} = 2\mu_x = 2(42) = 84$$

and the mean of $x_1 + x_2$ is

$$\mu_{x_1+x_2} = \mu_{x_1} + \mu_{x_2} = 42 + 42 = 84$$

However, the variance of $2x$ is

$$\sigma_{2x}^2 = (2)^2\sigma_x^2 = 4(16^2) = 1024$$

whereas the variance of $x_1 + x_2$ is

$$\sigma_{x_1+x_2}^2 = \sigma_{x_1}^2 + \sigma_{2x_2} = 16^2 + 16^2 = 256 + 256 = 512$$

This means that observations of $2x$ for randomly selected passengers will show more variability than observations of $x_1 + x_2$ for two randomly selected passengers. This may seem counterintuitive, but it is true! For example, each row of the first two columns of the accompanying table show simulated baggage weights for two passengers (an x_1 and an x_2 value) that were selected at random from a probability distribution with mean 42 and standard deviation 16. The third column of the table contains the value of $2x_1$ (two times the passenger 1 baggage weight) and the last column contains the values of $x_1 + x_2$ (the sum of the passenger 1 baggage weight and the passenger 2 baggage weight).

x_1	x_2	$2x_1$	$x_1 + x_2$	x_1	x_2	$2x_1$	$x_1 + x_2$
56	41	112	97	44	23	88	67
30	5	60	35	48	54	96	102
32	15	64	47	52	69	104	121
55	45	110	100	56	63	112	119
40	54	80	94	79	35	158	114
41	68	82	109	18	34	36	52
42	35	84	77	8	71	16	79
33	47	66	80	51	58	102	109
46	52	92	98	18	49	36	67
73	40	146	113	41	46	82	87

Figure 7.15 shows dotplots of the 20 values of $2x_1$ and the 20 values of $x_1 + x_2$. You can see that there is less variability in the $x_1 + x_2$ values than in the values of $2x_1$.

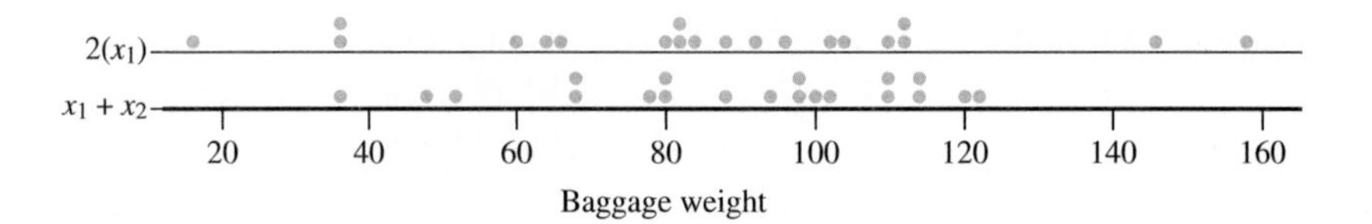

FIGURE 7.15
Dotplots of $2x_1$ and $x_1 + x_2$.

EXERCISES 7.27 - 7.42

7.27 Consider selecting a household in rural Thailand at random. Define the random variable x to be

x = number of individuals living in the selected household

Based on information in an article that appeared in the ***Journal of Applied Probability*** **(2011: 173–188)**, the probability distribution of x is as follows:

x	1	2	3	4	5	6	7	8	9	10
$p(x)$	.140	.175	.220	.260	.155	.025	.015	.005	.004	.001

Calculate the mean value of the random variable x. (Hint: See Example 7.9.)

7.28 The probability distribution of x, the number of defective tires on a randomly selected automobile checked at a certain inspection station, is given in the following table:

x	0	1	2	3	4
$p(x)$	.54	.16	.06	.04	.20

a. Calculate the mean value of x.

b. What is the probability that x exceeds its mean value?

7.29 Consider the following probability distribution for y = the number of broken eggs in a carton:

y	0	1	2	3	4
$p(y)$	.65	.20	.10	.04	.01

a. Calculate and interpret μ_y.

b. In the long run, for what percentage of cartons is the number of broken eggs less than μ_y? Does this surprise you?

c. Why doesn't $\mu_y = (0 + 1 + 2 + 3 + 4)/5 = 2.0$? Explain.

7.30 Referring to the previous exercise, use the result of Part (a) along with the fact that a carton contains 12 eggs to determine the mean value of z = the number of unbroken eggs. (Hint: z can be written as a linear function of y; see Example 7.15.)

7.31 Exercise 7.8 gave the following probability distribution for x = the number of courses for which a randomly selected student at a certain university is registered:

x	1	2	3	4	5	6	7
$p(x)$	.02	.03	.09	.25	.40	.16	.05

It can be easily verified that $\mu = 4.66$ and $\sigma = 1.20$.

a. Because $\mu - \sigma = 3.46$, the x values 1, 2, and 3 are more than 1 standard deviation below the mean. What is the probability that x is more than 1 standard deviation below its mean? (Hint: See Example 7.13.)

b. What x values are more than 2 standard deviations away from the mean value (either less than $\mu - 2\sigma$ or greater than $\mu + 2\sigma$)?

c. What is the probability that x is more than 2 standard deviations away from its mean value?

7.32 Example 7.11 gave the probability distributions of x = number of flaws in a randomly selected glass panel for two suppliers of glass used in the manufacture of flat screen TVs. If the manufacturer wanted to select a single supplier for glass panels, which of these two suppliers would you recommend? Justify your choice based on consideration of both center and variability. (Hint: See Example 7.11.)

7.33 Consider a large ferry that can accommodate cars and buses. The toll for cars is \$3, and the toll for buses is \$10. Let x and y denote the number of cars and buses, respectively, carried on a single trip. Cars and buses are accommodated on different levels of the ferry, so the number of buses accommodated on any trip is independent of the number of cars on the trip. Suppose that x and y have the following probability distributions:

x	0	1	2	3	4	5
$p(x)$	.05	.10	.25	.30	.20	.10

y	0	1	2
$p(y)$	.50	.30	.20

a. Compute the mean and standard deviation of x.

b. Compute the mean and standard deviation of y.

c. Compute the mean and variance of the total amount of money collected in tolls from cars.

d. Compute the mean and variance of the total amount of money collected in tolls from buses.

e. Compute the mean and variance of z = total number of vehicles (cars and buses) on the ferry.

f. Compute the mean and variance of w = total amount of money collected in tolls.

7.34 Suppose that for a given computer salesperson, the probability distribution of x = the number of systems sold in 1 month is given by the following table:

x	1	2	3	4	5	6	7	8
$p(x)$	.05	.10	.12	.30	.30	.11	.01	.01

a. Find the mean value of x (the mean number of systems sold).

b. Find the variance and standard deviation of x. How would you interpret these values?

c. What is the probability that the number of systems sold is within 1 standard deviation of its mean value?

d. What is the probability that the number of systems sold is more than 2 standard deviations from the mean? (Hint: See Example 7.13.)

7.35 A local television station sells 15-second, 30-second, and 60-second advertising spots. Let x denote the length of a randomly selected commercial appearing on this station, and suppose that the probability distribution of x is given by the following table:

x	15	30	60
$p(x)$	.1	.3	.6

a. Find the average length for commercials appearing on this station.

b. If a 15-second spot sells for \$500, a 30-second spot for \$800, and a 60-second spot for \$1000, find the average amount paid for commercials appearing on this station. (Hint: Consider a new variable, y = cost, and then find the probability distribution and mean value of y.)

7.36 An author has written a book and submitted it to a publisher. The publisher offers to print the book and gives the author the choice between a flat payment of \$10,000 and a royalty plan. Under the royalty plan the author would receive \$1 for each copy of the book sold. The author thinks that the following table gives the probability distribution of the variable x = the number of books that will be sold:

x	1000	5000	10,000	20,000
$p(x)$	.05	.30	.40	.25

Which payment plan should the author choose? Why?

7.37 A grocery store has an express line for customers purchasing five or fewer items. Let x be the number of items purchased by a randomly selected customer using this line. Give examples of two different assignments of probabilities such that the resulting distributions have the same mean but quite different standard deviations.

7.38 An appliance dealer sells three different models of upright freezers having 13.5, 15.9, and 19.1 cubic feet of storage space. Let x = the amount of storage space purchased by the next customer to buy a freezer. Suppose that x has the following probability distribution:

x	13.5	15.9	19.1
$p(x)$	.2	.5	.3

a. Calculate the mean and standard deviation of x.

b. If the price of the freezer depends on the size of the storage space, x, such that $Price = 25x - 8.5$, what is the mean price paid by the next customer?

c. What is the standard deviation of the price paid?

7.39 ▼ To assemble a piece of furniture, a wood peg must be inserted into a predrilled hole. Suppose that the diameter of a randomly selected peg is a random variable with mean 0.25 inch and standard deviation 0.006 inch and that the diameter of a randomly selected hole is a random variable with mean 0.253 inch and standard deviation 0.002 inch. Let x_1 = peg diameter, and let x_2 = denote hole diameter.

a. Why would the random variable y, defined as $y = x_2 - x_1$, be of interest to the furniture manufacturer?

b. What is the mean value of the random variable y?

c. Assuming that x_1 and x_2 are independent, what is the standard deviation of y?

d. Is it reasonable to think that x_1 and x_2 are independent? Explain.

e. Based on your answers to Parts (b) and (c), do you think that finding a peg that is too big to fit in the predrilled hole would be a relatively common or a relatively rare occurrence? Explain.

7.40 A multiple-choice exam consists of 50 questions. Each question has five choices, of which only one is correct. Suppose that the total score on the exam is computed as

$$y = x_1 - \frac{1}{4}x_2$$

where x_1 = number of correct responses and x_2 = number of incorrect responses. (Calculating a total score by subtracting a term based on the number of incorrect responses is known as a correction for guessing and is designed to discourage test takers from choosing answers at random.)

a. It can be shown that if a totally unprepared student answers all 50 questions by just selecting one of the five answers at random, then $\mu_{x_1} = 10$ and $\mu_{x_2} = 40$. What is the mean value of the total score, y? Does this surprise you? Explain. (Hint: See Example 7.16.)

b. Explain why it is unreasonable to use the formulas given in this section to compute the variance or standard deviation of y.

7.41 Consider a game in which a red die and a blue die are rolled. Let x_R denote the value showing on the uppermost face of the red die, and define x_B similarly for the blue die.

a. The probability distribution of x_R is

x_R	1	2	3	4	5	6
$p(x_R)$	1/6	1/6	1/6	1/6	1/6	1/6

Find the mean, variance, and standard deviation of x_R.

b. What are the values of the mean, variance, and standard deviation of x_B? (You should be able to answer this question without doing any additional calculations.)

c. Suppose that you are offered a choice of the following two games:

Game 1: Costs $7 to play, and you win y_1 dollars, where $y_1 = x_R + x_B$.

Game 2: Doesn't cost anything to play initially, but you "win" $3y_2$ dollars, where $y_2 = x_R - x_B$. If y_2 is negative, you must pay that amount; if it is positive, you receive that amount.

For Game 1, the net amount won in a game is $w_1 = y_1 - 7 = x_R + x_B - 7$. What are the mean and standard deviation of w_1?

d. For Game 2, the net amount won in a game is $w_2 = 3y_2 = 3(x_R - x_B)$. What are the mean and standard deviation of w_2?

e. Based on your answers to Parts (c) and (d), if you had to play, which game would you choose and why?

7.42 The states of Ohio, Iowa, and Idaho are often confused, probably because the names sound so similar. Each year, the State Tourism Directors of these three states drive to a meeting in one of the state capitals to discuss strategies for attracting tourists to their states so that the states will become better known.

The location of the meeting is selected at random from the three state capitals. The shortest highway distance from Boise, Idaho to Columbus, Ohio passes through Des Moines, Iowa. The highway distance from Boise to Des Moines is 1350 miles, and the distance from Des Moines to Columbus is 650 miles. Let d_1 represent the driving distance from Columbus to the meeting, with d_2 and d_3 representing the distances from Des Moines and Boise, respectively.

a. Find the probability distribution of d_1 and display it in a table.

b. What is the expected value of d_1?

c. What is the value of the standard deviation of d_1?

d. Consider the probability distributions of d_2 and d_3. Is either probability distribution the same as the probability distribution of d_1? Justify your answer.

e. Define a new random variable $t = d_1 + d_2$. Find the probability distribution of t.

f. For each of the following statements, indicate if the statement is true or false and provide statistical evidence to support your answer.

i. $E(t) = E(d_1) + E(d_2)$ (Hint: $E(t)$ is the expected value of t and another way of denoting the mean of t.)

ii. $\sigma_t^2 = \sigma_{d_1}^2 + \sigma_{d_2}^2$

Bold exercises answered in back ● Data set available online ▼ Video Solution available

7.5 Binomial and Geometric Distributions

In this section, we introduce two of the more commonly encountered discrete probability distributions: the binomial distribution and the geometric distribution. These distributions arise when the chance experiment of interest consists of making a sequence of dichotomous observations (two possible values for each observation). The process of making a single such observation is called a *trial*.

For example, one characteristic of blood type is Rh factor, which can be either positive or negative. We can think of a chance experiment that consists of noting the Rh factor for each of 25 blood donors as a sequence of 25 dichotomous trials, where each trial consists of observing the Rh factor (positive or negative) of a single donor.

We could also conduct a different chance experiment that consists of observing the Rh factor of blood donors until a donor who is Rh-negative is encountered. This second experiment can also be viewed as a sequence of dichotomous trials, but the total number of trials in this experiment is not predetermined, as it was in the previous example, where we knew in advance that there would be 25 trials. Experiments of the two types just described are typical of those leading to the binomial and the geometric probability distributions, respectively.

Binomial Distributions

Suppose that we decide to record the gender of each of the next 25 newborn children at a particular hospital. What is the chance that at least 15 are female? What is the chance that between 10 and 15 are female? How many among the 25 can we expect to be female? These and other similar questions can be answered by studying the **binomial probability distribution**.

The binomial distribution arises when the chance experiment of interest is a **binomial experiment**. Binomial experiments have the properties listed in the following box.

Properties of a Binomial Experiment

A binomial experiment consists of a sequence of trials with the following conditions:

1. There are a fixed number of trials.
2. Each trial can result in one of only two possible outcomes, labeled success (S) and failure (F).
3. Outcomes of different trials are independent.
4. The probability that a trial results in a success is the same for each trial.

The **binomial random variable x** is defined as

x = number of successes observed when a binomial experiment is performed

The probability distribution of x is called the **binomial probability distribution.**

The term *success* here does not necessarily have any of its usual connotations. Which of the two possible outcomes is labeled "success" is determined by the random variable of interest. For example, if the variable counts the number of female births among the next 25 births at a particular hospital, then a female birth would be labeled a success (because this is what the variable counts). If male births were counted instead, a male birth would be labeled a success and a female birth a failure.

One situation in which a binominal probability distribution arises was given in Example 7.5. In that example, we considered x = number among four customers who selected an energy efficient refrigerator (rather than a less expensive model). This is a binomial experiment with four trials, where the purchase of an energy efficient refrigerator is considered a success and $P(\text{success}) = P(E) = .4$. The 16 possible outcomes, along with the associated probabilities, were displayed in Table 7.1.

Consider now the case of five customers, a binomial experiment with five trials. The possible values of

$$x = \text{number who purchase an energy efficient refrigerator}$$

are 0, 1, 2, 3, 4, and 5. There are 32 possible outcomes of the binomial experiment, each one a sequence of five successes and failures. Five of these outcomes result in $x = 1$: *SFFFF*, *FSFFF*, *FFSFF*, *FFFSF*, and *FFFFS*.

Because the trials are independent, the first of these outcomes has probability

$$\begin{aligned} P(SFFFF) &= P(S)P(F)P(F)P(F)P(F) \\ &= (.4)(.6)(.6)(.6)(.6) \\ &= (.4)(.6)^4 \\ &= .05184 \end{aligned}$$

The probability calculation will be the same for any outcome with only one success ($x = 1$). It does not matter where in the sequence the single success occurs. It follows that

$$\begin{aligned} p(1) &= P(x = 1) \\ &= P(SFFFF \text{ or } FSFFF \text{ or } FFSFF \text{ or } FFFSF \text{ or } FFFFS) \\ &= .05184 + .05184 + .05184 + .05184 + .05184 \\ &= (5)(.05184) \\ &= .25920 \end{aligned}$$

Similarly, there are 10 outcomes for which $x = 2$, because there are 10 ways to select two from among the five trials to be the *S*'s: *SSFFF*, *SFSFF*, . . . , and *FFFSS*. The probability of each results from multiplying together (.4) two times and (.6) three times. For example,

$$\begin{aligned} P(SSFFF) &= (.4)(.4)(.6)(.6)(.6) \\ &= (.4)^2(.6)^3 \\ &= .03456 \end{aligned}$$

and so

$$\begin{aligned} p(2) &= P(x = 2) \\ &= P(SSFFF) + \cdots + P(FFFSS) \\ &= (10)(.4)^2(.6)^3 \\ &= .34560 \end{aligned}$$

The general form of the formula for calculating the probabilities associated with the different possible values of x is

$$\begin{aligned} p(x) &= P(x\ S\text{'s among the five trials}) \\ &= (\text{number of outcomes with } x\ S\text{'s}) \cdot (\text{probability of any given outcome with } x\ S\text{'s}) \\ &= (\text{number of outcomes with } x\ S\text{'s}) \cdot (.4)^x(.6)^{5-x} \end{aligned}$$

This form was seen previously where $p(2) = 10(.4)^2(.6)^3$.

The letter n is used to denote the number of trials in the binomial experiment. Then the number of outcomes with x *S*'s is the number of ways of selecting x from among the n trials to be the success trials. A simple expression for this quantity is

$$\text{number of outcomes with } x \text{ successes} = \frac{n!}{x!(n-x)!}$$

where, for any positive whole number m, the symbol $m!$ (read "m factorial") is defined by

$$m! = m(m-1)(m-2)\cdots(2)(1)$$

and $0! = 1$.

The Binomial Distribution

Notation:

n = number of independent trials in a binomial experiment

p = constant probability that any particular trial results in a success

Then

$$p(x) = P(x \text{ successes among } n \text{ trials})$$
$$= \frac{n!}{x!(n-x)!}p^x(1-p)^{n-x} \quad x = 0, 1, 2, \ldots, n$$

The expressions $\binom{n}{x}$ or ${}_nC_x$ are sometimes used in place of $\frac{n!}{x!(n-x)!}$. Both are read as "n choose x" and represent the number of ways of choosing x items from a set of n. Using this notation, the binomial probability function can also be written as

$$p(x) = \binom{n}{x}p^x(1-p)^{n-x} \quad x = 0, 1, 2, \ldots, n$$

or

$$p(x) = {}_nC_x p^x(1-p)^{n-x} \quad x = 0, 1, 2, \ldots, n$$

Notice that here the binomial probability distribution is specified using a formula that allows calculation of the various probabilities rather than by giving a table or a probability histogram.

EXAMPLE 7.18 Recognizing Your Roommate's Scent

Understand the context

An interesting experiment was described in the paper **"Sociochemosensory and Emotional Functions" (*Psychological Science* [2009]: 1118–1123)**. The authors of this paper wondered if college students could recognize their roommate by scent. They carried out an experiment in which female college students used fragrance-free soap, deodorant, shampoo, and laundry detergent for a period of time. Their bedding was also laundered using a fragrance-free detergent. Each person was then given a new t-shirt that she slept in for one night. The shirt was then collected and sealed in an airtight bag. Later, the roommate was presented with three identical t-shirts (one worn by her roommate and two worn by other women) and asked to pick the one that smelled most like her roommate. (Yes, hard to believe, but people really do research like this!)

This process was repeated a second time, with the shirts refolded and rearranged before the second trial. The researchers recorded how many times (0, 1, or 2) that the shirt worn by the roommate was correctly identified.

This can be viewed as a binomial experiment consisting of $n = 2$ trials. Each trial results in either a correct identification or an incorrect identification. Because the researchers counted the number of correct identifications, a correct identification is considered a success. We can then define

x = number of correct identifications

Formulate a plan ⟩

Suppose that a participant is not able to identify her roommate by smell. If this is the case, she is essentially just picking one of the three shirts at random and so the probability of success (picking the correct shirt) is 1/3. And, if a participant can't identify her roommate by smell, it is also reasonable to regard the two trials as independent. In this case, the experiment satisfies the conditions of a binomial experiment, and x is a binomial random variable with $n = 2$ and $p = 1/3$.

Do the work ⟩

We can use the binomial probability distribution formula to compute the probability associated with each of the possible x values as follows:

$$p(0) = \binom{2}{0}\left(\frac{1}{3}\right)^0\left(\frac{2}{3}\right)^2 = \frac{2!}{0!2!}\left(\frac{1}{3}\right)^0\left(\frac{2}{3}\right)^2 = (1)(1)\left(\frac{2}{3}\right)^2 = .4444$$

$$p(1) = \binom{2}{1}\left(\frac{1}{3}\right)^1\left(\frac{2}{3}\right)^1 = \frac{2!}{1!1!}\left(\frac{1}{3}\right)^1\left(\frac{2}{3}\right)^1 = (2)\left(\frac{1}{3}\right)^1\left(\frac{2}{3}\right)^1 = .4444$$

$$p(2) = \binom{2}{2}\left(\frac{1}{3}\right)^2\left(\frac{2}{3}\right)^0 = \frac{2!}{2!0!}\left(\frac{1}{3}\right)^2\left(\frac{2}{3}\right)^0 = (1)\left(\frac{1}{3}\right)^2(1) = .1111$$

Summarizing in table form gives

x	$p(x)$
0	.4444
1	.4444
2	.1111

Interpret the results ⟩

This means that about 44.4% of the time, a person who is just guessing would pick the correct shirt on neither trial, about 44.4% of the time the correct shirt would be identified on one of the two trials, and about 11.1% of the time the correct shirt would be identified on both trials.

The authors actually performed this experiment with 44 subjects. They reported that 47.7% of the subjects identified the correct shirt on neither trial, 22.7% identified the correct shirt on one trial, and 31.7% identified the correct shirt on both trials. The fact that these observed percentages differed quite a bit from what would have been expected if participants were just guessing (as specified by the binomial probabilities in the table above) was interpreted by the authors as evidence that some women could in fact identify their roommates by smell. ■

EXAMPLE 7.19 Computer Sales

Understand the context ⟩

Sixty percent of all computers sold by a large computer retailer are laptops and 40% are desktop models. The type of computer purchased by each of the next 12 customers will be noted. Define a random variable x as

x = number of computers among these 12 that are laptops

Formulate a plan ⟩

Because x counts the number of laptops, we use S to denote the sale of a laptop. Then x is a binomial random variable with $n = 12$ and $p = P(S) = .60$. The probability distribution of x is given by

$$p(x) = \frac{12!}{x!(12-x)!}(.6)^x(.4)^{12-x} \quad x = 0, 1, 2, \ldots, 12$$

Do the work ⟩

The probability that exactly four computers are laptops is

$$\begin{aligned} p(4) &= P(x = 4) \\ &= \frac{12!}{4!8!}(.6)^4(.4)^8 \\ &= (495)(.6)^4(.4)^8 \\ &= .042 \end{aligned}$$

Interpret the results ⟩

If group after group of 12 purchases is examined, the long-run percentage of those with exactly four laptops will be 4.2%.

The probability that between four and seven (inclusive) are laptops is

$$P(4 \le x \le 7) = P(x = 4 \textit{ or } x = 5 \textit{ or } x = 6 \textit{ or } x = 7)$$

Since these outcomes are mutually exclusive, this is equal to

$$\begin{aligned} P(4 \le x \le 7) &= p(4) + p(5) + p(6) + p(7) \\ &= \frac{12!}{4!8!}(.6)^4(.4)^8 + \cdots + \frac{12!}{7!5!}(.6)^7(.4)^5 \\ &= .042 + .101 + .177 + .227 \\ &= .547 \end{aligned}$$

Notice that

$$\begin{aligned} P(4 < x < 7) &= P(x = 5 \textit{ or } x = 6) \\ &= p(5) + p(6) \\ &= .278 \end{aligned}$$

so the probability depends on whether $<$ or $\le$ appears in the inequality. (This is typical of *discrete* random variables.) ■

The binomial distribution formula can be tedious to use unless n is small. Statistical software and most graphing calculators can compute binomial probabilities. If you don't have access to technology, Appendix Table 9 gives binomial probabilities for selected n in combination with various values of p.

Using Appendix Table 9

To find $p(x)$ for any particular value of x,

1. Locate the part of the table corresponding to your value of n (5, 10, 15, 20, or 25).
2. Move down to the row labeled with your value of x.
3. Go across to the column headed by the specified value of p.

The desired probability is at the intersection of the designated x row and p column. For example, when $n = 20$ and $p = .8$,

$$p(15) = P(x = 15) = (\text{entry at intersection of } n = 15 \text{ row and } p = .8 \text{ column}) = .175$$

Although $p(x)$ is positive for every possible x value, many probabilities are zero to three decimal places, so they appear as .000 in the table. More extensive binomial tables are available.

Sampling Without Replacement

Usually, sampling is carried out without replacement. This means that once an element has been selected for the sample, it is not a candidate for future selection. If sampling is done by selecting an element from the population, observing whether it is a success or a failure, and then returning it to the population before the next selection is made, the variable

$$x = \text{number of successes}$$

observed in the sample would fit all the requirements of a binomial random variable.

When sampling is done without replacement, the trials (individual selections) are not independent. In this case, the number of successes observed in the sample does not have a binomial distribution but rather a different type of distribution called a *hypergeometric distribution*. The probability calculations for this distribution are even more tedious than for the binomial distribution. Fortunately, when the sample size n is much smaller than N, the population size, probabilities calculated using the binomial distribution and the hypergeometric distribution are very close in value. They are so close, in fact, that statisticians often ignore the difference and use the binomial probabilities in place of the hypergeometric probabilities. You can use the following guideline for determining whether the binomial probability distribution is appropriate when sampling without replacement.

Let x denote the number of S's in a sample of size n selected without replacement from a population consisting of N individuals or objects. If $(n/N) \leq 0.05$ (that is, at most 5% of the population is sampled), then the binomial distribution gives a good approximation to the actual (hypergeometric) probability distribution of x.*

EXAMPLE 7.20 Online Security

Understand the context ❯

The 2009 National Cyber Security Alliance (NCSA) Symantec Online Safety Study found that only about 25% of adult Americans change passwords on online accounts at least quarterly as recommended by the NCSA (**"New Study Shows Need for Americans to Focus on Securing Online Accounts and Backing Up Critical Data," PRNewswire, October 29, 2009**). Suppose that exactly 25% of adult Americans change passwords quarterly. Consider a random sample of $n = 20$ adult Americans (much less than 5% of the population). Then

Formulate a plan ❯

$$x = \text{the number in the sample who change passwords quarterly}$$

has (approximately) a binomial distribution with $n = 20$ and $p = .25$. The probability that five of those sampled change passwords quarterly is

Do the work ❯

$$\begin{aligned} P(5) &= P(x = 5) \\ &= \text{entry in } x = 5 \text{ row and } p = .25 \text{ column in Appendix Table 9 for } n = 20 \\ &= .202 \end{aligned}$$

The probability that at least half of those in the sample (that is, 10 or more) change passwords quarterly is

$$\begin{aligned} P(x \geq 10) &= P(x = 10, 11, 12, \ldots, 20) \\ &= p(10) + p(11) + \cdots + p(20) \\ &= .010 + .003 + .001 + \cdots + .000 \\ &= .014 \end{aligned}$$

Interpret the results ❯

If, in fact, $p = .25$, only about 1.4% of all samples of size 20 would result in at least 10 people who change passwords quarterly. Because $P(x \geq 10)$ is so small when $p = .25$, if $x \geq 10$ were actually observed, we would have to wonder whether the reported value of $p = .25$ is correct.

Although it is possible that we would observe $x \geq 10$ when $p = .25$ (this would happen about 1.4% of the time in the long run), it might also be the case that p is actually greater than .25. In Chapter 10, we show how hypothesis-testing methods can be used to decide which of two contradictory claims about a population (such as $p = .25$ or $p > 25$) is more believable.

*In Chapter 8, we will see a different situation where a similar condition is introduced, but where the requirement is that at most 10% of the population is included in the sample. Be careful not to confuse these rules.

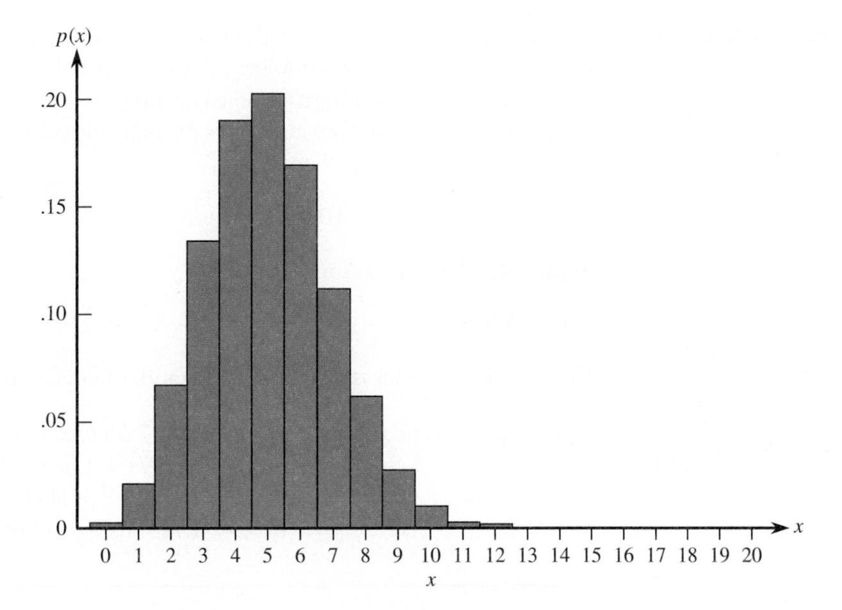

FIGURE 7.16
The binomial probability histogram when $n = 20$ and $p = .25$.

Technology, the binomial formula or tables can be used to compute each of the 21 probabilities $p(0), p(1), \ldots, p(20)$. Figure 7.16 shows the probability histogram for the binomial distribution with $n = 20$ and $p = .25$. Notice that the distribution is skewed to the right. (The binomial distribution is symmetric only when $p = .5$.) ■

Mean and Standard Deviation of a Binomial Random Variable

A binomial random variable x based on n trials has possible values $0, 1, 2, \ldots, n$, so the mean value is

$$\mu_x = \Sigma x p(x) = (0)p(0) + (1)p(1) + \cdots + (n)p(n)$$

and the variance of x is

$$\begin{aligned}\sigma_x^2 &= \Sigma(x - \mu_x)^2 \cdot p(x) \\ &= (0 - \mu_x)^2 p(0) + (1 - \mu_x)^2 p(1) + \cdots + (n - \mu_x)^2 p(n)\end{aligned}$$

These expressions would be very tedious to evaluate for any particular values of n and p. Fortunately, there are simple formulas for the mean and standard deviation of a binomial random variable.

The mean value and the standard deviation of a binomial random variable x are, respectively,

$$\mu_x = np \quad \text{and} \quad \sigma_x = \sqrt{np(1 - p)}$$

EXAMPLE 7.21 Credit Cards Paid in Full

Understand the context ⟩

The paper **"Debt Literacy, Financial Experiences and Overindebtedness" (*Social Science Research Network,* Working Paper w14808, 2008)** reported that 37% of credit card users pay their bills in full each month. This figure represents an average over different types of credit cards issued by many different banks.

Suppose that 40% of all individuals holding Visa cards issued by a certain bank pay in full each month. A random sample of $n = 25$ cardholders is to be selected. The bank is interested in the variable

$$x = \text{number in the sample who pay in full each month}$$

Formulate a plan ⟩ Even though sampling is done without replacement, the sample size $n = 25$ is most likely very small compared to the total number of credit card holders, so we can approximate the probability distribution of x using a binomial distribution with $n = 25$ and $p = .4$. We have defined "paid in full" as a success because this is the outcome counted by the random variable x.

Do the work ⟩ The mean value of x is then

$$\mu_x = np = 25(.40) = 10.0$$

and the standard deviation is

$$\sigma_x = \sqrt{np(1-p)} = \sqrt{25(.40)(.60)} = \sqrt{6} = 2.45$$

The probability that x is farther than 1 standard deviation from its mean value is

$$\begin{aligned} P(x < \mu_x - \sigma_x \text{ or } x > \mu_x + \sigma_x) &= P(x < 7.55 \text{ or } x > 12.45) \\ &= P(x \le 7) + P(x \ge 13) \\ &= p(0) + \cdots + p(7) + p(13) + \cdots + p(25) \\ &= .307 \quad \text{(using technology or Appendix Table 9)} \end{aligned}$$

■

The value of σ_x is 0 when $p = 0$ or $p = 1$. In these two cases, there is no uncertainty in x. We are sure to observe $x = 0$ when $p = 0$ and $x = n$ when $p = 1$. It is also easily verified that $p(1 - p)$ is largest when $p = .5$. This means that the binomial distribution spreads out the most when sampling from a 50–50 population. The farther p is from .5, the less spread out and the more skewed the binomial distribution.

Geometric Distributions

A binomial random variable is defined as the number of successes in n independent trials, where each trial can result in either a success or a failure and the probability of success is the same for each trial. Suppose, however, that we are not interested in the number of successes in a fixed number of trials but rather in the number of trials that must be carried out before a success occurs. Two examples are counting the number of boxes of cereal that must be purchased before finding one with a rare toy and counting the number of games that a professional bowler must play before achieving a score over 250.

The variable

x = number of trials to first success

is called a **geometric random variable,** and the probability distribution that describes its behavior is called a **geometric probability distribution.**

Suppose an experiment consists of a sequence of trials with the following conditions:

1. The trials are independent.
2. Each trial can result in one of two possible outcomes, success and failure.
3. The probability of success is the same for all trials.

A **geometric random variable** is defined as

x = number of trials until the first success is observed (including the success trial)

The probability distribution of x is called the **geometric probability distribution**.

For example, suppose that 40% of the students who drive to campus at your university carry jumper cables. Your car has a dead battery and you don't have jumper cables, so you decide to stop students who are headed to the parking lot and ask them whether they have a pair of jumper cables. You might be interested in the number of students you would have to stop before finding one who has jumper cables.

If we define success as a student with jumper cables, a trial would consist of asking an individual student for help. The random variable

x = number of students who must be stopped before finding one with jumper cables

is an example of a geometric random variable, because it can be viewed as the number of trials to the first success in a sequence of independent trials.

The probability distribution of a geometric random variable is easy to construct. We use p to denote the probability of success on any given trial. Possible outcomes can be denoted as follows:

Outcome	x = Number of Trials to First Success
S	1
FS	2
FFS	3
⋮	⋮
$FFFFFFS$	7
⋮	⋮

Each possible outcome consists of 0 or more failures followed by a single success. So,

$$\begin{aligned} p(x) &= P(x \text{ trials to first success}) \\ &= P(FF\ldots FS) \end{aligned}$$

$x - 1$ failures followed by a success on trial x

Because the probability of success is p for each trial, the probability of failure for each trial is $1 - p$. Because the trials are independent,

$$\begin{aligned} p(x) &= P(x \text{ trials to first success}) = P(FF\ldots FS) \\ &= P(F)P(F)\cdots P(F)P(S) \\ &= (1-p)(1-p)\cdots(1-p)p \\ &= (1-p)^{x-1}p \end{aligned}$$

This leads us to the formula for the geometric probability distribution.

Geometric Probability Distribution

If x is a geometric random variable with probability of success $= p$ for each trial, then

$$p(x) = (1-p)^{x-1}p \quad x = 1, 2, 3, \ldots$$

Greg Ceo/Stone/Getty Images

EXAMPLE 7.22 Jumper Cables

Consider the jumper cable problem described previously. For this problem, $p = .4$, because 40% of the students who drive to campus carry jumper cables. The probability distribution of

x = number of students who must be stopped before finding a student with jumper cables

is

$$p(x) = (.6)^{x-1}(.4) \qquad x = 1, 2, 3, \ldots$$

The probability distribution can now be used to compute various probabilities. For example, the probability that the first student stopped has jumper cables (that is, $x = 1$) is

$$p(1) = (.6)^{1-1}(.4) = (.6)^0(.4) = .4$$

The probability that three or fewer students must be stopped is

$$\begin{aligned} P(x \leq 3) &= p(1) + p(2) + p(3) \\ &= (.6)^0(.4) + (.6)^1(.4) + (.6)^2(.4) \\ &= .4 + .24 + .144 \\ &= .784 \end{aligned}$$

■

EXERCISES 7.43 - 7.62

7.43 NBC News reported that 1 in 20 children in the U.S. has a food allergy (May 2, 2013). Consider selecting 10 children at random. Define the random variable x as

x = number of children in the sample of 10 that have a food allergy

Find the following probabilities. (Hint: See Examples 7.19 and 7.20.)

a. $p(x < 3)$

b. $p(x \leq 3)$

c. $p(x \geq 4)$

d. $p(1 \leq x \leq 3)$

7.44 The article **"Should You Report That Fender-Bender?" (*Consumer Reports*, 2013:15)** reported that 7 in 10 auto accidents involve a single vehicle. Suppose 15 accidents are randomly selected. (Hint: See Examples 7.19 and 7.20.)

a. What is the probability that exactly four involve a single vehicle?

b. What is the probability that at most four involve a single vehicle?

c. What is the probability that exactly six involve multiple vehicles?

7.45 ▼ The ***Los Angeles Times* (December 13, 1992)** reported that what airline passengers like to do most on long flights is rest or sleep. In a survey of 3697 passengers, almost 80% did so. Suppose that for a particular route the actual percentage is exactly 80%, and consider randomly selecting six passengers. Then x, the number among the selected six who rested or slept, is a binomial random variable with $n = 6$ and $p = .8$.

a. Calculate $p(4)$, and interpret this probability.

b. Calculate $p(6)$, the probability that all six selected passengers rested or slept.

c. Determine $P(x \geq 4)$.

7.46 Refer to the previous exercise, and suppose that 10 rather than six passengers are selected ($n = 10$, $p = .8$). (Hint: Use technology or Appendix Table 9.)

a. What is $p(8)$?

b. Calculate $P(x \leq 7)$.

c. Calculate the probability that more than half of the selected passengers rested or slept.

7.47 Twenty-five percent of the customers of a grocery store use an express checkout. Consider five randomly selected customers, and let x denote the number among the five who use the express checkout.

a. What is $p(2)$, that is, $P(x = 2)$?

b. What is $P(x \leq 1)$?

c. What is $P(2 \leq x)$? (Hint: Make use of your answer from Part (b).)

d. What is $P(x \neq 2)$?

7.48 Example 7.18 described a study in which a person was asked to determine which of three t-shirts had been worn by her roommate by smelling the shirts **("Sociochemosensory and Emotional Functions," *Psychological Science* [2009]: 1118–1123)**. Suppose that instead of three shirts, each participant was asked to choose among four shirts and that the process was repeated five times. Then, assuming that the participant is choosing at random, x = number of correct identifications is a binomial random variable with $n = 5$ and $p = 1/4$.

a. What are the possible values of x?

b. For each possible value of x, find the associated probability $p(x)$ and display the possible x values and $p(x)$ values in a table.

c. Construct a probability histogram for the probability distribution of x.

7.49 In a press release dated **October 2, 2008, The National Cyber Security Alliance** reported that approximately 80% of adult Americans who own a computer claim

Bold exercises answered in back ● Data set available online ▼ Video Solution available

to have a firewall installed on their computer to prevent hackers from stealing personal information. This estimate was based on a survey of 3000 people. It was also reported that in a study of 400 computers, only about 40% actually had a firewall installed.

a. Suppose that the true proportion of computer owners who have a firewall installed is .80. If 20 computer owners are selected at random, what is the probability that more than 15 have a firewall installed?

b. Suppose that the true proportion of computer owners who have a firewall installed is .40. If 20 computer owners are selected at random, what is the probability that more than 15 have a firewall installed?

c. Suppose that a random sample of 20 computer owners is selected and that 14 have a firewall installed. Is it more likely that the true proportion of computer owners who have a firewall installed is .40 or .80? Justify your answer based on probability calculations.

7.50 A breeder of show dogs is interested in the number of female puppies in a litter. If a birth is equally likely to result in a male or a female puppy, give the probability distribution of the variable $x =$ number of female puppies in a litter of size 5.

7.51 The article **"FBI Says Fewer than 25 Failed Polygraph Test"** (***San Luis Obispo Tribune*, July 29, 2001**) states that false-positives in polygraph tests (tests in which an individual fails even though he or she is telling the truth) are relatively common and occur about 15% of the time. Suppose that such a test is given to 10 trustworthy individuals.

a. What is the probability that all 10 pass?

b. What is the probability that more than two fail, even though all are trustworthy?

c. The article indicated that 500 FBI agents were required to take a polygraph test. Consider the random variable $x =$ number of the 500 tested who fail. If all 500 agents tested are trustworthy, what are the mean and standard deviation of x?

d. The headline indicates that fewer than 25 of the 500 agents tested failed the test. Is this a surprising result if all 500 are trustworthy? Answer based on the values of the mean and standard deviation from Part (c).

7.52 Industrial quality control programs often include inspection of incoming materials from suppliers. If parts are purchased in large lots, a typical plan might be to select 20 parts at random from a lot and inspect them. A lot might be judged acceptable if one or fewer defective parts are found among those inspected. Otherwise, the lot is rejected and returned to the supplier. Use technology or Appendix Table 9 to find the probability of accepting lots that have each of the following (Hint: Identify success with a defective part):

a. 5% defective parts

b. 10% defective parts

c. 20% defective parts

7.53 Suppose that the probability is .1 that any given citrus tree will show measurable damage when the temperature falls to 30°F. (Hint: See Example 7.21.)

a. If the temperature does drop to 30°F, what is the expected number of citrus trees showing damage in orchards of 2000 trees?

b. What is the standard deviation of the number of trees that show damage?

7.54 Thirty percent of all automobiles undergoing an emissions inspection at a certain inspection station fail the inspection.

a. Among 15 randomly selected cars, what is the probability that at most five fail the inspection?

b. Among 15 randomly selected cars, what is the probability that between five and 10 (inclusive) fail to pass inspection?

c. Among 25 randomly selected cars, what is the mean value of the number that pass inspection, and what is the standard deviation of the number that pass inspection?

d. What is the probability that among 25 randomly selected cars, the number that pass is within 1 standard deviation of the mean value? (Hint: See Example 7.21.)

7.55 You are to take a multiple-choice exam consisting of 100 questions with five possible responses to each question. Suppose that you have not studied and so must guess (select one of the five answers in a completely random fashion) on each question. Let x represent the number of correct responses on the test.

a. What kind of probability distribution does x have?

b. What is your expected score on the exam? (Hint: Your expected score is the mean value of the x distribution.)

c. Compute the variance and standard deviation of x.

d. Based on your answers to Parts (b) and (c), is it likely that you would score over 50 on this exam? Explain the reasoning behind your answer.

7.56 Suppose that 20% of the 10,000 signatures on a certain recall petition are invalid. Would the number of invalid signatures in a sample of 2000 of these signatures have (approximately) a binomial distribution? Explain.

7.57 A city ordinance requires that a smoke detector be installed in all residential housing. There is concern that too many residences are still without detectors, so a costly inspection program is being contemplated. Let p be the proportion of all residences that have a detector. A random sample of 25 residences is selected. If the sample strongly suggests that $p < .80$ (less than 80% have detectors), as opposed to $p \geq .80$, the program will be implemented. Let x be the number of residences among the 25 that have a detector, and consider the following decision rule: Reject the claim that $p \geq .8$ and implement the program if $x \leq 15$.

a. What is the probability that the program is implemented when $p = .80$?

b. What is the probability that the program is not implemented if $p = .70$?

c. What is the probability that the program is not implemented if $p = .60$?

d. How do the "error probabilities" of Parts (b) and (c) change if the value 15 in the decision rule is changed to 14?

7.58 Suppose that 90% of all registered California voters favor banning the release of information from exit polls in presidential elections until after the polls in California close. A random sample of 25 registered California voters is to be selected.

a. What is the probability that more than 20 favor the ban?

b. What is the probability that at least 20 favor the ban?

c. What are the mean value and standard deviation of the number of voters in the sample who favor the ban?

d. If fewer than 20 voters in the sample favor the ban, is this at odds with the assertion that (at least) 90% of California registered voters favors the ban? (Hint: Consider $P(x < 20)$ when $p = .9$.)

7.59 Suppose a playlist on an MP3 music player consists of 100 songs, of which eight are by a particular artist. Suppose that songs are played by selecting a song at random (with replacement) from the playlist. The random variable x represents the number of songs played until a song by this artist is played.

a. Explain why the probability distribution of x is not binomial.

b. Find the following probabilities. (Hint: See Example 7.22.)

i. $p(4)$

ii. $P(x \leq 4)$

iii. $P(x > 4)$

iv. $P(x \geq 4)$

c. Interpret each of the probabilities in Part (b) and explain the difference between them.

7.60 Sophie is a dog that loves to play catch. Unfortunately, she isn't very good, and the probability that she catches a ball is only .1. Let x be the number of tosses required until Sophie catches a ball.

a. Does x have a binomial or a geometric distribution?

b. What is the probability that it will take exactly two tosses for Sophie to catch a ball?

c. What is the probability that more than three tosses will be required?

7.61 Suppose that 5% of cereal boxes contain a prize and the other 95% contain the message, "Sorry, try again." Consider the random variable x, where x = number of boxes purchased until a prize is found.

a. What is the probability that at most two boxes must be purchased?

b. What is the probability that exactly four boxes must be purchased?

c. What is the probability that more than four boxes must be purchased?

7.62 ▼ The article on polygraph testing of FBI agents referenced in Exercise 7.51 indicated that the probability of a false-positive (a trustworthy person who nonetheless fails the test) is .15. Let x be the number of trustworthy FBI agents tested until someone fails the test.

a. Describe the probability distribution of x?

b. What is the probability that the first false-positive will occur when the third person is tested?

c. What is the probability that fewer than four are tested before the first false-positive occurs?

d. What is the probability that more than three agents are tested before the first false-positive occurs?

7.6 Normal Distributions

Normal distributions formalize the notion of mound-shaped histograms introduced in Chapter 4. Normal distributions are widely used for two reasons. First, they provide a reasonable approximation to the distribution of many different variables. They also play a central role in many of the inferential procedures that will be discussed in later chapters.

Normal distributions are continuous probability distributions that are bell-shaped and symmetric, as shown in Figure 7.17. Normal distributions are sometimes referred to as *normal curves*.

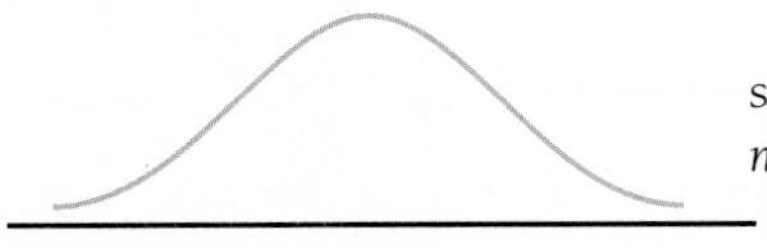

FIGURE 7.17
A normal distribution.

There are many different normal distributions, and they are distinguished from one another by their mean μ and standard deviation σ. The mean μ of a normal distribution describes where the corresponding curve is centered. The standard deviation σ describes how much the curve spreads out around that center. As with all continuous probability distributions, the total area under any normal curve is equal to 1.

Three normal distributions are shown in Figure 7.18. Notice that the smaller the standard deviation, the taller and narrower the corresponding curve. Remember that areas under a continuous probability distribution curve represent probabilities, so when the standard deviation is small, a larger area is concentrated near the center of the curve. This means that the chance of observing a value near the mean is much greater (because μ is at the center).

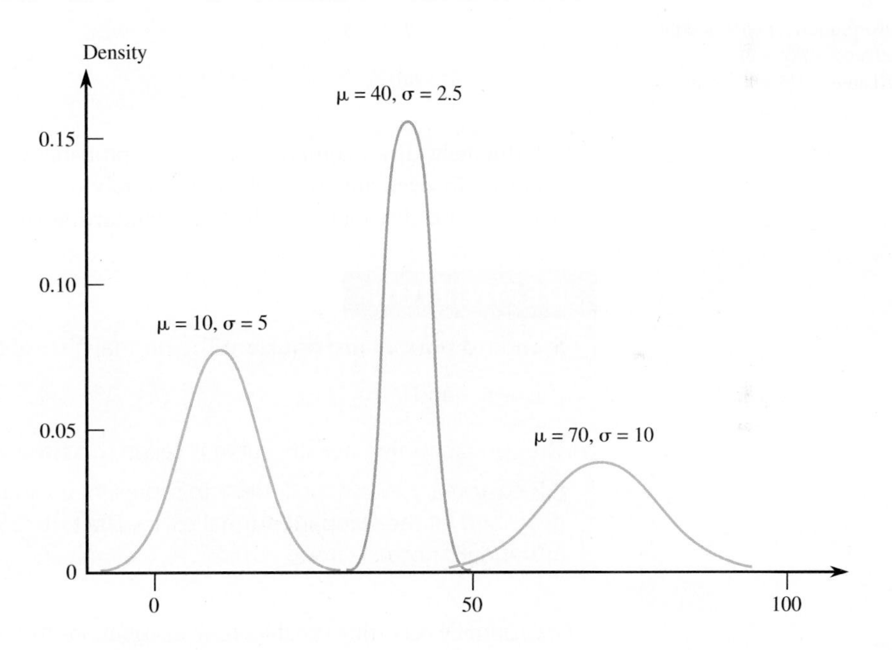

FIGURE 7.18
Three normal distributions.

The value of μ is the number on the measurement axis lying directly below the top of the curve. The value of σ can be approximated from a picture of the curve. Consider the normal curve in Figure 7.19. Starting at the top (above $\mu = 100$) and moving to the right, the curve turns downward until it is above the value 110. After that point, it continues to decrease in height but is turning upward rather than downward. Similarly, to the left of $\mu = 100$, the curve turns downward until it reaches 90 and then begins to turn upward. The curve changes from turning downward to turning upward at a distance of 10 on either side of μ. In general, σ is the distance to either side of μ at which a normal curve changes from turning downward to turning upward, so $\sigma = 10$ for the normal curve in Figure 7.19.

If a particular normal distribution is to be used to describe the behavior of a random variable, a mean and a standard deviation must be specified. For example, a normal distribution with mean 7 and standard deviation 1 might be used as a model for the distribution of $x =$ birth weight (in pounds). If this model is a reasonable description of the probability distribution, we could use areas under the normal curve with $\mu = 7$ and $\sigma = 1$ to approximate various probabilities related to birth weight.

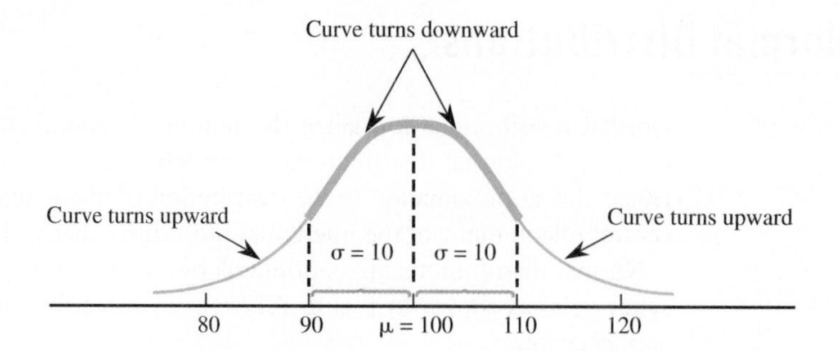

FIGURE 7.19
Mean μ and standard deviation σ for a normal curve.

The probability that a birth weight is over 8 pounds (expressed symbolically as $P(x > 8)$) corresponds to the shaded area in Figure 7.20(a). The shaded area in Figure 7.20(b) represents the probability of a birth weight between 6.5 and 8 pounds, $P(6.5 < x < 8)$.

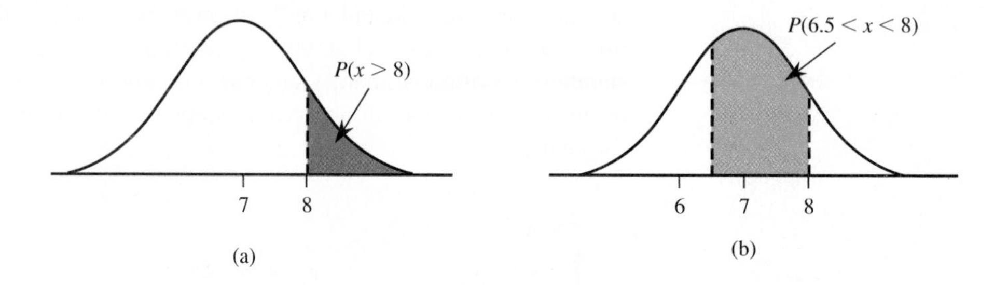

FIGURE 7.20
Normal distribution for birth weight: (a) shaded area = $P(x > 8)$; (b) shaded area = $P(6.5 < x < 8)$.

Unfortunately, direct computation of such probabilities (areas under a normal curve) is not simple. To overcome this difficulty, we rely on technology or a table of areas for a reference normal distribution, called the **standard normal distribution**.

DEFINITION

Standard normal distribution: The normal distribution with

$\mu = 0$ and $\sigma = 1$

The corresponding density curve is called the **standard normal curve.**

It is customary to use the letter z to represent a variable whose distribution is described by the standard normal curve. The term ***z* curve** is often used in place of standard normal curve.

Few naturally occurring variables have distributions that are well described by the standard normal distribution. However, this distribution is important because it is also used in probability calculations for other normal distributions. When we are interested in finding a probability based on some other normal curve, we either rely on technology or we first translate our problem into an equivalent problem that involves finding an area under the standard normal curve. A table for the standard normal distribution is then used to find the desired area. To be able to do this, we must first learn to work with the standard normal distribution.

The Standard Normal Distribution

In working with normal distributions, we need two general skills:

1. We must be able to use the normal distribution to compute probabilities, which are areas under a normal curve and above given intervals.
2. We must be able to characterize extreme values in the distribution, such as the largest 5%, the smallest 1%, and the most extreme 5% (which would include the largest 2.5% and the smallest 2.5%).

Let's begin by looking at how to accomplish these tasks when the distribution of interest is the standard normal distribution.

The standard normal or z curve is shown in Figure 7.21(a). It is centered at $\mu = 0$, and the standard deviation, $\sigma = 1$, is a measure of the extent to which it spreads out about its mean.

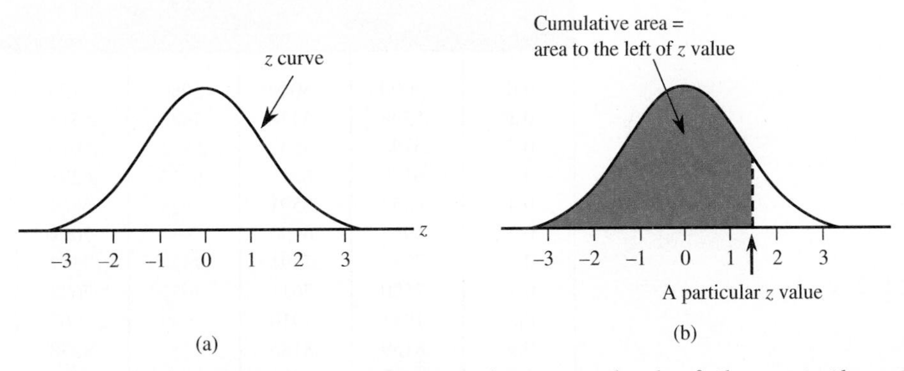

FIGURE 7.21
(a) A standard normal (z) curve; (b) a cumulative area.

Notice that this picture is consistent with the Empirical Rule of Chapter 4. About 95% of the area (probability) is associated with values that are within 2 standard deviations of the mean (between -2 and 2), and almost all of the area is associated with values that are within 3 standard deviations of the mean (between -3 and 3).

In the examples that follow, Appendix Table 2 is used. If you have access to technology (a statistics program or a graphing calculator), you can use technology to work these examples.

Appendix Table 2 tabulates cumulative z curve areas of the sort shown in Figure 7.21(b) for many different values of z. The smallest value for which the cumulative area is given is -3.89, a value far out in the lower tail of the z curve. The next smallest value for which the area appears is -3.88, then -3.87, then -3.86, and so on in increments of 0.01, terminating with the cumulative area to the left of 3.89.

Using the Table of Standard Normal Curve Areas

For any number z^* between -3.89 and 3.89 and rounded to two decimal places, Appendix Table 2 gives

$$(\text{area under } z \text{ curve to the left of } z^*) = P(z < z^*) = P(z \le z^*)$$

where the letter z is used to represent a random variable whose distribution is the standard normal distribution.

To find this probability using the table, locate the following:

1. The row labeled with the sign of z^* and the digit to either side of the decimal point (for example, -1.7 or 0.5)
2. The column identified with the second digit to the right of the decimal point in z^* (for example, .06 if $z^* = -1.76$)

The number at the intersection of this row and column is the desired probability, $P(z < z^*)$.

A portion of the table of standard normal curve areas appears in Figure 7.22. To find the area under the z curve to the left of 1.42, look in the row labeled 1.4 and the column labeled .02 (the highlighted row and column in Figure 7.22). From the table, the corresponding cumulative area is .9222. So

$$z \text{ curve area to the left of } 1.42 = .9222$$

We can also use the table to find the area to the right of a particular value. Because the total area under the z curve is 1, it follows that

$$(z \text{ curve area to the right of } 1.42) = 1 - (z \text{ curve area to the left of } 1.42)$$
$$= 1 - .9222$$
$$= .0778$$

z^*	.00	.01	.02	.03	.04	.05
0.0	.5000	.5040	.5080	.5120	.5160	.5199
0.1	.5398	.5438	.5478	.5517	.5557	.5596
0.2	.5793	.5832	.5871	.5910	.5948	.5987
0.3	.6179	.6217	.6255	.6293	.6331	.6368
0.4	.6554	.6591	.6628	.6664	.6700	.6736
0.5	.6915	.6950	.6985	.7019	.7054	.7088
0.6	.7257	.7291	.7324	.7357	.7389	.7422
0.7	.7580	.7611	.7642	.7673	.7704	.7734
0.8	.7881	.7910	.7939	.7967	.7995	.8023
0.9	.8159	.8186	.8212	.8238	.8264	.8289
1.0	.8413	.8438	.8461	.8485	.8508	.8531
1.1	.8643	.8665	.8686	.8708	.8729	.8749
1.2	.8849	.8869	.8888	.8907	.8925	.8944
1.3	.9032	.9049	.9066	.9082	.9099	.9115
1.4	.9192	.9207	.9222	.9236	.9251	.9265
1.5	.9332	.9345	.9357	.9370	.9382	.9394
1.6	.9452	.9463	.9474	.9484	.9495	.9505
1.7	.9554	.9564	.9573	.9582	.9591	.9599
1.8	.9641	.9649	.9656	.9664	.9671	.9678

$P(z < 1.42)$

FIGURE 7.22 Portion of the table of standard normal curve areas.

These probabilities can be interpreted to mean that in a long sequence of observations, approximately 92.22% of the observed z values will be smaller than 1.42, and 7.78% will be larger than 1.42.

EXAMPLE 7.23 Finding Standard Normal Curve Areas

The probability $P(z < -1.76)$ is found at the intersection of the -1.7 row and the .06 column of the z table. The result is

$$P(z < -1.76) = .0392$$

as shown in the following figure:

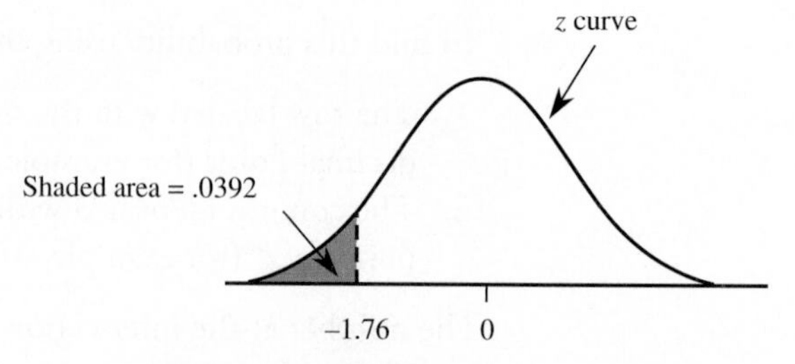

In other words, in a long sequence of observations, approximately 3.9% of the observed z values will be smaller than -1.76. Similarly,

$$P(z \leq 0.58) = \text{entry in 0.5 row and .08 column of Table 2} = .7190$$

as shown in the following figure:

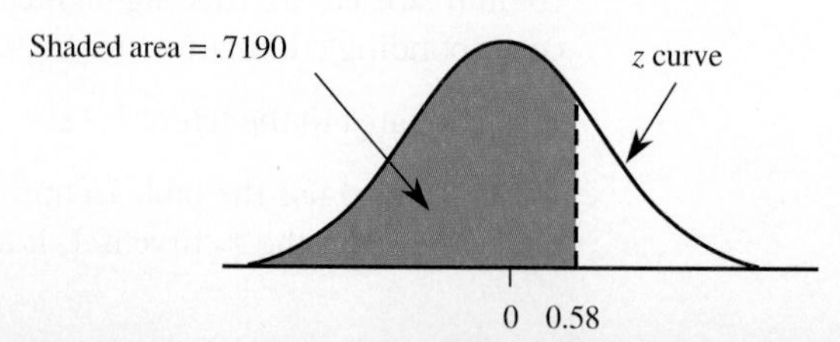

Now consider $P(z < -4.12)$. This probability does not appear in Appendix Table 2; there is no -4.1 row. However, it must be less than $P(z < -3.89)$, the smallest z value in the table, because -4.12 is farther out in the lower tail of the z curve. Since $P(z < -3.89) \approx .0000$ (that is, zero to four decimal places), it follows that

$$P(z < -4.12) \approx 0$$

Similarly,

$$P(z < 4.18) > P(z < 3.89) \approx 1.0000$$

from which we conclude that

$$P(z < 4.18) \approx 1$$

■

As illustrated in Example 7.23, we can use the cumulative areas tabulated in Appendix Table 2 to calculate other probabilities involving z. The probability that z is larger than a value c is

$$P(z > c) = \text{area under the } z \text{ curve to the right of } c = 1 - P(z \leq c)$$

In other words, the area to the right of a value (a right-tail area) is 1 minus the corresponding cumulative area. This is illustrated in Figure 7.23.

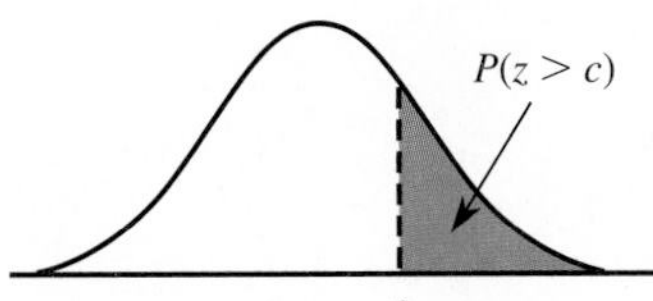

=

1

−

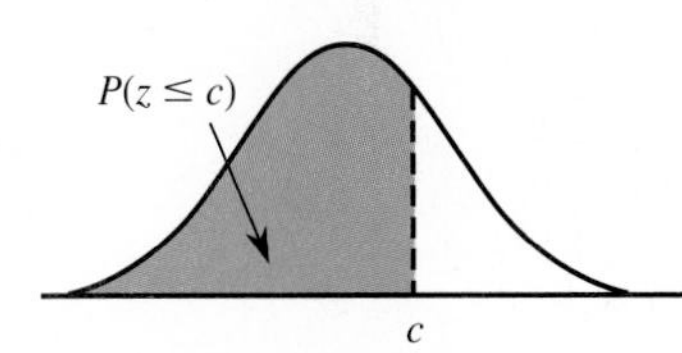

FIGURE 7.23
The relationship between an upper-tail area and a cumulative area.

Similarly, the probability that z falls in the interval between a lower limit a and an upper limit b is

$$\begin{aligned} P(a < z < b) &= \text{area under the } z \text{ curve and above the interval from } a \text{ to } b \\ &= P(z < b) - P(z < a) \end{aligned}$$

That is, $P(a < z < b)$ is the difference between two cumulative areas, as illustrated in Figure 7.24.

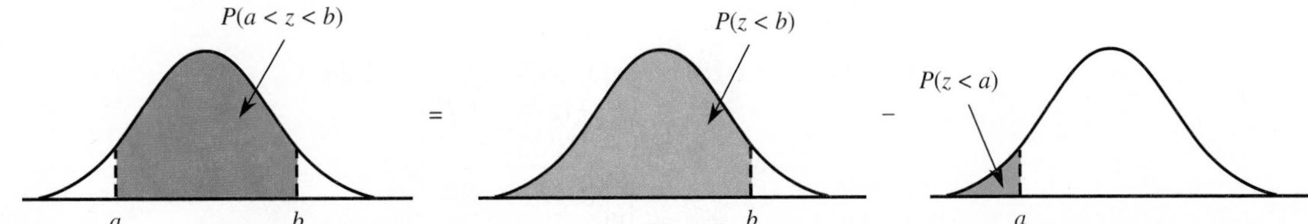

FIGURE 7.24
$P(a < z < b)$ as the difference between the two cumulative areas.

EXAMPLE 7.24 More About Standard Normal Curve Areas

The probability that z is between -1.76 and 0.58 is

$$\begin{aligned} P(-1.76 < z < 0.58) &= P(z < 0.58) - P(z < -1.76) \\ &= .7190 - .0392 \\ &= .6798 \end{aligned}$$

as shown in the following figure:

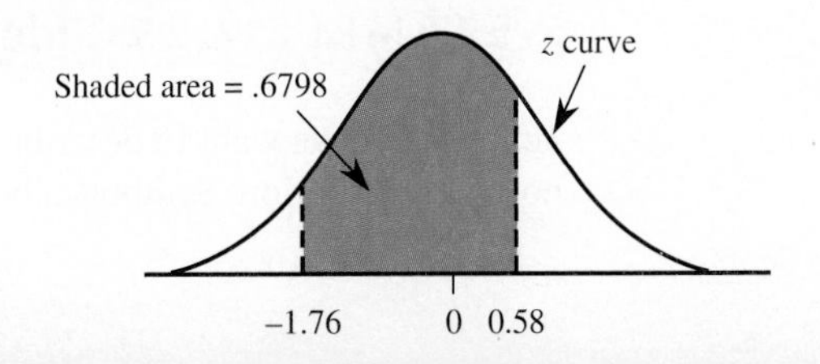

The probability that z is between -2 and $+2$ (within 2 standard deviations of its mean, since $\mu = 0$ and $\sigma = 1$) is

$$\begin{aligned} P(-2.00 < z < 2.00) &= P(z < 2.00) - P(z < -2.00) \\ &= .9772 - .0228 \\ &= .9544 \\ &\approx .95 \end{aligned}$$

as shown in the following figure:

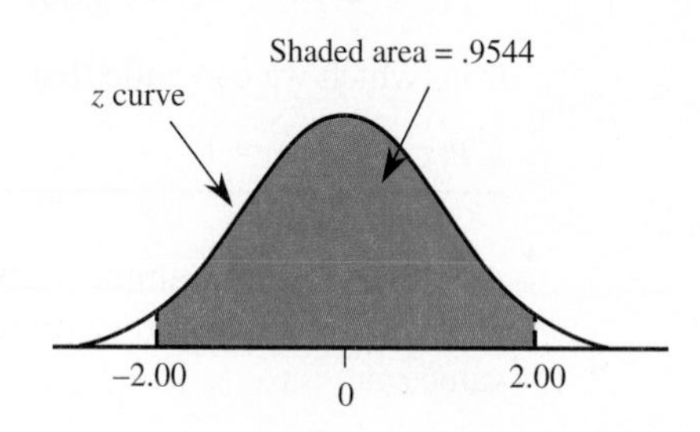

This last probability is the basis for one part of the Empirical Rule, which states that when a histogram is well approximated by a normal curve, approximately 95% of the values are within 2 standard deviations of the mean.

The probability that the value of z exceeds 1.96 is

$$\begin{aligned} P(z > 1.96) &= 1 - P(z < 1.96) \\ &= 1 - .9750 \\ &= .0250 \end{aligned}$$

as shown in the following figure:

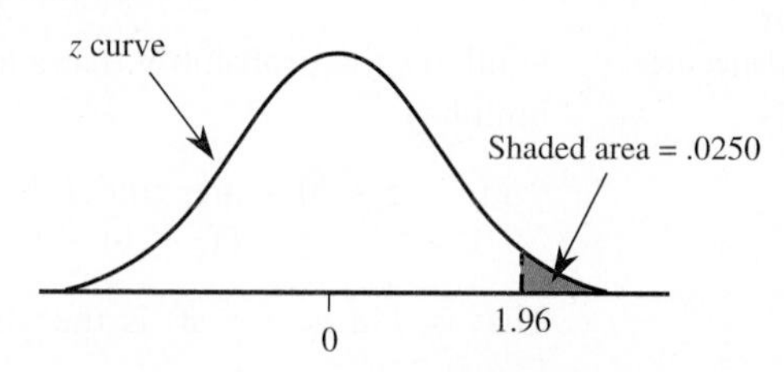

That is, 2.5% of the area under the z curve lies to the right of 1.96 in the upper tail.

Similarly,

$$\begin{aligned} P(z > -1.28) &= \text{area to the right of } -1.28 \\ &= 1 - P(z < -1.28) \\ &= 1 - .1003 \\ &= .8997 \\ &\approx .90 \end{aligned}$$

■

Identifying Extreme Values

Suppose that we want to describe the values included in the smallest 2% of a distribution or the values making up the most extreme 5% (which includes the largest 2.5% and the smallest 2.5%). Let's see how we can identify extreme values in the distribution by working through Examples 7.25 and 7.26.

EXAMPLE 7.25 Identifying Extreme Values

Suppose that we want to describe the values that make up the smallest 2% of the standard normal distribution. Symbolically, we are trying to find a value (call it z^*) such that

$$P(z < z^*) = .02$$

This is illustrated in Figure 7.25, which shows that the cumulative area for z^* is .02. Therefore, we look for a cumulative area of .0200 in the body of Appendix Table 2. The closest cumulative area in the table is .0202, in the -2.0 row and .05 column. We will use $z^* = -2.05$, the best approximation from the table. Variable values less than -2.05 make up the smallest 2% of the standard normal distribution.

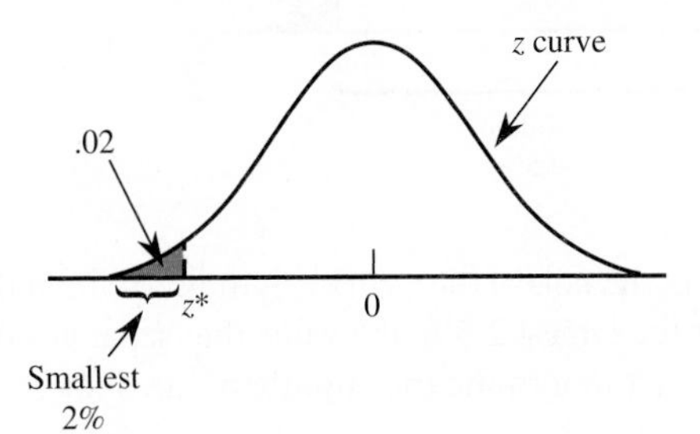

FIGURE 7.25
The smallest 2% of the standard normal distribution.

Now suppose that we had been interested in the largest 5% of all z values. We would then be trying to find a value of z^* for which

$$P(z > z^*) = .05$$

as illustrated in Figure 7.26. Because Appendix Table 2 always works with cumulative area (area to the left), the first step is to determine

$$\text{area to the left of } z^* = 1 - .05 = .95$$

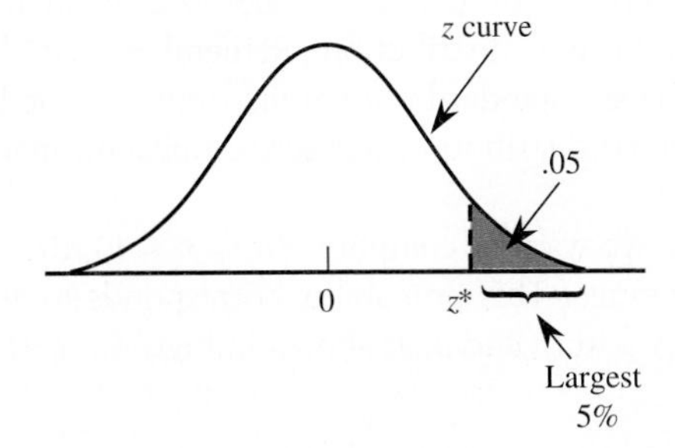

FIGURE 7.26
The largest 5% of the standard normal distribution.

Looking for the cumulative area closest to .95 in Appendix Table 2, we find that .95 falls exactly halfway between .9495 (corresponding to a z value of 1.64) and .9505 (corresponding to a z value of 1.65). Because .9500 is exactly halfway between the two areas, we use a z value that is halfway between 1.64 and 1.65. (If one value had been closer to .9500 than the other, we would just use the z value corresponding to the closest area.) This gives

$$z^* = \frac{1.64 + 1.65}{2} = 1.645$$

Values greater than 1.645 make up the largest 5% of the standard normal distribution. By symmetry, -1.645 separates the smallest 5% of all z values from the others. ■

EXAMPLE 7.26 More Extremes

Sometimes we are interested in identifying the most extreme (unusually large *or* small) values in a distribution. Consider describing the values that make up the most extreme 5% of the standard normal distribution. That is, we want to separate the middle 95% from the extreme 5%. This is illustrated in Figure 7.27.

Because the standard normal distribution is symmetric, the most extreme 5% is equally divided between the high side and the low side of the distribution, resulting in an area of

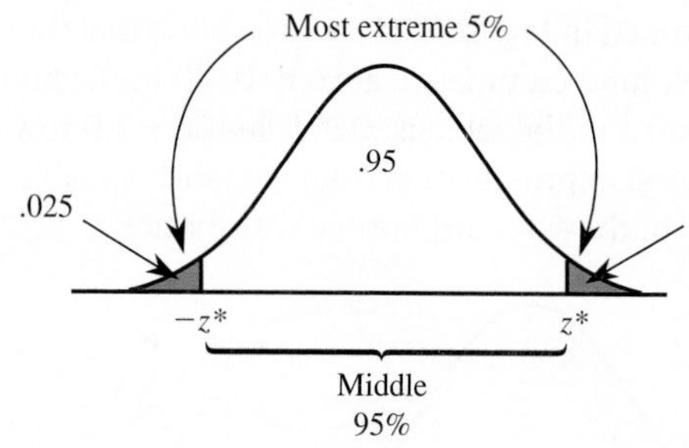

Most extreme 5%

.95

.025 .025

$-z^*$ z^*

Middle 95%

FIGURE 7.27
The most extreme 5% of the standard normal distribution.

.025 for each of the tails of the z curve. Symmetry about 0 implies that if z^* denotes the value that separates the largest 2.5%, the value that separates the smallest 2.5% is simply $-z^*$.

To find z^*, first determine the cumulative area for z^*, which is

$$\text{area to the left of } z^* = .95 + .025 = .975$$

The cumulative area .9750 appears in the 1.9 row and .06 column of Appendix Table 2, so $z^* = 1.96$. For the standard normal distribution, 95% of the variable values fall between -1.96 and 1.96; the most extreme 5% are those values that are either greater than 1.96 or less than -1.96. ■

Other Normal Distributions

We now show how z curve areas can be used to calculate probabilities and to describe values for any normal distribution. Remember that the letter z is reserved for those variables that have a standard normal distribution. The letter x is used more generally for any variable whose distribution is described by a normal curve with mean μ and standard deviation σ.

Suppose that we want to compute $P(a < x < b)$, the probability that the variable x lies in a particular range. This probability corresponds to an area under a normal curve and above the interval from a to b, as shown in Figure 7.28(a).

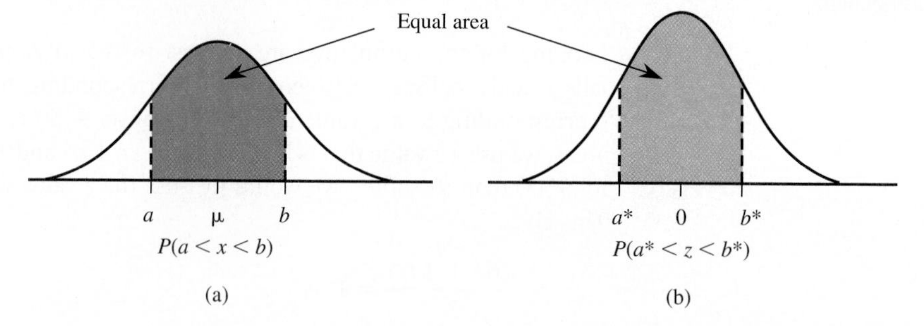

FIGURE 7.28
Equality of nonstandard and standard normal curve areas.

The strategy for obtaining this probability is to find an equivalent problem involving the standard normal distribution. Finding an equivalent problem means determining an interval (a^*, b^*) that has the same probability for z (same area under the z curve) as does the interval (a, b) in our original normal distribution (see Figure 7.28). The asterisk is used to distinguish a and b, the values for the original normal distribution with mean μ and standard deviation σ, from a^* and b^*, the values for the z curve.

To find a^* and b^*, we simply calculate z scores for the endpoints of the interval for which a probability is desired. This process is called *standardizing* the endpoints. For example, suppose that the variable x has a normal distribution with mean $\mu = 100$ and standard deviation $\sigma = 5$. To find

$$P(98 < x < 107)$$

we first translate this problem into an equivalent problem for the standard normal distribution.

Recall from Chapter 4 that a z score, or standardized score, tells how many standard deviations away from the mean a value lies. The z score is calculated by first subtracting the mean and then dividing by the standard deviation. Converting the lower endpoint $a = 98$ to a z score gives

$$a^* = \frac{98 - 100}{5} = \frac{-2}{5} = -.40$$

and converting the upper endpoint yields

$$b^* = \frac{107 - 100}{5} = \frac{7}{5} = 1.40$$

Then

$$P(98 < x < 107) = P(-.40 < z < 1.40)$$

The probability $P(-.40 < z < 1.40)$ can now be evaluated using technology or Appendix Table 2.

Finding Probabilities

To calculate probabilities for any normal distribution, standardize the relevant values and then use technology or the table of z curve areas. More specifically, if x is a variable whose behavior is described by a normal distribution with mean μ and standard deviation σ, then

$$P(x < b) = P(z < b^*)$$
$$P(x > a) = P(z > a^*)$$
$$P(a < x < b) = P(a^* < z < b^*)$$

where z is a variable whose distribution is standard normal and

$$a^* = \frac{a - \mu}{\sigma} \qquad b^* = \frac{b - \mu}{\sigma}$$

EXAMPLE 7.27 Newborn Birth Weights

Step-by-step technology instructions available online

Understand the context)

Data from the paper **"Fetal Growth Parameters and Birth Weight: Their Relationship to Neonatal Body Composition"** (***Ultrasound in Obstetrics and Gynecology*** **[2009]: 441–446)** suggest that a normal distribution with mean $\mu = 3500$ grams and standard deviation $\sigma = 600$ grams is a reasonable model for the probability distribution of the continuous numerical variable $x =$ birth weight of a randomly selected full-term baby. What proportion of birth weights are between 2900 and 4700 grams?

To answer this question, we must find

$$P(2900 < x < 4700)$$

Do the work)

First, we translate the interval endpoints to equivalent endpoints for the standard normal distribution:

$$a^* = \frac{a - \mu}{\sigma} = \frac{2900 - 3500}{600} = -1.00$$

$$b^* = \frac{b - \mu}{\sigma} = \frac{4700 - 3500}{600} = 2.00$$

Then

$$\begin{aligned} P(2900 < x < 4700) &= P(-1.00 < z < 2.00) \\ &= (z \text{ curve area to the left of } 2.00) \\ &\quad - (z \text{ curve area to the left of } -1.00) \\ &= .9772 - .1587 \\ &= .8185 \end{aligned}$$

Interpret the results) The probabilities for x and z are shown in Figure 7.29. If birth weights were observed for many babies from this population, about 82% of them would fall between 2900 and 4700 grams.

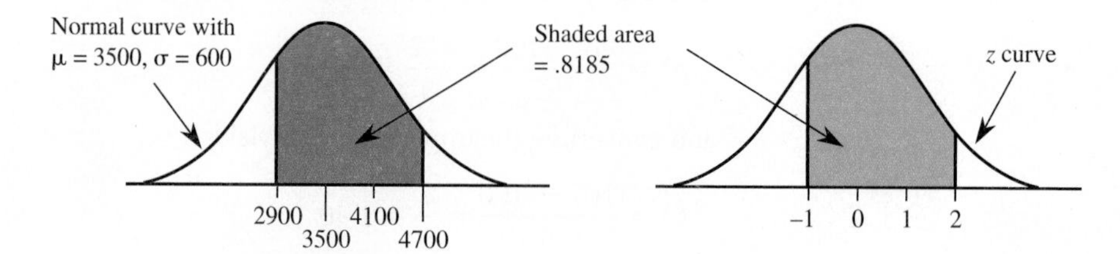

FIGURE 7.29
$P(2900 < x < 4700)$ and corresponding z curve area for the birth weight distribution of Example 7.27.

What is the probability that a randomly chosen baby will have a birth weight greater than 4500? To evaluate $P(x > 4500)$, we first compute

$$a^* = \frac{a - \mu}{\sigma} = \frac{4500 - 3500}{600} = 1.67$$

Then (see Figure 7.30)

$$\begin{aligned} P(x > 4500) &= P(z > 1.67) \\ &= z \text{ curve area to the right of } 1.67 \\ &= 1 - (z \text{ curve area to the left of } 1.67) \\ &= 1 - .9525 \\ &= .0475 \end{aligned}$$

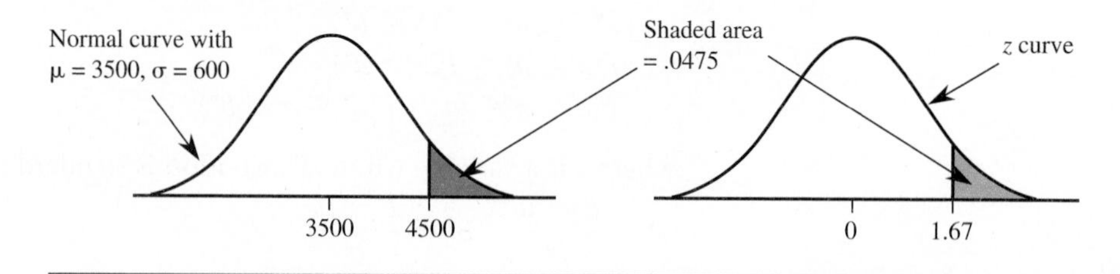

FIGURE 7.30
$P(x > 4500)$ and corresponding z curve area for the birth weight distribution of Example 7.27.

EXAMPLE 7.28 IQ Scores

Understand the context) Although there is some controversy regarding the appropriateness of IQ scores as a measure of intelligence, IQ scores are commonly used for a variety of purposes. One commonly used IQ scale (the Stanford-Binet) has a mean of 100 and a standard deviation of 15, and IQ scores are approximately normally distributed. (IQ score is actually a discrete variable [because it is based on the number of correct responses on a test], but its population distribution closely resembles a normal curve.) If we define the random variable

x = IQ score of a randomly selected individual

then x has approximately a normal distribution with $\mu = 100$ and $\sigma = 15$.

Do the work) One way to become eligible for membership in Mensa, an organization purportedly for those of high intelligence, is to have a Stanford–Binet IQ score above 130. What proportion of the population would qualify for Mensa membership? An answer to this question requires evaluating $P(x > 130)$. This probability is shown in Figure 7.31. With $a = 130$,

$$a^* = \frac{a - \mu}{\sigma} = \frac{130 - 100}{15} = 2.00$$

So (see Figure 7.32)

$$\begin{aligned} P(x > 130) &= P(z > 2.00) \\ &= z \text{ curve area to the right of } 2.00 \\ &= 1 - (z \text{ curve area to the left of } 2.00) \\ &= 1 - .9772 \\ &= .0228 \end{aligned}$$

Interpret the results Only 2.28% of the population would qualify for Mensa membership.

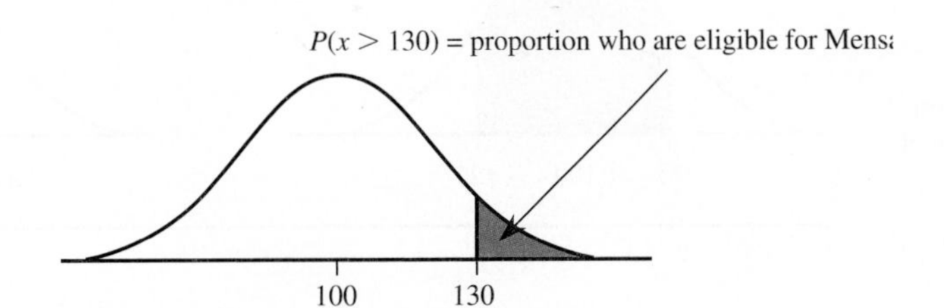

FIGURE 7.31
Normal distribution and desired proportion for Example 7.28.

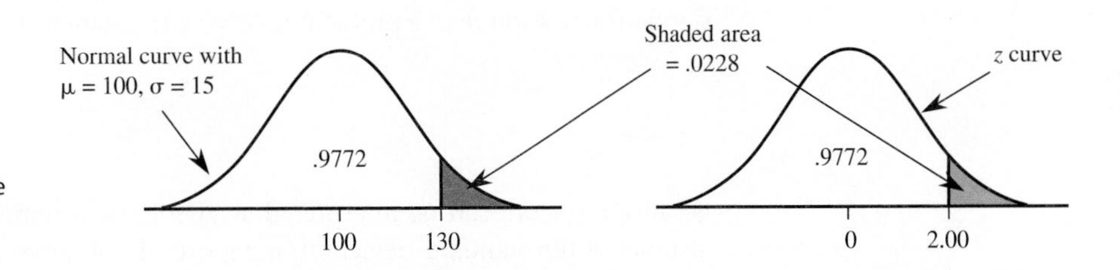

FIGURE 7.32
$P(x > 130)$ and corresponding z curve area for the IQ distribution of Example 7.28.

Suppose that we are interested in the proportion of the population with IQ scores below 80—that is, $P(x < 80)$. With $b = 80$,

$$b^* = \frac{b - \mu}{\sigma} = \frac{80 - 100}{15} = -1.33$$

So

$$\begin{aligned} P(x < 80) &= P(z < -1.33) \\ &= z \text{ curve area to the left of } -1.33 \\ &= .0918 \end{aligned}$$

as shown in Figure 7.33. This probability (.0918) tells us that just a little over 9% of the population has an IQ score below 80.

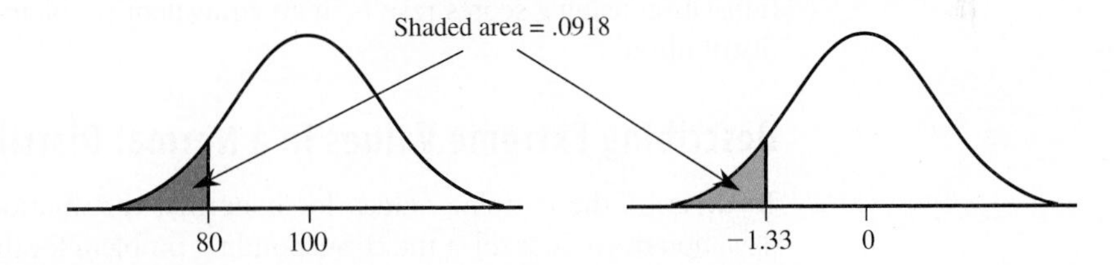

FIGURE 7.33
$P(x < 80)$ and corresponding z curve area for the IQ distribution of Example 7.28.

Now consider the proportion of the population with IQs between 75 and 125. Using $a = 75$ and $b = 125$, we obtain

$$a^* = \frac{75 - 100}{15} = -1.67 \qquad b^* = \frac{125 - 100}{15} = 1.67$$

so

$$\begin{aligned} P(75 < x < 125) &= P(-1.67 < z < 1.67) \\ &= z \text{ curve area between } -1.67 \text{ and } 1.67 \\ &= (z \text{ curve area to the left of } 1.67) \\ &\quad - (z \text{ curve area to the left of } -1.67) \\ &= .9525 - .0475 \\ &= .9050 \end{aligned}$$

This is illustrated in Figure 7.34. The calculation tells us that 90.5% of the population has an IQ score between 75 and 125. Of the 9.5% whose IQ score is not between 75 and 125, half of them (4.75%) have scores over 125, and the other half have scores below 75.

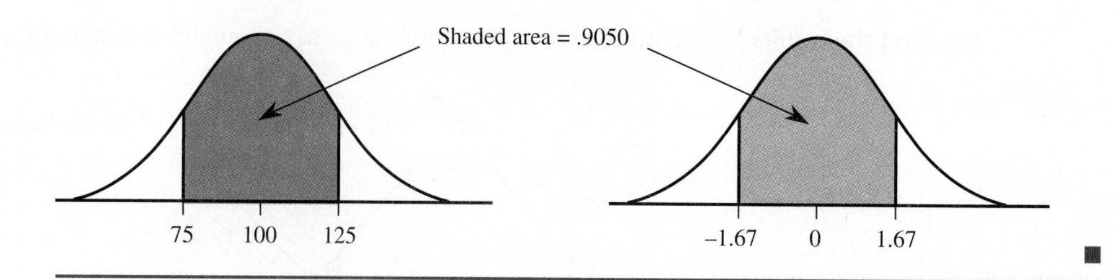

FIGURE 7.34
$P(75 < x < 125)$ and corresponding z curve area for the IQ distribution of Example 7.28.

■

When we translate from a problem involving a normal distribution with mean μ and standard deviation σ to a problem involving the standard normal distribution, we convert to z scores:

$$z = \frac{x - \mu}{\sigma}$$

Because a z score can be interpreted as giving the distance of an x value from the mean in units of the standard deviation, a z score of 1.4 corresponds to an x value that is 1.4 standard deviations above the mean, and a z score of -2.1 corresponds to an x value that is 2.1 standard deviations below the mean.

Suppose that we are trying to evaluate $P(x < 60)$ for a variable whose distribution is normal with $\mu = 50$ and $\sigma = 5$. Converting the endpoint 60 to a z score gives

$$z = \frac{60 - 50}{5} = 2$$

which tells us that the value 60 is 2 standard deviations above the mean. We then have

$$P(x < 60) = P(z < 2)$$

where z is a standard normal variable. Notice that for the standard normal distribution, the value 2 is also 2 standard deviations above the mean, because the mean is 0 and the standard deviation is 1. The value $z = 2$ is located the same distance (measured in standard deviations) from the mean of the standard normal distribution as is the value $x = 60$ from the mean in the normal distribution with $\mu = 50$ and $\sigma = 5$. This is why the translation using z scores results in an equivalent problem involving the standard normal distribution.

Describing Extreme Values in a Normal Distribution

To describe the extreme values for a normal distribution with mean μ and standard deviation σ, we first solve the corresponding problem for the standard normal distribution and then translate our answer into one for the normal distribution of interest. This process is illustrated in Example 7.29.

EXAMPLE 7.29 Registration Times

Understand the context ❭

Data on the length of time required to complete registration for classes using an online registration system suggest that the distribution of the variable

$$x = \text{time to register}$$

for students at a particular university can be well approximated by a normal distribution with mean $\mu = 12$ minutes and standard deviation $\sigma = 2$ minutes. (The normal distribution might not be an appropriate model for x = time to register at another university. Many factors influence the shape, center, and spread of such a distribution.)

Because some students do not log off properly, the university would like to log off students automatically after some amount of time has elapsed. This time will be chosen so that only 1% of the students are logged off while they are still attempting to register. To determine the amount of time that should be allowed before disconnecting a student, we need to describe the largest 1% of the distribution of time to register. These are the individuals who will be mistakenly disconnected. This is illustrated in Figure 7.35(a). To determine the value of x^*, we first solve the analogous problem for the standard normal distribution, as shown in Figure 7.35(b).

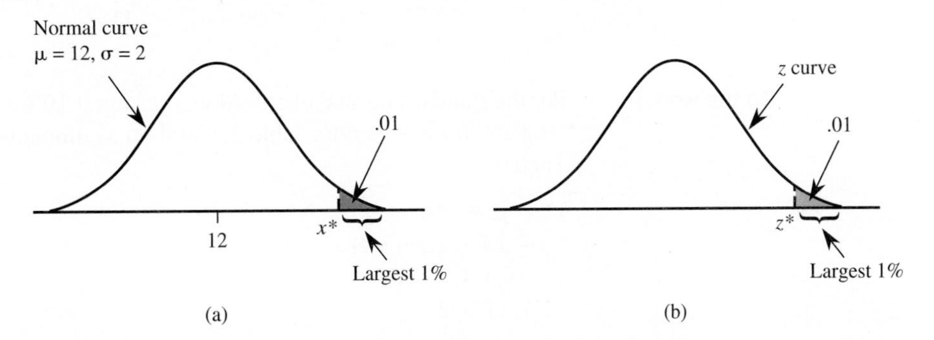

FIGURE 7.35
Capturing the largest 1% of the registration times

Do the work)

By looking at Appendix Table 2 for a cumulative area of .99, we find the closest entry (.9901) in the 2.3 row and the .03 column, from which $z^* = 2.33$. For the standard normal distribution, the largest 1% of the distribution is made up of those values greater than 2.33.

An equivalent statement is that the largest 1% are those with z scores greater than 2.33. This implies that in the distribution of x = time to register (or any other normal distribution), the largest 1% are those values with z scores greater than 2.33 or, equivalently, those x values more than 2.33 standard deviations above the mean. Here, the standard deviation is 2, so 2.33 standard deviations is 2.33(2), and it follows that

$$x^* = 12 + 2.33(2) = 12 + 4.66 = 16.66$$

Interpret the results)

The largest 1% of the distribution for time to register is made up of values that are greater than 16.66 minutes. If the university system was set to log off students after 16.66 minutes, only 1% of the students registering would be logged off before completing their registration. ■

A general formula for converting a z score back to an x value results from solving $z^* = \frac{x^* - \mu}{\sigma}$ for x^*, as shown in the accompanying box.

To convert a z score z^* back to an x value x^*, use

$$x^* = \mu + z^*\sigma$$

EXAMPLE 7.30 Garbage Truck Processing Times

Understand the context)

Garbage trucks entering a particular waste management facility are weighed and then they offload garbage into a landfill. Data from the paper **"Estimating Waste Transfer Station Delays Using GPS" (*Waste Management* [2008]: 1742–1750)** suggest that a normal distribution with mean $\mu = 13$ minutes and $\sigma = 3.9$ minutes is a reasonable model for the probability distribution of the random variable x = total processing time for a garbage truck at this waste management facility (total processing time includes waiting time as well as the time required to weigh the truck and offload the garbage).

Suppose that we want to describe the total processing times of the trucks making up the 10% with the longest processing times. These trucks would be the 10% with times corresponding to the shaded region in the accompanying illustration.

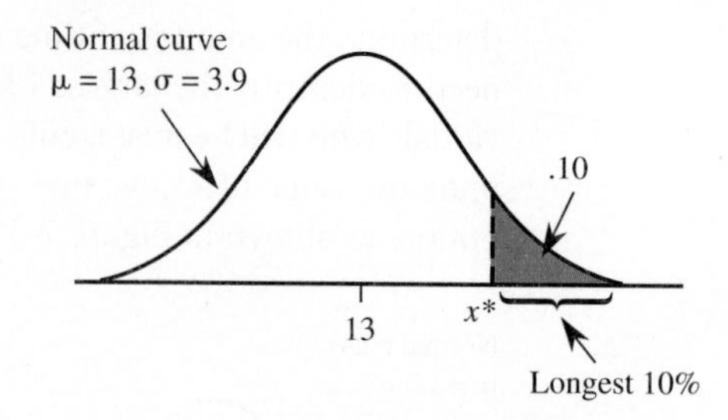

Do the work ⟩ For the standard normal distribution, the largest 10% are those with z values greater than $z^* = 1.28$ (from Appendix Table 2, based on a cumulative area of .90).
Then

$$\begin{aligned} x^* &= \mu + z^*\sigma \\ &= 13 + 1.28(3.9) \\ &= 13 + 4.992 \\ &= 17.992 \end{aligned}$$

Interpret the results ⟩ About 10% of the garbage trucks using this facility would have a total processing time of more than 17.992 minutes.

The 5% with the fastest processing times would be those with z values less than $z^* = -1.645$ (from Appendix Table 2, based on a cumulative area of .05). Then

$$\begin{aligned} x^* &= \mu + z^*\sigma \\ &= 13 + (-1.645)(3.9) \\ &= 13 - 6.416 \\ &= 6.584 \end{aligned}$$

About 5% of the garbage trucks processed at this facility will have total processing times of less than 6.584 minutes. ■

EXERCISES 7.63 - 7.82

7.63 Determine the following standard normal (z) curve areas. (Hint: See Examples 7.23 and 7.24.)

a. The area under the z curve to the left of 1.75

b. The area under the z curve to the left of -0.68

c. The area under the z curve to the right of 1.20

d. The area under the z curve to the right of -2.82

e. The area under the z curve between -2.22 and 0.53

f. The area under the z curve between -1 and 1

g. The area under the z curve between -4 and 4

7.64 Determine each of the following areas under the standard normal (z) curve:

a. To the left of -1.28

b. To the right of 1.28

c. Between -1 and 2

d. To the right of 0

e. To the right of -5

f. Between -1.6 and 2.5

g. To the left of 0.23

7.65 Let z denote a random variable that has a standard normal distribution. Determine each of the following probabilities:

a. $P(z < 2.36)$

b. $P(z \leq 2.36)$

c. $P(z < -1.23)$

d. $P(1.14 < z < 3.35)$

e. $P(-0.77 \leq z \leq -0.55)$

f. $P(z > 2)$

g. $P(z \geq -3.38)$

h. $P(z < 4.98)$

7.66 Let z denote a random variable having a normal distribution with $\mu = 0$ and $\sigma = 1$. Determine each of the following probabilities. (Hint: See Examples 7.27 and 7.28.)

a. $P(z < 0.10)$

b. $P(z < -0.10)$

c. $P(0.40 < z < 0.85)$

d. $P(-0.85 < z < -0.40)$

e. $P(-0.40 < z < 0.85)$

f. $P(z > -1.25)$

g. $P(z < -1.50$ or $z > 2.50)$

7.67 Let z denote a variable that has a standard normal distribution. Determine the value z^* to satisfy the following conditions. (Hint: See Example 7.25.)

a. $P(z < z^*) = .025$

b. $P(z < z^*) = .01$

c. $P(z < z^*) = .05$

d. $P(z > z^*) = .02$

e. $P(z > z^*) = .01$

f. $P(z > z^*$ or $z < -z^*) = .20$

7.68 Determine the value z^* that

a. Separates the largest 3% of all z values from the others (Hint: See Example 7.26.)

b. Separates the largest 1% of all z values from the others

c. Separates the smallest 4% of all z values from the others

d. Separates the smallest 10% of all z values from the others

7.69 Determine the value of z^* such that

a. $-z^*$ and z^* separate the middle 95% of all z values from the most extreme 5% (Hint: See Example 7.26.)

b. $-z^*$ and z^* separate the middle 90% of all z values from the most extreme 10%

c. $-z^*$ and z^* separate the middle 98% of all z values from the most extreme 2%

d. $-z^*$ and z^* separate the middle 92% of all z values from the most extreme 8%

7.70 Because $P(z < .44) = .67$, 67% of all z values are less than .44, and .44 is the 67th percentile of the standard normal distribution. Determine the value of each of the following percentiles for the standard normal distribution (Hint: If the cumulative area that you must look for does not appear in the z table, use the closest entry, see Example 7.27.):

a. The 91st percentile (Hint: Look for area .9100.)

b. The 77th percentile

c. The 50th percentile

d. The 9th percentile

e. What is the relationship between the 70th z percentile and the 30th z percentile?

7.71 Consider the population of all 1-gallon cans of dusty rose paint manufactured by a particular paint company. Suppose that a normal distribution with mean $\mu = 5$ ml and standard deviation $\sigma = 0.2$ ml is a reasonable model for the distribution of the variable x = amount of red dye in the paint mixture. Use the normal distribution model to calculate the following probabilities. (Hint: See Examaples 7.27 and 7.28.)

a. $P(x < 5.0)$

b. $P(x < 5.4)$

c. $P(x \leq 5.4)$

d. $P(4.6 < x < 5.2)$

e. $P(x > 4.5)$

f. $P(x > 4.0)$

7.72 Consider babies born in the "normal" range of 37–43 weeks gestational age. The paper referenced in Example 7.27 (**"Fetal Growth Parameters and Birth Weight: Their Relationship to Neonatal Body Composition,"** ***Ultrasound in Obstetrics and Gynecology*** **[2009]: 441–446)** suggests that a normal distribution with mean $\mu = 3500$ grams and standard deviation $\sigma = 600$ grams is a reasonable model for the probability distribution of the continuous numerical variable x = birth weight of a randomly selected full-term baby.

a. What is the probability that the birth weight of a randomly selected full-term baby exceeds 4000 g?

b. What is the probability that the birth weight of a randomly selected full-term baby is between 3000 and 4000 g?

c. What is the probability that the birth weight of a randomly selected full-term baby is either less than 2000 g or greater than 5000 g?

d. What is the probability that the birth weight of a randomly selected full-term baby exceeds 7 pounds? (Hint: 1 lb = 453.59 g.)

e. How would you characterize the most extreme 0.1% of all full-term baby birth weights?

f. If x is a random variable with a normal distribution and a is a numerical constant ($a \neq 0$), then $y = ax$ also has a normal distribution. Use this formula to determine the distribution of full-term baby birth weight expressed in pounds (shape, mean, and standard deviation), and then recalculate the probability from Part (d). How does this compare to your previous answer?

7.73 Emissions of nitrogen oxides, which are major constituents of smog, can be modeled using a normal distribution. Let x denote the amount of this pollutant emitted by a randomly selected vehicle (in parts per billion). The distribution of x can be described by a normal distribution with $\mu = 1.6$ and $\sigma = 0.4$. Suppose that the EPA wants to offer some sort of incentive to get the worst polluters off the road.

What emission levels constitute the worst 10% of the vehicles?

7.74 The paper referenced in Example 7.30 (**"Estimating Waste Transfer Station Delays Using GPS,"** ***Waste Management*** **[2008]: 1742–1750)** describing processing times for garbage trucks also provided information on processing times at a second facility. At this second facility, the mean total processing time was 9.9 minutes and the standard deviation of the processing times was 6.2 minutes. Explain why a normal distribution with mean 9.9 and standard deviation 6.2 would not be an appropriate model for the probability distribution of the variable x = total processing time of a randomly selected truck entering this facility.

7.75 The size of the left upper chamber of the heart is one measure of cardiovascular health. When the upper left chamber is enlarged, the risk of heart problems is increased. The paper **"Left Atrial Size Increases with Body Mass Index in Children"** (***International Journal of Cardiology*** **[2009]: 1–7)** described a study in which the left atrial size was measured for a large number of children age 5 to 15 years. Based on this data, the authors concluded that for healthy children, left atrial diameter was approximately normally distributed with a mean of 26.4 mm and a standard deviation of 4.2 mm.

a. Approximately what proportion of healthy children have left atrial diameters less than 24 mm?

b. Approximately what proportion of healthy children have left atrial diameters greater than 32 mm?

c. Approximately what proportion of healthy children have left atrial diameters between 25 and 30 mm?

d. For healthy children, what is the value for which only about 20% have a larger left atrial diameter?

7.76 The paper referenced in the previous exercise also included data on left atrial diameter for children who were considered overweight. For these children, left atrial diameter was approximately normally distributed with a mean of 28 mm and a standard deviation of 4.7 mm.

a. Approximately what proportion of overweight children have left atrial diameters less than 25 mm?

b. Approximately what proportion of overweight children have left atrial diameters greater than 32 mm?

c. Approximately what proportion of overweight children have left atrial diameters between 25 and 30 mm?

d. What proportion of overweight children has left atrial diameters greater than the mean for healthy children?

7.77 According to the paper **"Commuters' Exposure to Particulate Matter and Carbon Monoxide in Hanoi, Vietnam"** (***Transportation Research*** **[2008]: 206–211)**, the carbon monoxide exposure of someone riding a motorbike for 5 km on a highway in Hanoi is approximately normally distributed with a mean of 18.6 ppm. Suppose that the standard deviation of carbon monoxide exposure is 5.7 ppm. Approximately what proportion of those who ride a motorbike for 5 km on a Hanoi highway will experience a carbon monoxide exposure of

a. more than 20 ppm?

b. more than 25 ppm?

7.78 A machine that cuts corks for wine bottles operates in such a way that the distribution of the diameter for the corks produced is well approximated by a normal distribution with mean 3 cm and standard deviation 0.1 cm. The specifications call for corks with diameters between 2.9 and 3.1 cm. A cork not meeting the specifications is considered defective. (A cork that is too small leaks and causes the wine to deteriorate. A cork that is too large doesn't fit in the bottle.) What proportion of corks produced by this machine are defective?

7.79 Refer to the previous exercise. Suppose that there are two machines available for cutting corks. The machine described in the preceding problem produces corks with diameters that are approximately normally distributed with mean 3 cm and standard deviation 0.1 cm. The second machine produces corks with diameters that are approximately normally distributed with mean 3.05 cm and standard deviation 0.01 cm. Which machine would you recommend? (Hint: Which machine would produce fewer defective corks?)

7.80 A gasoline tank for a certain car is designed to hold 15 gallons of gas. Suppose that the variable x = actual capacity of a randomly selected tank has a distribution that is well approximated by a normal curve with mean 15.0 gallons and standard deviation 0.1 gallon.

a. What is the probability that a randomly selected tank will hold at most 14.8 gallons?

b. What is the probability that a randomly selected tank will hold between 14.7 and 15.1 gallons?

c. If two such tanks are independently selected, what is the probability that both hold at most 15 gallons?

7.81 ▼ The time that it takes a randomly selected job applicant to perform a certain task has a distribution that can be approximated by a normal distribution with a mean value of 120 seconds and a standard deviation of 20 seconds. The fastest 10% are to be given advanced training. What task times qualify individuals for such training?

7.82 ▼ Suppose that the distribution of typing speed in words per minute (wpm) for experienced typists using a new type of split keyboard can be approximated by a normal curve with mean 60 wpm and standard deviation 15 wpm (**"The Effects of Split Keyboard Geometry on Upper body Postures,"** ***Ergonomics*** **[2009]: 104–111**).

a. What is the probability that a randomly selected typist's speed is at most 60 wpm?

b. What is the probability that a randomly selected typist's speed is less than 60 wpm?

c. What is the probability that a randomly selected typist's speed is between 45 and 90 wpm?

d. Would you be surprised to find a typist in this population whose speed exceeded 105 wpm?

e. Suppose that two typists are independently selected. What is the probability that both their typing speeds exceed 75 wpm?

f. Suppose that special training is to be made available to the slowest 20% of the typists. What typing speeds would qualify individuals for this training?

Bold exercises answered in back ● Data set available online ▼ Video Solution available

7.7 Checking for Normality and Normalizing Transformations

Some of the most frequently used statistical methods are valid only when a sample $x_1, x_2, \ldots, x_n$ has come from a population distribution that is at least approximately normal. One way to see whether an assumption of population normality is plausible is to construct a **normal probability plot** of the data.

One version of this plot uses quantities called **normal scores.** The values of the normal scores depend on the sample size n. For example, the normal scores when $n = 10$ are as follows:

−1.539	−1.001	−.656	−.376	−.123
.123	.376	.656	1.001	1.539

To interpret these numbers, think of selecting sample after sample from a standard normal distribution, each one consisting of $n = 10$ observations. Then −1.539 is the long-run average of the smallest observation from each sample, −1.001 is the long-run average of the second smallest observation from each sample, and so on. In other words, −1.539 is the mean value of the smallest observation in a sample of size 10 from the z distribution and −1.001 is the mean value of the second smallest observation, and so on.

Tables of normal scores for many different sample sizes are available. Alternatively, many software packages (such as Minitab and JMP) and some graphing calculators can compute these scores on request and then construct a normal probability plot. Not all calculators and software packages use the same algorithm to compute normal scores. However, these differences in normal scores do not change the overall character of a normal probability plot, so either the tabulated values or those given by the computer or calculator can be used.

After the sample observations are ordered from smallest to largest, the smallest normal score is paired with the smallest observation, the second smallest normal score with the second smallest observation, and so on. The first number in a pair is the normal score, and the second number in the pair is the observed data value. A normal probability plot is just a scatterplot of these (normal score, observed value) pairs.

If the sample has been selected from a *standard* normal distribution, the second number in each pair should be reasonably close to the first number (ordered observation ≈ corresponding mean value). Then the n plotted points will fall near a line with slope equal to 1 (a 45° line) passing through (0, 0). When the sample has been obtained from *some* normal population distribution (but not necessarily the standard normal distribution), the plotted points should be close to *some* straight line (but not necessarily one with slope 1 and intercept 0).

DEFINITION

Normal probability plot: A scatterplot of the (normal score, observed value) pairs. A strong linear pattern in a normal probability plot suggests it is plausible that the population distribution is normal. On the other hand, systematic departure from a straight-line pattern (such as curvature in the plot) indicates that it is not reasonable to assume that the population distribution is normal.

EXAMPLE 7.31 Egg Weights

Understand the context ❯

The following data represent egg weights (in grams) for a sample of 10 eggs. These data are consistent with summary quantities in the paper **"Evaluation of Egg Quality Traits of Chickens Reared under Backyard System in Western Uttar Pradesh"** (***Indian Journal of Poultry Science*, 2009**).

53.04 53.50 52.53 53.00 53.07 52.86 52.66 53.23 53.26 53.16

Do the work ❯

Arranging the sample observations in order from smallest to largest results in

52.53 52.66 52.86 53.00 53.04 53.07 53.16 53.23 53.26 53.50

Pairing these ordered observations with the normal scores for a sample of size 10 (previously given) results in the following 10 pairs that can be used to construct the normal probability plot:

(−1.539, 52.53)	(−1.001, 52.66)
(−0.656, 52.86)	(−0.376, 53.00)
(−0.123, 53.04)	(0.123, 53.07)
(0.376, 53.16)	(0.656, 53.23)
(1.001, 53.26)	(1.539, 53.50)

Interpret the results ❯

The normal probability plot is shown in Figure 7.36. The linear pattern in the plot indicates that it is plausible that the egg-weight distribution from which these observations were selected is normal.

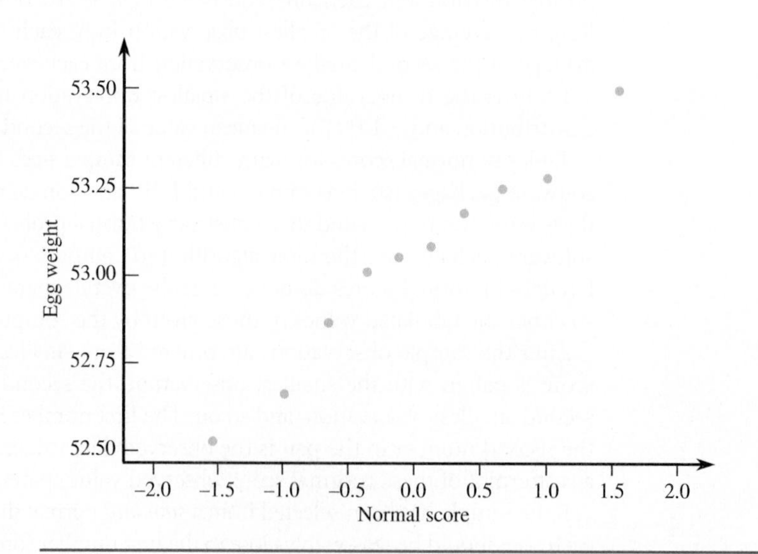

FIGURE 7.36
A normal probability plot for data of Example 7.31.

■

The decision as to whether a normal probability plot shows a strong linear pattern is somewhat subjective. Particularly when n is small, normality should not be ruled out unless the departure from linearity is clear-cut. Figure 7.37 displays several plots that suggest a nonnormal population distribution.

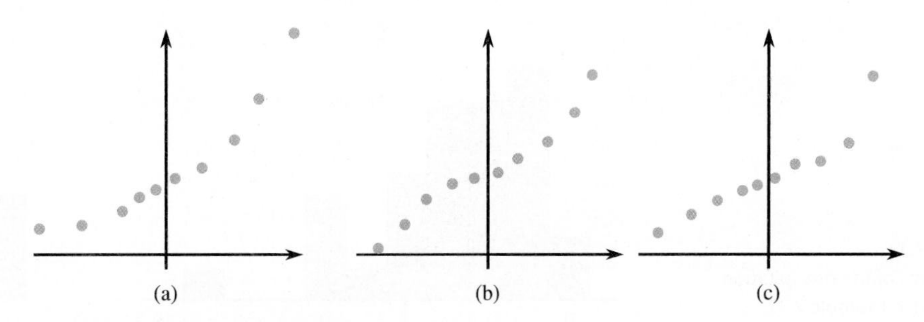

FIGURE 7.37
Plots suggesting nonnormality:
(a) indication that the population distribution is skewed;
(b) indication that the population distribution has heavier tails than a normal curve;
(c) presence of an outlier.

Transforming Data to Obtain a Distribution That Is Approximately Normal (Optional)

Many of the most frequently used statistical methods are valid only when the sample is selected at random from a population whose distribution is at least approximately normal. When a sample histogram shows a distinctly nonnormal shape, it is common to use a transformation or reexpression of the data.

By *transforming* data, we mean applying some specified mathematical function (such as the square root, logarithm, or reciprocal) to each data value to produce a set of transformed data. We can then study and summarize the distribution of these transformed values using methods that require normality.

We saw in Chapter 5 that, with bivariate data, one or both of the variables can be transformed in an attempt to find two variables that are linearly related. With univariate data, a transformation is usually chosen to produce a distribution of transformed values that is more symmetric and more closely approximated by a normal curve than the original distribution.

EXAMPLE 7.32 Rainfall Data

Understand the context

● Data that have been used by several investigators to introduce the concept of transformation consist of values of March precipitation for Minneapolis–St. Paul over a period of 30 years. These values are given in Table 7.2, along with the square root of each value. Histograms of both the original and the transformed data appear in Figure 7.38.

The distribution of the original data is clearly skewed, with a long upper tail. The square-root transformation results in a substantially more symmetric distribution, with a typical value of around 1.1.

TABLE 7.2 Original and Square-Root-Transformed Values of March Precipitation in Minneapolis–St. Paul over a 30-year Period

Year	Precipitation	$\sqrt{\text{Precipitation}}$	Year	Precipitation	$\sqrt{\text{Precipitation}}$
1	.77	.88	16	1.62	1.27
2	1.74	1.32	17	1.31	1.14
3	.81	.90	18	.32	.57
4	1.20	1.10	19	.59	.77
5	1.95	1.40	20	.81	.90
6	1.20	1.10	21	2.81	1.68
7	.47	.69	22	1.87	1.37
8	1.43	1.20	23	1.18	1.09
9	3.37	1.84	24	1.35	1.16
10	2.20	1.48	25	4.75	2.18
11	3.00	1.73	26	2.48	1.57
12	3.09	1.76	27	.96	.98
13	1.51	1.23	28	1.89	1.37
14	2.10	1.45	29	.90	.95
15	.52	.72	30	2.05	1.43

● Data set available online

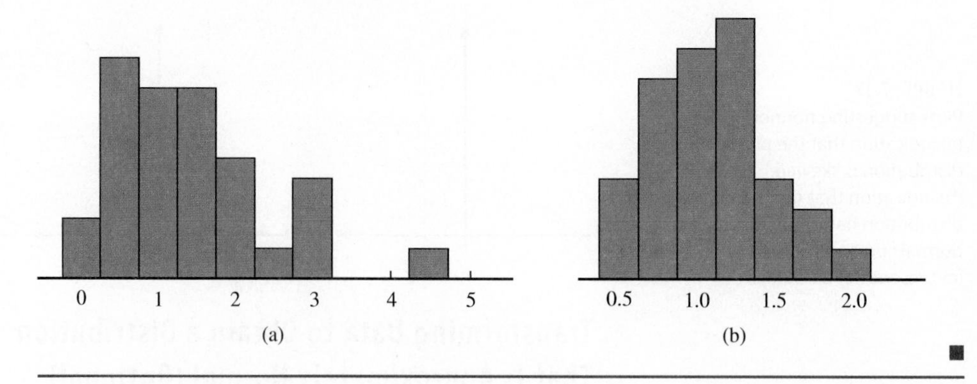

FIGURE 7.38
Histograms of the precipitation data used in Example 7.32:
(a) untransformed data;
(b) square-root transformed data.

Logarithmic transformations are also common and, as with bivariate data, either the natural logarithm or the base 10 logarithm can be used. A logarithmic transformation is usually applied to data that are positively skewed (a long upper tail). This affects values in the upper tail substantially more than values in the lower tail, yielding a more symmetric—and often more nearly normal—distribution.

EXAMPLE 7.33 Markers for Kidney Disease

Two measures of kidney function are the levels of a substance called AGT found in blood and urine. The paper **"Urinary Angiotensinogen as a Potential Biomarker of Severity of Chronic Kidney Diseases"** (***Journal of the American Society of Hypertension*** **[2008]: 349–354)** describes a study in which blood plasma AGT levels and urinary AGT levels were measured for a sample of adults with chronic kidney disease. Representative data (consistent with summary quantities and descriptions given in the paper) for 40 patients are given in Table 7.3.

TABLE 7.3 Plasma and Urinary AGT Levels

Plasma AGT	Plasma AGT	Urinary AGT	Urinary AGT
21.0	16.7	56.2	41.7
36.0	20.2	288.4	29.5
22.9	24.5	45.7	208.9
8.0	18.5	426.6	229.1
27.3	40.2	190.6	186.2
32.4	18.8	616.6	29.5
17.2	28.1	97.7	229.1
30.9	26.8	66.1	13.5
27.2	24.1	2.6	407.4
30.0	14.1	74.1	1122.0
35.1	18.9	14.5	66.1
21.6	25.6	56.2	7.4
22.7	10.2	812.8	177.8
2.5	29.2	11.5	6.2
30.2	29.5	346.7	67.6
27.3	24.3	9.6	20.0
19.6	22.3	288.4	28.8
19.0	16.5	147.9	186.2
13.4	25.6	17.0	141.3
18.0	23.0	575.4	724.4

The authors of the paper stated that the distribution of plasma AGT levels was approximately normal. Minitab was used to construct the histogram and normal probability plot for the plasma AGT levels shown in Figure 7.39. The histogram is reasonably symmetric and the normal probability plot shows a strong linear pattern. This is consistent with the authors' statement about the approximate normality of the plasma AGT levels.

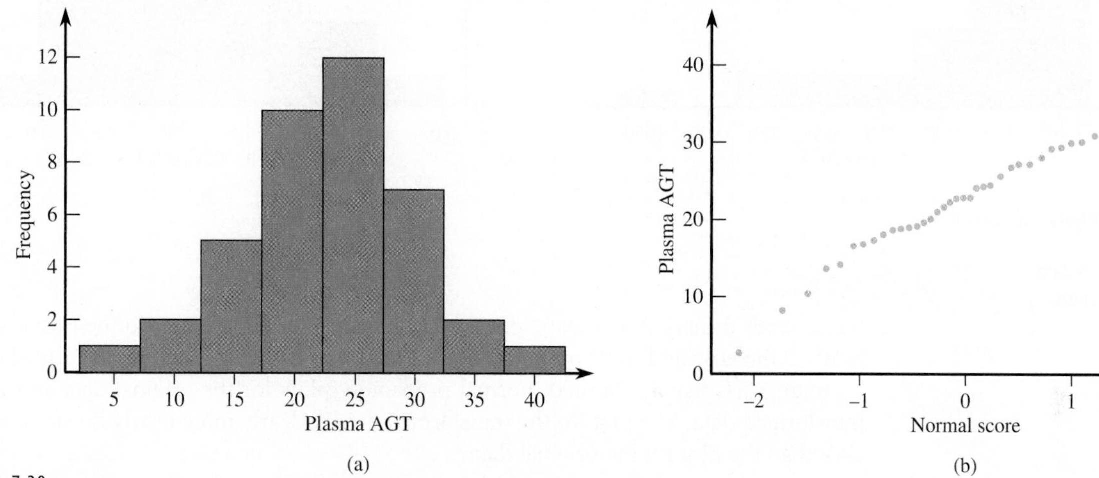

FIGURE 7.39
Graphical displays for the plasma AGT data of Example 7.33:
(a) histogram;
(b) normal probability plot.

When the authors considered urinary AGT levels, they found that the distribution of the sample data was skewed, and they used a log transformation in order to obtain a distribution that was more approximately normal. Table 7.4 gives the urinary AGT levels along with the log-transformed data. Figure 7.40 shows histograms of the original urinary AGT data and the

TABLE 7.4 Urinary AGT Levels and Log-Transformed Levels

Urinary AGT	Log(Urinary AGT)	Urinary AGT	Log(Urinary AGT)
56.2	1.75	41.7	1.62
288.4	2.46	29.5	1.47
45.7	1.66	208.9	2.32
426.6	2.63	229.1	2.36
190.6	2.28	186.2	2.27
616.6	2.79	29.5	1.47
97.7	1.99	229.1	2.36
66.1	1.82	13.5	1.13
2.6	0.41	407.4	2.61
74.1	1.87	1122.0	3.05
14.5	1.16	66.1	1.82
56.2	1.75	7.4	0.87
812.8	2.91	177.8	2.25
11.5	1.06	6.2	0.79
346.7	2.54	67.6	1.83
9.6	0.98	20.0	1.30
288.4	2.46	28.8	1.46
147.9	2.17	186.2	2.27
17.0	1.23	141.3	2.15
575.4	2.76	724.4	2.86

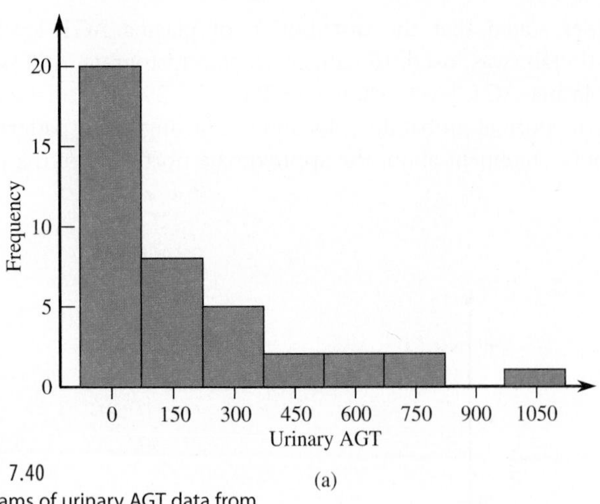

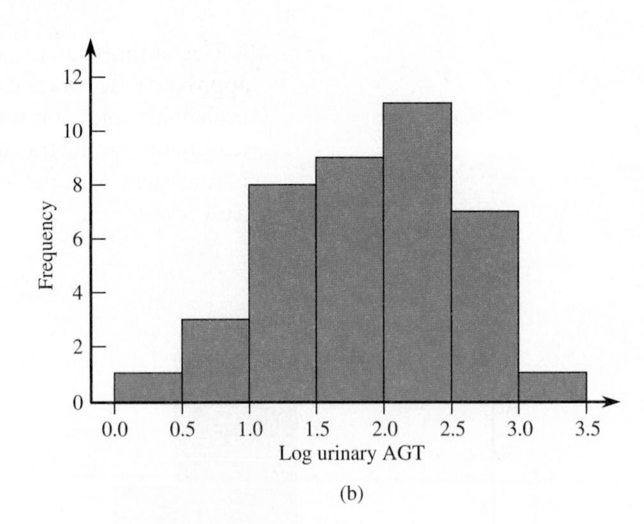

FIGURE 7.40
Histograms of urinary AGT data from Example 7.33:
(a) untransformed data;
(b) transformed data.

transformed urinary AGT data. Notice that the histogram for the transformed data is more nearly symmetric and more mound shaped than the histogram of the untransformed data.

Figure 7.41 displays Minitab normal probability plots for the original data and for the transformed data. The plot for the transformed data is clearly more nearly linear in appearance than the plot for the original data.

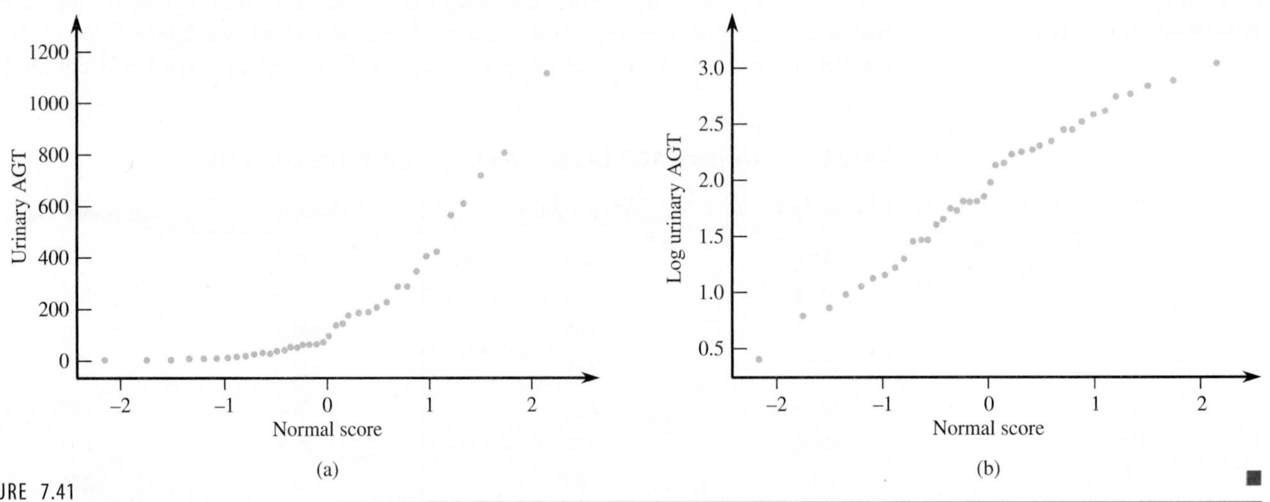

FIGURE 7.41
Minitab normal probability plots for the urinary AGT data of
(a) original data;
(b) transformed data.

Selecting a Transformation

Occasionally, a particular transformation can be dictated by some theoretical argument, but often this is not the case and you may want to try several different transformations to find one that is satisfactory. Figure 7.42, from the article **"Distribution of Sperm Counts in Suspected Infertile Men"** (***Journal of Reproduction and Fertility*** **[1983]: 91–96**), shows what can result from such a search. Other investigators in this field had previously used all three of the transformations illustrated, but the log transformation shown in Figure 7.42(b) appears to be the best choice.

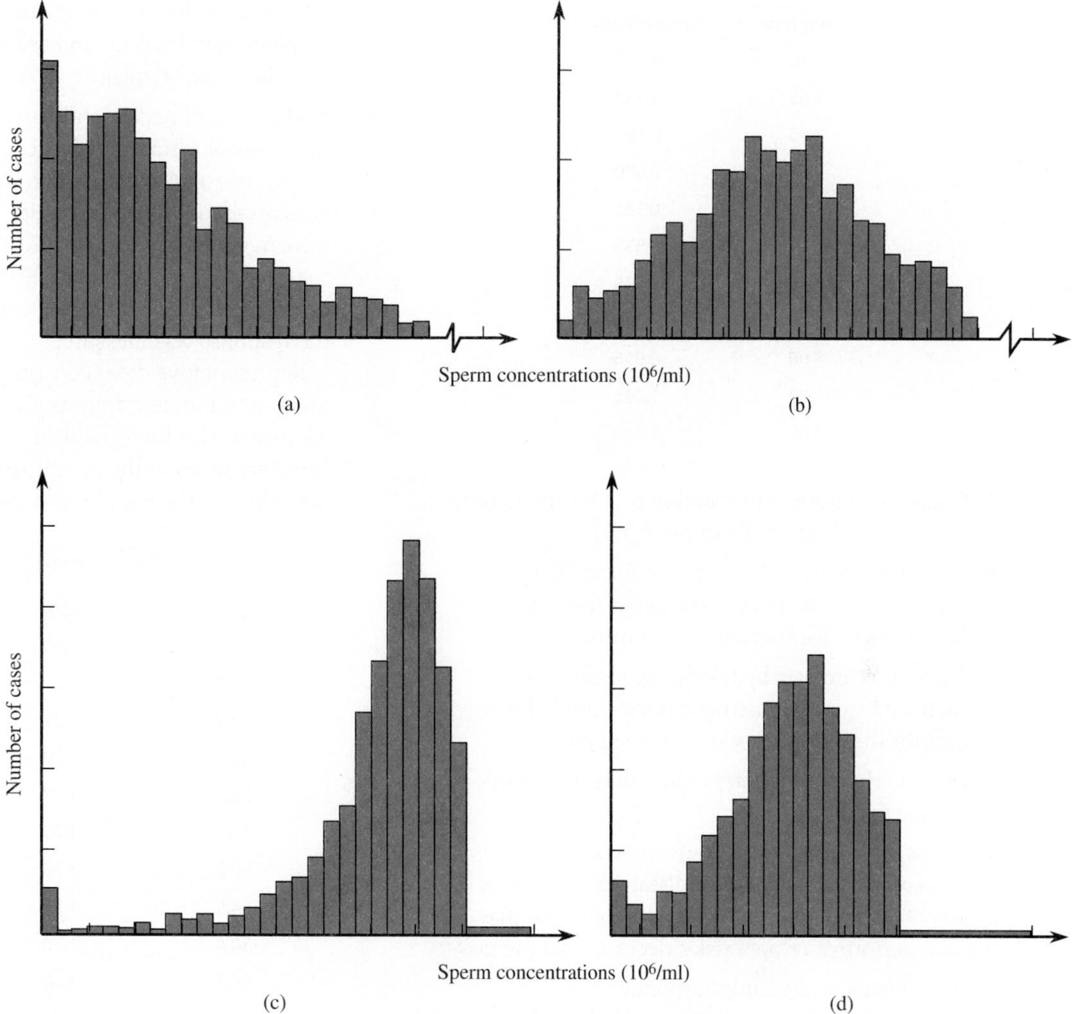

FIGURE 7.42
Histograms of sperm concentrations for 1711 suspected infertile men:
(a) untransformed data (highly skewed);
(b) log-transformed data (reasonably symmetric);
(c) square-root-transformed data;
(d) cube-root-transformed data.

EXERCISES 7.83 - 7.93

7.83 ● The authors of the paper **"Development of Nutritionally At-Risk Young Children is Predicted by Malaria, Anemia, and Stunting in Pemba, Zanzibar" (*The Journal of Nutrition* [2009]:763–772)** studied factors that might be related to dietary deficiencies in children. Children were observed for a length of time and the time spent in various activities was recorded. One variable of interest was the length of time (in minutes) a child spent fussing.

The authors comment that the distribution of fussing times was skewed and that they used a square root transformation to create a distribution that was more approximately normal. Data consistent with summary quantities in the paper for 15 children are given in the accompanying table. Normal scores for a samples size of 15 are also given.

Fussing Time	Normal Score
0.05	−1.739
0.10	−1.245
0.15	−0.946
0.40	−0.714

continued

Fussing Time	Normal Score
0.70	−0.515
1.05	−0.333
1.95	−0.165
2.15	0.000
3.70	0.165
3.90	0.335
4.50	0.515
6.00	0.714
8.00	0.946
11.00	1.245
14.00	1.739

a. Construct a normal probability plot for the fussing time data. (Hint: See Example 7.31.)

b. Does the plot from Part (a) look linear? Do you agree with the authors of the paper that the fussing time distribution is not normal?

c. Transform the data by taking the square root of each data value. Construct a normal probability plot for the square root transformed data.

d. How do the normal probability plots from Parts (a) and (c) compare?

7.84 ● The paper **"Risk Behavior, Decision Making, and Music Genre in Adolescent Males" (Marshall University, May 2009)** examined the effect of type of music playing and performance on a risky, decision-making task.

a. Participants in the study responded to a questionnaire that was used to assign a risk behavior score. Risk behavior scores (read from a graph that appeared in the paper) for 15 participants follow. Use these data to construct a normal probability plot (the normal scores for a sample of size 15 appear in the previous exercise).

102	105	113	120	125	127	134	135
139	141	144	145	149	150	160	

b. Participants also completed a positive and negative affect scale (PANAS) designed to measure emotional response to music. PANAS values (read from a graph that appeared in the paper) for 15 participants follow. Use these data to construct a normal probability plot (the normal scores for a sample of size 15 appear in the previous exercise).

36	40	45	47	48	49	50	52
53	54	56	59	61	62	70	

c. The author of the paper states that he believes that it is reasonable to consider both risk behavior scores and PANAS scores to be approximately normally distributed. Do the normal probability plots from Parts (a) and (b) support this conclusion? Explain.

7.85 ● Measures of nerve conductivity are used in the diagnosis of certain medical conditions. The paper **"Effects of Age, Gender, Height, and Weight on Late Responses and Nerve Conduction Study Parameters" (*Acta Neurologica Taiwanica* [2009]: 242–249)** describes a study in which the ulnar nerve was stimulated in healthy patients and the amplitude and velocity of the response was measured.

Representative data (consistent with summary quantities and descriptions given in the paper) for 30 patients for the variable x = response velocity (m/s) are given in the accompanying table. Also given are values of the log of x and the square root of x.

x	log(x)	sqrt(x)
60.1	1.78	7.75
48.7	1.69	6.98
51.7	1.71	7.19
52.9	1.72	7.27
50.5	1.70	7.11
58.5	1.77	7.65
53.6	1.73	7.32
60.3	1.78	7.77
64.5	1.81	8.03
50.4	1.70	7.10
56.5	1.75	7.52
55.5	1.74	7.45
53.0	1.72	7.28
50.5	1.70	7.11
54.0	1.73	7.35
53.6	1.73	7.32
55.2	1.74	7.43
57.9	1.76	7.61
61.5	1.79	7.84
58.0	1.76	7.62
57.6	1.76	7.59
67.1	1.83	8.19
56.2	1.75	7.50
53.8	1.73	7.33
55.7	1.75	7.46
52.9	1.72	7.27
54.0	1.73	7.35
52.6	1.72	7.25
61.8	1.79	7.86
62.8	1.80	7.92

a. Construct a histogram of the untransformed data.

b. Does the distribution of x appear to be approximately normal? Explain.

c. Construct a histogram of the log transformed data.

d. Is the histogram of the transformed data more symmetric than the histogram of the untransformed data?

e. Construct a histogram of the square root transformed data.

f. Do either of the two transformations (square root or log) result in a histogram that is more nearly normal in shape? (Hint: See Example 7.33.)

7.86 ● Macular degeneration is the most common cause of blindness in people older than 60 years. One variable thought to be related to a type of inflammation associated with this disease is level of a substance called soluble Fas ligand (sFasL) in the blood.

The accompanying table contains representative data on x = sFasL level for 10 patients with age-related macular degeneration. These data are consistent with summary quantities and descriptions of the data given in the paper **"Associations of Plasma-Soluble Fas Ligand with Aging and Age-Related Macular Degeneration"** (***Investigative Ophthalmology & Visual Science*** **[2008]: 1345–1349**).

The authors of the paper noted that the distribution of sFasL level was skewed and recommended a cube-root transformation. The cube-root values and the normal scores for a sample size of 10 are also given in the accompanying table.

x	Cube Root of x	Normal Score
0.069	0.41	−1.539
0.074	0.42	−1.001
0.176	0.56	−0.656
0.185	0.57	−0.376
0.216	0.60	−0.123
0.287	0.66	0.123
0.343	0.70	0.376
0.343	0.70	0.656
0.512	0.80	1.001
0.729	0.90	1.539

a. Construct a normal probability plot using the untransformed data.

b. Does the normal probability plot appear linear or curved?

c. Construct a normal probability plot using the transformed data. Does the normal probability plot appear more nearly linear than the plot for the untransformed data?

7.87 The following normal probability plot was constructed using part of the data appearing in the paper **"Trace Metals in Sea Scallops"** (***Environmental Concentration and Toxicology*** **19: 1326–1334**).

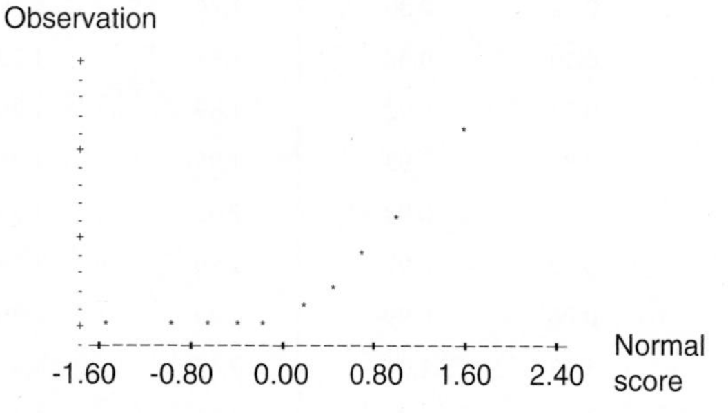

The variable under study was the amount of cadmium in North Atlantic scallops. Do the sample data suggest that the cadmium concentration distribution is not normal? Explain.

7.88 ● Consider the following 10 observations on the lifetime (in hours) for a certain type of power supply: 152.7, 172.0, 172.5, 173.3, 193.0, 204.7, 216.5, 234.9, 262.6, and 422.6. Construct a normal probability plot, and comment on the plausibility of a normal distribution as a model for power supply lifetime. (The normal scores for a sample of size 10 are −1.539, −1.001, −.656, −.376, −.123, .123, .376, .656, 1.001, and 1.539.)

7.89 ● Consider the following sample of 25 observations on the diameter x (in centimeters) of a disk used in a certain system:

16.01	16.08	16.13	15.94	16.05	16.27	15.89
15.84	15.95	16.10	15.92	16.04	15.82	16.15
16.06	15.66	15.78	15.99	16.29	16.15	16.19
16.22	16.07	16.13	16.11			

The 13 largest normal scores for a sample of size 25 are 1.965, 1.524, 1.263, 1.067, 0.905, 0.764, 0.637, 0.519, 0.409, 0.303, 0.200, 0.100, and 0. The 12 smallest scores result from placing a negative sign in front of each of the given nonzero scores. Construct a normal probability plot. Does it appear plausible that disk diameter is normally distributed? Explain.

7.90 ● Example 7.32 examined rainfall data for Minneapolis–St. Paul. The square-root transformation was used to obtain a distribution of values that was more nearly symmetric than the distribution of the original data. Another transformation that has been suggested by meteorologists is the cube root: transformed value = (original value)$^{1/3}$. The original values and their cube roots (the transformed values) are given in the following table:

Original	Transformed	Original	Transformed
0.32	0.68	1.51	1.15
0.47	0.78	1.62	1.17
0.52	0.80	1.74	1.20
0.59	0.84	1.87	1.23
0.77	0.92	1.89	1.24
0.81	0.93	1.95	1.25
0.81	0.93	2.05	1.27
0.90	0.97	2.10	1.28
0.96	0.99	2.20	1.30
1.18	1.06	2.48	1.35
1.20	1.06	2.81	1.41
1.20	1.06	3.00	1.44
1.31	1.09	3.09	1.46
1.35	1.11	3.37	1.50
1.43	1.13	4.75	1.68

Construct a histogram of the transformed data. Compare your histogram to those given in Figure 7.38. Which of the cube-root and square-root transformations appear to result in a histogram that is more nearly symmetric?

7.91 The article **"The Distribution of Buying Frequency Rates" (*Journal of Marketing Research* [1980]: 210–216)** reported the results of a $3\frac{1}{2}$-year study of toothpaste purchases. The investigators conducted their research using a national sample of 2071 households and recorded the number of toothpaste purchases for each household participating in the study. The results are given in the following frequency distribution:

Number of Purchases	Number of Households (Frequency)
10 to <20	904
20 to <30	500
30 to <40	258
40 to <50	167
50 to <60	94
60 to <70	56
70 to <80	26
80 to <90	20
90 to <100	13
100 to <110	9
110 to <120	7
120 to <130	6
130 to <140	6
140 to <150	3
150 to <160	0
160 to <170	2

a. Draw a histogram for this frequency distribution. Would you describe the histogram as positively or negatively skewed?

b. Does the square-root transformation result in a histogram that is more nearly symmetric than that of the original data? (Be careful! This one is a bit tricky because you don't have the raw data; transforming the endpoints of the class intervals will result in class intervals that are not necessarily of equal widths, so the histogram of the transformed values will have to be drawn with this in mind.)

7.92 ● The paper **"Temperature and the Northern Distributions of Wintering Birds" (*Ecology* [1991]: 2274–2285)** gave the following body masses (in grams) for 50 different bird species:

7.7	10.1	21.6	8.6	12.0	11.4	16.6	9.4
11.5	9.0	8.2	20.2	48.5	21.6	26.1	6.2
19.1	21.0	28.1	10.6	31.6	6.7	5.0	68.8
23.9	19.8	20.1	6.0	99.6	19.8	16.5	9.0
448.0	21.3	17.4	36.9	34.0	41.0	15.9	12.5
10.2	31.0	21.5	11.9	32.5	9.8	93.9	10.9
19.6	14.5						

a. Draw a histogram based on class intervals 5 to <10, 10 to <15, 15 to <20, 20 to <25, 25 to <30, 30 to <40, 40 to <50, 50 to <100, and 100 to <500. Is a transformation of the data desirable? Explain.

b. Use a calculator or statistical computer package to calculate logarithms of these observations, and construct a histogram. Is the log transformation successful in producing a more symmetric distribution?

c. Consider transformed value $= \dfrac{1}{\sqrt{\text{original value}}}$ and construct a histogram of the transformed data. Does the histogram appear to resemble a normal curve?

7.93 The following figure appeared in the paper **"EDTA-Extractable Copper, Zinc, and Manganese in Soils of the Canterbury Plains" (*New Zealand Journal of Agricultural Research* [1984]: 207–217)**. A large number of topsoil samples were analyzed for manganese (Mn), zinc (Zn), and copper (Cu), and the resulting data were summarized using histograms.

The investigators transformed each data set using logarithms in an effort to obtain more nearly symmetric distributions of values. Do you think the transformations were successful? Explain.

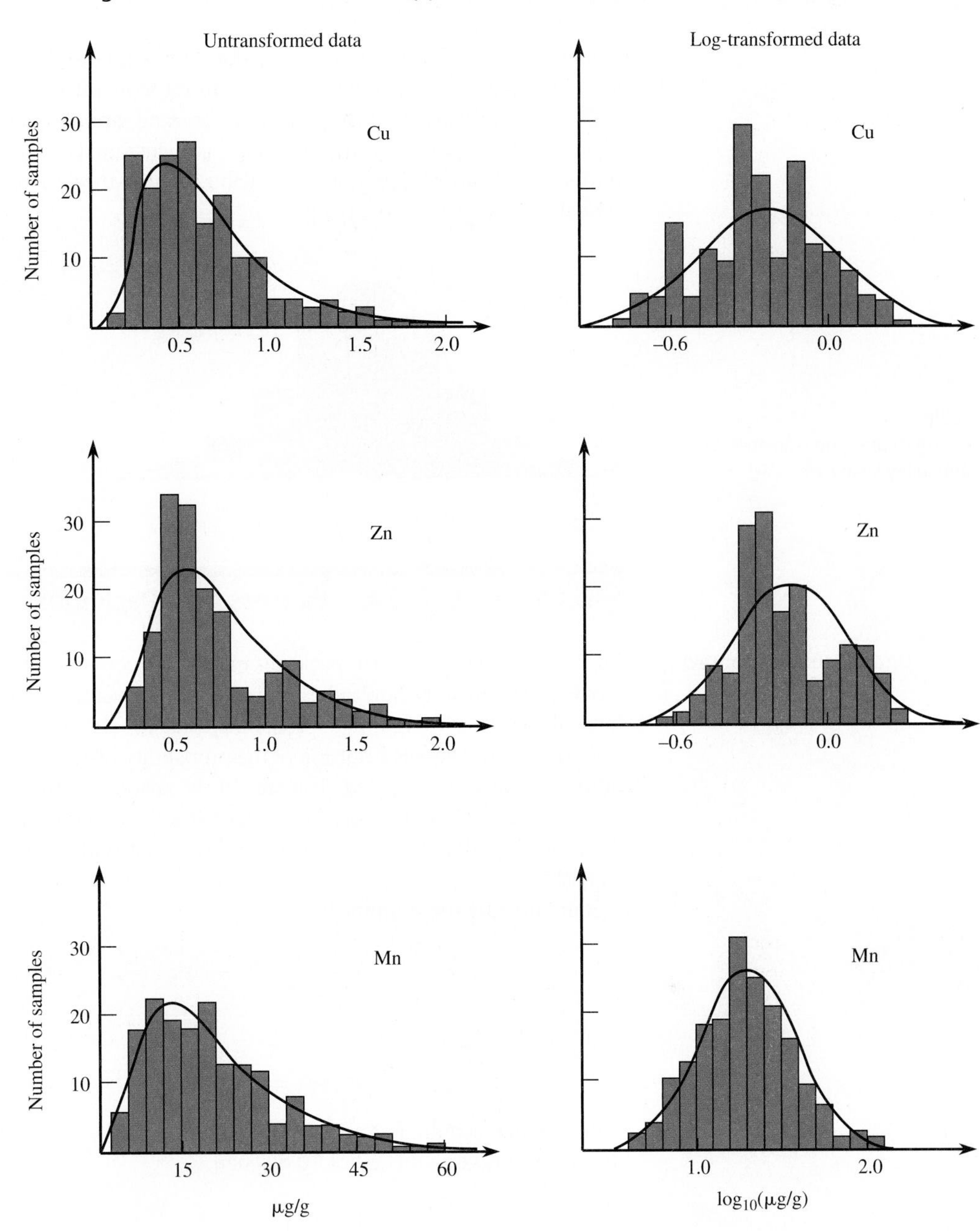

7.8 Using the Normal Distribution to Approximate a Discrete Distribution (Optional)

The distributions of many random variables can be approximated by a carefully chosen normal distribution. In this section, we show how probabilities for some discrete random variables can be approximated using a normal curve. The most important case of this is the approximation of binomial probabilities.

The Normal Curve and Discrete Variables

The probability distribution of a discrete random variable x is represented graphically by a probability histogram. The probability of a particular value is the area of the rectangle centered at that value. Possible values of x are isolated points on the number line, usually whole numbers. For example, if x = the IQ of a randomly selected 8-year-old child, then x is a discrete random variable, because an IQ score must be a whole number.

Often, a probability histogram can be well approximated by a normal curve, as illustrated in Figure 7.43. In such cases, it is customary to say that x has approximately a normal distribution. The normal distribution can then be used to calculate approximate probabilities of events involving x.

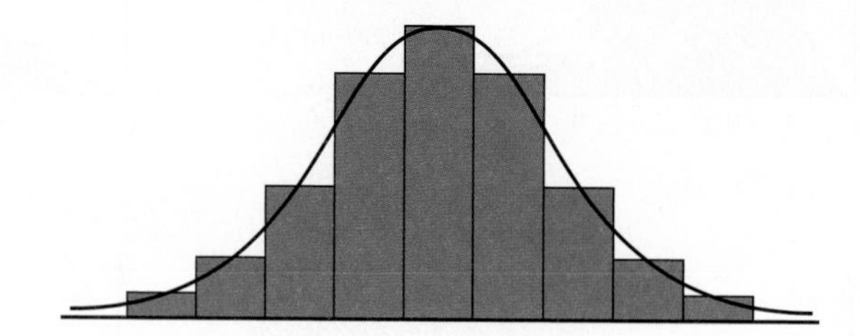

FIGURE 7.43
A normal curve approximation to a probability histogram.

EXAMPLE 7.34 Express Mail Packages

Understand the context)

The number of express mail packages mailed at a certain post office on a randomly selected day is approximately normally distributed with mean 18 and standard deviation 6. Let's first calculate the approximate probability that $x = 20$.

Figure 7.44(a) shows a portion of the probability histogram for x with the approximating normal curve superimposed. The area of the shaded rectangle is $P(x = 20)$. The left edge of this rectangle is at 19.5 on the horizontal scale, and the right edge is at 20.5. Therefore, the desired probability is approximately the area under the normal curve between 19.5 and 20.5.

Standardizing these limits gives

$$\frac{20.5 - 18}{6} = .42 \qquad \frac{19.5 - 18}{6} = .25$$

from which we get

$$P(x = 20) \approx P(.25 < z < .42) = .6628 - .5987 = .0641$$

In a similar fashion, Figure 7.44(b) shows that $P(x \leq 10)$ is approximately the area under the normal curve to the left of 10.5. Then

$$P(x \leq 10) \approx P\left(z \leq \frac{10.5 - 18}{6}\right) = P(z \leq -1.25)$$
$$= .1056$$

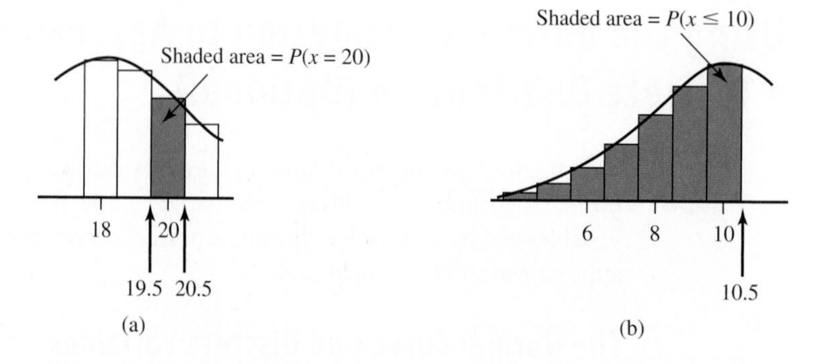

FIGURE 7.44
The normal approximation for Example 7.34.

The calculation of probabilities in Example 7.34 illustrates the use of what is known as a **continuity correction.** Because the rectangle for $x = 10$ extends to 10.5 on the right, we use the normal curve area to the left of 10.5 rather than 10. In general, if possible x values are consecutive whole numbers, then $P(a \le x \le b)$ will be approximately the normal curve area between limits $a - \frac{1}{2}$ and $b + \frac{1}{2}$.

Normal Approximation to a Binomial Distribution

Figure 7.45 shows the probability histograms for two binomial distributions, one with $n = 25$, $p = .4$, and the other with $n = 25$, $p = .1$. For each distribution, we computed $\mu = np$ and $\sigma = \sqrt{np(1-p)}$ and then we superimposed a normal curve with this μ and σ on the corresponding probability histogram.

A normal curve fits the probability histogram well in the first case (Figure 7.45(a)). When this happens, binomial probabilities can be accurately approximated by areas under the normal curve. Because of this, statisticians say that both x (the number of successes) and x/n (the proportion of successes) are approximately normally distributed.

In the second case (Figure 7.45(b)), the normal curve does not give a good approximation because the probability histogram is skewed, whereas the normal curve is symmetric.

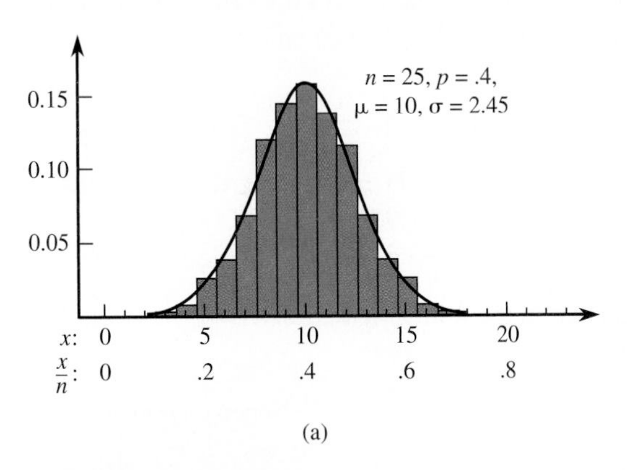

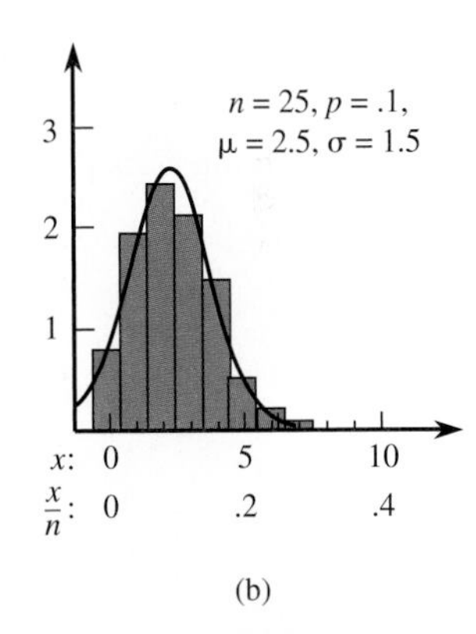

FIGURE 7.45
Normal approximations to binomial distributions.

Let x be a binomial random variable based on n trials and success probability p, so that

$$\mu = np \text{ and } \sigma = \sqrt{np(1-p)}$$

If n and p are such that

$$np \ge 10 \text{ and } n(1-p) \ge 10$$

then the distribution of x is approximately normal.

Combining this result with the continuity correction implies that

$$P(a \le x \le b) = P\left(\frac{a - \frac{1}{2} - \mu}{\sigma} \le z \le \frac{b + \frac{1}{2} - \mu}{\sigma}\right)$$

That is, the probability that x is between a and b inclusive is approximately the area under the approximating normal curve between $a - \frac{1}{2}$ and $b + \frac{1}{2}$. Similarly,

$$P(x \le b) \approx P\left(z \le \frac{b + \frac{1}{2} - \mu}{\sigma}\right) \qquad P(a \le x) \approx P\left(\frac{a - \frac{1}{2} - \mu}{\sigma} \le z\right)$$

When either $np < 10$ or $n(1-p) < 10$, the binomial distribution is too skewed for the normal approximation to give reasonably accurate probability estimates.

EXAMPLE 7.35 Premature Babies

Understand the context

Premature babies are those born before 37 weeks, and those born before 34 weeks are most at risk. The paper **"Some Thoughts on the True Value of Ultrasound"** (***Ultrasound in Obstetrics and Gynecology*** **[2007]: 671–674)** reported that 2% of births in the United States occur before 34 weeks.

Suppose that 1000 births are randomly selected and that the number of these births that occurred prior to 34 weeks, x, is to be determined. Because

$$np = 1000(.02) = 20 \geq 10$$
$$n(1-p) = 1000(.98) = 980 \geq 10$$

the distribution of x is approximately normal with

$$\mu = np = 1000(.02) = 20$$
$$\sigma = \sqrt{np(1-p)} = \sqrt{1000(.02)(.98)} = \sqrt{19.60} = 4.427$$

The probability that the number of babies in a sample of 1000 born prior to 34 weeks will be between 10 and 25 (inclusive) is

Do the work

$$\begin{aligned} P(10 \leq x \leq 25) &= P\left(\frac{9.5-20}{4.427} \leq z \leq \frac{25.5-20}{4.427}\right) \\ &= P(-2.37 \leq z \leq 1.29) \\ &= .9015 - .0089 \\ &= .8926 \end{aligned}$$

as shown in the following figure:

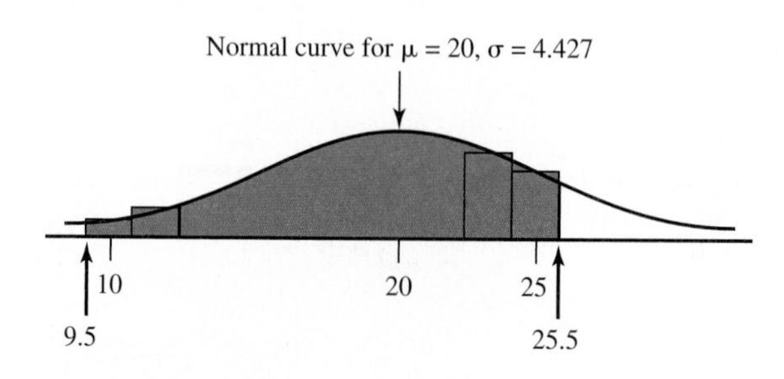

■

EXERCISES 7.94 - 7.104

7.94 Let x denote the IQ of an individual selected at random from a certain population. The value of x must be a whole number. Suppose that the distribution of x can be approximated by a normal distribution with mean value 100 and standard deviation 15. Approximate the following probabilities. (Hint: See Example 7.34.)

a. $P(x = 100)$

b. $P(x \leq 110)$

c. $P(x < 110)$ (Hint: $x < 110$ is the same as $x \leq 109$.)

d. $P(75 \leq x \leq 125)$

7.95 Suppose that the distribution of

x = the number of items produced by an assembly line during an 8-hour shift

can be approximated by a normal distribution with mean value 150 and standard deviation 10.

a. What is the approximate probability that the number of items produced is at most 120?

b. What is the approximate probability that at least 125 items are produced?

c. What is the approximate probability that between 135 and 160 (inclusive) items are produced?

7.96 The number of vehicles leaving a turnpike at a certain exit during a particular time period has approximately a normal distribution with mean value 500 and standard deviation 75. What is the approximate probability that the number of cars exiting during this period is

a. at least 650?

b. strictly between 400 and 550? (*Strictly* means that the values 400 and 550 are not included.)

c. between 400 and 550 (inclusive)?

7.97 Suppose that x has a binomial distribution with $n = 50$ and $p = .6$, so that $\mu = np = 30$ and $\sigma = \sqrt{np(1-p)} = 3.4641$. Approximate the following probabilities using the normal approximation with the continuity correction. (Hint: See Example 7.35.)

a. $P(x = 30)$

b. $P(x = 25)$

c. $P(x \leq 25)$

d. $P(25 \leq x \leq 40)$

e. $P(25 < x < 40)$ (Hint: $25 < x < 40$ is the same as $26 \leq x \leq 39$.)

7.98 Symptom validity tests (SVTs) are sometimes used to confirm diagnosis of psychiatric disorders. The paper **"Developing a Symptom Validity Test for Post-Traumatic Stress Disorder: Application of the Binomial Distribution" (*Journal of Anxiety Disorders* [2008]: 1297–1302)** investigated the use of SVTs in the diagnosis of post-traumatic stress disorder.

One SVT proposed is a 60-item test (called the MENT test), where each item has only a correct or incorrect response. The MENT test is designed so that responses to the individual questions can be considered independent of one another. For this reason, the authors of the paper believe that the score on the MENT test can be viewed as a binomial random variable with $n = 60$.

The MENT test is designed to help in distinguishing fictitious claims of post-traumatic stress disorder. The items on the MENT test are written so that the correct response to an item should be relatively obvious, even to people suffering from stress disorders. Researchers have found that a patient with a fictitious claim of stress disorder will try to "fake" the test, and that the probability of a correct response to an item for these patients is .7 (compared to .96 for other patients).

The authors used a normal approximation to the binomial distribution with $n = 60$ and $p = .70$ to approximate various probabilities of interest, where

x = number of correct responses on the MENT test for a patient who is trying to fake the test.

a. Verify that it is appropriate to use a normal approximation to the binomial distribution in this situation.

b. Approximate the following probabilities:

i. $P(x = 42)$

ii. $P(x < 42)$

iii. $P(x \leq 42)$

c. Explain why the probabilities computed in Part (b) are not all equal.

d. The authors computed the exact binomial probability of a score of 42 or less for someone who is not faking the test. Using $p = .96$, they found

$p(x \leq 42) = .000000000013$

Explain why the authors computed this probability using the binomial formula rather than using a normal approximation.

e. The authors propose that someone who scores 42 or less on the MENT exam is faking the test. Explain why this is reasonable, using some of the probabilities from Parts (b) and (d) as justification.

7.99 Studies have found that women diagnosed with cancer in one breast also sometimes have cancer in the other breast that was not initially detected by mammogram or physical examination (**"MRI Evaluation of the Contralateral Breast in Women with Recently Diagnosed Breast Cancer," *The New England Journal of Medicine* [2007]: 1295–1303**).

To determine if magnetic resonance imaging (MRI) could detect missed tumors in the other breast, 969 women diagnosed with cancer in one breast had an MRI exam. The MRI detected tumors in the other breast in 30 of these women.

a. Use $p = \frac{30}{969} = .031$ as an estimate of the probability that a woman diagnosed with cancer in one breast has an undetected tumor in the other breast. Consider a random sample of 500 women diagnosed with cancer in one breast. Explain why it is reasonable to think that the random variable

x = number in the sample who have an undetected tumor in the other breast has a binomial distribution with $n = 500$ and $p = .031$.

b. Is it reasonable to use the normal distribution to approximate probabilities for the random variable x defined in Part (a)? Explain why or why not.

c. Approximate the following probabilities:

i. $P(x < 10)$

ii. $P(10 \leq x \leq 25)$

iii. $P(x > 20)$

d. For each of the probabilities in Part (c), write a sentence interpreting the probability.

7.100 Seventy percent of the bicycles sold by a certain store are mountain bikes. Among 100 randomly selected bike purchases, what is the approximate probability that

a. At most 75 are mountain bikes?

b. Between 60 and 75 (inclusive) are mountain bikes?

c. More than 80 are mountain bikes?

d. At most 30 are not mountain bikes?

7.101 Suppose that 25% of the fire alarms in a large city are false alarms. Let x denote the number of false alarms in a random sample of 100 alarms. Approximate the following probabilities:

a. $P(20 \leq x \leq 30)$

b. $P(20 < x < 30)$

c. $P(x \geq 35)$

d. The probability that x is farther than 2 standard deviations from its mean value

7.102 Suppose that 65% of all registered voters in a certain area favor a 7-day waiting period before purchase of a handgun. Among 225 randomly selected registered voters, what is the approximate probability that

a. At least 150 favor such a waiting period?

b. More than 150 favor such a waiting period?

c. Fewer than 125 favor such a waiting period?

7.103 Flashlight bulbs manufactured by a certain company are sometimes defective.

a. If 5% of all such bulbs are defective, could the techniques of this section be used to approximate the probability that at least five of the bulbs in a random sample of size 50 are defective? If so, calculate this probability; if not, explain why not.

b. Reconsider the question posed in Part (a) for the probability that at least 20 bulbs in a random sample of size 500 are defective.

7.104 A company that manufactures mufflers for cars offers a lifetime warranty on its products, provided that ownership of the car does not change. Suppose that only 20% of its mufflers are replaced under this warranty.

a. In a random sample of 400 purchases, what is the approximate probability that between 75 and 100 (inclusive) mufflers are replaced under warranty?

b. Among 400 randomly selected purchases, what is the approximate probability that at most 70 mufflers are ultimately replaced under warranty?

c. If you were told that fewer than 50 among 400 randomly selected purchases were ever replaced under warranty, would you question the 20% figure? Explain.

Bold exercises answered in back ● Data set available online ▼ Video Solution available

ACTIVITY 7.1 Is It Real?

Background: Three students were asked to complete an assignment that asked them to do the following:

a. Flip a coin 30 times, and note the number of heads observed in the 30 flips.

b. Repeat Step (a) 100 times, to obtain 100 observations of the random variable x = number of heads in 30 flips.

c. Construct a dotplot of the 100 x values.

Because this was a tedious assignment, one or more of the did not really carry out the coin flipping and just made up 100 values of x that they thought would look "real." The dotplots produced by these three students are shown at the top of the next page.

1. Do you think that any of the three students made up the x values shown in their dotplot? If so, which ones, and what about the dotplot makes you think the student did not actually do the coin flipping?

2. Working as a group, each student in your class should flip a coin 30 times and note the number of heads in the 30 tosses. If there are fewer than 50 students in the class, each student should repeat this process until there are a total of at least 50 observations of x = number of heads in 30 flips. Using the data from the entire class, construct a dotplot of the x values.

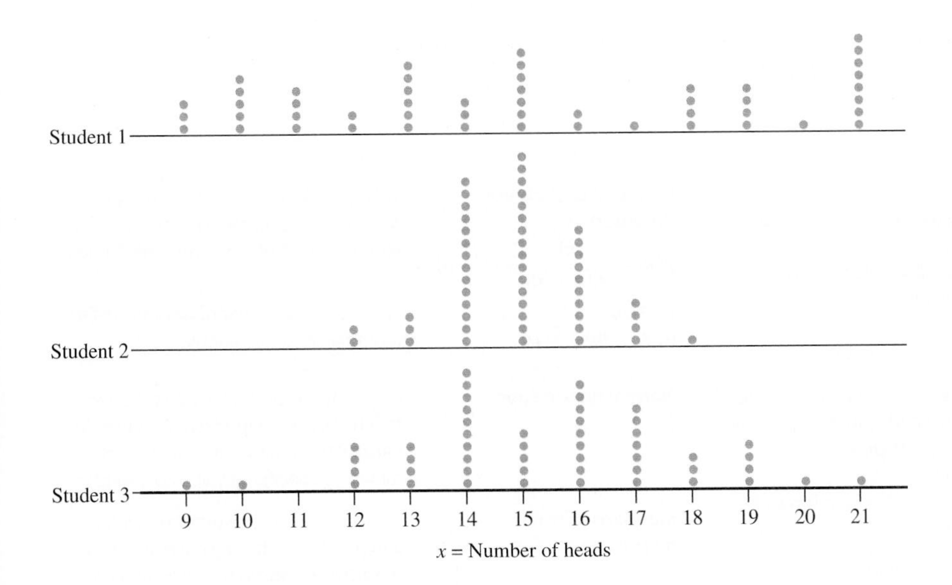

3. After looking at the dotplot in Step 2 that resulted from actually flipping a coin 30 times and observing number of heads, reconsider your answers in Question 1. For each of the three students, explain why you now think that he or she did or did not actually do the coin flipping.

ACTIVITY 7.2 Rotten Eggs?

Background: ***The Salt Lake Tribune*** **(October 11, 2002)** printed the following account of an exchange between a restaurant manager and a health inspector:

> The recipe calls for four fresh eggs for each quiche. A Salt Lake County Health Department inspector paid a visit recently and pointed out that research by the Food and Drug Administration indicates that one in four eggs carries salmonella bacterium, so restaurants should never use more than three eggs when preparing quiche. The manager on duty wondered aloud if simply throwing out three eggs from each dozen and using the remaining nine in four-egg quiches would serve the same purpose.

1. Working in a group or as a class, discuss the folly of the above statement!
2. Suppose the following argument is made for three-egg quiches rather than four-egg quiches: Let $x \leq$ number of eggs that carry salmonella. Then

   $$p(0) = p(x = 0) = (0.75)^3 = .422$$

   for three-egg quiches and

   $$p(0) = p(x = 0) = (0.75)^4 = .316$$

   for four-egg quiches. What assumption must be made to justify these probability calculations? Do you think this is reasonable or not? Explain.
3. Suppose that a carton of one dozen eggs does happen to have exactly three eggs that carry salmonella and that the manager does as he proposes: selects three eggs at random and throws them out, then uses the remaining nine eggs in four-egg quiches. Let x = number of eggs that carry salmonella among four eggs selected at random from the remaining nine.

   Working with a partner, conduct a simulation to approximate the distribution of x by carrying out the following sequence of steps:

   a. Take 12 identical slips of paper and write "Good" on nine of them and "Bad" on the remaining three. Place the slips of paper in a paper bag or some other container.
   b. Mix the slips and then select three at random and remove them from the bag.
   c. Mix the remaining slips and select four "eggs" from the bags.
   d. Note the number of bad eggs among the four selected. (This is an observed x value.)
   e. Replace all slips, so that the bag now contains all 12 "eggs."
   f. Repeat Steps (b)–(d) at least 10 times, each time recording the observed x value.
4. Combine the observations from your group with those from the other groups. Use the resulting data to approximate the distribution of x. Comment on the resulting distribution in the context of the risk of salmonella exposure if the manager's proposed procedure is used.

SUMMARY Key Concepts and Formulas

TERM OR FORMULA	COMMENT
Random variable: discrete or continuous	A numerical variable with a value determined by the outcome of a chance experiment. A random variable is discrete if its possible values are isolated points along the number line and continuous if its possible values form an entire interval on the number line.
Probability distribution of a discrete random variable *x*	A formula, table, or graph that gives the probability associated with each possible *x* value. Conditions on $p(x)$ are (1) $p(x) \geq 0$, and (2) $\Sigma p(x) = 1$, where the sum is over all possible *x* values.
Probability distribution of a continuous random variable *x*	Specified by a smooth (density) curve for which the total area under the curve is 1. The probability $P(a < x < b)$ is the area under the curve and above the interval from *a* to *b*; this is also $P(a \leq x \leq b)$.
μ_x **and** σ_x	The mean and standard deviation, respectively, of a random variable *x*. These quantities describe the center and extent of spread about the center of the variable's probability distribution.
$\mu_x = \Sigma xp(x)$	The mean value of a discrete random variable *x*. It locates the center of the variable's probability distribution.
$\sigma_x^2 = \Sigma(x - \mu_x)^2 p(x)$ $\sigma_x = \sqrt{\sigma_x^2}$	The variance and standard deviation, respectively, of a discrete random variable. These are measures of the extent to which the variable's distribution spreads out about the mean μ_x.
Binomial probability distribution $p(x) = \frac{n!}{x!(n-x)!} p^x(1-p)^{n-x}$	This formula gives the probability of observing *x* successes ($x = 0, 1, \ldots, n$) among *n* trials of a binomial experiment.
$\mu_x = np$ $\sigma_x = \sqrt{np(1-p)}$	The mean and standard deviation of a binomial random variable.
Normal distribution	A continuous probability distribution that has a bell-shaped density curve. A particular normal distribution is determined by specifying values of μ and σ.
Standard normal distribution	This is the normal distribution with $\mu = 0$ and $\sigma = 1$. The density curve is called the *z* curve, and *z* is the letter commonly used to denote a variable having this distribution. Areas under the *z* curve to the left of various values are given in Appendix Table 2.
$z = \frac{x - \mu}{\sigma}$	*z* is obtained by "standardizing": subtracting the mean and then dividing by the standard deviation. When *x* has a normal distribution, *z* has a standard normal distribution.
Normal probability plot	A graph used to judge the plausibility of the assumption that a sample has been selected from a normal population distribution. If the plot is reasonably straight, this assumption is reasonable.
Normal approximation to the binomial distribution	When both $np \geq 10$ and $n(1-p) \geq 10$, binomial probabilities are well approximated by corresponding areas under a normal curve with $\mu = np$ and $\sigma = \sqrt{np(1-p)}$.

CHAPTER REVIEW Exercises 7.105 - 7.126

7.105 Let *x* denote the duration of a randomly selected pregnancy (the time elapsed between conception and birth). Accepted values for the mean value and standard deviation of *x* are 266 days and 16 days, respectively. Suppose that the probability distribution of *x* is (approximately) normal.

a. What is the probability that the duration of pregnancy is between 250 and 300 days?

b. What is the probability that the duration of pregnancy is at most 240 days?

c. What is the probability that the duration of pregnancy is within 16 days of the mean duration?

d. A "Dear Abby" column dated January 20, 1973, contained a letter from a woman who stated that the duration of her pregnancy was exactly 310 days. (She wrote that the last visit with her husband, who was in the Navy, occurred 310 days before birth.) What is the probability that duration of pregnancy is 310 days or more? Does this probability make you a bit skeptical of the claim?

e. Some insurance companies will pay the medical expenses associated with childbirth only if the insurance has been in effect for more than 9 months (275 days). This restriction is designed to ensure that the insurance company pays benefits for only those pregnancies for which conception occurred during coverage. Suppose that conception occurred 2 weeks after coverage began. What is the probability that the insurance company will refuse to pay benefits because of the 275-day insurance requirement?

7.106 A soft-drink machine dispenses only regular Coke and Diet Coke. Sixty percent of all purchases from this machine are diet drinks. The machine currently has 10 cans of each type. If 15 customers want to purchase drinks before the machine is restocked, what is the probability that each of the 15 is able to purchase the type of drink desired? (Hint: Let x denote the number among the 15 who want a diet drink. For which possible values of x is everyone satisfied?)

7.107 A business has six customer service telephone lines. Let x denote the number of lines in use at a specified time. Suppose that the probability distribution of x is as follows:

x	0	1	2	3	4	5	6
$p(x)$	.10	.15	.20	.25	.20	.06	.04

Write each of the following events in terms of x, and then calculate the probability of each one:

a. At most three lines are in use

b. Fewer than three lines are in use

c. At least three lines are in use

d. Between two and five lines (inclusive) are in use

e. Between two and four lines (inclusive) are not in use

f. At least four lines are not in use

7.108 Refer to the probability distribution of the previous exercise.

a. Calculate the mean value and standard deviation of x.

b. What is the probability that the number of lines in use is farther than 3 standard deviations from the mean value?

7.109 A new battery's voltage may be acceptable (A) or unacceptable (U). A certain flashlight requires two batteries, so batteries will be independently selected and tested until two acceptable ones have been found. Suppose that 80% of all batteries have acceptable voltages, and let y denote the number of batteries that must be tested.

a. What is $p(2)$, that is, $P(y = 2)$?

b. What is $p(3)$? (Hint: There are two different outcomes that result in $y = 3$.)

c. In order to have $y = 5$, what must be true of the fifth battery selected? List the four outcomes for which $y = 5$, and then determine $p(5)$.

d. Use the pattern in your answers for Parts (a)–(c) to obtain a general formula for $p(y)$.

7.110 A pizza company advertises that it puts 0.5 pounds of real mozzarella cheese on its medium pizzas. Suppose that the amount of cheese on a randomly selected medium pizza is normally distributed with a mean value of 0.5 pounds and a standard deviation of 0.025 pounds.

a. What is the probability that the amount of cheese on a medium pizza is between 0.525 and 0.550 pounds?

b. What is the probability that the amount of cheese on a medium pizza exceeds the mean value by more than 2 standard deviations?

c. What is the probability that three randomly selected medium pizzas all have at least 0.475 pounds of cheese?

7.111 Suppose that fuel efficiency for a particular model car under specified conditions is normally distributed with a mean value of 30.0 mpg and a standard deviation of 1.2 mpg.

a. What is the probability that the fuel efficiency for a randomly selected car of this type is between 29 and 31 mpg?

b. Would it surprise you to find that the efficiency of a randomly selected car of this model is less than 25 mpg?

c. If three cars of this model are randomly selected, what is the probability that all three have efficiencies exceeding 32 mpg?

d. Find a number c^* such that 95% of all cars of this model have efficiencies exceeding c^* (i.e., $P(x > c^*) = .95$).

7.112 A coin is flipped 25 times. Let x be the number of flips that result in heads (H). Consider the following rule for deciding whether or not the coin is fair:

Judge the coin fair if $8 \leq x \leq 17$.
Judge the coin biased if either $x \leq 7$ or $x \geq 18$.

a. What is the probability of judging the coin biased when it is actually fair?

b. What is the probability of judging the coin fair when $P(H) = .9$, so that there is a substantial bias? Repeat for $P(H) = .1$.

c. What is the probability of judging the coin fair when P(H) = .6? when P(H) = .4? Why are these probabilities so large compared to the probabilities in Part (b)?

d. What happens to the "error probabilities" of Parts (a) and (b) if the decision rule is changed so that the coin is judged fair if $7 \leq x \leq 18$ and unfair otherwise? Is this a better rule than the one first proposed? Explain.

7.113 The probability distribution of x, the number of defective tires on a randomly selected automobile checked at a certain inspection station, is given in the following table:

x	0	1	2	3	4
$p(x)$	.54	.16	.06	.04	.20

The mean value of x is $\mu_x = 1.2$. Calculate the values of σ_x^2 and σ_x.

7.114 The amount of time spent by a statistical consultant with a client at their first meeting is a random variable having a normal distribution with a mean value of 60 minutes and a standard deviation of 10 minutes.

a. What is the probability that more than 45 minutes is spent at the first meeting?

b. What amount of time is exceeded by only 10% of all clients at a first meeting?

c. If the consultant assesses a fixed charge of $10 (for overhead) and then charges $50 per hour, what is the mean revenue from a client's first meeting?

7.115 The lifetime of a certain brand of battery is normally distributed with a mean value of 6 hours and a standard deviation of 0.8 hours when it is used in a particular DVD player. Suppose that two new batteries are independently selected and put into the player. The player ceases to function as soon as one of the batteries fails.

a. What is the probability that the player functions for at least 4 hours?

b. What is the probability that the DVD player works for at most 7 hours?

c. Find a number c^* such that only 5% of all DVD players will function without battery replacement for more than c^* hours.

7.116 A machine producing vitamin E capsules operates so that the actual amount of vitamin E in each capsule is normally distributed with a mean of 5 mg and a standard deviation of 0.05 mg. What is the probability that a randomly selected capsule contains less than 4.9 mg of vitamin E? at least 5.2 mg?

7.117 The ***Wall Street Journal* (February 15, 1972)** reported that General Electric was sued in Texas for sex discrimination over a minimum height requirement of 5 ft. 7 in. The suit claimed that this restriction eliminated more than 94% of adult females from consideration. Let x represent the height of a randomly selected adult woman. Suppose that x is approximately normally distributed with mean 66 inches (5 ft. 6 in.) and standard deviation 2 inches.

a. Is the claim that 94% of all women are shorter than 5 ft. 7 in. correct?

b. What proportion of adult women would be excluded from employment as a result of the height restriction?

7.118 The longest "run" of S's in the sequence *SSFSSSSFFS* has length 4, corresponding to the S's on the fourth, fifth, sixth, and seventh positions. Consider a binomial experiment with $n = 4$, and let y be the length in the longest run of S's.

a. When $p = .5$, the 16 possible outcomes are equally likely. Determine the probability distribution of y in this case (first list all outcomes and the y value for each one). Then calculate μ_y.

b. Repeat Part (a) for the case $p = .6$.

c. Let z denote the longest run of either S's or F's. Determine the probability distribution of z when $p = .5$.

7.119 Two sisters, Allison and Teri, have agreed to meet between 1 and 6 P.M. on a particular day. In fact, Allison is equally likely to arrive at exactly 1 P.M., 2 P.M., 3 P.M., 4 P.M., 5 P.M., or 6 P.M. Teri is also equally likely to arrive at each of these six times, and Allison's and Teri's arrival times are independent of one another. Thus there are 36 equally likely (Allison, Teri) arrival-time pairs, for example, (2, 3) or (6, 1). Suppose that the first person to arrive waits until the second person arrives; let w be the amount of time the first person has to wait.

a. What is the probability distribution of w?

b. On average, how much time do you expect to elapse between the two arrivals?

7.120 Four people—a, b, c, and d—are waiting to give blood. Of these four, a and b have type AB blood, whereas c and d do not. An emergency call has just come in for some type AB blood. If blood donations are taken one by one from the four people in random order and x is the number of donations needed to obtain an AB individual (so possible x values are 1, 2, and 3), what is the probability distribution of x?

7.121 Kyle and Lygia are going to play a series of Trivial Pursuit games. The first person to win four games will be declared the winner. Suppose that outcomes of successive games are independent and that the probability of Lygia winning any particular game is .6. Define a random variable x as the number of games played in the series.

a. What is $p(4)$? (Hint: Either Kyle or Lygia could win four straight games.)

b. What is $p(5)$? (Hint: For Lygia to win in exactly five games, what has to happen in the first four games and in Game 5?)

c. Determine the probability distribution of x.

d. On average, how many games will the series last?

7.122 Refer to the previous exercise, and let y be the number of games won by the series loser. Determine the probability distribution of y.

7.123 Suppose that your statistics professor tells you that the scores on a midterm exam were approximately normally distributed with a mean of 78 and a standard deviation of 7. The top 15% of all scores have been designated A's. Your score is 89. Did you receive an A? Explain.

7.124 Suppose that the pH of soil samples taken from a certain geographic region is normally distributed with a mean pH of 6.00 and a standard deviation of 0.10. If the pH of a randomly selected soil sample from this region is determined, answer the following questions about it:

a. What is the probability that the resulting pH is between 5.90 and 6.15?

b. What is the probability that the resulting pH exceeds 6.10?

c. What is the probability that the resulting pH is at most 5.95?

d. What value will be exceeded by only 5% of all such pH values?

7.125 The lightbulbs used to provide exterior lighting for a large office building have an average lifetime of 700 hours. If length of life is approximately normally distributed with a standard deviation of 50 hours, how often should all the bulbs be replaced so that no more than 20% of the bulbs will have already burned out?

7.126 There are approximately 40,000 travel agencies in the United States, of which 11,000 are members of the American Society of Travel Agents (booking a tour through an ASTA member increases the likelihood of a refund in the event of cancellation).

a. If x is the number of ASTA members among 5000 randomly selected agencies, could you use the methods of Section 7.8 to approximate $P(1200 < x < 1400)$? Why or why not?

b. In a random sample of 100 agencies, what are the mean value and standard deviation of the number of ASTA members?

c. If the sample size in Part (b) is doubled, does the standard deviation double? Explain.

Bold exercises answered in back ● Data set available online ▼ Video Solution available

TECHNOLOGY NOTES

Finding Normal Probabilities

JMP

1. After opening a new data table, click **Rows** then select **Add rows**
2. Type **1** in the box next to **How many rows to add:**
3. Click **OK**
4. Double-click on the **Column 1** heading
5. Click **Column Properties** and select **Formula**
6. Click **Edit Formula**
7. In the box under **Functions (grouped)** click **Probability** then select **Normal Distribution**
8. In the white box at the bottom half of the screen, double-click in the red box around x

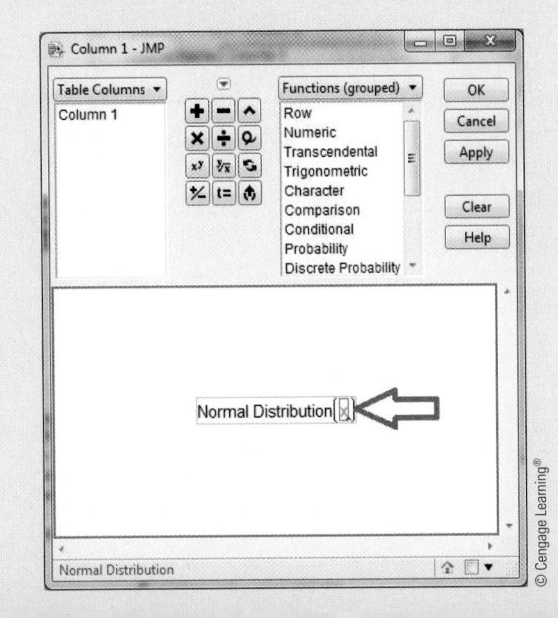

9. Type the value that you would like to find a probability for and click **OK**
10. Click **OK**

Note: This procedure outputs the value for $P(X \leq x)$. If you want to find the value for $P(X \geq x)$ you will need to subtract the output from one.

Minitab

1. Click **Calc** then click **Probability Distributions** then click **Normal...**
2. In the box next to **Mean:** type the mean of the normal distribution that you are working with
3. In the box next to **Standard deviation:** type the standard deviation of the normal distribution that you are working with
4. Click the radio button next to **Input constant:**
5. Click in the box next to **Input constant:** and type the value that you are finding the probability for
6. Click **OK**

Note: This procedure outputs the value for $P(X \leq x)$. If you want to find the value for $P(X \geq x)$ you will need to subtract the output from one.

Note: You may also type a column of values for which you would like to find probabilities (these must ALL be from the SAME distribution) and use the **Input Column** option to find probabilities for each value in the column selected.

SPSS

1. Type in the values that you would like to find the probabilities for in one column (this column will be automatically titled VAR00001)
2. Click **Transform** and select **Compute Variable...**
3. In the box under **Target Variable:** type NormProb (this will be the title of a new column that lists the cumulative probabilities for each value in VAR00001)
4. Under **Function group:** select **CDF & Noncentral CDF**
5. Under **Functions and Special Variables:** double-click **Cdf.Normal**
6. In the box under **Numeric Expression:** you should see CDF.NORMAL(?,?,?)
7. Highlight the first ? and double-click VAR00001
8. Highlight the second ? and type the mean of the normal distribution that you are working with
9. Highlight the third ? and type the standard deviation of the normal distribution that you are working with

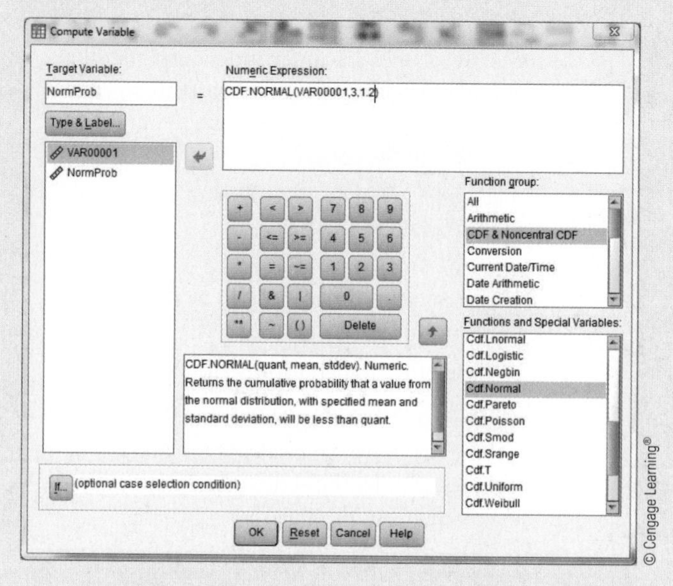

10. Click **OK**

Note: This procedure outputs the value for $P(X \leq x)$. If you want to find the value for $P(X \geq x)$ you will need to subtract the output from one.

Excel

1. Click in an empty cell
2. Click the **Formulas** ribbon
3. Select **Insert Function**
4. Select **Statistical** from the drop-down menu for category
5. Select **NORMDIST** from the **Select a function:** box
6. Click **OK**
7. Click in the box next to **X** and type the data value that you would like to find the probability for
8. Click in the box next to **Mean** and type the mean of the Normal distribution that you are working with
9. Click in the box next to **Standard_dev** and type the standard deviation of the Normal distribution that you are working with
10. Click in the box next to **Cumulative** and type **TRUE**
11. Click **OK**

Note: This procedure outputs the value for $P(X \leq x)$. If you want to find the value for $P(X \geq x)$ you will need to subtract the output from one.

TI-83/84

$P(x_1 < X < x_2)$

1. Press the **2nd** key then press the **VARS** key
2. Highlight **normalcdf(** and press **ENTER**
3. Type in the lower bound x_1
4. Type a comma
5. Type in the upper bound x_2
6. Type a comma
7. Type in the value of the mean
8. Type a comma
9. Type in the value of the standard deviation

10. Type a right parenthesis
11. Press **ENTER**

Note: If you are looking for $P(X < x)$, use a lower bound of −10,000,000. If you are looking for $P(X > x)$, use an upper bound of 10,000,000.

TI-Nspire

$P(x_1 < X < x_2)$

1. Enter the Scratchpad
2. Press the **menu** key and select **5:Probablity** then **5:Distributions** then **2:Normal Cdf…** then press **enter**
3. In the box next to **Lower Bound** type the value for x_1
4. In the box next to **Upper Bound** type the value for x_2
5. In the box next to μ type the value of the mean
6. In the box next to σ type the value of the standard deviation
7. Press **OK**

Note: If you are looking for $P(X < x)$, use a lower bound of −10,000,000. If you are looking for $P(X > x)$, use an upper bound of 10,000,000.

Finding Binomial Probabilities

JMP

$P(X \leq x)$

1. After opening a new data table, click **Rows** then select **Add rows**
2. Type **1** in the box next to **How many rows to add:**
3. Click **OK**
4. Double-click on the **Column 1** heading
5. Click **Column Properties** and select **Formula**
6. Click **Edit Formula**
7. In the box under **Functions (grouped)** click **Discrete Probability** then select **Binomial Distribution**
8. In the white box at the bottom half of the screen, double-click in the red box around p and type the value for the success probability, **p**, and press **enter** on your keyboard
9. Double-click in the red box around n and type the value for the number of trials, **n**, and press **enter** on your keyboard
10. Double-click in the red box around **k** and type the value for the number of successes for which you would like to find the probability
11. Click **OK**
12. Click **OK**

$P(X = x)$

1. After opening a new data table, click **Rows** then select **Add rows**
2. Type **1** in the box next to **How many rows to add:**
3. Click **OK**
4. Double-click on the **Column 1** heading
5. Click **Column Properties** and select **Formula**
6. Click **Edit Formula**
7. In the box under **Functions (grouped)** click **Discrete Probability** then select **Binomial Probability**
8. In the white box at the bottom half of the screen, double-click in the red box around **p** and type the value for the success probability, p, and press **enter** on your keyboard
9. Double-click in the red box around **n** and type the value for the number of trials, n, and press **enter** on your keyboard
10. Double-click in the red box around **k** and type the value for the number of successes for which you would like to find the probability
11. Click **OK**
12. Click **OK**

Minitab

$P(X \leq x)$

1. Click **Calc** then click **Probability Distributions** then click **Binomial…**
2. In the box next to **Number of Trials:** input the value for n, the total number of trials
3. In the box next to **Probability of Success:** input the value for p, the success probability
4. Click the radio button next to **Input Constant**
5. In the box next to **Input Constant:** type the value that you want to find the probability for
6. Click **OK**

Note: You may also type a column of values for which you would like to find probabilities (these must ALL be from the SAME distribution) and use the **Input Column** option to find probabilities for each value in the column selected.

$P(X = x)$

1. Click **Calc** then click **Probability Distributions** then click **Binomial…**
2. Click the radio button next to **Probability**
3. In the box next to **Number of Trials:** input the value for n, the total number of trials
4. In the box next to **Probability of Success:** input the value for p, the success probability
5. Click the radio button next to **Input Constant**
6. In the box next to **Input Constant:** type the value that you want to find the probability for
7. Click **OK**

Note: You may also type a column of values for which you would like to find probabilities (these must ALL be from the SAME distribution) and use the **Input Column** option to find probabilities for each value in the column selected.

SPSS

$P(X = x)$

1. Type in the values that you would like to find the probabilities for in one column (this column will be automatically titled VAR00001)
2. Click **Transform** and select **Compute Variable…**
3. In the box under **Target Variable:** type **BinomProb** (this will be the title of a new column that lists the cumulative probabilities for each value in VAR00001)

4. Under **Function group:** select **CDF & Noncentral CDF**
5. Under **Functions and Special Variables:** double-click **Cdf.Binom**
6. In the box under **Numeric Expression:** you should see CDF.BINOM(?,?,?)
7. Highlight the first ? and double-click VAR00001
8. Highlight the second ? and type the value for n, the total number of trials
9. Highlight the third ? and type success probability, p
10. Click **OK**

$P(X = x)$

1. Type in the values that you would like to find the probabilities for in one column (this column will be automatically titled VAR00001)
2. Click **Transform** and select **Compute Variable…**
3. In the box under **Target Variable:** type BinProb (this will be the title of a new column that lists the cumulative probabilities for each value in VAR00001)
4. Under **Function group:** select **PDF & Noncentral PDF**
5. Under **Functions and Special Variables:** double-click Pdf.Binom
6. In the box under **Numeric Expression:** you should see PDF.BINOM(?,?,?)
7. Highlight the first ? and double-click VAR00001
8. Highlight the second ? and type the value for n, the total number of trials
9. Highlight the third ? and type success probability, p
10. Click **OK**

Excel

1. Click in an empty cell
2. Click the **Formulas** ribbon
3. Select **Insert Function**
4. Select **Statistical** from the drop-down menu for category
5. Select **BINOMDIST** from the **Select a function:** box
6. Click **OK**
7. Click in the box next to **Number_s** and type the number of successes that you are finding a probability for
8. Click in the box next to **Trials** and type the value for n, the total number of trials
9. Click in the box next to **Probability_s** and type success probability, p
10. Click in the box next to **Cumulative** and type **TRUE** if you are finding $P(X \leq x)$ or type **FALSE** if you are finding $P(X = x)$
11. Click **OK**

Note: This procedure outputs the value for $P(X \leq x)$ when **TRUE** is used as input for **Cumulative.** If you want to find the value for $P(X \geq x)$ you will need to subtract this output from one.

TI-83/84

$P(X \leq x)$

1. Press the 2nd key then press the **VARS** key
2. Highlight the **binomcdf(** option and press the **ENTER** key
3. Type in the number of trials, n
4. Type a comma
5. Type in the success probability, p
6. Type a comma
7. Type the value for x
8. Type a right parenthesis
9. Press **ENTER**

$P(X = x)$

1. Press the **2nd** key then press the **VARS** key
2. Highlight the **binompdf(** option and press the **ENTER** key
3. Type in the number of trials, n
4. Type a comma
5. Type in the success probability, p
6. Type a comma
7. Type the value for x
8. Type a right parenthesis
9. Press **ENTER**

TI-Nspire

$P(x_1 \leq X \leq x_2)$

1. Enter the Scratchpad
2. Press the **menu** key and select **5:Probability** then select **5:Distributions** then select **E:Binomial Cdf…** and press the **enter** key
3. In the box next to **Num Trials, n** type in the number of trials, n
4. In the box next to **Prob Success, p** type in the success probability, p
5. In the box next to **Lower Bound** type in the value for x_1
6. In the box next to **Upper Bound** type in the value for x_2
7. Press **OK**

Note: In order to find $P(X \leq x)$ input a 0 for the lower bound.

$P(X = x)$

1. Enter the Calculate Scratchpad
2. Press the **menu** key and select **5:Probability** then select **5:Distributions** then select **D:Binomial Pdf…** and press the **enter** key
3. In the box next to **Num Trials, n** type in the number of trials, n
4. In the box next to **Prob Success, p** type in the success probability, p
5. In the box next to **X Value** type the value for x
6. Press **OK**

Normal Probability Plots

JMP

1. Enter the raw data into a column
2. Click **Analyze** then select **Distribution**
3. Click and drag the column name containing the data from the box under **Select Columns** to the box next to **Y, Columns**
4. Click **OK**
5. Click the red arrow next to the column name
6. Select **Normal Quantile Plot**

Minitab

1. Input the data for which you would like to check Normality into a column
2. Click **Graph** then click **Probability Plot...**
3. Highlight the **Single** plot
4. Click **OK**
5. Double click the column name for the column that contains your data to move it into the **Graph Variables:** box
6. Click **OK**

SPSS

1. Input the data for which you would like to check Normality into a column
2. Click **Analyze** then select **Descriptive Statistics** then select **Q-Q Plots...**
3. Highlight the column name for the variable
4. Click the arrow to move the variable to the **Variables:** box
5. Click **OK**

Note: The normal probability plot is output with several other plots and statistics. This plot can be found under the title **Normal Q-Q Plot.**

Excel

Excel does not have the functionality to automatically produce Normal Probability Plots. However, Excel can produce Normal Probability Plots in the course of running a regression using the Analysis ToolPak.

TI-83/84

1. Input the raw data into **L1** (In order to access lists press the **STAT** key, highlight the option called **Edit...** then press **ENTER**)
2. Press the **2nd** key then press the **Y =** key
3. Highlight **Plot1** and press **ENTER**
4. Highlight **On** and press **ENTER**
5. Highlight the plot type on the second row, third column and press **ENTER**
6. Press **GRAPH**

TI-Nspire

1. Enter the data into a data list (In order to access data lists select the spreadsheet option and press **enter)**

Note: Be sure to title the list by selecting the top row of the column and typing a title.

2. Press **menu** and select **3:Data** then **6:QuickGraph** then press **enter**
3. Press **menu** and select **1:Plot Type** then select **4:Normal Probability Plot** and press **enter**

CUMULATIVE REVIEW EXERCISES CR7.1 - CR7.21

CR7.1 The paper **"The Psychology of Security" (*Communications of the ACM* [2008]: 36–41)** states that people are more likely to gamble for a loss than to accept a guaranteed loss. For example, the authors indicate that when presented with a scenario that allows people to choose between a guaranteed gain of $5 and a gain of $10 if the flip of a coin results in a head, more people will choose the guaranteed gain. However, when presented with a scenario that involves choosing between a guaranteed loss of $5 and a loss of $10 if the flip of a coin results in a head, more people will choose the gambling option. Describe how you would design an experiment that would allow you to test this theory.

CR7.2 ● The article **"Water Consumption in the Bay Area" (*San Jose Mercury News,* January 16, 2010)** reported the accompanying values for water consumption (gallons per person per day) for residential customers of 27 San Francisco Bay area water agencies:

49 52 53 59 59 61 69 70 70 73 79 80
85 87 89 90 92 93 93 95 96 97 114 149
208 318 334

a. Use the given water consumption values to construct a boxplot. Describe any interesting features of the boxplot. Are there any outliers in the data set?
b. Compute the mean and standard deviation for this data set.
c. If the two largest values were deleted from the data set, would the standard deviation of this new data set be larger or smaller than the standard deviation you computed for the entire data set?
d. How do the values of the mean and median of the entire data set compare? Is this consistent with the shape of the boxplot from Part (a)? Explain.

CR7.3 Red-light cameras are used in many places to deter drivers from running red lights. The following graphical display is similar to one that appeared in the article **"Communities Put a Halt to Red-Light Cameras" (*USA Today,* January 18, 2010)**. Based on this graph, would it be correct to conclude that there were fewer red-light cameras in 2009 than in 2002? Explain.

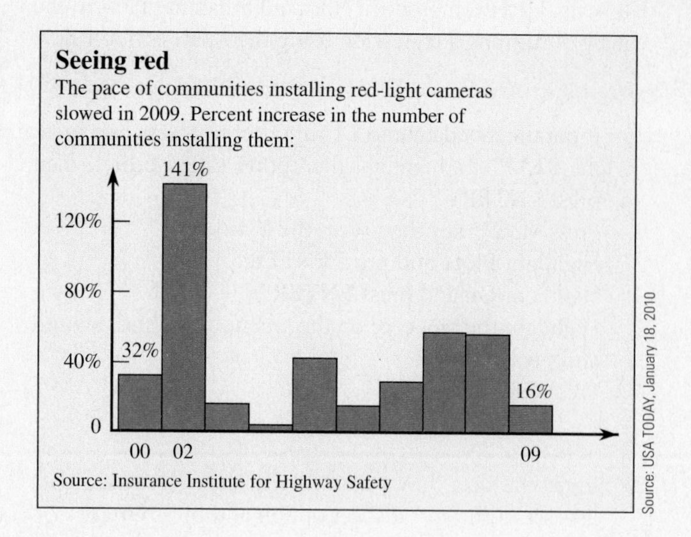

CR7.4 The following quote is from ***USA Today* (January 21, 2010)**:

> Most Americans think the Census is important, and the majority say they will participate in the count this spring—although Hispanics, younger people and the less educated are not as enthusiastic, a Pew Research Center survey found. The poll of 1504 adults found that perceptions differ: 74% of blacks and 72% of Hispanics rate the Census "very important" vs. 57% of whites. Believing it's important doesn't necessarily mean participating: Fewer than half of Hispanics, 57% of blacks and 61% of whites say they will fill out and mail the forms. Age is the greatest factor in participation: Only 36% of respondents who are 30 or younger say they will definitely participate.

Suppose that it is reasonable to regard survey participants as representative of adult Americans.

a. Suppose that an adult American is to be selected at random and consider the following events:

I = the event that the selected individual thinks that the Census is very important

W = the event that the selected individual is white

Are the events I and W independent or dependent events? Explain.

b. Give an example of two census-related events different than the ones defined in Part (a) and indicate whether they are independent or dependent events. Explain your reasoning.

CR7.5 Beverly is building a fireplace. She is using special decorative 8-inch bricks for the bottom row of the fireplace. Although the bricks are advertised as 8-inch bricks, the actual length of a brick is variable, and the brick lengths are approximately normally distributed with a mean of 8 inches and a standard deviation of 0.2 inches.

Unfortunately, Beverly didn't buy enough of the 8-inch bricks, and needs to get two more bricks to try to fill the remaining 15.9 inches. The two additional bricks cannot have a combined length of more than 15.9 inches and must fit in the 15.9-inch space with no more than .1 inch of space left over.

What is the probability that two randomly selected bricks will allow Beverly to complete the bottom row of the fireplace? (Hint: Define b_1 = length of the first brick and b_2 = length of the second brick. A variable that is a linear combination of independent variables that have normal distributions also has a normal distribution.)

CR7.6 ● Manatees are large marine mammals found along the coast of Florida. Data on the total number of manatee deaths per year reported by the **Florida Marine Research Center** is given in the accompanying table. Construct a time series plot of these data and comment on any trends over time.

Year	Number of Manatee Deaths	Year	Number of Manatee Deaths
1974	7	1985	112
1975	28	1986	117
1976	58	1987	109
1977	109	1988	127
1978	77	1989	164
1979	72	1990	199
1980	56	1991	166
1981	112	1992	156
1982	111	1993	137
1983	71	1994	177
1984	119		

CR7.7 Two shipping services offer overnight delivery of parcels, and both promise delivery before 10 A.M. A mail-order catalog company ships 30% of its overnight packages using service 1 and 70% using service 2. Service 1 fails to meet the 10 A.M. delivery promise 10% of the time, whereas Service 2 fails to deliver by 10 A.M. 8% of the time.

Suppose that you made a purchase from this company and were expecting your package by 10 A.M., but it is late. Which shipping service was more likely to have been used?

CR7.8 The paper **"The Effect of Temperature and Humidity on Size of Segregated Traffic Exhaust Particle Emissions" (*Atmospheric Environment* [2008]: 2369–2382)** gave the following summary quantities for a measure of traffic flow (vehicles/second) during peak traffic hours at a particular location recorded daily over a long sequence of days:

Mean = 0.41 Standard Deviation = 0.26 Median = 0.45
5th percentile = 0.03 Lower quartile = 0.18
Upper quartile = 0.57 95th percentile 2 0.86

Based on these summary quantities, do you think that the distribution of the measure of traffic flow could have been approximately normal? Explain your reasoning.

CR7.9 The article **"Doctors Misdiagnose More Women, Blacks" (*San Luis Obispo Tribune,* April 20, 2000)** gave the following information, which is based on a large study of more than 10,000 patients treated in emergency rooms in the eastern and midwestern United States:

1. Doctors misdiagnosed heart attacks in 2.1% of all patients.
2. Doctors misdiagnosed heart attacks in 4.3% of black patients.
3. Doctors misdiagnosed heart attacks in 7% of women under 55 years old.

Use the following event definitions:

M = event that a heart attack is misdiagnosed,
B = event that a patient is black, and
W = event that a patient is a woman under 55 years old.

Translate each of the three given statements into probability notation.

CR7.10 The ***Cedar Rapids Gazette* (November 20, 1999)** reported the following information on compliance with child restraint laws for cities in Iowa:

City	Number of Children Observed	Number Properly Restrained
Cedar Falls	210	173
Cedar Rapids	231	206
Dubuque	182	135
Iowa City (city)	175	140
Iowa City (interstate)	63	47

a. Use the information provided to estimate the following probabilities:
   i. The probability that a randomly selected child is properly restrained given that the child is observed in Dubuque.
   ii. The probability that a randomly selected child is properly restrained given that the child is observed in a city that has "Cedar" in its name.

b. Suppose that you are observing children in the Iowa City area. Use a tree diagram to illustrate the possible outcomes of an observation that considers both the location of the observation (city or interstate) and whether the child observed was properly restrained.

CR7.11 According to a study conducted by a risk assessment firm **(*Associated Press,* December 8, 2005)**, drivers residing within 1 mile of a restaurant are 30% more likely to be in an accident in a given policy year. Consider the following two events:

A = event that a driver has an accident during a policy year
R = event that a driver lives within 1 mile of a restaurant

Which of the following four probability statements is consistent with the findings of this survey? Justify your choice.

i. $P(A|R) = .3$

ii. $P(A|R^C) = .3$

iii. $\dfrac{P(A|R)}{P(A|R^C)} = .3$

iv. $\dfrac{P(A|R) - P(A|R^C)}{P(A|R^C)} = .3$

CR7.12 The article **"Men, Women at Odds on Gun Control" (*Cedar Rapids Gazette,* September 8, 1999)** included the following statement: "The survey found that 56% of American adults favored stricter gun control laws. Sixty-six percent of women favored the tougher laws, compared with 45% of men." These figures are based on a large telephone survey conducted by Associated Press Polls. If an adult is selected at random, are the events *selected adult is female* and *selected adult favors stricter gun control* independent events? Explain.

CR7.13 Suppose that a new Internet company Mumble.com requires all employees to take a drug test. Mumble.com can afford only the inexpensive drug test—the one with a 5% false-positive rate and a 10% false-negative rate. (That means that 5% of those who are not using drugs will incorrectly test positive and that 10% of those who are actually using drugs will test negative.) Suppose that 10% of those who work for Mumble.com are using the drugs for which Mumble is checking. (Hint: It may be helpful to draw a tree diagram to answer the questions that follow.)

a. If one employee is chosen at random, what is the probability that the employee both uses drugs and tests positive?

b. If one employee is chosen at random, what is the probability that the employee does not use drugs but tests positive anyway?

c. If one employee is chosen at random, what is the probability that the employee tests positive?

d. If we know that a randomly chosen employee has tested positive, what is the probability that he or she uses drugs?

CR7.14 Refer to the previous exercise. Suppose that because of the high rate of false-positives for the drug test, Mumble.com has instituted a mandatory independent second test for those who test positive on the first test.

a. If one employee is selected at random, what is the probability that the selected employee uses drugs and tests positive twice?

b. If one employee is selected at random, what is the probability that the employee tests positive twice?

c. If we know that the randomly chosen employee has tested positive twice, what is the probability that he or she uses drugs?

d. What is the chance that an individual who does use drugs doesn't test positive twice (either this employee tests negative on the first round and doesn't need a retest, or this employee tests positive the first time and that result is followed by a negative result on the retest)?

e. Discuss the benefits and drawbacks of using a retest scheme such as the one proposed in this question.

CR7.15 A chemical supply company currently has in stock 100 pounds of a certain chemical, which it sells to customers in 5-pound lots. Let x = the number of lots ordered by a randomly chosen customer. The probability distribution of x is as follows:

x	1	2	3	4
$p(x)$	.2	.4	.3	.1

a. Calculate the mean value of x.
b. Calculate the variance and standard deviation of x.

CR7.16 Return to the previous exercise, and let y denote the amount of material (in pounds) left after the next customer's order is shipped. Find the mean and variance of y. (Hint: y is a linear function of x.)

CR7.17 An experiment was conducted to investigate whether a graphologist (a handwriting analyst) could distinguish a normal person's handwriting from that of a psychotic. A well-known expert was given 10 files, each containing handwriting samples from a normal person and from a person diagnosed as psychotic, and asked to identify the psychotic's handwriting.

The graphologist made correct identifications in 6 of the 10 trials (data taken from ***Statistics in the Real World*, by R. J. Larsen and D. F. Stroup [New York: Macmillan, 1976]**). Does this evidence indicate that the graphologist has an ability to distinguish the handwriting of psychotics? (Hint: What is the probability of correctly guessing six or more times out of 10? Your answer should depend on whether this probability is relatively small or relatively large.)

CR7.18 A machine that produces ball bearings has initially been set so that the true average diameter of the bearings it produces is 0.500 inches. A bearing is acceptable if its diameter is within 0.004 inches of this target value. Suppose, however, that the setting has changed during the course of production, so that the distribution of the diameters produced is well approximated by a normal distribution with mean 0.499 inches and standard deviation 0.002 inches. What percentage of the bearings produced will not be acceptable?

CR7.19 Consider the variable

x = time required for a college student to complete a standardized exam

Suppose that for the population of students at a particular university, the distribution of x is well approximated by a normal distribution with mean 45 minutes and standard deviation 5 minutes.

a. If 50 minutes is allowed for the exam, what proportion of students at this university would be unable to finish in the allotted time?
b. How much time should be allowed for the exam if we wanted 90% of the students taking the test to be able to finish in the allotted time?
c. How much time is required for the fastest 25% of all students to complete the exam?

CR7.20 ● The paper **"The Load-Life Relationship for M50 Bearings with Silicon Nitride Ceramic Balls" (*Lubrication Engineering* [1984]: 153–159)** reported the accompanying data on bearing load life (in millions of revolutions). The corresponding normal scores are also given.

x	Normal Score	x	Normal Score
47.1	−1.867	240.0	0.062
68.1	−1.408	240.0	0.187
68.1	−1.131	278.0	0.315
90.8	−0.921	278.0	0.448
103.6	−0.745	289.0	0.590
106.0	−0.590	289.0	0.745
115.0	−0.448	367.0	0.921
126.0	−0.315	385.9	1.131
146.6	−0.187	392.0	1.408
229.0	−0.062	395.0	1.867

Construct a normal probability plot. Is it plausible that load-bearing life is a normal distribution?

CR7.21 ● The following data are a sample of survival times (days from diagnosis) for patients suffering from chronic leukemia of a certain type:

7	47	58	74	177	232	273	285
317	429	440	445	455	468	495	497
532	571	579	581	650	702	715	779
881	900	930	968	1077	1109	1314	1334
1367	1534	1712	1784	1877	1886	2045	2056
2260	2429	2509					

a. Construct a relative frequency distribution for this data set, and draw the corresponding histogram.
b. Would you describe this histogram as having a positive or a negative skew?
c. Would you recommend transforming the data? Explain.

CHAPTER 8

Sampling Variability and Sampling Distributions

Christian Petersen/Getty Images Sport/Getty Images

The inferential methods presented in Chapters 9–15 use information contained in a sample to reach conclusions about one or more characteristics of the population from which the sample was selected. For example, let μ denote the true mean fat content of quarter-pound hamburgers marketed by a national fast-food chain. To learn something about μ, we might obtain a sample of $n = 50$ hamburgers and determine the fat content for each one. The sample data might produce a mean of $\bar{x} = 28.4$ grams. How close is this sample mean to the population mean, μ? If we selected another sample of 50 quarter-pound burgers and then determine the sample mean fat content, would this second value be near 28.4, or might it be quite different?

These questions can be addressed by studying what is called the *sampling distribution* of $\bar{x}$. Just as the probability distribution of a numerical variable describes its long-run behavior, the sampling distribution of $\bar{x}$ provides information about the long-run behavior of $\bar{x}$ when sample after sample is selected.

In this chapter, we also consider the sampling distribution of a sample proportion (the fraction of individuals or objects in a sample that have some characteristic of interest). The sampling distribution of a sample proportion, $\hat{p}$, provides information about the long-run behavior of the sample proportion. This knowledge allows us to make inferences about a population proportion.

Chapter 8: Learning Objectives

STUDENTS WILL UNDERSTAND:

- that the value of a sample statistic varies from sample to sample.
- that sampling distribution describes sample-to-sample variability in the values of a statistic.
- how the standard deviations of the sampling distributions of $\bar{x}$ and $\hat{p}$ are related to sample size.

STUDENTS WILL BE ABLE TO:

- describe general properties of the sampling distribution of $\bar{x}$.
- describe general properties of the sampling distribution of $\hat{p}$.
- determine when the sampling distribution of $\bar{x}$ or $\hat{p}$ is approximately normal.
- use properties of the sampling distribution to reason informally about the value of a population mean or a population proportion.

8.1 Statistics and Sampling Variability

A quantity computed from the values in a sample is called a **statistic**. Values of statistics such as the sample mean $\bar{x}$, the sample median, the sample standard deviation s, or the proportion of individuals in a sample that possess a particular property $\hat{p}$, are our primary sources of information about various population characteristics.

The usual way to obtain information about the value of a population characteristic is by selecting a sample from the population. For example, to gain insight about the mean credit card balance for students at a particular university, we might select a sample of 50 students at the university. Each student would be asked about his or her credit card balance to yield a value of x = current balance.

We could construct a histogram of the 50 sample x values, and we could view this histogram as a rough approximation of the population distribution of x. In a similar way, we can view the sample mean $\bar{x}$ (the mean of a sample of n values) as an approximation of μ, the mean of the population distribution.

It would be nice if the value of $\bar{x}$ were equal to the value of the population mean μ, but this is not usually the case. Not only will the value of $\bar{x}$ for a particular sample from a population usually differ from μ, but the $\bar{x}$ values from different samples also usually differ from one another. For example, two different samples of 50 student credit card balances will usually result in different $\bar{x}$ values. This sample-to-sample variability makes it challenging to generalize from a sample to the population from which it was selected. An understanding of sample-to-sample variability enables us to meet this challenge.

DEFINITION

Statistic: A quantity computed from values in a sample.

Sampling variability: The observed value of a statistic depends on the particular sample selected from the population and it will vary from sample to sample. This variability is called **sampling variability**.

EXAMPLE 8.1 Exploring Sampling Variability

Consider a small population consisting of the 20 students enrolled in an upper division class. The amount of money (in dollars) each of the 20 students spent on textbooks for the current semester is shown in the following table:

Student	Amount Spent on Books	Student	Amount Spent on Books	Student	Amount Spent on Books
1	367	5	375	9	378
2	358	6	395	10	268
3	442	7	322	11	419
4	361	8	370	12	363

continued

Student	Amount Spent on Books	Student	Amount Spent on Books	Student	Amount Spent on Books
13	365	16	284	19	330
14	362	17	331	20	423
15	433	18	259		

For this population,

$$\mu = \frac{367 + 358 + \cdots + 423}{20} = 360.25$$

Suppose we don't know the value of the population mean, so we decide to estimate μ by taking a random sample of five students and computing $\bar{x}$, the sample mean amount spent on textbooks. Is this a reasonable thing to do? Is the estimate that results likely to be close to the value of μ, the population mean?

To answer these questions, consider a simple experiment that allows us to examine the behavior of the statistic $\bar{x}$ when random samples of size 5 are repeatedly selected. (This scenario is not realistic. If a population consisted of only 20 individuals, we would probably conduct a census rather than select a sample. However, this small population size is easier to work with as we develop the idea of sampling variability.)

Let's first select a random sample of size 5 from this population. This can be done by writing the numbers from 1 to 20 on otherwise identical slips of paper, mixing them well, and then selecting 5 slips without replacement. The numbers on the slips selected identify which of the 20 students will be included in our sample. Alternatively, either a table of random digits or a random number generator can be used to determine which 5 students should be selected.

We used Minitab to obtain five random numbers between 1 and 20, resulting in 17, 20, 7, 11, and 9. This results in the following sample of amounts spent on books:

331 423 322 419 378

For this sample,

$$\bar{x} = \frac{1873}{5} = 374.60$$

The sample mean is larger than the population mean of \$360.25 by about \$15. Is this difference typical, or is this particular sample mean unusually far away from μ? Taking more samples will provide some additional insight.

Four more random samples (Samples 2–5) from this same population are shown here.

Sample 2		Sample 3		Sample 4		Sample 5	
Student	x	Student	x	Student	x	Student	x
4	361	15	433	20	423	18	259
15	433	12	363	16	284	8	370
12	363	3	442	19	330	9	378
1	367	7	322	1	367	7	322
18	259	18	259	8	370	14	362
$\bar{x}$	**356.60**	$\bar{x}$	**363.80**	$\bar{x}$	**354.80**	$\bar{x}$	**338.20**

Because $\mu = 360.25$, we can see the following:

1. The value of $\bar{x}$ varies from one random sample to another (sampling variability).
2. Some samples produced $\bar{x}$ values larger than μ (Samples 1 and 3), whereas others produced values smaller than μ (Samples 2, 4, and 5).
3. Samples 2, 3, and 4 produced $\bar{x}$ values that were fairly close to the population mean, but Sample 5 resulted in a value that was \$22 below the population mean.

Continuing with the experiment, we selected 45 additional random samples of size $n = 5$. The resulting sample means are as follows:

Sample	$\bar{x}$	Sample	$\bar{x}$	Sample	$\bar{x}$
6	374.6	21	355.0	36	353.4
7	356.6	22	407.2	37	379.6
8	363.8	23	380.0	38	352.6
9	354.8	24	377.4	39	342.2
10	338.2	25	341.2	40	362.6
11	375.6	26	316.0	41	315.4
12	379.2	27	370.0	42	366.2
13	341.6	28	401.0	43	361.4
14	355.4	29	347.0	44	375
15	363.8	30	373.8	45	401.4
16	339.6	31	382.8	46	337
17	348.2	32	320.4	47	387.4
18	430.8	33	313.6	48	349.2
19	388.8	34	387.6	49	336.8
20	352.8	35	314.8	50	364.6

Figure 8.1 is a density histogram of the 50 sample means. It provides insight about the behavior of $\bar{x}$. Most samples resulted in $\bar{x}$ values that are reasonably near $\mu = 360.25$, falling between 335 and 395. A few samples, however, produced values that were far from μ. If we were to take a sample of size 5 from this population and use $\bar{x}$ as an estimate of the population mean μ, we should *not* necessarily expect $\bar{x}$ to be close to μ.

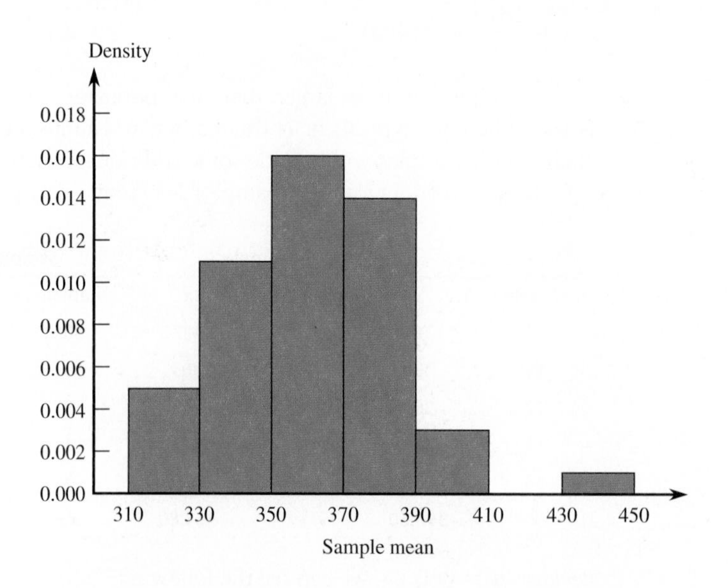

FIGURE 8.1
Density histogram of $\bar{x}$ values from the 50 random samples of Example 8.1.

The density histogram in Figure 8.1 visually conveys information about the sampling variability in the statistic $\bar{x}$. It provides an approximation to the distribution of $\bar{x}$ values that would have been observed if we had considered every different possible sample of size 5 from this population. ■

In the example just considered, we obtained the approximate sampling distribution of the statistic $\bar{x}$ by considering just 50 different samples. The actual **sampling distribution** comes from considering *all* possible samples of size n.

DEFINITION

Sampling distribution: The distribution that would be formed by considering the value of a sample statistic for every possible different sample of a given size from a population.

The sampling distribution of a statistic, such as $\bar{x}$, provides important information about variation in the values of the statistic. The density histogram of Figure 8.1 is an *approximation* of the sampling distribution of the statistic $\bar{x}$ for samples of size 5 from the population described in Example 8.1.

We could have determined the actual sampling distribution of $\bar{x}$ by considering every possible sample of size 5 from the population of 20 students, computing the mean for each sample, and then constructing a density histogram of the $\bar{x}$ values. But this would have been a lot of work—there are 15,504 different possible samples of size 5. And, for more realistic situations with larger population and sample sizes, the situation becomes even worse because there are so many possible samples that must be considered.

Fortunately, as we look at a few more examples in the sections that follow, patterns emerge that enable us to describe some important aspects of the sampling distributions for some statistics without actually having to look at all possible samples.

EXERCISES 8.1 - 8.9

8.1 Explain the difference between a population characteristic and a statistic.

8.2 What is the difference between $\bar{x}$ and μ? between s and σ?

8.3 ▼ For each of the following statements, identify the number that appears in boldface type as the value of either a population characteristic or a statistic:

a. A department store reports that **84%** of all customers who use the store's credit plan pay their bills on time.

b. A sample of 100 students at a large university had a mean age of **24.1** years.

c. The Department of Motor Vehicles reports that **22%** of all vehicles registered in a particular state are imports.

d. A hospital reports that based on the 10 most recent cases, the mean length of stay for surgical patients is **6.4** days.

e. A consumer group, after testing 100 batteries of a certain brand, reported an average life of **63** hours of use.

8.4 Consider a population consisting of the following five values, which represent the number of DVD rentals during the academic year for each of five housemates:

8 14 16 10 11

a. Compute the mean of this population.

b. Select a random sample of size 2 by writing the five numbers in this population on slips of paper, mixing them, and then selecting two. Compute the mean for your sample.

c. Repeatedly select random samples of size 2, and compute the $\bar{x}$ value for each sample until you have the $\bar{x}$ values for 25 samples.

d. Construct a density histogram using the 25 $\bar{x}$ values. Are most of the $\bar{x}$ values near the population mean? Do the $\bar{x}$ values differ a lot from sample to sample, or do they tend to be similar? (Hint: See Example 8.1.)

8.5 Select 10 additional random samples of size 5 from the population of 20 students given in Example 8.1, and compute the mean amount spent on books for each of the 10 samples. Are the $\bar{x}$ values consistent with the results of the sampling experiment summarized in Figure 8.1?

8.6 Suppose that the sampling experiment described in Example 8.1 had used samples of size 10 rather than size 5.

a. If 50 samples of size 10 were selected, the $\bar{x}$ value for each sample computed, and a density histogram constructed, how do you think this histogram would differ from the density histogram constructed for samples of size 5 (Figure 8.1)?

b. In what way do you think the density histograms would be similar?

8.7 ▼ Consider the following population: {1, 2, 3, 4}. For this population the mean is

$$\mu = \frac{1 + 2 + 3 + 4}{4} = 2.5$$

Suppose that a random sample of size 2 is to be selected without replacement from this population. There are 12 possible samples (provided that the order in which observations are selected is taken into account):

1, 2 1, 3 1, 4 2, 1 2, 3 2, 4
3, 1 3, 2 3, 4 4, 1 4, 2 4, 3

a. Compute the sample mean for each of the 12 possible samples.

b. Use the sample means to construct the sampling distribution of $\bar{x}$. Display the sampling distribution as a density histogram.

c. Suppose that a random sample of size 2 is to be selected, but this time sampling will be done with replacement. Using a method similar to that of Part (a), construct the sampling distribution of $\bar{x}$. (Hint: There are 16 different possible samples in this case.)

d. In what ways are the two sampling distributions of Parts (b) and (c) similar? In what ways are they different?

8.8 Simulate sampling from the population of Exercise 8.7 by using four slips of paper individually marked 1, 2, 3, and 4. Select a sample of size 2 without replacement, and compute $\bar{x}$. Repeat this process 50 times, and construct a density histogram of the 50 $\bar{x}$ values. How does this sampling distribution compare to the actual sampling distribution of $\bar{x}$ from Exercise 8.7, Part (b)?

8.9 Consider the following population: {2, 3, 3, 4, 4}. The value of μ is 3.2, but suppose that this is not known to an investigator. Three possible statistics for estimating μ are

Statistic 1: the sample mean, $\bar{x}$
Statistic 2: the sample median
Statistic 3: the average of the largest and the smallest values in the sample

A random sample of size 3 will be selected without replacement. Provided that we disregard the order in which the observations are selected, there are 10 possible samples that might result (writing 3 and 3*, 4 and 4* to distinguish the two 3's and the two 4's in the population):

2, 3, 3* 2, 3, 4 2, 3, 4* 2, 3*, 4 2, 3*, 4*
2, 4, 4* 3, 3*, 4 3, 3*, 4* 3, 4, 4* 3*, 4, 4*

a. For each of these 10 samples, compute Statistics 1, 2, and 3.

b. Construct the sampling distribution of each of these statistics.

c. Which statistic would you recommend for estimating μ and why?

Bold exercises answered in back ● Data set available online ▼ Video Solution available

8.2 The Sampling Distribution of a Sample Mean

When the objective of a statistical investigation is to learn about the value of a population mean μ, it is natural to consider the sample mean $\bar{x}$. To understand how inferential procedures based on $\bar{x}$ work, we must first study how sampling variability causes $\bar{x}$ to vary in value from one sample to another.

The behavior of $\bar{x}$ is described by its sampling distribution. The sample size n and characteristics of the population—its shape, mean value μ, and standard deviation σ—are important in determining properties of the sampling distribution of $\bar{x}$.

It is helpful first to consider the results of some sampling experiments. In Examples 8.2 and 8.3, we start with a specified population distribution, fix a sample size n, and select 500 different random samples of this size. We then compute $\bar{x}$ for each sample and construct a sample histogram of these 500 $\bar{x}$ values. Because 500 is reasonably large (a reasonably long sequence of samples), the density histogram of the $\bar{x}$ values should closely resemble the actual sampling distribution of $\bar{x}$ (which would be obtained from an unending sequence of $\bar{x}$ values).

We then repeat the experiment for several different values of n to see how the choice of sample size affects the sampling distribution. This enables us to identify some patterns that will be helpful in understanding important properties of the sampling distribution of $\bar{x}$.

EXAMPLE 8.2 Blood Platelet Volume

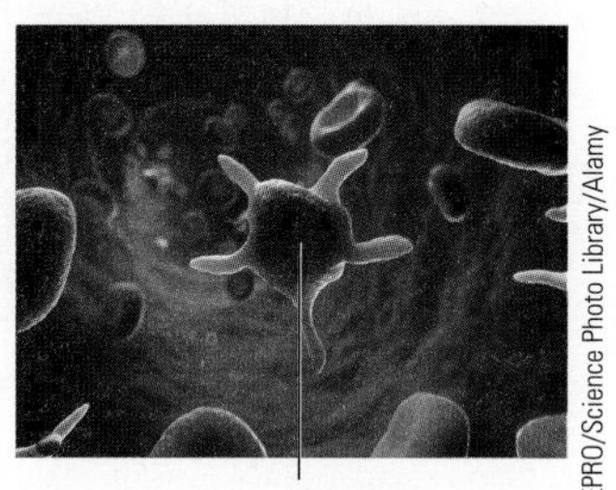

SCIEPRO/Science Photo Library/Alamy

The paper **"Mean Platelet Volume in Patients with Metabolic Syndrome and Its Relationship with Coronary Artery Disease" (*Thrombosis Research* [2007]: 245–250)** includes data that suggest that the distribution of platelet volume for patients who do not have metabolic syndrome (a combination of factors that indicates a high risk of heart disease) is approximately normal with mean $\mu = 8.25$ and standard deviation $\sigma = 0.75$.

Figure 8.2 shows a normal curve centered at 8.25, the mean value of platelet volume. The value of the population standard deviation, 0.75, determines the extent to which the x distribution spreads out about its mean value.

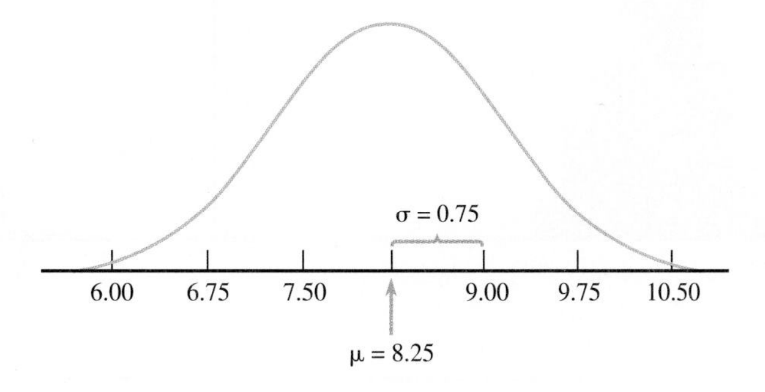

FIGURE 8.2
Normal distribution of platelet size ($\mu = 8.25$ and $\sigma = 0.75$).

We first used Minitab to select 500 random samples from this population distribution, with each sample consisting of $n = 5$ observations. A density histogram of the resulting 500 $\bar{x}$ values appears in Figure 8.3(a). This procedure was repeated for samples of size $n = 10$, $n = 20$, and $n = 30$. The resulting density histograms of the $\bar{x}$ values are displayed in Figure 8.3(b)–(d).

The first thing to notice about the histograms is their shape. Each of the four histograms is approximately normal in shape. The resemblance would be even more striking if each histogram had been based on many more than 500 $\bar{x}$ values.

Second, notice that each histogram is centered approximately at 8.25, the mean of the population being sampled. Had the histograms been based on an unending sequence of $\bar{x}$ values, their centers would have been exactly equal to the population mean, 8.25.

The final aspect of the histograms to note is their spread relative to one another. The smaller the value of n, the greater the extent to which the sampling distribution spreads out around the population mean value. This is why the histograms for $n = 20$ and $n = 30$ are based on narrower class intervals than those for the two smaller sample sizes. For the larger sample sizes, most of the $\bar{x}$ values are quite close to 8.25. This is the effect of averaging. When n is small, a single unusual x value can result in an $\bar{x}$ value far from the center. With a larger sample size, any unusual x values, when averaged with the other sample values, still tend to yield an $\bar{x}$ value close to μ. Combining these insights yields a result that should appeal to your intuition: *$\bar{x}$ based on a large sample will tend to be closer to μ than $\bar{x}$ based on a small sample.*

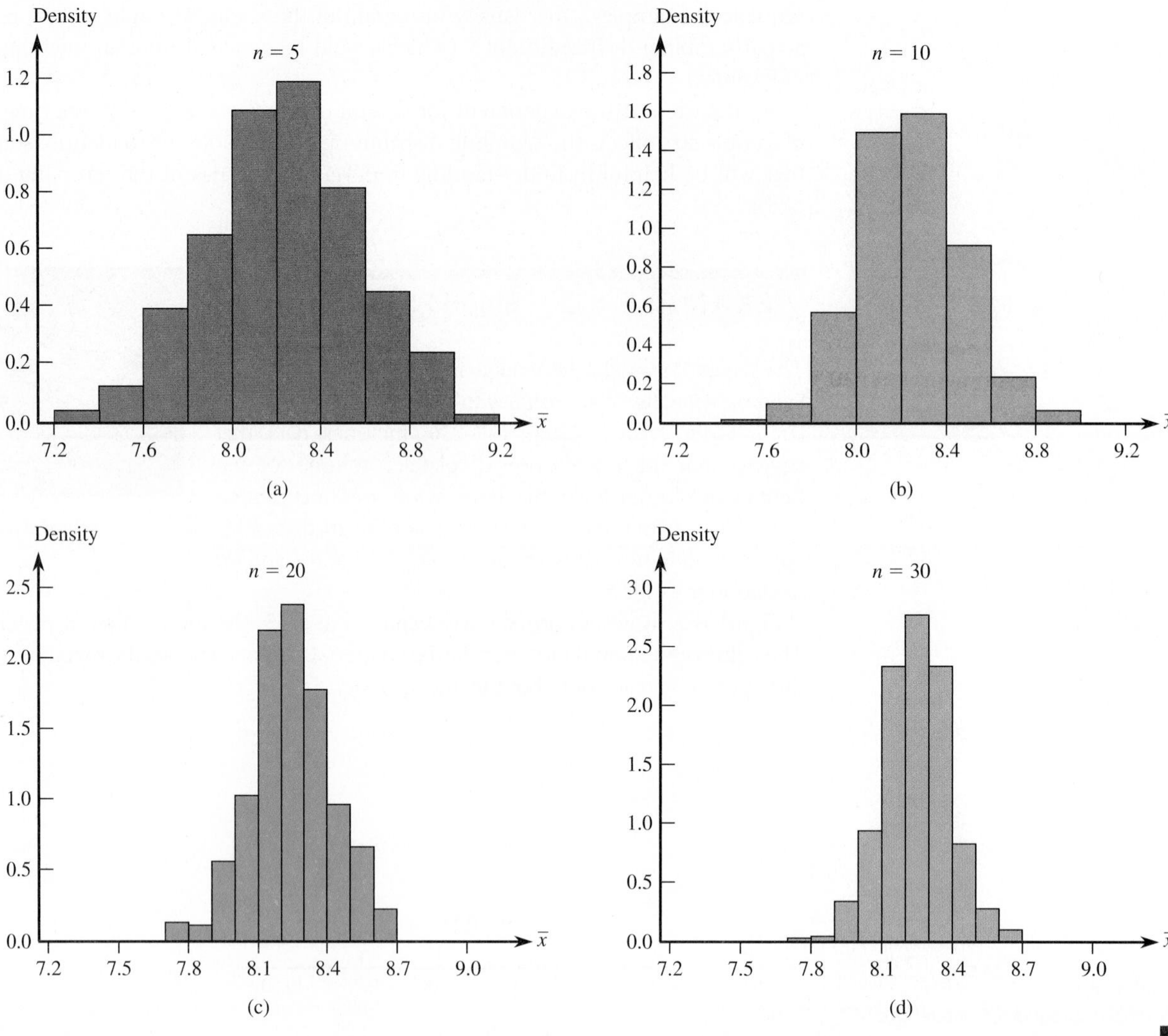

FIGURE 8.3
Density histograms for $\bar{x}$ based on 500 samples, each consisting of n observations, for Example 8.2: (a) $n = 5$; (b) $n = 10$; (c) $n = 20$; (d) $n = 30$.

EXAMPLE 8.3 Time to First Goal in Hockey

Now consider properties of the $\bar{x}$ distribution when the population is quite skewed (and thus very unlike a normal distribution). The paper **"Is the Overtime Period in an NHL Game Long Enough?"** **(*American Statistician* [2008]: 151–154)** gave data on the time (in minutes) from the start of the game to the first goal scored for the 281 regular season games from the 2005–2006 season that went into overtime.

Figure 8.4 displays a density histogram of the data (read from a graph that appeared in the paper). The histogram has a long upper tail, indicating that the first goal is scored in the first 20 minutes of most games, but that for some games, the first goal is not scored until much later in the game.

If we think of the 281 values as a population, the histogram in Figure 8.4 shows the distribution of values in that population. The skewed shape makes identification of the mean value from the picture more difficult than for a normal distribution, but we computed the average of the 281 values to be $\mu = 13$ minutes.

For each of the sample sizes $n = 5$, 10, 20, and 30, we selected 500 random samples of size n. This was done with replacement to approximate more nearly the usual situation, in which the sample size n is only a small fraction of the population size. We then constructed a histogram of the 500 $\bar{x}$ values for each of the four sample sizes. These histograms are displayed in Figure 8.5.

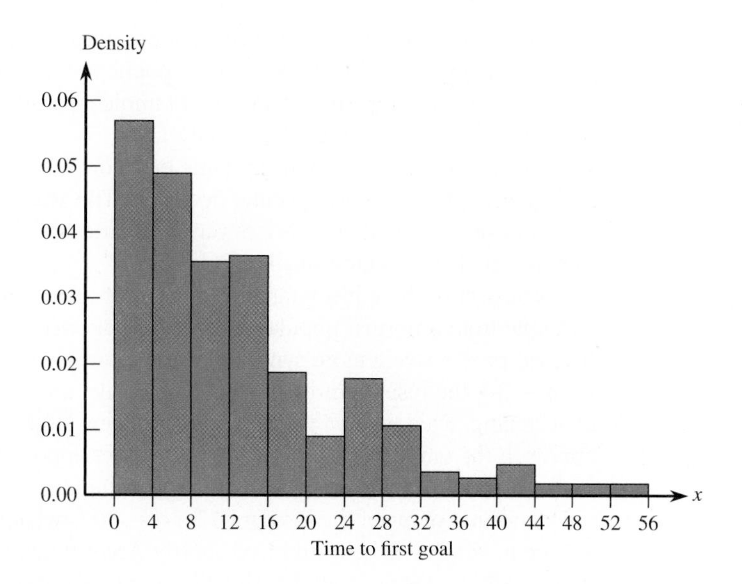

FIGURE 8.4
The population distribution for Example 8.3 ($\mu = 13$).

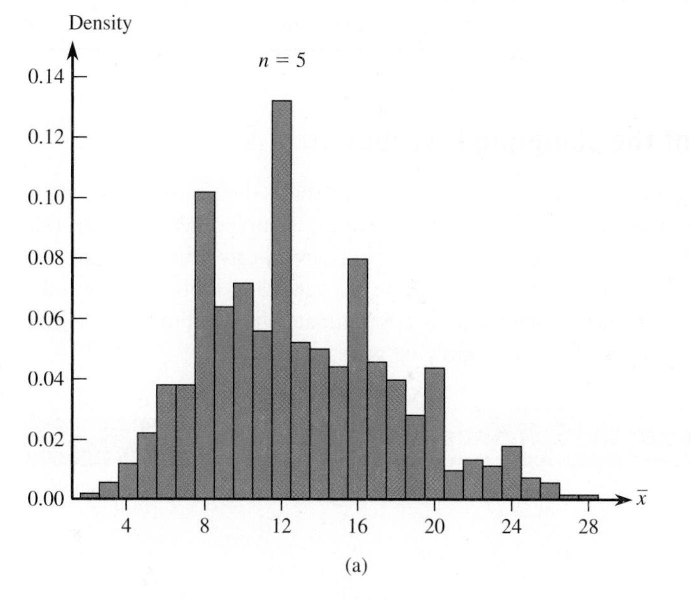

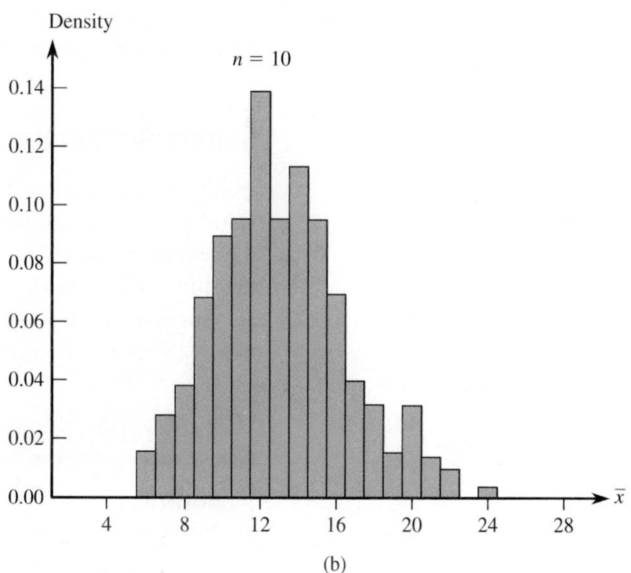

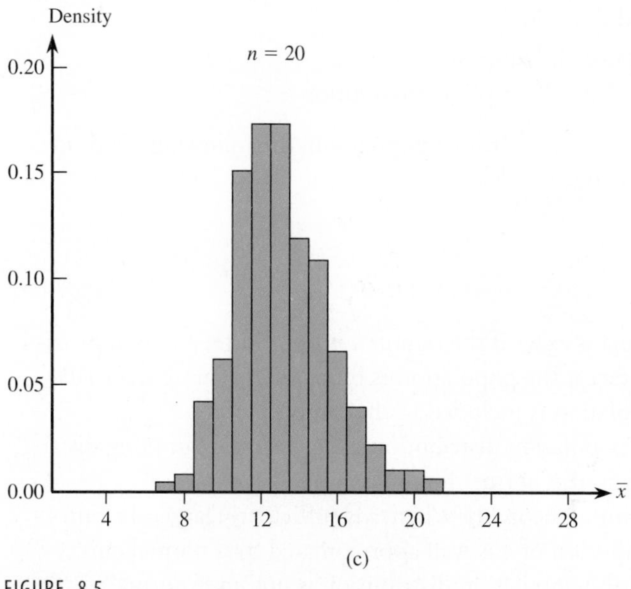

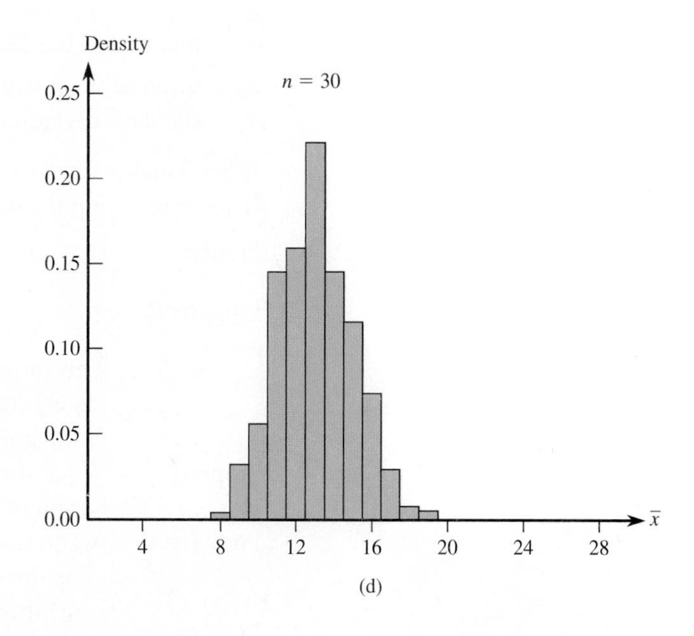

FIGURE 8.5
Four histograms of 500 $\bar{x}$ values for Example 8.3: (a) $n = 5$; (b) $n = 10$; (c) $n = 20$; (d) $n = 30$.

As with samples from a normal population, the averages of the 500 $\bar{x}$ values for the four different sample sizes are all close to the population mean $\mu = 13$. If each histogram had been based on an unending sequence of sample means rather than just 500 of them, each one would have been centered at exactly 13.

Comparison of the four $\bar{x}$ histograms in Figure 8.5 also shows that as n increases, the histogram's spread around its center decreases. This was also true of increasing sample sizes from a normal population: $\bar{x}$ is less variable (varies less from sample to sample) for a large sample size than it is for a small sample size.

One aspect of these histograms distinguishes them from the distribution of $\bar{x}$ based on a sample from a normal population. They are skewed and differ in shape more, but they become progressively more symmetric as the sample size increases. We can also see that for $n = 30$, the histogram has a shape much like a normal curve. Again this is the effect of averaging. Even when n is large, one of the few large x values in the population doesn't appear in the sample very often. When one does appear, its contribution to $\bar{x}$ is swamped by the contributions of more typical sample values.

The normal shape of the histogram for $n = 30$ is what is predicted by the Central Limit Theorem, which will be introduced shortly. According to this theorem, even if the population distribution is not well described by a normal curve, the sampling distribution of $\bar{x}$ is approximately normal in shape when the sample size n is reasonably large. ■

General Properties of the Sampling Distribution of $\bar{x}$

Examples 8.2 and 8.3 suggest that for any n, the center of the $\bar{x}$ distribution (the mean value of $\bar{x}$) coincides with the mean of the population being sampled. We also see that the spread of the $\bar{x}$ distribution decreases as n increases. This indicates that the standard deviation of $\bar{x}$ is smaller for large n than for small n. The histograms in Figures 8.3 and 8.5 also suggest that in some cases, the $\bar{x}$ distribution is approximately normal in shape. These observations are stated more formally in the following general rules.

General Properties of the Sampling Distribution of $\bar{x}$

Notation

$\bar{x}$ sample mean
n sample size
μ population mean
σ population standard deviation
$\mu_{\bar{x}}$ mean of the $\bar{x}$ sampling distribution
$\sigma_{\bar{x}}$ standard deviation of the $\bar{x}$ sampling distribution

When random samples are selected from a population, the following are properties of the sampling distribution of $\bar{x}$:

Property 1. $\mu_{\bar{x}} = \mu$.

Property 2. $\sigma_{\bar{x}} = \dfrac{\sigma}{\sqrt{n}}$.

This formula is exact if the population is infinite, and is approximately correct if the population is finite and no more than 10% of the population is included in the sample.

Property 3. When the population distribution is normal, the sampling distribution of $\bar{x}$ is also normal for any sample size n.

Property 4. (Central Limit Theorem) When n is sufficiently large, the sampling distribution of $\bar{x}$ is well approximated by a normal curve, even when the population distribution is not itself normal.

Property 1, $\mu_{\bar{x}} = \mu$, states that the sampling distribution of $\bar{x}$ is always centered at the mean of the population sampled.

Property 2, $\sigma_{\bar{x}} = \frac{\sigma}{\sqrt{n}}$, not only states that the spread of the sampling distribution of $\bar{x}$ decreases as n increases, but also gives a precise relationship between the standard deviation of the $\bar{x}$ distribution and the population standard deviation and sample size. For example, when $n = 4$,

$$\sigma_{\bar{x}} = \frac{\sigma}{\sqrt{n}} = \frac{\sigma}{\sqrt{4}} = \frac{\sigma}{2}$$

so the $\bar{x}$ distribution has a standard deviation only half as large as the population standard deviation.

Properties 3 and 4 specify when the $\bar{x}$ distribution is normal (when the population is normal) or approximately normal (when the sample size is large).

Figure 8.6 illustrates these rules by showing several $\bar{x}$ distributions superimposed over a graph of the population distribution.

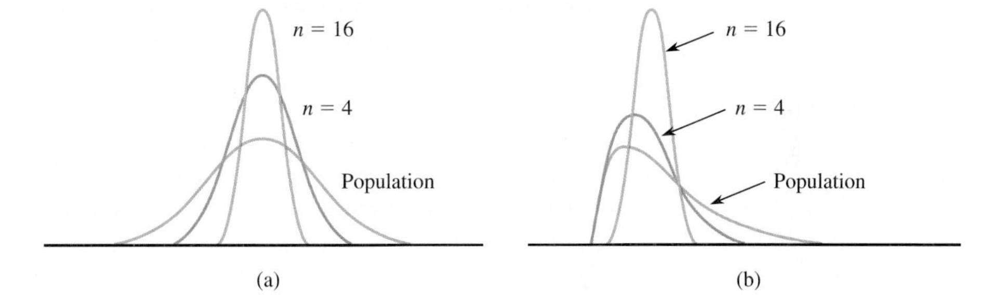

FIGURE 8.6
Population distribution and sampling distributions of $\bar{x}$:
(a) symmetric population;
(b) skewed population.

The Central Limit Theorem, Property 4, states that when n is sufficiently large, the $\bar{x}$ distribution is approximately normal, no matter what the shape of the population distribution. This result has enabled statisticians to develop procedures for making inferences about a population mean μ using a large sample, even when the shape of the population distribution is unknown.

Recall that a variable is standardized by subtracting the mean value and then dividing by its standard deviation. Using Properties 1 and 2 to standardize $\bar{x}$ gives an important consequence of the last two properties.

If n is large or the population distribution is normal, the standardized variable

$$z = \frac{\bar{x} - \mu_{\bar{x}}}{\sigma_{\bar{x}}} = \frac{\bar{x} - \mu}{\frac{\sigma}{\sqrt{n}}}$$

has (at least approximately) a standard normal (z) distribution.

Application of the Central Limit Theorem in specific situations requires a rule of thumb for deciding when n is sufficiently large. Such a rule is not as easy to come by as one might think. Look back at Figure 8.5, which shows the approximate sampling distribution of $\bar{x}$ for $n = 5$, 10, 20, and 30 when the population distribution is quite skewed. Certainly the histogram for $n = 5$ is not well described by a normal curve, and this is still true of the histogram for $n = 10$, particularly in the tails of the histogram. Among the four histograms, only the histogram for $n = 30$ has a reasonably normal shape.

On the other hand, when the population distribution is normal, the sampling distribution of $\bar{x}$ is normal for any n. If the population distribution is somewhat skewed but not to the extent of Figure 8.4, we might expect the $\bar{x}$ sampling distribution to be a bit skewed for $n = 5$ but quite well described by a normal curve for n as small as 10 or 15.

How large n must be for the $\bar{x}$ distribution to be approximately normal depends on how much the population distribution differs from a normal distribution. The closer the population distribution is to being normal, the smaller the value of n necessary for the Central Limit Theorem approximation to be accurate.

Many statisticians recommend the following conservative rule:

> The Central Limit Theorem can safely be applied if n is greater than or equal to 30.

If the population distribution is believed to be reasonably close to a normal distribution, an n of 15 or 20 is often large enough for $\bar{x}$ to have approximately a normal distribution. At the other extreme, we can imagine a distribution with a much longer tail than that of Figure 8.4, in which case even $n = 40$ or 50 would not suffice for approximate normality of $\bar{x}$. In practice, however, a sample size of 30 or more is usually sufficient.

EXAMPLE 8.4 Courting Scorpion Flies

Understand the context ❯

The authors of the paper **"Should I Stay or Should I Go? Condition- and Status-Dependent Courtship Decisions in the Scorpion Fly *Panorpa Cognate*" (*Animal Behaviour* [2009]: 491–497)** studied the courtship behavior of mating scorpion flies. One variable of interest was x = courtship time, which was defined as the time from the beginning of a female-male interaction until mating.

Data from the paper suggest that it is reasonable to think that the mean and standard deviation of x are $\mu = 117.1$ minutes and $\sigma = 109.1$ minutes. Notice that the distribution of courtship times cannot be normal because for a normal distribution centered at 117.1 and with such a large standard deviation, it would not be uncommon to observe negative values, but courtship time can't have a negative value.

The sampling distribution of

$\bar{x}$ = mean courtship time for a random sample of 20 scorpion fly mating pairs

would have mean

$$\mu_{\bar{x}} = \mu = 117.1 \text{ minutes}$$

So the sampling distribution is centered at 117.1. The standard deviation of $\bar{x}$ is

$$\sigma_{\bar{x}} = \frac{\sigma}{\sqrt{n}} = \frac{109.1}{\sqrt{20}} = 24.40$$

which is smaller than the population standard deviation σ. Because the population distribution is not normal and because the sample size is smaller than 30, we cannot assume that the sampling distribution of $\bar{x}$ is approximately normal in shape. ■

EXAMPLE 8.5 Soda Volumes

Understand the context ❯

A soft-drink bottler claims that, on average, cans contain 12 ounces of soda. Let x denote the actual volume of soda in a randomly selected can. Suppose that x is normally distributed with $\sigma = 0.16$ ounces. Sixteen cans are to be selected, and the soda volume will be determined for each one. Let $\bar{x}$ denote the resulting sample mean soda volume. Because the x distribution is normal, the sampling distribution of $\bar{x}$ is also normal. *If the bottler's claim is correct,* the sampling distribution of $\bar{x}$ has a mean value of $\mu_{\bar{x}} = \mu = 12$ and a standard deviation of

$$\sigma_{\bar{x}} = \frac{\sigma}{\sqrt{n}} = \frac{.16}{\sqrt{16}} = .04$$

Formulate a plan ⟩

To calculate a probability involving $\bar{x}$, we standardize by subtracting the mean value, 12, and dividing by the standard deviation (of $\bar{x}$), which is 0.04. For example, the probability that the sample mean soda volume is between 11.96 ounces and 12.08 ounces is the area between 11.96 and 12.08 under the normal curve with mean 12 and standard deviation 0.04, as shown in the following figure.

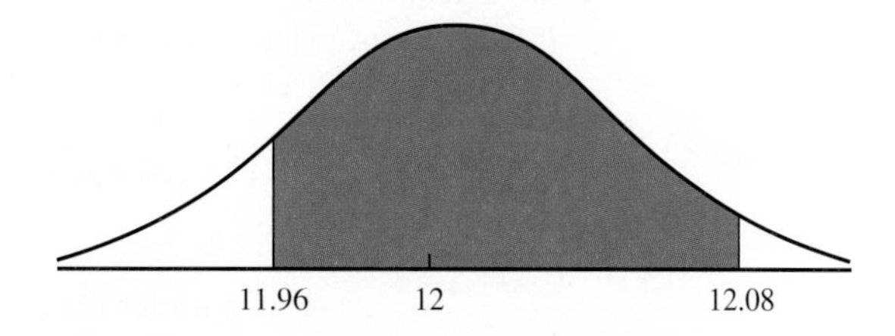

This area is calculated by first standardizing the interval limits:

Do the work ⟩

$$\text{Lower limit } a^* = \frac{11.96 - 12}{.04} = -1.0$$

$$\text{Upper limit } b^* = \frac{12.08 - 12}{.04} = 2.0$$

Then (using Appendix Table 2)

$$\begin{aligned} P(11.96 \le \bar{x} \le 12.08) &= \text{area under the } z \text{ curve between } -1.0 \text{ and } 2.0 \\ &= (\text{area to the left of } 2.0) - (\text{area to the left of } -1.0) \\ &= .9772 - .1587 \\ &= .8185 \end{aligned}$$

The probability that the sample mean soda volume is at most 11.9 ounces is

$$\begin{aligned} P(\bar{x} \le 11.9) &= P\left(z \le \frac{11.9 - 12}{.04} = -2.5\right) \\ &= (\text{area under the z curve to the left of } -2.5) = .0062 \end{aligned}$$

Interpret the results ⟩

If the x distribution is as described and the bottler's claim is correct, a sample mean soda volume based on a random sample of 16 observations is less than 11.9 ounces for fewer than 1% of all such samples. If we observed an $\bar{x}$ value that is smaller than 11.9 ounces, this would cast doubt on the bottler's claim that the average soda volume is 12 ounces. ∎

EXAMPLE 8.6 Fat Content of Hot Dogs

Understand the context ⟩

A hot dog manufacturer claims that one of its brands of hot dogs has an average fat content of $\mu = 18$ grams per hot dog. Consumers of this brand would probably not be disturbed if the mean is less than 18 but would be unhappy if it exceeds 18. Let x denote the fat content of a randomly selected hot dog, and suppose that σ, the standard deviation of the x distribution, is 1.

An independent testing organization is asked to analyze a random sample of 36 hot dogs. Let $\bar{x}$ be the average fat content for this sample. The sample size, $n = 36$, is large enough to rely on the Central Limit Theorem and to regard the $\bar{x}$ distribution as approximately normal. The standard deviation of the $\bar{x}$ distribution is

$$\sigma_{\bar{x}} = \frac{\sigma}{\sqrt{n}} = \frac{1}{\sqrt{36}} = .1667$$

If the manufacturer's claim is correct, we know that $\mu_{\bar{x}} = \mu = 18$ grams.

Suppose that the sample resulted in a mean of $\bar{x} = 18.4$ grams. Does this result suggest that the manufacturer's claim is incorrect?

Formulate a plan

We can answer this question by looking at the sampling distribution of $\bar{x}$. Because of sampling variability, even if $\mu = 18$, we know that $\bar{x}$ will not usually be exactly 18. But, is it likely that we would see a sample mean at least as large as 18.4 when the population mean is really 18?

If the company's claim is correct,

Do the work

$$P(\bar{x} \geq 18.4) \approx P\left(z \geq \frac{18.4 - 18}{.1667}\right)$$
$$= P(z \geq 2.4)$$
$$= \text{area under the } z \text{ curve to the right of } 2.4$$
$$= 1 - .9918 = .0082$$

Interpret the results

Values of $\bar{x}$ at least as large as 18.4 will be observed only about 0.82% of the time when a random sample of size 36 is taken from a population with mean 18 and standard deviation 1. The value $\bar{x} = 18.4$ is enough greater than 18 that we would question the manufacturer's claim. ■

Other Cases

We now know a great deal about the sampling distribution of $\bar{x}$ in two cases: for a normal population distribution and for a large sample size. What happens when the population distribution is not normal and n is small? Although it is still true that $\mu_{\bar{x}} = \mu$ and $\sigma_{\bar{x}} = \frac{\sigma}{\sqrt{n}}$, unfortunately there is no general result about the shape of the distribution.

When the objective is to make an inference about the center of such a population, one way to proceed is to make an assumption about the shape of the distribution. Statisticians have proposed and studied a number of such models. Theoretical methods or simulation can be used to describe the $\bar{x}$ distribution corresponding to the assumed model.

An alternative strategy is to use one of the transformations presented in Chapter 7 to create a data set that more closely resembles a sample from a normal population and then to base inferences on the transformed data. Yet another path is to use an inferential procedure based on a statistic other than $\bar{x}$. Consult a statistician or a more advanced text for more information.

EXERCISES 8.10 - 8.22

8.10 A random sample is selected from a population with mean $\mu = 100$ and standard deviation $\sigma = 10$. Determine the mean and standard deviation of the $\bar{x}$ sampling distribution for each of the following sample sizes:

a. $n = 9$
b. $n = 15$
c. $n = 36$
d. $n = 50$
e. $n = 100$
f. $n = 400$

8.11 For which of the sample sizes given in the previous exercise would it be reasonable to think that the $\bar{x}$ sampling distribution is approximately normal in shape?

8.12 Explain the difference between σ and $\sigma_{\bar{x}}$ and between μ and $\mu_{\bar{x}}$.

8.13 ▼ Suppose that a random sample of size 64 is to be selected from a population with mean 40 and standard deviation 5.

a. What are the mean and standard deviation of the $\bar{x}$ sampling distribution? Describe the shape of the $\bar{x}$ sampling distribution.

Bold exercises answered in back ● Data set available online ▼ Video Solution available

b. What is the approximate probability that $\bar{x}$ will be within 0.5 of the population mean μ? (Hint: See Examples 8.5 and 8.6.)

c. What is the approximate probability that $\bar{x}$ will differ from μ by more than 0.7?

8.14 The time that a randomly selected individual waits for an elevator in an office building has a uniform distribution over the interval from 0 to 1 minute. For this distribution $\mu = 0.5$ and $\sigma = 0.289$.

a. Let $\bar{x}$ be the sample mean waiting time for a random sample of 16 individuals. What are the mean and standard deviation of the sampling distribution of $\bar{x}$?

b. Answer Part (a) for a random sample of 50 individuals.

c. Draw a picture of a good approximation to the sampling distribution of $\bar{x}$ when $n = 50$.

8.15 Let x denote the time (in minutes) that it takes a fifth-grade student to read a certain passage. Suppose that the mean value and standard deviation of x are $\mu = 2$ minutes and $\sigma = 0.8$ minutes, respectively.

a. If $\bar{x}$ is the sample mean time for a random sample of $n = 9$ students, where is the $\bar{x}$ distribution centered, and how much does it spread out around the center (as described by its standard deviation)?

b. Repeat Part (a) for a sample of size of $n = 20$ and again for a sample of size $n = 100$. How do the centers and spreads of the three $\bar{x}$ distributions compare to one another?

c. Which of the sample sizes in Part (b) would be most likely to result in an $\bar{x}$ value close to μ, and why?

8.16 In the library on a university campus, there is a sign in the elevator that indicates a limit of 16 persons. In addition, there is a weight limit of 2500 pounds. Assume that the average weight of students, faculty, and staff on campus is 150 pounds, that the standard deviation is 27 pounds, and that the distribution of weights of individuals on campus is approximately normal. Suppose a random sample of 16 persons from the campus will be selected.

a. What is the mean of the $\bar{x}$ sampling distribution?

b. What is the standard deviation of the $\bar{x}$ sampling distribution?

c. What average weights for a sample of 16 people will result in the total weight exceeding the weight limit of 2500 pounds?

d. What is the chance that a random sample of 16 people will exceed the weight limit?

8.17 Suppose that the mean value of interpupillary distance (the distance between the pupils of the left and right eyes) for adult males is 65 mm and that the population standard deviation is 5 mm.

a. If the distribution of interpupillary distance is normal and a random sample of $n = 25$ adult males is to be selected, what is the probability that the sample mean distance $\bar{x}$ for these 25 will be between 64 and 67 mm? at least 68 mm? (Hint: See Examples 8.5 and 8.6.)

b. Suppose that a random sample of 100 adult males is to be obtained. Without assuming that interpupillary distance is normally distributed, what is the approximate probability that the sample mean distance will be between 64 and 67 mm? at least 68 mm?

8.18 Suppose that a sample of size 100 is to be drawn from a population with standard deviation 10.

a. What is the probability that the sample mean will be within 1 of the value of μ?

b. For this example ($n = 100$, $\sigma = 10$), complete each of the following statements by computing the appropriate value:

i. Approximately 95% of the time, $\bar{x}$ will be within _____ of μ.

ii. Approximately 0.3% of the time, $\bar{x}$ will be farther than _____ from μ.

8.19 A manufacturing process is designed to produce bolts with a 0.5-inch diameter. Once each day, a random sample of 36 bolts is selected and the bolt diameters are recorded. If the resulting sample mean is less than 0.49 inches or greater than 0.51 inches, the process is shut down for adjustment. The standard deviation for diameter is 0.02 inches.

What is the probability that the manufacturing line will be shut down unnecessarily? (Hint: Find the probability of observing an $\bar{x}$ in the shut-down range when the true process mean really is 0.5 inches.)

8.20 College students with checking accounts typically write relatively few checks in any given month, whereas adults who are not students typically write many more checks during a month. Suppose that 50% of a bank's accounts are held by students and that 50% are held by adults who are not students. Let x denote the number of checks written in a given month by a randomly selected bank customer.

a. Give a sketch of what the probability distribution of x might look like.

b. Suppose that the mean value of x is 22.0 and that the standard deviation is 16.5. A random sample of $n = 100$ customers is to be selected. Where is the sampling distribution of $\bar{x}$ centered, and what is the standard deviation of the $\bar{x}$ distribution? Sketch a rough picture of the sampling distribution.

c. Referring to Part (b), what is the approximate probability that $\bar{x}$ is at most 20? at least 25?

8.21 ▼ An airplane with room for 100 passengers has a total baggage limit of 6000 pounds. Suppose that the total weight of the baggage checked by an individual passenger is a random variable x with a mean value of 50 pounds and a standard deviation of 20 pounds. If 100 passengers will board a flight, what is the approximate probability that the total weight of their baggage will exceed the limit? (Hint: With $n = 100$, the total weight exceeds the limit when the average weight $\bar{x}$ exceeds 6000/100.)

8.22 The thickness (in millimeters) of the coating applied to disk drives is one characteristic that determines the usefulness of the product. When no unusual circumstances are present, the thickness (x) has a normal distribution with a mean of 2 mm and a standard deviation of 0.05 mm. Suppose that the process will be monitored by selecting a random sample of 16 drives from each shift's production and determining $\bar{x}$, the mean coating thickness for the sample.

a. Describe the sampling distribution of $\bar{x}$ for a random sample of size 16.

b. When no unusual circumstances are present, we expect $\bar{x}$ to be within $3\sigma_{\bar{x}}$ of 2 mm, the desired value. An $\bar{x}$ value farther from 2 than $3\sigma_{\bar{x}}$ is interpreted as an indication of a problem that needs attention. Compute $2 \pm 3\sigma_{\bar{x}}$.

c. Referring to Part (b), what is the probability that a sample mean will be outside $2 \pm 3\sigma_{\bar{x}}$ just by chance (that is, when there are no unusual circumstances)?

d. Suppose that a machine used to apply the coating is out of adjustment, resulting in a mean coating thickness of 2.05 mm. What is the probability that a problem will be detected when the next sample is taken? (Hint: This will occur if $\bar{x} > 2 + 3\sigma_{\bar{x}}$ or $\bar{x} < 2 - 3\sigma_{\bar{x}}$ when $\mu = 2.05$.)

Bold exercises answered in back ● Data set available online ▼ Video Solution available

8.3 The Sampling Distribution of a Sample Proportion

The objective of many statistical investigations is to draw a conclusion about the proportion of individuals or objects in a population that possess a specified property—for example, cell phones that don't require service during the warranty period or coffee drinkers who regularly drink decaffeinated coffee. Traditionally, any individual or object that possesses the property of interest is labeled a success (S), and one that does not possess the property is termed a failure (F).

The letter p is used to denote the proportion of successes in the population. The value of p is a number between 0 and 1, and $100p$ is the percentage of successes in the population. For example, if $p = .75$, 75% of the population members are successes, and if $p = .01$, the population contains only 1% successes and 99% failures.

The value of p is usually unknown. When a random sample of size n is selected from this type of population, some of the individuals in the sample are successes, and the rest are failures. The statistic that provides a basis for making inferences about p is $\hat{p}$, the **sample proportion of successes**:

$$\hat{p} = \frac{\text{number of S's in the sample}}{n}$$

For example, if there are three successes in a sample of size 5, then $\hat{p} = 3/5 = .6$.

Just as making inferences about μ requires knowing something about the sampling distribution of the statistic $\bar{x}$, making inferences about p requires first learning about properties of the sampling distribution of the statistic $\hat{p}$. For example, when $n = 5$, the six possible values of $\hat{p}$ are 0, .2 (from 1/5), .4, .6, .8, and 1. The sampling distribution of $\hat{p}$ gives the probability of each of these six possible values. These probabilities are the long-run proportions of the time that these values would occur if random samples with $n = 5$ were selected over and over again.

As we did for the sample mean, we will look at some simulation experiments to develop an intuitive understanding of the distribution of the sample proportion before stating general properties. In each example, 500 random samples of size n are selected from a population having a specified value of p. We compute $\hat{p}$ for each sample and then construct a density histogram of the 500 values.

EXAMPLE 8.7 Gender of College Students

In the fall of 2008, there were 18,516 students enrolled at California Polytechnic State University, San Luis Obispo. Of these students, 8091 (43.7%) were female. To illustrate properties of the sampling distribution of a sample proportion, we will simulate sampling from this Cal Poly student population.

With S denoting a female student and F a male student, the proportion of S's in the population is $p = .437$. A statistical software package was used to select 500 samples of size $n = 10$, then 500 samples of size $n = 25$, then 500 samples with $n = 50$, and finally 500 samples with $n = 100$. Histograms of the 500 values of $\hat{p}$ for each of the four sample sizes are displayed in Figure 8.7.

The most noticeable feature of the histogram shapes is that all are approximately symmetric. All four histograms appear to be centered at roughly .437, the value of p for the population sampled. Had the histograms been based on an unending sequence of samples, each histogram would have been centered at exactly .437. Finally, as was the case with the sampling distribution of $\bar{x}$, the histograms spread out more for small sample sizes than for large sample sizes. Not surprisingly, the value of $\hat{p}$ based on a large sample size tends to be closer to p, the population proportion of successes, than does $\hat{p}$ from a small sample.

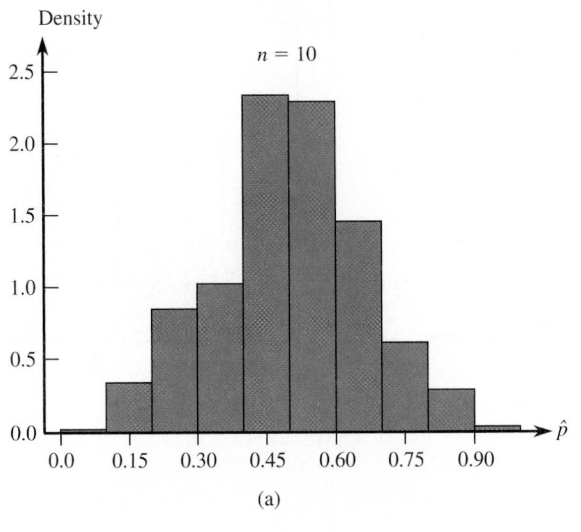

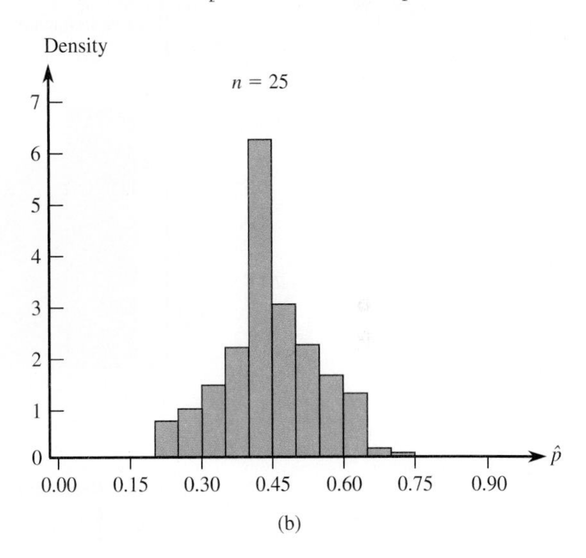

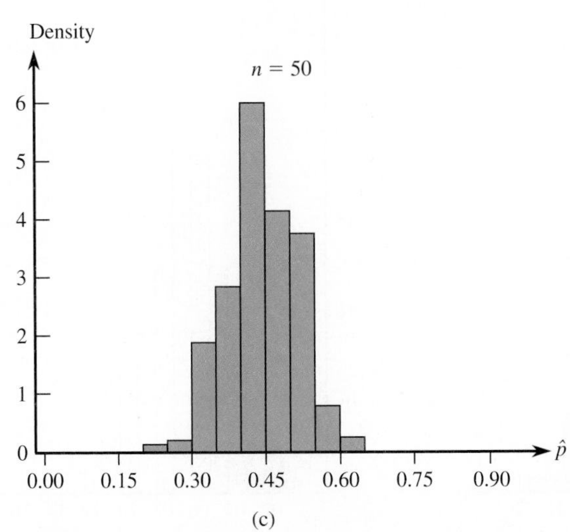

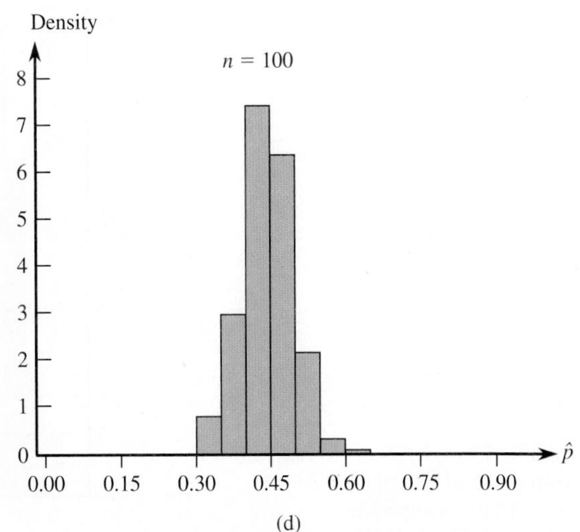

FIGURE 8.7
Histograms for 500 values of $\hat{p}$ ($p = .437$) for Example 8.7: (a) $n = 10$; (b) $n = 25$; (c) $n = 50$; (d) $n = 100$.

■

EXAMPLE 8.8 Contracting Hepatitis from Blood Transfusion

The development of viral hepatitis after a blood transfusion can cause serious complications for a patient. The article **"Lack of Awareness Results in Poor Autologous Blood Transfusion" (*Health Care Management*, May 15, 2003)** reported that hepatitis occurs in 7% of patients who receive blood transfusions during heart surgery. Here, we simulate sampling from the population of blood recipients, with S denoting a recipient who contracts hepatitis (not the sort of characteristic one usually thinks of as a success, but the S–F labeling is arbitrary), so $p = .07$. Figure 8.8 displays histograms of 500 values of $\hat{p}$ for the four sample sizes $n = 10, 25, 50$, and 100.

As was the case in Example 8.7, all four histograms are centered at approximately the value of p for the population being sampled. (The average of the $\hat{p}$ values for these simulations are .0690, .0677, .0707, and .0694.) If the histograms had been based on an unending sequence of samples, they would all have been centered at exactly $p = .07$.

The spread of a histogram based on a large n is smaller than the spread of a histogram resulting from a small sample size. The larger the value of n, the closer the sample proportion $\hat{p}$ tends to be to the value of the population proportion p.

Also notice that there is a progression toward the shape of a normal curve as n increases. The histograms for $n = 10$ and $n = 25$ exhibit substantial skew, and the skew of the histogram for $n = 50$ is still moderate (compare Figure 8.8(c) with Figure 8.8(d)). Only the

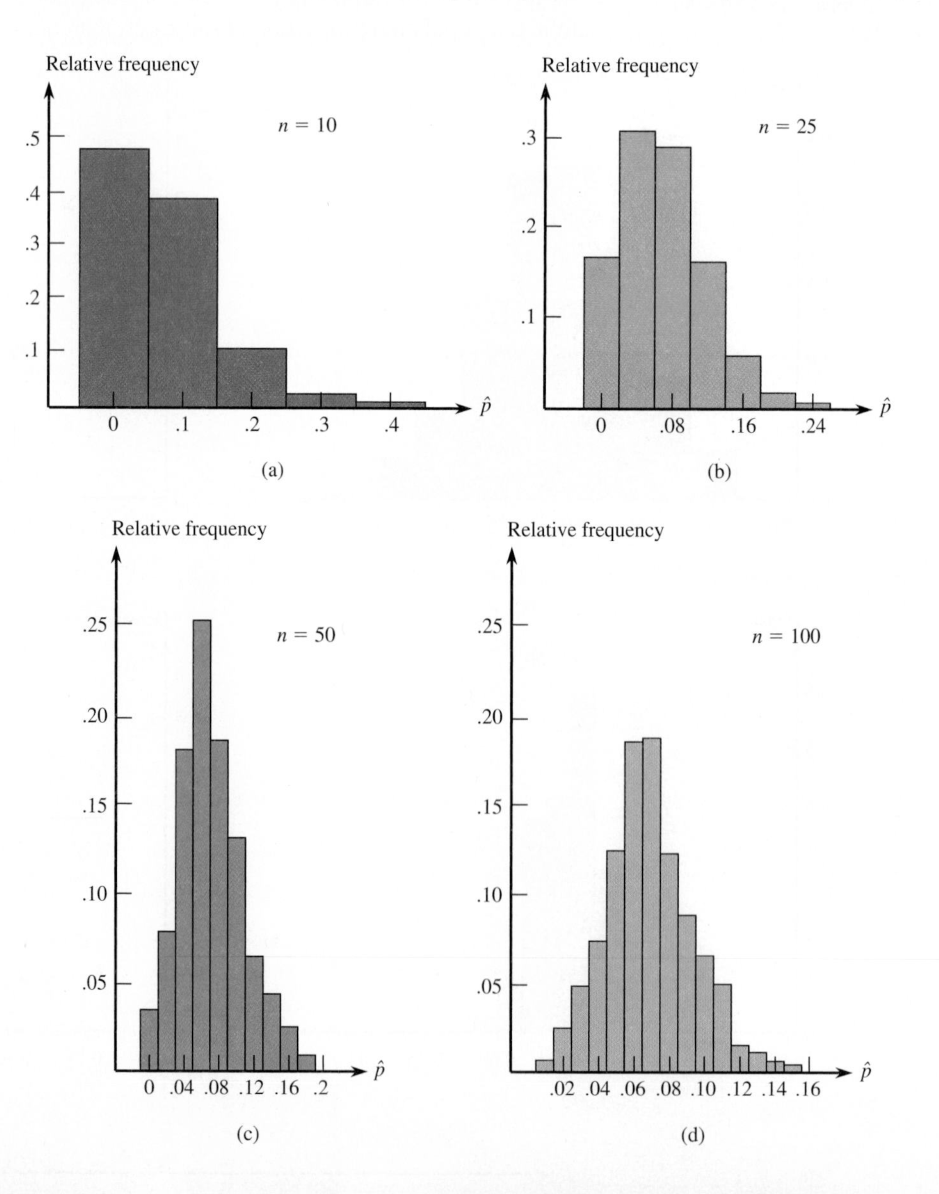

FIGURE 8.8
Histograms of 500 values of $\hat{p}$ ($p = .07$) for Example 8.8: (a) $n = 10$; (b) $n = 25$; (c) $n = 50$; (d) $n = 100$.

histogram for $n = 100$ is described reasonably well by a normal curve. It appears that whether a normal curve provides a good approximation to the sampling distribution of $\hat{p}$ depends on the values of both n and p. Knowing only that $n = 50$ is not enough to guarantee that the shape of the histogram is approximately normal. ■

General Properties of the Sampling Distribution of $\hat{p}$

Examples 8.7 and 8.8 suggest that the sampling distribution of $\hat{p}$ depends on both n, the sample size, and p, the proportion of successes in the population. Key results are stated more formally in the following general rules.

General Properties of the Sampling Distribution of $\hat{p}$

Notation

$\hat{p}$ sample proportion
n sample size
p population proportion
$\mu_{\hat{p}}$ mean of the $\hat{p}$ sampling distribution
$\sigma_{\hat{p}}$ standard deviation of the $\hat{p}$ sampling distribution

When random samples are selected from a population, the following are properties of the sampling distribution of $\hat{p}$:

Property 1. $\mu_{\hat{p}} = p$.

Property 2. $\sigma_{\hat{p}} = \sqrt{\frac{p(1-p)}{n}}$.

This formula is exact if the population is infinite, and is approximately correct if the population is finite and no more than 10% of the population is included in the sample.

Property 3. When n is large and p is not too near 0 or 1, the sampling distribution of $\hat{p}$ is approximately normal.

The sampling distribution of $\hat{p}$ is always centered at the value of the population success proportion p, and the extent to which the distribution spreads out about p decreases as the sample size n increases.

Examples 8.7 and 8.8 indicate that both p and n must be considered in judging whether the sampling distribution of $\hat{p}$ is approximately normal.

The farther the value of p is from .5, the larger n must be in order for the sampling distribution of $\hat{p}$ to be approximately normal.

A conservative rule of thumb is that if both

$np \geq 10$ and $n(1 - p) \geq 10$,

then a normal distribution provides a reasonable approximation to the sampling distribution of $\hat{p}$.

A sample size of $n = 100$ is not by itself sufficient to justify the use of a normal approximation. If $p = .01$, the distribution of $\hat{p}$ is positively skewed even when $n = 100$, so a bell-shaped curve does not give a good approximation. Similarly, if $n = 100$ and $p = .99$ (so that $n(1 - p) = 1 < 10$), the distribution of $\hat{p}$ has a substantial negative skew. The conditions $np \geq 10$ and $n(1 - p) \geq 10$ ensure that the sampling distribution of $\hat{p}$ is not too skewed. If $p = .5$, the normal approximation can be used for n as small as 20, whereas for $p = .05$ or .95, n should be at least 200.

EXAMPLE 8.9 Blood Transfusions Continued

Understand the context

The proportion of all cardiac patients receiving blood transfusions who contract hepatitis was given as .07 in the article referenced in Example 8.8. Suppose that a new blood screening procedure is believed to reduce the incidence rate of hepatitis. Blood screened using this procedure is given to $n = 200$ blood recipients. Only 6 of the 200 patients contract hepatitis.

This appears to be a favorable result, because $\hat{p} = 6/200 = .03$. The question of interest to medical researchers is, Does this result indicate that the true (long-run) proportion of patients who contract hepatitis when the new screening procedure is used is less than .07, or could this result be plausibly attributed to sampling variability (that is, to the fact that $\hat{p}$ typically differs from the population proportion, p)?

If the screening procedure is not effective and so $p = .07$,

$$\mu_{\hat{p}} = 0.7$$

$$\sigma_{\hat{p}} = \sqrt{\frac{p(1-p)}{200}} = \sqrt{\frac{(.07)(.93)}{200}} = .018$$

Because

$$np = 200(.07) = 14 \geq 10$$

and

$$n(1-p) = 200(.93) = 186 \geq 10$$

the sampling distribution of $\hat{p}$ is approximately normal.

Do the work

Then, if the screening procedure is not effective,

$$P(\hat{p} \leq .03) = P\left(z \leq \frac{.03 - .07}{.018}\right)$$
$$= P(z \leq -2.22)$$
$$= .0132$$

Interpret the results

This small probability tells us that it is unlikely that a sample proportion .03 or smaller would be observed if the screening procedure was not effective. The new screening procedure appears to yield a smaller incidence rate for hepatitis. ■

EXERCISES 8.23 - 8.31

8.23 A random sample is to be selected from a population that has a proportion of successes $p = .65$. Determine the mean and standard deviation of the sampling distribution of $\hat{p}$ for each of the following sample sizes:

a. $n = 10$
b. $n = 20$
c. $n = 30$
d. $n = 50$
e. $n = 100$
f. $n = 200$

8.24 a. For which of the sample sizes given in the previous exercise would the sampling distribution of $\hat{p}$ be approximately normal if $p = .65$?

b. For which of the sample sizes given in the previous exercise would the sampling distribution be approximately normal if $p = .2$?

8.25 ▼ The article **"Unmarried Couples More Likely to Be Interracial"** (***San Luis Obispo Tribune*, March 13, 2002**) reported that 7% of married couples in the United States are mixed racially or ethnically. Consider the population consisting of all married couples in the United States. (Hint: See Example 8.9.)

a. A random sample of $n = 100$ couples will be selected from this population and $\hat{p}$, the proportion of couples that are mixed racially or ethnically, will be computed. What are the mean and standard deviation of the sampling distribution of $\hat{p}$?

b. Is it reasonable to assume that the sampling distribution of $\hat{p}$ is approximately normal for random samples of size $n = 100$? Explain.

c. Suppose that the sample size is $n = 200$ rather than $n = 100$, as in Part (b). Does the change in sample size change the mean and standard deviation of the sampling distribution of $\hat{p}$? If so, what are the new values for the mean and standard deviation? If not, explain why not.

d. Is it reasonable to assume that the sampling distribution of $\hat{p}$ is approximately normal for random samples of size $n = 200$? Explain.

e. When $n = 200$, what is the probability that the proportion of couples in the sample who are racially or ethnically mixed will be greater than .10?

8.26 The article referenced in the previous exercise reported that for unmarried couples living together, the proportion that are racially or ethnically mixed is .15. Answer the questions posed in Parts (a)–(e) of the previous exercise for the population of unmarried couples living together.

8.27 ▼ A certain chromosome defect occurs in only 1 in 200 adult Caucasian males. A random sample of $n = 100$ adult Caucasian males is to be obtained.

a. What is the mean value of the sample proportion $\hat{p}$, and what is the standard deviation of the sample proportion?

b. Does $\hat{p}$ have approximately a normal distribution in this case? Explain.

c. What is the smallest value of n for which the sampling distribution of $\hat{p}$ is approximately normal?

8.28 The article **"Should Pregnant Women Move? Linking Risks for Birth Defects with Proximity to Toxic Waste Sites" (*Chance* [1992]: 40–45)** reported that in a large study carried out in the state of New York, approximately 30% of the study subjects lived within 1 mile of a hazardous waste site. Let p denote the proportion of all New York residents who live within 1 mile of such a site, and suppose that $p = .3$.

a. Would $\hat{p}$ based on a random sample of only 10 residents have approximately a normal distribution? Explain why or why not.

b. What are the mean value and standard deviation of $\hat{p}$ based on a random sample of size 400?

c. When $n = 400$, what is $P(.25 \leq \hat{p} \leq .35)$? (Hint: See Example 8.9.)

d. Is the probability calculated in Part (c) larger or smaller than would be the case if $n = 500$? Answer without actually calculating this probability.

8.29 The article **"Thrillers" (*Newsweek*, April 22, 1985)** stated, "Surveys tell us that more than half of America's college graduates are avid readers of mystery novels." Let p denote the actual proportion of college graduates who are avid readers of mystery novels. Consider a sample proportion $\hat{p}$ that is based on a random sample of 225 college graduates.

a. If $p = .5$, what are the mean value and standard deviation of $\hat{p}$? Answer this question for $p = .6$. Does $\hat{p}$ have approximately a normal distribution in both cases? Explain.

b. Calculate $P(\hat{p} \geq .6)$ for both $p = .5$ and $p = .6$.

c. Without doing any calculations, how do you think the probabilities in Part (b) would change if n were 400 rather than 225?

8.30 Suppose that a particular candidate for public office is in fact favored by 48% of all registered voters in the district. A polling organization will take a random sample of 500 voters and will use $\hat{p}$, the sample proportion, to estimate p. What is the approximate probability that $\hat{p}$ will be greater than .5, causing the polling organization to incorrectly predict the result of the upcoming election? (Hint: See Example 8.9.)

8.31 ▼ A manufacturer of computer printers purchases plastic ink cartridges from a vendor. When a large shipment is received, a random sample of 200 cartridges is selected, and each cartridge is inspected. If the sample proportion of defective cartridges is more than .02, the entire shipment is returned to the vendor.

a. What is the approximate probability that a shipment will be returned if the true proportion of defective cartridges in the shipment is .05?

b. What is the approximate probability that a shipment will not be returned if the true proportion of defective cartridges in the shipment is .10?

ACTIVITY 8.1 Do Students Who Take the SATs Multiple Times Have an Advantage in College Admissions?

Technology activity: Requires use of a computer or a graphing calculator.

Background: ***The Chronicle of Higher Education*** **(January 29, 2003)** summarized an article that appeared on the ***American Prospect*** web site titled **"College Try: Why Universities Should Stop Encouraging Applicants to Take the SATs Over and Over Again."** This paper argued that current college admission policies that permit applicants to take the SAT exam multiple times and then use the highest score for consideration of admission favor students from families with higher incomes (who can afford to take the exam many times). The author proposed two alternatives that he believes would be fairer than using the highest score: (1) Use the average of all test scores, or (2) use only the most recent score.

In this activity, you will investigate the differences between the three possibilities by looking at the sampling distributions of three statistics for a test taker who takes the exam twice and for a test taker who takes the exam five times. The three statistics are

Max = maximum score
Mean = average score
Recent = most recent score

An individual's score on the SAT exam fluctuates between test administrations. Suppose that a particular student's "true ability" is reflected by an SAT score of 1200 but, because of chance fluctuations, the test score on any particular administration of the exam can be considered a random variable that has a distribution that is approximately normal with mean 1200 and standard deviation 30. If we select a sample from this normal distribution, the resulting set of observations can be viewed as a collection of test scores that might have been obtained by this student.

Part 1: Begin by considering what happens if this student takes the exam twice. You will use simulation to generate samples of two test scores, Score1 and Score2, for this student. Then you will compute the values of Max, Mean, and Recent for each pair of scores. The resulting values of Max, Mean, and Recent will be used to construct approximations to the sampling distributions of the three statistics.

The instructions that follow assume the use of Minitab. If you are using a different software package or a graphing calculator, your instructor will provide alternative instructions.

a. Obtain 500 sets of two test scores by generating observations from a normal distribution with mean 1200 and standard deviation 30.

Minitab: Calc → Random Data → Normal
Enter 500 in the Generate box (to get 500 sets of scores)
Enter C1-C2 in the Store in Columns box (to get two test scores in each set)
Enter 1200 in the Mean box (because we want scores from a normal distribution with mean 1200)
Enter 30 in the Standard Deviation box (because we want scores from a normal distribution with standard deviation 30)
Click on OK

b. Looking at the Minitab worksheet, you should now see 500 rows of values in each of the first two columns. The two values in any particular row can be regarded as the test scores that might be observed when the student takes the test twice. For each pair of test scores, we now calculate the values of Max, Mean, and Recent.

i. Recent is just the last test score, so the values in C2 are the values of Recent. Name this column recent2 by typing the name into the gray box at the top of C2.

ii. Compute the maximum test score (Max) for each pair of scores, and store the values in C3, as follows:

Minitab: Calc → Row statistics
Click the button for maximum
Enter C1-C2 in the Input variables box
Enter C3 in the Store Result In box.
Click on OK

You should now see the maximum value for each pair in C3. Name this column max2.

iii. Compute the average test score (Mean) for each pair of scores, and store the values in C4, as follows:

Minitab: Calc → Row statistics
Click the button for mean
Enter C1-C2 in the Input Variables box
Enter C4 in the Store Result In box.
Click on OK

You should now see the average for each pair in C4. Name this column mean2.

c. Construct density histograms for each of the three statistics (these density histograms approximate the sampling distributions of the three statistics), as follows:

Minitab: Graph → Histogram
Enter max2, mean2, and recent2 into the first three rows of the Graph Variables box
Click on the Options button. Select Density. Click on OK. (This will produce histograms that use the density scale rather than the frequency scale.)
Click on the Frame drop-down menu, and select Multiple Graphs. Select Same X and Same Y. (This will cause Minitab to use the same scales for all three histograms, so that they can be easily compared.)
Click on OK.

Part 2: Now you will produce approximate sampling distributions for these same three statistics, but for the case of a student who takes the exam five times. Follow the same steps as in Part 1, with the following modifications:

a. Obtain 500 sets of five test scores, and store these values in columns C11– C15.

b. Recent will just be the values in C15; name this column recent5. Compute the Max and Mean values, and store them in columns C16 and C17. Name these columns max5 and mean5.

c. Construct density histograms for max5, mean5, and recent5.

Part 3: Now use the approximate sampling distributions constructed in Parts 1 and 2 to answer the following questions.

a. The statistic that is the average of the test scores is just a sample mean (for a sample of size 2 in Part 1 and for a sample of size 5 in Part 2). How do the sampling distributions of mean2 and mean5 compare to what is expected based on the general properties of the $\bar{x}$ distribution given in Section 8.2? Explain.

b. Based on the three distributions from Part 1, for a two-time test taker, describe the advantage of using the maximum score compared to using either the average score or the most recent score.

c. Now consider the approximate sampling distributions of the maximum score for two-time and for five-time test takers. How do these two distributions compare?

d. Does a student who takes the exam five times have a big advantage over a student of equal ability who takes the exam only twice if the maximum score is used for college admission decisions? Explain.

e. If you were writing admission procedures for a selective university, would you recommend using the maximum test score, the average test score, or the most recent test score in making admission decisions? Write a paragraph explaining your choice.

SUMMARY Key Concepts and Formulas

TERM OR FORMULA	COMMENT
Statistic	Any quantity whose value is computed from sample data.
Sampling distribution	The probability distribution of a statistic. The sampling distribution describes the long-run behavior of the statistic.
Sampling distribution of $\bar{x}$	The probability distribution of the sample mean $\bar{x}$ based on a random sample of size n. Properties of the $\bar{x}$ sampling distribution: $\mu_{\bar{x}} = \mu$ and $\sigma_{\bar{x}} = \frac{\sigma}{\sqrt{n}}$ (where μ and σ are the population mean and standard deviation, respectively). In addition, when the population distribution is normal or the sample size is large, the sampling distribution of $\bar{x}$ is (approximately) normal.
Central Limit Theorem	This important theorem states that when n is sufficiently large, the $\bar{x}$ distribution will be approximately normal. The standard rule of thumb is that the theorem can safely be applied when n is greater than or equal to 30.
Sampling distribution of $\hat{p}$	The probability distribution of the sample proportion $\hat{p}$, based on a random sample of size n. When the sample size is sufficiently large, the sampling distribution of $\hat{p}$ is approximately normal, with $\mu_{\hat{p}} = p$ and $\sigma_{\hat{p}} = \sqrt{\frac{p(1-p)}{n}}$ where p is the value of the population proportion.

CHAPTER REVIEW Exercises 8.32 - 8.37

8.32 The nicotine content in a single cigarette of a particular brand has a distribution with mean 0.8 mg and standard deviation 0.1 mg. If 100 randomly selected cigarettes of this brand are analyzed, what is the probability that the resulting sample mean nicotine content will be less than 0.79? less than 0.77?

8.33 Let $x_1, x_2, \ldots, x_{100}$ denote the actual net weights (in pounds) of 100 randomly selected bags of fertilizer. Suppose that the weight of a randomly selected bag has a distribution with mean 50 pounds and variance 1 pound^2. Let $\bar{x}$ be the sample mean weight ($n = 100$).

a. Describe the sampling distribution of $\bar{x}$.

b. What is the probability that the sample mean is between 49.75 pounds and 50.25 pounds?

c. What is the probability that the sample mean is less than 50 pounds?

8.34 Suppose that 20% of the subscribers of a cable television company watch the shopping channel at least once a week. The cable company is trying to decide whether to replace this channel with a new local station. A survey of 100 subscribers will be undertaken. The cable company has decided to keep the shopping channel if the sample proportion is greater than .25.

What is the approximate probability that the cable company will keep the shopping channel, even though the proportion of all subscribers who watch it is only .20?

8.35 Water permeability of concrete can be measured by letting water flow across the surface and determining the amount lost (in inches per hour). Suppose that the permeability index x for a randomly selected concrete specimen of a particular type is normally distributed with mean value 1000 and standard deviation 150.

a. How likely is it that a single randomly selected specimen will have a permeability index between 850 and 1300?

b. If the permeability index is to be determined for each specimen in a random sample of size 10, how likely is it that the sample mean permeability index will be

i. between 950 and 1100?

ii. between 850 and 1300?

8.36 Suppose that 40% of all U.S. employees contribute to a retirement plan ($p = .40$).

a. In a random sample of 100 employees, what is the approximate probability that at least half of those in the sample contribute to a retirement plan?

b. Suppose you were told that at least 60 of the 100 employees in a sample from your state contribute to a retirement plan. Would you think $p = .40$ for your state? Explain.

8.37 The amount of money spent by a customer at a discount store has a mean of \$100 and a standard deviation of \$30. What is the probability that a randomly selected group of 50 shoppers will spend a total of more than \$5300? (Hint: The total will be more than \$5300 when the sample mean exceeds what value?)

CHAPTER 9

Estimation Using a Single Sample

George Hall/Documentary Value/Corbis

Most American college students make use of the Internet for both academic and social purposes. The authors of the paper "U.S. College Students' Internet Use: Race, Gender and Digital Divides" (*Journal of Computer-Mediated Communication* [2009]: 244–264) describe the results of a survey of 7421 students at 40 colleges and universities. The sample was selected in a way that the authors believed would result in a sample that reflected general demographics of college students in the U.S.

The authors wanted to use the sample data to estimate the proportion of college students who spend more than 3 hours a day on the Internet. The methods introduced in this chapter can be used to estimate this proportion. Because the estimate will be based only on a sample rather than on a census of all U.S. college students, it is important that this estimate be constructed in a way that also conveys information about the anticipated accuracy.

Chapter 9: Learning Objectives

STUDENTS WILL UNDERSTAND:

- what it means for a statistic to be an unbiased estimator of a population characteristic.
- the relationships between sample size, margin of error, and the width of a confidence interval.
- the factors that affect the width of a confidence interval.
- the meaning of the confidence level associated with a confidence interval.

STUDENTS WILL BE ABLE TO:

- construct and interpret a confidence interval for a population proportion.
- construct and interpret a confidence interval for a population mean.
- determine the sample size necessary to achieve a desired bound on error when estimating a population proportion or a population mean.

9.1 Point Estimation

The simplest approach to estimating a population characteristic involves using sample data to compute a single number that can be regarded as a plausible value of the population characteristic. For example, sample data might suggest that 1000 hours is a plausible value for μ, the mean lifetime for all lightbulbs of a particular brand. A sample survey of students at a particular university might lead to the statement that .41 is a plausible value for p, the proportion of all students at the university who favor a fee for recreational facilities.

DEFINITION

Point estimate: A single number that is based on sample data and represents a plausible value of a population characteristic.

A point estimate is obtained by first selecting an appropriate statistic. The value of the statistic for a given sample is then used as the point estimate. For example, the computed value of the sample mean is one point estimate of a population mean μ, and the sample proportion is a point estimate of a population proportion, p.

EXAMPLE 9.1 Internet Use by College Students

Understand the context)

One of the purposes of the survey described in the chapter introduction was to estimate the proportion of college students who spend more than 3 hours a day on the Internet. Based on information given in the paper, 2998 of the 7421 students surveyed reported Internet use of more than 3 hours per day. We can use this information to estimate p, where p is the proportion of all U.S. college students who use the Internet more than 3 hours a day.

With a success identified as a student who uses the Internet more than 3 hours a day, p is the population proportion of successes. The statistic

$$\hat{p} = \frac{\text{number of successes in the sample}}{n}$$

which is the sample proportion of successes, is an obvious choice for obtaining a point estimate of p. Using the reported information, the point estimate of p is

Do the work)

$$\hat{p} = \frac{2998}{7421} = .404$$

Interpret the results)

Based on this random sample, we estimate that 40.4% of college students in the United States spend more than 3 hours a day on the Internet. ■

For purposes of estimating a population proportion p, there is no obvious alternative to the statistic $\hat{p}$. In other situations, such as the one illustrated in Example 9.2, there may be several statistics that might be used to obtain an estimate.

EXAMPLE 9.2 Academic Reading

Understand the context)

The paper **"The Impact of Internet and Television Use on the Reading Habits and Practices of College Students"** (***Journal of Adolescent and Adult Literacy*** **[2009]: 609–619)** investigates the reading habits of college students. The authors distinguished between recreational reading and academic reading and asked students to keep track of time spent reading. The following observations represent the number of hours spent on academic reading in 1 week by

20 college students (these data are compatible with summary values given in the paper and have been arranged in order from smallest to largest):

Consider the data ⟩

1.7	3.8	4.7	9.6	11.7	12.3	12.3	12.4	12.6	13.4
14.1	14.2	15.8	15.9	18.7	19.4	21.2	21.9	23.3	28.2

A dotplot of the data is shown here:

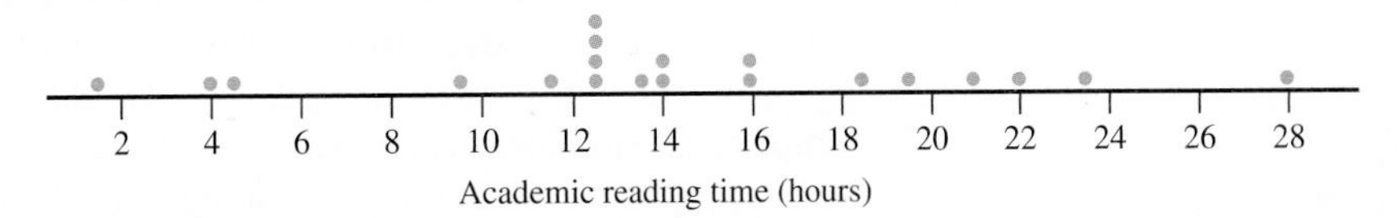

From the dotplot, we can see that the distribution of academic reading time is approximately symmetric.

Suppose a point estimate of μ, the mean academic reading time per week for all college students, is desired. An obvious choice of a statistic for estimating μ is the sample mean, $\bar{x}$. However, there are other possibilities. We might consider using the sample median, because the data set exhibits some symmetry. (If the corresponding population distribution is symmetric, the population mean μ and the population median are equal).

The estimates of μ based on the mean and the median for this sample are

Do the work ⟩

$$\text{sample mean} = \bar{x} = \frac{\Sigma x}{n} = \frac{287.2}{20} = 14.36$$

$$\text{sample median} = \frac{13.4 + 14.1}{2} = 13.75$$

The estimates of the mean academic reading time per week for college students differ somewhat from one another. The choice between them should depend on which statistic tends, on average, to produce an estimate closest to the actual value of μ. The following subsection discusses criteria for choosing among competing statistics. ■

Choosing a Statistic for Computing an Estimate

As illustrated in Example 9.2, there may be more than one statistic that is reasonable to use as a point estimate of a specified population characteristic. We would like to use a statistic that tends to produce an accurate estimate—that is, an estimate close to the value of the population characteristic. Information about the accuracy of estimation for a particular statistic is provided by the statistic's sampling distribution.

Figure 9.1 displays the sampling distributions of three different statistics. The value of the population characteristic, which is denoted by *true value* in the figure, is marked on the measurement axis. The distribution in Figure 9.1(a) is that of a statistic unlikely to yield an

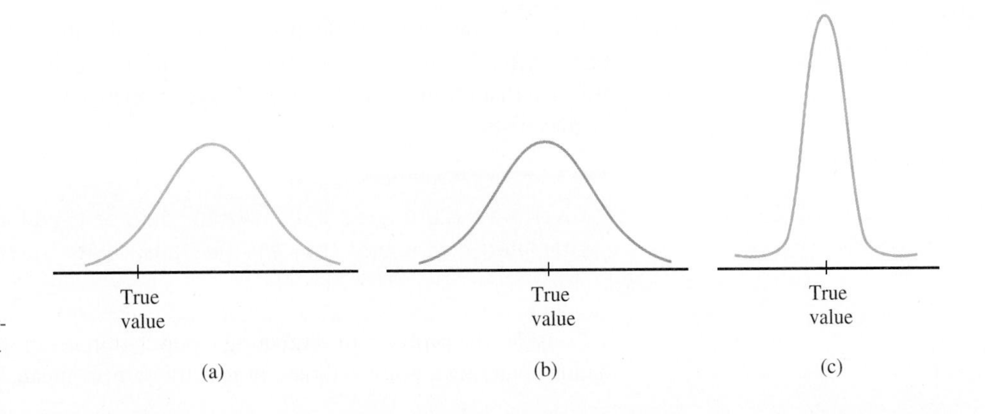

FIGURE 9.1
Sampling distributions of three different statistics for estimating a population characteristic.

estimate close to the true value. The distribution is centered to the right of the true value, making it very likely that an estimate (a value of the statistic for a particular sample) will be larger than the true value. If this statistic is used to compute an estimate based on a first sample, then another estimate based on a second sample, and another estimate based on a third sample, and so on, the long-run average value of these estimates will be greater than the true value.

The sampling distribution of Figure 9.1(b) is centered at the true value. This means that although one estimate may be smaller than the true value and another may be larger, when this statistic is used many times over with different random samples, there will be no long-run tendency to over- or underestimate the true value. Notice that even though the sampling distribution is correctly centered, it spreads out quite a bit around the true value. Because of this, some estimates resulting from the use of this statistic will be far above or far below the true value, even though there is no systematic tendency to underestimate or overestimate the true value.

In contrast, the mean value of the statistic with the distribution shown in Figure 9.1(c) is equal to the true value of the population characteristic (implying no systematic tendency to over- or underestimate the actual value). The statistic's standard deviation is relatively small. This means that estimates based on this third statistic will almost always be quite close to the true value—certainly more often than estimates resulting from the statistic with the sampling distribution shown in Figure 9.1(b).

DEFINITION

Unbiased statistic: A statistic whose mean value is equal to the value of the population characteristic being estimated.

Biased statistic: A statistic that is not unbiased.

As an example of a statistic that is biased, consider using the sample range as an estimate of the population range. Because the range of a population is defined as the difference between the largest value in the population and the smallest value, the range for a sample tends to underestimate the population range. This is because the largest value in a sample must be less than or equal to the largest value in the population and the smallest sample value must be greater than or equal to the smallest value in the population. The sample range equals the population range *only* if the sample happens to include both the largest and the smallest values in the population. In all other instances, the value of the sample range is smaller than the population range. This means that $\mu_{\text{sample range}}$ is less than the population range, implying bias.

One of the general results concerning the sampling distribution of $\bar{x}$, the sample mean, is that $\mu_{\bar{x}} = \mu$. This result says that the $\bar{x}$ values from all possible random samples of size n center around μ, the population mean. For example, if $\mu = 100$, the $\bar{x}$ distribution is centered at 100, whereas if $\mu = 5200$, then the $\bar{x}$ distribution is centered at 5200. This means that $\bar{x}$ is an unbiased statistic for estimating μ. Similarly, because the sampling distribution of $\hat{p}$ is centered at p, it follows that $\hat{p}$ is an unbiased statistic for estimating a population proportion.

Using an unbiased statistic that also has a small standard deviation ensures that there will be no systematic tendency to under- or overestimate the value of the population characteristic *and* that estimates will almost always be relatively close to the value of the population characteristic.

Given several unbiased statistics that could be used for estimating a population characteristic, the best choice is the statistic with the smallest standard deviation.

Consider the problem of estimating a population mean, μ. The obvious choice of statistic for obtaining a point estimate of μ is the sample mean, $\bar{x}$, an unbiased statistic for this purpose. However, when the population distribution is symmetric, $\bar{x}$ is not the only choice. Other unbiased statistics for estimating μ in this case include the sample median. Which statistic should be used? If the population distribution is normal, then $\bar{x}$ has a smaller standard deviation than any other unbiased statistic for estimating μ. This means that if the population distribution is normal, the sample mean would be the best choice.

Now consider estimating another population characteristic, the population variance, σ^2. The sample variance

$$s^2 = \frac{\Sigma(x - \bar{x})^2}{n - 1}$$

is a good choice for obtaining a point estimate of the population variance, σ^2. It can be shown that s^2 is an unbiased statistic for estimating σ^2. This means that whatever the value of σ^2, the sampling distribution of s^2 is centered at that value. This is the reason that the divisor $(n - 1)$ is used. An alternative statistic is the average squared deviation

$$\frac{\Sigma(x - \bar{x})^2}{n}$$

which has a more natural divisor than s^2. However, the average squared deviation is biased, with its values tending to be smaller, on average, than the value of σ^2.

EXAMPLE 9.3 Airborne Times for Flights from San Francisco to Washington, D.C.

Understand the context ❯

The **Bureau of Transportation Statistics** provides data on U.S. airline flights. The airborne times (in minutes) for nonstop flights from San Francisco to Washington Dulles airport for 10 randomly selected flights in June 2009 are:

270 256 267 285 274 275 266 258 271 281

For these data $\Sigma x = 2703$, $\Sigma x^2 = 731{,}373$, $n = 10$, and

$$\sum(x - \bar{x})^2 = \sum x^2 - \frac{(\Sigma x)^2}{n}$$

$$= 731{,}373 - \frac{(2703)^2}{10}$$

$$= 752.1$$

Do the work ❯

We use σ^2 to denote the actual variance in airborne time for June 2009 nonstop flights from San Francisco to Washington Dulles airport. Using the sample variance s^2 to provide a point estimate of σ^2 yields

$$s^2 = \frac{\Sigma(x - \bar{x})^2}{n - 1} = \frac{752.1}{9} = 83.57$$

Using the average squared deviation (with divisor $n = 10$), the resulting point estimate is

$$\frac{\Sigma(x - \bar{x})^2}{n} = \frac{752.1}{10} = 75.21$$

Interpret the results ❯

Because s^2 is an unbiased statistic for estimating σ^2, most statisticians would recommend using the point estimate 83.57. ■

An obvious choice of a statistic for estimating the population standard deviation σ is the sample standard deviation s. For the data given in Example 9.3,

$$s = \sqrt{83.57} = 9.14$$

Unfortunately, the fact that s^2 is an unbiased statistic for estimating σ^2 does not imply that s is an unbiased statistic for estimating σ. The sample standard deviation tends to underestimate slightly the true value of σ. However, unbiasedness is not the only criterion by which a statistic can be judged, and there are other good reasons for using s to estimate σ. In what follows, whenever we need to estimate σ based on a single random sample, we will use the statistic s to obtain a point estimate.

EXERCISES 9.1 - 9.9

9.1 ▼ Three different statistics are being considered for estimating a population characteristic. The sampling distributions of the three statistics are shown in the following illustration:

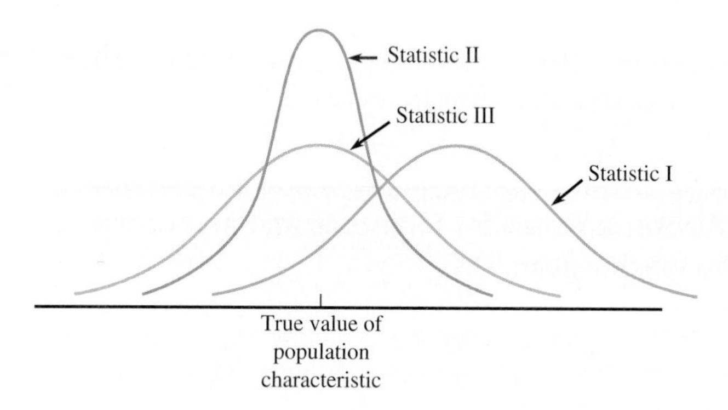

Which statistic would you recommend? Explain your choice. (Hint: See the section on choosing a statistic.)

9.2 a. Why is an unbiased statistic generally preferred over a biased statistic for estimating a population characteristic?

b. Does unbiasedness alone guarantee that the estimate will be close to the true value? Explain.

c. Under what circumstances might you choose a biased statistic over an unbiased statistic if two statistics are available for estimating a population characteristic?

9.3 Consumption of fast food is a topic of interest to researchers in the field of nutrition. The article **"Effects of Fast-Food Consumption on Energy Intake and Diet Quality Among Children"** (***Pediatrics*** **[2004]: 112–118)** reported that 1720 of those in a random sample of 6212 U.S. children indicated that on a typical day, they ate fast food. Estimate p, the proportion of children in the United States who eat fast food on a typical day.

9.4 ● Data consistent with summary quantities in the article referenced in the previous exercise on total calorie consumption on a particular day are given for a sample of children who did not eat fast food on that day and for a sample of children who did eat fast food on that day. Assume that it is reasonable to regard these samples as representative of the population of children in the United States.

No Fast Food

2331	1918	1009	1730	1469	2053	2143	1981
1852	1777	1765	1827	1648	1506	2669	

Fast Food

2523	1758	934	2328	2434	2267	2526	1195
890	1511	875	2207	1811	1250	2117	

a. Use the given information to estimate the mean calorie intake for children in the United States on a day when no fast food is consumed.

b. Use the given information to estimate the mean calorie intake for children in the United States on a day when fast food is consumed.

c. Use the given information to produce estimates of the standard deviations of calorie intake for days when no fast food is consumed and for days when fast food is consumed.

9.5 Each person in a random sample of 20 students at a particular university was asked whether he or she is registered to vote. The responses (R = registered, N = not registered) are given here:

R R N R N N R R R N R R R R R N R R R N

Use these data to estimate p, the proportion of all students at the university who are registered to vote.

9.6 Suppose that each of 935 smokers received a nicotine patch, which delivers nicotine to the bloodstream but at a much slower rate than cigarettes do. Dosage was decreased to 0 over a 12-week period. Suppose that 245 of the subjects were still not smoking 6 months after treatment. Assuming it is reasonable to regard this sample as representative of all smokers, estimate the percentage of all smokers who, when given this treatment, would refrain from smoking for at least 6 months.

9.7 ● Given below are the sodium contents (in mg) for seven brands of hot dogs rated as "very good" by ***Consumer Reports*** **(www.consumerreports.org)**:

420 470 350 360 270 550 530

a. Use the given data to produce a point estimate of μ, the true mean sodium content for hot dogs.

b. Use the given data to produce a point estimate of σ^2, the variance of sodium content for hot dogs.

c. Use the given data to produce an estimate of σ, the standard deviation of sodium content. Is the statistic you used to produce your estimate unbiased? (Hint: See the discussion following Example 9.3.)

9.8 ● A random sample of $n = 12$ four-year-old red pine trees was selected, and the diameter (in inches) of each tree's main stem was measured. The resulting observations are as follows:

11.3 10.7 12.4 15.2 10.1 12.1 16.2 10.5
11.4 11.0 10.7 12.0

a. Compute a point estimate of σ, the population standard deviation of main stem diameter. What statistic did you use to obtain your estimate?

b. Suppose that the diameter distribution is normal. Then the 90th percentile of the diameter distribution is $\mu + 1.28\sigma$ (so 90% of all trees have diameters less than this value). Compute a point estimate for this percentile. (Hint: First compute an estimate of μ and then use it along with your estimate of σ from Part (a).)

9.9 ● A random sample of 10 houses heated with natural gas in a particular area is selected, and the amount of gas (in therms) used during the month of January is determined for each house. The resulting observations are as follows:

103 156 118 89 125 147 122 109 138 99

a. Let μ_J denote the average gas usage during January by all houses in this area. Compute a point estimate of μ_J.

b. Suppose that 10,000 houses in this area use natural gas for heating. Let τ denote the total amount of gas used by all of these houses during January. Estimate τ using the given data. What statistic did you use in computing your estimate?

c. Use the data in Part (a) to estimate p, the proportion of all houses that used at least 100 therms.

d. Give a point estimate of the population median usage based on the sample of Part (a). Which statistic did you use?

Bold exercises answered in back ● Data set available online ▼ Video Solution available

9.2 Large-Sample Confidence Interval for a Population Proportion

In Section 9.1, we saw how to use a statistic to produce a point estimate of a population characteristic. The value of a point estimate depends on which sample, out of all the possible samples, happens to be selected. Different samples usually produce different estimates as a result of chance differences from one sample to another. Because of sampling variability, rarely is the point estimate from a sample exactly equal to the actual value of the population characteristic. We hope that the chosen statistic produces an estimate that is close, on average, to the true value.

Although a point estimate may represent our best single-number estimate for the value of the population characteristic, it is not the only plausible value. As an alternative to a point estimate, we can use the sample data to report an interval of plausible values for the population characteristic. For example, we might be confident that for all text messages sent from cell phones, the proportion p of messages that are longer than 50 characters is in the interval from .53 to .57. The narrowness of this interval implies that we have rather precise information about the value of p. If, with the same high degree of confidence, we could only state that p was between .32 and .74, it would be clear that we had relatively imprecise knowledge of the value of p.

DEFINITION

Confidence interval: An interval of plausible values for a population characteristic.

A confidence interval is constructed so that we have a chosen degree of confidence that the actual value of the population characteristic will be between the lower and upper endpoints of the interval.

Associated with each confidence interval is a **confidence level**. The confidence level provides information on how much "confidence" we can have in the *method* used to construct the interval estimate (*not* our confidence in any one particular interval).

Usual choices for confidence levels are 90%, 95%, and 99%, although other confidence levels are also possible. If we were to construct a 95% confidence interval using the method to be described shortly, we would be using a method that is "successful" 95% of the time. This means that if this method was used to generate an interval estimate over and over again with different random samples, in the long run 95% of the resulting intervals would include the actual value of the characteristic being estimated. Similarly, a 99% confidence interval is one that is constructed using a method that is, in the long run, successful in capturing the actual value of the population characteristic 99% of the time.

DEFINITION

Confidence level: The success rate of the *method* used to construct a confidence interval.

One goal of many statistical studies is to estimate the proportion of individuals or objects in a population that possess a particular property of interest (whatever we have defined as a "success"). For example, a university administrator might be interested in the proportion of students who prefer a new registration system to the previous registration method. In a different setting, a quality control engineer might be concerned about the proportion of defective parts produced by a particular manufacturing process.

Recall that p denotes the proportion of the population that possess the property of interest (successes). Previously, we used the sample proportion

$$\hat{p} = \frac{\text{number in the sample that possess the property of interest}}{n}$$

to calculate a point estimate of p. We can also use $\hat{p}$ to construct a confidence interval for p.

Although a small-sample confidence interval for p can be obtained, our focus is on the large-sample case. The construction of the large-sample interval is based on properties of the sampling distribution of the statistic $\hat{p}$:

1. The sampling distribution of $\hat{p}$ is centered at p. This means that $\mu_{\hat{p}} = p$, so $\hat{p}$ is an unbiased statistic for estimating p.
2. As long as the sample size is less than 10% of the population size, the standard deviation of $\hat{p}$ is well approximated by $\sigma_{\hat{p}} = \sqrt{\frac{p(1-p)}{n}}$
3. As long as n is large ($np \geq 10$ and $n(1 - p) \geq 10$), the sampling distribution of $\hat{p}$ is well approximated by a normal curve.

The accompanying box summarizes these properties.

When n is large and the sample size is less than 10% of the population size, the statistic $\hat{p}$ has a sampling distribution that is approximately normal with mean p and standard deviation $\sqrt{\frac{p(1-p)}{n}}$.

The development of a confidence interval for p is easier to follow if we begin by selecting a particular confidence level. For a confidence level of 95%, the table of standard normal (z) curve areas (Appendix Table 2) can be used to determine a value z^* such that a central area of .95 falls between $-z^*$ and z^*. In this case, the remaining area of .05 is divided equally between the two tails, as shown in Figure 9.2. The total area to the left of the desired z^* is .975 (.95 central area + .025 area below $-z^*$). By locating .9750 in the body of Appendix Table 2, we find that the corresponding z critical value is $z^* = 1.96$.

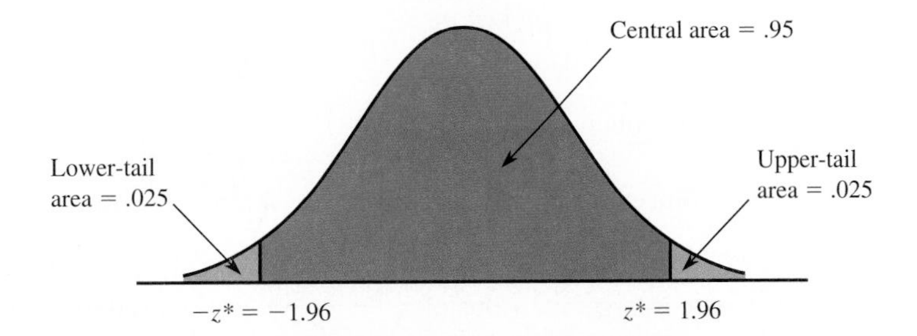

FIGURE 9.2
Capturing a central area of .95 under the z curve.

Generalizing this result to normal distributions other than the standard normal distribution tells us that for *any* normal distribution, about 95% of the values are within 1.96 standard deviations of the mean. For large random samples, the sampling distribution of $\hat{p}$ is approximately normal with mean $\mu_{\hat{p}} = p$ and standard deviation $\sigma_{\hat{p}} = \sqrt{\frac{p(1-p)}{n}}$, and we get the following result.

When n is large, approximately 95% of all random samples of size n will result in a value of $\hat{p}$ that is within $1.96\sigma_{\hat{p}} = 1.96\sqrt{\frac{p(1-p)}{n}}$ of the actual value of the population proportion p.

If $\hat{p}$ is within $1.96\sqrt{\frac{p(1-p)}{n}}$ of p, this means the interval

$$\hat{p} - 1.96\sqrt{\frac{p(1-p)}{n}} \text{ to } \hat{p} + 1.96\sqrt{\frac{p(1-p)}{n}}$$

will capture p. This will happen for 95% of all possible samples. However, if $\hat{p}$ is farther away from p than $1.96\sqrt{\frac{p(1-p)}{n}}$ (which will happen for about 5% of all possible samples), the interval will not include the value of p. This is shown in Figure 9.3.

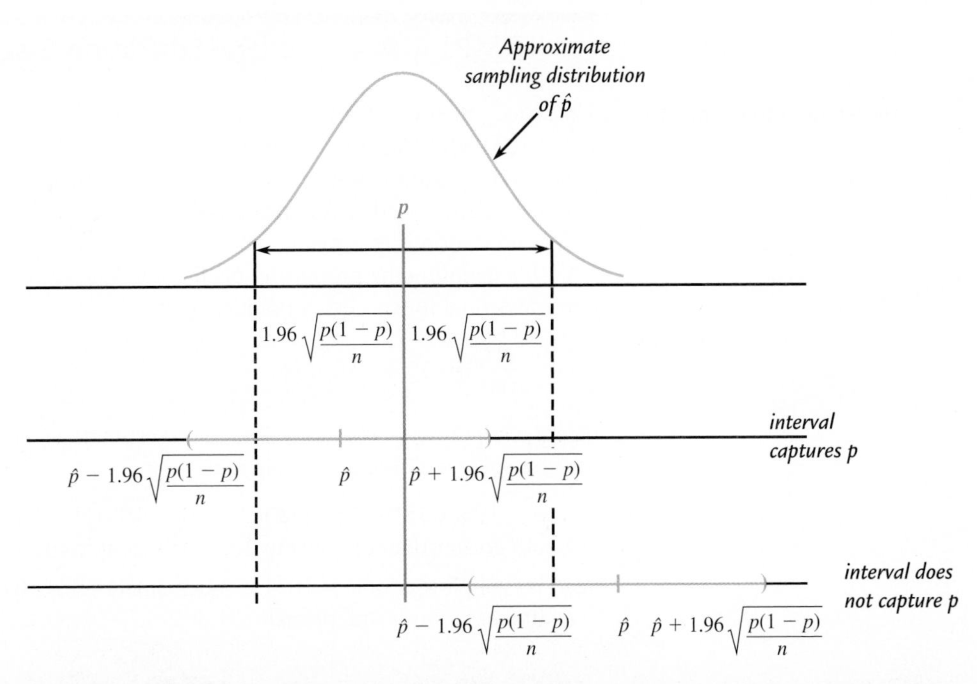

FIGURE 9.3
The population proportion p is captured in the interval from $\hat{p} - 1.96\sqrt{\frac{p(1-p)}{n}}$ to $\hat{p} + 1.96\sqrt{\frac{p(1-p)}{n}}$ when $\hat{p}$ is within $1.96\sqrt{\frac{p(1-p)}{n}}$ of p.

Because $\hat{p}$ is within $1.96\sigma_{\hat{p}}$ of p for 95% of all possible random samples, this means that in repeated random sampling, 95% of the intervals

$$\hat{p} - 1.96\sqrt{\frac{p(1-p)}{n}} \text{ to } \hat{p} + 1.96\sqrt{\frac{p(1-p)}{n}}$$

will contain p.

Since the value of p is unknown, $\sqrt{\frac{p(1-p)}{n}}$ must be estimated. As long as the sample size is large, the value of $\sqrt{\frac{\hat{p}(1-\hat{p})}{n}}$ can be used in place of $\sqrt{\frac{p(1-p)}{n}}$.

When n is large, a **95% confidence interval for p** is

$$\left(\hat{p} - 1.96\sqrt{\frac{\hat{p}(1-\hat{p})}{n}}, \hat{p} + 1.96\sqrt{\frac{\hat{p}(1-\hat{p})}{n}}\right)$$

An abbreviated formula for the interval is

$$\hat{p} \pm 1.96\sqrt{\frac{\hat{p}(1-\hat{p})}{n}}$$

where $\hat{p} + 1.96\sqrt{\frac{\hat{p}(1-\hat{p})}{n}}$ gives the upper endpoint of the interval and $\hat{p} - 1.96\sqrt{\frac{\hat{p}(1-\hat{p})}{n}}$ gives the lower endpoint of the interval.

The interval can be used as long as

1. $n\hat{p} \geq 10$ and $n(1 - \hat{p}) \geq 10$,
2. the sample size is less than 10% of the population size if sampling is without replacement, and
3. the sample can be regarded as a random sample from the population of interest.

EXAMPLE 9.4 College Education Essential for Success?

Understand the context ❯

The article **"How Well Are U.S. Colleges Run?" (*USA Today*, February 17, 2010)** describes a survey of 1031 adult Americans. The survey was carried out by the National Center for Public Policy and the sample was selected in a way that makes it reasonable to regard the sample as representative of adult Americans. Of those surveyed, 567 indicated that they believed a college education is essential for success.

With p denoting the proportion of all adult Americans who believe that a college education is essential for success, a point estimate of p is

$$\hat{p} = \frac{567}{1031} = .55$$

Before computing a confidence interval to estimate p, we should check to make sure that the three necessary conditions are met:

1. $n\hat{p} = 1031(.55) = 567$ and $n(1 - \hat{p}) = 1031(1 - .55) = 1031(.45) = 364$ are both greater than or equal to 10, so the sample size is large enough to proceed.
2. The sample size of $n = 1031$ is much smaller than 10% of the population size (the number of adult Americans).

3. The sample was selected in a way designed to produce a representative sample. So, it is reasonable to regard the sample as a random sample from the population.

Formulate a plan ⟩ Because all three conditions are met, it is appropriate to use the sample data to construct a 95% confidence interval for p.

Do the work ⟩ A 95% confidence interval for p is

$$\hat{p} \pm 1.96\sqrt{\frac{\hat{p}(1-\hat{p})}{n}} = .55 \pm 1.96\sqrt{\frac{(.55)(1-.55)}{1031}}$$
$$= .55 \pm (1.96)(.015)$$
$$= .55 \pm .029$$
$$= (.521, .579)$$

Interpret the results ⟩ Based on this sample, we can be 95% confident that p, the proportion of adult Americans who believe a college education is essential for success, is between .521 and .579. We used a *method* to construct this estimate that in the long run will successfully capture the actual value of p 95% of the time. ■

The 95% confidence interval for p calculated in Example 9.4 is (.521, .579). It is tempting to say that there is a "probability" of .95 that p is between .521 and .579. *Do not yield to this temptation!* The 95% refers to the percentage of *all* possible samples resulting in an interval that includes the value of p. In other words, if we take random sample after random sample from the population and use each one separately to compute a 95% confidence interval, in the long run roughly 95% of these intervals will capture the value of p.

Figure 9.4 illustrates this concept for intervals generated from 100 different random samples. In this particular set of 100 intervals, 93 include p, whereas 7 do not. Any specific interval, and our interval (.521, .579) in particular, either includes p or it does not (remember, the value of p is fixed but not known to us). We cannot make a chance (probability) statement concerning this particular interval. *The confidence level 95% refers to the*

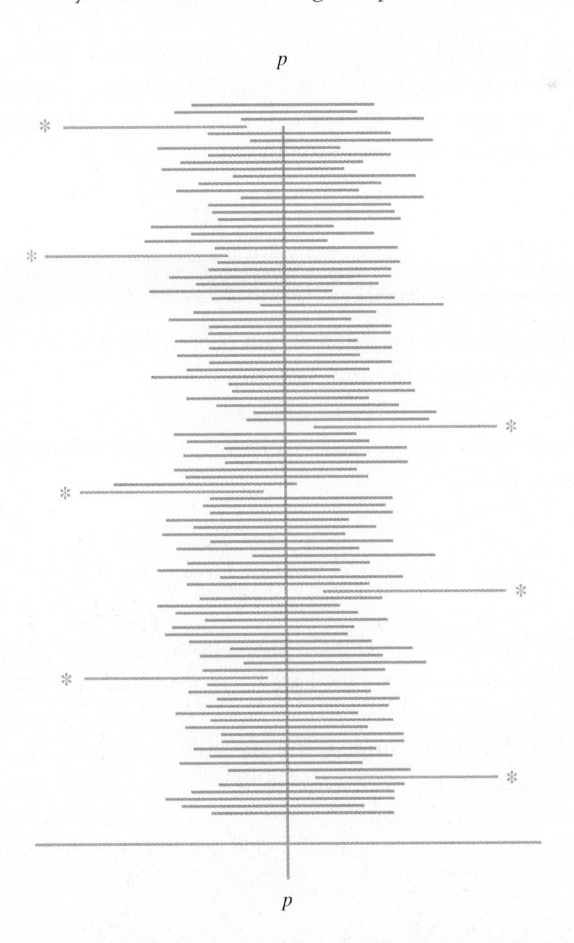

FIGURE 9.4
One hundred 95% confidence intervals for p computed from 100 different random samples (asterisks identify intervals that do not include p).

method used to construct the interval rather than to any particular interval, such as the one we obtained.

The formula given for a 95% confidence interval can easily be adapted for other confidence levels. The choice of a 95% confidence level led to the use of the z value 1.96 (chosen to capture a central area of .95 under the standard normal curve) in the confidence interval formula. Any other confidence level can be obtained by using an appropriate z critical value in place of 1.96.

For example, suppose that we wanted to achieve a confidence level of 99%. To obtain a central area of .99, the appropriate z critical value would have a cumulative area (area to the left) of .995, as illustrated in Figure 9.5. From Appendix Table 2, we find that the corresponding z critical value is $z = 2.58$. A 99% confidence interval for p is then obtained by using 2.58 in place of 1.96 in the formula for the 95% confidence interval.

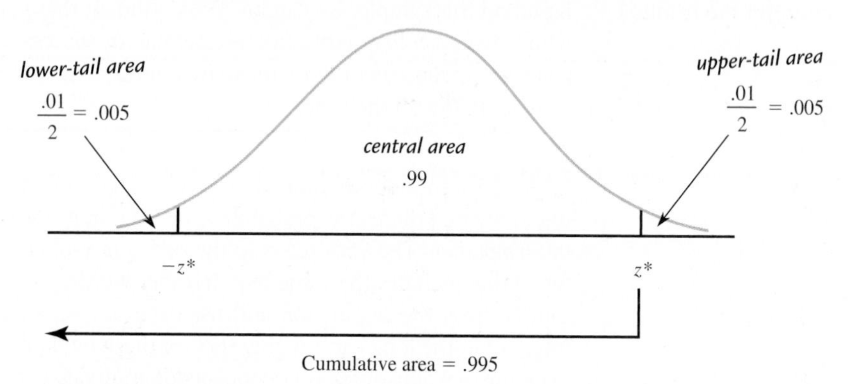

FIGURE 9.5
Finding the z critical value for a 99% confidence level.

Why settle for 95% confidence when 99% confidence is possible? Because the higher confidence level comes at a price. The resulting interval is wider than the 95% interval. The width of the 95% interval is $2\left(1.96\sqrt{\frac{\hat{p}(1-\hat{p})}{n}}\right)$, whereas the 99% interval has width $2\left(2.58\sqrt{\frac{\hat{p}(1-\hat{p})}{n}}\right)$. The higher *reliability* of the 99% interval (where "reliability" is specified by the confidence level) entails a loss in precision (as indicated by the wider interval). In the opinion of many investigators, a 95% confidence interval produces a reasonable compromise between reliability and precision.

The Large-Sample Confidence Interval for p

The general formula for a confidence interval for a population proportion p when

1. $\hat{p}$ is the sample proportion from a **simple random sample,**
2. the sample size **n is large** ($n\hat{p} \geq 10$ and $n(1 - \hat{p}) \geq 10$), and
3. if the sample is selected without replacement, **the sample size is small relative to the population size** (n is at most 10% of the population size)*

is

$$\hat{p} \pm (z \text{ critical value})\sqrt{\frac{\hat{p}(1-\hat{p})}{n}}$$

The desired confidence level determines which z critical value is used. The three most commonly used confidence levels, 90%, 95%, and 99%, use z critical values 1.645, 1.96, and 2.58, respectively.

Note: This interval is not appropriate for small samples. It is possible to construct a confidence interval in the small-sample case, but this is beyond the scope of this textbook.

*In Chapter 7, we saw a different situation where a similar condition is introduced, but where the requirement was that at most 5% of the population is included in the sample. Be careful not to confuse these two rules.

EXAMPLE 9.5 Dangerous Driving

Wilfried Krecichwost /The Image Bank/Getty Images

Understand the context ❯

The article **"Nine Out of Ten Drivers Admit in Survey to Having Done Something Dangerous"** (***Knight Ridder Newspapers,*** **July 8, 2005**) reported the results of a survey of 1100 drivers. Of those surveyed, 990 admitted to careless or aggressive driving during the previous 6 months. Assuming that it is reasonable to regard this sample of 1100 as representative of the population of drivers, we can use this information to construct an estimate of p, the proportion of all drivers who have engaged in careless or aggressive driving in the past 6 months.

For this sample

$$\hat{p} = \frac{990}{1100} = .900$$

Formulate a plan ❯

Because the sample size is less than 10% of the population size and $n\hat{p} = 990$ and $n(1 - \hat{p}) = 110$ are both greater than or equal to 10, the conditions necessary for appropriate use of the formula for a large-sample confidence interval are met.

A 90% confidence interval for p is then

Do the work ❯

$$\begin{aligned}\hat{p} \pm (z \text{ critical value})\sqrt{\frac{\hat{p}(1-\hat{p})}{n}} &= .900 \pm 1.645\sqrt{\frac{(.900)(.100)}{1100}} \\ &= .900 \pm (1.645)(.009) \\ &= .900 \pm .015 \\ &= (.885, .915)\end{aligned}$$

Interpret the results ❯

Based on these sample data, we can be 90% confident that the proportion of all drivers who have engaged in careless or aggressive driving in the past 6 months is between .885 and .915. We have used a *method* to construct this interval estimate that has a 10% error rate. ■

The confidence level for the z confidence interval for a population proportion is only approximate. When we report a 95% confidence interval for a population proportion, the 95% confidence level implies that we have used a method that produces an interval that includes the actual value of the population proportion 95% of the time in repeated random sampling. In fact, because the normal distribution is only an approximation to the sampling distribution of $\hat{p}$, the actual confidence level may differ somewhat from the reported value. If the conditions (1) $n\hat{p} \geq 10$ and $n(1 - \hat{p}) \geq 10$ and (2) n is at most 10% of the population size if sampling without replacement are met, the normal approximation is reasonable and the actual confidence level is usually not too different from the reported level. This is why it is important to check these conditions before computing and reporting a large-sample z confidence interval for a population proportion.

What should you do if these conditions are not met? If the sample size is too small to satisfy the $n\hat{p}$ and $n(1 - \hat{p})$ greater than or equal to 10 condition, an alternative procedure can be used. Consult a statistician or a more advanced textbook in this case. If the condition that the sample size is less than 10% of the population size when sampling without replacement is not satisfied, the large-sample z confidence interval tends to be conservative (that is, it tends to be wider than is necessary to achieve the desired confidence level). In this case, a finite population correction factor can be used to obtain a more precise interval. Again, it would be wise to consult a statistician or a more advanced textbook.

An Alternative to the Large-Sample z Interval

Investigators have shown that in some instances, even when the sample size conditions of the large-sample z confidence interval for a population proportion are met, the actual confidence level associated with the method may be noticeably different from the reported confidence level.

A modified interval that has an actual confidence level that is closer to the reported confidence level is based on a modified sample proportion, $\hat{p}_{mod}$, the proportion of successes after adding two successes and two failures to the sample. Then $\hat{p}_{mod}$ is

$$\hat{p}_{mod} = \frac{\text{number of successes} + 2}{n + 4}$$

$\hat{p}_{mod}$ is used in place of $\hat{p}$ in the usual confidence interval formula. Properties of this modified confidence interval are investigated in Activity 9.2 at the end of the chapter.

General Form of a Confidence Interval

Many confidence intervals have the same general form as the large-sample z confidence interval for p just considered. We started with a statistic $\hat{p}$, from which a point estimate for p was obtained. The standard deviation of this statistic is $\sqrt{p(1 - p)/n}$. This resulted in a confidence interval of the form

$$\begin{pmatrix}\text{point estimate using}\\ \text{a specified statistic}\end{pmatrix} \pm (\text{critical value})\begin{pmatrix}\text{standard deviation}\\ \text{of the statistic}\end{pmatrix}$$

Because p was unknown, we estimated the standard deviation of the statistic by $\sqrt{\hat{p}(1 - \hat{p})/n}$, which yielded an interval of the form

$$\begin{pmatrix}\text{point estimate using}\\ \text{a specified statistic}\end{pmatrix} \pm (\text{critical value})\begin{pmatrix}\text{estimated}\\ \text{standard deviation}\\ \text{of the statistic}\end{pmatrix}$$

For a population characteristic other than p, a statistic for estimating the characteristic is selected. Then (drawing on statistical theory) a formula for the standard deviation of the statistic is given. In practice, it is almost always necessary to estimate this standard deviation, so that the interval

$$\begin{pmatrix}\text{point estimate using}\\ \text{a specified statistic}\end{pmatrix} \pm (\text{critical value})\begin{pmatrix}\text{estimated}\\ \text{standard deviation}\\ \text{of the statistic}\end{pmatrix}$$

is the prototype confidence interval. It is common practice to refer to both the standard deviation of a statistic and the *estimated* standard deviation of a statistic as the **standard error**. In this textbook, when we use the term standard error, we mean the estimated standard deviation of a statistic.

DEFINITION

Standard error: The estimated standard deviation of a statistic.

The 95% confidence interval for p is based on the fact that, for approximately 95% of all random samples, $\hat{p}$ is within $1.96\sqrt{\frac{p(1-p)}{n}}$ of p. The quantity $1.96\sqrt{\frac{p(1-p)}{n}}$ is sometimes called the **bound on the error of estimation** associated with a 95% confidence level. This means that we have 95% confidence that the point estimate $\hat{p}$ is no farther than this quantity from p.

DEFINITION

Bound on error: If the sampling distribution of a statistic is (at least approximately) normal, the **bound on error of estimation, *B*,** associated with a 95% confidence interval is (1.96)·(standard error of the statistic).

Choosing the Sample Size

Before collecting any data, an investigator may wish to determine a sample size for which a particular value of the bound on the error is achieved. For example, with p representing the actual proportion of students at a university who purchase textbooks online, the objective of an investigation may be to estimate p to within .05 with 95% confidence.

The value of n necessary to achieve this is obtained by setting $1.96\sqrt{\frac{p(1-p)}{n}}$ equal to .05 and solving for n.

In general, suppose that we wish to estimate p to within an amount B (the specified bound on the error of estimation) with 95% confidence. To find the necessary sample size, consider the equation

$$B = 1.96\sqrt{\frac{p(1-p)}{n}}$$

Solving this equation for n results in

$$n = p(1-p)\left(\frac{1.96}{B}\right)^2$$

Unfortunately, the use of this formula requires the value of p, which is unknown. One possible way to proceed is to carry out a preliminary study and use the resulting data to get a rough estimate of p. In other cases, prior knowledge may suggest a reasonable estimate of p.

If there is no reasonable basis for estimating p and a preliminary study is not feasible, a conservative solution follows from the observation that $p(1-p)$ is never larger than .25 (its value when $p = .5$). Replacing $p(1-p)$ with .25, the maximum value, yields

$$n = .25\left(\frac{1.96}{B}\right)^2$$

Using this formula to obtain n gives us a sample size for which we can be 95% confident that $\hat{p}$ will be within B of p, no matter what the value of p.

The sample size required to estimate a population proportion p to within an amount B with 95% confidence is

$$n = p(1-p)\left(\frac{1.96}{B}\right)^2$$

The value of p may be estimated using prior information. In the absence of any such information, using $p = .5$ in this formula gives a conservatively large value for the required sample size (this value of p gives a larger n than would any other value).

EXAMPLE 9.6 Sniffing Out Cancer

Understand the context ❯

Researchers have found biochemical markers of cancer in the exhaled breath of cancer patients, but chemical analysis of breath specimens has not yet proven effective in clinical diagnosis. The authors of the paper **"Diagnostic Accuracy of Canine Scent Detection in Early- and Late-Stage Lung and Breast Cancers"** (***Integrative Cancer Therapies*** **[2006]: 1–10)** describe a study to investigate whether dogs can be trained to identify the presence or absence of cancer by sniffing breath specimens.

Suppose we want to collect data that would allow us to estimate the long-run proportion of accurate identifications for a particular dog that has completed training. The dog has been trained to lie down when presented with a breath specimen from a cancer patient and to remain standing when presented with a specimen from a person who does not have cancer. How many different breath specimens should be used if we want to estimate the long-run proportion of correct identifications for this dog to within .05 with 95% confidence?

Do the work ⟩ Using a conservative value of $p = .5$ in the formula for required sample size gives

$$n = p(1 - p)\left(\frac{1.96}{B}\right)^2 = (.5)(.5)\left(\frac{1.96}{.05}\right)^2 = 384.16$$

Interpret the results ⟩ Thus, a sample of at least 385 breath specimens should be used. Notice that in sample size calculations, we always round up. ■

EXERCISES 9.10 - 9.33

9.10 ▼ For each of the following choices, explain which would result in a wider large-sample confidence interval for p. (Hint: Consider the confidence interval formula.)

a. 90% confidence level or 95% confidence level

b. $n = 100$ or $n = 400$

9.11 The formula used to compute a large-sample confidence interval for p is

$$\hat{p} \pm (z \text{ critical value})\sqrt{\frac{\hat{p}(1 - \hat{p})}{n}}$$

What is the appropriate z critical value for each of the following confidence levels?

a. 95%

b. 90%

c. 99%

d. 80%

e. 85%

9.12 The use of the interval

$$\hat{p} \pm (z \text{ critical value})\sqrt{\frac{\hat{p}(1 - \hat{p})}{n}}$$

requires a large sample. For each of the following combinations of n and $\hat{p}$, indicate whether the sample size is large enough for use of this interval to be appropriate.

a. $n = 50$ and $\hat{p} = .30$

b. $n = 50$ and $\hat{p} = .05$

c. $n = 15$ and $\hat{p} = .45$

d. $n = 100$ and $\hat{p} = .01$

e. $n = 100$ and $\hat{p} = .70$

f. $n = 40$ and $\hat{p} = .25$

g. $n = 60$ and $\hat{p} = .25$

h. $n = 80$ and $\hat{p} = .10$

9.13 Discuss how each of the following factors affects the width of the confidence interval for p. (Hint: Consider the confidence interval formula.)

a. The confidence level

b. The sample size

c. The value of $\hat{p}$

9.14 The article **"Career Expert Provides DOs and DON'Ts for Job Seekers on Social Networking" (CareerBuilder.com, August 19, 2009)** included data from a survey of 2667 hiring managers and human resource professionals. The article noted that many employers are using social networks to screen job applicants and that this practice is becoming more common. Of the 2667 people who participated in the survey, 1200 indicated that they use social networking sites (such as Facebook, MySpace, and LinkedIn) to research job applicants.

For the purposes of this exercise, assume that the sample is representative of hiring managers and human resource professionals. Construct and interpret a 95% confidence interval for the proportion of hiring managers and human resource professionals who use social networking sites to research job applicants. (Hint: See Example 9.4.)

9.15 Based on data from a survey of 1200 randomly selected Facebook users **(*USA Today*, March 24, 2010)**, the following is a 90% confidence interval for the proportion of all Facebook users who say it is not OK to "friend" someone who reports to you at work: (0.60, 0.64). What is the meaning of the 90% confidence level associated with this interval?

9.16 In a survey on supernatural experiences, 722 of 4013 adult Americans surveyed reported that they had seen or been with a ghost **("What Supernatural Experiences We've Had," *USA Today*, February 8, 2010)**.

a. What assumption must be made in order for it to be appropriate to use the formula of this section to construct a confidence interval to estimate the proportion of all adult Americans who have seen or been with a ghost?

b. Construct and interpret a 90% confidence interval for the proportion of all adult Americans who have seen or been with a ghost. (Hint: See Example 9.5.)

c. Would a 99% confidence interval be narrower or wider than the interval computed in Part (b)? Justify your answer.

9.17 If a hurricane was headed your way, would you evacuate? The headline of a press release issued **January 21, 2009** by the survey research company **International Communications Research (icrsurvey.com)** states, "Thirty-one Percent of People on High-Risk Coast Will Refuse Evacuation Order, Survey of Hurricane Preparedness Finds." This headline was based on a survey of 5046 adults who live within 20 miles of the coast in high hurricane risk counties of eight southern states. In selecting the sample, care was taken to ensure that the sample would be representative of the population of coastal residents in these states.

a. Use this information to estimate the proportion of coastal residents who would evacuate using a 98% confidence interval.

b. Write a few sentences interpreting the interval and the confidence level associated with the interval.

9.18 The study **"Digital Footprints" (Pew Internet & American Life Project, www.pewinternet.org, 2007)** reported that 47% of Internet users have searched for information about themselves online. The 47% figure was based on a random sample of Internet users. For purposes of this exercise, suppose that the sample size was $n = 300$ (the actual sample size was much larger).

Construct and interpret a 90% confidence interval for the proportion of Internet users who have searched online for information about themselves.

9.19 The article **"Kids Digital Day: Almost 8 Hours" (*USA Today*, January 20, 2010)** summarized results from a national survey of 2002 Americans age 8 to 18. The sample was selected in a way that was expected to result in a sample representative of Americans in this age group.

a. Of those surveyed, 1321 reported owning a cell phone. Use this information to construct and interpret a 90% confidence interval estimate of the proportion of all Americans age 8 to 18 who own a cell phone.

b. Of those surveyed, 1522 reported owning an MP3 music player. Use this information to construct and interpret a 90% confidence interval estimate of the proportion of all Americans age 8 to 18 who own an MP3 music player.

c. Explain why the confidence interval from Part (b) is narrower than the confidence interval from Part (a) even though the confidence level and the sample size used to compute the two intervals was the same.

9.20 The article **"Students Increasingly Turn to Credit Cards" (*San Luis Obispo Tribune*, July 21, 2006)** reported that 37% of college freshmen and 48% of college seniors carry a credit card balance from month to month. Suppose that the reported percentages were based on random samples of 1000 college freshmen and 1000 college seniors.

a. Construct a 90% confidence interval for the proportion of college freshmen who carry a credit card balance from month to month.

b. Construct a 90% confidence interval for the proportion of college seniors who carry a credit card balance from month to month.

c. Explain why the two 90% confidence intervals from Parts (a) and (b) are not the same width.

9.21 ▼ The article **"CSI Effect Has Juries Wanting More Evidence" (*USA Today*, August 5, 2004)** examines how the popularity of crime-scene investigation television shows is influencing jurors' expectations of what evidence should be produced at a trial. In a survey of 500 potential jurors, one study found that 350 were regular watchers of at least one crime-scene forensics television series.

a. Assuming that it is reasonable to regard this sample of 500 potential jurors as representative of potential jurors in the United States, use the given information to construct and interpret a 95% confidence interval for the proportion of all potential jurors who regularly watch at least one crime-scene investigation series.

b. Would a 99% confidence interval be wider or narrower than the 95% confidence interval from Part (a)?

9.22 In a survey of 1000 randomly selected adults in the United States, participants were asked what their most favorite and what their least favorite subject was when they were in school **(Associated Press, August 17, 2005)**. In what might seem like a contradiction, math was chosen more often than any other subject in both categories! Math was chosen by 230 of the 1000 as the favorite subject, and it was also chosen by 370 of the 1000 as the least favorite subject.

a. Construct a 95% confidence interval for the proportion of U.S. adults for whom math was the favorite subject in school.

b. Construct a 95% confidence interval for the proportion of U.S. adults for whom math was the least favorite subject.

9.23 The report **"2005 Electronic Monitoring & Surveillance Survey: Many Companies Monitoring, Recording, Videotaping—and Firing—Employees" (American Management Association, 2005)** summarized the results of a survey of 526 U.S. businesses. The report stated that 137 of the 526 businesses had fired workers for misuse of the Internet and 131 had fired workers for e-mail misuse. For purposes of this exercise, assume that it is reasonable to regard this sample as representative of businesses in the United States.

a. Construct and interpret a 95% confidence interval for the proportion of U.S. businesses that have fired workers for misuse of the Internet.

b. What are two reasons why a 90% confidence interval for the proportion of U.S. businesses that have fired workers for misuse of e-mail would be narrower than the 95% confidence interval computed in Part (a)?

9.24 In an AP-AOL sports poll **(Associated Press, December 18, 2005)**, 394 of 1000 randomly selected U.S. adults indicated that they considered themselves to be baseball fans. Of the 394 baseball fans, 272 stated that they thought the designated hitter rule should either be expanded to both baseball leagues or eliminated.

a. Construct a 95% confidence interval for the proportion of U.S. adults who consider themselves to be baseball fans.

b. Construct a 95% confidence interval for the proportion of those who consider themselves to be baseball fans who think the designated hitter rule should be expanded to both leagues or eliminated.

c. Explain why the confidence intervals of Parts (a) and (b) are not the same width even though they both have a confidence level of 95%.

9.25 The article **"Viewers Speak Out Against Reality TV" (Associated Press, September 12, 2005)** included the following statement: "Few people believe there's much reality in reality TV: a total of 82% said the shows are either 'totally made up' or 'mostly distorted.'" This statement was based on a survey of 1002 randomly selected adults. Compute and interpret a bound on the error of estimation for the reported proportion of .82. (Hint: See the definition of bound on error.)

9.26 One thousand randomly selected adult Americans participated in a survey conducted by the **Associated Press (June 2006)**. When asked "Do you think it is sometimes justified to lie or do you think lying is never justified?" 52% responded that lying was never justified. When asked about lying to avoid hurting someone's feelings, 650 responded that this was often or sometimes okay.

a. Construct a 90% confidence interval for the proportion of adult Americans who think lying is never justified.

b. Construct a 90% confidence interval for the proportion of adult Americans who think that it is often or sometimes okay to lie to avoid hurting someone's feelings.

c. Comment on the apparent inconsistency in the responses given by the individuals in this sample.

9.27 ***USA Today*** **(October 14, 2002)** reported that 36% of adult drivers admit that they often or sometimes talk on a cell phone when driving. This estimate was based on data from a sample of 1004 adult drivers, and a bound on the error of estimation of 3.1% was reported. Assuming a 95% confidence level, do you agree with the reported bound on the error? Explain.

9.28 The Gallup Organization conducts an annual survey on crime. It was reported that 25% of all households experienced some sort of crime during the past year. This estimate was based on a sample of 1002 randomly selected households. The report states, "One can say with 95% confidence that the margin of sampling error is ± 3 percentage points." Explain how this statement can be justified.

9.29 The article **"Hospitals Dispute Medtronic Data on Wires" (*The Wall Street Journal*, February 4, 2010)** describes several studies of the failure rate of defibrillators used in the treatment of heart problems. In one study conducted by the Mayo Clinic, it was reported that failures were experienced within the first 2 years by 18 of 89 patients under 50 years old and 13 of 362 patients age 50 and older who received a particular type of defibrillator. Assume it is reasonable to regard these two samples as representative of patients in the two age groups who receive this type of defibrillator.

a. Construct and interpret a 95% confidence interval for the proportion of patients under 50 years old who experience a failure within the first 2 years after receiving this type of defibrillator.

b. Construct and interpret a 99% confidence interval for the proportion of patients age 50 and older who experience a failure within the first 2 years after receiving this type of defibrillator.

c. Suppose that the researchers wanted to estimate the proportion of patients under 50 years old who experience a failure within the first 2 years after receiving this type of defibrillator to within .03 with 95% confidence. How large a sample should be used? Use the results of the study as a preliminary estimate of the population proportion. (Hint: See Example 9.6.)

9.30 Based on a representative sample of 511 U.S. teenagers age 12 to 17, International Communications Research estimated that the proportion of teens who support keeping the legal drinking age at 21 is $\hat{p} = 0.64$ (64%). The press release titled **"Majority of Teens (Still) Favor the Legal Drinking Age" (www.icrsurvey.com, January 21, 2009)** also reported a margin of error of 0.04 (4%) for this estimate. Show how the reported value for the margin of error was computed.

9.31 A discussion of digital ethics appears in the article **"Academic Cheating, Aided by Cell Phones or Web, Shown to be Common" (*Los Angeles Times*, June 17, 2009)**. One question posed in the article is: What proportion of college students have used cell phones to cheat on an exam? Suppose you have been asked to estimate this proportion for students enrolled at a large university. How many students should you include in your sample if you want to estimate this proportion to within .02 with 95% confidence?

9.32 In spite of the potential safety hazards, some people would like to have an Internet connection in their car. A preliminary survey of adult Americans has estimated this proportion to be somewhere around .30 **(*USA Today*, May 1, 2009)**.

a. Use the given preliminary estimate to determine the sample size required to estimate the proportion of adult Americans who would like an Internet connection in their car to within .02 with 95% confidence.

b. The formula for determining sample size given in this section corresponds to a confidence level of 95%. How would you modify this formula if a 99% confidence level was desired?

c. Use the given preliminary estimate to determine the sample size required to estimate the proportion of adult Americans who would like an Internet connection in their car to within .02 with 99% confidence.

9.33 Data from a survey of a representative sample was used to estimate that 32% of computer users in 2011 had tried to get on a Wi-Fi network that was not their own in order to save money **(*USA Today*, May 16, 2011)**. Suppose you decide to conduct a survey to estimate this proportion for the current year. What is the required sample size if you want to estimate this proportion with a margin of error of 0.05? Compute the required sample size first using 0.32 as a preliminary estimate of p and then using the conservative value of 0.5. How do the two sample sizes compare? What sample size would you recommend for this study?

Bold exercises answered in back ● Data set available online ▼ Video Solution available

9.3 Confidence Interval for a Population Mean

In this section, we consider how to use information from a random sample to construct a confidence interval estimate of a population mean, μ. We begin by considering the case in which σ, the population standard deviation, is known and the sample size n is large enough for the Central Limit Theorem to apply. In this case, the following three properties about the sampling distribution of $\bar{x}$ hold:

1. The sampling distribution of $\bar{x}$ is centered at μ, so $\bar{x}$ is an unbiased statistic for estimating μ ($\mu_{\bar{x}} = \mu$).
2. The standard deviation of $\bar{x}$ is $\sigma_{\bar{x}} = \dfrac{\sigma}{\sqrt{n}}$.
3. As long as n is large (generally $n \geq 30$), the sampling distribution of $\bar{x}$ is approximately normal, even when the population distribution itself is not normal.

The same reasoning that was used to develop the large-sample confidence interval for a population proportion p can be used to obtain a confidence interval estimate for μ.

The One-Sample z Confidence Interval for μ

The general formula for a confidence interval for a population mean μ when
1. $\bar{x}$ is the sample mean from a **simple random sample**,
2. the **sample size n is large** (generally $n \geq 30$), and
3. **σ, the population standard deviation, is known**

is

$$\bar{x} \pm (z \text{ critical value})\left(\frac{\sigma}{\sqrt{n}}\right)$$

EXAMPLE 9.7 Cosmic Radiation

Understand the context

Cosmic radiation levels rise with increasing altitude, prompting researchers to consider how pilots and flight crews might be affected by increased exposure to cosmic radiation. The paper **"Estimated Cosmic Radiation Doses for Flight Personnel"** (***Space Medicine and Medical Engineering*** **[2002]: 265–269)** reported a mean annual cosmic radiation dose of 219 mrems for a sample of flight personnel of Xinjiang Airlines. Suppose that this mean was based on a random sample of 100 flight crew members.

Let μ denote the mean annual cosmic radiation exposure for all Xinjiang Airlines flight crew members. Although σ, the true population standard deviation, is not usually known, suppose for illustrative purposes that $\sigma = 35$ mrem is known. Because the sample size is large and σ is known, a 95% confidence interval for μ is

Do the work

$$\begin{aligned}\bar{x} \pm (z \text{ critical value})\left(\frac{\sigma}{\sqrt{n}}\right) &= 219 \pm (1.96)\left(\frac{35}{\sqrt{100}}\right)\\ &= 219 \pm 6.86\\ &= (212.14, 225.86)\end{aligned}$$

Interpret the results

Based on this sample, *plausible* values of μ, the actual mean annual cosmic radiation exposure for Xinjaing Airlines flight crew members, are between 212.14 and 225.86 mrem. A 95% confidence level is associated with the method used to produce this interval estimate. ■

The confidence interval just introduced is appropriate when σ is known and n is large, and it can be used regardless of the shape of the population distribution. This is because this confidence interval is based on the Central Limit Theorem, which says that when n is sufficiently large, the sampling distribution of $\bar{x}$ is approximately normal for any population distribution.

When n is small, the Central Limit Theorem cannot be used to justify the normality of the $\bar{x}$ sampling distribution, so the z confidence interval can not be used. One way to proceed in the small-sample case is to make a specific assumption about the shape of the population distribution and then to use a method that is valid under this assumption.

One instance where this is easy to do is when it is reasonable to believe that the population distribution is normal in shape. Recall that for a normal population distribution the sampling distribution of $\bar{x}$ is normal even for small sample sizes. So, if n is small but the population distribution is normal, the same confidence interval formula just introduced can still be used.

If it is reasonable to believe that the distribution of values in the population is normal, a confidence interval for μ (when σ is known) is

$$\bar{x} \pm (z \text{ critical value})\left(\frac{\sigma}{\sqrt{n}}\right)$$

This interval is appropriate even when n is small, as long as it is reasonable to think that the population distribution is normal in shape.

There are several ways that sample data can be used to assess the plausibility of normality. Two common ways are to look at a normal probability plot of the sample data (looking for a plot that is reasonably straight) or to construct a dotplot or a boxplot of the data (looking for approximate symmetry and no outliers).

Confidence Interval for μ When σ Is Unknown

The confidence interval just developed has an obvious drawback: To compute the interval endpoints, σ must be known. Unfortunately, this is rarely the case in practice. We now turn our attention to the situation when σ is unknown. The development of the confidence interval in this instance depends on the assumption that the population distribution is normal. This assumption is not critical if the sample size is large, but it is important when the sample size is small.

To understand the derivation of this confidence interval, begin by taking another look at the previous 95% confidence interval. We know that $\mu_{\bar{x}} = \mu$ and $\sigma_{\bar{x}} = \frac{\sigma}{\sqrt{n}}$. Also, when the population distribution is normal, the $\bar{x}$ distribution is normal. These facts imply that the standardized variable

$$z = \frac{\bar{x} - \mu}{\frac{\sigma}{\sqrt{n}}}$$

has approximately a standard normal distribution. Because the interval from -1.96 to 1.96 captures an area of .95 under the z curve, approximately 95% of all samples result in an $\bar{x}$ value that satisfies

$$-1.96 < \frac{\bar{x} - \mu}{\frac{\sigma}{\sqrt{n}}} < 1.96$$

Manipulating these inequalities to isolate μ in the middle results in the equivalent inequalities:

$$\bar{x} - 1.96\left(\frac{\sigma}{\sqrt{n}}\right) < \mu < \bar{x} + 1.96\left(\frac{\sigma}{\sqrt{n}}\right)$$

The term $\bar{x} - 1.96\left(\frac{\sigma}{\sqrt{n}}\right)$ is the lower endpoint of the 95% large-sample confidence interval for μ, and $\bar{x} + 1.96\left(\frac{\sigma}{\sqrt{n}}\right)$ is the upper endpoint.

If σ is unknown, we must use the sample data to estimate σ. If we use the sample standard deviation as our estimate, the result is a different standardized variable denoted by t:

$$t = \frac{\bar{x} - \mu}{\frac{s}{\sqrt{n}}}$$

The value of s may not be all that close to σ, especially when n is small. As a consequence, the use of s in place of σ introduces extra variability. The value of z varies from sample to sample, because different samples generally result in different $\bar{x}$ values. There is even more variability in t, because different samples may result in different values of both $\bar{x}$ and s. Because of this, the distribution of t is more spread out than the standard normal (z) distribution.

To develop an appropriate confidence interval, we must investigate the probability distribution of the standardized variable t for a sample from a normal population. This requires that we first learn about probability distributions called *t distributions*.

t Distributions

Just as there are many different normal distributions, there are also many different t distributions. While normal distributions are distinguished from one another by their mean μ and standard deviation σ, t distributions are distinguished by a positive whole number called the number of *degrees of freedom* (df). There is a t distribution with 1 df, another with 2 df, and so on.

Important Properties of *t* Distributions

1. The *t* distribution corresponding to any particular number of degrees of freedom is bell shaped and centered at zero (just like the standard normal (z) distribution).
2. Each *t* distribution is more spread out than the standard normal (z) distribution.
3. As the number of degrees of freedom increases, the spread of the corresponding *t* distribution decreases.
4. As the number of degrees of freedom increases, the corresponding sequence of *t* distributions approaches the standard normal (z) distribution.

The properties discussed in the preceding box are illustrated in Figure 9.6, which shows two *t* curves along with the *z* curve.

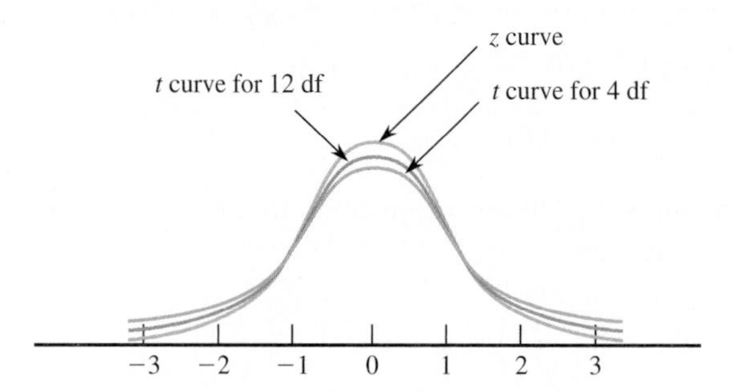

FIGURE 9.6
Comparison of the *z* curve and *t* curves for 12 df and 4 df.

Appendix Table 3 gives selected critical values for various *t* distributions. The central areas for which values are tabulated are .80, .90, .95, .98, .99, .998, and .999. To find a particular critical value, go down the left margin of the table to the row labeled with the desired number of degrees of freedom. Then move over in that row to the column headed by the desired central area.

For example, the value in the 12-df row under the column corresponding to central area .95 is 2.18, so 95% of the area under the *t* curve with 12 df lies between −2.18 and 2.18. Moving over two columns, we find the critical value for central area .99 (still with 12 df) to be 3.06 (see Figure 9.7). Moving down the .99 column to the 20-df row, we see the critical value is 2.85, so the area between −2.85 and 2.85 under the *t* curve with 20 df is .99.

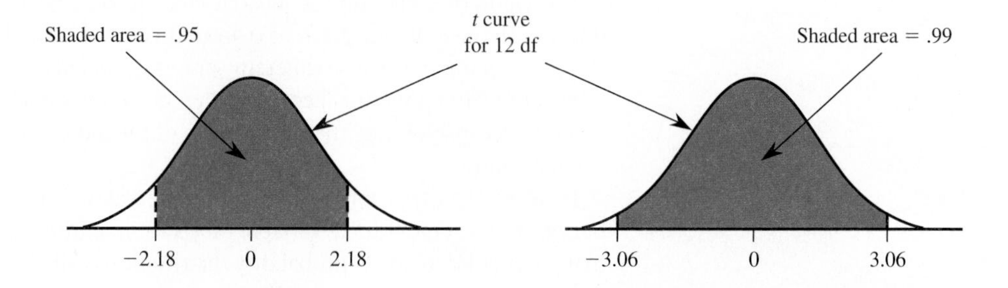

FIGURE 9.7
t critical values illustrated.

Notice that the critical values increase from left to right in each row of Appendix Table 3. This makes sense because as we move to the right, we capture larger central areas. In each column, the critical values decrease as we move downward, reflecting decreasing spread for *t* distributions with larger degrees of freedom.

The larger the number of degrees of freedom, the more closely the *t* curve resembles the *z* curve. To emphasize this, we have included the *z* critical values as the last row of the *t* table. Furthermore, once the number of degrees of freedom exceeds 30, the critical values change little as the number of degrees of freedom increases. For this reason, Appendix Table 3 jumps from 30 df to 40 df, then to 60 df, then to 120 df, and finally to the row of z critical values.

If we need a critical value for a number of degrees of freedom between those tabulated, we just use the critical value for the closest df. For df > 120, we use the z critical values. Many graphing calculators calculate t critical values for any number of degrees of freedom, so if you are using such a calculator, it is not necessary to approximate the t critical values as described.

One-Sample *t* Confidence Interval

The fact that the sampling distribution of $\frac{\bar{x} - \mu}{(\sigma/\sqrt{n})}$ is approximately the z (standard normal) distribution when n is large led to the z confidence interval when σ is known. In the same way, the following proposition allows us to obtain a confidence interval when the population distribution is normal but σ is unknown.

If $\bar{x}$ and s are the mean and standard deviation of a random sample from a normal population distribution, then the probability distribution of the standardized variable

$$t = \frac{\bar{x} - \mu}{\frac{s}{\sqrt{n}}}$$

is the t distribution with df $= n - 1$.

To see how this result leads to the desired confidence interval, consider the case $n = 25$. We use the t distribution with df $= n - 1 = 24$. From Appendix Table 3, the interval between -2.06 and 2.06 captures a central area of .95 under the t curve with 24 df. This means that 95% of all samples of size $n = 25$ from a normal population result in values of $\bar{x}$ and s for which

$$-2.06 < \frac{\bar{x} - \mu}{\frac{s}{\sqrt{n}}} < 2.06$$

Algebraically manipulating these inequalities to isolate μ yields

$$\bar{x} - 2.06\left(\frac{s}{\sqrt{25}}\right) < \mu < \bar{x} + 2.06\left(\frac{s}{\sqrt{25}}\right)$$

The 95% confidence interval for μ in this situation extends from the lower endpoint $\bar{x} - 2.06\left(\frac{s}{\sqrt{25}}\right)$ to the upper endpoint $\bar{x} + 2.06\left(\frac{s}{\sqrt{25}}\right)$. This interval can also be written

$$\bar{x} \pm 2.06\left(\frac{s}{\sqrt{25}}\right)$$

The differences between this interval and the interval when σ is known are the use of the t critical value 2.06 rather than the z critical value 1.96 and the use of the sample standard deviation as an estimate of σ. The extra uncertainty that results from estimating σ causes the t interval to be wider than the z interval.

If the sample size is something other than 25 or if the desired confidence level is something other than 95%, a different t critical value (obtained from Appendix Table 3 or by using technology) is used in place of 2.06.

The One-Sample t Confidence Interval for μ

The general formula for a confidence interval for a population mean μ based on a sample of size n when

1. $\bar{x}$ is the sample mean from a **simple random sample,**
2. the **population distribution is normal,** *or* **the sample size n is large** (generally $n \geq 30$), and
3. σ**, the population standard deviation, is unknown**

is

$$\bar{x} \pm (t \text{ critical value})\left(\frac{s}{\sqrt{n}}\right)$$

where the t critical value is based on df $= n - 1$. Appendix Table 3 gives critical values appropriate for each of the confidence levels 90%, 95%, and 99%, as well as several other less frequently used confidence levels.

If n is large (generally $n \geq 30$), the normality of the population distribution is not critical. *However, this confidence interval is appropriate for small n only when the population distribution is (at least approximately) normal.* If this is not the case, as might be suggested by a normal probability plot or boxplot, another estimation method should be used.

EXAMPLE 9.8 Drive-Through Medicine

Understand the context ❯

During a flu outbreak, many people visit emergency rooms, where they often must wait in crowded waiting rooms where other patients may be exposed. The paper **"Drive-Through Medicine: A Novel Proposal for Rapid Evaluation of Patients during an Influenza Pandemic"** (***Annals of Emergency Medicine*** **[2010]: 268–273)** describes an interesting study of the feasibility of a drive-through model where flu patients are evaluated while they remain in their cars. One of the interesting observations from this study was that not only were patients kept relatively isolated and away from other patients, but the time to process a patient was shorter because delays related to turning over examination rooms were eliminated.

In the experiment, 38 volunteers were each given a scenario from a randomly selected set of flu cases seen in the emergency room. The scenarios provided the volunteer with a medical history and a description of symptoms that would allow the volunteer to respond to questions from the examining physician. These volunteer patients were then processed using a drive-through procedure that was implemented in the parking structure of Stanford University Hospital. The time to process each case from admission to discharge was recorded.

Consider the data ❯

Data read from a graph that appears in the paper was used to compute the following summary statistics for admission-to-discharge processing times (in minutes):

$n = 38 \qquad \bar{x} = 26 \qquad s = 1.57$

Formulate a plan ❯

A boxplot of the 38 processing times did show a couple of outliers on the high end, corresponding to unusually long processing times, suggesting that it is probably not reasonable to think of the population distribution of drive-through processing times as being approximately normal. However, because the sample size is greater than 30 and the distribution of sample processing times was not extremely skewed, it is appropriate to consider using the t confidence interval to estimate the mean admission-to-discharge processing time for flu patients using the drive-through procedure. Because the 38 flu scenarios were thought to be representative of the population of flu patients seen in emergency rooms and the sample size is large, we can use the formula for the t confidence interval to compute a 95% confidence interval.

For this example, $n = 38$, df $= 37$, and the appropriate t critical value is 2.02 (from the 40-df row of Appendix Table 3). The 95% confidence interval is then

Do the work ⟩

$$\bar{x} \pm (t \text{ critical value})\left(\frac{s}{\sqrt{n}}\right) = 26 \pm (2.02)\left(\frac{1.57}{\sqrt{38}}\right)$$
$$= 26 \pm .514$$
$$= (25.486, 26.514)$$

Interpret the results ⟩

Based on the sample data, we believe that the actual mean admission-to-discharge processing time for flu patients processed using the drive-through procedure is between 25.486 minutes and 25.514 minutes. We used a method that has a 5% error rate to construct this interval. The authors of the paper indicated that the average processing time for flu patients seen in the emergency room was about 90 minutes, so it appears that the drive-through procedure has promise both in terms of keeping flu patients isolated and also in reducing processing time. ■

EXAMPLE 9.9 Waiting for Surgery

Understand the context ⟩

The Cardiac Care Network in Ontario, Canada, collected information on the time between the date a patient was recommended for heart surgery and the surgery date for cardiac patients in Ontario **("Wait Times Data Guide," Ministry of Health and Long-Term Care, Ontario, Canada, 2006)**. The reported mean wait times (in days) for samples of patients for two cardiac procedures are given in the accompanying table. (The standard deviations in the table were estimated from information on wait-time variability included in the report.)

Consider the data ⟩

Surgical Procedure	Sample Size	Mean Wait Time	Standard Deviation
Bypass	539	19	10
Angiography	847	18	9

Formulate a plan ⟩

If we had access to the raw data (the 539 + 847 = 1386 individual wait-time observations), we might begin by looking at boxplots. Data consistent with the given summary quantities were used to generate the boxplots of Figure 9.8. The boxplots for the two surgical procedures are similar. There are outliers in both data sets, which might cause us to question the normality of the two wait-time distributions, but because the sample sizes are large, it is still appropriate to use the t confidence interval.

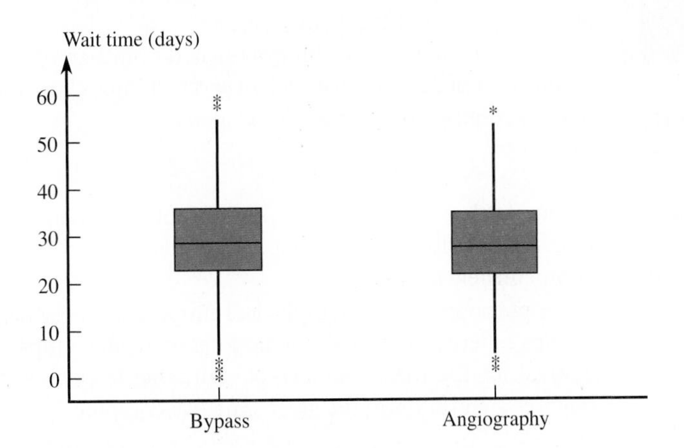

FIGURE 9.8
Boxplots for Example 9.9.

As a next step, we can use the confidence interval of this section to estimate the actual mean wait time for each of the two procedures. Let's first focus on the sample of bypass patients. For this group,

$$\text{sample size} = n = 539$$
$$\text{sample mean wait time} = \bar{x} = 19$$
$$\text{sample standard deviation} = s = 10$$

The report referenced here indicated that it is reasonable to regard these data as representative of the Ontario population. So, with μ denoting the mean wait time for bypass surgery in Ontario, we can estimate μ using a 90% confidence interval.

Do the work ⟩

From Appendix Table 3, we use t critical value $= 1.645$ (from the z critical value row because df $= n - 1 = 538 > 120$, the largest number of degrees of freedom in the table). The 90% confidence interval for μ is

$$\bar{x} \pm (t \text{ critical value})\left(\frac{s}{\sqrt{n}}\right) = 19 \pm (1.645)\left(\frac{10}{\sqrt{539}}\right)$$
$$= 19 \pm .709$$
$$= (18.291, 19.709)$$

Interpret the results ⟩

Based on this sample, we are 90% confident that μ is between 18.291 days and 19.709 days. This interval is fairly narrow, indicating that our information about the value of μ is relatively precise.

A graphing calculator or any of the commercially available statistical computing packages can produce t confidence intervals. Confidence interval output from Minitab for the angiography data is shown here.

One-Sample T

N	Mean	StDev	SE Mean	90% CI
847	18.0000	9.0000	0.3092	(17.4908, 18.5092)

The 90% confidence interval for mean wait time for angiography extends from 17.4908 days to 18.5092 days. This interval is narrower than the 90% interval for bypass surgery wait time for two reasons: the sample size is larger (847 rather than 539) and the sample standard deviation is smaller (9 rather than 10). ■

Alan and Sandy Carey /Photodisc/Getty Images

EXAMPLE 9.10 Selfish Chimps?

● The article **"Chimps Aren't Charitable" (*Newsday*, November 2, 2005)** summarized the results of a research study published in the journal *Nature*. In this study, chimpanzees learned to use an apparatus that dispensed food when either of two ropes was pulled. When one of the ropes was pulled, only the chimp controlling the apparatus received food. When the other rope was pulled, food was dispensed both to the chimp controlling the apparatus and also to a chimp in the adjoining cage.

Understand the context ⟩

The accompanying data (approximated from a graph in the paper) represent the number of times out of 36 trials that each of seven chimps chose the option that would provide food to both chimps (the "charitable" response).

23 22 21 24 19 20 20

Formulate a plan ⟩

Figure 9.9 is a normal probability plot of these data. The plot is reasonably straight, so it seems plausible that the population distribution of number of charitable responses is approximately normal.

For purposes of this example, let's suppose it is reasonable to regard this sample of seven chimps as representative of the population of all chimpanzees. Calculation of a confidence interval for the mean number of charitable responses for the population of all chimps requires $\bar{x}$ and s. From the given data, we compute

Step-by-step technology instructions available online

$\bar{x} = 21.29 \quad s = 1.80$

The t critical value for a 99% confidence interval based on 6 df is 3.71. The interval is

Do the work ⟩

$$\bar{x} \pm (t \text{ critical value})\left(\frac{s}{\sqrt{n}}\right) = 21.29 \pm (3.71)\left(\frac{1.80}{\sqrt{7}}\right)$$
$$= 21.29 \pm 2.52$$
$$= (18.77, 23.81)$$

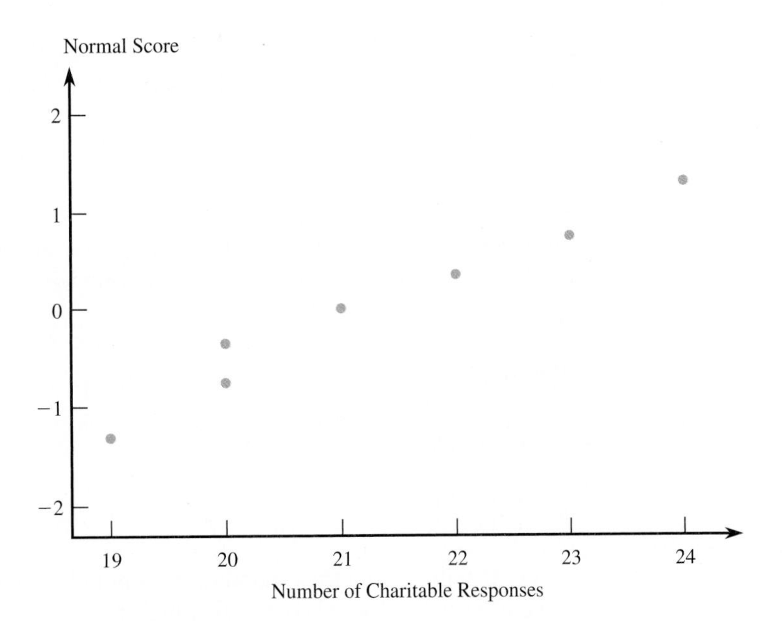

FIGURE 9.9
Normal probability plot for data of Example 9.10

A statistical software package could also have been used to compute the 99% confidence interval. The following is output from SPSS. The slight discrepancy between the hand-calculated interval and the one reported by SPSS occurs because SPSS uses more decimal accuracy in $\bar{x}$, s, and t critical values.

One-Sample Statistics

	N	Mean	Std. Deviation	Std. Error Mean
CharitableResponses	7	21.2857	1.79947	.68014

One-Sample

	99% Confidence Interval	
	Lower	**Upper**
CharitableResponses	18.7642	23.8073

Interpret the results ⟩

With 99% confidence, we estimate the population mean number of charitable responses (out of 36 trials) to be between 18.77 and 23.81. Remember that the 99% confidence level implies that if the same formula is used to calculate intervals for sample after sample randomly selected from the population of chimps, in the long run 99% of these intervals will capture the value of μ between the lower and upper confidence limits.

Notice that based on this interval, we would conclude that, on average, chimps choose the charitable option more than half the time (more than 18 out of 36 trials). The ***Newsday*** headline **"Chimps Aren't Charitable"** was based on additional data from the study indicating that chimps' charitable behavior was no different when there was another chimp in the adjacent cage than when the adjacent cage was empty. We will revisit this study in Chapter 11 to investigate this further. ■

Choosing the Sample Size

When estimating μ using a large sample or using a small sample from a normal population, the bound B on the error of estimation associated with a 95% confidence interval is

$$B = 1.96\left(\frac{\sigma}{\sqrt{n}}\right)$$

Before collecting any data, an investigator may wish to determine a sample size for which a particular value of the bound is achieved. For example, with μ representing the average fuel efficiency (in miles per gallon, mpg) for all cars of a certain type, the objective of an investigation may be to estimate μ to within 1 mpg with 95% confidence. The value of n necessary to achieve this is obtained by setting $B = 1$ and then solving

$$1 = 1.96\left(\frac{\sigma}{\sqrt{n}}\right)$$

for n.

In general, suppose that we wish to estimate μ to within an amount B (the specified bound on the error of estimation) with 95% confidence. Finding the necessary sample size requires solving the equation

$$B = 1.96\left(\frac{\sigma}{\sqrt{n}}\right)$$

for n. The result is

$$n = \left(\frac{1.96\sigma}{B}\right)^2$$

Notice that the greater the variability in the population (larger σ), the greater the required sample size will be. And, of course, the smaller the desired bound on error is, the larger the required sample size will be.

Use of the sample-size formula requires that σ be known, but this is rarely the case in practice. One possible strategy for estimating σ is to carry out a preliminary study and use the resulting sample standard deviation (or a somewhat larger value, to be conservative) to determine n for the main part of the study. Another possibility is simply to make an educated guess about the value of σ and to use that value to calculate n. For a population distribution that is not too skewed, dividing the anticipated range (the difference between the largest and the smallest values) by 4 often gives a rough idea of the value of the standard deviation.

The sample size required to estimate a population mean μ to within an amount B with 95% confidence is

$$n = \left(\frac{1.96\sigma}{B}\right)^2$$

If σ is unknown, it may be estimated based on previous information or, for a population that is not too skewed, by using (range)/4.

If the desired confidence level is something other than 95%, 1.96 is replaced by the appropriate z critical value (for example, 2.58 for 99% confidence).

EXAMPLE 9.11 Cost of Textbooks

The financial aid office wishes to estimate the mean cost of textbooks per quarter for students at a particular university. For the estimate to be useful, it should be within \$20 of the true population mean. How large a sample should be used to be 95% confident of achieving this level of accuracy?

To determine the required sample size, we must have a value for σ. The financial aid office is pretty sure that the amount spent on books varies widely, with most values between \$150 and \$550. A reasonable estimate of σ is then

$$\frac{range}{4} = \frac{550 - 150}{4} = \frac{400}{4} = 100$$

The required sample size is

$$n = \left(\frac{1.96\sigma}{B}\right)^2 = \left(\frac{(1.96)(100)}{20}\right)^2 = (9.8)^2 = 96.04$$

Rounding up, a sample size of 97 or larger is recommended. ■

EXERCISES 9.34 - 9.52

9.34 Given a variable that has a t distribution with the specified degrees of freedom, what percentage of the time will its value fall in the indicated region?

a. 10 df, between -1.81 and 1.81

b. 10 df, between -2.23 and 2.23

c. 24 df, between -2.06 and 2.06

d. 24 df, between -2.80 and 2.80

e. 24 df, outside the interval from -2.80 to 2.80

f. 24 df, to the right of 2.80

g. 10 df, to the left of -1.81

9.35 The formula used to compute a confidence interval for the mean of a normal population when n is small is

$$\bar{x} \pm (t \text{ critical value})\frac{s}{\sqrt{n}}$$

What is the appropriate t critical value for each of the following confidence levels and sample sizes?

a. 95% confidence, $n = 17$

b. 90% confidence, $n = 12$

c. 99% confidence, $n = 24$

d. 90% confidence, $n = 25$

e. 90% confidence, $n = 13$

f. 95% confidence, $n = 10$

9.36 The two intervals (114.4, 115.6) and (114.1, 115.9) are confidence intervals (computed using the same sample data) for μ = true average resonance frequency (in hertz) for all tennis rackets of a certain type.

a. What is the value of the sample mean resonance frequency? (Hint: Where is the confidence interval centered?)

b. The confidence level for one of these intervals is 90% and for the other it is 99%. Which is which, and how can you tell?

9.37 ▼ Samples of two different models of cars were selected, and the actual speed for each car was determined when the speedometer registered 50 mph. The resulting 95% confidence intervals for mean actual speed were (51.3, 52.7) and (49.4, 50.6). Assuming that the two sample standard deviations are equal, which confidence interval is based on the larger sample size? Explain your reasoning.

9.38 The authors of the paper **"Deception and Design: The Impact of Communication Technology on Lying Behavior"** (***Proceedings of Computer Human Interaction*** **[2004]**) asked 30 students in an upper division communications course at a large university to keep a journal for 7 days, recording each social interaction and whether or not they told any lies during that interaction. A lie was defined as "any time you intentionally try to mislead someone." The paper reported that the mean number of lies per day for the 30 students was 1.58 and the standard deviation of number of lies per day was 1.02.

a. What assumption must be made in order for the t confidence interval of this section to be an appropriate method for estimating μ, the mean number of lies per day for all students at this university?

b. Would you recommend using the t confidence interval to construct an estimate of μ as defined in Part (a)? Explain why or why not.

9.39 In a study of academic procrastination, the authors of the paper **"Correlates and Consequences of Behavioral Procrastination"** (***Procrastination, Current Issues and New Directions*** **[2000]**) reported that for a sample of 411 undergraduate students at a midsize public university preparing for a final exam in an introductory psychology course, the mean time spent studying for the exam was 7.74 hours and the standard deviation of study times was 3.40 hours. For purposes of this exercise, assume that it is reasonable to regard this sample as representative of students taking introductory psychology at this university.

a. Construct a 95% confidence interval to estimate μ, the mean time spent studying for the final exam for students taking introductory psychology at this university. (Hint: See Example 9.8.)

b. The paper also gave the following sample statistics for the percentage of study time that occurred in the 24 hours prior to the exam:

$n = 411 \qquad \bar{x} = 43.18 \qquad s = 21.46$

Construct and interpret a 90% confidence interval for the mean percentage of study time that occurs in the 24 hours prior to the exam.

9.40 Medical research has shown that repeated wrist extensions beyond 20 degrees increase the risk of wrist and hand injuries. Each of 24 students at Cornell University used a proposed new computer mouse design, and while using the mouse, each student's wrist extension was recorded. Data consistent with summary values given in the paper "Comparative Study of Two Computer Mouse Designs" **(Cornell Human Factors Laboratory Technical Report RP7992)** are given here.

27	28	24	26	27	25	25	24	24	24	25	28
22	25	24	28	27	26	31	25	28	27	27	25

a. Use these data to estimate the mean wrist extension for people using this new mouse design using a 90% confidence interval.

b. Are any assumptions required in order for it to be appropriate to generalize your estimate to the population of Cornell students? To the population of all university students?

c. Based on your interval from Part (a), do you think there is reason to believe that the mean wrist extension for people using the new mouse design is greater than 20 degrees? Explain why or why not.

9.41 In **June 2009**, Harris Interactive conducted its **Great Schools Survey**. In this survey, the sample consisted of 1086 adults who were parents of school-aged children. The sample was selected in a way that makes it reasonable to regard it as representative of the population of parents of school-aged children. One question on the survey asked respondents how much time (in hours) per month they spent volunteering at their children's school during the previous school year. The following summary statistics for time volunteered per month were given:

$n = 1086 \qquad \bar{x} = 5.6 \qquad \text{median} = 1$

a. What does the fact that the mean is so much larger than the median tell you about the distribution of time spent volunteering per month?

b. Based on your answer to Part (a), explain why it is not reasonable to assume that the population distribution of time spent volunteering is approximately normal.

c. Explain why it is appropriate to use the t confidence interval to estimate the mean time spent volunteering for the population of parents of school-aged children even though the population distribution is not approximately normal.

d. Suppose that the sample standard deviation was $s = 5.2$. Compute and interpret a 98% confidence interval for μ, the mean time spent volunteering for the population of parents of school-aged children.

9.42 The authors of the paper **"Driven to Distraction"** (***Psychological Science*** **[2001]: 462–466)** describe an experiment to evaluate the effect of using a cell phone on reaction time. Subjects were asked to perform a simulated driving task while talking on a cell phone. While performing this task, occasional red and green lights flashed on the computer screen. If a green light flashed, subjects were to continue driving, but if a red light flashed, subjects were to brake as quickly as possible. The reaction time (in msec) was recorded. The following summary statistics are based on a graph that appeared in the paper:

$n = 48 \qquad \bar{x} = 530 \qquad s = 70$

a. Construct and interpret a 95% confidence interval for μ, the mean time to react to a red light while talking on a cell phone.

b. What assumption must be made in order to generalize this confidence interval to the population of all drivers?

c. Suppose that the researchers wanted to estimate the mean reaction time to within 5 msec with 95% confidence. Using the sample standard deviation from the study described as a preliminary estimate of the standard deviation of reaction times, compute the required sample size. (Hint: See Example 9.11.)

9.43 Suppose that a random sample of 50 bottles of a particular brand of cough medicine is selected and the alcohol content of each bottle is determined. Let μ denote the mean alcohol content (in percent) for the population of all bottles of the brand under study. Suppose that the sample of 50 results in a 95% confidence interval for μ of (7.8, 9.4).

a. Would a 90% confidence interval have been narrower or wider than the given interval? Explain your answer.

b. Consider the following statement: There is a 95% chance that μ is between 7.8 and 9.4. Is this statement correct? Why or why not?

c. Consider the following statement: If the process of selecting a random sample of size 50 and then computing the corresponding 95% confidence interval is repeated 100 times, 95 of the resulting intervals will include μ. Is this statement correct? Why or why not?

9.44 ▼ Acrylic bone cement is sometimes used in hip and knee replacements to fix an artificial joint in place. The force required to break an acrylic bone cement bond was measured for six specimens under specified conditions, and the resulting mean and standard deviation were 306.09 Newtons and 41.97 Newtons, respectively. Assuming that it is reasonable to believe that breaking force under these conditions has a distribution that is approximately normal, estimate the mean breaking force for acrylic bone cement under the specified conditions using a 95% confidence interval.

9.45 The article **"The Association Between Television Viewing and Irregular Sleep Schedules Among Children Less Than 3 Years of Age"** **(*Pediatrics*** **[2005]: 851–856)** reported the accompanying 95% confidence intervals for average TV viewing time (in hours per day) for three different age groups.

Age Group	95% Confidence Interval
Less than 12 months	(0.8, 1.0)
12 to 23 months	(1.4, 1.8)
24 to 35 months	(2.1, 2.5)

a. Suppose that the sample sizes for each of the three age group samples were equal. Based on the given confidence intervals, which of the age group samples had the greatest variability in TV viewing time? Explain your choice. (Hint: Consider the formula for the confidence interval.)

b. Now suppose that the sample standard deviations for the three age group samples were equal, but that the three sample sizes might have been different. Which of the three age-group samples had the largest sample size? Explain your choice.

c. The interval (.768, 1.032) is either a 90% confidence interval or a 99% confidence interval for the mean TV viewing time computed using the sample data for children less than 12 months old. Is the confidence level for this interval 90% or 99%? Explain your choice.

9.46 The article **"Most Canadians Plan to Buy Treats, Many Will Buy Pumpkins, Decorations and/or Costumes" (Ipsos-Reid, October 24, 2005)** summarized results from a survey of 1000 randomly selected Canadian residents. Each individual in the sample was asked how much he or she anticipated spending on Halloween during 2005. The resulting sample mean and standard deviation were \$46.65 and \$83.70, respectively.

a. Explain how it could be possible for the standard deviation of the anticipated Halloween expense to be larger than the mean anticipated expense.

b. Is it reasonable to think that the distribution of the variable *anticipated Halloween expense* is approximately normal? Explain why or why not.

c. Is it appropriate to use the t confidence interval to estimate the mean anticipated Halloween expense for Canadian residents? Explain why or why not.

d. If appropriate, construct and interpret a 99% confidence interval for the mean anticipated Halloween expense for Canadian residents.

9.47 Because of safety considerations, in May 2003 the Federal Aviation Administration (FAA) changed its guidelines for how small commuter airlines must estimate passenger weights. Under the old rule, airlines used 180 pounds as a typical passenger weight (including carry-on luggage) in warm months and 185 pounds as a typical weight in cold months.

The ***Alaska Journal of Commerce*** **(May 25, 2003)** reported that Frontier Airlines conducted a study to estimate average passenger plus carry-on weights. They found an average summer weight of 183 pounds and a winter average of 190 pounds. Suppose that each of these estimates was based on a random sample of 100 passengers and that the sample standard deviations were 20 pounds for the summer weights and 23 pounds for the winter weights.

a. Construct and interpret a 95% confidence interval for the mean summer weight (including carry-on luggage) of Frontier Airlines passengers.

b. Construct and interpret a 95% confidence interval for the mean winter weight (including carry-on luggage) of Frontier Airlines passengers.

c. The new FAA recommendations are 190 pounds for summer and 195 pounds for winter. Comment on these recommendations in light of the confidence interval estimates from Parts (a) and (b).

9.48 ● Example 9.3 gave the following airborne times (in minutes) for 10 randomly selected flights from San Francisco to Washington Dulles airport:

270 256 267 285 274 275 266 258 271 281

a. Compute and interpret a 90% confidence interval for the mean airborne time for flights from San Francisco to Washington Dulles. (Hint: See Example 9.10.)

b. Give an interpretation of the 90% confidence level associated with the interval estimate in Part (a).

c. If a flight from San Francisco to Washington Dulles is scheduled to depart at 10 A.M., what would you recommend for the published arrival time? Explain.

9.49 ● Fat content (in grams) for seven randomly selected hot dogs that were rated as very good by **Consumer Reports (www.consumerreports.org)** is shown below. Is it reasonable to use this data and the t confidence interval of this section to construct a confidence interval for the mean fat content of hot dogs rated as very good by Consumer Reports? Explain why or why not. (Hint: See Example 9.10.)

14 15 11 10 6 15 16

9.50 ● Five students visiting the student health center for a free dental examination during National Dental Hygiene Month were asked how many months had passed since their last visit to a dentist. Their responses were as follows:

6 17 11 22 29

Assuming that these five students can be considered a random sample of all students participating in the free checkup program, construct a 95% confidence interval for the mean number of months elapsed since the last visit to a dentist for the population of students participating in the program.

9.51 The Bureau of Alcohol, Tobacco, and Firearms (BATF) has been concerned about lead levels in California wines. In a previous testing of wine specimens, lead levels ranging from 50 to 700 parts per billion were recorded. How many wine specimens should be tested if the BATF wishes to estimate the true mean lead level for California wines to within 10 parts per billion with 95% confidence? (Hint: See Example 9.11.)

9.52 The formula described in this section for determining sample size corresponds to a confidence level of 95%. What would be the appropriate formula for determining sample size when the desired confidence level is 90%? 98%?

Bold exercises answered in back ● Data set available online ▼ Video Solution available

9.4 Interpreting and Communicating the Results of Statistical Analyses

The purpose of most surveys and many research studies is to produce estimates of population characteristics. One way of providing such an estimate is to construct and report a confidence interval for the population characteristic of interest.

Communicating the Results of Statistical Analyses

When using sample data to estimate a population characteristic, a point estimate or a confidence interval estimate might be used. Confidence intervals are generally preferred because a point estimate by itself does not convey any information about the accuracy of the estimate. For this reason, whenever you report the value of a point estimate, it is a good idea to also include an estimate of the bound on the error of estimation.

Reporting and interpreting a confidence interval estimate requires a bit of care. First, always report both the confidence interval and the confidence level associated with the method used to produce the interval. Then, remember that both the confidence interval and the confidence level should be interpreted.

A good strategy is to begin with an interpretation of the confidence interval in the context of the problem and then to follow that with an interpretation of the confidence level. For example, if a 90% confidence interval for p, the proportion of students at a particular university who own a laptop computer, is (.56, .78), we might say

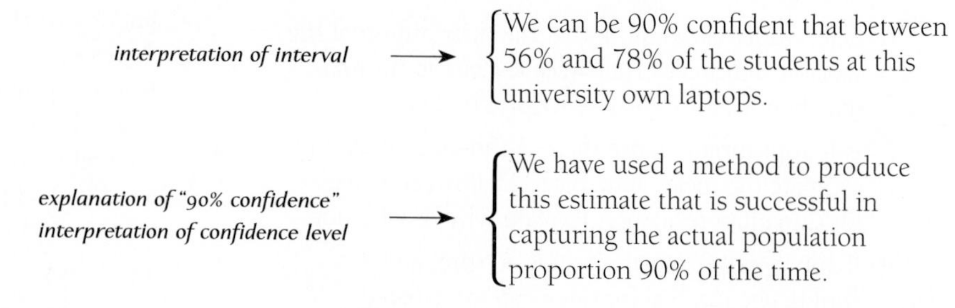

interpretation of interval
We can be 90% confident that between 56% and 78% of the students at this university own laptops.
explanation of "90% confidence"
interpretation of confidence level
We have used a method to produce this estimate that is successful in capturing the actual population proportion 90% of the time.

When providing an interpretation of a confidence interval, remember that the interval is an estimate of a population characteristic and be careful *not* to say that the interval applies to individual values in the population or to the values of sample statistics. For example, if a 99% confidence interval for μ, the mean amount of ketchup in bottles labeled as 12 ounces, is (11.94, 11.98), this does *not* tell us that 99% of 12-ounce ketchup bottles contain between 11.94 and 11.98 ounces of ketchup. Nor does it tell us that 99% of samples of the same size would have sample means in this particular range. The confidence interval is an estimate of the *mean* for *all* bottles in the *population* of interest.

Interpreting the Results of Statistical Analyses

Unfortunately, there is no customary way of reporting the estimates of population characteristics in published sources. Possibilities include

confidence interval
estimate $\pm$ bound on error
estimate $\pm$ standard error

If the population characteristic being estimated is a population mean, then you may also see

sample mean $\pm$ sample standard deviation

If the interval reported is described as a confidence interval, a confidence level should accompany it. These intervals can be interpreted just as we have interpreted the confidence intervals in this chapter. The confidence level specifies the long-run error rate associated with the method used to construct the interval. For example, a 95% confidence level specifies a 5% long-run error rate.

A form particularly common in news articles is estimate $\pm$ bound on error, where the bound on error is also sometimes called the **margin of error**. The bound on error reported is usually two times the standard deviation of the estimate. This method of reporting is a little less formal than a confidence interval and, if the sample size is reasonably large, is roughly equivalent to reporting a 95% confidence interval. You can interpret these intervals as you would a confidence interval with approximate confidence level of 95%.

You must use care in interpreting intervals reported in the form of an estimate $\pm$ standard error. Recall from Section 9.2 that the general form of a confidence interval is

estimate $\pm$ (critical value)(standard deviation of the estimate)

In journal articles, the estimated standard deviation of the estimate is usually referred to as the *standard error*. The critical value in the confidence interval formula was determined by the form of the sampling distribution of the estimate and by the confidence level. Notice that the reported form, estimate $\pm$ standard error, is equivalent to a confidence interval with the critical value set equal to 1. For a statistic whose sampling distribution is approximately normal (such as the mean of a large sample or a large-sample proportion), a critical value of 1 corresponds to an approximate confidence level of about 68%. Because a confidence level of 68% is rather low, you may want to use the given information and the confidence interval formula to convert to an interval with a higher confidence level.

When researchers are trying to estimate a population mean, they sometimes report sample mean $\pm$ sample standard deviation. Be particularly careful here. To convert this information into a useful interval estimate of the population mean, you must first convert the sample standard deviation to the standard error of the sample mean (by dividing by $\sqrt{n}$) and then use the standard error and an appropriate critical value to construct a confidence interval.

For example, suppose that a random sample of size 100 is used to estimate the population mean. If the sample resulted in a sample mean of 500 and a sample standard deviation of 20, you might find the published results summarized in any of the following ways:

95% confidence interval for the population mean: (496.08, 503.92)
mean $\pm$ bound on error: 500 $\pm$ 4
mean $\pm$ standard error: 500 $\pm$ 2
mean $\pm$ standard deviation: 500 $\pm$ 20

What to Look For in Published Data

Here are some questions to ask when you encounter interval estimates in published reports.

- Is the reported interval a confidence interval, mean $\pm$ bound on error, mean $\pm$ standard error, or mean $\pm$ standard deviation? If the reported interval is not a confidence interval, you may want to construct a confidence interval from the given information.

- What confidence level is associated with the given interval? Is the choice of confidence level reasonable? What does the confidence level say about the long-run error rate of the method used to construct the interval?
- Is the reported interval relatively narrow or relatively wide? Has the population characteristic been estimated precisely?

For example, the article **"Use of a Cast Compared with a Functional Ankle Brace After Operative Treatment of an Ankle Fracture"** (***Journal of Bone and Joint Surgery*** **[2003]: 205–211**) compared two different methods of immobilizing an ankle after surgery to repair damage from a fracture. The article includes the following statement:

> The mean duration (and standard deviation) between the operation and return to work was 63±13 days (median, sixty-three days; range, thirty three to ninety-eight days) for the cast group and 65±19 days (median, sixty-two days; range, eight to 131 days) for the brace group; the difference was not significant.

This is an example of a case where we must be careful—the reported intervals are of the form estimate ± standard deviation. We can use this information to construct a confidence interval for the mean time between surgery and return to work for each method of immobilization. One hundred patients participated in the study, with 50 wearing a cast after surgery and 50 wearing an ankle brace (random assignment was used to assign patients to treatment groups). Because the sample sizes are both large, we can use the *t* confidence interval formula

$$\text{mean} \pm (t \text{ critical value})\left(\frac{s}{\sqrt{n}}\right)$$

Each sample has df = 50 − 1 = 49. The closest df value in Appendix Table 3 is for df = 40, and the corresponding *t* critical value for a 95% confidence level is 2.02. The corresponding intervals are

$$\text{Cast: } 63 \pm 2.02\left(\frac{13}{\sqrt{50}}\right) = 63 \pm 3.71 = (59.29, 66.71)$$

$$\text{Brace: } 65 \pm 2.02\left(\frac{19}{\sqrt{50}}\right) = 65 \pm 5.43 = (59.57, 70.43)$$

The chosen confidence level of 95% implies that the method used to construct each of the intervals has a 5% long-run error rate. Assuming that it is reasonable to view these samples as representative of the patient population, we can interpret these intervals as follows: We can be 95% confident that the mean return-to-work time for those treated with a cast is between 59.29 and 66.71 days. We can be 95% confident that the mean return-to-work time for those treated with an ankle brace is between 59.57 and 70.43 days.

These intervals are relatively wide, indicating that the values of the treatment means have not been estimated as precisely as we might like. This is not surprising, given the sample sizes and the variability in each sample. Note that the two intervals overlap. This supports the statement that the difference between the two immobilization methods was not significant. Formal methods for directly comparing two groups, covered in Chapter 11, could be used to further investigate this issue.

A Word to the Wise: Cautions and Limitations

When working with point and confidence interval estimates, here are a few things you need to keep in mind:

1. In order for an estimate to be useful, we must know something about accuracy. You should beware of point estimates that are not accompanied by a bound on error or some other measure of accuracy.
2. A confidence interval estimate that is wide indicates that we don't have very precise information about the population characteristics being estimated. Don't be

fooled by a high confidence level if the resulting interval is wide. High confidence, while desirable, is not the same thing as saying we have precise information about the value of a population characteristic.

The width of a confidence interval is affected by the confidence level, the sample size, and the standard deviation of the statistic used as the basis for constructing the interval. The best strategy for decreasing the width of a confidence interval is to take a larger sample. It is far better to think about this before collecting data and to use the required sample size formulas to determine a sample size that will result in a confidence interval estimate that is narrow enough to provide useful information.

3. The accuracy of estimates depends on the sample size, not the population size. This may be counter to intuition, but as long as the sample size is small relative to the population size (n less than 10% of the population size), the bound on error for estimating a population proportion with 95% confidence is approximately $2\sqrt{\frac{\hat{p}(1-\hat{p})}{n}}$ and for estimating a population mean with 95% confidence is approximately $2\frac{s}{\sqrt{n}}$.

   Notice that each of these involves the sample size n, and both bounds decrease as the sample size increases. Neither approximate bound on error depends on the population size.

   The size of the population does need to be considered if sampling is without replacement and the sample size is more than 10% of the population size. In this case, a **finite population correction factor** $\sqrt{\frac{N-n}{N-1}}$ is used to adjust the bound on error (the given bound is multiplied by the correction factor). Since this correction factor is always less than 1, the adjusted bound on error is smaller.

4. Assumptions and "plausibility" conditions are important. The confidence interval procedures of this chapter require certain assumptions. If these assumptions are met, the confidence intervals provide us with a method for using sample data to estimate population characteristics with confidence. When the assumptions associated with a confidence interval procedure are in fact true, the confidence level specifies a correct success rate for the method. However, assumptions (such as the assumption of a normal population distribution) are rarely exactly met in practice. Fortunately, in most cases, as long as the assumptions are approximately met, the confidence interval procedures still work well.

   In general, we can only determine if assumptions are "plausible" or approximately met, and that we are in the situation where we expect the inferential procedure to work reasonably well. This is usually confirmed by knowledge of the data collection process and by using the sample data to check certain "plausibility conditions."

   The formal assumptions for the large-sample z confidence interval for a population proportion are

   1. The sample is a random sample from the population of interest.
   2. The sample size is large enough for the sampling distribution of $\hat{p}$ to be approximately normal.
   3. Sampling is without replacement.

   Whether the random sample assumption is plausible will depend on how the sample was selected and the intended population. Plausibility conditions for the other two assumptions are the following:

   $n\hat{p} \geq 10$ and $n(1-\hat{p}) \geq 10$ (so the sampling distribution of $\hat{p}$ is approximately normal), and

   n is less than 10% of the population size (so that the formula for the standard deviation of $\hat{p}$ provides a good approximation to the actual standard deviation).

The formal assumptions for the t confidence interval for a population mean are

1. The sample is a random sample from the population of interest.
2. The population distribution is normal, so that the distribution of $t = \frac{\bar{x} - \mu}{s/\sqrt{n}}$ has a t distribution.

As was the case for proportions, the plausibility of the random sample assumption will depend on how the sample was selected and the population of interest. The plausibility conditions for the normal population distribution assumption are the following:

A normal probability plot of the data is reasonably straight (indicating that the population distribution is approximately normal)

or

The data distribution is approximately symmetric and there are no outliers. This may be confirmed by looking at a dotplot, boxplot, stem-and-leaf display, or histogram of the data.

Alternatively, if n is large ($n \geq 30$), the sampling distribution of $\bar{x}$ will be approximately normal even for nonnormal population distributions. This implies that use of the t interval is appropriate even if population normality is not plausible.

In the end, you must decide that the assumptions are met or that they are "plausible" and that the inferential method used will provide reasonable results. This is also true for the inferential methods introduced in the chapters that follow.

5. Watch out for the "±" when reading published reports. Don't fall into the trap of thinking confidence interval every time you see a ± in an expression. As was discussed earlier in this section, published reports are not consistent, and in addition to confidence intervals, it is common to see estimate ± standard error and estimate ± sample standard deviation reported.

EXERCISES 9.53 - 9.55

9.53 The following quote is from the article **"Credit Card Debt Rises Faster for Seniors" (*USA Today*, July 28, 2009)**:

> The study, which will be released today by Demos, a liberal public policy group, shows that low- and middle-income consumers 65 and older carried $10,235 in average credit card debt last year.

What additional information would you want in order to evaluate the accuracy of this estimate? Explain.

9.54 Authors of the news release titled **"Major Gaps Still Exist Between the Perception and the Reality of Americans' Internet Security Protections, Study Finds" (The National Cyber Security Alliance)** estimated the proportion of Americans who claim to have a firewall installed on their computer to protect them from computer hackers to be .80 based on a survey conducted by the Zogby market research firm. They also estimated the proportion of those who actually have a firewall installed to be .42, based on checkups performed by Norton's PC Help software. The following quote is from the news release:

> For the study, NCSA commissioned a Zogby survey of more than 3000 Americans and Symantec conducted checkups of 400 Americans' personal computers performed by PC Help by Norton (www.norton.com/tuneup). The Zogby poll has a margin of error of +/− 1.6% and the checkup has a margin of error of +/− 5%.

Explain why the margins of error for the two estimated proportions are different.

9.55 The paper **"The Curious Promiscuity of Queen Honey Bees (*Apis mellifera*): Evolutionary and Behavioral Mechanisms" (*Annals of Zoology Fennici* [2001]:255–265)** describes a study of the mating behavior of queen honeybees. The following quote is from the paper:

> Queens flew for an average of 24.2 ± 9.21 minutes on their mating flights, which is consistent with previous findings. On those flights, queens effectively mated with 4.6 ± 3.47 males (mean ± SD).

a. The intervals reported in the quote from the paper were based on data from the mating flights of $n = 30$ queen honeybees. One of the two intervals reported is stated to be a confidence interval for a population mean. Which interval is this? Justify your choice.

b. Use the given information to construct a 95% confidence interval for the mean number of partners on a mating flight for queen honeybees. For purposes of this exercise, assume that it is reasonable to consider these 30 queen honeybees as representative of the population of queen honeybees.

Bold exercises answered in back ● Data set available online ▼ Video Solution available

ACTIVITY 9.1 Getting a Feel for Confidence Level

Technology Activity (Applet): Open the applet (available at www.cengagebrain.com) called ConfidenceIntervals. You should see a screen like the one shown here.

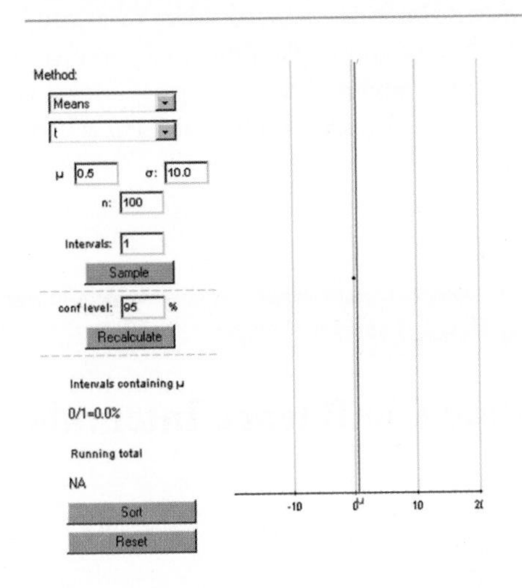

Getting Started: If the "Method" box does not say "Means," use the drop-down menu to select Means. In the box just below, select "t' from the drop-down menu. This applet will select a random sample from a specified normal population distribution and then use the sample to construct a confidence interval for the population mean. The interval is then plotted on the display, and you can see if the resulting interval contains the actual value of the population mean.

For purposes of this activity, we will sample from a normal population with mean 100 and standard deviation 5. We will begin with a sample size of $n = 10$. In the applet window, set $\mu = 100$, $\sigma = 5$, and $n = 10$. Leave the conf-level box set at 95%. Click the "Recalculate" button to rescale the picture on the right. Now click on the sample button. You should see a confidence interval appear on the display on the right-hand side. If the interval contains the actual mean of 100, the interval is drawn in green; if 100 is not in the confidence interval, the interval is shown in red. Your screen should look something like the following.

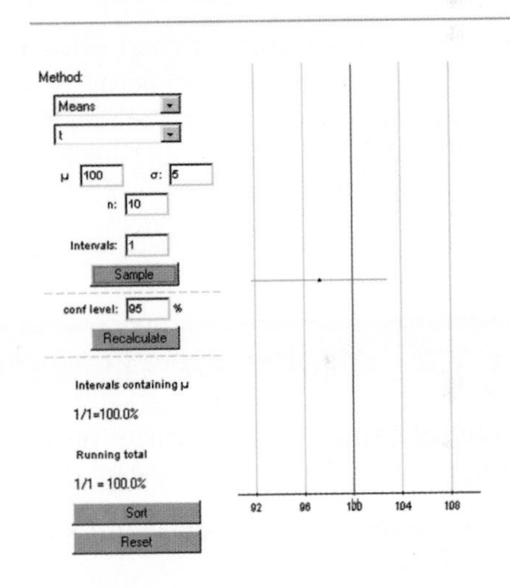

Part 1: Click on the "Sample" button several more times, and notice how the confidence interval estimate changes from sample to sample. Also notice that at the bottom of the left-hand side of the display, the applet is keeping track of the proportion of all the intervals calculated so far that include the actual value of μ. If we were to construct a large number of intervals, this proportion should closely approximate the capture rate for the confidence interval method.

To look at more than one interval at a time, change the "Intervals" box from 1 to 100, and then click the sample button. You should see a screen similar to the one at the top left of the next page, with 100 intervals in the display on the right-hand side. Again, intervals containing 100 (the value of μ in this case) will be green and those that do not contain 100 will be red. Also note that the capture proportion on the left-hand side has also been updated to reflect what happened with the 100 newly generated intervals.

Simulating Confidence Intervals

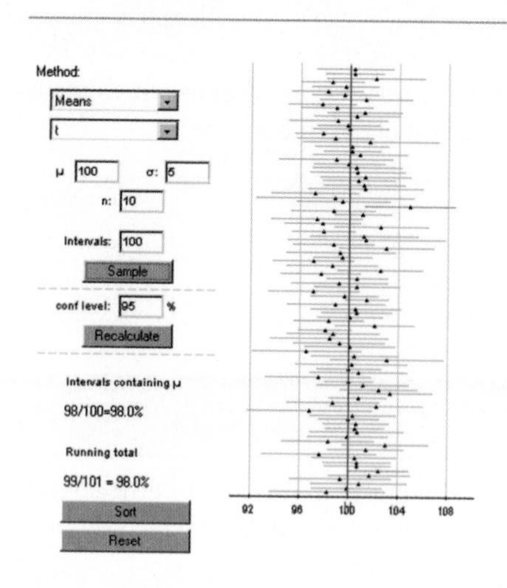

Continue generating intervals until you have seen at least 1000 intervals, and then answer the following question:

a. How does the proportion of intervals constructed that contain $\mu = 100$ compare to the stated confidence level of 95%? On how many intervals was your proportion based? (Note—if you followed the instructions, this should be at least 1000.)

Experiment with three other confidence levels of your choice, and then answer the following question:

b. In general, is the proportion of computed t confidence intervals that contain $\mu = 100$ close to the stated confidence level?

Part 2: When the population is normal but σ is unknown, we construct a confidence interval for a population mean using a t critical value rather than a z critical value. How important is this distinction?

Let's investigate. Use the drop-down menu to change the box just below the method box that says "Means" from "t" to "z with s." The applet will now construct intervals using the sample standard deviation, but will use a z critical value rather than the t critical value.

Use the applet to construct at least 1000 95% intervals, and then answer the following question:

c. Comment on how the proportion of the computed intervals that include the actual value of the population mean compares to the stated confidence level of 95%. Is this surprising? Explain why or why not.

Now experiment with some different samples sizes. What happens when $n = 20$? $n = 50$? $n = 100$? Use what you have learned to write a paragraph explaining what these simulations tell you about the advisability of using a z critical value in the construction of a confidence interval for μ when σ is unknown.

ACTIVITY 9.2 An Alternative Confidence Interval for a Population Proportion

Technology Activity (Applet): This activity presumes that you have already worked through Activity 9.1.

Background: In Section 9.2, it was suggested that a confidence interval of the form

$$\hat{p}_{\text{mod}} \pm (z \text{ critical value})\sqrt{\frac{\hat{p}_{\text{mod}}(1 - \hat{p}_{\text{mod}})}{n}}$$

where $\hat{p}_{\text{mod}} = \dfrac{\text{successes} + 2}{n + 4}$ is an alternative to the usual large-sample z confidence interval. This alternative interval is preferred by many statisticians because, in repeated sampling, the proportion of intervals constructed that include the actual value of the population proportion, p, tends to be closer to the stated confidence level. In this activity, we will explore how the "capture rates" for the two different interval estimation methods compare.

Open the applet (available at www.cengagebrain.com) called ConfidenceIntervals. You should see a screen like the one shown.

Simulating Confidence Intervals

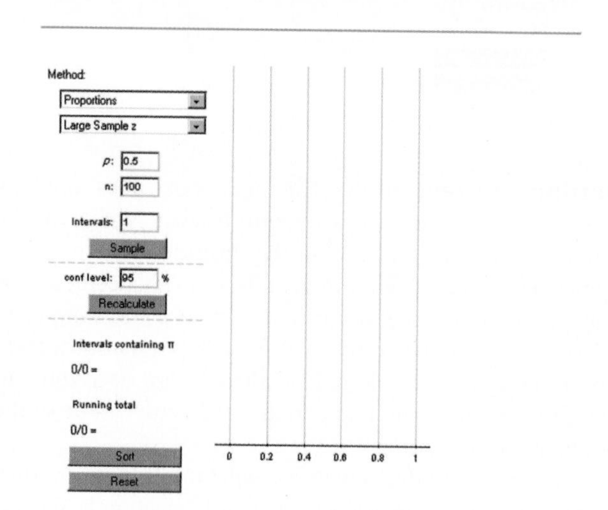

Select "Proportion" from the Method box drop-down menu, and then select "Large Sample z" from the drop-down menu of the second box. We will consider sampling from a population with $p = .3$ using a sample size of 40. In the applet window, enter $p = .3$ and $n = 40$. Note that $n = 40$ is large enough to satisfy $np \geq 10$ and $n(1 - p) \geq 10$.

Set the "Intervals" box to 100, and then use the applet to construct a large number (at least 1000) of 95% confidence intervals.

1. How does the proportion of intervals constructed that include $p = .3$, the population proportion, compare to .95? Does this surprise you? Explain.

   Now use the drop-down menu to change "Large Sample z" to "Modified." Now the applet will construct the alternative confidence interval that is based on $\hat{p}_{\text{mod}}$. Use the applet to construct a large number (at least 1000) of 95% confidence intervals.
2. How does the proportion of intervals constructed that include $p = .3$, the population proportion, compare to .95? Is this proportion closer to .95 than was the case for the large-sample z interval?
3. Experiment with different combinations of values of sample size and population proportion p. Can you find a combination for which the large sample z interval has a capture rate that is close to 95%? Can you find a combination for which it has a capture rate that is even farther from 95% than it was for $n = 40$ and $p = .3$? How does the modified interval perform in each of these cases?

ACTIVITY 9.3 Verifying Signatures on a Recall Petition

Background: In 2003, petitions were submitted to the California Secretary of State calling for the recall of Governor Gray Davis. Each of California's 58 counties then had to report the number of valid signatures on the petitions from that county so that the State could determine whether there were enough valid signatures to certify the recall and set a date for the recall election. The following paragraph appeared in the ***San Luis Obispo Tribune*** **(July 23, 2003)**:

> In the campaign to recall Gov. Gray Davis, the secretary of state is reporting 16,000 verified signatures from San Luis Obispo County. In all, the County Clerk's Office received 18,866 signatures on recall petitions and was instructed by the state to check a random sample of 567. Out of those, 84.48% were good. The verification process includes checking whether the signer is a registered voter and whether the address and signature on the recall petition match the voter registration.

1. Use the data from the random sample of 567 San Luis Obispo County signatures to construct a 95% confidence interval for the proportion of petition signatures that are valid.
2. How do you think that the reported figure of 16,000 verified signature for San Luis Obispo County was obtained?
3. Based on your confidence interval from Step 1, explain why you think that the reported figure of 16,000 verified signatures is or is not reasonable.

ACTIVITY 9.4 A Meaningful Paragraph

Write a meaningful paragraph that includes the following six terms: **sample**, **population**, **confidence level**, **estimate**, **mean**, **margin of error**.

A "meaningful paragraph" is a coherent piece writing in an appropriate context that uses all of the listed words. The paragraph should show that you understand the meanings of the terms and their relationship to one another. A sequence of sentences that just define the terms is *not* a meaningful paragraph. When choosing a context, think carefully about the terms you need to use. Choosing a good context will make writing a meaningful paragraph easier.

SUMMARY Key Concepts and Formulas

TERM OR FORMULA	COMMENT
Point estimate	A single number, based on sample data, that represents a plausible value of a population characteristic.
Unbiased statistic	A statistic that has a sampling distribution with a mean equal to the value of the population characteristic to be estimated.
Confidence interval	An interval computed from sample data that provides a range of plausible values for a population characteristic.
Confidence level	A number that provides information on how much "confidence" we can have in the method used to construct a confidence interval estimate. The confidence level specifies the percentage of all possible samples that will produce an interval containing the value of the population characteristic.
$\hat{p} \pm (z \text{ critical value})\sqrt{\frac{\hat{p}(1-\hat{p})}{n}}$	A formula used to construct a confidence interval for p when the sample size is large.
$n = p(1-p)\left(\frac{1.96}{B}\right)^2$	A formula used to compute the sample size necessary for estimating p to within an amount B with 95% confidence. (For other confidence levels, replace 1.96 with an appropriate z critical value.)
$\bar{x} \pm (z \text{ critical value})\frac{\sigma}{\sqrt{n}}$	A formula used to construct a confidence interval for μ when σ is known and either the sample size is large or the population distribution is normal.
$\bar{x} \pm (t \text{ critical value})\frac{s}{\sqrt{n}}$	A formula used to construct a confidence interval for μ when σ is unknown and either the sample size is large or the population distribution is normal.
$n = \left(\frac{1.96\sigma}{B}\right)^2$	A formula used to compute the sample size necessary for estimating μ to within an amount B with 95% confidence. (For other confidence levels, replace 1.96 with an appropriate z critical value.)

CHAPTER REVIEW Exercises 9.56 - 9.69

9.56 How much money do people spend on graduation gifts? In 2007, the **National Retail Federation (www.nrf.com)** surveyed 2815 consumers who reported that they bought one or more graduation gifts that year. The sample was selected in a way designed to produce a sample representative of adult Americans who purchased graduation gifts in 2007. For this sample, the mean amount spent per gift was \$55.05. Suppose that the sample standard deviation was \$20. Construct and interpret a 98% confidence interval for the mean amount of money spent per graduation gift in 2007.

9.57 ▼ **"Tongue Piercing May Speed Tooth Loss, Researchers Say"** is the headline of an article that appeared in the ***San Luis Obispo Tribune* (June 5, 2002)**. The article describes a study of 52 young adults with pierced tongues. The researchers found receding gums, which can lead to tooth loss, in 18 of the participants.

a. Construct a 95% confidence interval for the proportion of young adults with pierced tongues who have receding gums.

b. What assumptions must be made for use of the z confidence interval to be appropriate?

9.58 In a study of 1710 schoolchildren in Australia (***Herald Sun*, October 27, 1994**), 1060 children indicated that they normally watch TV before school in the morning. (Interestingly, only 35% of the parents said their children watched TV before school!)

a. Construct a 95% confidence interval for the true proportion of Australian children who say they watch TV before school.

b. What assumption about the sample must be true for the method used to construct the interval to be valid?

9.59 The authors of the paper **"Short-Term Health and Economic Benefits of Smoking Cessation: Low Birth**

Weight" (*Pediatrics* [1999]: 1312–1320) investigated the medical cost associated with babies born to mothers who smoke. The paper included estimates of mean medical cost for low-birth-weight babies for different ethnic groups. For a sample of 654 Hispanic low-birth-weight babies, the mean medical cost was \$55,007 and the standard error ($s/\sqrt{n}$) was \$3011. For a sample of 13 Native American low-birth-weight babies, the mean and standard error were \$73,418 and \$29,577, respectively. Explain why the two standard errors are so different.

9.60 The article **"Consumers Show Increased Liking for Diesel Autos" (*USA Today*, January 29, 2003)** reported that 27% of U.S. consumers would opt for a diesel car if it ran as cleanly and performed as well as a car with a gas engine. Suppose that you suspect that the proportion might be different in your area and that you want to conduct a survey to estimate this proportion for the adult residents of your city.

a. What is the required sample size if you want to estimate this proportion to within .05 with 95% confidence? Compute the required sample size first using .27 as a preliminary estimate of p and then using the conservative value of .5.

b. How do the two sample sizes compare? What sample size would you recommend for this study?

9.61 Seventy-seven students at the University of Virginia were asked to keep a diary of conversations with their mothers, recording any lies they told during these conversations **(*San Luis Obispo Telegram-Tribune*, August 16, 1995)**. It was reported that the mean number of lies per conversation was 0.5. Suppose that the standard deviation (which was not reported) was 0.4.

a. Suppose that this group of 77 is a random sample from the population of students at this university. Construct a 95% confidence interval for the mean number of lies per conversation for this population.

b. The interval in Part (a) does not include 0. Does this imply that all students lie to their mothers? Explain.

9.62 A manufacturer of small appliances purchases plastic handles for coffeepots from an outside vendor. If a handle is cracked, it is considered defective and must be discarded. A large shipment of plastic handles is received. The proportion of defective handles p is of interest. How many handles from the shipment should be inspected to estimate p to within 0.1 with 95% confidence?

9.63 An article in the ***Chicago Tribune* (August 29, 1999)** reported that in a poll of residents of the Chicago suburbs, 43% felt that their financial situation had improved during the past year. The following statement is from the article: "The findings of this Tribune poll are based on interviews with 930 randomly selected suburban residents. The sample included suburban Cook County plus DuPage, Kane, Lake, McHenry, and Will Counties. In a sample of this size, one can say with 95% certainty that results will differ by no more than 3%t from results obtained if all residents had been included in the poll."

Comment on this statement. Give a statistical argument to justify the claim that the estimate of 43% is within 3% of the true proportion of residents who feel that their financial situation has improved.

9.64 A manufacturer of college textbooks is interested in estimating the strength of the bindings produced by a particular binding machine. Strength can be measured by recording the force required to pull the pages from the binding. If this force is measured in pounds, how many books should be tested to estimate the mean force required to break the binding to within 0.1 pounds with 95% conficence? Assume that σ is known to be 0.8 pound.

9.65 High-profile legal cases have many people reevaluating the jury system. Many believe that juries in criminal trials should be able to convict on less than a unanimous vote. To assess support for this idea, investigators asked each individual in a random sample of Californians whether they favored allowing conviction by a 10–2 verdict in criminal cases not involving the death penalty. The Associated Press **(*San Luis Obispo Telegram-Tribune*, September 13, 1995)** reported that 71% supported the 10–2 verdict. Suppose that the sample size for this survey was $n = 900$. Compute and interpret a 99% confidence interval for the proportion of Californians who favor the 10–2 verdict.

9.66 The confidence intervals presented in this chapter give both lower and upper bounds on plausible values for the population characteristic being estimated. In some instances, only an upper bound or only a lower bound is appropriate. Using the same reasoning that gave the large sample interval in Section 9.3, we can say that when n is large, 99% of all samples have

$$\mu < \bar{x} + 2.33\frac{s}{\sqrt{n}}$$

(because the area under the z curve to the left of 2.33 is .99). Thus, $\bar{x} + 2.33\frac{s}{\sqrt{n}}$ is a 99% upper confidence bound for μ. Use the data of Example 9.9 to calculate the 99% upper confidence bound for the true mean wait time for bypass patients in Ontario.

9.67 When n is large, the statistic s is approximately unbiased for estimating σ and has approximately a normal distribution. The standard deviation of this statistic when the population distribution is normal is $\sigma_s \approx \frac{\sigma}{\sqrt{2n}}$ which can be estimated by $\frac{s}{\sqrt{2n}}$. An approximate large-sample confidence interval for the population standard deviation σ is then

$$s \pm (z \text{ critical value})\frac{s}{\sqrt{2n}}$$

Use the data of Example 9.9 to obtain a 95% confidence interval for the true standard deviation of waiting time for angiography.

9.68 The interval from −2.33 to 1.75 captures an area of .95 under the z curve. This implies that another large-sample 95% confidence interval for μ has lower limit $\bar{x} - 2.33\frac{\sigma}{\sqrt{n}}$ and upper limit $\bar{x} + 1.75\frac{\sigma}{\sqrt{n}}$. Would you recommend using this 95% interval over the 95% interval $\bar{x} \pm 1.96\frac{\sigma}{\sqrt{n}}$ discussed in the text? Explain. (Hint: Look at the width of each interval.)

9.69 The eating habits of 12 bats were examined in the article **"Foraging Behavior of the Indian False Vampire Bat" (*Biotropica* [1991]: 63–67)**. These bats consume insects and frogs. For these 12 bats, the mean time to consume a frog was $\bar{x} = 21.9$ minutes. Suppose that the standard deviation was $s = 7.7$ minutes. Construct and interpret a 90% confidence interval for the mean suppertime of a vampire bat whose meal consists of a frog. What assumptions must be reasonable for the one-sample t interval to be appropriate?

Bold exercises answered in back ● Data set available online ▼ Video Solution available

TECHNOLOGY NOTES

Confidence Intervals for Proportions

JMP

Summarized data

1. Enter the data table into the JMP data table with categories in the first column and counts in the second column

Untitled - JMP [2]

File Edit Tables Rows Cols DOE Analyze Graph Tools View Window

	Column 1	Column 2
1	Yes	130
2	No	45

2. Click **Analyze** and select **Distribution**
3. Click and drag the first column name from the box under **Select Columns** to the box next to **Y, Columns**
4. Click and drag the second column name from the box under **Select Columns** to the box next to **Freq**
5. Click **OK**
6. Click the red arrow next to the column name and click **Confidence Interval** then select the appropriate level or select **Other** to input a level that is not listed

Raw Data

1. Enter the raw data into a column
2. Click **Analyze** and select **Distribution**
3. Click and drag the first column name from the box under **Select Columns** to the box next to **Y, Columns**
4. Click **OK**
5. Click the red arrow next to the column name and click **Confidence Interval** then select the appropriate level or select **Other** to input a level that is not listed

Minitab

Summarized data

1. Click Stat then click Basic Statistics then click 1 Proportion...
2. Click the radio button next to **Summarized data**
3. In the box next to **Number of Trials:** type the value for n, the total number of trials
4. In the box next to **Number of events:** type the value for the number of successes
5. Click **Options...**
6. Input the appropriate confidence level in the box next to **Confidence Level**
7. Check the box next to **Use test and interval based on normal distribution**
8. Click **OK**
9. Click **OK**

Raw data

1. Input the raw data into a column
2. Click Stat then click Basic Statistics then click 1 Proportion...

3. Click in the box under **Samples in columns:**
4. Double click the column name where the raw data is stored
5. Click **Options...**
6. Input the appropriate confidence level in the box next to **Confidence Level**
7. Check the box next to **Use test and interval based on normal distribution**
8. Click **OK**
9. Click **OK**

SPSS

SPSS does not have the functionality to automatically produce confidence intervals for a population proportion.

Excel

Excel does not have the functionality to automatically produce confidence intervals for a population proportion. However, you can manually type in the formula for the lower and upper limits separately into two different cells to have Excel calculate the result for you.

TI-83/84

1. Press the **STAT** key
2. Highlight **TESTS**
3. Highlight **1-PropZInterval** and press **ENTER**
4. Next to ***x*** type the number of successes
5. Next to ***n*** type the number of trials, *n*
6. Next to **C-Level** type the value for the confidence level
7. Highlight **Calculate** and press **ENTER**

TI-Nspire

1. Enter the Calculate Scratchpad
2. Press the **menu** key then select **6:Statistics** then select **6:Confidence Intervals** then **5:1-Prop *z* Interval...** then press **enter**
3. In the box next to **Successes, *x*** type the number of successes
4. In the box next to ***n*** type the number of trials, *n*
5. In the box next to **C Level** type the confidence level
6. Press **OK**

Confidence Interval for μ Based on *t*-distribution

JMP

1. Input the data into a column
2. Click **Analyze** and select **Distribution**
3. Click and drag the column name from the box under **Select Columns** to the box next to **Y, Response**
4. Click **OK**
5. Click the red arrow next to the column name and select **Confidence Interval** then select the appropriate confidence level or click **Other** to specify a level

Minitab

Summarized data

1. Click **Stat** then click **Basic Statistics** then click **1-sample t...**
2. Click the radio button next to **Summarized data**
3. In the box next to **Sample size:** type the value for *n*, the sample size
4. In the box next to **Mean:** type the value for the sample mean
5. In the box next to **Standard deviation:** type the value for the sample standard deviation
6. Click **Options...**
7. Input the appropriate confidence level in the box next to **Confidence Level**
8. Click **OK**
9. Click **OK**

Raw data

1. Input the raw data into a column
2. Click **Stat** then click **Basic Statistics** then click **1-sample t...**
3. Click in the box under **Samples in columns:**
4. Double click the column name where the raw data are stored
5. Click **Options...**
6. Input the appropriate confidence level in the box next to **Confidence Level**
7. Click **OK**
8. Click **OK**

SPSS

1. Input the data into a column
2. Click **Analyze** then select **Compare Means** then select **One-Sample T Test...**
3. Highlight the column name for the variable
4. Click the arrow to move the variable to the **Test Variable(s):** box
5. Click **Options...**
6. Input the appropriate confidence level in the box next to **Confidence Interval Percentage:**
7. Click **Continue**
8. Click **OK**

Note: The confidence interval for the population means is in the **One-Sample Test** table.

Excel

Note: Excel does not have the functionality to produce a confidence interval for a single population mean automatically. However, you can use Excel to produce the estimate for the sample mean and the margin of error for the confidence interval using the steps below.

1. Input the raw data into a column
2. Click the **Data** ribbon
3. Click **Data Analysis** in the **Analysis** group

   Note: If you do not see **Data Analysis** listed on the Ribbon, see the Technology Notes for Chapter 2 for instructions on installing this add-on.

4. Select **Descriptive Statistics** from the dialog box and click **OK**

5. Click in the box next to **Input Range:** and select the data (if you used a column title, check the box next to **Labels in first row**)
6. Click the box next to **Confidence Level for Mean**:
7. In the box to the right of **Confidence Level for Mean:** type in the confidence level you are using
8. Click **OK**

Note: The margin of error can be found in the row titled Confidence Level. In order to find the lower limit of the confidence interval, subtract this from the mean shown in the output. To find the upper limit of the confidence interval, add this to the mean shown in the output.

TI-83/84

Summarized data

1. Press **STAT**
2. Highlight **TESTS**
3. Highlight **TInterval…** and press **ENTER**
4. Highlight **Stats** and press **ENTER**
5. Next to $\bar{x}$ input the value for the sample mean
6. Next to **sx** input the value for the sample standard deviation
7. Next to **n** input the value for the sample size
8. Next to **C-Level** input the appropriate confidence level
9. Highlight **Calculate** and press **ENTER**

Raw data

1. Enter the data into **L1** (In order to access lists press the **STAT** key, highlight the option called **Edit…** then press **ENTER**)
2. Press **STAT**
3. Highlight **TESTS**
4. Highlight **TInterval…** and press **ENTER**
5. Highlight **Data** and press **ENTER**
6. Next to **C-Level** input the appropriate confidence level
7. Highlight **Calculate** and press **ENTER**

TI-Nspire

Summarized data

1. Enter the Calculate Scratchpad
2. Press the **menu** key and select **6:Statistics** then **6:Confidence Intervals** then **2:t Interval…** and press **enter**
3. From the drop-down menu select **Stats**
4. Press **OK**
5. Next to $\bar{x}$ input the value for the sample mean
6. Next to **sx** input the value for the sample standard deviation
7. Next to **n** input the value for the sample size
8. Next to **C Level** input the appropriate confidence level
9. Press **OK**

Raw data

1. Enter the data into a data list (In order to access data lists select the spreadsheet option and press **enter**)

   Note: Be sure to title the list by selecting the top row of the column and typing a title.

2. Press the **menu** key and select **4:Statistics** then **3:Confidence Intervals** then **2:t Interval…** and press **enter**
3. From the drop-down menu select **Data**
4. Press **OK**
5. Next to **List** select the list containing your data
6. Next to **C-Level** input the appropriate confidence level
7. Press **OK**

CHAPTER 10

Hypothesis Testing Using a Single Sample

PictureNet/Spirit/Corbis

In Chapter 9, we considered situations in which the primary goal was to estimate the unknown value of some population characteristic. Sample data can also be used to decide whether some claim or *hypothesis* about a population characteristic is plausible.

For example, sharing of prescription drugs is a practice that has many associated risks. Is this a common practice among teens? Is there evidence that more than 10% of teens have shared prescription medications with a friend? The article **"Many Teens Share Prescription Drugs" (*Calgary Herald,* August 3, 2009)** summarized the results of a survey of a representative sample of 592 U.S. teens age 12 to 17 and reported that 118 of those surveyed admitted to having shared a prescription drug with a friend. With p representing this proportion for all U.S. teens age 12 to 17, we can use the hypothesis testing methods of this chapter to decide whether the sample data provide convincing evidence that p is greater than .10.

As another example, a report released by the National Association of Colleges and Employers stated that the average starting salary for students graduating with a bachelor's degree in 2010 is \$48,351 **("Winter 2010 Salary Survey," www.naceweb.org)**. Suppose that you are interested in investigating whether the mean starting salary for students graduating from your university this year is greater than the 2010 average of \$48,351. You select a random sample of $n = 40$ graduates from the current graduating class of your university and determine the starting salary of each one.

If this sample produced a mean starting salary of \$49,958 and a standard deviation of \$1214, is it reasonable to conclude that μ, the mean starting salary for all graduates in the current graduating class at your university, is greater than \$48,351? We will see in this chapter how the sample data can be analyzed to decide whether $\mu > 48,351$ is a reasonable conclusion.

Chapter 10: Learning Objectives

STUDENTS WILL UNDERSTAND:

- that rejecting the null hypothesis implies strong support for the alternative hypothesis.
- why failing to reject the null hypothesis does not imply strong support for the null hypothesis.
- the reasoning used to reach a decision in a hypothesis test.

STUDENTS WILL BE ABLE TO:

- translate a research question into null and alternative hypotheses.
- describe Type I and Type II errors in context.
- carry out a large-sample *z* test for a population proportion and interpret the results in context.
- carry out a *t* test for a population mean and interpret the results in context.
- describe the effect of the significance level and the sample size on the power of a test.

10.1 Hypotheses and Test Procedures

A hypothesis is a claim or statement about the value of a single population characteristic or the values of several population characteristics. The following are examples of legitimate hypotheses:

$\mu = 1000$, where μ is the mean number of characters in an e-mail message
$p < .01$, where p is the proportion of e-mail messages that are undeliverable

In contrast, the statements $\bar{x} = 1000$ and $\hat{p} = .01$ are *not* hypotheses, because $\bar{x}$ and $\hat{p}$ are *sample* characteristics.

A test of hypotheses is a method that uses sample data to decide between two competing claims (hypotheses) about a population characteristic. One hypothesis might be $\mu = 1000$ and the other $\mu \neq 1000$ or one hypothesis might be $p = .01$ and the other $p < .01$. If it were possible to carry out a census of the entire population, we would know which of the two hypotheses is correct, but usually we must decide between them using information from a sample.

A criminal trial is a familiar situation in which a choice between two contradictory claims must be made. The person accused of the crime must be judged either guilty or not guilty. Under the U.S. system of justice, the individual on trial is initially presumed not guilty. Only strong evidence to the contrary causes the not guilty claim to be rejected in favor of a guilty verdict. The burden is thus put on the prosecution to prove the guilty claim. The French perspective in criminal proceedings is the opposite. Once enough evidence has been presented to justify bringing an individual to trial, the initial assumption is that the accused is guilty. The burden of proof then falls on the accused to establish otherwise.

As in a judicial proceeding, we initially assume that a particular hypothesis, called the **null hypothesis,** is the correct one. We then consider the evidence (the sample data) and reject the null hypothesis in favor of the competing hypothesis, called the **alternative hypothesis,** only if there is *convincing* evidence against the null hypothesis.

DEFINITION

Null hypothesis: A claim about a population characteristic that is initially assumed to be true. The null hypothesis is denoted by H_0.

Alternative hypothesis: A competing claim about a population characteristic. The alternative hypothesis is denoted by H_a.

In carrying out a test of H_0 versus H_a, the null hypothesis H_0 will be rejected in favor of H_a only if sample evidence strongly suggests that H_0 is false.

If the sample does not provide such evidence, H_0 will not be rejected.

The two possible conclusions are *reject* H_0 or *fail to reject* H_0.

EXAMPLE 10.1 Tennis Ball Diameters

Because of variation in the manufacturing process, tennis balls produced by a particular machine do not have identical diameters. Let μ denote the mean diameter for all tennis balls currently being produced. Suppose that the machine was initially calibrated to achieve the design specification $\mu = 3$ inches. However, the manufacturer is now concerned that the diameters no longer conform to this specification. If sample evidence suggests that $\mu \neq 3$ inches, the production process will be halted while the machine is recalibrated. Because stopping production is costly, the manufacturer wants to be quite sure that $\mu \neq 3$ inches before undertaking recalibration. Under these circumstances, a sensible choice of hypotheses is

H_0: $\mu = 3$ (the specification is being met, so recalibration is unnecessary)
H_a: $\mu \neq 3$ (the specification is not being met, so recalibration is necessary)

H_0 would be rejected in favor of H_a only if the sample provides convincing evidence against the null hypothesis. ■

EXAMPLE 10.2 Compact Florescent Lightbulb Lifetimes

Understand the context)

Compact fluorescent (cfl) lightbulbs are much more energy efficient than standard incandescent light bulbs. Ecobulb brand 60-watt cfl lightbulbs state on the package "Average life 8,000 hours." Let μ denote the true mean life of Ecobulb 60-watt cfl lightbulbs. Then the advertised claim is $\mu = 8000$ hours. People who purchase this brand would be unhappy if μ is actually less than the advertised value.

Suppose that a sample of Ecobulb cfl lightbulbs is selected and tested. The lifetime for each bulb in the sample is recorded. The sample results can then be used to test the hypothesis $\mu = 8000$ hours against the hypothesis $\mu < 8000$ hours. The accusation that the company is overstating the mean lifetime is a serious one, and it is reasonable to require convincing evidence before concluding that $\mu < 8000$. This suggests that the claim $\mu = 8000$ should be selected as the null hypothesis and that $\mu < 8000$ should be selected as the alternative hypothesis. Then

H_0: $\mu = 8000$

would be rejected in favor of

H_a: $\mu < 8000$

only if sample evidence strongly suggests that the initial assumption, $\mu = 8000$ hours, is not plausible. ■

Because the alternative hypothesis in Example 10.2 asserted that $\mu < 8000$ (the actual mean lifetime is less than the advertised value), it might have seemed sensible to state H_0 as the inequality $\mu \geq 8000$. The assertion $\mu \geq 8000$ is in fact the *implicit* null hypothesis, but we will state H_0 explicitly as a claim of equality. There are several reasons for this. First, the development of a decision rule is most easily understood if there is only a single hypothesized value of μ (or p or whatever other population characteristic is under consideration). Second, suppose that the sample data provided compelling evidence that H_0: $\mu = 8000$ should be rejected in favor of H_a: $\mu < 8000$. This means that we were convinced by the sample data that the population mean was smaller than 8000. It follows that we would have also been convinced that the population mean could not have been 8001 or 8010 or any other value that was larger than 8000. As a consequence, the conclusion when testing H_0: $\mu = 8000$ versus H_a: $\mu < 8000$ is always the same as the conclusion for a test where the null hypothesis is H_0: $\mu \geq 8000$. For these reasons, it is customary to state the null hypothesis H_0 as a claim of equality.

The form of a null hypothesis is

H_0: population characteristic = hypothesized value

where the hypothesized value is a specific number determined by the problem context.

The alternative hypothesis will have one of the following three forms:

H_a: population characteristic > hypothesized value
H_a: population characteristic < hypothesized value
H_a: population characteristic ≠ hypothesized value

Notice that we can test H_0: $p = .1$ versus H_a: $p < .1$, but we can't test H_0: $\mu = 50$ versus H_a: $\mu > 100$. The number appearing in the alternative hypothesis must be identical to the hypothesized value in H_0.

Example 10.3 illustrates how the selection of H_0 (the claim initially assumed to be true) and H_a depends on the objectives of a study.

EXAMPLE 10.3 Evaluating a New Medical Treatment

Understand the context)

A medical research team has been given the task of evaluating a new laser treatment for certain types of tumors. Consider the following two scenarios:

Scenario 1: The current standard treatment is considered reasonable and safe by the medical community, has no major side effects, and has a known success rate of 0.85 (85%).

Scenario 2: The current standard treatment sometimes has serious side effects, is costly, and has a known success rate of 0.30 (30%).

In the first scenario, research efforts would probably be directed toward determining whether the new treatment has a higher success rate than the standard treatment. Unless convincing evidence of this is presented, it is unlikely that current medical practice would be changed. With p representing the true proportion of successes for the laser treatment, the following hypotheses would be tested:

H_0: $p = .85$ versus H_a: $p > .85$

In this case, rejection of the null hypothesis indicates convincing evidence that the success rate is higher for the new treatment.

In the second scenario, the current standard treatment does not have much to recommend it. The new laser treatment may be considered preferable because of cost or because it has fewer or less serious side effects, as long as the success rate for the new procedure is no worse than that of the standard treatment. Here, researchers might decide to test the hypothesis

H_0: $p = .30$ versus H_a: $p < .30$

If the null hypothesis is rejected, the new treatment will not be put forward as an alternative to the standard treatment, because there is strong evidence that the laser method has a lower success rate.

If the null hypothesis is not rejected, we are able to conclude only that there is not convincing evidence that the success rate for the laser treatment is lower than that for the standard. This is *not* the same as saying that we have evidence that the laser treatment is as good as the standard treatment. If medical practice were to embrace the new procedure, it would not be because it has a higher success rate but rather because it costs less or has fewer side effects, and there is not strong evidence that it has a lower success rate than the standard treatment. ■

A statistical hypothesis test is only capable of demonstrating strong support for the alternative hypothesis (by rejection of the null hypothesis). When the null hypothesis is not rejected, it does not mean strong support for H_0—only lack of strong evidence against it.

In the lightbulb scenario of Example 10.2, if H_0: $\mu = 8000$ is rejected in favor of H_a: $\mu < 8000$, it is because we have strong evidence that actual mean lifetime is less than the advertised value. However, nonrejection of H_0 does not necessarily provide strong support for the advertised claim. If the objective is to demonstrate that the average lifetime is greater than 8000 hours, the hypotheses that would be tested are H_0: $\mu = 8000$ versus H_a: $\mu > 8000$. Then rejection of H_0 indicates strong evidence that $\mu > 8000$. *When deciding which alternative hypothesis to use, keep the research objectives in mind.*

EXERCISES 10.1 - 10.11

10.1 Explain why the statement $\bar{x} = 50$ is not a legitimate hypothesis.

10.2 For the following pairs, indicate which do not comply with the rules for setting up hypotheses, and explain why:

a. H_0: $\mu = 15$, H_a: $\mu = 15$

b. H_0: $p = .4$, H_a: $p > .6$

c. H_0: $\mu = 123$, H_a: $\mu < 123$

d. H_0: $\mu = 123$, H_a: $\mu = 125$

e. H_0: $\hat{p} = .1$, H_a: $\hat{p} \neq .1$

10.3 To determine whether the pipe welds in a nuclear power plant meet specifications, a random sample of welds is selected and tests are conducted on each weld in the sample. Weld strength is measured as the force required to break the weld. Suppose that the specifications state that the mean strength of welds should exceed 100 lb/in^2. The inspection team decides to test H_0: $\mu = 100$ versus H_a: $\mu > 100$. Explain why this alternative hypothesis was chosen rather than $\mu < 100$.

10.4 Do state laws that allow private citizens to carry concealed weapons result in a reduced crime rate? The author of a study carried out by the Brookings Institution is reported as saying, "The strongest thing I could say is that I don't see any strong evidence that they are reducing crime" (**San Luis Obispo Tribune, January 23, 2003**).

a. Is this conclusion consistent with testing

H_0: concealed weapons laws reduce crime

versus

H_a: concealed weapons laws do not reduce crime

or with testing

H_0: concealed weapons laws do not reduce crime

versus

H_a: concealed weapons laws reduce crime

Explain.

b. Does the stated conclusion indicate that the null hypothesis was rejected or not rejected? Explain.

10.5 The MMR vaccine is a measles-mumps-rubella vaccine. Consider the following quote from the article **"Review Finds No Link Between Vaccine and Autism" (San Luis Obispo Tribune, October 19, 2005)**: "'We found no evidence that giving MMR causes Crohn's disease and/or autism in the children that get the MMR,' said Tom Jefferson, one of the authors of *The Cochrane Review*. 'That does not mean it doesn't cause it. It means we could find no evidence of it.'"

In the context of a hypothesis test with the null hypothesis being that MMR vaccine does not cause autism, explain why the author could not conclude that the MMR vaccine does not cause autism. (Hint: See discussion following Example 10.3.)

10.6 CareerBuilder.com conducted a survey to learn about the proportion of employers who had ever sent an employee home because they were dressed inappropriately (**June 17, 2008, www.careerbuilder.com**). Suppose you are interested in determining if the resulting data provided strong evidence in support of the claim that more than one-third of employers have sent an employee home to change clothes. What pair of hypotheses should you test in order to answer this question?

10.7 A study sponsored by the American Savings Education Council (**"Preparing for Their Future: A Look at the Financial State of Gen X and Gen Y," March 2008**) included data from a survey of 1752 people age 19 to 36. One of the survey questions asked participants how satisfied they were with their current financial situation. Suppose you want to determine if the survey data provides convincing evidence that fewer than 10% of adults age 19 to 39 are very satisfied with their current financial situation. With $p =$ the proportion of adults age 19 to 39 who are very satisfied with their financial situation, what hypotheses should you test?

10.8 A researcher speculates that because of differences in diet, Japanese children may have a lower mean blood cholesterol level than U.S. children do. Suppose that the mean level for U.S. children is known to be 170. Let μ represent the mean blood cholesterol level for all Japanese children. What hypotheses should the researcher test?

10.9 A county commissioner must vote on a resolution that would commit substantial resources to the construction of a sewer in an outlying residential area. Her fiscal decisions have been criticized in the past, so she decides to take a survey of constituents to find out whether they favor spending money for a sewer system. She will vote to appropriate funds only if she is convinced that a majority of the people in her district favor the measure. What hypotheses should she test?

10.10 A cruise ship charges passengers $3 for a can of soda. Because of passenger complaints, the ship manager has decided to try out a plan with a lower price. He thinks that with a lower price, more cans will be sold, which would mean that the ship would still make a reasonable total profit. With the old pricing, the mean number of cans sold per passenger for a 10-day trip was 10.3 cans.

Suppose μ represents the mean number of cans per passenger for the new pricing. What hypotheses should the ship manager test if he wants to determine if the mean number of cans sold is higher for the new pricing plan?

10.11 The report **"How Teens Use Media" (Nielsen, June 2009)** says that 83% of U.S. teens use text messaging. Suppose you plan to select a random sample of 400 students at the local high school and will ask each one if he or she uses text messaging. You plan to use the resulting data to decide if there is evidence that the proportion of students at the high school who use text messaging differs from the national figure given in the Nielsen report. What hypotheses should you test?

Bold exercises answered in back ● Data set available online ▼ Video Solution available

10.2 Errors in Hypothesis Testing

Once hypotheses have been formulated, a **test procedure** uses sample data to determine whether H_0 should be rejected. Just as a jury may reach the wrong verdict in a trial, there is some chance that using a test procedure with sample data may lead us to the wrong conclusion about a population characteristic. In this section, we discuss the kinds of errors that can occur and consider how the choice of a test procedure influences the chances of these errors.

One erroneous conclusion in a criminal trial is for a jury to convict an innocent person, and another is for a guilty person to be set free. Similarly, there are two different types of errors that might be made when making a decision in a hypothesis testing problem. One type of error involves rejecting H_0 even though the null hypothesis is true. The second type of error results from failing to reject H_0 when it is false. These errors are known as Type I and Type II errors, respectively.

DEFINITION

Type I error: The error of rejecting H_0 when H_0 is true

Type II error: The error of failing to reject H_0 when H_0 is false

The only way to guarantee that neither type of error occurs is to base the decision on a census of the entire population. Risk of error is the price paid for basing the decision on sample data.

EXAMPLE 10.4 On-Time Arrivals

Understand the context ⟩

The **U.S. Bureau of Transportation Statistics** reports that for the 12 months ending in July 2012, 83.7% of all domestic passenger flights arrived on time (meaning within 15 minutes of the scheduled arrival). Suppose that an airline with a poor on-time record decides to offer its employees a bonus if, in an upcoming month, the airline's proportion of on-time flights exceeds the overall industry rate of .837. We can use p to represent the actual proportion of the airline's flights that are on time during the month of interest. A random sample of flights might be selected and used as a basis for choosing between

H_0: $p = .837$ and H_a: $p > .837$

In this context, a Type I error (rejecting a true H_0) results in the airline rewarding its employees when in fact the actual proportion of on-time flights did not exceed .837. A Type II error (not rejecting a false H_0) results in the airline employees *not* receiving a reward that they deserved. ■

EXAMPLE 10.5 Slowing the Growth of Tumors

Understand the context ⟩

In 2004, Vertex Pharmaceuticals, a biotechnology company, issued a press release announcing that it had filed an application with the Food and Drug Administration to begin clinical trials of an experimental drug VX-680 that had been found to reduce the growth rate of pancreatic and colon cancer tumors in animal studies (***New York Times*, February 24, 2004**).

Let μ denote the true mean growth rate of tumors for patients receiving the experimental drug. Data resulting from the planned clinical trials can be used to test

H_0: μ = mean growth rate of tumors without the experimental drug

versus

H_a: μ < mean growth rate of tumors without the experimental drug

The null hypothesis states that the experimental drug is not effective—that the mean growth rate of tumors for patients receiving the experimental drug is the same as for patients who do not take the experimental drug. The alternative hypothesis states that the experimental drug is effective in reducing the mean growth rate of tumors.

In this context, a Type I error consists of incorrectly concluding that the experimental drug is effective in slowing the growth rate of tumors. A potential consequence of making a Type I error would be that the company would continue to devote resources to the development of the drug when it really is not effective.

A Type II error consists of not concluding that the experimental drug is effective when in fact the mean growth rate of tumors is reduced. A potential consequence of making a Type II error is that the company might abandon development of a drug that was effective. ■

Examples 10.4 and 10.5 illustrate the two different types of error that might occur when testing hypotheses. The associated consequences of making a Type I and a Type II error are quite different. The accompanying box introduces the terminology and notation used to describe error probabilities.

The **probability of a Type I error** is denoted by α and is called the **significance level** of the test. For example, a test with $\alpha = .01$ is said to have a significance level of .01.

The **probability of a Type II error** is denoted by β.

EXAMPLE 10.6 Blood Test for Ovarian Cancer

Understand the context)

Women with ovarian cancer usually are not diagnosed until the disease is in an advanced stage, when it is most difficult to treat. The paper **"Diagnostic Markers for Early Detection of Ovarian Cancer" (*Clinical Cancer Research* [2008]: 1065–1072)** describes a new approach to diagnosing ovarian cancer that is based on using six different blood biomarkers (a blood biomarker is a biochemical characteristic that is measured in laboratory testing). The authors report the following results using the six biomarkers:

- For 156 women known to have ovarian cancer, the biomarkers correctly identified 151 as having ovarian cancer.
- For 362 women known not to have ovarian cancer, the biomarkers correctly identified 360 of them as being ovarian cancer free.

We can think of using this blood test to choose between two hypotheses:

H_0: woman has ovarian cancer
H_a: woman does not have ovarian cancer

Notice that although these are not "statistical hypotheses" (statements about a population characteristic), the possible decision errors are analogous to Type I and Type II errors.

In this situation, believing that a woman with ovarian cancer is cancer free would be a Type I error—rejecting the hypothesis of ovarian cancer when it is in fact true. Believing that a woman who is actually cancer free does have ovarian cancer is a Type II error—not rejecting the null hypothesis when it is in fact false.

Based on the study results, we can estimate the error probabilities. The probability of a Type I error, α, is approximately $5/156 = .032$. The probability of a Type II error, β, is approximately $2/362 = .006$. ■

The ideal test procedure would result in both $\alpha = 0$ and $\beta = 0$. However, if we must base our decision on incomplete information—a sample rather than a census—it is impossible to achieve this ideal. The standard test procedures allow us to control α, but they provide no direct control over β. Because α represents the probability of rejecting a true null hypothesis, selecting a significance level $\alpha = .05$ results in a test procedure that, used over and over with different random samples, rejects a *true* H_0 about 5 times in 100. Selecting $\alpha = .01$ results in a test procedure with a Type I error rate of 1% in long-term repeated use. Choosing a small value for α implies that the user wants a procedure for which the risk of a Type I error is quite small.

One question arises naturally at this point: If we can select α, the probability of making a Type I error, why would we ever select $\alpha = .05$ rather than $\alpha = .01$? Why not always select a very small value for α? To achieve a small probability of making a Type I error, we would need the corresponding test procedure to require the evidence against H_0 to be very strong before the null hypothesis can be rejected. Although this makes a Type I error unlikely, it increases the risk of a Type II error (*not* rejecting H_0 when it should have been rejected). The choice will depend on the consequences of Type I and Type II errors. If a Type II error has serious consequences, it may be a good idea to select a somewhat larger value for α.

In general, there is a compromise between small α and small β, leading to the following widely accepted principle for specifying a test procedure.

After assessing the consequences of Type I and Type II errors, identify the largest α that is tolerable for the problem. Then use a test procedure with this maximum acceptable value as the level of significance (because using a smaller α increases β). In other words, use the largest acceptable value for α.

EXAMPLE 10.7 Lead in Tap Water

Understand the context

The Environmental Protection Agency (EPA) has adopted what is known as the Lead and Copper Rule, which defines drinking water as unsafe if the concentration of lead is 15 parts per billion (ppb) or greater or if the concentration of copper is 1.3 parts per million (ppm) or greater.

With μ denoting the mean concentration of lead, the manager of a community water system might use lead level measurements from a sample of water specimens to test

$$H_0: \mu = 15 \quad \text{versus} \quad H_a: \mu < 15$$

The null hypothesis (which also implicitly includes the $\mu > 15$ case) states that the mean lead concentration is excessive by EPA standards. The alternative hypothesis states that the mean lead concentration is at an acceptable level and that the water system meets EPA standards for lead.

In this context, a Type I error leads to the conclusion that a water source meets EPA standards for lead when in fact it does not. Possible consequences of this type of error include health risks associated with excessive lead consumption (for example, increased blood pressure, hearing loss, and, in severe cases, anemia and kidney damage).

A Type II error is to conclude that the water does not meet EPA standards for lead when in fact it actually does. Possible consequences of a Type II error include elimination of a community water source. Because a Type I error might result in potentially serious public health risks, a small value of α (Type I error probability), such as $\alpha = .01$, could be selected. Of course, selecting a small value for α increases the risk of a Type II error. If the community has only one water source, a Type II error could also have very serious consequences for the community, and we might want to rethink the choice of α. ■

EXERCISES 10.12 - 10.22

10.12 Researchers at the University of Washington and Harvard University analyzed records of breast cancer screening and diagnostic evaluations (**"Mammogram Cancer Scares More Frequent than Thought,"** ***USA Today,*** **April 16, 1998**). Discussing the benefits and downsides of the screening process, the article states that, although the rate of false-positives is higher than previously thought, if radiologists were less aggressive in following up on suspicious tests, the rate of false-positives would fall but the rate of missed cancers would rise.

Suppose that such a screening test is used to decide between a null hypothesis of

H_0: no cancer is present

and an alternative hypothesis of

H_a: cancer is present.

(Although these are not hypotheses about a population characteristic, this exercise illustrates the definitions of Type I and Type II errors.) (Hint: See Example 10.6.)

a. Would a false-positive (thinking that cancer is present when in fact it is not) be a Type I error or a Type II error?

b. Describe a Type I error in the context of this problem, and discuss the possible consequences of making a Type I error.

c. Describe a Type II error in the context of this problem, and discuss the possible consequences of making a Type II error.

d. What aspect of the relationship between the probability of Type I and Type II errors is being described by the statement in the article that if radiologists were less aggressive in following up on suspicious tests, the rate of false-positives would fall but the rate of missed cancers would rise?

10.13 The paper **"MRI Evaluation of the Contralateral Breast in Women with Recently Diagnosed Breast Cancer"** (***New England Journal of Medicine*** **[2007]: 1295–1303**) describes a study of the use of MRI (Magnetic Resonance Imaging) exams in the diagnosis of breast cancer. The purpose of the study was to determine if MRI exams do a better job than mammograms of determining if women who have recently been diagnosed with cancer in one breast have cancer in the other breast. The study participants were 969 women who had been diagnosed with cancer in one breast and for whom a mammogram did not detect cancer in the other breast.

These women had an MRI exam of the other breast, and 121 of those exams indicated possible cancer. After undergoing biopsies, it was determined that 30 of the 121 did in fact have cancer in the other breast, whereas 91 did not. The women were all followed for one year, and three of the women for whom the MRI exam did not indicate cancer in the other breast were subsequently diagnosed with cancer that the MRI did not detect. The accompanying table summarizes this information.

	Cancer Present	Cancer Not Present	Total
MRI Positive for Cancer	30	91	**121**
MRI Negative for Cancer	3	845	**848**
Total	**33**	**936**	**969**

Suppose that for women recently diagnosed with cancer in only one breast, the MRI is used to decide between the two "hypotheses"

H_0: woman has cancer in the other breast
H_a: woman does not have cancer in the other breast

(Although these are not hypotheses about a population characteristic, this exercise illustrates the definitions of Type I and Type II errors.) (Hint: See Example 10.7.)

a. One possible error would be deciding that a woman who does have cancer in the other breast is cancer-free. Is this a Type I or a Type II error? Use the information in the table to approximate the probability of this type of error.

b. There is a second type of error that is possible in this setting. Describe this error and use the information in the given table to approximate the probability of this type of error.

10.14 Medical personnel are required to report suspected cases of child abuse. Because some diseases have symptoms that mimic those of child abuse, doctors who see a child with these symptoms must decide between two competing hypotheses:

H_0: symptoms are due to child abuse
H_a: symptoms are due to disease

(Although these are not hypotheses about a population characteristic, this exercise illustrates the definitions of Type I and Type II errors.) The article **"Blurred Line Between Illness, Abuse Creates Problem for Authorities" (*Macon Telegraph*, February 28, 2000)** included the following quote from a doctor in Atlanta regarding the consequences of making an incorrect decision: "If it's disease, the worst you have is an angry family. If it is abuse, the other kids (in the family) are in deadly danger."

a. For the given hypotheses, describe Type I and Type II errors.

b. Based on the quote regarding consequences of the two kinds of error, which type of error does the doctor quoted consider more serious? Explain.

10.15 Ann Landers, in her advice column of October 24, 1994 (***San Luis Obispo Telegram-Tribune***), described the reliability of DNA paternity testing as follows: "To get a completely accurate result, you would have to be tested, and so would (the man) and your mother. The test is 100% accurate if the man is *not* the father and 99.9% accurate if he is."

a. Consider using the results of DNA paternity testing to decide between the following two hypotheses:

H_0: a particular man is the father
H_a: a particular man is not the father

In the context of this problem, describe Type I and Type II errors. (Although these are not hypotheses about a population characteristic, this exercise illustrates the definitions of Type I and Type II errors.)

b. Based on the information given, what are the values of α, the probability of a Type I error, and β, the probability of a Type II error?

c. Ann Landers also stated, "If the mother is not tested, there is a 0.8% chance of a false positive." For the hypotheses given in Part (a), what is the value of β if the decision is based on DNA testing in which the mother is not tested?

10.16 A television manufacturer claims that (at least) 90% of its TV sets will not need service during the first 3 years of operation. A consumer agency wishes to check this claim, so it obtains a random sample of $n = 100$ purchasers and asks each whether the set purchased needed repair during the first 3 years after purchase. Let $\hat{p}$ be the sample proportion of responses indicating no repair (so that no repair is identified with a success). Let p denote the actual proportion of successes for all sets made by this manufacturer.

The agency does not want to claim false advertising unless sample evidence strongly suggests that $p < .9$. The appropriate hypotheses are then H_0: $p = .9$ versus H_a: $p < .9$.

a. In the context of this problem, describe Type I and Type II errors, and discuss the possible consequences of each.

b. Would you recommend a test procedure that uses $\alpha = .10$ or one that uses $\alpha = .01$? Explain.

10.17 A manufacturer of hand-held calculators receives large shipments of printed circuits from a supplier. It is too costly and time-consuming to inspect all incoming circuits, so when each shipment arrives, a sample is selected for inspection. Information from the sample is then used to test H_0: $p = .01$ versus H_a: $p > .01$, where p is the actual proportion of defective circuits in the shipment.

If the null hypothesis is not rejected, the shipment is accepted, and the circuits are used in the production of calculators. If the null hypothesis is rejected, the entire shipment is returned to the supplier because of inferior quality. (A shipment is defined to be of inferior quality if it contains more than 1% defective circuits.)

a. In this context, define Type I and Type II errors.

b. From the calculator manufacturer's point of view, which type of error is considered more serious?

c. From the printed circuit supplier's point of view, which type of error is considered more serious?

10.18 Water specimens are taken from water used for cooling as it is being discharged from a power plant into a river. It has been determined that as long as the mean temperature of the discharged water is at most 150°F, there will be no negative effects on the river's ecosystem. To investigate whether the plant is in compliance with regulations that prohibit a mean discharge water temperature above 150°F, researchers will take 50 water specimens at randomly selected times and record the temperature of each specimen. The resulting data will be used to test the hypotheses H_0: $\mu = 150°F$ versus H_a: $\mu > 150°F$.

a. In the context of this example, describe Type I and Type II errors.

b. Which type of error would you consider more serious? Explain.

10.19 Suppose that for a particular hypothesis test, the consequences of a Type I error are very serious. Would you want to carry out the test using a small significance level α (such as 0.01) or a larger significance level (such as 0.10)? Explain the reason for your choice. (Hint: See discussion just before Example 10.7.)

10.20 Suppose that you are an inspector for the Fish and Game Department and that you are given the task of determining whether to prohibit fishing along part of the Oregon coast. You will close an area to fishing if it is determined that fish in that region have an unacceptably high mercury content.

a. Assuming that a mercury concentration of 5 ppm is considered the maximum safe concentration, which of the following pairs of hypotheses would you test:

$$H_0: \mu = 5 \quad \text{versus} \quad H_a: \mu > 5$$

or

$$H_0: \mu = 5 \quad \text{versus} \quad H_a: \mu < 5$$

Give the reasons for your choice.

b. Would you prefer a significance level of .1 or .01 for your test? Explain. (Hint: See discussion just before Example 10.7.)

10.21 The National Cancer Institute conducted a 2-year study to determine whether cancer death rates for areas near nuclear power plants are higher than for areas without nuclear facilities **(*San Luis Obispo Telegram-Tribune*, September 17, 1990)**. A spokesperson for the Cancer Institute said, "From the data at hand, there was no convincing evidence of any increased risk of death from any of the cancers surveyed due to living near nuclear facilities. However, no study can prove the absence of an effect."

a. Let p denote the proportion of the population in areas near nuclear power plants who die of cancer during a given year. The researchers at the Cancer Institute might have considered the two rival hypotheses of the form

H_0: p = value for areas without nuclear facilities
H_a: $p >$ value for areas without nuclear facilities

Did the researchers reject H_0 or fail to reject H_0?

b. If the Cancer Institute researchers were incorrect in their conclusion that there is no increased cancer risk associated with living near a nuclear power plant, are they making a Type I or a Type II error? Explain.

c. Comment on the spokesperson's last statement that no study can *prove* the absence of an effect. Do you agree with this statement?

10.22 An automobile manufacturer is considering using robots for part of its assembly process. Converting to robots is an expensive process, so it will be undertaken only if there is strong evidence that the proportion of defective installations is lower for the robots than for human assemblers. Let p denote the proportion of defective installations for the robots. It is known that human assemblers have a defect proportion of .02.

a. Which of the following pairs of hypotheses should the manufacturer test:

H_0: $p = .02$ versus H_a: $p < .02$

or

H_0: $p = .02$ versus H_a: $p > .02$

Explain your answer.

b. In the context of this exercise, describe Type I and Type II errors.

c. Would you prefer a test with $\alpha = .01$ or $\alpha = .1$? Explain your reasoning.

Bold exercises answered in back ● Data set available online ▼ Video Solution available

10.3 Large-Sample Hypothesis Tests for a Population Proportion

Now that the basic concepts of hypothesis testing have been introduced, we are ready to consider how to use sample data to decide between a null and an alternative hypothesis. In a hypothesis test, there are two possible conclusions: We either reject H_0 or we fail to reject H_0. The fundamental idea behind hypothesis-testing procedures is this: *We reject the null hypothesis if the observed sample is very unlikely to have occurred when H_0 is true.*

In this section, we consider testing hypotheses about a population proportion when the sample size n is large. As before, p denotes the proportion of individuals or objects in a specified population that possess a certain property. A random sample of n individuals or objects is selected from the population. The sample proportion

$$\hat{p} = \frac{\text{number in the sample that possess the property}}{n}$$

serves as the basis for testing hypotheses about p.

The large-sample test procedure is based on the same properties of the sampling distribution of $\hat{p}$ that were used previously to obtain a confidence interval for p:

1. $\mu_{\hat{p}} = p$
2. $\sigma_{\hat{p}} = \sqrt{\dfrac{p(1-p)}{n}}$
3. When n is large, the sampling distribution of $\hat{p}$ is approximately normal.

These three results imply that the standardized variable

$$z = \frac{\hat{p} - p}{\sqrt{\dfrac{p(1-p)}{n}}}$$

has approximately a standard normal distribution when n is large. Example 10.8 shows how this information allows us to make a decision in a hypothesis test.

EXAMPLE 10.8 Impact of Food Labels

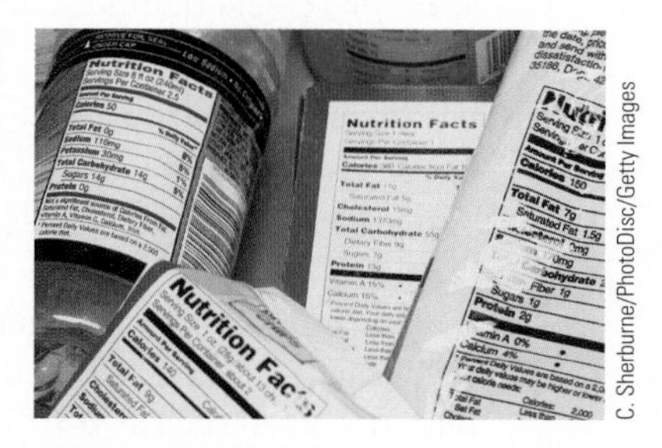
C. Sherburne/PhotoDisc/Getty Images

Understand the context ❯

In **June 2006**, an **Associated Press** survey was conducted to investigate how people use the nutritional information provided on food package labels. Interviews were conducted with 1003 randomly selected adult Americans, and each participant was asked a series of questions, including the following two:

Question 1: When purchasing packaged food, how often do you check the nutrition labeling on the package?

Question 2: How often do you purchase foods that are bad for you, even after you've checked the nutrition labels?

It was reported that 582 responded "frequently" to the question about checking labels and 441 responded very often or somewhat often to the question about purchasing "bad" foods even after checking the label.

Let's start by looking at the responses to the first question. Based on these data, is it reasonable to conclude that a majority of adult Americans frequently check the nutritional labels when purchasing packaged foods? We can answer this question by considering

p = actual proportion of adult Americans who frequently check nutritional labels

and testing the following hypotheses:

H_0: $p = .5$
H_a: $p > .5$ (The proportion of adult Americans who frequently check nutritional labels is greater than .5. That is, more than half (a majority) frequently check nutritional labels.)

Recall that in a hypothesis test, the null hypothesis is rejected only if there is convincing evidence against it—in this case, convincing evidence that $p > .5$. If H_0 is rejected, there is strong support for the claim that a majority of adult Americans frequently check nutritional labels when purchasing packaged foods.

For this sample,

$$\hat{p} = \frac{582}{1003} = .58$$

The observed sample proportion is certainly greater than .5, but this could just be due to sampling variability. That is, when $p = .5$ (meaning H_0 is true), the sample proportion $\hat{p}$ usually differs somewhat from .5 simply because of chance variation from one sample to another. Is it plausible that a sample proportion of $\hat{p} = .58$ occurred as a result of this chance variation, or is it unusual to observe a sample proportion this large when $p = .5$?

To answer this question, we form a *test statistic,* the quantity used as a basis for making a decision between H_0 and H_a. Creating a test statistic involves replacing p with the hypothesized value in the z variable $z = \dfrac{\hat{p} - p}{\sqrt{p(1 - p)/n}}$ to obtain

$$z = \frac{\hat{p} - .5}{\sqrt{\dfrac{(.5)(.5)}{n}}}$$

If the null hypothesis is true and the sample size is large, this statistic should have approximately a standard normal distribution, because in this case

1. $\mu_{\hat{p}} = .5$
2. $\sigma_{\hat{p}} = \sqrt{\dfrac{(.5)(.5)}{n}}$
3. $\hat{p}$ has approximately a normal distribution.

The calculated value of z expresses the distance between $\hat{p}$ and the hypothesized value of p as a number of standard deviations. For example, if $z = 3$, then the value of $\hat{p}$ that came from the sample is 3 standard deviations (of $\hat{p}$) greater than what we would have expected if the null hypothesis were true. How likely is it that a z value at least this inconsistent with H_0 would be observed if in fact H_0 is true? If H_0 is true, the test statistic has (approximately) a standard normal distribution. This means that

$$P(z \geq 3 \text{ when } H_0 \text{ is true}) = \text{area under the } z \text{ curve to the right of } 3.00 = .0013$$

It follows that if H_0 is true, fewer than 1% of all samples produce a value of z at least as inconsistent with H_0 as $z = 3$. Because this z value is in the most extreme 1% of the z distribution, it is sensible to reject H_0.

For our data,

Do the work

$$z = \frac{\hat{p} - .5}{\sqrt{\frac{(.5)(.5)}{n}}} = \frac{.58 - .5}{\sqrt{\frac{(.5)(.5)}{1003}}} = \frac{.08}{.016} = 5.00$$

That is, $\hat{p} = .58$ is 5 standard deviations greater than what we would expect it to be if the null hypothesis H_0: $p = .5$ was true. The sample data appear to be much more consistent with the alternative hypothesis, H_a: $p > .5$. In particular,

$$\begin{aligned} &P(\text{value of } z \text{ is at least as contradictory to } H_0 \text{ as } 5.00 \text{ when } H_0 \text{ is true}) \\ &\quad = P(z \geq 5.00 \text{ when } H_0 \text{ is true}) \\ &\quad = \text{area under the } z \text{ curve to the right of } 5.00 \\ &\quad \approx 0 \end{aligned}$$

Interpret the results

There is virtually no chance of seeing a sample proportion and corresponding z value this extreme as a result of chance variation alone when H_0 is true. If $\hat{p}$ is 5 standard deviations or more away from .5, how can we believe that $p = .5$? The evidence for rejecting H_0 in favor of H_a is very compelling.

Interestingly, in spite of the fact that there is strong evidence that a majority of adult Americans frequently check nutritional labels, the data on responses to the second question suggest that the percentage of people who then ignore the information on the label and purchase "bad" foods anyway is not small—the sample proportion who responded very often or somewhat often was .44. ■

The preceding example illustrates the reasoning behind large-sample procedures for testing hypotheses about p (and other test procedures as well). We begin by assuming that the null hypothesis is true. The sample is then examined in light of this assumption. If the observed sample proportion would not be unusual when H_0 is true, then chance variability from one sample to another is a plausible explanation for what has been observed, and H_0 should not be rejected. On the other hand, if the observed sample proportion would have been quite unlikely when H_0 is true, we would take the sample as convincing evidence against the null hypothesis and we should reject H_0. We base a decision to reject or to fail to reject the null hypothesis on an assessment of how extreme or unlikely the observed sample is if H_0 is true.

The assessment of how inconsistent the observed data are with H_0 is based on first computing the value of the test statistic

$$z = \frac{\hat{p} - \text{hypothesized value}}{\sqrt{\frac{(\text{hypothesized value})(1 - \text{hypothesized value})}{n}}}$$

We then calculate the **P-value**, the probability, assuming that H_0 is true, of obtaining a z value at least as inconsistent with H_0 as what was actually observed.

DEFINITION

Test statistic: A value computed using sample data. It is the value used to make the decision to reject or fail to reject H_0.

***P*-value:** A measure of inconsistency between the hypothesized value for a population characteristic and the observed sample. It is the probability, assuming that H_0 is true, of obtaining a test statistic value at least as inconsistent with H_0 as what was observed. The *P*-value is also sometimes called the **observed significance level.**

EXAMPLE 10.9 Detecting Plagiarism

Understand the context ⟩

Plagiarism is a growing concern among college and university faculty members, and many universities are now using software tools to detect student work that is not original.

Researchers at an Australian university that introduced the use of plagiarism detection software in a number of courses surveyed 171 students enrolled in those courses (**"Student and Staff Perceptions of the Effectiveness of Plagiarism Detection Software,"** ***Australian Journal of Educational Technology*** **[2008]: 222–240**). In the survey, 58 of the 171 students indicated that they believed that the use of plagiarism-detection software unfairly targeted students.

Assuming it is reasonable to regard the sample as representative of students at this university, does the sample provide convincing evidence that more than one-third of the students at the university believe that the use of plagiarism-detection software unfairly targets students?

With

p = proportion of all students at the university who believe that the use of plagiarism-detection software unfairly targets students

the relevant hypotheses are

$$H_0\text{: } p = \frac{1}{3} = .33$$

$$H_a\text{: } p > .33$$

The sample proportion is $\hat{p} = \dfrac{58}{171} = .34$.

Formulate a plan ⟩

Is the value of $\hat{p}$ enough greater than one-third that we shold reject H_0?

Because the sample size is large, the statistic

$$z = \frac{\hat{p} - .33}{\sqrt{\dfrac{(.33)(1 - .33)}{n}}}$$

has approximately a standard normal distribution when H_0 is true. The calculated value of the test statistic is

Do the work ⟩

$$z = \frac{.34 - .33}{\sqrt{\dfrac{(.33)(1 - .33)}{171}}} = \frac{.01}{.036} = 0.28$$

The probability that a z value at least this inconsistent with H_0 would be observed if in fact H_0 is true is

$$\begin{aligned} P\text{-value} &= P(z \geq 0.28 \text{ when } H_0 \text{ is true}) \\ &= \text{area under the } z \text{ curve to the right of } 0.28 \\ &= 1 - .6103 \\ &= .3897 \end{aligned}$$

Interpret the results ⟩

This probability indicates that when $p = .33$, it would not be unusual to observe a sample proportion as large as .34. When H_0 is true, roughly 40% of all samples would have a sample proportion as large as or larger than .34. This means that a sample proportion of .34 is reasonably consistent with the null hypothesis. Although .34 is larger than the hypothesized value of $p = .33$, chance variation from sample to sample is a plausible explanation for what was observed. There is not strong evidence that the proportion of students who believe that the use of plagiarism detection software unfairly targets students is greater than one-third. ■

As illustrated by Examples 10.8 and 10.9, small *P*-values indicate that sample results are inconsistent with H_0, whereas larger *P*-values are interpreted as meaning that the data are consistent with H_0 and that sampling variability alone is a plausible explanation for what was observed in the sample. As you probably noticed, the two cases examined ($P\text{-value} \approx 0$ and $P\text{-value} = .3897$) were such that a decision between rejecting or not rejecting H_0 was clear-cut. A decision in other cases might not be so obvious. For example, what if the sample had resulted in a *P*-value of .04? Is this unusual enough to warrant rejection of H_0? How small must the *P*-value be before H_0 should be rejected?

The answers to these questions depend on the significance level, α (the probability of a Type I error), selected for the test. For example, suppose that we set $\alpha = .05$. This implies that the probability of rejecting a true null hypothesis is .05. To obtain a test procedure with this probability of Type I error, we would reject the null hypothesis if the sample result is among the most unusual 5% of all samples when H_0 is true. We would reject H_0 if the computed $P\text{-value} \le .05$. If we had selected $\alpha = .01$, H_0 would be rejected only if we observed a sample result so extreme that it would be among the most unusual 1% if H_0 is true (which occurs when $P\text{-value} \le .01$).

A decision to reject or to fail to reject H_0 results from comparing the *P*-value to the chosen α:

Reject H_0 if $P\text{-value} \le \alpha$.

Fail to reject H_0 if $P\text{-value} > \alpha$.

Suppose, for example, that the $P\text{-value} = .0352$ and that a significance level of .05 is chosen. Then, because

$$P\text{-value} = .0352 \le .05 = \alpha$$

H_0 would be rejected. This would not be the case, though, for $\alpha = .01$, because then $P\text{-value} > \alpha$.

Computing a *P*-Value for a Large-Sample Test Concerning *p*

The computation of the *P*-value depends on the form of the inequality in the alternative hypothesis, H_a. Suppose, for example, that we wish to test

$$H_0\text{: } p = .6 \qquad \text{versus} \qquad H_a\text{: } p > .6$$

based on a large sample. The appropriate test statistic is

$$z = \frac{\hat{p} - .6}{\sqrt{\dfrac{(.6)(1 - .6)}{n}}}$$

Values of $\hat{p}$ inconsistent with H_0 and much more consistent with H_a are those much *greater* than .6 (because $p = .6$ when H_0 is true and $p > .6$ when H_0 is false and H_a is true). Such values of $\hat{p}$ correspond to z values considerably greater than 0.

If $n = 400$ and $\hat{p} = .679$, then

$$z = \frac{.679 - .6}{\sqrt{\dfrac{(.6)(1 - .6)}{400}}} = \frac{.079}{.025} = 3.16$$

The value $\hat{p} = .679$ is more than 3 standard deviations larger than what we would have expected if H_0 were true. Then,

$$\begin{aligned}
P\text{-value} &= P(z \text{ at least as inconsistent with } H_0 \text{ as } 3.16 \text{ when } H_0 \text{ is true}) \\
&= P(z \geq 3.16) \\
&= \text{area under the } z \text{ curve to the right of } 3.16 \\
&= 1 - .9992 \\
&= .0008
\end{aligned}$$

This P-value is illustrated in Figure 10.1. If H_0 is true, in the long run, only 8 out of 10,000 samples would result in a z value as or more extreme than what actually resulted. Most people would consider such a z quite unusual. Using a significance level of .01, we would reject the null hypothesis because P-value $= .0008 \leq .01 = \alpha$.

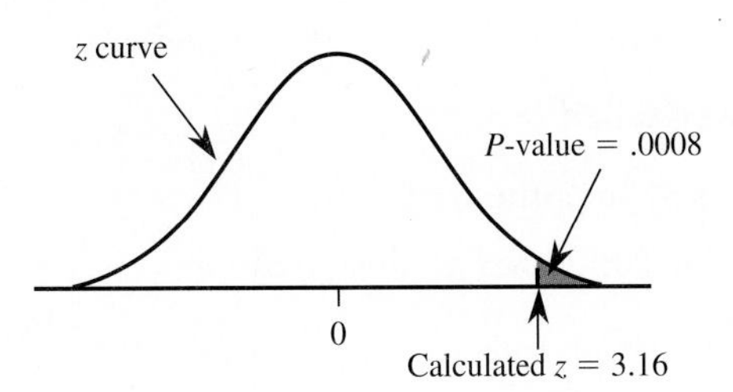

FIGURE 10.1
Calculating a P-value.

Now consider testing H_0: $p = .3$ versus H_a: $p \neq .3$. A value of $\hat{p}$ *either* much greater than .3 *or* much less than .3 is inconsistent with H_0 and provides support for H_a. Such a $\hat{p}$ corresponds to a z value far out in *either* tail of the z curve. If

$$z = \frac{\hat{p} - .3}{\sqrt{\dfrac{(.3)(1 - .3)}{n}}} = 1.75$$

then (as shown in Figure 10.2)

$$\begin{aligned}
P\text{-value} &= P(z \text{ value at least as inconsistent with } H_0 \text{ as } 1.75 \text{ when } H_0 \text{ is true}) \\
&= P(z \geq 1.75 \text{ or } z \leq -1.75) \\
&= (z \text{ curve area to the right of } 1.75) + (z \text{ curve area to the left of } -1.75) \\
&= (1 - .9599) + .0401 \\
&= .0802
\end{aligned}$$

If $z = -1.75$, the P-value in this situation is also .0802 because 1.75 and -1.75 are equally inconsistent with H_0.

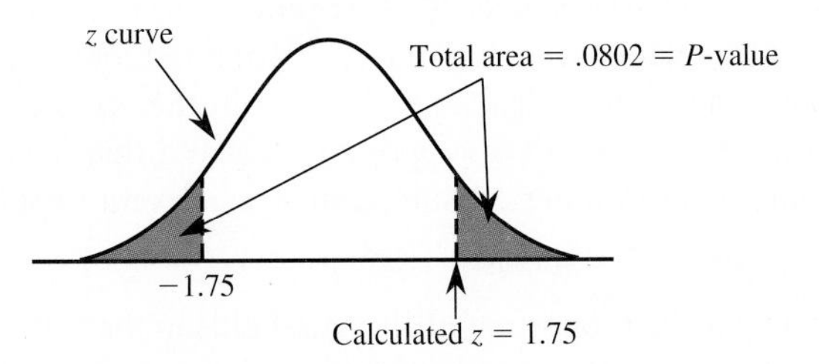

FIGURE 10.2
P-value as the sum of two tail areas.

The symmetry of the z curve implies that when the test is two-tailed (the "not equal" alternative), it is not necessary to add two curve areas. Instead,

If z is positive, P-value $= 2$(area to the right of z).
If z is negative, P-value $= 2$(area to the left of z).

Determination of the *P*-Value When the Test Statistic Is *z*

1. **Upper-tailed test:**

   H_a: $p >$ hypothesized value

   P-value computed as illustrated:

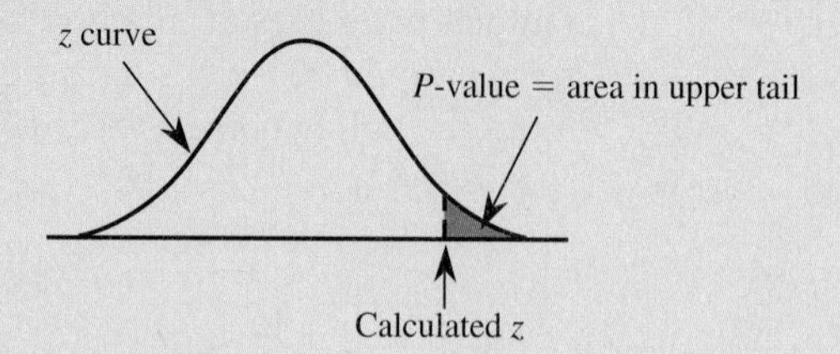

2. **Lower-tailed test:**

   H_a: $p <$ hypothesized value

   P-value computed as illustrated:

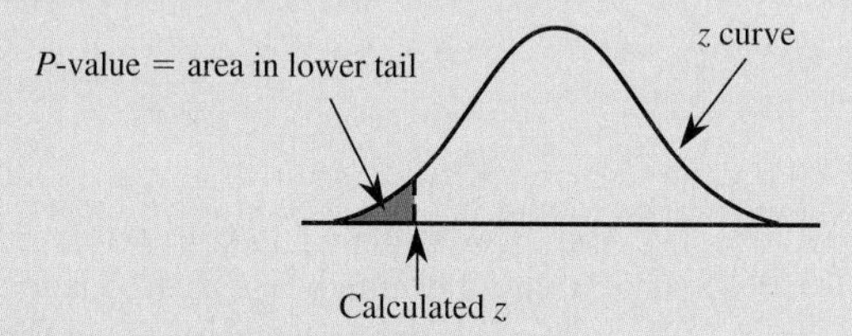

3. **Two-tailed test:**

   H_a: $p \neq$ hypothesized value

   P-value computed as illustrated:

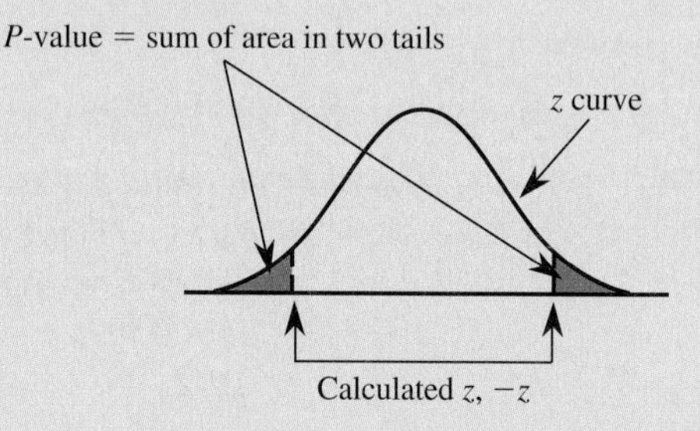

EXAMPLE 10.10 Water Conservation

Understand the context ❯

Suppose that in December 2014, a county-wide water conservation campaign was conducted in a particular county. In January 2015, a random sample of 500 homes is selected, and water usage is recorded for each home in the sample. The county supervisors wanted to know whether the data supported the claim that fewer than half the households in the county reduced water consumption. The relevant hypotheses are

$$H_0\text{: } p = .5 \qquad \text{versus} \qquad H_a\text{: } p < .5$$

where p is the proportion of all households in the county with reduced water usage.

Formulate a plan ❯

Suppose that the sample results were $n = 500$ and $\hat{p} = .440$. Because the sample size is large and this is a lower-tailed test, we can compute the *P*-value by first calculating the value of the z test statistic

$$z = \frac{\hat{p} - 5}{\sqrt{\frac{(.5)(1 - .5)}{n}}}$$

and then finding the area under the z curve to the left of this z.

Based on the observed sample data,

Do the work)

$$z = \frac{.440 - .5}{\sqrt{\frac{(.5)(1 - .5)}{500}}} = \frac{-.060}{.0224} = -2.68$$

The P-value is then equal to the area under the z curve and to the left of -2.68. From the entry in the -2.6 row and .08 column of Appendix Table 2, we find that

P-value = .0037

Interpret the results)

Using a .01 significance level, we reject H_0 (because $.0037 \leq .01$), leading us to conclude that there is convincing evidence that the proportion with reduced water usage was less than .5. Notice that rejection of H_0 would not be justified if a *very* small significance level, such as .001, had been selected. ■

Example 10.10 illustrates the calculation of a P-value for a lower-tailed test. The use of P-values in upper-tailed and two-tailed tests is illustrated in Examples 10.11 and 10.12. But first we summarize large-sample tests of hypotheses about a population proportion and introduce a step-by-step procedure for carrying out a hypothesis test.

Summary of Large-Sample z Test for p

Null hypothesis: H_0: p = hypothesized value

Test statistic: $$z = \frac{\hat{p} - \text{hypothesized value}}{\sqrt{\frac{(\text{hypothesized value})(1 - \text{hypothesized value})}{n}}}$$

Alternative Hypothesis:	***P*-Value:**
H_a: $p >$ hypothesized value	Area under z curve to right of calculated z
H_a: $p <$ hypothesized value	Area under z curve to left of calculated z
H_a: $p \neq$ hypothesized value	(1) 2(area to right of z) if z is positive, or (2) 2(area to left of z) if z is negative

Assumptions:
1. $\hat{p}$ is the sample proportion from a *random sample.*
2. The *sample size is large.* This test can be used if n satisfies both n(hypothesized value) ≥ 10 and n(1 − hypothesized value) ≥ 10.
3. If sampling is without replacement, the sample size is no more than 10% of the population size.

We recommend that the following sequence of steps be used when carrying out a hypothesis test.

Steps in a Hypothesis Test

1. Describe the population characteristic about which hypotheses are to be tested.
2. State the null hypothesis H_0.
3. State the alternative hypothesis H_a.
4. Select the significance level α for the test.
5. Display the test statistic to be used, with substitution of the hypothesized value identified in Step 2 but without any computation at this point.
6. Check to make sure that any assumptions required for the test are reasonable.

continued

7. Compute all quantities appearing in the test statistic and then the value of the test statistic itself.
8. Determine the P-value associated with the observed value of the test statistic.
9. State the conclusion (which is reject H_0 if P-value $\leq \alpha$ and fail to reject H_0 otherwise). The conclusion should then be stated in the context of the problem, and the level of significance should be included.

Steps 1–4 constitute a statement of the problem, Steps 5–8 give the analysis that leads to a decision, and Step 9 provides the conclusion.

Step-by-step technology instructions available online

EXAMPLE 10.11 Unfit Teens

Understand the context ❯

The article **"7 Million U.S. Teens Would Flunk Treadmill Tests" (Associated Press, December 11, 2005)** summarized the results of a study in which 2205 adolescents age 12 to 19 took a cardiovascular treadmill test. The researchers conducting the study indicated that the sample was selected in such a way that it could be regarded as representative of adolescents nationwide. Of the 2205 adolescents tested, 750 showed a low level of cardiovascular fitness. Does this sample provide support for the claim that more than 30% of adolescents have a low level of cardiovascular fitness? We answer this question by following the nine steps for carrying out a hypothesis test. We will use a .05 significance level for this example.

1. Population characteristic of interest:

   p = proportion of all adolescents who have a low level of cardiovascular fitness

2. Null hypothesis: H_0: $p = .3$
3. Alternative hypothesis: H_a: $p > .3$ (the percentage of adolescents with a low fitness level is greater than 30%)
4. Significance level: $\alpha = .05$

Formulate a plan ❯

5. Test statistic:

$$z = \frac{\hat{p} - \text{hypothesized value}}{\sqrt{\frac{(\text{hypothesized value})(1 - \text{hypothesized value})}{n}}} = \frac{\hat{p} - .3}{\sqrt{\frac{(.3)(1 - .3)}{n}}}$$

6. Assumptions: This test requires a random sample and a large sample size. The given sample was considered to be representative of adolescents nationwide, and if this is the case, it is reasonable to regard the sample as if it were a random sample. The sample size was $n = 2205$. Since $2205(.3) \geq 10$ and $2205(1 - .3) \geq 10$, the large-sample test is appropriate. The population is all adolescents age 12 to 19, so the sample size is small compared to the population size.

Do the work ❯

7. Computations: $n = 2205$ and $\hat{p} = 750/2205 = .34$, so

$$z = \frac{.34 - .3}{\sqrt{\frac{(.3)(1 - .3)}{2205}}} = \frac{.04}{.010} = 4.00$$

8. P-value: This is an upper-tailed test (the inequality in H_a is >), so the P-value is the area to the right of the computed z value. Since $z = 4.00$ is so far out in the upper tail of the standard normal distribution, the area to its right is negligible, and

   P-value ≈ 0

Interpret the results ❯

9. Conclusion: Since P-value $\leq \alpha$ ($0 \leq .05$), H_0 is rejected at the .05 level of significance. We conclude that the proportion of adolescents who have a low level of

cardiovascular fitness is greater than .3. That is, the sample provides convincing evidence in support of the claim that more than 30% of adolescents have a low fitness level. ■

EXAMPLE 10.12 College Attendance

Understand the context ❯

The report **"California's Education Skills Gap: Modest Improvements Could Yield Big Gains" (Public Policy Institute of California, April 16, 2008, www.ppic.org)** states that nationwide, 61% of high school graduates go on to attend a two-year or four-year college the year after graduation. The college-going rate for high school graduates in California was estimated to be 55%. Suppose that the estimate of 55% was based on a random sample of 1500 California high school graduates in 2009.

Can we reasonably conclude that the proportion of California high school graduates in 2009 who attended college the year after graduation is different from the national figure? We will use the nine-step hypothesis testing procedure and a significance level of $\alpha = .01$ to answer this question.

1. p = proportion of all 2009 high school graduates who attended college the year after graduation
2. H_0: $p = .61$
3. H_a: $p \neq .61$ (differs from the national proportion)
4. Significance level: $\alpha = .01$

Formulate a plan ❯

5. Test statistic:

$$z = \frac{\hat{p} - \text{hypothesized value}}{\sqrt{\frac{(\text{hypothesized value})(1 - \text{hypothesized value})}{n}}} = \frac{\hat{p} - .61}{\sqrt{\frac{(.61)(.39)}{n}}}$$

6. Assumptions: This test requires a random sample and a large sample size. The given sample was a random sample, the population size is much larger than the sample size, and the sample size was $n = 1500$. Because $1500(.61) \geq 10$ and $1500(.39) \geq 10$, the large-sample test is appropriate.

Do the work ❯

7. Computations: $\hat{p} = .55$, so

$$z = \frac{.55 - .61}{\sqrt{\frac{(.61)(.39)}{1500}}} = \frac{-.06}{.013} = -4.62$$

8. *P*-value: The area under the z curve to the left of -4.62 is approximately 0, so *P*-value $\approx 2(0) = 0$.

Interpret the results ❯

9. Conclusion: At significance level .01, we reject H_0 because *P*-value $\approx 0 < .01 = \alpha$. The data provide convincing evidence that the proportion of 2009 California high school graduates who attended college during the year after graduation differs from the nationwide proportion. ■

Most statistical computer packages and graphing calculators can calculate and report *P*-values for a variety of hypothesis-testing situations, including the large sample test for a proportion. Minitab was used to carry out the test of Example 10.10, and the resulting computer output follows:

Test and Confidence Interval for One Proportion

Test of p = 0.5 vs p < 0.5

Sample	X	N	Sample p	95.0 % CI	Z-Value	P-Value
1	220	500	0.440000	(0.396491, 0.483509)	−2.68	0.004

From the Minitab output, $z = -2.68$, and the associated P-value is .004. The small difference between the P-value here and the one computed in Example 10.10 (.0037) is the result of rounding.

It is also possible to compute the value of the z test statistic and then use a statistical computer package or graphing calculator to determine the corresponding P-value as an area under the standard normal curve. For example, the user can specify a value and Minitab will determine the area to the left of this value for any particular normal distribution. Because of this, the computer can be used in place of Appendix Table 2. In Example 10.10, the computed z was -2.68. Using Minitab gives the following output:

```
Normal with mean = 0 and standard deviation = 1.00000
        x        P(X ≤ x)
  -2.6800        0.0037
```

From the output, we learn that the area to the left of $-2.68 = .0037$, which agrees with the value obtained by using the tables.

EXERCISES 10.23 - 10.41

10.23 Use the definition of the P-value to explain the following:

a. Why H_0 would be rejected if P-value = .0003

b. Why H_0 would not be rejected if P-value = .350

10.24 For which of the following P-values will the null hypothesis be rejected when performing a test with a significance level of .05:

a. .001
b. .021
c. .078
d. .047
e. .148

10.25 Pairs of P-values and significance levels, α, are given. For each pair, state whether you would reject H_0 at the given significance level.

a. P-value = .084, $\alpha = .05$

b. P-value = .003, $\alpha = .001$

c. P-value = .498, $\alpha = .05$

d. P-value = .084, $\alpha = .10$

e. P-value = .039, $\alpha = .01$

f. P-value = .218, $\alpha = .10$

10.26 Let p denote the proportion of students at a particular university that use the fitness center on campus on a regular basis. For a large-sample z test of H_0: $p = .5$ versus H_a: $p > .5$, find the P-value associated with each of the given values of the test statistic:

a. 1.40
b. 0.93
c. 1.96
d. 2.45
e. -0.17

10.27 Assuming a random sample from a large population, for which of the following null hypotheses and sample sizes n is the large-sample z test appropriate:

a. H_0: $p = .2$, $n = 25$

b. H_0: $p = .6$, $n = 210$

c. H_0: $p = .9$, $n = 100$

d. H_0: $p = .05$, $n = 75$

10.28 In a survey conducted by CareerBuilder.com, employers were asked if they had ever sent an employee home because they were dressed inappropriately (**June 17, 2008, www.careerbuilder.com**). A total of 2765 employers responded to the survey, with 968 saying that they had sent an employee home for inappropriate attire. In a press release, CareerBuilder makes the claim that more than one-third of employers have sent an employee home to change clothes.

Do the sample data provide convincing evidence in support of this claim? Test the relevant hypotheses using $\alpha = .05$. For purposes of this exercise, assume that it is reasonable to regard the sample as representative of employers in the United States. (Hint: See Example 10.11.)

10.29 In a survey of 1000 women age 22 to 35 who work full time, 540 indicated that they would be willing to give up some personal time in order to make more money (***USA Today*, March 4, 2010**). The sample was selected in a way that was designed to produce a sample that was representative of women in the targeted age group.

a. Do the sample data provide convincing evidence that the majority of women age 22 to 35 who work full-time would be willing to give up some personal time for more money? Test the relevant hypotheses using $\alpha = .01$.

b. Would it be reasonable to generalize the conclusion from Part (a) to all working women? Explain why or why not.

10.30 The paper **"Debt Literacy, Financial Experiences and Over-Indebtedness" (*Social Science Research Network*, Working paper W14808, 2008)** included analysis of data from a national sample of 1000 Americans. One question on the survey was:

> "You owe $3000 on your credit card. You pay a minimum payment of $30 each month. At an Annual Percentage Rate of 12% (or 1% per month), how many years would it take to eliminate your credit card debt if you made no additional charges?"

Answer options for this question were: (a) less than 5 years; (b) between 5 and 10 years; (c) between 10 and 15 years; (d) never—you will continue to be in debt; (e) don't know; and (f) prefer not to answer.

a. Only 354 of the 1000 respondents chose the correct answer of never. For purposes of this exercise, you can assume that the sample is representative of adult Americans. Is there convincing evidence that the proportion of adult Americans who can answer this question correctly is less than .40 (40%)? Use $\alpha = .05$ to test the appropriate hypotheses. (Hint: See Example 10.10.)

b. The paper also reported that 37.8% of those in the sample chose one of the wrong answers (a, b, and c) as their response to this question. Is it reasonable to conclude that more than one-third of adult Americans would select a wrong answer to this question? Use $\alpha = .05$.

10.31 **"Most Like it Hot"** is the title of a press release issued by the **Pew Research Center (March 18, 2009, www.pewsocialtrends.org)**. The press release states that "by an overwhelming margin, Americans want to live in a sunny place." This statement is based on data from a nationally representative sample of 2260 adult Americans. Of those surveyed, 1288 indicated that they would prefer to live in a hot climate rather than a cold climate.

Do the sample data provide convincing evidence that a majority of all adult Americans prefer a hot climate over a cold climate? Use the nine-step hypothesis testing process with $\alpha = .01$ to answer this question.

10.32 In a survey of 1005 adult Americans, 46% indicated that they were somewhat interested or very interested in having web access in their cars **(*USA Today*, May 1, 2009)**. Suppose that the marketing manager of a car manufacturer claims that the 46% is based only on a sample and that 46% is close to half, so there is no reason to believe that the proportion of all adult Americans who want car web access is less than .50. Is the marketing manager correct in his claim? Provide statistical evidence to support your answer. For purposes of this exercise, assume that the sample can be considered as representative of adult Americans.

10.33 The paper **"Teens and Distracted Driving" (Pew Internet & American Life Project, 2009)** reported that in a representative sample of 283 American teens age 16 to 17, there were 74 who indicated that they had sent a text message while driving. For purposes of this exercise, assume that this sample is a random sample of 16- to 17-year-old Americans. Do these data provide convincing evidence that more than a quarter of Americans age 16 to 17 have sent a text message while driving? Test the appropriate hypotheses using a significance level of 0.01.

10.34 The report **"How Teens Use Media" (Nielsen, June 2009)** says that 37% of teens in the U.S. access the Internet from a mobile phone. Suppose you plan to select a random sample of 500 students at the local high school and will ask each student in the sample if he or she accesses the Internet from a mobile phone. You plan to use the resulting data to decide if there is evidence that the proportion of students at the high school who access the web using a mobile phone differs from the national figure given in the Nielsen report. What hypotheses should you test?

10.35 The article **"Irritated by Spam? Get Ready for Spit" (*USA Today*, November 10, 2004)** predicts that "spit," spam that is delivered via Internet phone lines and cell phones, will be a growing problem as more people turn to web-based phone services. In a 2004 poll of 5500 cell phone users conducted by the Yankee Group, 20% indicated that they had received commercial messages or ads on their cell phones. Is there sufficient evidence to conclude that the proportion of cell phone users who have received commercial messages or ads in 2004 was greater than the proportion of .13 reported for the previous year?

10.36 According to a ***Washington Post-ABC News*** poll, 331 of 502 randomly selected U.S. adults interviewed said they would not be bothered if the National Security Agency collected records of personal telephone calls they had made. Is there sufficient evidence to conclude that a majority of U.S. adults feel this way? Test the appropriate hypotheses using a .01 significance level.

10.37 According to a survey of 1000 adult Americans conducted by **Opinion Research Corporation**, 210 of those surveyed said playing the lottery would be the most practical way for them to accumulate

$200,000 in net wealth in their lifetime (**"One in Five Believe Path to Riches Is the Lottery," *San Luis Obispo Tribune*, January 11, 2006**). Although the article does not describe how the sample was selected, for purposes of this exercise, assume that the sample can be regarded as a random sample of adult Americans. Is there convincing evidence that more than 20% of adult Americans believe that playing the lottery is the best strategy for accumulating $200,000 in net wealth?

10.38 The article **"Cops Get Screened for Digital Dirt" (*USA Today*, Nov. 12, 2010)** summarizes a report of law enforcement agencies regarding the use of social media to screen applicants for employment. The report was based on a survey of 728 law enforcement agencies. One question on the survey asked if the agency routinely reviewed applicant's social media activity during background checks. For purposes of this exercise, suppose that the 728 agencies were selected at random, and that you want to use the survey data to decide if there is convincing evidence that more than 25% of law enforcement agencies review applicants' social media activity as part of routine background checks.

a. The sampling distribution of $\hat{p}$ describes the behavior of $\hat{p}$ when random samples are selected from a particular population. Describe the shape, center, and spread of the sampling distribution of $\hat{p}$ for samples of size 728 if the null hypothesis H_0: $p = 0.25$ is true.

b. Would you be surprised to observe a sample proportion of $\hat{p} = 0.27$ for a sample of size 728 if the null hypothesis H_0: $p = 0.25$ is true? Explain why or why not.

c. Would you be surprised to observe a sample proportion of $\hat{p} = 0.31$ for a sample of size 728 if the null hypothesis H_0: $p = 0.25$ is true? Explain why or why not.

10.39 Refer back to the previous exercise. The actual sample proportion observed in the study was $\hat{p} = 0.33$. Based on this sample proportion, is there convincing evidence that more than 25% of law enforcement agencies review social media activity as part of background checks, or is this sample proportion consistent with what you would expect to see when the null hypothesis is true?

10.40 The report **"2007 Electronic Monitoring & Surveillance Survey: Many Companies Monitoring, Recording, Videotaping—and Firing—Employees" (American Management Association, 2007)** summarized the results of a survey of 304 U.S. businesses. Of these companies, 201 indicated that they monitor employees' web site visits. For purposes of this exercise, assume that it is reasonable to regard this sample as representative of businesses in the United States.

a. Is there sufficient evidence to conclude that more than 60% of U.S. businesses monitor employees' web site visits? Test the appropriate hypotheses using a significance level of .01.

b. Is there sufficient evidence to conclude that a majority of U.S. businesses monitor employees' web site visits? Test the appropriate hypotheses using a significance level of .01.

10.41 The article **"Fewer Parolees Land Back Behind Bars" (*Associated Press*, April 11, 2006)** includes the following statement: "Just over 38% of all felons who were released from prison in 2003 landed back behind bars by the end of the following year, the lowest rate since 1979." Explain why it would not be necessary to carry out a hypothesis test to determine if the proportion of felons released in 2003 was less than .40.

Bold exercises answered in back ● Data set available online ▼ Video Solution available

10.4 Hypothesis Tests for a Population Mean

We now turn our attention to developing a method for testing hypotheses about a population mean. The test procedures in this case are based on the same two results that led to the z and t confidence intervals in Chapter 9:

1. When n is large or the population distribution is approximately normal, then

$$z = \frac{\bar{x} - \mu}{\frac{\sigma}{\sqrt{n}}}$$

has approximately a standard normal distribution.

2. When n is large or the population distribution is approximately normal, then

$$t = \frac{\bar{x} - \mu}{\frac{s}{\sqrt{n}}}$$

has approximately a t distribution with $\text{df} = n - 1$.

This means that if we are interested in testing a null hypothesis of the form

H_0: μ = hypothesized value

and n is large or the population distribution is approximately normal, one of the following z or t test statistics can be used:

Case 1: σ known

Test statistic: $z = \frac{\bar{x} - \text{hypothesized value}}{\frac{\sigma}{\sqrt{n}}}$

P-value: Computed as an area under the z curve

Case 2: σ unknown

Test statistic: $t = \frac{\bar{x} - \text{hypothesized value}}{\frac{s}{\sqrt{n}}}$

P-value: Computed as an area under the t curve with $\text{df} = n - 1$

Because it is rarely the case that σ, the population standard deviation, is known, we focus our attention on the test procedure for the case in which σ is unknown.

When testing a hypothesis about a population mean, the null hypothesis specifies a particular hypothesized value for μ. We write the null hypothesis as H_0: μ = hypothesized value. The alternative hypothesis has one of the following three forms, depending on the research question being addressed:

H_a: $\mu >$ hypothesized value
H_a: $\mu <$ hypothesized value
H_a: $\mu \neq$ hypothesized value

If n is large or if the population distribution is approximately normal, the test statistic

$$t = \frac{\bar{x} - \text{hypothesized value}}{\frac{s}{\sqrt{n}}}$$

can be used. For example, if the null hypothesis to be tested is H_0: $\mu = 100$, the test statistic becomes

$$t = \frac{\bar{x} - 100}{\frac{s}{\sqrt{n}}}$$

Consider the alternative hypothesis H_a: $\mu > 100$, and suppose that a sample of size $n = 24$ results in $\bar{x} = 104.20$ and $s = 8.23$. Then the test statistic value is

$$t = \frac{104.20 - 100}{\frac{8.23}{\sqrt{24}}} = \frac{4.20}{1.6799} = 2.50$$

Because this is an upper-tailed test, the *P*-value is the area under an appropriate t curve (here with df $= 24 - 1 = 23$) to the right of 2.50.

Appendix Table 4 is a tabulation of t curve tail areas. Each column of the table is for a different number of degrees of freedom: 1, 2, 3, . . . , 30, 35, 40, 60, 120, and a last column for df $= \infty$, which is the same as for the z curve. The table gives the area under each t curve to the right of values ranging from 0.0 to 4.0 in increments of 0.1. Part of this table appears in Figure 10.3. For example,

area under the 23-df t curve to the right of 2.5 = .010
= *P*-value for an upper-tailed t test

Suppose that $t = -2.7$ for a lower-tailed test based on 23 df. Then, because each t curve is symmetric about 0,

P-value = area to the left of -2.7 = area to the right of 2.7 = .006

As is the case for z tests, we double the tail area to obtain the *P*-value for two-tailed t tests. This means that if $t = 2.6$ or if $t = -2.6$ for a two-tailed t test with 23 df, then

P-value = 2(.008) = .016

Once past 30 df, the tail areas change very little, so the last column (∞) in Appendix Table 4 provides a good approximation.

t \ df	1	2	. . .	22	23	24	. . .	60	120
0.0									
0.1									
⋮				⋮	⋮	⋮			
2.5			. . .	.010	.010	.010	. . .		
2.6			. . .	.008	.008	.008	. . .		
2.7			. . .	.007	.006	.006	. . .		
2.8			. . .	.005	.005	.005	. . .		
⋮				⋮	⋮	⋮			
4.0									

Area under 23-df t curve to right of 2.7

FIGURE 10.3
Part of Appendix Table 4: t curve tail areas.

A graphing calculator or a statistics package can also be used to compute *P*-values for t tests.

The following two boxes show how the *P*-value is obtained as a t curve area and give a general description of the test procedure.

Finding *P*-Values for a *t* Test

1. **Upper-tailed test:**

   H_a: $\mu >$ hypothesized value

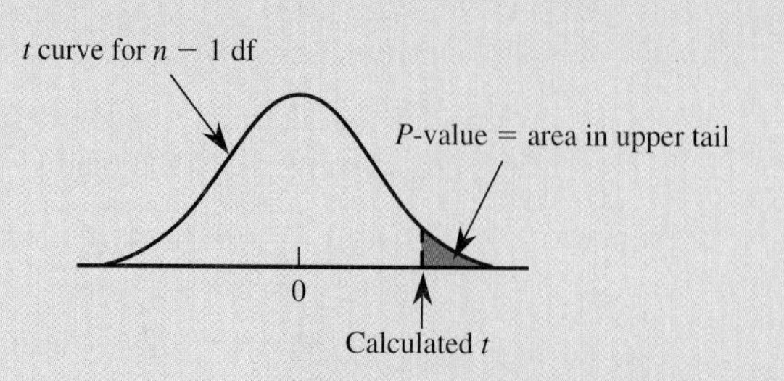

continued

2. **Lower-tailed test:**

   H_a: $\mu <$ hypothesized value

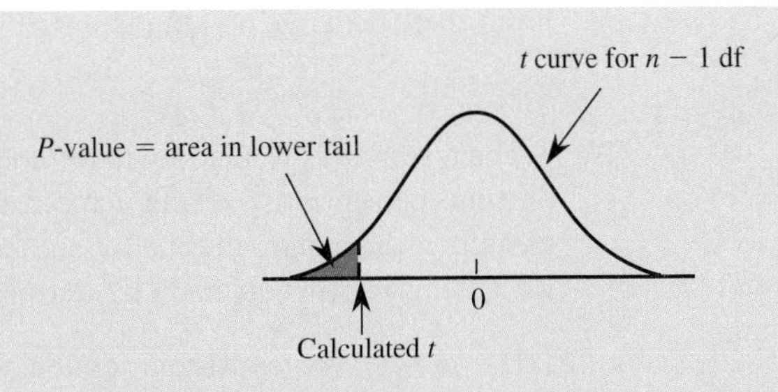

3. **Two-tailed test:**

   H_a: $\mu \neq$ hypothesized value

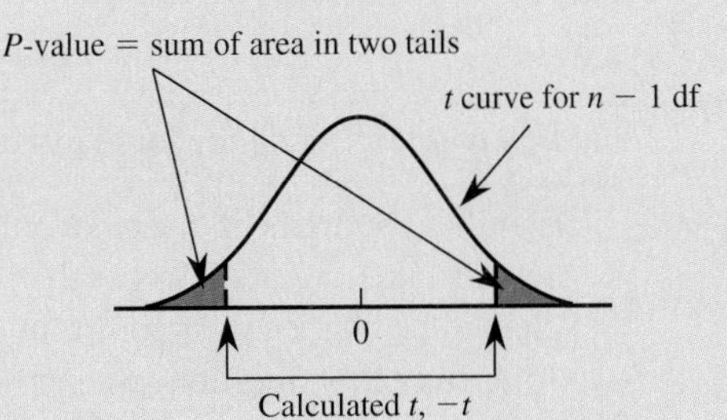

Appendix Table 4 gives upper-tail t curve areas to the right of values 0.0, 0.1, . . . , 4.0. These areas are P-values for upper-tailed tests and, by symmetry, also for lower-tailed tests. Doubling an area gives the P-value for a two-tailed test.

The One-Sample t Test for a Population Mean

Null hypothesis: H_0: $\mu =$ hypothesized value

Test statistic: $t = \dfrac{\bar{x} - \text{hypothesized value}}{\dfrac{s}{\sqrt{n}}}$

Alternative Hypothesis:	***P*-Value:**
H_a: $\mu >$ hypothesized value	Area to the right of calculated t under t curve with df $= n - 1$
H_a: $\mu <$ hypothesized value	Area to the left of calculated t under t curve with df $= n - 1$
H_a: $\mu \neq$ hypothesized value	(1) 2(area to the right of t) if t is positive, or (2) 2(area to the left of t) if t is negative

Assumptions: 1. $\bar{x}$ and s are the sample mean and sample standard deviation from a *random sample.*
2. The *sample size is large* (generally $n \geq 30$) or *the population distribution is at least approximately normal.*

EXAMPLE 10.13 Time Stands Still (or So It Seems)

● A study conducted by researchers at Pennsylvania State University investigated whether time perception, an indication of a person's ability to concentrate, is impaired during nicotine withdrawal. The study results were presented in the paper **"Smoking Abstinence Impairs Time Estimation Accuracy in Cigarette Smokers"** (***Psychopharmacology Bulletin*** **[2003]: 90–95**).

After a 24-hour smoking abstinence, 20 smokers were asked to estimate how much time had passed during a 45-second period. Suppose the resulting data on perceived elapsed time (in seconds) were as follows (these data are artificial but are consistent with summary quantities given in the paper):

69	65	72	73	59	55	39	52	67	57
56	50	70	47	56	45	70	64	67	53

● Data set available online

From these data, we obtain

$n = 20 \qquad \bar{x} = 59.30 \qquad s = 9.84$

The researchers wanted to determine whether smoking abstinence had a negative impact on time perception, causing elapsed time to be overestimated. With μ representing the mean perceived elapsed time for smokers who have abstained from smoking for 24 hours, we can answer this question by testing

H_0: $\mu = 45$ (no consistent tendency to overestimate the time elapsed)

versus

H_a: $\mu > 45$ (tendency for elapsed time to be overestimated)

The null hypothesis is rejected only if there is convincing evidence that $\mu > 45$. The observed value, 59.30, is larger than 45, but can a sample mean as large as this be plausibly explained by chance variation from one sample to another when $\mu = 45$?

To answer this question, we carry out a hypothesis test with a significance level of .05 using the nine-step procedure described in Section 10.3.

1. Population characteristic of interest:

Understand the context

μ = mean perceived elapsed time for smokers who have abstained from smoking for 24 hours

2. Null hypothesis: H_0: $\mu = 45$
3. Alternative hypothesis: H_a: $\mu > 45$
4. Significance level: $\alpha = .05$

Formulate a plan

5. Test statistic: $t = \dfrac{\bar{x} - \text{hypothesized value}}{\dfrac{s}{\sqrt{n}}} = \dfrac{\bar{x} - 45}{\dfrac{s}{\sqrt{n}}}$

6. Assumptions: This test requires a random sample and either a large sample size or a normal population distribution. The authors of the paper believed that it was reasonable to consider this sample as representative of smokers in general, and if this is the case, it is reasonable to regard it as if it were a random sample. Because the sample size is only 20, for the t test to be appropriate, we must be willing to assume that the population distribution of perceived elapsed times is at least approximately normal. Is this reasonable? The following graph gives a boxplot of the data:

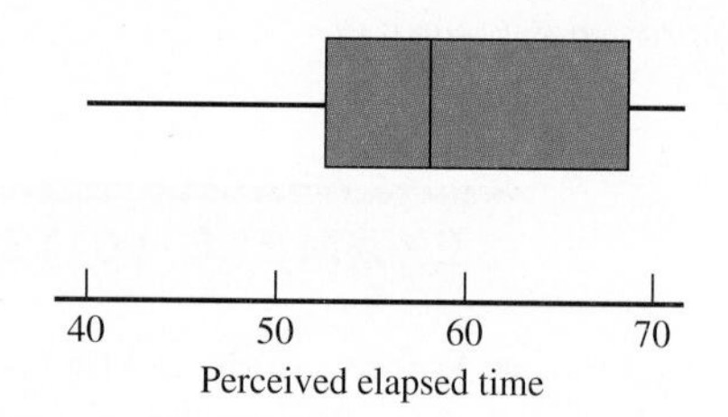

Although the boxplot is not perfectly symmetric, it does not appear to be too skewed and there are no outliers, so the use of the t test is reasonable.

7. Computations: $n = 20$, $\bar{x} = 59.30$, and $s = 9.84$, so

Do the work

$$t = \frac{59.30 - 45}{\dfrac{9.84}{\sqrt{20}}} = \frac{14.30}{2.20} = 6.50$$

8. *P*-value: This is an upper-tailed test (the inequality in H_a is "greater than"), so the *P*-value is the area to the right of the computed t value. Because df $= 20 - 1 = 19$, we can use the df $= 19$ column of Appendix Table 4 to find the *P*-value. With $t = 6.50$, we obtain *P*-value $=$ area to the right of $6.50 \approx 0$ (because 6.50 is greater than 4.0, the largest tabulated value).

Interpret the results ⟩

9. Conclusion: Because *P*-value $\leq \alpha$, we reject H_0 at the .05 level of significance. There is almost no chance of seeing a sample mean this extreme as a result of just chance variation when H_0 is true. There is convincing evidence that the mean perceived time elapsed is greater than the actual time elapsed of 45 seconds.

This paper also looked at perception of elapsed time for a sample of nonsmokers and for a sample of smokers who had not abstained from smoking. The investigators found that the null hypothesis of $\mu = 45$ could not be rejected for either of these groups. ■

EXAMPLE 10.14 Goofing Off at Work

Step-by-step technology instructions available online

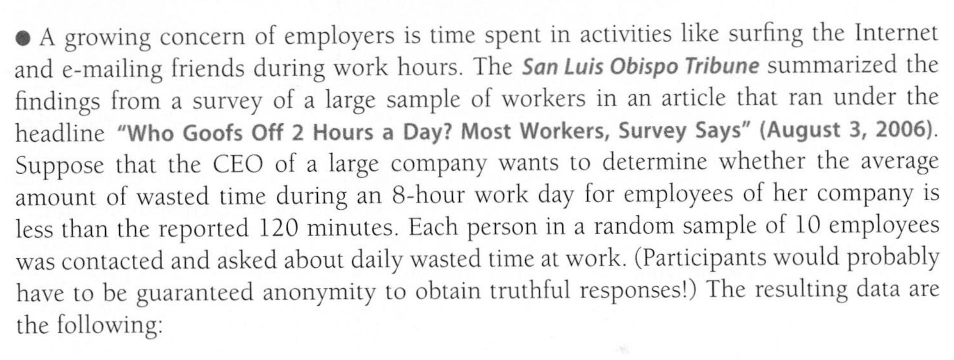

● A growing concern of employers is time spent in activities like surfing the Internet and e-mailing friends during work hours. The ***San Luis Obispo Tribune*** summarized the findings from a survey of a large sample of workers in an article that ran under the headline **"Who Goofs Off 2 Hours a Day? Most Workers, Survey Says" (August 3, 2006)**. Suppose that the CEO of a large company wants to determine whether the average amount of wasted time during an 8-hour work day for employees of her company is less than the reported 120 minutes. Each person in a random sample of 10 employees was contacted and asked about daily wasted time at work. (Participants would probably have to be guaranteed anonymity to obtain truthful responses!) The resulting data are the following:

108 112 117 130 111 131 113 113 105 128

Summary quantities are $n = 10$, $\bar{x} = 116.80$, and $s = 9.45$.

Do these data provide evidence that the mean wasted time for this company is less than 120 minutes? To answer this question, let's carry out a hypothesis test with $\alpha = .05$.

Understand the context ⟩

1. $\mu =$ mean daily wasted time for employees of this company
2. H_0: $\mu = 120$
3. H_a: $\mu < 120$
4. $\alpha = .05$

Formulate a plan ⟩

5. $t = \dfrac{\bar{x} - \text{hypothesized value}}{\frac{s}{\sqrt{n}}} = \dfrac{\bar{x} - 120}{\frac{s}{\sqrt{n}}}$

6. This test requires a random sample and either a large sample or a normal population distribution. The given sample was a random sample of employees. Because the sample size is small, we must be willing to assume that the population distribution of times is at least approximately normal. The accompanying normal probability plot appears to be reasonably straight, and although the normal probability plot and the boxplot reveal some skewness in the sample, there are no outliers.

● Data set available online

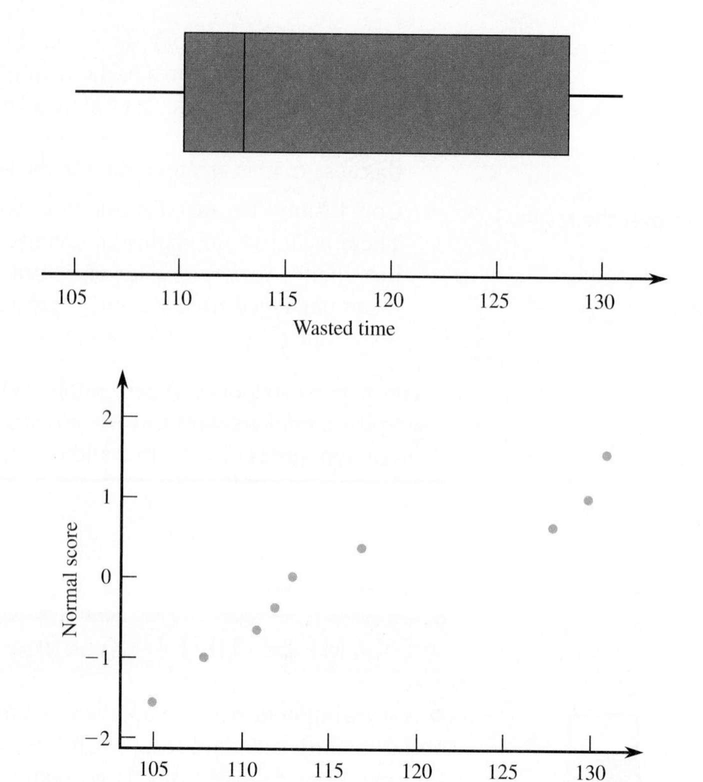

Based on these observations, it is plausible that the population distribution is approximately normal, so we proceed with the t test.

Do the work ⟩

7. Test statistic: $t = \dfrac{116.80 - 120}{\dfrac{9.45}{\sqrt{10}}} = -1.07$

8. From the df = 9 column of Appendix Table 4 and by rounding the test statistic value to −1.1, we get

   P-value = area to the left of −1.1 = area to the right of 1.1 = .150

as shown:

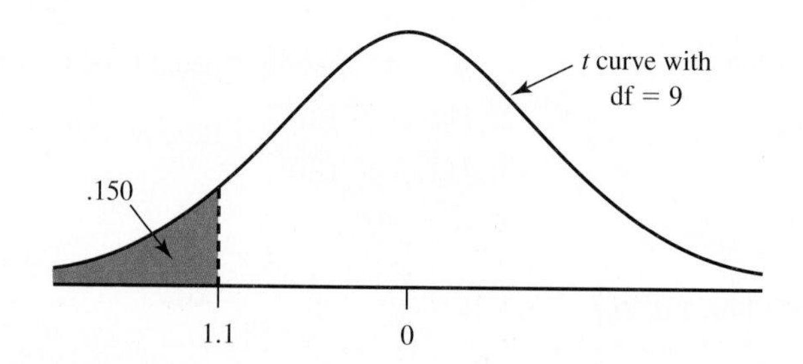

Interpret the results ⟩

9. Because the P-value $> \alpha$, we fail to reject H_0. There is not sufficient evidence to conclude that the mean wasted time per 8-hour work day for employees at this company is less than 120 minutes.

Minitab could also have been used to carry out the test, as shown in the output below.

One-Sample T: Wasted Time

Test of mu = 120 vs < 120

Variable	N	Mean	StDev	SE Mean	95% Upper Bound	T	P
Wasted Time	10	116.800	9.449	2.988	122.278	−1.07	0.156

Although we had to round the computed t value to -1.1 to use Appendix Table 4, Minitab was able to compute the P-value corresponding to the actual value of the test statistic, which was P-value $= 0.156$. ■

EXAMPLE 10.15 Cricket Love

Dynamic Graphics Group/Creatas/Alamy

The article **"Well-Fed Crickets Bowl Maidens Over"** (***Nature Science Update*, February 11, 1999**) reported that female field crickets are attracted to males that have high chirp rates and hypothesized that chirp rate is related to nutritional status. The usual chirp rate for male field crickets was reported to vary around a mean of 60 chirps per second. To investigate whether chirp rate was related to nutritional status, investigators fed male crickets a high protein diet for 8 days, after which chirp rate was measured.

The mean chirp rate for the crickets on the high protein diet was reported to be 109 chirps per second. Is this convincing evidence that the mean chirp rate for crickets on a high protein diet is greater than 60 (which would then imply an advantage in attracting the ladies)?

Suppose that the sample size and sample standard deviation are $n = 32$ and $s = 40$. Let's test the relevant hypotheses with $\alpha = .01$.

Understand the context ⟩

1. μ = mean chirp rate for crickets on a high protein diet
2. H_0: $\mu = 60$
3. H_a: $\mu > 60$
4. $\alpha = .01$

Formulate a plan ⟩

5. $t = \dfrac{\bar{x} - \text{hypothesized value}}{\dfrac{s}{\sqrt{n}}} = \dfrac{\bar{x} - 60}{\dfrac{s}{\sqrt{n}}}$

6. This test requires a random sample and either a large sample or a normal population distribution. Because the sample size is large ($n = 32$), it is reasonable to proceed with the t test as long as we are willing to consider the 32 male field crickets in this study as if they were a random sample from the population of male field crickets.

Do the work ⟩

7. Test statistic: $t = \dfrac{109 - 60}{\dfrac{40}{\sqrt{32}}} = \dfrac{49}{7.07} = 6.93$

8. This is an upper-tailed test, so the P-value is the area under the t curve with df $= 31$ and to the right of 6.93. From Appendix Table 4, P-value ≈ 0.

Interpret the results ⟩

9. Because P-value ≈ 0, which is less than the significance level, α, we reject H_0. There is convincing evidence that the mean chirp rate is higher for male field crickets that eat a high protein diet. ■

Statistical Versus Practical Significance

Carrying out a hypothesis test amounts to deciding whether the value obtained for the test statistic could plausibly have resulted when H_0 is true. When the value of the test statistic leads to rejection of H_0, it is customary to say that the result is **statistically significant** at the chosen significance level α. The finding of statistical significance means that, in the investigator's opinion, the observed deviation from what was expected under H_0 cannot reasonably be attributed to only chance variation. However, statistical significance is not the same as concluding that the true situation differs from what the null hypothesis states in any practical sense. That is, even after H_0 has been rejected, the data may suggest that there is no *practical* difference between the actual value of the population characteristic and what the null hypothesis states that value to be. This is illustrated in Example 10.16.

EXAMPLE 10.16 "Significant" but Unimpressive Test Score Improvement

Understand the context) Let μ denote the average score on a standardized test for all children in a certain region of the United States. The average score for all children in the United States is 100. Regional education authorities are interested in testing H_0: $\mu = 100$ versus H_a: $\mu > 100$ using a significance level of .001.

A sample of 2500 children resulted in the values $n = 2500$, $\bar{x} = 101.0$, and $s = 15.0$. Then

$$t = \frac{101.10 - 100}{\frac{15}{\sqrt{2500}}} = 3.3$$

This is an upper-tailed test, so (using the z column of Appendix Table 4 because df = 2499) P-value = area to the right of $3.33 \approx .000$. Because P-value $< .001$, we reject H_0. There is evidence that the mean score for this region is greater than 100.

Interpret the results) However, with $n = 2500$, the point estimate $\bar{x} = 101.0$ is likely to be very close to the actual value of μ. This means that it looks as though H_0 was rejected because $\mu \approx 101$ rather than 100. From a practical point of view, a 1-point difference is most probably of no practical importance. ■

EXERCISES 10.42 - 10.58

10.42 Give as much information as you can about the P-value of a t test in each of the following situations:

a. Upper-tailed test, df = 8, $t = 2.0$

b. Upper-tailed test, $n = 14$, $t = 3.2$

c. Lower-tailed test, df = 10, $t = -2.4$

d. Lower-tailed test, $n = 22$, $t = -4.2$

e. Two-tailed test, df = 15, $t = -1.6$

f. Two-tailed test, $n = 16$, $t = 1.6$

g. Two-tailed test, $n = 16$, $t = 6.3$

10.43 Give as much information as you can about the P-value of a t test in each of the following situations:

a. Two-tailed test, df = 9, $t = 0.73$

b. Upper-tailed test, df = 10, $t = -0.5$

c. Lower-tailed test, $n = 20$, $t = -2.1$

d. Lower-tailed test, $n = 20$, $t = -5.1$

e. Two-tailed test, $n = 40$, $t = 1.7$

10.44 Paint used to paint lines on roads must reflect enough light to be clearly visible at night. Let μ denote the mean reflectometer reading for a new type of paint under consideration. A test of H_0: $\mu = 20$ versus H_a: $\mu > 20$ based on a sample of 15 observations gave $t = 3.2$. What conclusion is appropriate at each of the following significance levels?

a. $\alpha = .05$

b. $\alpha = .01$

c. $\alpha = .001$

10.45 A certain pen has been designed so that true average writing lifetime under controlled conditions (involving the use of a writing machine) is at least 10 hours. A random sample of 18 pens is selected, the writing lifetime of each is determined, and a normal probability plot of the resulting data supports the use of a one-sample t test. The relevant hypotheses are H_0: $\mu = 10$ versus H_a: $\mu < 10$.

a. If $t = -2.3$ and $\alpha = .05$ is selected, what conclusion is appropriate?

b. If $t = -1.83$ and $\alpha = .01$ is selected, what conclusion is appropriate?

c. If $t = 0.47$, what conclusion is appropriate?

10.46 The true average diameter of ball bearings of a certain type is supposed to be 0.5 inch. What conclusion is appropriate when testing H_0: $\mu = 0.5$ versus H_a: $\mu \neq 0.5$ inch each of the following situations:

a. $n = 13, t = 1.6, \alpha = .05$

b. $n = 13, t = -1.6, \alpha = .05$

c. $n = 25, t = -2.6, \alpha = .01$

d. $n = 25, t = -3.6$

10.47 The paper **"Playing Active Video Games Increases Energy Expenditure in Children" (*Pediatrics* [2009]: 534–539)** describes an interesting investigation of the possible cardiovascular benefits of active video games. Mean heart rate for healthy boys age 10 to 13 after walking on a treadmill at 2.6 km/hour for 6 minutes is 98 beats per minute (bpm). For each of 14 boys, heart rate was measured after 15 minutes of playing Wii Bowling. The resulting sample mean and standard deviation were 101 bpm and 15 bpm, respectively. For purposes of this exercise, assume that it is reasonable to regard the sample of boys as representative of boys age 10 to 13 and that the distribution of heart rates after 15 minutes of Wii Bowling is approximately normal.

a. Does the sample provide convincing evidence that the mean heart rate after 15 minutes of Wii Bowling is different from the known mean heart rate after 6 minutes walking on the treadmill? Carry out a hypothesis test using $\alpha = .01$.

b. The known resting mean heart rate for boys in this age group is 66 bpm. Is there convincing evidence that the mean heart rate after Wii Bowling for 15 minutes is higher than the known mean resting heart rate for boys of this age? Use $\alpha = .01$. (Hint: See Example 10.15.)

c. Based on the outcomes of the tests in Parts (a) and (b), write a paragraph comparing the benefits of treadmill walking and Wii Bowling in terms of raising heart rate over the resting heart rate.

10.48 A study of fast-food intake is described in the paper **"What People Buy From Fast-Food Restaurants" (*Obesity* [2009]: 1369–1374)**. Adult customers at three hamburger chains (McDonald's, Burger King, and Wendy's) at lunchtime in New York City were approached as they entered the restaurant and asked to provide their receipt when exiting. The receipts were then used to determine what was purchased and the number of calories consumed was determined. In all, 3857 people participated in the study. The sample mean number of calories consumed was 857 and the sample standard deviation was 677.

a. The sample standard deviation is quite large. What does this tell you about number of calories consumed in a hamburger-chain lunchtime fast-food purchase in New York City?

b. Given the values of the sample mean and standard deviation and the fact that the number of calories consumed can't be negative, explain why it is *not* reasonable to assume that the distribution of calories consumed is normal.

c. Based on a recommended daily intake of 2000 calories, the online **Healthy Dining Finder (www.healthydiningfinder.com)** recommends a target of 750 calories for lunch. Assuming that it is reasonable to regard the sample of 3857 fast-food purchases as representative of all hamburger-chain lunchtime purchases in New York City, carry out a hypothesis test to determine if the sample provides convincing evidence that the mean number of calories in a New York City hamburger-chain lunchtime purchase is greater than the lunch recommendation of 750 calories. Use $\alpha = .01$. (Hint: See Example 10.15.)

d. Would it be reasonable to generalize the conclusion of the test in Part (c) to the lunchtime fast-food purchases of all adult Americans? Explain why or why not.

e. Explain why it is better to use the customer receipt to determine what was ordered rather than just asking a customer leaving the restaurant what he or she purchased.

f. Do you think that asking customers before they order to provide a receipt when they leave could have introduced a potential bias? Explain.

10.49 The report **"Highest Paying Jobs for 2009–10 Bachelor's Degree Graduates" (National Association of Colleges and Employers, February 2010)** states that the mean yearly salary offer for students graduating with a degree in accounting in 2010 is \$48,722. Suppose that a random sample of 50 accounting graduates at a large university who received job offers resulted in a mean offer of \$49,850 and a standard deviation of \$3300. Do the sample data provide strong support for the claim that the mean salary offer for accounting graduates of this university is higher than the 2010 national average of \$48,722? Test the relevant hypotheses using $\alpha = .05$.

10.50 ● ***The Economist*** collects data each year on the price of a Big Mac in various countries around the world. The price of a Big Mac for a sample of McDonald's restaurants in Europe in January 2014 resulted in the following Big Mac prices (after conversion to U.S. dollars):

5.18 4.95 4.07 4.68 5.22 4.67
4.14 4.98 5.15 5.56 5.36 4.60

The mean price of a Big Mac in the U.S. in January 2014 was \$4.62. For purposes of this exercise, assume it is reasonable to regard the sample as representative of European McDonald's restaurants. Does the sample provide convincing evidence that the mean January 2014 price of a Big Mac in Europe is greater than the reported U.S. price? Test the relevant hypotheses using $\alpha = .05$.

10.51 A credit bureau analysis of undergraduate students' credit records found that the average number of credit cards in an undergraduate's wallet was 4.09 (**"Undergraduate Students and Credit Cards in 2004," Nellie Mae, May 2005**). It was also reported that in a random sample of 132 undergraduates, the sample mean number of credit cards that the students said they carried was 2.6. The sample standard deviation was not reported, but for purposes of this exercise, suppose that it was 1.2. Is there convincing evidence that the mean number of credit cards that undergraduates report carrying is less than the credit bureau's figure of 4.09? (Hint: See Example 10.14.)

10.52 ● Medical research has shown that repeated wrist extension beyond 20 degrees increases the risk of wrist and hand injuries. Each of 24 students at Cornell University used a proposed new computer mouse design. While using the mouse, each student's wrist extension was recorded. Data consistent with summary values given in the paper **"Comparative Study of Two Computer Mouse Designs" (Cornell Human Factors Laboratory Technical Report RP7992)** are given. Use these data to test the hypothesis that the mean wrist extension for people using this new mouse design is greater than 20 degrees. Are any assumptions required in order for it to be appropriate to generalize the results of your test to the population of Cornell students? To the population of all university students? (Hint: See Example 10.13.)

27 28 24 26 27 25 25 24 24 24 25 28
22 25 24 28 27 26 31 25 28 27 27 25

10.53 The international polling organization Ipsos reported data from a survey of 2000 randomly selected Canadians who carry debit cards (**Canadian Account Habits Survey, July 24, 2006**). Participants in this survey were asked what they considered the minimum purchase amount for which it would be acceptable to use a debit card. Suppose that the sample mean and standard deviation were \$9.15 and \$7.60, respectively. (These values are consistent with a histogram of the sample data that appears in the report.)

Do these data provide convincing evidence that the mean minimum purchase amount for which Canadians consider the use of a debit card to be appropriate is less than \$10? Carry out a hypothesis test with a significance level of .01.

10.54 A comprehensive study conducted by the National Institute of Child Health and Human Development tracked more than 1000 children from an early age through elementary school (***New York Times*, November 1, 2005**). The study concluded that children who spent more than 30 hours a week in child care before entering school tended to score higher in math and reading when they were in the third grade. The researchers cautioned that the findings should not be a cause for alarm because the effects of child care were found to be small.

Explain how the difference between the sample mean math score for third graders who spent long hours in child care and the known overall mean for third graders could be small but the researchers could still reach the conclusion that the mean for the child care group is significantly higher than the overall mean for third graders. (Hint: See discussion of statistical versus practical significance.)

10.55 In a study of computer use, 1000 randomly selected Canadian Internet users were asked how much time they spend using the Internet in a typical week (**Ipsos Reid, August 9, 2005**). The mean of the sample observations was 12.7 hours.

a. The sample standard deviation was not reported, but suppose that it was 5 hours. Carry out a hypothesis test with a significance level of .05 to decide if there is convincing evidence that the mean time spent using the Internet by Canadians is greater than 12.5 hours.

b. Now suppose that the sample standard deviation was 2 hours. Carry out a hypothesis test with a significance level of .05 to decide if there is convincing evidence that the mean time spent using the Internet by Canadians is greater than 12.5 hours.

c. Explain why the null hypothesis was rejected in the test of Part (b) but not in the test of Part (a).

10.56 The paper titled **"Music for Pain Relief" (*The Cochrane Database of Systematic Reviews*, April 19, 2006)** concluded, based on a review of 51 studies of the effect of music on pain intensity, that

> "Listening to music reduces pain intensity levels . . . However, the magnitude of these positive effects is small, the clinical relevance of music for pain relief in clinical practice is unclear."

Are the authors of this paper claiming that the pain reduction attributable to listening to music is not statistically significant, not practically significant, or neither statistically nor practically significant? Explain.

10.57 ● ▼ Many consumers pay careful attention to stated nutritional contents on packaged foods when making purchases. It is therefore important that the information on packages be accurate. A random sample of $n = 12$ frozen dinners of a certain type was selected from production during a particular period, and the calorie content of each one was determined. Here are the resulting observations, along with a boxplot and normal probability plot:

255 244 239 242 265 245 259 248
225 226 251 233

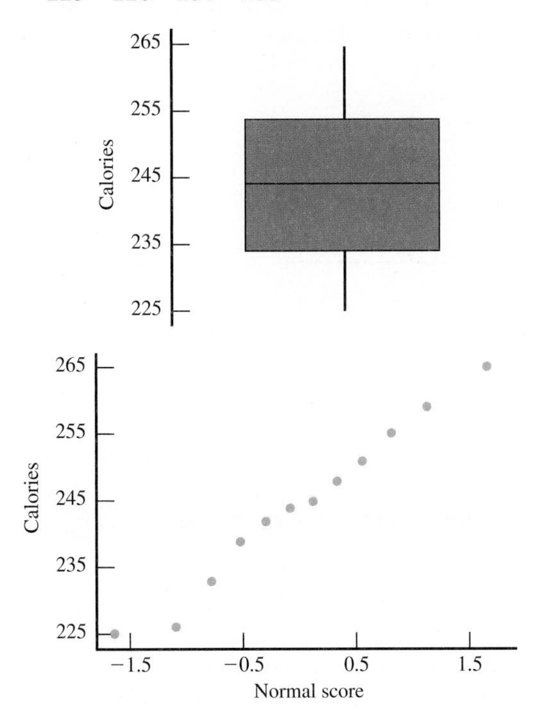

a. Is it reasonable to test hypotheses about mean calorie content μ by using a t test? Explain why or why not.

b. The stated calorie content is 240. Does the boxplot suggest that actual average content differs from the stated value? Explain your reasoning.

c. Carry out a formal test of the hypotheses suggested in Part (b).

10.58 ● Much concern has been expressed regarding the practice of using nitrates as meat preservatives. In one study involving possible effects of these chemicals, bacteria cultures were grown in a medium containing nitrates. The rate of uptake of radio-labeled amino acid (in dpm, disintegrations per minute) was then determined for each culture, yielding the following observations:

7251 6871 9632 6866 9094 5849 8957 7978
7064 7494 7883 8178 7523 8724 7468

Suppose that it is known that the mean rate of uptake for cultures without nitrates is 8000.

Do the data suggest that the addition of nitrates results in a decrease in the mean rate of uptake? Test the appropriate hypotheses using a significance level of .10.

Bold exercises answered in back ● Data set available online ▼ Video Solution available

10.5 Power and Probability of Type II Error

In this chapter, we have introduced test procedures for testing hypotheses about population characteristics, such as μ and p. What characterizes a "good" test procedure? It makes sense to think that a good test procedure is one that has both a small probability of rejecting H_0 when it is true (a Type I error) and a high probability of rejecting H_0 when it is false.

The test procedures presented in this chapter allow us to directly control the probability of rejecting a true H_0 by our choice of the significance level α. But what about the probability of rejecting H_0 when it is false? As we will see, several factors influence this probability. Let's begin by considering an example.

Suppose that the student body president at a university is interested in studying the amount of money that students spend on textbooks each semester. The director of the financial aid office believes that the average amount spent on books is \$500 per semester and uses this figure to determine the amount of financial aid for which a student is eligible. The student body president plans to ask each individual in a random sample of students how much he or she spent on books this semester and has decided to use the resulting data to test

$$H_0: \mu = 500 \quad \text{versus} \quad H_a: \mu > 500$$

using a significance level of .05. If the actual mean is 500 (or less than 500), the correct decision is to fail to reject the null hypothesis. Incorrectly rejecting the null hypothesis is a Type I error. On the other hand, if the actual mean is 525 or 505 or even 501, the correct decision is to reject the null hypothesis. Not rejecting the null hypothesis is a Type II error. How likely is it that the null hypothesis will in fact be rejected?

If the actual mean is 501, the probability that we reject H_0: $\mu = 500$ is not very great. This is because when we carry out the test, we are essentially looking at the sample mean and asking, if it looks like what we would expect to see if the population mean were 500. As illustrated in Figure 10.4, if the actual mean is greater than but very close to 500, chances are that the sample mean will look pretty much like what we would expect to see if the population mean were 500, and we will not be convinced that the null hypothesis should be rejected. If the true mean is 525, it is less likely that the sample will be mistaken for a sample from a population with mean 500 because sample means will tend to cluster around 525. In this case, it is more likely that we will correctly reject H_0. If the actual mean is 550, rejection of H_0 is even more likely.

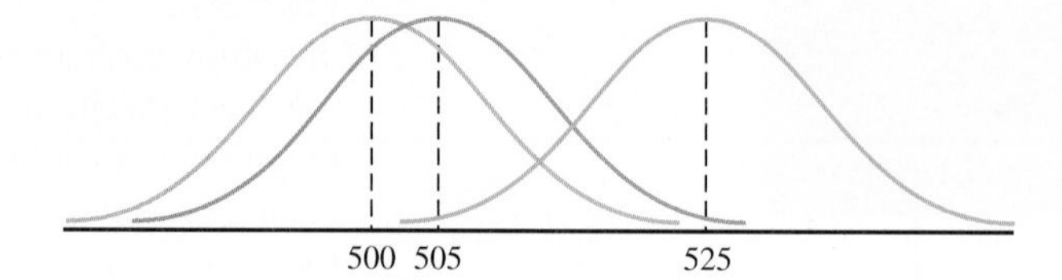

FIGURE 10.4
Sampling distribution of $\bar{x}$ for $\mu = 500, 505, 525$.

When we consider the probability of rejecting the null hypothesis, we are looking at what statisticians refer to as the **power** of the test.

DEFINITION

Power of a test: The probability of rejecting the null hypothesis.

From the previous discussion, it should be apparent that when a hypothesis about a population mean is being tested, the power of the test depends on the true value of the population mean, μ. Because the actual value of μ is unknown (if we knew the value of μ we wouldn't be doing the hypothesis test!), we cannot know what the power is for the actual value of μ. However, it is possible to gain some insight into the power of a test by looking at a number of "what if" scenarios. For example, we might ask, What is the power if the actual mean is 525? or What is the power if the actual mean is 505? and so on. We can determine the power at $\mu = 525$, the power at $\mu = 505$, and the power at any other value of interest. Although it is technically possible to consider power when the null hypothesis is true, an investigator is usually concerned about the power only at values for which the null hypothesis is false.

In general, when testing a hypothesis about a population characteristic, there are three factors that influence the power of the test:

1. The size of the difference between the actual value of the population characteristic and the hypothesized value (the value that appears in the null hypothesis).
2. The choice of significance level, α.
3. The sample size.

Effect of Various Factors on the Power of a Test

1. The larger the size of the discrepancy between the hypothesized value and the actual value of the population characteristic, the greater the power.
2. The larger the significance level, α, the greater the power of the test.
3. The larger the sample size, the greater the power of the test.

Let's consider each of the statements in the box above. The first statement has already been discussed in the context of the textbook example. Because power is the probability of rejecting the null hypothesis, it makes sense that the power will be higher when the actual value of a population characteristic is quite different from the hypothesized value than when it is close to the hypothesized value.

The effect of significance level on power is not quite as obvious. To understand the relationship between power and significance level, it helps to consider the relationship between power and β, the probability of a Type II error.

When H_0 is false, power $= 1 - \beta$.

This relationship follows from the definitions of power and Type II error. A Type II error results from *not* rejecting a false H_0. Because power is the probability of rejecting H_0, it follows that *when H_0 is false*

$$\begin{aligned} \text{power} &= \text{probability of rejecting a false } H_0 \\ &= 1 - \text{probability of not rejecting a false } H_0 \\ &= 1 - \beta \end{aligned}$$

Recall from Section 10.2 that the choice of α, the Type I error probability, affects the value of β, the Type II error probability. Choosing a larger value for α results in a smaller value for β (and therefore a larger value for $1 - \beta$). In terms of power, this means that choosing a larger value for α results in a larger value for the power of the test. The larger the Type I error probability we are willing to tolerate, the more likely it is that the test will be able to detect any particular departure from H_0.

The third factor that affects the power of a test is the sample size. When H_0 is false, the power of a test is the probability that we will "detect" that H_0 is false and, based on the observed sample, reject H_0. Intuition suggests that we will be more likely to detect a departure from H_0 with a large sample than with a small sample. This is in fact the case—the larger the sample size, the higher the power.

For example, consider testing the hypotheses presented previously:

$$H_0\!: \mu = 500 \quad \text{versus} \quad H_a\!: \mu > 500$$

The observations about power imply the following:

1. For any value of μ exceeding 500, the power of a test based on a sample of size 100 is greater than the power of a test based on a sample of size 75 (assuming the same significance level).
2. For any value of μ exceeding 500, the power of a test using a significance level of .05 is greater than the power of a test using a significance level of .01 (assuming the same sample size).
3. For any value of μ exceeding 500, the power of the test is greater if the actual mean is 550 than if the actual mean is 525 (assuming the same sample size and significance level).

As was mentioned previously in this section, it is impossible to calculate the *exact* power of a test because in practice we do not know the values of population characteristics. However, we can evaluate the power at a selected alternative value which would tell us whether the power would be high or low if this alternative value is the actual value.

The following optional subsection shows how Type II error probabilities and power can be evaluated for selected tests.

Calculating Power and Type II Error Probabilities for Selected Tests (Optional)

The test procedures presented in this chapter are designed to control the probability of a Type I error (rejecting H_0 when H_0 is true) at the desired significance level α. However, little has been said so far about calculating the value of β, the probability of a Type II error (not rejecting H_0 when H_0 is false). Here, we consider the calculation of β and power for the hypothesis tests previously introduced.

When we carry out a hypothesis test, we specify the desired value of α, the probability of a Type I error. The probability of a Type II error, β, is the probability of not rejecting H_0 even though it is false. Suppose that we are testing

$$H_0\text{: } \mu = 1.5 \quad \text{versus} \quad H_a\text{: } \mu > 1.5$$

Because we do not know the actual value of μ, we cannot calculate the actual value of β. However, the probability of Type II error can be investigated by calculating β for several different potential values of μ, such as $\mu = 1.55$, $\mu = 1.6$, and $\mu = 1.7$. Once a value of β has been determined, the power of the test at the corresponding alternative value is just $1 - \beta$.

EXAMPLE 10.17 Calculating Power

An airline claims that the mean time on hold for callers to its customer service phone line is 1.5 minutes. We might investigate this claim by testing

$$H_0\text{: } \mu = 1.5 \quad \text{versus} \quad H_a\text{: } \mu > 1.5$$

where μ is the actual mean customer hold time. A random sample of $n = 36$ calls is to be selected, and the resulting data will be used to reach a conclusion.

Suppose that the standard deviation of hold time (σ) is known to be 0.20 minutes and that a significance level of .01 is to be used. Our test statistic is

$$z = \frac{\bar{x} - 1.5}{\frac{.20}{\sqrt{n}}} = \frac{\bar{x} - 1.5}{\frac{.20}{\sqrt{36}}} = \frac{\bar{x} - 1.5}{.0333}$$

The inequality in H_a implies that

$$P\text{-value} = \text{area under } z \text{ curve to the right of calculated } z$$

From Appendix Table 2, it is easily verified that the z critical value 2.33 captures an upper-tail z curve area of .01. This means that P-value $\le .01$ only when $z \ge 2.33$. This is equivalent to the decision rule

$$\text{reject } H_0 \text{ if calculated } z \ge 2.33$$

which becomes

$$\text{reject } H_0 \text{ if } \frac{\bar{x} - 1.5}{.0333} \ge 2.33$$

Solving this inequality for $\bar{x}$ we get

$$\bar{x} \ge 1.5 + 2.33(.0333)$$

or

$$\bar{x} \ge 1.578$$

So if $\bar{x} \ge 1.578$, we will reject H_0, and if $\bar{x} < 1.578$, we will fail to reject H_0. This decision rule corresponds to $\alpha = .01$.

Suppose now that $\mu = 1.6$ (so that H_0 is false). A Type II error will then occur if $\bar{x} < 1.578$. What is the probability that this occurs? If $\mu = 1.6$, the sampling distribution of $\bar{x}$ is approximately normal, centered at 1.6, and has a standard deviation of .0333. The probability of observing an $\bar{x}$ value less than 1.578 can then be determined by finding an area under a normal curve with mean 1.6 and standard deviation .0333, as illustrated in Figure 10.5.

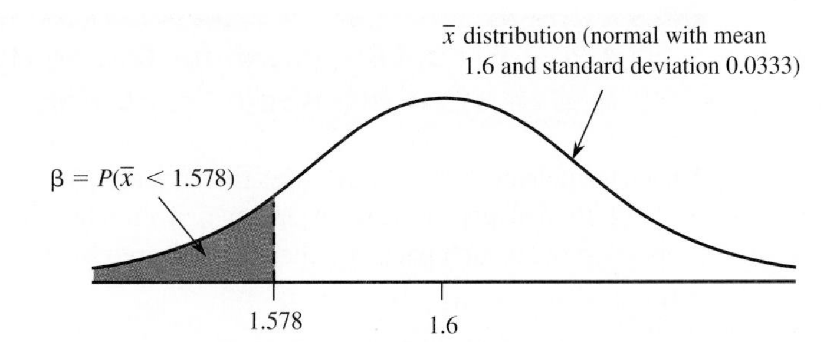

FIGURE 10.5
β for $\mu = 1.6$ in Example 10.17.

Because the curve in Figure 10.5 is not the standard normal (z) curve, we must first convert to a z score before using Appendix Table 2 or technology to find the area. Here,

$$z \text{ score for } 1.578 = \frac{1.578 - \mu_{\bar{x}}}{\sigma_{\bar{x}}} = \frac{1.578 - 1.6}{.0333} = -.66$$

and

area under z curve to left of $-0.66 = .2546$

So, if $\mu = 1.6$, $\beta = .2546$. This means that if μ is 1.6, about 25% of all samples would still result in $\bar{x}$ values less than 1.578, and we would not reject H_0.

The power of the test at $\mu = 1.6$ is then

$$\begin{aligned}(\text{power at } \mu = 1.6) &= 1 - (\beta \text{ when } \mu \text{ is } 1.6)\\ &= 1 - .2546\\ &= .7454\end{aligned}$$

This means that if the actual mean is 1.6, the probability of rejecting H_0: $\mu = 1.5$ in favor of H_a: $\mu > 1.5$ is .7454. If μ is 1.6 and the test is used repeatedly with random samples selected from the population, in the long run about 75% of the samples will result in the correct conclusion to reject H_0.

Now consider β and power when $\mu = 1.65$. The normal curve in Figure 10.5 would then be centered at 1.65. Because β is the area to the left of 1.578 and the curve has shifted to the right, β decreases. Converting 1.578 to a z score and using Appendix Table 2 gives $\beta = .0154$. Also,

$$(\text{power at } \mu = 1.65) = 1 - .0154 = .9846$$

As expected, the power at $\mu = 1.65$ is greater than the power at $\mu = 1.6$ because 1.65 is farther from the hypothesized value of 1.5. ■

Statistical software and graphing calcultors can calculate the power for specified values of σ, α, n, and the difference between the actual and hypothesized values of μ. The following Minitab output shows power calculations corresponding to those in Example 10.17:

```
1-Sample Z Test
Testing mean = null (versus > null)
Alpha = 0.01   Sigma = 0.2   Sample Size = 36
Difference   Power
0.10         0.7497
0.15         0.9851
```

The slight differences between the power values computed by Minitab and those previously obtained are due to rounding in Example 10.17.

The probability of a Type II error and the power for z tests concerning a population proportion are calculated in a similar manner.

EXAMPLE 10.18 Power for Testing Hypotheses About Proportions

A package delivery service advertises that at least 90% of all packages brought to its office by 9 A.M. for delivery in the same city are delivered by noon that day. Let p denote the actual proportion of all such packages that are delivered by noon. The hypotheses of interest are

$$H_0\text{: } p = .9 \quad \text{versus} \quad H_a\text{: } p < .9$$

where the alternative hypothesis states that the company's claim is untrue.

The value $p = .8$ represents a substantial departure from the company's claim. If the hypotheses are tested using significance level .01 and a sample of $n = 225$ packages, what is the probability that the departure from H_0 represented by this alternative value will go undetected?

At significance level .01, H_0 is rejected if P-value $\leq .01$. For the case of a lower-tailed test, this is the same as rejecting H_0 if

$$z = \frac{\hat{p} - \mu_{\hat{p}}}{\sigma_{\hat{p}}} = \frac{\hat{p} - .9}{\sqrt{\frac{(.9)(.1)}{225}}} = \frac{\hat{p} - .9}{.02} \leq -2.33$$

(Because -2.33 captures a lower-tail z curve area of .01, the smallest 1% of all z values satisfy $z \leq -2.33$.) This inequality is equivalent to $\hat{p} \leq .853$, so H_0 is *not* rejected if $\hat{p} > .853$. When $p = .8$, $\hat{p}$ has approximately a normal distribution with

$$\mu_{\hat{p}} = .8$$

$$\sigma_{\hat{p}} = \sqrt{\frac{(.8)(.2)}{225}} = .0267$$

Then β is the probability of obtaining a sample proportion greater than .853, as illustrated in Figure 10.6.

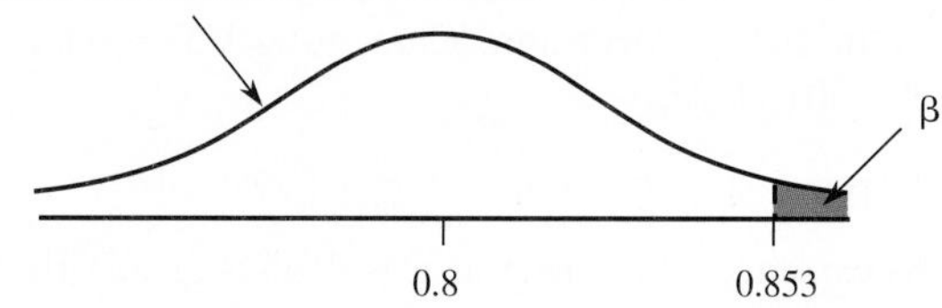

FIGURE 10.6
β for $p = .8$ in Example 10.18.

Converting to a z score results in

$$z = \frac{.853 - .8}{.0267} = 1.99$$

and Appendix Table 2 gives

$$\beta = 1 - .9767 = .0233$$

When $p = .8$ and a significance level of .01 is used, less than 3% of all samples of size $n = 225$ will result in a Type II error. The power of the test at $p = .8$ is $1 - .0233 = .9767$. This means that the probability of rejecting H_0: $p = .9$ in favor of H_a: $p < .9$ when p is really .8 is .9767, which is quite high. ■

β and Power for the t Test (Optional)

The power and β values for t tests can be approximated by using a set of curves specially constructed for this purpose or by using appropriate software. As with the z test, the value of β depends not only on the actual value of μ but also on the selected significance level α. β increases as α is made smaller. In addition, β depends on the number of degrees of freedom, $n - 1$. For any fixed significance level α, it should be easier for the test to detect a specific departure from H_0 when n is large than when n is small. For a fixed alternative value, β decreases as $n - 1$ increases.

Unfortunately, there is one other quantity on which β depends: the population standard deviation σ. As σ increases, so does $\sigma_{\bar{x}}$. This in turn makes it more likely that an $\bar{x}$ value far from μ will be observed just by chance, resulting in an incorrect conclusion. Once α is specified and n is fixed, the determination of β at a particular alternative value of μ requires that a value of σ be chosen, because each different value of σ yields a different value of β. (This did not present a problem with the z test because when using a z test, the value of σ is known.) If the investigator can specify a range of plausible values for σ, then using the largest such value will give a pessimistic β (one on the high side) and a pessimistic value of power (one on the low side).

Figure 10.7 shows three different β curves for a one-tailed t test (appropriate for H_a: $\mu >$ hypothesized value or for H_a: $\mu <$ hypothesized value). A more complete set of curves for both one- and two-tailed tests when $\alpha = .05$ and when $\alpha = .01$ appears in Appendix Table 5. To determine β, first compute the quantity

$$d = \frac{|\text{alternative value} - \text{hypothesized value}|}{\sigma}$$

Then locate d on the horizontal axis, move directly up to the curve for $n - 1$ df, and move over to the vertical axis to find β.

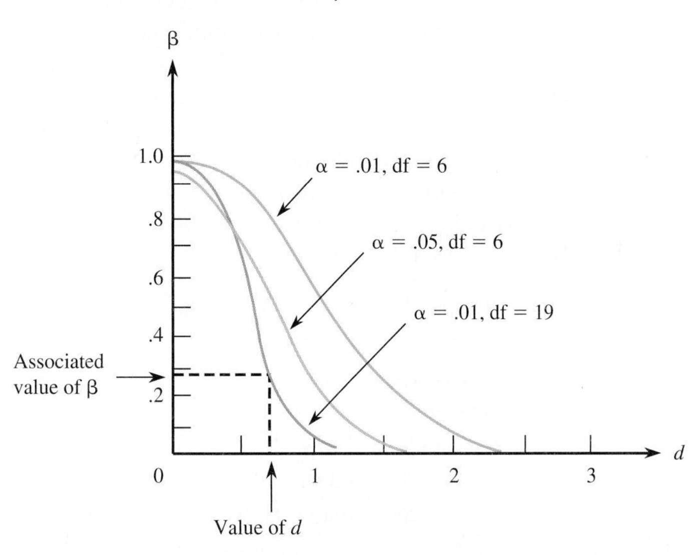

FIGURE 10.7
β curves for the one-tailed t test.

EXAMPLE 10.19 β and Power for t Tests

Consider testing

H_0: $\mu = 100$ versus H_a: $\mu > 100$

and focus on the alternative value $\mu = 110$. Suppose that $\sigma = 10$, the sample size is $n = 7$, and a significance level of .01 has been selected. For $\sigma = 10$,

$$d = \frac{|110 - 100|}{10} = \frac{10}{10} = 1$$

Figure 10.7 (using df $= 7 - 1 = 6$) gives $\beta \approx .6$. This means that if $\sigma = 10$, the significance level is .01 and $n = 7$, when $\mu = 110$, roughly 60% of all samples result in an incorrect decision to not reject H_0! Equivalently, the power of the test at $\mu = 110$ is only $1 - .6 = .4$. The probability of rejecting H_0 when $\mu = 110$ is not very large.

If a .05 significance level is used instead, then $\beta \approx .3$, which is still rather large. Using a .01 significance level with $n = 20$ (df = 19) yields, from Figure 10.7, $\beta \approx .05$. At the alternative value $\mu = 110$, for $\sigma = 10$, a significance level of .01, and $n = 20$, β is smaller than for a significance level of .05 with $n = 7$. Substantially increasing n counterbalances using the smaller α.

Now consider the alternative $\mu = 105$, again with $\sigma = 10$, so that

$$d = \frac{|105 - 100|}{10} = \frac{5}{10} = .5$$

Then, from Figure 10.7, $\beta = .95$ when $\alpha = .01$, $n = 7$; $\beta = .7$ when $\alpha = .05$, $n = 7$; and $\beta = .65$ when $\alpha = .01$, $n = 20$. These values of β are all quite large. With $\sigma = 10$, $\mu = 105$ is too close to the hypothesized value of 100 for any of these three tests to have a good chance of detecting such a departure from H_0. A substantial decrease in β would require using a much larger sample size. For example, from Appendix Table 5, $\beta = .08$ when $\alpha = .05$ and $n = 40$.

The curves in Figure 10.7 also give β when testing H_0: $\mu = 100$ versus H_a: $\mu < 100$. If the alternative value $\mu = 90$ is of interest and $\sigma = 10$,

$$d = \frac{|90 - 100|}{10} = \frac{10}{10} = 1$$

and values of β are the same as those given in the first paragraph of this example. ■

Because curves for only selected degrees of freedom appear in Appendix Table 5, other degrees of freedom require a visual approximation. For example, the 27-df curve (for $n = 28$) lies between the 19-df and 29-df curves, which do appear, and it is closer to the 29-df curve. This type of approximation is adequate because it is the general magnitude of β—large, small, or moderate—that is of primary concern.

Minitab can also evaluate power for the t test. For example, the following output shows Minitab calculations for power at $\mu = 110$ for samples of size 7 and 20 when $\alpha = .01$. The corresponding approximate values from Appendix Table 5 found in Example 10.19 are fairly close to the Minitab values.

```
1-Sample t Test
Testing mean = null (versus > null)
Calculating power for mean = null + 10
Alpha = 0.01      Sigma = 10
 Sample Size      Power
           7      0.3968
          20      0.9653
```

The β curves in Appendix Table 5 are those for t tests. When the alternative value in H_a corresponds to a value of d relatively close to 0, β for a t test may be rather large. One might wonder whether there is another type of test that has the same level of significance α as does the t test and smaller values of β. The following result provides the answer to this question.

When the population distribution is normal, the t test for testing hypotheses about μ has smaller β than does any other test procedure that has the same level of significance α.

Stated another way, among all tests with level of significance α, the t test makes β as small as it can possibly be when the population distribution is normal. In this sense, the t test is a best test. Statisticians have also shown that when the population distribution is not too far from a normal distribution, no test procedure can improve on the t test by very much (i.e., no test procedure can have the same α and substantially smaller β).

However, when the population distribution is believed to be strongly nonnormal (heavy-tailed, highly skewed, or multimodal), the t test should not be used. Then it's time to consult your friendly neighborhood statistician, who can provide you with alternative methods of analysis.

EXERCISES 10.59 - 10.65

10.59 The power of a test is influenced by the sample size and the choice of significance level.

a. Explain how increasing the sample size affects the power (when significance level is held fixed).

b. Explain how increasing the significance level affects the power (when sample size is held fixed).

10.60 Water specimens are taken from water used for cooling as it is being discharged from a power plant into a river. It has been determined that as long as the mean temperature of the discharged water is at most 150°F, there will be no negative effects on the river's ecosystem. To investigate whether the plant is in compliance with regulations that prohibit a mean discharge water temperature above 150°F, a scientist will take 50 water specimens at randomly selected times and will record the water temperature of each specimen. She will then use a z statistic

$$z = \frac{\bar{x} - 150}{\frac{\sigma}{\sqrt{n}}}$$

to decide between the hypotheses H_0: $\mu = 150$ and H_a: $\mu > 150$, where μ is the mean temperature of discharged water. Assume that σ is known to be 10.

a. Explain why use of the z statistic is appropriate in this setting.

b. Describe Type I and Type II errors in this context.

c. The rejection of H_0 when $z \geq 1.8$ corresponds to what value of α? (That is, what is the area under the z curve to the right of 1.8?)

d. Suppose that the actual value for μ is 153 and that H_0 is to be rejected if $z \geq 1.8$. Draw a sketch (similar to that of Figure 10.5) of the sampling distribution of $\bar{x}$, and shade the region that would represent β, the probability of making a Type II error.

e. For the hypotheses and test procedure described, compute the value of β when $\mu = 153$. (Hint: See Example 10.17.)

f. For the hypotheses and test procedure described, what is the value of β if $\mu = 160$?

g. What would be the conclusion of the test if H_0 is rejected when $z \geq 1.8$ and $\bar{x} = 152.4$? What type of error might have been made in reaching this conclusion?

10.61 ▼ Let μ denote the true average lifetime (in hours) for a certain type of battery under controlled laboratory conditions. A test of H_0: $\mu = 10$ versus H_a: $\mu < 10$ will be based on a sample of size 36. Suppose that σ is known to be 0.6, from which $\sigma_{\bar{x}} = .1$. The appropriate test statistic is then

$$z = \frac{\bar{x} - 10}{0.1}$$

a. What is α for the test procedure that rejects H_0 if $z \leq -1.28$?

b. If the test procedure of Part (a) is used, calculate β when $\mu = 9.8$, and interpret this error probability.

c. Without doing any calculation, explain how β when $\mu = 9.5$ compares to β when $\mu = 9.8$. Then check your assertion by computing β when $\mu = 9.5$.

d. What is the power of the test when $\mu = 9.8$? when $\mu = 9.5$?

10.62 The city council in a large city has become concerned about the trend toward exclusion of renters with children in apartments within the city. The housing coordinator has decided to select a random sample of 125 apartments and determine for each whether children are permitted. Let p be the proportion of all apartments that prohibit children. If the city council is convinced that p is greater than 0.75, it will consider appropriate legislation.

a. If 102 of the 125 sampled apartments exclude renters with children, would a level .05 test lead you to the conclusion that more than 75% of all apartments exclude children?

b. What is the power of the test when $p = .8$ and $\alpha = .05$? (Hint: See Example 10.18.)

10.63 The amount of shaft wear after a fixed mileage was determined for each of seven randomly selected internal combustion engines, resulting in a mean of 0.0372 inch and a standard deviation of 0.0125 inch.

a. Assuming that the distribution of shaft wear is normal, Use $\alpha = .05$ to test the hypotheses H_0: $\mu = .035$ versus H_a: $\mu > .035$.

b. Using $\sigma = 0.0125$, $\alpha = .05$, and Appendix Table 5, what is the approximate value of β, the probability of a Type II error, when $\mu = .04$? (Hint: See Example 10.19.)

c. What is the approximate power of the test when $\mu = .04$ and $\alpha = .05$?

10.64 Optical fibers are used in telecommunications to transmit light. Suppose current technology allows production of fibers that transmit light about 50 km. Researchers are trying to develop a new type of glass fiber that will increase this distance. In evaluating a new fiber, it is of interest to test H_0: $\mu = 50$ versus H_a: $\mu > 50$, with μ denoting the mean transmission distance for the new optical fiber.

a. Assuming $\sigma = 10$ and $n = 10$, use Appendix Table 5 to approximate β, the probability of a Type II error, for each of the given alternative values of μ when a test with significance level .05 is employed:

i. 52 ii. 55 iii. 60 iv. 70

b. What happens to β in each of the cases in Part (a) if σ is actually larger than 10? Explain your reasoning.

10.65 Let μ denote the mean diameter for bearings of a certain type. A test of H_0: $\mu = 0.5$ versus H_a: $\mu \neq 0.5$ will be based on a sample of n bearings. The diameter distribution is believed to be normal. Determine the value of β in each of the following cases:

a. $n = 15, \alpha = .05, \sigma = 0.02, \mu = 0.52$

b. $n = 15, \alpha = .05, \sigma = 0.02, \mu = 0.48$

c. $n = 15, \alpha = .01, \sigma = 0.02, \mu = 0.52$

d. $n = 15, \alpha = .05, \sigma = 0.02, \mu = 0.54$

e. $n = 15, \alpha = .05, \sigma = 0.04, \mu = 0.54$

f. $n = 20, \alpha = .05, \sigma = 0.04, \mu = 0.54$

g. Is the way in which β changes as n, α, σ, and μ vary consistent with your intuition? Explain.

Bold exercises answered in back ● Data set available online ▼ Video Solution available

10.6 Interpreting and Communicating the Results of Statistical Analyses

The nine-step procedure that we have proposed for testing hypotheses provides a systematic approach for carrying out a complete test. However, you rarely see the results of a hypothesis test reported in publications in such a complete way.

Communicating the Results of Statistical Analyses

When summarizing the results of a hypothesis test, it is important that you include several things in the summary in order to provide all the relevant information. These are:

1. *Hypotheses.* Whether specified in symbols or described in words, it is important that both the null and the alternative hypotheses be clearly stated. If you are using symbols to define the hypotheses, be sure to describe them in the context of the problem at hand (for example, μ = population mean calorie intake).
2. *Test procedure.* You should be clear about what test procedure was used (for example, large-sample z test for proportions) and why you think it was reasonable to use this procedure. The plausibility of any required assumptions should be satisfactorily addressed.
3. *Test statistic.* Be sure to include the value of the test statistic and the P-value. Including the P-value allows a reader who may have chosen a different significance level to see whether she would have reached the same or a different conclusion.
4. *Conclusion in context.* Never end the report of a hypothesis test with the statement "I rejected (or did not reject) H_0." Always provide a conclusion that is in the

context of the problem and that answers the original research question which the hypothesis test was designed to answer. Be sure also to indicate the level of significance used as a basis for the decision.

Interpreting the Results of Statistical Analyses

When the results of a hypothesis test are reported in a journal article or other published source, it is common to find only the value of the test statistic and the associated *P*-value accompanying the discussion of conclusions drawn from the data. Often, especially in newspaper articles, only sample summary statistics are given, with the conclusion immediately following. You may have to fill in some of the intermediate steps for yourself to see whether or not the conclusion is justified.

For example, the article **"Physicians' Knowledge of Herbal Toxicities and Adverse Herb-Drug Interactions" (*European Journal of Emergency Medicine,* August 2004)** summarizes the results of a study to assess doctors' familiarity with adverse effects of herbal remedies as follows:

> "A total of 142 surveys and quizzes were completed by 59 attending physicians, 57 resident physicians, and 26 medical students. The mean subject score on the quiz was only slightly higher than would have occurred from random guessing."

The quiz consisted of 16 multiple-choice questions. If each question had four possible choices, the statement that the mean quiz score was only slightly higher than would have occurred from random guessing suggests that the researchers considered the hypotheses H_0: $\mu = 4$ and H_a: $\mu > 4$, where μ represents the mean score for the population of all physicians and medical students and the null hypothesis corresponds to the expected number of correct choices for someone who is guessing. Assuming that it is reasonable to regard this sample as representative of the population of interest, the data from the sample could be used to carry out a test of these hypotheses.

What to Look For in Published Data

Here are some questions to consider when you are reading a report that contains the results of a hypothesis test:

- What hypotheses are being tested? Are the hypotheses about a population mean, a population proportion, or some other population characteristic?
- Was the appropriate test used? Does the validity of the test depend on any assumptions about the sample or about the population from which the sample was selected? If so, are the assumptions reasonable?
- What is the *P*-value associated with the test? Was a significance level reported (as opposed to simply reporting the *P*-value)? Is the chosen significance level reasonable?
- Are the conclusions drawn consistent with the results of the hypothesis test?

For example, consider the following statement from the paper **"Didgeridoo Playing as Alternative Treatment for Obstructive Sleep Apnoea Syndrome" (*British Medical Journal* [2006]: 266–270)**: "We found that four months of training of the upper airways by didgeridoo playing reduces daytime sleepiness in people with snoring and obstructive apnoea syndrome." This statement was supported by data on a measure of daytime sleepiness called the Epworth scale. For the 14 participants in the study, the mean improvement in Epworth scale was 4.4 and the standard deviation was 3.7.

The paper does not indicate what test was performed or what the value of the test statistic was. It appears that the hypotheses of interest are H_0: $\mu = 0$ (no improvement) versus H_a: $\mu > 0$, where μ represents the mean improvement in Epworth score after four months of didgeridoo playing for all people with snoring and obstructive sleep apnoea. Because the sample size is not large, the one-sample t test would be appropriate if the sample can be considered a random sample and the distribution of Epworth scale improvement scores is approximately normal. If these assumptions are reasonable (something that was not addressed in the paper), the t test results in $t = 4.45$ and an associated P-value of .000. Because the reported P-value is so small H_0 would be rejected, supporting the conclusion in the paper that didgeridoo playing is an effective treatment. (In case you are wondering, a didgeridoo is an Australian Aboriginal woodwind instrument.)

A Word to the Wise: Cautions and Limitations

There are several things you should watch for when conducting a hypothesis test or when evaluating a written summary of a hypothesis test.

1. The result of a hypothesis test can never show strong support for the null hypothesis. Make sure that you don't confuse "There is no reason to believe the null hypothesis is not true" with the statement "There is convincing evidence that the null hypothesis is true." These are very different statements!
2. If you have complete information for the population, don't carry out a hypothesis test! It should be obvious that no test is needed to answer questions about a population if you have complete information and don't need to generalize from a sample, but people sometimes forget this fact. For example, in an article on growth in the number of prisoners by state, the ***San Luis Obispo Tribune* (August 13, 2001)** reported "California's numbers showed a statistically insignificant change, with 66 fewer prisoners at the end of 2000." The use of the term "statistically insignificant" implies some sort of statistical inference, which is not appropriate when a complete accounting of the entire prison population is known. Perhaps the author confused statistical and practical significance. Which brings us to . . .
3. Don't confuse statistical significance with practical significance. When statistical significance has been declared, be sure to step back and evaluate the result in light of its practical importance. For example, we may be convinced that the proportion who respond favorably to a proposed medical treatment is greater than .4, the known proportion that responds favorably for the currently recommended treatments. But if our estimate of this proportion for the proposed treatment is .405, is this of any practical interest? It might be if the proposed treatment is less costly or has fewer side effects, but in other cases it may not be of any real interest. Results must always be interpreted in context.

EXERCISES 10.66 - 10.67

10.66 In 2006, Boston Scientific sought approval for a new heart stent (a medical device used to open clogged arteries) called the Liberte. This stent was being proposed as an alternative to a stent called the Express that was already on the market. The following excerpt is from an article that appeared in ***The Wall Street Journal* (August 14, 2008)**:

> Boston Scientific wasn't required to prove that the Liberte was 'superior' than a previous treatment, the agency decided—only that it wasn't 'inferior' to Express. Boston Scientific proposed—and the FDA okayed—a benchmark in which Liberte could be up to three percentage points worse than Express—meaning that if 6% of Express patients' arteries reclog, Boston Scientific would have to prove that Liberte's rate of reclogging was less than 9%. Anything more would be considered 'inferior.' . . . In the end, after nine months, the Atlas study found that 85 of the patients suffered reclogging. In comparison, historical data on 991 patients implanted with the Express stent show a 7% rate. Boston Scientific then had to answer this question: Could the study have gotten such results if the Liberte were truly inferior to Express?"

Assume a 7% reclogging rate for the Express stent. Explain why it would be appropriate for Boston

Scientific to carry out a hypothesis test using the following hypotheses:

H_0: $p = .10$

H_a: $p < .10$

where p is the proportion of patients receiving Liberte stents that suffer reclogging. Be sure to address both the choice of the hypothesized value and the form of the alternative hypothesis in your explanation.

10.67 The article **"Boy or Girl: Which Gender Baby Would You Pick?" (*LiveScience*, March 23, 2005, www.livescience.com)** summarized the findings of a study that was published in ***Fertility and Sterility***. The *LiveScience* article makes the following statements: "When given the opportunity to choose the sex of their baby, women are just as likely to choose pink socks as blue, a new study shows" and "Of the 561 women who participated in the study, 229 said they would like to choose the sex of a future child. Among these 229, there was no greater demand for boys or girls." These statements are equivalent to the claim that for women who would like to choose the baby's sex, the proportion who would choose a girl is 0.50 or 50%.

a. The journal article on which the *LiveScience* summary was based **("Preimplantation Sex-Selection Demand and Preferences in an Infertility Population," *Fertility and Sterility* [2005]: 649–658)** states that of the 229 women who wanted to select the baby's sex, 89 wanted a boy and 140 wanted a girl. Does this provide convincing evidence against the statement of no preference in the *LiveScience* summary? Test the relevant hypotheses using $\alpha = .05$. Be sure to state any assumptions you must make about the way the sample was selected in order for your test to be appropriate.

b. The journal article also provided the following information about the study:

- A survey with 19 questions was mailed to 1385 women who had visited the Center for Reproductive Medicine at Brigham and Women's Hospital.
- 561 women returned the survey.

Do you think it is reasonable to generalize the results from this survey to a larger population? Do you have any concerns about the way the sample was selected or about potential sources of bias? Explain.

Bold exercises answered in back ● Data set available online ▼ Video Solution available

ACTIVITY 10.1 Comparing the *t* and *z* Distributions

Technology Activity: Requires use of a computer or a graphing calculator.

The instructions that follow assume the use of Minitab. If you are using a different software package or a graphing calculator, your instructor will provide alternative instructions.

Background: Suppose a random sample will be selected from a population that is known to have a normal distribution. Then the statistic

$$z = \frac{\bar{x} - \mu}{\frac{\sigma}{\sqrt{n}}}$$

has a standard normal (z) distribution. Since it is rarely the case that σ is known, inferences for population means are usually based on the statistic $t = \dfrac{\bar{x} - \mu}{(s/\sqrt{n})}$, which has a t distribution rather than a z distribution. The informal justification for this was that the use of s to estimate σ introduces additional variability, resulting in a statistic whose distribution is more spread out than is the z distribution.

In this activity, you will use simulation to sample from a known normal population and then investigate how the behavior of $t = \dfrac{\bar{x} - \mu}{s/(\sqrt{n})}$ compares with the behavior of $z = \dfrac{\bar{x} - \mu}{\sigma/(\sqrt{n})}$.

1. Generate 200 random samples of size 5 from a normal population with mean 100 and standard deviation 10.

   Using Minitab, go to the Calc Menu. Then

```
Calc→ Random Data → Normal
In the "Generate" box, enter 200
In the "Store in columns" box, enter c1-c5
In the mean box, enter 100
In the standard deviation box, enter 10
Click on OK
```

   You should now see 200 rows of data in each of the first 5 columns of the Minitab worksheet.

2. Each row contains five values that have been randomly selected from a normal population with mean 100 and standard deviation 10. Viewing each row as a sample of size 5 from this population, calculate the mean and standard deviation for each of the

200 samples (the 200 rows) by using Minitab's row statistics functions, which can also be found under the Calc menu:

Calc→ Row statistics
Choose the "Mean" button
In the "Input Variables" box, enter c1-c5
In the "Store result in" box, enter c7
Click on OK

You should now see the 200 sample means in column 7 of the Minitab worksheet. Name this column "x-bar" by typing the name in the gray box at the top of c7.

Now follow a similar process to compute the 200 sample standard deviations, and store them in c8. Name c8 "s."

3. Next, calculate the value of the z statistic for each of the 200 samples. We can calculate z in this example because we know that the samples were selected from a population for which $\sigma = 10$. Use the calculator function of Minitab to

compute $z = \dfrac{\bar{x} - \mu}{(\sigma/\sqrt{n})} = \dfrac{\bar{x} - 100}{(10/\sqrt{5})}$ as follows:

Calc → Calculator
In the "Store results in" box, enter c10
In the "Expression box" type in the following: (c7-100)/(10/sqrt(5))
Click on OK

You should now see the z values for the 200 samples in c11. Name c11 "z."

4. Now calculate the value of the t statistic for each of the 200 samples. Use the calculator function of Minitab to compute $t = \dfrac{\bar{x} - \mu}{(s/\sqrt{n})} = \dfrac{\bar{x} - 100}{(s/\sqrt{5})}$ as follows:

Calc → Calculator
In the "Store results in" box, enter c11
In the "Expression box" type in the following: (c7-100)/(c8/sqrt(5))
Click on OK

You should now see the t values for the 200 samples in c10. Name c10 "t."

5. Graphs, at last! Now construct histograms of the 200 z values and the 200 t values. These two graphical displays will provide insight about how each of these two statistics behaves in repeated sampling. Use the same scale for the two histograms so that it will be easier to compare the two distributions.

Graph → Histogram
In the "Graph variables" box, enter c10 for graph 1 and c11 for graph 2
Click the Frame dropdown menu and select multiple graphs. Then under the scale choices, select "Same X and same Y."

6. Now use the histograms from Step 5 to answer the following questions:
   a. Write a brief description of the shape, center, and spread for the histogram of the z values. Is what you see in the histogram consistent with what you would have expected to see? Explain. (Hint: In theory, what is the distribution of the z statistic?)
   b. How does the histogram of the t values compare to the z histogram? Be sure to comment on center, shape, and spread.
   c. Is your answer to Part (b) consistent with what would be expected for a statistic that has a t distribution? Explain.
   d. The z and t histograms are based on only 200 samples, and they only approximate the corresponding sampling distributions. The 5th percentile for the standard normal distribution is -1.645 and the 95th percentile is $+1.645$. For a t distribution with df $= 5 - 1 = 4$, the 5th and 95th percentiles are -2.13 and $+2.13$, respectively. How do these percentiles compare to those of the distributions displayed in the histograms? (Hint: Sort the 200 z values—in Minitab, choose "Sort" from the Manip menu. Once the values are sorted, percentiles from the histogram can be found by counting in 10 [which is 5% of 200] values from either end of the sorted list. Then repeat this with the t values.)
   e. Are the results of your simulation and analysis consistent with the statement that the statistic $z = \dfrac{\bar{x} - \mu}{(\sigma/\sqrt{n})}$ has a standard normal (z) distribution and the statistic $t = \dfrac{\bar{x} - \mu}{(s/\sqrt{n})}$ has a t distribution? Explain.

ACTIVITY 10.2 A Meaningful Paragraph

Write a meaningful paragraph that includes the following six terms: **hypotheses, P-value, reject H_0, Type I error, statistical significance, practical significance.**

A "meaningful paragraph" is a coherent piece of writing in an appropriate context that uses all of the listed words. The paragraph should show that you understand the meaning of the terms and their relationship to one another. A sequence of sentences that just define the terms is *not* a meaningful paragraph. When choosing a context, think carefully about the terms you need to use. Choosing a good context will make writing a meaningful paragraph easier.

SUMMARY Key Concepts and Formulas

TERM OR FORMULA	COMMENT
Hypothesis	A claim about the value of a population characteristic.
Null hypothesis, H_0	The hypothesis initially assumed to be true. It has the form H_0: population characteristic = hypothesized value.
Alternative hypothesis, H_a	A hypothesis that specifies a claim that is contradictory to H_0 and is judged the more plausible claim when H_0 is rejected.
Type I error	Rejecting H_0 when H_0 is true. The probability of a Type I error is denoted by α and is referred to as the significance level for the test.
Type II error	Not rejecting H_0 when H_0 is false. The probability of a Type II error is denoted by β.
Test statistic	A value computed from sample data that is then used as the basis for making a decision between H_0 and H_a.
P-value	The probability, computed assuming H_0 is true, of obtaining a value of the test statistic at least as contradictory to H_0 as what actually resulted. H_0 is rejected if P-value $\leq \alpha$ and not rejected if P-value $> \alpha$, where α is the chosen significance level.
$z = \dfrac{\hat{p} - \text{hypothesized value}}{\sqrt{\dfrac{(\text{hyp. val})(1 - \text{hyp. val})}{n}}}$	A test statistic for testing H_0: p = hypothesized value when the sample size is large. The P-value is determined as an area under the z curve.
$z = \dfrac{\bar{x} - \text{hypothesized value}}{\dfrac{\sigma}{\sqrt{n}}}$	A test statistic for testing H_0: μ = hypothesized value when σ is known and either the population distribution is normal or the sample size is large. The P-value is determined as an area under the z curve.
$t = \dfrac{\bar{x} - \text{hypothesized value}}{\dfrac{s}{\sqrt{n}}}$	A test statistic for testing H_0: μ = hypothesized value when σ is unknown and either the population distribution is normal or the sample size is large. The P-value is determined from the t curve with df = $n - 1$.
Power	The power of a test is the probability of rejecting the null hypothesis. Power is affected by the size of the difference between the hypothesized value and the actual value, the sample size, and the significance level.

CHAPTER REVIEW Exercises 10.68 - 10.82

10.68 In a representative sample of 1000 adult Americans, only 430 could name at least one justice who is currently serving on the U.S. Supreme Court **(Ipsos, January 10, 2006)**. Using a significance level of .01, carry out a hypothesis test to determine if there is convincing evidence to support the claim that fewer than half of adult Americans can name at least one justice currently serving on the Supreme Court.

10.69 ▼ In a national survey of 2013 adults, 1590 responded that lack of respect and courtesy in American society is a serious problem, and 1283 indicated that they believe that rudeness is a more serious problem than in past years **(Associated Press, April 3, 2002)**. Is there convincing evidence that less than three-quarters of U.S. adults believe that rudeness is a worsening problem? Test the relevant hypotheses using a significance level of .05.

10.70 Students at the Akademia Podlaka conducted an experiment to determine whether the Belgium-minted Euro coin was equally likely to land heads up or tails up. Coins were spun on a smooth surface, and in 250 spins, 140 landed with the heads side up **(*New Scientist*, January 4, 2002)**.

a. Should the students interpret this result as convincing evidence that the proportion of the time the coin would land heads up is not .5? Test the relevant hypotheses using $\alpha = .01$.

b. Would your conclusion be different if a significance level of .05 had been used? Explain.

10.71 An article titled **"Teen Boys Forget Whatever It Was"** appeared in the Australian newspaper ***The Mercury* (April 21, 1997)**. It described a study of academic performance and attention span and reported

Bold exercises answered in back ● Data set available online ▼ Video Solution available

that the mean time to distraction for teenage boys working on an independent task was 4 minutes. Although the sample size was not given in the article, suppose that this mean was based on a random sample of 50 teenage Australian boys and that the sample standard deviation was 1.4 minutes.

Is there convincing evidence that the average attention span for teenage boys is less than 5 minutes? Test the relevant hypotheses using $\alpha = .01$.

10.72 The authors of the article **"Perceived Risks of Heart Disease and Cancer Among Cigarette Smokers" (*Journal of the American Medical Association* [1999]: 1019–1021)** expressed the concern that a majority of smokers do not view themselves as being at increased risk of heart disease or cancer. A study of 737 current smokers selected at random from U.S. households with telephones found that of the 737 smokers surveyed, 295 indicated that they believed they have a higher than average risk of cancer.

Do these data suggest that p, the proportion of all smokers who view themselves as being at increased risk of cancer is in fact less than .5, as claimed by the authors of the paper? Test the relevant hypotheses using $\alpha = .05$.

10.73 A number of initiatives on the topic of legalized gambling have appeared on state ballots. A political candidate has decided to support legalization of casino gambling if he is convinced that more than two-thirds of U.S. adults approve of casino gambling. Suppose that 1523 adults selected at random from households with telephones were asked whether they approved of casino gambling. The number in the sample who approved was 1035. Does the sample provide convincing evidence that more than two-thirds of U.S. adults approve?

10.74 Although arsenic is known to be a poison, it also has some beneficial medicinal uses. In one study of the use of arsenic to treat acute promyelocytic leukemia (APL), a rare type of blood cell cancer, APL patients were given an arsenic compound as part of their treatment. Of those receiving arsenic, 42% were in remission and showed no signs of leukemia in a subsequent examination **(*Washington Post*, November 5, 1998)**. It is known that 15% of APL patients go into remission after the conventional treatment.

Suppose that the study had included 100 randomly selected patients (the actual number in the study was much smaller). Is there sufficient evidence to conclude that the proportion in remission for the arsenic treatment is greater than .15, the remission proportion for the conventional treatment? Test the relevant hypotheses using a .01 significance level.

10.75 Many people have misconceptions about how profitable small, consistent investments can be. In a survey of 1010 randomly selected U.S. adults **(Associated Press, October 29, 1999)**, only 374 responded that they thought that an investment of $25 per week over 40 years with a 7% annual return would result in a sum of over $100,000 (the correct amount is $286,640). Is there sufficient evidence to conclude that less than 40% of U.S. adults are aware that such an investment would result in a sum of over $100,000? Test the relevant hypotheses using $\alpha = .05$.

10.76 The same survey described in the previous exercise also asked the individuals in the sample what they thought was their best chance to obtain more than $500,000 in their lifetime. Twenty-eight percent responded "win a lottery or sweepstakes." Does this provide convincing evidence that more than one-fourth of U.S. adults see a lottery or sweepstakes win as their best chance of accumulating $500,000? Carry out a test using a significance level of .01.

10.77 Speed, size, and strength are thought to be important factors in football performance. The article **"Physical and Performance Characteristics of NCAA Division I Football Players" (*Research Quarterly for Exercise and Sport* [1990]: 395–401)** reported on physical characteristics of Division I starting football players in the 1988 football season. Information for teams ranked in the top 20 was obtained, and it was reported that the mean weight of starters on top-20 teams was 105 kg. A random sample of 33 starting players (various positions were represented) from Division I teams that were not ranked in the top 20 resulted in a sample mean weight of 103.3 kg and a sample standard deviation of 16.3 kg. Is there sufficient evidence to conclude that the mean weight for non-top-20 starters is less than 105, the known value for top-20 teams?

10.78 Duck hunting in populated areas faces opposition on the basis of safety and environmental issues. In a survey to assess public opinion regarding duck hunting on Morro Bay (located along the central coast of California), a random sample of 750 local residents included 560 who strongly opposed hunting on the bay. Does this sample provide sufficient evidence to conclude that the majority of local residents oppose hunting on Morro Bay? Test the relevant hypotheses using $\alpha = .01$.

10.79 Past experience has indicated that the true response rate is 40% when individuals are approached with a request to fill out a questionnaire and return it in a stamped and addressed envelope. An investigator believes that if the person distributing the questionnaire is stigmatized in some obvious way, potential respondents would feel sorry for the distributor and thus tend to respond at a rate higher than 40%. To investigate this theory, a distributor is fitted with an eye patch. Of the 200 questionnaires distributed by this individual, 109 were returned. Does this provide convincing evidence that the response rate in this situation exceeds the rate in the past? State and test the appropriate hypotheses at significance level .05.

10.80 ● An automobile manufacturer who wishes to advertise that one of its models achieves 30 mpg (miles per gallon) decides to carry out a fuel efficiency test. Six nonprofessional drivers were selected, and each one drove a car from Phoenix to Los Angeles. The resulting fuel efficiencies (in miles per gallon) are:

27.2 29.3 31.2 28.4 30.3 29.6

Assuming that fuel efficiency is normally distributed under these circumstances, do the data contradict the claim that true average fuel efficiency is (at least) 30 mpg?

10.81 A student organization uses the proceeds from a particular soft-drink dispensing machine to finance its activities. The price per can had been \$0.75 for a long time, and the average daily revenue during that period was \$75.00. The price was recently increased to \$1.00 per can. A random sample of $n = 20$ days after the price increase yielded a sample mean daily revenue and sample standard deviation of \$70.00 and \$4.20, respectively.

Does this information suggest that the mean daily revenue has decreased from its value before the price increase? Test the appropriate hypotheses using $\alpha = .05$.

10.82 A hot tub manufacturer advertises that with its heating equipment, a temperature of 100°F can be achieved on average in 15 minutes or less. A random sample of 25 tubs is selected, and the time necessary to achieve a 100°F temperature is determined for each tub. The sample mean time and sample standard deviation are 17.5 minutes and 2.2 minutes, respectively. Does this information cast doubt on the company's claim? Carry out a test of hypotheses using significance level .05.

Bold exercises answered in back ● Data set available online ▼ Video Solution available

TECHNOLOGY NOTES

z Test for Proportions

JMP

Summarized data

1. Enter the data into the JMP data table with categories in the first column and counts in the second column
2. Click **Analyze** and select **Distribution**
3. Click and drag the first column name from the box under **Select Columns** to the box next to **Y, Columns**
4. Click and drag the second column name from the box under **Select Columns** to the box next to **Freq**
5. Click **OK**
6. Click the red arrow next to the column name and click **Test Probabilities**
7. Under the **Test Probabilities** section that appears in the output, click in the box across from **Yes** under the **Hypoth Prob** and type the hypothesized value for p
8. Select the appropriate option for alternative
9. Click **Done**

Note: In the two-sided case, JMP uses the square of the z test statistic, called the Chi-Square test statistic. The two methods are mathematically identical.

Note: In the one-sided cases, JMP uses the exact binomial test rather than the z test.

Raw data

1. Enter the raw data into a column

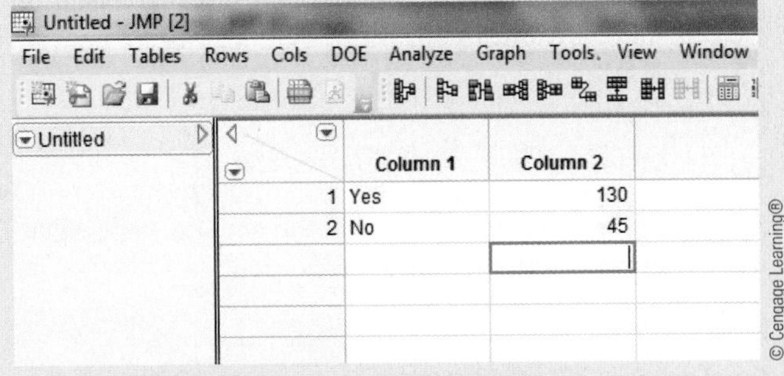

2. Click **Analyze** and select **Distribution**
3. Click and drag the first column name from the box under **Select Columns** to the box next to **Y, Columns**
4. Click **OK**
5. Click the red arrow next to the column name and click **Test Probabilities**
6. Under the **Test Probabilities** section that appears in the output, click in the box across from **Yes** under the **Hypoth Prob** and type the hypothesized value for p

7. Select the appropriate option for alternative
8. Click **Done**

Note: In the two-sided case, JMP uses the square of the z test statistic, called the Chi-Square test statistic. The two methods are mathematically identical.

Note: In the one-sided cases, JMP uses the exact binomial test rather than the z test.

Minitab

Summarized data

1. Click **Stat** then click **Basic Statistics** then click **1 Proportion...**
2. Click the radio button next to **Summarized data**
3. In the box next to **Number of Trials:** type the value for n, the sample size
4. In the box next to **Number of events:** type the value for the number of successes
5. Click **Options...**
6. Input the appropriate hypothesized value in the box next to **Test proportion:**
7. Select the appropriate alternative hypothesis from the drop-down menu next to **Alternative**
8. Check the box next to **Use test and interval based on normal distribution**
9. Click **OK**
10. Click **OK**

Raw data

1. Input the raw data into a column
2. Click **Stat** then click **Basic Statistics** then click **1 Proportion...**
3. Click in the box under **Samples in columns:**
4. Double click the column name where the raw data are stored
5. Click **Options...**
6. Select the appropriate alternative hypothesis from the drop-down menu next to **Alternative**
7. Check the box next to **Use test and interval based on normal distribution**
8. Click **OK**
9. Click **OK**

SPSS

SPSS does not have the functionality to automatically calculate a z test for a testing a single proportion.

Excel

Excel does not have the functionality to automatically calculate a z test for a testing a single population proportion. You may type in the formula by hand into a cell to have Excel calculate the value of the test statistic for you. Then use methods from Chapter 6 to find the P-value using the Normal Distribution.

TI-83/84

1. Press the **STAT** key
2. Highlight **TESTS**
3. Highlight **1-PropZTest...** and press **ENTER**
4. Next to **p0** type the hypothesized value for p
5. Next to **x** type the number of successes
6. Next to **n** type the sample size, n
7. Next to **prop**, highlight the appropriate alternative hypothesis
8. Highlight **Calculate** and press **ENTER**

TI-Nspire

1. Enter the Calculate Scratchpad
2. Press the **menu** key then select **6:Statistics** then select **7:Stat Tests** then **5:1-Prop z Test...** then press **enter**
3. In the box next to **p0** type the hypothesized value for p
4. In the box next to **Successes, x** type the number of successes
5. In the box next to **n** type the sample size, n
6. In the box next to **Alternate Hyp** choose the appropriate alternative hypothesis from the drop-down menu
7. Press **OK**

t Test for population mean, μ

JMP

1. Input the data into a column
2. Click **Analyze** and select **Distribution**
3. Click and drag the column name from the box under **Select Columns** to the box next to **Y, Response**
4. Click **OK**
5. Click the red arrow next to the column name and select **Test Mean**
6. In the box next to **Specify Hypothesized Mean**, type the hypothesized value of the mean, μ_0
7. Click **OK**

Note: The output provides results for all three possible alternative hypotheses.

Minitab

Summarized data

1. Click **Stat** then click **Basic Statistics** then click **1-sample t...**
2. Click the radio button next to **Summarized data**
3. In the box next to **Sample size:** type the value for n, the sample size
4. In the box next to **Mean:** type the value for the sample mean
5. In the box next to **Standard deviation:** type the value for the sample standard deviation
6. In the box next to **Test mean:** type the hypothesized value of the population mean
7. Click **Options...**
8. Select the appropriate alternative hypothesis from the drop-down menu next to **Alternative:**
9. Click **OK**
10. Click **OK**

Raw data

1. Input the raw data into a column
2. Click **Stat** then click **Basic Statistics** then click **1-sample t...**
3. Click in the box under **Samples in columns:**
4. Double click the column name where the raw data are stored
5. In the box next to **Test mean:** type the hypothesized value of the population mean
6. Click **Options...**
7. Select the appropriate alternative hypothesis from the drop-down menu next to **Alternative:**
8. Click **OK**
9. Click **OK**

SPSS

1. Input the data into a column
2. Click **Analyze** then select **Compare Means** then select **One-Sample T Test...**
3. Highlight the column name for the variable
4. Click the arrow to move the variable to the **Test Variable(s):** box
5. In the box next to **Test Value:** input the hypothesized test value
6. Click **OK**

Note: This procedure produces a two-sided *P*-value.

Excel

Excel does not have the functionality to automatically produce a *t* test for a single population mean. However, you may type the formula into an empty cell manually to have Excel calculate the value of the test statistic for you. You can then use the steps below to find a *P*-value for the test statistic.

1. Select an empty cell
2. Click on **Formulas**
3. Click **Insert Function**
4. Select **Statistical** from the drop-down box for category
5. Select **TDIST** and click **OK**
6. Click in the box next to **X** and select the cell containing your test statistic or type it manually
7. Click in the box next to **Deg_freedom** and type the number of degrees of freedom (n-1)
8. Click in the box next to **Tails** and type 1 for a one-tailed *P*-value or 2 for a two-tailed *P*-value
9. Click **OK**

Note: Choosing a one-tailed distribution in Step 8 will result in returning $P(X \geq x)$.

TI-83/84

Summarized data

1. Press **STAT**
2. Highlight **TESTS**
3. Highlight **T-Test...**
4. Highlight **Stats** and press **ENTER**
5. Next to μ_0 type the hypothesized value for the population mean
6. Next to $\bar{x}$ input the value for the sample mean
7. Next to **sx** input the value for the sample standard deviation
8. Next to **n** input the value for the sample size
9. Next to μ highlight the appropriate alternative hypothesis and press **ENTER**
10. Highlight **Calculate** and press **ENTER**

Raw data

1. Enter the data into **L1** (In order to access lists press the **STAT** key, highlight the option called **Edit...** then press **ENTER**)
2. Press **STAT**
3. Highlight **TESTS**
4. Highlight **T-Test...**
5. Highlight **Data** and press **ENTER**
6. Next to μ_0 type the hypothesized value for the population mean
7. Next to μ highlight the appropriate alternative hypothesis and press **ENTER**
8. Highlight **Calculate** and press **ENTER**

TI-Nspire

Summarized data

1. Enter the Calculate Scratchpad
2. Press the **menu** key and select **6:Statistics** then **7:Stat Tests** then **2:t test...** and press **enter**
3. From the drop-down menu select **Stats**
4. Press **OK**
5. Next to μ_0 type the hypothesized value for the population mean
6. Next to $\bar{x}$ input the value for the sample mean
7. Next to **sx** input the value for the sample standard deviation
8. Next to **n** input the value for the sample size
9. Next to **Alternate Hyp** select the appropriate alternative hypothesis from the drop-down menu
10. Press **OK**

Raw data

1. Enter the data into a data list (In order to access data lists select the spreadsheet option and press **enter**)
   Note: Be sure to title the list by selecting the top row of the column and typing a title.
2. Press the **menu** key and select **4:Statistics** then **7:Stat Tests** then **2:t test...** and press **enter**
3. From the drop-down menu select **Data**
4. Press **OK**
5. Next to μ_0 input the hypothesized value for the population mean
6. Next to **List** select the list containing your data
7. Next to **Alternate Hyp** select the appropriate alternative hypothesis from the drop-down menu
8. Press **OK**

CUMULATIVE REVIEW EXERCISES CR10.1 - CR10.16

CR10.1 The ***AARP Bulletin* (March 2010)** included the following short news brief: "Older adults who did 1 hour of tai chi twice weekly cut their pain from knee osteoarthritis considerably in a 12-week study conducted at Tufts University School of Medicine." Suppose you were asked to design a study to investigate this claim. Describe an experiment that would allow comparison of the reduction in knee pain for those who did 1 hour of tai chi twice weekly to the reduction in knee pain for those who did not do tai chi. Include a discussion of how study participants would be selected, how pain reduction would be measured, and how participants would be assigned to experimental groups.

CR10.2 The following graphical display is similar to one that appeared in ***USA Today* (June 3, 2009)**. Write a few sentences critiquing this graphical display. Do you think it does a good job of creating a visual representation of the three percentages in the display?

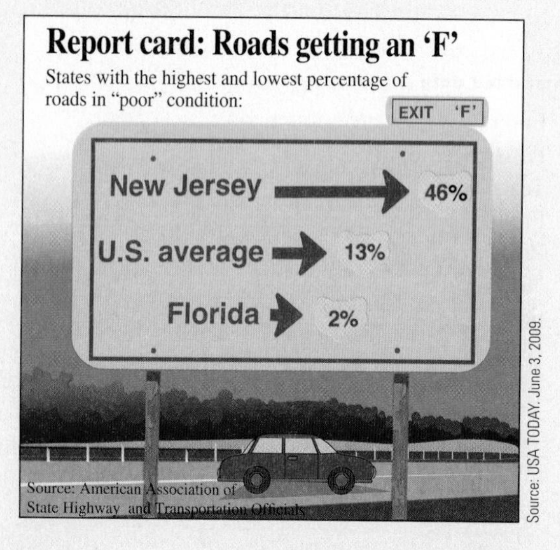

CR10.3 ● The article **"Flyers Trapped on Tarmac Push for Rules on Release" (*USA Today*, July 28, 2009)** included the accompanying data on the number of flights with a tarmac delay of more than 3 hours between October 2008 and May 2009 for U.S. airlines.

Airline	Number of Flights	Rate per 100,000 Flights
AirTran	7	0.4
Alaska	0	0.0
American	48	1.3
American Eagle	44	1.6
Atlantic Southeast	11	0.6
Comair	29	2.7
Continental	72	4.1
Delta	81	2.8
ExpressJet	93	4.9
Frontier	5	0.9
Hawaiian	0	0.0
JetBlue	18	1.4
Mesa	17	1.1
Northwest	24	1.2
Pinnacle	13	0.7
SkyWest	29	0.8
Southwest	11	0.1
United	29	1.1
US Airways	46	1.6

continued

a. Construct a dotplot of the data on number of flights delayed for more than 3 hours. Are there any unusual observations that stand out in the dot plot? What airlines appear to be the worst in terms of number of flights delayed on the tarmac for more than 3 hours?

b. Construct a dotplot of the data on rate per 100,000 flights. Write a few sentences describing the interesting features of this plot.

c. If you wanted to compare airlines on the basis of tarmac delays, would you recommend using the data on number of flights delayed or on rate per 100,000 flights? Explain the reason for your choice.

CR10.4 ● The article **"Wait Times on Rise to See Doctor" (*USA Today*, June 4, 2009)** gave the accompanying data on average wait times in days to get an appointment with a medical specialist in 15 U.S. cities. Construct a boxplot of the average wait-time data. Are there any outliers in the data set?

City	Average Appointment Wait Time
Atlanta	11.2
Boston	49.6
Dallas	19.2
Denver	15.4
Detroit	12.0
Houston	23.4
Los Angeles	24.2
Miami	15.4
Minneapolis	19.8

continued

City	Average Appointment Wait Time
New York	19.2
Philadelphia	27.0
Portland	14.4
San Diego	20.2
Seattle	14.2
Washington, D.C.	22.6

CR10.5 The report **"New Study Shows Need for Americans to Focus on Securing Online Accounts and Backing up Critical Data" (PRNewswire, October 29, 2009)** states that only 25% of Americans change computer passwords quarterly, in spite of a recommendation from the National Cyber Security Alliance that passwords be changed at least once every 90 days. For purposes of this exercise, assume that the 25% figure is correct for the population of adult Americans.

a. If a random sample of 20 adult Americans is selected, what is the probability that exactly 3 of them change passwords quarterly?

b. What is the probability that more than 8 people in a random sample of 20 adult Americans change passwords quarterly?

c. What is the mean and standard deviation of the variable x = number of people in a random sample of 100 adult Americans who change passwords quarterly?

d. Find the approximate probability that the number of people who change passwords quarterly in a random sample of 100 adult Americans is less than 20.

CR10.6 The article **"Should Canada Allow Direct-to-Consumer Advertising of Prescription Drugs?" (*Canadian Family Physician* [2009]: 130–131)** calls for the legalization of advertising of prescription drugs in Canada. Suppose you wanted to conduct a survey to estimate the proportion of Canadians who would support allowing this type of advertising. How large a random sample would be required to estimate this proportion to within .02 with 95% confidence?

CR10.7 The National Association of Colleges and Employers carries out a student survey each year. A summary of data from the 2009 survey included the following information:

- 26% of students graduating in 2009 intended to go on to graduate or professional school.
- Only 40% of those who graduated in 2009 received at least one job offer prior to graduation.
- Of those who received a job offer, only 45% had accepted an offer by the time they graduated.

Consider the following events:

O = event that a randomly selected 2009 graduate received at least one job offer

A = event that a randomly selected 2009 graduate accepted a job offer prior to graduation

G = event that a randomly selected 2009 graduate plans to attend graduate or professional school

Compute the following probabilities.

a. $P(O)$

b. $P(A)$

c. $P(G)$

d. $P(A|O)$

e. $P(O|A)$

f. $P(A \cap O)$

CR10.8 It probably wouldn't surprise you to know that Valentine's Day means big business for florists, jewelry stores, and restaurants. But would it surprise you to know that it is also a big day for pet stores? In January 2010, the National Retail Federation conducted a survey of consumers who they believed were selected in a way that would produce a sample representative of the population of adults in the United States **("This Valentine's Day, Couples Cut Back on Gifts to Each Other, According to NRF Survey," www.nrf.com)**. One of the questions in the survey asked if the respondent planned to spend money on a Valentine's Day gift for his or her pet this year.

a. The proportion who responded that they did plan to purchase a gift for their pet was .173. Suppose that the sample size for this survey was $n = 200$. Construct and interpret a 95% confidence interval for the proportion of all U.S. adults who planned to purchase a Valentine's Day gift for their pet in 2010.

b. The actual sample size for the survey was much larger than 200. Would a 95% confidence interval computed using the actual sample size have been narrower or wider than the confidence interval computed in Part (a)?

c. Still assuming a sample size of $n = 200$, carry out a hypothesis test to determine if the data provides convincing evidence that the proportion who planned to buy a Valentine's Day gift for their pet in 2010 was greater than .15. Use a significance level of .05.

CR10.9 The article **"Doctors Cite Burnout in Mistakes" (*San Luis Obispo Tribune*, March 5, 2002)** reported that many doctors who are completing their residency have financial struggles that could interfere with training. In a sample of 115 residents, 38 reported that they worked moonlighting jobs and 22 reported a credit card debt of more than $3000. Suppose that it is reasonable to consider this sample of 115 as a random sample of all medical residents in the United States.

a. Construct and interpret a 95% confidence interval for the proportion of U.S. medical residents who work moonlighting jobs.

b. Construct and interpret a 90% confidence interval for the proportion of U.S. medical residents who have a credit card debt of more than $3000.
c. Give two reasons why the confidence interval in Part (a) is wider than the confidence interval in Part (b).

CR10.10 The National Geographic Society conducted a study that included 3000 respondents, age 18 to 24, in nine different countries (*San Luis Obispo Tribune*, **November 21, 2002**). The society found that 10% of the participants could not identify their own country on a blank world map.

a. Construct a 90% confidence interval for the proportion who can identify their own country on a blank world map.
b. What assumptions are necessary for the confidence interval in Part (a) to be valid?
c. To what population would it be reasonable to generalize the confidence interval estimate from Part (a)?

CR10.11 **"Heinz Plays Catch-up After Under-Filling Ketchup Containers"** is the headline of an article that appeared on CNN.com **(November 30, 2000)**. The article stated that Heinz had agreed to put an extra 1% of ketchup into each ketchup container sold in California for a 1-year period. Suppose that you want to make sure that Heinz is in fact fulfilling its end of the agreement. You plan to take a sample of 20-oz bottles shipped to California, measure the amount of ketchup in each bottle, and then use the resulting data to estimate the mean amount of ketchup in 20-oz bottles. A small pilot study showed that the amount of ketchup in 20-oz bottles varied from 19.9 to 20.3 oz. How many bottles should be included in the sample if you want to estimate the true mean amount of ketchup to within 0.1 oz with 95% confidence?

CR10.12 In a survey conducted by Yahoo Small Business, 1432 of 1813 adults surveyed said that they would alter their shopping habits if gas prices remain high **(Associated Press, November 30, 2005)**. The article did not say how the sample was selected, but for purposes of this exercise, assume that it is reasonable to regard this sample as representative of adult Americans. Based on these survey data, is it reasonable to conclude that more than three-quarters of adult Americans plan to alter their shopping habits if gas prices remain high?

CR10.13 In an AP-AOL sports poll **(Associated Press, December 18, 2005)**, 272 of 394 randomly selected baseball fans stated that they thought the designated hitter rule should either be expanded to both baseball leagues or eliminated. Based on the given information, is there sufficient evidence to conclude that a majority of baseball fans feel this way?

CR10.14 ▼ The article **"Americans Seek Spiritual Guidance on Web"** (*San Luis Obispo Tribune*, **October 12, 2002**) reported that 68% of the general population belong to a religious community. In a survey on Internet use, 84% of "religion surfers" (defined as those who seek spiritual help online or who have used the web to search for prayer and devotional resources) belong to a religious community. Suppose that this result was based on a sample of 512 religion surfers. Is there convincing evidence that the proportion of religion surfers who belong to a religious community is different from .68, the proportion for the general population? Use $\alpha = .05$.

CR10.15 A survey of teenagers and parents in Canada conducted by the polling organization Ipsos **("Untangling the Web: The Facts About Kids and the Internet," January 25, 2006)** included questions about Internet use. It was reported that for a sample of 534 randomly selected teens, the mean number of hours per week spent online was 14.6 and the standard deviation was 11.6.

a. What does the large standard deviation, 11.6 hours, tell you about the distribution of online times for this sample of teens?
b. Do the sample data provide convincing evidence that the mean number of hours that teens spend online is greater than 10 hours per week?

CR10.16 The same survey referenced in the previous exercise reported that for a random sample of 676 parents of Canadian teens, the mean number of hours parents thought their teens spent online was 6.5 and the sample standard deviation was 8.6.

a. Do the sample data provide convincing evidence that the mean number of hours that parents think their teens spend online is less than 10 hours per week?
b. Write a few sentences commenting on the results of the test in Part (a) and of the test in Part (b) of the previous exercise.

Bold exercises answered in back ● Data set available online ▼ Video Solution available

Appendix A

APPENDIX **A**

Statistical Tables 760

Table 1 Random Numbers 760
Table 2 Standard Normal Probabilities (Cumulative z Curve Areas) 762
Table 3 t Critical Values 764
Table 4 Tail Areas for t Curves 765
Table 5 Curves of $\beta = P$(Type II Error) for t Tests 768
Table 6 Values That Capture Specified Upper-Tail F Curve Areas 769
Table 7 Critical Values of q for the Studentized Range Distribution 773
Table 8 Upper-Tail Areas for Chi-Square Distributions 774
Table 9 Binomial Probabilities 776

Appendix A: Statistical Tables

TABLE 1 Random Numbers

Row																				
1	4	5	1	8	5	0	3	3	7	1	2	8	4	5	1	1	0	9	5	7
2	4	2	5	5	8	0	4	5	7	0	7	0	3	6	6	1	3	1	3	1
3	8	9	9	3	4	3	5	0	6	3	9	1	1	8	2	6	9	2	0	9
4	8	9	0	7	2	9	9	0	4	7	6	7	4	7	1	3	4	3	5	3
5	5	7	3	1	0	3	7	4	7	8	5	2	0	1	3	7	7	6	3	6
6	0	9	3	8	7	6	7	9	9	5	6	2	5	6	5	8	4	2	6	4
7	4	1	0	1	0	2	2	0	4	7	5	1	1	9	4	7	9	7	5	1
8	6	4	7	3	6	3	4	5	1	2	3	1	1	8	0	0	4	8	2	0
9	8	0	2	8	7	9	3	8	4	0	4	2	0	8	9	1	2	3	3	2
10	9	4	6	0	6	9	7	8	8	2	5	2	9	6	0	1	4	6	0	5
11	6	6	9	5	7	4	4	6	3	2	0	6	0	8	9	1	3	6	1	8
12	0	7	1	7	7	7	2	9	7	8	7	5	8	8	6	9	8	4	1	0
13	6	1	3	0	9	7	3	3	6	6	0	4	1	8	3	2	6	7	6	8
14	2	2	3	6	2	1	3	0	2	2	6	6	9	7	0	2	1	2	5	8
15	0	7	1	7	4	2	0	0	0	1	3	1	2	0	4	7	8	4	1	0
16	6	6	5	1	6	1	8	1	5	5	2	6	2	0	1	1	5	2	3	6
17	9	9	6	2	5	3	5	9	8	3	7	5	0	1	3	9	3	8	0	8
18	9	9	9	6	1	2	9	3	4	6	5	6	4	6	5	8	2	7	4	0
19	2	5	6	3	1	9	8	1	1	0	3	5	6	7	9	1	4	5	2	0
20	5	1	1	9	8	1	2	1	1	6	9	8	1	8	1	9	9	1	2	0
21	1	9	8	0	7	4	6	8	4	0	3	0	8	1	1	0	6	2	3	2
22	9	7	0	9	6	3	8	9	9	7	0	6	5	4	3	6	5	0	3	2
23	1	7	6	4	8	2	0	3	9	6	3	6	2	1	0	7	7	3	1	7
24	6	2	5	8	2	0	7	8	6	4	6	6	8	9	2	0	6	9	0	4
25	1	5	7	1	1	1	9	5	1	4	5	2	8	3	4	3	0	7	3	5
26	1	4	6	6	5	6	0	1	9	4	0	5	2	7	6	4	3	6	8	8
27	1	8	5	0	2	1	6	8	0	7	7	2	6	2	6	7	5	4	8	7
28	7	8	7	4	6	5	4	3	7	9	3	9	2	7	9	5	4	2	3	1
29	1	6	3	2	8	3	7	3	0	7	2	4	8	0	9	9	9	4	7	0
30	2	8	9	0	8	1	6	8	1	7	3	1	3	0	9	7	2	5	7	9
31	0	7	8	8	6	5	7	5	5	4	0	0	3	4	1	2	7	3	7	9
32	8	4	0	1	4	5	1	9	1	1	2	1	5	3	2	8	5	5	7	5
33	7	3	5	9	7	0	4	9	1	2	1	3	2	5	1	9	3	3	8	3
34	4	7	2	6	7	6	9	9	2	7	8	7	5	5	5	2	4	4	3	4
35	9	3	3	7	0	7	0	5	7	5	6	9	5	4	3	1	4	6	6	8
36	0	2	4	9	7	8	1	6	3	8	7	8	0	5	6	7	2	7	5	0
37	7	1	0	1	8	4	7	1	2	9	3	8	0	0	8	7	9	2	8	6
38	9	7	9	4	4	5	3	1	9	3	4	5	0	6	3	5	9	6	9	8
39	0	4	2	5	0	0	9	9	6	4	0	6	9	0	3	8	3	5	7	2
40	0	7	1	2	3	6	1	7	9	3	9	5	4	6	8	4	8	8	0	6
41	3	5	6	6	2	4	4	5	6	3	7	8	7	6	5	2	0	4	3	2
42	6	6	8	5	5	2	9	7	9	3	3	1	6	9	5	9	7	1	1	2
43	9	5	0	4	3	1	1	7	3	9	2	7	7	4	7	0	3	1	2	8
44	5	1	7	8	9	4	7	2	9	2	8	9	9	8	0	6	3	7	2	1
45	1	6	3	9	4	1	3	2	1	1	8	5	6	3	4	1	9	3	1	7
46	4	4	8	6	4	0	3	8	3	8	3	5	9	5	9	4	8	3	9	4
47	7	7	6	6	4	5	4	4	8	4	4	0	3	9	8	5	2	0	2	3
48	2	5	6	6	3	7	0	6	5	6	9	0	1	9	5	2	6	9	1	2

TABLE 1 Random Numbers (*Continued*)

Row																				
49	9	4	0	4	7	5	3	2	8	7	2	7	4	9	3	9	6	5	5	6
50	7	3	1	5	6	6	5	0	3	5	3	7	2	8	6	2	4	1	8	7
51	7	5	8	2	8	8	8	7	6	4	1	1	0	2	3	1	9	3	6	0
52	3	3	6	0	9	1	1	0	3	2	7	8	2	0	5	3	4	8	9	8
53	0	2	9	6	9	8	9	3	8	1	5	3	9	9	7	0	7	7	1	6
54	8	5	9	6	2	9	6	8	2	1	2	4	7	0	6	8	3	4	6	1
55	5	4	7	6	1	0	0	1	0	4	6	1	4	1	5	0	9	6	5	5
56	5	0	3	6	4	1	9	8	4	4	1	2	0	2	5	1	8	1	2	1
57	0	2	6	3	7	5	1	1	6	6	0	5	8	1	2	3	3	6	1	3
58	3	8	1	6	3	8	1	4	5	2	9	4	2	5	7	3	2	3	1	8
59	9	1	5	6	0	6	5	6	6	3	6	2	3	0	0	0	1	8	5	9
60	5	3	5	6	3	9	5	4	7	3	6	6	7	5	0	1	5	6	7	3
61	9	6	6	4	5	7	7	6	1	5	4	4	8	0	6	5	7	6	3	0
62	6	3	0	6	7	9	5	5	4	6	2	2	8	4	4	0	0	9	9	8
63	8	5	8	3	5	2	0	6	6	0	0	6	0	6	3	0	1	7	0	5
64	3	8	2	4	9	0	9	2	6	2	9	5	1	9	1	9	0	8	3	3
65	1	4	4	1	1	7	4	6	3	6	5	6	5	5	7	7	0	3	5	8
66	5	9	9	5	3	7	2	5	1	7	1	1	0	7	1	0	9	2	8	8
67	8	7	1	7	5	2	5	6	8	7	9	9	1	3	9	6	4	9	3	0
68	6	7	2	3	1	4	9	2	1	7	0	8	6	7	8	9	9	4	7	4
69	2	3	2	8	7	0	9	7	1	1	1	2	8	2	9	1	0	6	7	7
70	2	9	5	7	8	4	7	9	0	3	6	9	2	0	6	0	6	2	6	8
71	4	8	9	8	3	2	7	6	9	1	9	8	6	9	5	2	4	9	9	9
72	1	5	6	5	7	7	5	4	3	4	3	8	1	8	9	9	4	4	1	1
73	1	8	1	1	7	2	8	5	5	8	9	9	9	6	2	0	1	6	6	7
74	5	7	7	0	9	5	5	6	8	6	8	2	2	6	0	5	5	1	8	7
75	1	8	6	0	5	4	8	3	4	5	3	5	8	7	7	7	8	5	7	0
76	2	6	6	7	9	4	2	2	8	7	4	3	4	9	6	1	9	4	3	9
77	3	6	6	4	5	7	8	3	0	2	8	4	6	7	2	1	4	5	2	3
78	0	7	8	0	1	2	1	1	3	4	2	1	6	9	3	3	5	4	0	4
79	8	3	6	0	5	7	7	9	1	5	8	8	4	9	5	7	2	2	7	6
80	5	3	6	9	0	6	3	8	7	5	9	5	9	7	4	2	5	6	2	9
81	0	9	3	7	7	2	8	6	4	3	2	9	4	8	2	9	9	6	9	9
82	9	4	7	4	0	0	0	3	5	4	6	6	2	6	2	3	6	1	1	4
83	5	5	4	1	7	8	6	4	2	3	2	9	8	4	6	3	8	3	0	5
84	5	3	0	0	5	4	8	0	7	4	7	6	2	1	1	2	1	2	6	9
85	3	3	0	9	3	2	9	4	0	5	5	4	8	7	5	7	5	3	8	8
86	3	0	5	7	1	9	5	8	0	0	4	5	3	0	3	0	2	7	6	7
87	5	0	8	6	0	8	1	6	2	0	8	6	5	4	0	7	2	9	1	0
88	3	6	4	7	8	2	3	5	7	9	8	5	2	7	6	9	0	2	4	9
89	9	0	4	4	9	1	6	8	5	2	8	9	0	7	5	7	2	5	1	8
90	9	5	2	6	9	3	9	6	5	1	8	8	7	8	2	0	4	4	7	9
91	9	4	5	7	0	3	4	6	4	2	5	4	8	6	1	1	9	1	8	8
92	8	1	1	8	0	5	4	2	8	5	3	3	3	0	1	1	4	4	8	3
93	6	9	4	7	8	3	3	9	1	2	5	0	1	2	3	0	1	1	2	5
94	0	0	6	8	8	7	2	4	4	7	6	6	0	3	4	7	5	6	8	2
95	5	3	3	9	3	8	4	9	1	9	1	7	8	4	5	2	2	5	4	4
96	2	5	6	2	7	6	0	3	8	1	4	4	2	6	8	3	6	3	2	8
97	7	4	3	7	9	6	8	6	2	8	3	8	4	2	2	0	7	0	5	3
98	1	9	0	8	8	0	1	2	2	2	7	5	6	5	5	7	8	7	2	6
99	2	4	8	0	2	5	2	7	0	5	9	6	6	1	5	8	7	9	7	5
100	4	1	7	8	6	7	1	1	5	8	9	4	8	9	8	3	0	9	0	7

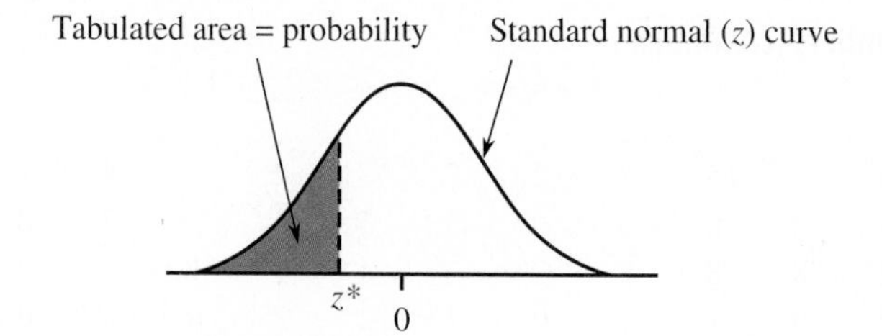

TABLE 2 **Standard Normal Probabilities (Cumulative *z* Curve Areas)**

z^*	.00	.01	.02	.03	.04	.05	.06	.07	.08	.09
−3.8	.0001	.0001	.0001	.0001	.0001	.0001	.0001	.0001	.0001	.0000
−3.7	.0001	.0001	.0001	.0001	.0001	.0001	.0001	.0001	.0001	.0001
−3.6	.0002	.0002	.0001	.0001	.0001	.0001	.0001	.0001	.0001	.0001
−3.5	.0002	.0002	.0002	.0002	.0002	.0002	.0002	.0002	.0002	.0002
−3.4	.0003	.0003	.0003	.0003	.0003	.0003	.0003	.0003	.0003	.0002
−3.3	.0005	.0005	.0005	.0004	.0004	.0004	.0004	.0004	.0004	.0003
−3.2	.0007	.0007	.0006	.0006	.0006	.0006	.0006	.0005	.0005	.0005
−3.1	.0010	.0009	.0009	.0009	.0008	.0008	.0008	.0008	.0007	.0007
−3.0	.0013	.0013	.0013	.0012	.0012	.0011	.0011	.0011	.0010	.0010
−2.9	.0019	.0018	.0018	.0017	.0016	.0016	.0015	.0015	.0014	.0014
−2.8	.0026	.0025	.0024	.0023	.0023	.0022	.0021	.0021	.0020	.0019
−2.7	.0035	.0034	.0033	.0032	.0031	.0030	.0029	.0028	.0027	.0026
−2.6	.0047	.0045	.0044	.0043	.0041	.0040	.0039	.0038	.0037	.0036
−2.5	.0062	.0060	.0059	.0057	.0055	.0054	.0052	.0051	.0049	.0048
−2.4	.0082	.0080	.0078	.0075	.0073	.0071	.0069	.0068	.0066	.0064
−2.3	.0107	.0104	.0102	.0099	.0096	.0094	.0091	.0089	.0087	.0084
−2.2	.0139	.0136	.0132	.0129	.0125	.0122	.0119	.0116	.0113	.0110
−2.1	.0179	.0174	.0170	.0166	.0162	.0158	.0154	.0150	.0146	.0143
−2.0	.0228	.0222	.0217	.0212	.0207	.0202	.0197	.0192	.0188	.0183
−1.9	.0287	.0281	.0274	.0268	.0262	.0256	.0250	.0244	.0239	.0233
−1.8	.0359	.0351	.0344	.0336	.0329	.0322	.0314	.0307	.0301	.0294
−1.7	.0446	.0436	.0427	.0418	.0409	.0401	.0392	.0384	.0375	.0367
−1.6	.0548	.0537	.0526	.0516	.0505	.0495	.0485	.0475	.0465	.0455
−1.5	.0668	.0655	.0643	.0630	.0618	.0606	.0594	.0582	.0571	.0559
−1.4	.0808	.0793	.0778	.0764	.0749	.0735	.0721	.0708	.0694	.0681
−1.3	.0968	.0951	.0934	.0918	.0901	.0885	.0869	.0853	.0838	.0823
−1.2	.1151	.1131	.1112	.1093	.1075	.1056	.1038	.1020	.1003	.0985
−1.1	.1357	.1335	.1314	.1292	.1271	.1251	.1230	.1210	.1190	.1170
−1.0	.1587	.1562	.1539	.1515	.1492	.1469	.1446	.1423	.1401	.1379
−0.9	.1841	.1814	.1788	.1762	.1736	.1711	.1685	.1660	.1635	.1611
−0.8	.2119	.2090	.2061	.2033	.2005	.1977	.1949	.1922	.1894	.1867
−0.7	.2420	.2389	.2358	.2327	.2296	.2266	.2236	.2206	.2177	.2148
−0.6	.2743	.2709	.2676	.2643	.2611	.2578	.2546	.2514	.2483	.2451
−0.5	.3085	.3050	.3015	.2981	.2946	.2912	.2877	.2843	.2810	.2776
−0.4	.3446	.3409	.3372	.3336	.3300	.3264	.3228	.3192	.3156	.3121
−0.3	.3821	.3783	.3745	.3707	.3669	.3632	.3594	.3557	.3520	.3483
−0.2	.4207	.4168	.4129	.4090	.4052	.4013	.3974	.3936	.3897	.3859
−0.1	.4602	.4562	.4522	.4483	.4443	.4404	.4364	.4325	.4286	.4247
−0.0	.5000	.4960	.4920	.4880	.4840	.4801	.4761	.4721	.4681	.4641
0.0	.5000	.5040	.5080	.5120	.5160	.5199	.5239	.5279	.5319	.5359
0.1	.5398	.5438	.5478	.5517	.5557	.5596	.5636	.5675	.5714	.5753
0.2	.5793	.5832	.5871	.5910	.5948	.5987	.6026	.6064	.6103	.6141
0.3	.6179	.6217	.6255	.6293	.6331	.6368	.6406	.6443	.6480	.6517
0.4	.6554	.6591	.6628	.6664	.6700	.6736	.6772	.6808	.6844	.6879

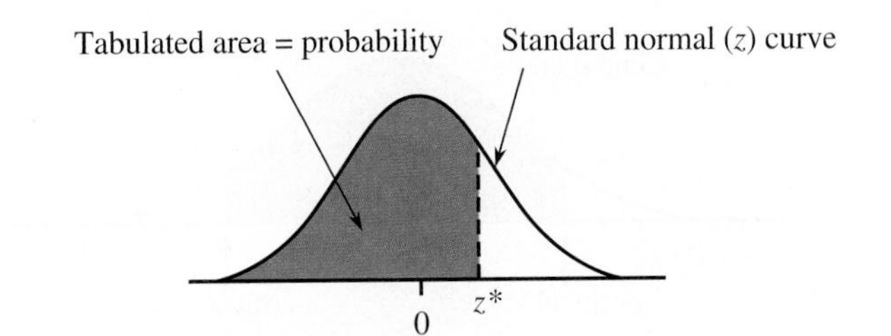

TABLE 2 Standard Normal Probabilities (Cumulative z Curve Areas) (*Continued*)

z^*	.00	.01	.02	.03	.04	.05	.06	.07	.08	.09
0.5	.6915	.6950	.6985	.7019	.7054	.7088	.7123	.7157	.7190	.7224
0.6	.7257	.7291	.7324	.7357	.7389	.7422	.7454	.7486	.7517	.7549
0.7	.7580	.7611	.7642	.7673	.7704	.7734	.7764	.7794	.7823	.7852
0.8	.7881	.7910	.7939	.7967	.7995	.8023	.8051	.8078	.8106	.8133
0.9	.8159	.8186	.8212	.8238	.8264	.8289	.8315	.8340	.8365	.8389
1.0	.8413	.8438	.8461	.8485	.8508	.8531	.8554	.8577	.8599	.8621
1.1	.8643	.8665	.8686	.8708	.8729	.8749	.8770	.8790	.8810	.8830
1.2	.8849	.8869	.8888	.8907	.8925	.8944	.8962	.8980	.8997	.9015
1.3	.9032	.9049	.9066	.9082	.9099	.9115	.9131	.9147	.9162	.9177
1.4	.9192	.9207	.9222	.9236	.9251	.9265	.9279	.9292	.9306	.9319
1.5	.9332	.9345	.9357	.9370	.9382	.9394	.9406	.9418	.9429	.9441
1.6	.9452	.9463	.9474	.9484	.9495	.9505	.9515	.9525	.9535	.9545
1.7	.9554	.9564	.9573	.9582	.9591	.9599	.9608	.9616	.9625	.9633
1.8	.9641	.9649	.9656	.9664	.9671	.9678	.9686	.9693	.9699	.9706
1.9	.9713	.9719	.9726	.9732	.9738	.9744	.9750	.9756	.9761	.9767
2.0	.9772	.9778	.9783	.9788	.9793	.9798	.9803	.9808	.9812	.9817
2.1	.9821	.9826	.9830	.9834	.9838	.9842	.9846	.9850	.9854	.9857
2.2	.9861	.9864	.9868	.9871	.9875	.9878	.9881	.9884	.9887	.9890
2.3	.9893	.9896	.9898	.9901	.9904	.9906	.9909	.9911	.9913	.9916
2.4	.9918	.9920	.9922	.9925	.9927	.9929	.9931	.9932	.9934	.9936
2.5	.9938	.9940	.9941	.9943	.9945	.9946	.9948	.9949	.9951	.9952
2.6	.9953	.9955	.9956	.9957	.9959	.9960	.9961	.9962	.9963	.9964
2.7	.9965	.9966	.9967	.9968	.9969	.9970	.9971	.9972	.9973	.9974
2.8	.9974	.9975	.9976	.9977	.9977	.9978	.9979	.9979	.9980	.9981
2.9	.9981	.9982	.9982	.9983	.9984	.9984	.9985	.9985	.9986	.9986
3.0	.9987	.9987	.9987	.9988	.9988	.9989	.9989	.9989	.9990	.9990
3.1	.9990	.9991	.9991	.9991	.9992	.9992	.9992	.9992	.9993	.9993
3.2	.9993	.9993	.9994	.9994	.9994	.9994	.9994	.9995	.9995	.9995
3.3	.9995	.9995	.9995	.9996	.9996	.9996	.9996	.9996	.9996	.9997
3.4	.9997	.9997	.9997	.9997	.9997	.9997	.9997	.9997	.9997	.9998
3.5	.9998	.9998	.9998	.9998	.9998	.9998	.9998	.9998	.9998	.9998
3.6	.9998	.9998	.9999	.9999	.9999	.9999	.9999	.9999	.9999	.9999
3.7	.9999	.9999	.9999	.9999	.9999	.9999	.9999	.9999	.9999	.9999
3.8	.9999	.9999	.9999	.9999	.9999	.9999	.9999	.9999	.9999	1.0000

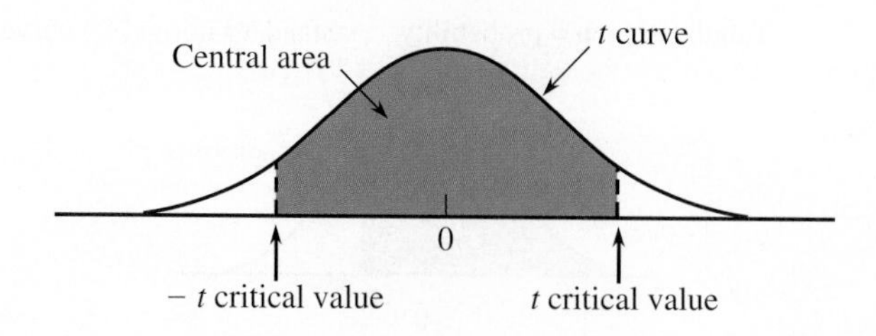

TABLE 3 *t* Critical Values

Central area captured: Confidence level:		.80 80%	.90 90%	.95 95%	.98 98%	.99 99%	.998 99.8%	.999 99.9%
Degrees of freedom	1	3.08	6.31	12.71	31.82	63.66	318.31	636.62
	2	1.89	2.92	4.30	6.97	9.93	23.33	31.60
	3	1.64	2.35	3.18	4.54	5.84	10.21	12.92
	4	1.53	2.13	2.78	3.75	4.60	7.17	8.61
	5	1.48	2.02	2.57	3.37	4.03	5.89	6.86
	6	1.44	1.94	2.45	3.14	3.71	5.21	5.96
	7	1.42	1.90	2.37	3.00	3.50	4.79	5.41
	8	1.40	1.86	2.31	2.90	3.36	4.50	5.04
	9	1.38	1.83	2.26	2.82	3.25	4.30	4.78
	10	1.37	1.81	2.23	2.76	3.17	4.14	4.59
	11	1.36	1.80	2.20	2.72	3.11	4.03	4.44
	12	1.36	1.78	2.18	2.68	3.06	3.93	4.32
	13	1.35	1.77	2.16	2.65	3.01	3.85	4.22
	14	1.35	1.76	2.15	2.62	2.98	3.79	4.14
	15	1.34	1.75	2.13	2.60	2.95	3.73	4.07
	16	1.34	1.75	2.12	2.58	2.92	3.69	4.02
	17	1.33	1.74	2.11	2.57	2.90	3.65	3.97
	18	1.33	1.73	2.10	2.55	2.88	3.61	3.92
	19	1.33	1.73	2.09	2.54	2.86	3.58	3.88
	20	1.33	1.73	2.09	2.53	2.85	3.55	3.85
	21	1.32	1.72	2.08	2.52	2.83	3.53	3.82
	22	1.32	1.72	2.07	2.51	2.82	3.51	3.79
	23	1.32	1.71	2.07	2.50	2.81	3.49	3.77
	24	1.32	1.71	2.06	2.49	2.80	3.47	3.75
	25	1.32	1.71	2.06	2.49	2.79	3.45	3.73
	26	1.32	1.71	2.06	2.48	2.78	3.44	3.71
	27	1.31	1.70	2.05	2.47	2.77	3.42	3.69
	28	1.31	1.70	2.05	2.47	2.76	3.41	3.67
	29	1.31	1.70	2.05	2.46	2.76	3.40	3.66
	30	1.31	1.70	2.04	2.46	2.75	3.39	3.65
	40	1.30	1.68	2.02	2.42	2.70	3.31	3.55
	60	1.30	1.67	2.00	2.39	2.66	3.23	3.46
	120	1.29	1.66	1.98	2.36	2.62	3.16	3.37
z critical values	∞	1.28	1.645	1.96	2.33	2.58	3.09	3.29

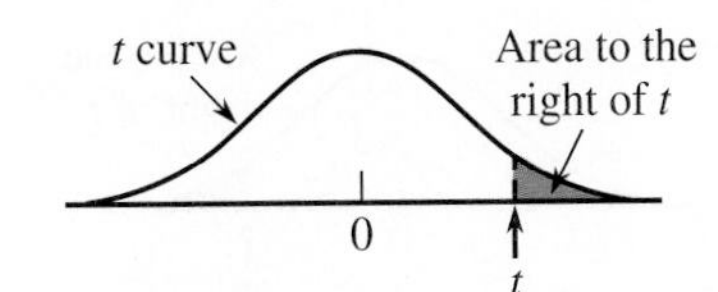

TABLE 4 Tail Areas for *t* Curves

t \ df	1	2	3	4	5	6	7	8	9	10	11	12
0.0	.500	.500	.500	.500	.500	.500	.500	.500	.500	.500	.500	.500
0.1	.468	.465	.463	.463	.462	.462	.462	.461	.461	.461	.461	.461
0.2	.437	.430	.427	.426	.425	.424	.424	.423	.423	.423	.423	.422
0.3	.407	.396	.392	.390	.388	.387	.386	.386	.386	.385	.385	.385
0.4	.379	.364	.358	.355	.353	.352	.351	.350	.349	.349	.348	.348
0.5	.352	.333	.326	.322	.319	.317	.316	.315	.315	.314	.313	.313
0.6	.328	.305	.295	.290	.287	.285	.284	.283	.282	.281	.280	.280
0.7	.306	.278	.267	.261	.258	.255	.253	.252	.251	.250	.249	.249
0.8	.285	.254	.241	.234	.230	.227	.225	.223	.222	.221	.220	.220
0.9	.267	.232	.217	.210	.205	.201	.199	.197	.196	.195	.194	.193
1.0	.250	.211	.196	.187	.182	.178	.175	.173	.172	.170	.169	.169
1.1	.235	.193	.176	.167	.162	.157	.154	.152	.150	.149	.147	.146
1.2	.221	.177	.158	.148	.142	.138	.135	.132	.130	.129	.128	.127
1.3	.209	.162	.142	.132	.125	.121	.117	.115	.113	.111	.110	.109
1.4	.197	.148	.128	.117	.110	.106	.102	.100	.098	.096	.095	.093
1.5	.187	.136	.115	.104	.097	.092	.089	.086	.084	.082	.081	.080
1.6	.178	.125	.104	.092	.085	.080	.077	.074	.072	.070	.069	.068
1.7	.169	.116	.094	.082	.075	.070	.066	.064	.062	.060	.059	.057
1.8	.161	.107	.085	.073	.066	.061	.057	.055	.053	.051	.050	.049
1.9	.154	.099	.077	.065	.058	.053	.050	.047	.045	.043	.042	.041
2.0	.148	.092	.070	.058	.051	.046	.043	.040	.038	.037	.035	.034
2.1	.141	.085	.063	.052	.045	.040	.037	.034	.033	.031	.030	.029
2.2	.136	.079	.058	.046	.040	.035	.032	.029	.028	.026	.025	.024
2.3	.131	.074	.052	.041	.035	.031	.027	.025	.023	.022	.021	.020
2.4	.126	.069	.048	.037	.031	.027	.024	.022	.020	.019	.018	.017
2.5	.121	.065	.044	.033	.027	.023	.020	.018	.017	.016	.015	.014
2.6	.117	.061	.040	.030	.024	.020	.018	.016	.014	.013	.012	.012
2.7	.113	.057	.037	.027	.021	.018	.015	.014	.012	.011	.010	.010
2.8	.109	.054	.034	.024	.019	.016	.013	.012	.010	.009	.009	.008
2.9	.106	.051	.031	.022	.017	.014	.011	.010	.009	.008	.007	.007
3.0	.102	.048	.029	.020	.015	.012	.010	.009	.007	.007	.006	.006
3.1	.099	.045	.027	.018	.013	.011	.009	.007	.006	.006	.005	.005
3.2	.096	.043	.025	.016	.012	.009	.008	.006	.005	.005	.004	.004
3.3	.094	.040	.023	.015	.011	.008	.007	.005	.005	.004	.004	.003
3.4	.091	.038	.021	.014	.010	.007	.006	.005	.004	.003	.003	.003
3.5	.089	.036	.020	.012	.009	.006	.005	.004	.003	.003	.002	.002
3.6	.086	.035	.018	.011	.008	.006	.004	.004	.003	.002	.002	.002
3.7	.084	.033	.017	.010	.007	.005	.004	.003	.002	.002	.002	.002
3.8	.082	.031	.016	.010	.006	.004	.003	.003	.002	.002	.001	.001
3.9	.080	.030	.015	.009	.006	.004	.003	.002	.002	.001	.001	.001
4.0	.078	.029	.014	.008	.005	.004	.003	.002	.002	.001	.001	.001

(continued)

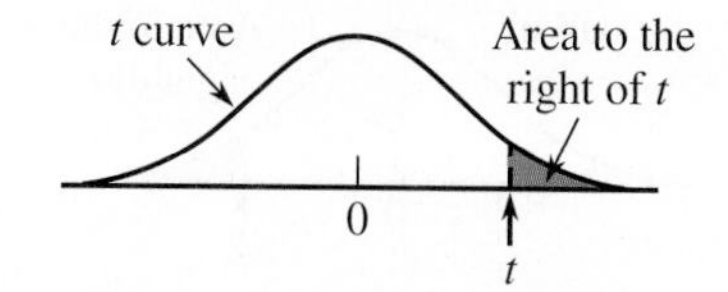

TABLE 4 Tail Areas for *t* Curves (*Continued*)

t \ df	13	14	15	16	17	18	19	20	21	22	23	24
0.0	.500	.500	.500	.500	.500	.500	.500	.500	.500	.500	.500	.500
0.1	.461	.461	.461	.461	.461	.461	.461	.461	.461	.461	.461	.461
0.2	.422	.422	.422	.422	.422	.422	.422	.422	.422	.422	.422	.422
0.3	.384	.384	.384	.384	.384	.384	.384	.384	.384	.383	.383	.383
0.4	.348	.347	.347	.347	.347	.347	.347	.347	.347	.347	.346	.346
0.5	.313	.312	.312	.312	.312	.312	.311	.311	.311	.311	.311	.311
0.6	.279	.279	.279	.278	.278	.278	.278	.278	.278	.277	.277	.277
0.7	.248	.247	.247	.247	.247	.246	.246	.246	.246	.246	.245	.245
0.8	.219	.218	.218	.218	.217	.217	.217	.217	.216	.216	.216	.216
0.9	.192	.191	.191	.191	.190	.190	.190	.189	.189	.189	.189	.189
1.0	.168	.167	.167	.166	.166	.165	.165	.165	.164	.164	.164	.164
1.1	.146	.144	.144	.144	.143	.143	.143	.142	.142	.142	.141	.141
1.2	.126	.124	.124	.124	.123	.123	.122	.122	.122	.121	.121	.121
1.3	.108	.107	.107	.106	.105	.105	.105	.104	.104	.104	.103	.103
1.4	.092	.091	.091	.090	.090	.089	.089	.089	.088	.088	.087	.087
1.5	.079	.077	.077	.077	.076	.075	.075	.075	.074	.074	.074	.073
1.6	.067	.065	.065	.065	.064	.064	.063	.063	.062	.062	.062	.061
1.7	.056	.055	.055	.054	.054	.053	.053	.052	.052	.052	.051	.051
1.8	.048	.046	.046	.045	.045	.044	.044	.043	.043	.043	.042	.042
1.9	.040	.038	.038	.038	.037	.037	.036	.036	.036	.035	.035	.035
2.0	.033	.032	.032	.031	.031	.030	.030	.030	.029	.029	.029	.028
2.1	.028	.027	.027	.026	.025	.025	.025	.024	.024	.024	.023	.023
2.2	.023	.022	.022	.021	.021	.021	.020	.020	.020	.019	.019	.019
2.3	.019	.018	.018	.018	.017	.017	.016	.016	.016	.016	.015	.015
2.4	.016	.015	.015	.014	.014	.014	.013	.013	.013	.013	.012	.012
2.5	.013	.012	.012	.012	.011	.011	.011	.011	.010	.010	.010	.010
2.6	.011	.010	.010	.010	.009	.009	.009	.009	.008	.008	.008	.008
2.7	.009	.008	.008	.008	.008	.007	.007	.007	.007	.007	.006	.006
2.8	.008	.007	.007	.006	.006	.006	.006	.006	.005	.005	.005	.005
2.9	.006	.005	.005	.005	.005	.005	.005	.004	.004	.004	.004	.004
3.0	.005	.004	.004	.004	.004	.004	.004	.004	.003	.003	.003	.003
3.1	.004	.004	.004	.003	.003	.003	.003	.003	.003	.003	.003	.002
3.2	.003	.003	.003	.003	.003	.002	.002	.002	.002	.002	.002	.002
3.3	.003	.002	.002	.002	.002	.002	.002	.002	.002	.002	.002	.001
3.4	.002	.002	.002	.002	.002	.002	.002	.001	.001	.001	.001	.001
3.5	.002	.002	.002	.001	.001	.001	.001	.001	.001	.001	.001	.001
3.6	.002	.001	.001	.001	.001	.001	.001	.001	.001	.001	.001	.001
3.7	.001	.001	.001	.001	.001	.001	.001	.001	.001	.001	.001	.001
3.8	.001	.001	.001	.001	.001	.001	.001	.001	.001	.000	.000	.000
3.9	.001	.001	.001	.001	.001	.001	.000	.000	.000	.000	.000	.000
4.0	.001	.001	.001	.001	.000	.000	.000	.000	.000	.000	.000	.000

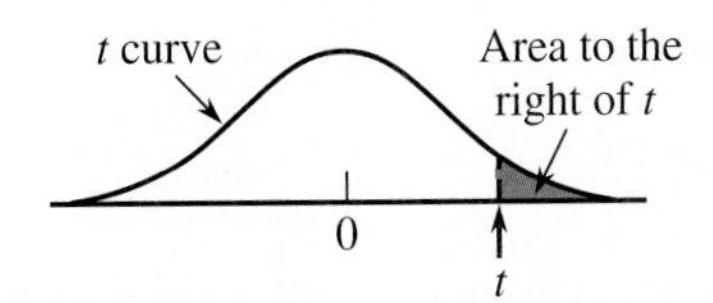

TABLE 4 Tail Areas for *t* Curves (*Continued*)

t \ df	25	26	27	28	29	30	35	40	60	120	∞(=z)
0.0	.500	.500	.500	.500	.500	.500	.500	.500	.500	.500	.500
0.1	.461	.461	.461	.461	.461	.461	.460	.460	.460	.460	.460
0.2	.422	.422	.421	.421	.421	.421	.421	.421	.421	.421	.421
0.3	.383	.383	.383	.383	.383	.383	.383	.383	.383	.382	.382
0.4	.346	.346	.346	.346	.346	.346	.346	.346	.345	.345	.345
0.5	.311	.311	.311	.310	.310	.310	.310	.310	.309	.309	.309
0.6	.277	.277	.277	.277	.277	.277	.276	.276	.275	.275	.274
0.7	.245	.245	.245	.245	.245	.245	.244	.244	.243	.243	.242
0.8	.216	.215	.215	.215	.215	.215	.215	.214	.213	.213	.212
0.9	.188	.188	.188	.188	.188	.188	.187	.187	.186	.185	.184
1.0	.163	.163	.163	.163	.163	.163	.162	.162	.161	.160	.159
1.1	.141	.141	.141	.140	.140	.140	.139	.139	.138	.137	.136
1.2	.121	.120	.120	.120	.120	.120	.119	.119	.117	.116	.115
1.3	.103	.103	.102	.102	.102	.102	.101	.101	.099	.098	.097
1.4	.087	.087	.086	.086	.086	.086	.085	.085	.083	.082	.081
1.5	.073	.073	.073	.072	.072	.072	.071	.071	.069	.068	.067
1.6	.061	.061	.061	.060	.060	.060	.059	.059	.057	.056	.055
1.7	.051	.051	.050	.050	.050	.050	.049	.048	.047	.046	.045
1.8	.042	.042	.042	.041	.041	.041	.040	.040	.038	.037	.036
1.9	.035	.034	.034	.034	.034	.034	.033	.032	.031	.030	.029
2.0	.028	.028	.028	.028	.027	.027	.027	.026	.025	.024	.023
2.1	.023	.023	.023	.022	.022	.022	.022	.021	.020	.019	.018
2.2	.019	.018	.018	.018	.018	.018	.017	.017	.016	.015	.014
2.3	.015	.015	.015	.015	.014	.014	.014	.013	.012	.012	.011
2.4	.012	.012	.012	.012	.012	.011	.011	.011	.010	.009	.008
2.5	.010	.010	.009	.009	.009	.009	.009	.008	.008	.007	.006
2.6	.008	.008	.007	.007	.007	.007	.007	.007	.006	.005	.005
2.7	.006	.006	.006	.006	.006	.006	.005	.005	.004	.004	.003
2.8	.005	.005	.005	.005	.005	.004	.004	.004	.003	.003	.003
2.9	.004	.004	.004	.004	.004	.003	.003	.003	.003	.002	.002
3.0	.003	.003	.003	.003	.003	.003	.002	.002	.002	.002	.001
3.1	.002	.002	.002	.002	.002	.002	.002	.002	.001	.001	.001
3.2	.002	.002	.002	.002	.002	.002	.001	.001	.001	.001	.001
3.3	.001	.001	.001	.001	.001	.001	.001	.001	.001	.001	.000
3.4	.001	.001	.001	.001	.001	.001	.001	.001	.001	.000	.000
3.5	.001	.001	.001	.001	.001	.001	.001	.001	.000	.000	.000
3.6	.001	.001	.001	.001	.001	.001	.000	.000	.000	.000	.000
3.7	.001	.001	.000	.000	.000	.000	.000	.000	.000	.000	.000
3.8	.000	.000	.000	.000	.000	.000	.000	.000	.000	.000	.000
3.9	.000	.000	.000	.000	.000	.000	.000	.000	.000	.000	.000
4.0	.000	.000	.000	.000	.000	.000	.000	.000	.000	.000	.000

TABLE 5 Curves of $\beta = P(\text{Type II Error})$ for t Tests

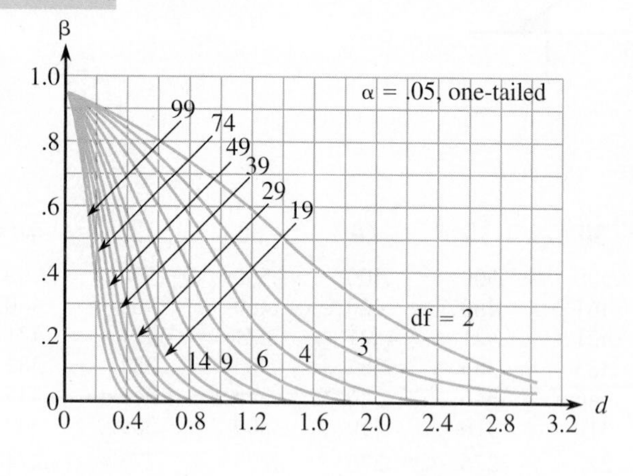

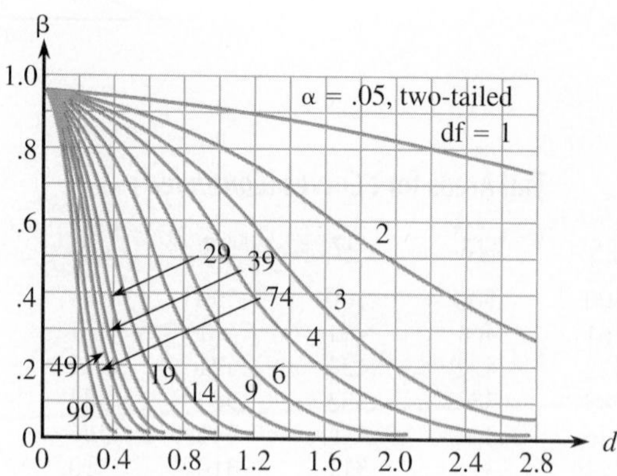

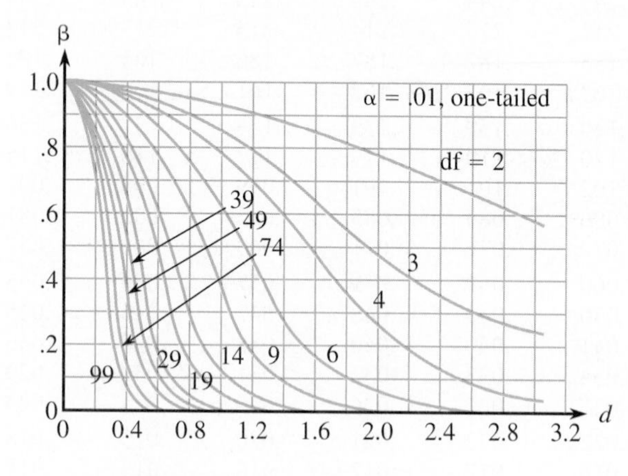

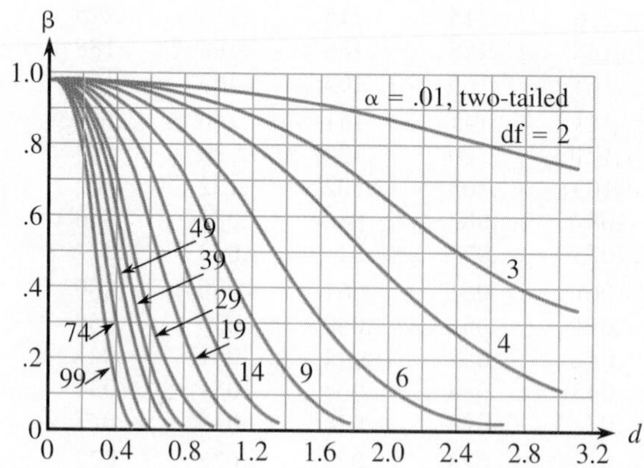

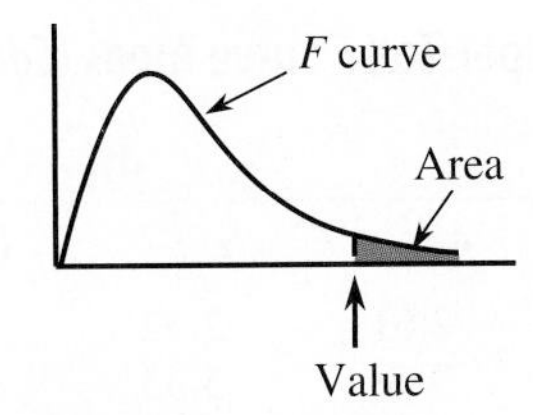

TABLE 6 **Values That Capture Specified Upper-Tail F Curve Areas**

		df_1									
df_2	Area	1	2	3	4	5	6	7	8	9	10
1	.10	39.86	49.50	53.59	55.83	57.24	58.20	58.91	59.44	59.86	60.19
	.05	161.40	199.50	215.70	224.60	230.20	234.00	236.80	238.90	240.50	241.90
	.01	4052.00	5000.00	5403.00	5625.00	5764.00	5859.00	5928.00	5981.00	6022.00	6056.00
2	.10	8.53	9.00	9.16	9.24	9.29	9.33	9.35	9.37	9.38	9.39
	.05	18.51	19.00	19.16	19.25	19.30	19.33	19.35	19.37	19.38	19.40
	.01	98.50	99.00	99.17	99.25	99.30	99.33	99.36	99.37	99.39	99.40
	.001	998.50	999.00	999.20	999.20	999.30	999.30	999.40	999.40	999.40	999.40
3	.10	5.54	5.46	5.39	5.34	5.31	5.28	5.27	5.25	5.24	5.23
	.05	10.13	9.55	9.28	9.12	9.01	8.94	8.89	8.85	8.81	8.79
	.01	34.12	30.82	29.46	28.71	28.24	27.91	27.67	27.49	27.35	27.23
	.001	167.00	148.50	141.10	137.10	134.60	132.80	131.60	130.60	129.90	129.20
4	.10	4.54	4.32	4.19	4.11	4.05	4.01	3.98	3.95	3.94	3.92
	.05	7.71	6.94	6.59	6.39	6.26	6.16	6.09	6.04	6.00	5.96
	.01	21.20	18.00	16.69	15.98	15.52	15.21	14.98	14.80	14.66	14.55
	.001	74.14	61.25	56.18	53.44	51.71	50.53	49.66	49.00	48.47	48.05
5	.10	4.06	3.78	3.62	3.52	3.45	3.40	3.37	3.34	3.32	3.30
	.05	6.61	5.79	5.41	5.19	5.05	4.95	4.88	4.82	4.77	4.74
	.01	16.26	13.27	12.06	11.39	10.97	10.67	10.46	10.29	10.16	10.05
	.001	47.18	37.12	33.20	31.09	29.75	28.83	28.16	27.65	27.24	26.92
6	.10	3.78	3.46	3.29	3.18	3.11	3.05	3.01	2.98	2.96	2.94
	.05	5.99	5.14	4.76	4.53	4.39	4.28	4.21	4.15	4.10	4.06
	.01	13.75	10.92	9.78	9.15	8.75	8.47	8.26	8.10	7.98	7.87
	.001	35.51	27.00	23.70	21.92	20.80	20.03	19.46	19.03	18.69	18.41
7	.10	3.59	3.26	3.07	2.96	2.88	2.83	2.78	2.75	2.72	2.70
	.05	5.59	4.74	4.35	4.12	3.97	3.87	3.79	3.73	3.68	3.64
	.01	12.25	9.55	8.45	7.85	7.46	7.19	6.99	6.84	6.72	6.62
	.001	29.25	21.69	18.77	17.20	16.21	15.52	15.02	14.63	14.33	14.08
8	.10	3.46	3.11	2.92	2.81	2.73	2.67	2.62	2.59	2.56	2.54
	.05	5.32	4.46	4.07	3.84	3.69	3.58	3.50	3.44	3.39	3.35
	.01	11.26	8.65	7.59	7.01	6.63	6.37	6.18	6.03	5.91	5.81
	.001	25.41	18.49	15.83	14.39	13.48	12.86	12.40	12.05	11.77	11.54
9	.10	3.36	3.01	2.81	2.69	2.61	2.55	2.51	2.47	2.44	2.42
	.05	5.12	4.26	3.86	3.63	3.48	3.37	3.29	3.23	3.18	3.14
	.01	10.56	8.02	6.99	6.42	6.06	5.80	5.61	5.47	5.35	5.26
	.001	22.86	16.39	13.90	12.56	11.71	11.13	10.70	10.37	10.11	9.89

(continued)

TABLE 6 **Values That Capture Specified Upper-Tail *F* Curve Areas (*Continued*)**

		df_1									
df_2	Area	1	2	3	4	5	6	7	8	9	10
10	.10	3.29	2.92	2.73	2.61	2.52	2.46	2.41	2.38	2.35	2.32
	.05	4.96	4.10	3.71	3.48	3.33	3.22	3.14	3.07	3.02	2.98
	.01	10.04	7.56	6.55	5.99	5.64	5.39	5.20	5.06	4.94	4.85
	.001	21.04	14.91	12.55	11.28	10.48	9.93	9.52	9.20	8.96	8.75
11	.10	3.23	2.86	2.66	2.54	2.45	2.39	2.34	2.30	2.27	2.25
	.05	4.84	3.98	3.59	3.36	3.20	3.09	3.01	2.95	2.90	2.85
	.01	9.65	7.21	6.22	5.67	5.32	5.07	4.89	4.74	4.63	4.54
	.001	19.69	13.81	11.56	10.35	9.58	9.05	8.66	8.35	8.12	7.92
12	.10	3.18	2.81	2.61	2.48	2.39	2.33	2.28	2.24	2.21	2.19
	.05	4.75	3.89	3.49	3.26	3.11	3.00	2.91	2.85	2.80	2.75
	.01	9.33	6.93	5.95	5.41	5.06	4.82	4.64	4.50	4.39	4.30
	.001	18.64	12.97	10.80	9.63	8.89	8.38	8.00	7.71	7.48	7.29
13	.10	3.14	2.76	2.56	2.43	2.35	2.28	2.23	2.20	2.16	2.14
	.05	4.67	3.81	3.41	3.18	3.03	2.92	2.83	2.77	2.71	2.67
	.01	9.07	6.70	5.74	5.21	4.86	4.62	4.44	4.30	4.19	4.10
	.001	17.82	12.31	10.21	9.07	8.35	7.86	7.49	7.21	6.98	6.80
14	.10	3.10	2.73	2.52	2.39	2.31	2.24	2.19	2.15	2.12	2.10
	.05	4.60	3.74	3.34	3.11	2.96	2.85	2.76	2.70	2.65	2.60
	.01	8.86	6.51	5.56	5.04	4.69	4.46	4.28	4.14	4.03	3.94
	.001	17.14	11.78	9.73	8.62	7.92	7.44	7.08	6.80	6.58	6.40
15	.10	3.07	2.70	2.49	2.36	2.27	2.21	2.16	2.12	2.09	2.06
	.05	4.54	3.68	3.29	3.06	2.90	2.79	2.71	2.64	2.59	2.54
	.01	8.68	6.36	5.42	4.89	4.56	4.32	4.14	4.00	3.89	3.80
	.001	16.59	11.34	9.34	8.25	7.57	7.09	6.74	6.47	6.26	6.08
16	.10	3.05	2.67	2.46	2.33	2.24	2.18	2.13	2.09	2.06	2.03
	.05	4.49	3.63	3.24	3.01	2.85	2.74	2.66	2.59	2.54	2.49
	.01	8.53	6.23	5.29	4.77	4.44	4.20	4.03	3.89	3.78	3.69
	.001	16.12	10.97	9.01	7.94	7.27	6.80	6.46	6.19	5.98	5.81
17	.10	3.03	2.64	2.44	2.31	2.22	2.15	2.10	2.06	2.03	2.00
	.05	4.45	3.59	3.20	2.96	2.81	2.70	2.61	2.55	2.49	2.45
	.01	8.40	6.11	5.18	4.67	4.34	4.10	3.93	3.79	3.68	3.59
	.001	15.72	10.66	8.73	7.68	7.02	6.56	6.22	5.96	5.75	5.58
18	.10	3.01	2.62	2.42	2.29	2.20	2.13	2.08	2.04	2.00	1.98
	.05	4.41	3.55	3.16	2.93	2.77	2.66	2.58	2.51	2.46	2.41
	.01	8.29	6.01	5.09	4.58	4.25	4.01	3.84	3.71	3.60	3.51
	.001	15.38	10.39	8.49	7.46	6.81	6.35	6.02	5.76	5.56	5.39
19	.10	2.99	2.61	2.40	2.27	2.18	2.11	2.06	2.02	1.98	1.96
	.05	4.38	3.52	3.13	2.90	2.74	2.63	2.54	2.48	2.42	2.38
	.01	8.18	5.93	5.01	4.50	4.17	3.94	3.77	3.63	3.52	3.43
	.001	15.08	10.16	8.28	7.27	6.62	6.18	5.85	5.59	5.39	5.22

TABLE 6 Values That Capture Specified Upper-Tail *F* Curve Areas (*Continued*)

		df_1									
df_2	Area	1	2	3	4	5	6	7	8	9	10
20	.10	2.97	2.59	2.38	2.25	2.16	2.09	2.04	2.00	1.96	1.94
	.05	4.35	3.49	3.10	2.87	2.71	2.60	2.51	2.45	2.39	2.35
	.01	8.10	5.85	4.94	4.43	4.10	3.87	3.70	3.56	3.46	3.37
	.001	14.82	9.95	8.10	7.10	6.46	6.02	5.69	5.44	5.24	5.08
21	.10	2.96	2.57	2.36	2.23	2.14	2.08	2.02	1.98	1.95	1.92
	.05	4.32	3.47	3.07	2.84	2.68	2.57	2.49	2.42	2.37	2.32
	.01	8.02	5.78	4.87	4.37	4.04	3.81	3.64	3.51	3.40	3.31
	.001	14.59	9.77	7.94	6.95	6.32	5.88	5.56	5.31	5.11	4.95
22	.10	2.95	2.56	2.35	2.22	2.13	2.06	2.01	1.97	1.93	1.90
	.05	4.30	3.44	3.05	2.82	2.66	2.55	2.46	2.40	2.34	2.30
	.01	7.95	5.72	4.82	4.31	3.99	3.76	3.59	3.45	3.35	3.26
	.001	14.38	9.61	7.80	6.81	6.19	5.76	5.44	5.19	4.99	4.83
23	.10	2.94	2.55	2.34	2.21	2.11	2.05	1.99	1.95	1.92	1.89
	.05	4.28	3.42	3.03	2.80	2.64	2.53	2.44	2.37	2.32	2.27
	.01	7.88	5.66	4.76	4.26	3.94	3.71	3.54	3.41	3.30	3.21
	.001	14.20	9.47	7.67	6.70	6.08	5.65	5.33	5.09	4.89	4.73
24	.10	2.93	2.54	2.33	2.19	2.10	2.04	1.98	1.94	1.91	1.88
	.05	4.26	3.40	3.01	2.78	2.62	2.51	2.42	2.36	2.30	2.25
	.01	7.82	5.61	4.72	4.22	3.90	3.67	3.50	3.36	3.26	3.17
	.001	14.03	9.34	7.55	6.59	5.98	5.55	5.23	4.99	4.80	4.64
25	.10	2.92	2.53	2.32	2.18	2.09	2.02	1.97	1.93	1.89	1.87
	.05	4.24	3.39	2.99	2.76	2.60	2.49	2.40	2.34	2.28	2.24
	.01	7.77	5.57	4.68	4.18	3.85	3.63	3.46	3.32	3.22	3.13
	.001	13.88	9.22	7.45	6.49	5.89	5.46	5.15	4.91	4.71	4.56
26	.10	2.91	2.52	2.31	2.17	2.08	2.01	1.96	1.92	1.88	1.86
	.05	4.23	3.37	2.98	2.74	2.59	2.47	2.39	2.32	2.27	2.22
	.01	7.72	5.53	4.64	4.14	3.82	3.59	3.42	3.29	3.18	3.09
	.001	13.74	9.12	7.36	6.41	5.80	5.38	5.07	4.83	4.64	4.48
27	.10	2.90	2.51	2.30	2.17	2.07	2.00	1.95	1.91	1.87	1.85
	.05	4.21	3.35	2.96	2.73	2.57	2.46	2.37	2.31	2.25	2.20
	.01	7.68	5.49	4.60	4.11	3.78	3.56	3.39	3.26	3.15	3.06
	.001	13.61	9.02	7.27	6.33	5.73	5.31	5.00	4.76	4.57	4.41
28	.10	2.89	2.50	2.29	2.16	2.06	2.00	1.94	1.90	1.87	1.84
	.05	4.20	3.34	2.95	2.71	2.56	2.45	2.36	2.29	2.24	2.19
	.01	7.64	5.45	4.57	4.07	3.75	3.53	3.36	3.23	3.12	3.03
	.001	13.50	8.93	7.19	6.25	5.66	5.24	4.93	4.69	4.50	4.35
29	.10	2.89	2.50	2.28	2.15	2.06	1.99	1.93	1.89	1.86	1.83
	.05	4.18	3.33	2.93	2.70	2.55	2.43	2.35	2.28	2.22	2.18
	.01	7.60	5.42	4.54	4.04	3.73	3.50	3.33	3.20	3.09	3.00
	.001	13.39	8.85	7.12	6.19	5.59	5.18	4.87	4.64	4.45	4.29

(continued)

TABLE 6 Values That Capture Specified Upper-Tail *F* Curve Areas (*Continued*)

		df_1									
df_2	Area	1	2	3	4	5	6	7	8	9	10
30	.10	2.88	2.49	2.28	2.14	2.05	1.98	1.93	1.88	1.85	1.82
	.05	4.17	3.32	2.92	2.69	2.53	2.42	2.33	2.27	2.21	2.16
	.01	7.56	5.39	4.51	4.02	3.70	3.47	3.30	3.17	3.07	2.98
	.001	13.29	8.77	7.05	6.12	5.53	5.12	4.82	4.58	4.39	4.24
40	.10	2.84	2.44	2.23	2.09	2.00	1.93	1.87	1.83	1.79	1.76
	.05	4.08	3.23	2.84	2.61	2.45	2.34	2.25	2.18	2.12	2.08
	.01	7.31	5.18	4.31	3.83	3.51	3.29	3.12	2.99	2.89	2.80
	.001	12.61	8.25	6.59	5.70	5.13	4.73	4.44	4.21	4.02	3.87
60	.10	2.79	2.39	2.18	2.04	1.95	1.87	1.82	1.77	1.74	1.71
	.05	4.00	3.15	2.76	2.53	2.37	2.25	2.17	2.10	2.04	1.99
	.01	7.08	4.98	4.13	3.65	3.34	3.12	2.95	2.82	2.72	2.63
	.001	11.97	7.77	6.17	5.31	4.76	4.37	4.09	3.86	3.69	3.54
90	.10	2.76	2.36	2.15	2.01	1.91	1.84	1.78	1.74	1.70	1.67
	.05	3.95	3.10	2.71	2.47	2.32	2.20	2.11	2.04	1.99	1.94
	.01	6.93	4.85	4.01	3.53	3.23	3.01	2.84	2.72	2.61	2.52
	.001	11.57	7.47	5.91	5.06	4.53	4.15	3.87	3.65	3.48	3.34
120	.10	2.75	2.35	2.13	1.99	1.90	1.82	1.77	1.72	1.68	1.65
	.05	3.92	3.07	2.68	2.45	2.29	2.18	2.09	2.02	1.96	1.91
	.01	6.85	4.79	3.95	3.48	3.17	2.96	2.79	2.66	2.56	2.47
	.001	11.38	7.32	5.78	4.95	4.42	4.04	3.77	3.55	3.38	3.24
240	.10	2.73	2.32	2.10	1.97	1.87	1.80	1.74	1.70	1.65	1.63
	.05	3.88	3.03	2.64	2.41	2.25	2.14	2.04	1.98	1.92	1.87
	.01	6.74	4.69	3.86	3.40	3.09	2.88	2.71	2.59	2.48	2.40
	.001	11.10	7.11	5.60	4.78	4.25	3.89	3.62	3.41	3.24	3.09
∞	.10	2.71	2.30	2.08	1.94	1.85	1.77	1.72	1.67	1.63	1.60
	.05	3.84	3.00	2.60	2.37	2.21	2.10	2.01	1.94	1.88	1.83
	.01	6.63	4.61	3.78	3.32	3.02	2.80	2.64	2.51	2.41	2.32
	.001	10.83	6.91	5.42	4.62	4.10	3.74	3.47	3.27	3.10	2.96

TABLE 7 Critical Values of q for the Studentized Range Distribution

Error df	Confidence level	Number of populations, treatments, or levels being compared							
		3	4	5	6	7	8	9	10
5	95%	4.60	5.22	5.67	6.03	6.33	6.58	6.80	6.99
	99%	6.98	7.80	8.42	8.91	9.32	9.67	9.97	10.24
6	95%	4.34	4.90	5.30	5.63	5.90	6.12	6.32	6.49
	99%	6.33	7.03	7.56	7.97	8.32	8.61	8.87	9.10
7	95%	4.16	4.68	5.06	5.36	5.61	5.82	6.00	6.16
	99%	5.92	6.54	7.01	7.37	7.68	7.94	8.17	8.37
8	95%	4.04	4.53	4.89	5.17	5.40	5.60	5.77	5.92
	99%	5.64	6.20	6.62	6.96	7.24	7.47	7.68	7.86
9	95%	3.95	4.41	4.76	5.02	5.24	5.43	5.59	5.74
	99%	5.43	5.96	6.35	6.66	6.91	7.13	7.33	7.49
10	95%	3.88	4.33	4.65	4.91	5.12	5.30	5.46	5.60
	99%	5.27	5.77	6.14	6.43	6.67	6.87	7.05	7.21
11	95%	3.82	4.26	4.57	4.82	5.03	5.20	5.35	5.49
	99%	5.15	5.62	5.97	6.25	6.48	6.67	6.84	6.99
12	95%	3.77	4.20	4.51	4.75	4.95	5.12	5.27	5.39
	99%	5.05	5.50	5.84	6.10	6.32	6.51	6.67	6.81
13	95%	3.73	4.15	4.45	4.69	4.88	5.05	5.19	5.32
	99%	4.96	5.40	5.73	5.98	6.19	6.37	6.53	6.67
14	95%	3.70	4.11	4.41	4.64	4.83	4.99	5.13	5.25
	99%	4.89	5.32	5.63	5.88	6.08	6.26	6.41	6.54
15	95%	3.67	4.08	4.37	4.59	4.78	4.94	5.08	5.20
	99%	4.84	5.25	5.56	5.80	5.99	6.16	6.31	6.44
16	95%	3.65	4.05	4.33	4.56	4.74	4.90	5.03	5.15
	99%	4.79	5.19	5.49	5.72	5.92	6.08	6.22	6.35
17	95%	3.63	4.02	4.30	4.52	4.70	4.86	4.99	5.11
	99%	4.74	5.14	5.43	5.66	5.85	6.01	6.15	6.27
18	95%	3.61	4.00	4.28	4.49	4.67	4.82	4.96	5.07
	99%	4.70	5.09	5.38	5.60	5.79	5.94	6.08	6.20
19	95%	3.59	3.98	4.25	4.47	4.65	4.79	4.92	5.04
	99%	4.67	5.05	5.33	5.55	5.73	5.89	6.02	6.14
20	95%	3.58	3.96	4.23	4.45	4.62	4.77	4.90	5.01
	99%	4.64	5.02	5.29	5.51	5.69	5.84	5.97	6.09
24	95%	3.53	3.90	4.17	4.37	4.54	4.68	4.81	4.92
	99%	4.55	4.91	5.17	5.37	5.54	5.69	5.81	5.92
30	95%	3.49	3.85	4.10	4.30	4.46	4.60	4.72	4.82
	99%	4.45	4.80	5.05	5.24	5.40	5.54	5.65	5.76
40	95%	3.44	3.79	4.04	4.23	4.39	4.52	4.63	4.73
	99%	4.37	4.70	4.93	5.11	5.26	5.39	5.50	5.60
60	95%	3.40	3.74	3.98	4.16	4.31	4.44	4.55	4.65
	99%	4.28	4.59	4.82	4.99	5.13	5.25	5.36	5.45
120	95%	3.36	3.68	3.92	4.10	4.24	4.36	4.47	4.56
	99%	4.20	4.50	4.71	4.87	5.01	5.12	5.21	5.30
∞	95%	3.31	3.63	3.86	4.03	4.17	4.29	4.39	4.47
	99%	4.12	4.40	4.60	4.76	4.88	4.99	5.08	5.16

TABLE 8 **Upper-Tail Areas for Chi-Square Distributions**

Right-tail area	df = 1	df = 2	df = 3	df = 4	df = 5
>0.100	<2.70	<4.60	<6.25	<7.77	<9.23
0.100	2.70	4.60	6.25	7.77	9.23
0.095	2.78	4.70	6.36	7.90	9.37
0.090	2.87	4.81	6.49	8.04	9.52
0.085	2.96	4.93	6.62	8.18	9.67
0.080	3.06	5.05	6.75	8.33	9.83
0.075	3.17	5.18	6.90	8.49	10.00
0.070	3.28	5.31	7.06	8.66	10.19
0.065	3.40	5.46	7.22	8.84	10.38
0.060	3.53	5.62	7.40	9.04	10.59
0.055	3.68	5.80	7.60	9.25	10.82
0.050	3.84	5.99	7.81	9.48	11.07
0.045	4.01	6.20	8.04	9.74	11.34
0.040	4.21	6.43	8.31	10.02	11.64
0.035	4.44	6.70	8.60	10.34	11.98
0.030	4.70	7.01	8.94	10.71	12.37
0.025	5.02	7.37	9.34	11.14	12.83
0.020	5.41	7.82	9.83	11.66	13.38
0.015	5.91	8.39	10.46	12.33	14.09
0.010	6.63	9.21	11.34	13.27	15.08
0.005	7.87	10.59	12.83	14.86	16.74
0.001	10.82	13.81	16.26	18.46	20.51
<0.001	>10.82	>13.81	>16.26	>18.46	>20.51

Right-tail area	df = 6	df = 7	df = 8	df = 9	df = 10
>0.100	<10.64	<12.01	<13.36	<14.68	<15.98
0.100	10.64	12.01	13.36	14.68	15.98
0.095	10.79	12.17	13.52	14.85	16.16
0.090	10.94	12.33	13.69	15.03	16.35
0.085	11.11	12.50	13.87	15.22	16.54
0.080	11.28	12.69	14.06	15.42	16.75
0.075	11.46	12.88	14.26	15.63	16.97
0.070	11.65	13.08	14.48	15.85	17.20
0.065	11.86	13.30	14.71	16.09	17.44
0.060	12.08	13.53	14.95	16.34	17.71
0.055	12.33	13.79	15.22	16.62	17.99
0.050	12.59	14.06	15.50	16.91	18.30
0.045	12.87	14.36	15.82	17.24	18.64
0.040	13.19	14.70	16.17	17.60	19.02
0.035	13.55	15.07	16.56	18.01	19.44
0.030	13.96	15.50	17.01	18.47	19.92
0.025	14.44	16.01	17.53	19.02	20.48
0.020	15.03	16.62	18.16	19.67	21.16
0.015	15.77	17.39	18.97	20.51	22.02
0.010	16.81	18.47	20.09	21.66	23.20
0.005	18.54	20.27	21.95	23.58	25.18
0.001	22.45	24.32	26.12	27.87	29.58
<0.001	>22.45	>24.32	>26.12	>27.87	>29.58

TABLE 8 Upper-Tail Areas for Chi-Square Distributions (*Continued*)

Right-tail area	df = 11	df = 12	df = 13	df = 14	df = 15
>0.100	<17.27	<18.54	<19.81	<21.06	<22.30
0.100	17.27	18.54	19.81	21.06	22.30
0.095	17.45	18.74	20.00	21.26	22.51
0.090	17.65	18.93	20.21	21.47	22.73
0.085	17.85	19.14	20.42	21.69	22.95
0.080	18.06	19.36	20.65	21.93	23.19
0.075	18.29	19.60	20.89	22.17	23.45
0.070	18.53	19.84	21.15	22.44	23.72
0.065	18.78	20.11	21.42	22.71	24.00
0.060	19.06	20.39	21.71	23.01	24.31
0.055	19.35	20.69	22.02	23.33	24.63
0.050	19.67	21.02	22.36	23.68	24.99
0.045	20.02	21.38	22.73	24.06	25.38
0.040	20.41	21.78	23.14	24.48	25.81
0.035	20.84	22.23	23.60	24.95	26.29
0.030	21.34	22.74	24.12	25.49	26.84
0.025	21.92	23.33	24.73	26.11	27.48
0.020	22.61	24.05	25.47	26.87	28.25
0.015	23.50	24.96	26.40	27.82	29.23
0.010	24.72	26.21	27.68	29.14	30.57
0.005	26.75	28.29	29.81	31.31	32.80
0.001	31.26	32.90	34.52	36.12	37.69
<0.001	>31.26	>32.90	>34.52	>36.12	>37.69

Right-tail area	df = 16	df = 17	df = 18	df = 19	df = 20
>0.100	<23.54	<24.77	<25.98	<27.20	<28.41
0.100	23.54	24.76	25.98	27.20	28.41
0.095	23.75	24.98	26.21	27.43	28.64
0.090	23.97	25.21	26.44	27.66	28.88
0.085	24.21	25.45	26.68	27.91	29.14
0.080	24.45	25.70	26.94	28.18	29.40
0.075	24.71	25.97	27.21	28.45	29.69
0.070	24.99	26.25	27.50	28.75	29.99
0.065	25.28	26.55	27.81	29.06	30.30
0.060	25.59	26.87	28.13	29.39	30.64
0.055	25.93	27.21	28.48	29.75	31.01
0.050	26.29	27.58	28.86	30.14	31.41
0.045	26.69	27.99	29.28	30.56	31.84
0.040	27.13	28.44	29.74	31.03	32.32
0.035	27.62	28.94	30.25	31.56	32.85
0.030	28.19	29.52	30.84	32.15	33.46
0.025	28.84	30.19	31.52	32.85	34.16
0.020	29.63	30.99	32.34	33.68	35.01
0.015	30.62	32.01	33.38	34.74	36.09
0.010	32.00	33.40	34.80	36.19	37.56
0.005	34.26	35.71	37.15	38.58	39.99
0.001	39.25	40.78	42.31	43.81	45.31
<0.001	>39.25	>40.78	>42.31	>43.81	>45.31

TABLE 9 Binomial Probabilities

$n = 5$

	p												
x	0.05	0.1	0.2	0.25	0.3	0.4	0.5	0.6	0.7	0.75	0.8	0.9	0.95
0	.774	.590	.328	.237	.168	.078	.031	.010	.002	.001	.000	.000	.000
1	.204	.328	.410	.396	.360	.259	.156	.077	.028	.015	.006	.000	.000
2	.021	.073	.205	.264	.309	.346	.313	.230	.132	.088	.051	.008	.001
3	.001	.008	.051	.088	.132	.230	.313	.346	.309	.264	.205	.073	.021
4	.000	.000	.006	.015	.028	.077	.156	.259	.360	.396	.410	.328	.204
5	.000	.000	.000	.001	.002	.010	.031	.078	.168	.237	.328	.590	.774

$n = 10$

	p												
x	0.05	0.1	0.2	0.25	0.3	0.4	0.5	0.6	0.7	0.75	0.8	0.9	0.95
0	.599	.349	.107	.056	.028	.006	.001	.000	.000	.000	.000	.000	.000
1	.315	.387	.268	.188	.121	.040	.010	.002	.000	.000	.000	.000	.000
2	.075	.194	.302	.282	.233	.121	.044	.011	.001	.000	.000	.000	.000
3	.010	.057	.201	.250	.267	.215	.117	.042	.009	.003	.001	.000	.000
4	.001	.011	.088	.146	.200	.251	.205	.111	.037	.016	.006	.000	.000
5	.000	.001	.026	.058	.103	.201	.246	.201	.103	.058	.026	.001	.000
6	.000	.000	.006	.016	.037	.111	.205	.251	.200	.146	.088	.011	.001
7	.000	.000	.001	.003	.009	.042	.117	.215	.267	.250	.201	.057	.010
8	.000	.000	.000	.000	.001	.011	.044	.121	.233	.282	.302	.194	.075
9	.000	.000	.000	.000	.000	.002	.010	.040	.121	.188	.268	.387	.315
10	.000	.000	.000	.000	.000	.000	.001	.006	.028	.056	.107	.349	.599

TABLE 9 Binomial Probabilities (*Continued*)

$n = 15$

	p												
x	0.05	0.1	0.2	0.25	0.3	0.4	0.5	0.6	0.7	0.75	0.8	0.9	0.95
0	.463	.206	.035	.013	.005	.000	.000	.000	.000	.000	.000	.000	.000
1	.366	.343	.132	.067	.031	.005	.000	.000	.000	.000	.000	.000	.000
2	.135	.267	.231	.156	.092	.022	.003	.000	.000	.000	.000	.000	.000
3	.031	.129	.250	.225	.170	.063	.014	.002	.000	.000	.000	.000	.000
4	.005	.043	.188	.225	.219	.127	.042	.007	.001	.000	.000	.000	.000
5	.001	.010	.103	.165	.206	.186	.092	.024	.003	.001	.000	.000	.000
6	.000	.002	.043	.092	.147	.207	.153	.061	.012	.003	.001	.000	.000
7	.000	.000	.014	.039	.081	.177	.196	.118	.035	.013	.003	.000	.000
8	.000	.000	.003	.013	.035	.118	.196	.177	.081	.039	.014	.000	.000
9	.000	.000	.001	.003	.012	.061	.153	.207	.147	.092	.043	.002	.000
10	.000	.000	.000	.001	.003	.024	.092	.186	.206	.165	.103	.010	.001
11	.000	.000	.000	.000	.001	.007	.042	.127	.219	.225	.188	.043	.005
12	.000	.000	.000	.000	.000	.002	.014	.063	.170	.225	.250	.129	.031
13	.000	.000	.000	.000	.000	.000	.003	.022	.092	.156	.231	.267	.135
14	.000	.000	.000	.000	.000	.000	.000	.005	.031	.067	.132	.343	.366
15	.000	.000	.000	.000	.000	.000	.000	.000	.005	.013	.035	.206	.463

$n = 20$

	p												
x	0.05	0.1	0.2	0.25	0.3	0.4	0.5	0.6	0.7	0.75	0.8	0.9	0.95
0	.358	.122	.012	.003	.001	.000	.000	.000	.000	.000	.000	.000	.000
1	.377	.270	.058	.021	.007	.000	.000	.000	.000	.000	.000	.000	.000
2	.189	.285	.137	.067	.028	.003	.000	.000	.000	.000	.000	.000	.000
3	.060	.190	.205	.134	.072	.012	.001	.000	.000	.000	.000	.000	.000
4	.013	.090	.218	.190	.130	.035	.005	.000	.000	.000	.000	.000	.000
5	.002	.032	.175	.202	.179	.075	.015	.001	.000	.000	.000	.000	.000
6	.000	.009	.109	.169	.192	.124	.037	.005	.000	.000	.000	.000	.000
7	.000	.002	.055	.112	.164	.166	.074	.015	.001	.000	.000	.000	.000
8	.000	.000	.022	.061	.114	.180	.120	.035	.004	.001	.000	.000	.000
9	.000	.000	.007	.027	.065	.160	.160	.071	.012	.003	.000	.000	.000
10	.000	.000	.002	.010	.031	.117	.176	.117	.031	.010	.002	.000	.000
11	.000	.000	.000	.003	.012	.071	.160	.160	.065	.027	.007	.000	.000
12	.000	.000	.000	.001	.004	.035	.120	.180	.114	.061	.022	.000	.000
13	.000	.000	.000	.000	.001	.015	.074	.166	.164	.112	.055	.002	.000
14	.000	.000	.000	.000	.000	.005	.037	.124	.192	.169	.109	.009	.000
15	.000	.000	.000	.000	.000	.001	.015	.075	.179	.202	.175	.032	.002
16	.000	.000	.000	.000	.000	.000	.005	.035	.130	.190	.218	.090	.013
17	.000	.000	.000	.000	.000	.000	.001	.012	.072	.134	.205	.190	.060
18	.000	.000	.000	.000	.000	.000	.000	.003	.028	.067	.137	.285	.189
19	.000	.000	.000	.000	.000	.000	.000	.000	.007	.021	.058	.270	.377
20	.000	.000	.000	.000	.000	.000	.000	.000	.001	.003	.012	.122	.358

(continued)

TABLE 9 Binomial Probabilities (*Continued*)

$n = 25$

	p												
x	0.05	0.1	0.2	0.25	0.3	0.4	0.5	0.6	0.7	0.75	0.8	0.9	0.95
0	.277	.072	.004	.001	.000	.000	.000	.000	.000	.000	.000	.000	.000
1	.365	.199	.024	.006	.001	.000	.000	.000	.000	.000	.000	.000	.000
2	.231	.266	.071	.025	.007	.000	.000	.000	.000	.000	.000	.000	.000
3	.093	.226	.136	.064	.024	.002	.000	.000	.000	.000	.000	.000	.000
4	.027	.138	.187	.118	.057	.007	.000	.000	.000	.000	.000	.000	.000
5	.006	.065	.196	.165	.103	.020	.002	.000	.000	.000	.000	.000	.000
6	.001	.024	.163	.183	.147	.044	.005	.000	.000	.000	.000	.000	.000
7	.000	.007	.111	.165	.171	.080	.014	.001	.000	.000	.000	.000	.000
8	.000	.002	.062	.124	.165	.120	.032	.003	.000	.000	.000	.000	.000
9	.000	.000	.029	.078	.134	.151	.061	.009	.000	.000	.000	.000	.000
10	.000	.000	.012	.042	.092	.161	.097	.021	.001	.000	.000	.000	.000
11	.000	.000	.004	.019	.054	.147	.133	.043	.004	.001	.000	.000	.000
12	.000	.000	.001	.007	.027	.114	.155	.076	.011	.002	.000	.000	.000
13	.000	.000	.000	.002	.011	.076	.155	.114	.027	.007	.001	.000	.000
14	.000	.000	.000	.001	.004	.043	.133	.147	.054	.019	.004	.000	.000
15	.000	.000	.000	.000	.001	.021	.097	.161	.092	.042	.012	.000	.000
16	.000	.000	.000	.000	.000	.009	.061	.151	.134	.078	.029	.000	.000
17	.000	.000	.000	.000	.000	.003	.032	.120	.165	.124	.062	.002	.000
18	.000	.000	.000	.000	.000	.001	.014	.080	.171	.165	.111	.007	.000
19	.000	.000	.000	.000	.000	.000	.005	.044	.147	.183	.163	.024	.001
20	.000	.000	.000	.000	.000	.000	.002	.020	.103	.165	.196	.065	.006
21	.000	.000	.000	.000	.000	.000	.000	.007	.057	.118	.187	.138	.027
22	.000	.000	.000	.000	.000	.000	.000	.002	.024	.064	.136	.226	.093
23	.000	.000	.000	.000	.000	.000	.000	.000	.007	.025	.071	.266	.231
24	.000	.000	.000	.000	.000	.000	.000	.000	.002	.006	.024	.199	.365
25	.000	.000	.000	.000	.000	.000	.000	.000	.000	.001	.004	.072	.277

Appendix B References

Chapter 1

Hock, Roger R. *Forty Studies That Changed Psychology: Exploration into the History of Psychological Research.* Prentice Hall, 1995.

Moore, David, and William Notz. *Statistics: Concepts and Controversies*, 8th ed. W. H. Freeman, 2012. (A nice, informal survey of statistical concepts and reasoning.)

Peck, Roxy, ed. *Statistics: A Guide to the Unknown*, 4th ed. Cengage Learning, 2006. (Short, nontechnical articles by a number of well-known statisticians and users of statistics on the application of statistics in various disciplines and subject areas.)

Utts, Jessica. *Seeing Through Statistics,* 4th ed. Cengage Learning, 2014. (A nice introduction to the fundamental ideas of statistical reasoning.)

Chapter 2

Cobb, George. *Introduction to the Design and Analysis of Experiments.* Wiley, 2008. (An interesting and thorough introduction to the design of experiments.)

Freedman, David, Robert Pisani, and Roger Purves. *Statistics*, 4th ed. W. W. Norton, 2007. (The first two chapters contain some interesting examples of both well-designed and poorly designed experimental studies.)

Lohr, Sharon. *Sampling Design and Analysis,* 2nd ed. Cengage Learning, 2010. (A nice discussion of sampling and sources of bias at an accessible level.)

Moore, David, and William Notz. *Statistics: Concepts and Controversies*, 8th ed. W. H. Freeman, 2012. (Contains an excellent chapter on the advantages and pitfalls of experimentation and another chapter in a similar vein on sample surveys and polls.)

Scheaffer, Richard L., William Mendenhall, and Lyman Ott. *Elementary Survey Sampling*, 7th ed. Cengage Learning, 2011. (An accessible yet thorough treatment of the subject.)

Sudman, Seymour, and Norman Bradburn. *Asking Questions: A Practical Guide to Questionnaire Design.* Jossey-Bass, 1982. (A good discussion of the art of questionnaire design.)

Chapter 3

Chambers, John, William Cleveland, Beat Kleiner, and Paul Tukey. *Graphical Methods for Data Analysis.* Wadsworth, 1983. (This is an excellent survey of methods, illustrated with numerous interesting examples.)

Cleveland, William. *The Elements of Graphing Data*, 2nd ed. Hobart Press, 1994. (An informal and informative introduction to various aspects of graphical analysis.)

Freedman, David, Robert Pisani, and Roger Purves. *Statistics*, 4th ed. W. W. Norton, 2007. (An excellent, informal introduction to concepts, with some insightful cautionary examples concerning misuses of statistical methods.)

Moore, David, and William Notz. *Statistics: Concepts and Controversies*, 8th ed. W. H. Freeman, 2012. (A nonmathematical yet highly entertaining introduction to our discipline. Two thumbs up!)

Chapter 4

Chambers, John, William Cleveland, Beat Kleiner, and Paul Tukey. *Graphical Methods for Data Analysis.* Wadsworth, 1983. (This is an excellent survey of methods, illustrated with numerous interesting examples.)

Cleveland, William. *The Elements of Graphing Data*, 2nd ed. Hobart Press, 1994. (An informal and informative introduction to various aspects of graphical analysis.)

Freedman, David, Robert Pisani, and Roger Purves. *Statistics*, 4th ed. W. W. Norton, 2007. (An excellent, informal introduction to concepts, with some insightful cautionary examples concerning misuses of statistical methods.)

Moore, David, and William Notz. *Statistics: Concepts and Controversies*, 8th ed. W. H. Freeman, 2012. (A nonmathematical yet highly entertaining introduction to our discipline. Two thumbs up!)

Chapter 5

Neter, John, William Wasserman, and Michael Kutner. *Applied Linear Statistical Models*, 5th ed. McGraw-Hill, 2005. (The first half of this book gives a comprehensive treatment of regression analysis without overindulging in mathematical development; a highly recommended reference.)

Chapter 6

Devore, Jay L. *Probability and Statistics for Engineering and the Sciences*, 8th ed. Cengage Learning, 2011. (The treatment of probability in this source is more comprehensive and at a somewhat higher mathematical level than ours is in this textbook.)

Mosteller, Frederick, Robert Rourke, and George Thomas. *Probability with Statistical Applications*. Addison-Wesley, 1970. (A good introduction to probability at a modest mathematical level.)

Chapter 8

Freedman, David, Robert Pisani, and Roger Purves. *Statistics*, 4th ed. W. W. Norton, 2007. (This book gives an excellent informal discussion of sampling distributions.)

Chapter 9

Devore, Jay L. *Probability and Statistics for Engineering and the Sciences*, 8th ed. Cengage Learning, 2011. (This book gives a somewhat general introduction to confidence intervals.)

Freedman, David, Robert Pisani, and Roger Purves, *Statistics*, 4th ed. W. W. Norton, 2007. (This book contains an informal discussion of confidence intervals.)

Chapter 10

The books by Freedman et al. and Moore listed in previous chapter references are excellent sources. Their orientation is primarily conceptual, with a minimum of mathematical development, and both sources offer many valuable insights.

Chapter 11

Devore, Jay. *Probability and Statistics for Engineering and the Sciences*, 8th ed. Cengage Learning, 2011. (Contains a somewhat more comprehensive treatment of the inferential material presented in this and the two previous chapters, although the notation is a bit more mathematical than that of the present textbook.)

Chapter 12

Agresti, Alan, and B. Finlay. *Statistical Methods for the Social Sciences*, 4th ed. Prentice Hall, 2009. (This book includes a good discussion of measures of association for two-way frequency tables.)

Everitt, B. S. *The Analysis of Contingency Tables*. Halsted Press, 1977. (A compact but informative survey of methods for analyzing categorical data.)

Mosteller, Frederick, and Robert Rourke. *Sturdy Statistics*. Addison-Wesley, 1973. (Contains several readable chapters on the varied uses of the chi-square statistic.)

Chapter 13

Neter, John, William Wasserman, and Michael Kutner. *Applied Linear Statistical Models*, 5th ed. McGraw-Hill, 2005. (The first half of this book gives a comprehensive treatment of regression analysis without overindulging in mathematical development; a highly recommended reference.)

Chapter 14

Kerlinger, Fred N., and Elazar J. Pedhazur, *Multiple Regression in Behavioral Research*, 3rd ed. Holt, Rinehart & Winston, 1997. (A readable introduction to multiple regression.)

Neter, John, William Wasserman, and Michael Kutner. *Applied Linear Statistical Models*, 4th ed. McGraw-Hill, 1996. (The first half of this book gives a comprehensive treatment of regression analysis without overindulging in mathematical development; a highly recommended reference.)

Chapter 15

Miller, Rupert. *Beyond ANOVA: The Basics of Applied Statistics*. Wiley, 1986. (This book contains a wealth of information concerning violations of basic assumptions and alternative methods of analysis.)

Winer, G. J., D. R. Brown, and K. M. Michels, *Statistical Principles in Experimental Design*, 3rd ed. McGraw-Hill, 1991. (This book contains extended discussion of ANOVA with many examples worked out in great detail.)

Chapter 16

Conover, W. J. *Practical Nonparametric Statistics*, 3rd ed. Wiley, 1999. (An accessible presentation of distribution-free methods.)

Daniel, Wayne. *Applied Nonparametric Statistics*, 2nd ed. PWS-Kent, 1990. (An elementary presentation of distribution-free methods, including the rank-sum test discussed in Section 16.1.)

Mosteller, Frederick, and Richard Rourke. *Sturdy Statistics*. Addison-Wesley, 1973. (A readable, intuitive development of distribution-free methods, including those based on ranks.)

Answers to Selected Odd-Numbered Exercises

Chapter 1

1.1 *Descriptive statistics* is the branch of statistics that involves the organization and summary of the values in a data set. *Inferential statistics* is the branch of statistics concerned with reaching conclusions about a population based on the information provided by a sample.
1.3 Sample
1.5 **a.** The population of interest is the set of all 15,000 students at the university. **b.** The sample is the 200 students who are interviewed.
1.7 The population is the set of all 7000 property owners. The sample is the group of 500 owners included in the survey.
1.9 The population is the set of 5000 used bricks. The sample is the set of 100 bricks she checks.
1.11 **a.** The researchers wanted to know if taking a garlic supplement reduces the risk of getting a cold. **b.** We would want to know how many people participated in the study and how the people were assigned to the two groups.
1.13 **a.** Categorical **b.** Categorical **c.** Numerical (discrete) **d.** Numerical (continuous) **e.** Categorical **f.** Numerical (continuous)
1.15 **a.** Continuous **b.** Continuous **c.** Continuous **d.** Discrete
1.17 **a.** Gender of purchaser, brand of motorcycle, telephone area code **b.** Number of previous motorcycles **c.** Bar chart **d.** Dotplot
1.19 **b.** Meat and poultry items appear to be relatively low cost sources of protein.
1.21 **b.** The most common reason was financial, accounting for 30.2% of students who left for non-academic reasons. The next two most common reasons were health and other personal reasons, accounting for 19.0% and 15.9%, respectively, of the students who left for non-academic reasons.
1.23 **a.** There were two sites that received far greater numbers of visits than the remaining 23 sites. Also, the distribution of the number of visits has the greatest density of points for the smaller numbers of visits, with the density decreasing as the number of visits increases. **b.** There were two sites that were used by far greater numbers of individuals (unique visitors) than the remaining 23 sites. However, these two sites are less far above the others in terms of the number of unique visitors than they are in terms of the total number of visits. **c.** The statistic "visits per unique visitor" tells us how heavily the individuals are using the sites.
1.25 **b.** Eastern states have, on average, lower wireless percents than states in the other two regions. The West and Middle states regions have, on average, roughly equal wireless percents.
1.27 **a.** *Rate per 10,000 flights*
1.29 **b.** By far the largest number of students purchased books at the campus bookstore. Many also used an online bookstore or an off-campus bookstore. Relatively few students fell into the other response categories.
1.31

Type of Household	Relative Frequency
Nonfamily	0.29
Married with children	0.27
Married without children	0.29
Single parent	0.15

1.33 **a.** Categorical **c.** No
1.35 **b.** The most frequently occurring violation categories were security (43%) and maintenance (39%). The least frequently occurring violation categories were flight operations (6%) and hazardous materials (3%).

Chapter 2

2.1 Observational study, because the researchers did not assign participants to the length of stay groups.
2.3 Experiment, because the professor determined which students were in each of the two experimental groups.
2.5 **a.** Experiment, because researchers decided which participants would receive which treatments. **b.** Yes, because the participants were randomly assigned to the treatments.
2.7 **a.** Experiment. **b.** Yes, because the participants were randomly assigned to the treatments.
2.9 We are told that moderate drinkers, as a group, tended to be better educated, wealthier, and more active than nondrinkers. It is possible the observed reduction in the risk of heart disease among moderate drinkers is caused by one of these attributes and not by the moderate drinking.
2.11 **a.** The data would need to be collected from a simple random sample of affluent Americans. **b.** No
2.13 Number the names on the list from 1 to n, where n is the number of students at the college. Then use a random number generator to select 100 different numbers between 1 and n. Students corresponding to these 100 numbers would be included in the sample.
2.15 Number names on list, use random number generator to select 30 numbers between 1 and 500. Signatures corresponding to the 30 selected numbers constitute the random sample.
2.17 **a.** all American women **b.** No, only women for three states were included in the sample. **c.** no **d.** The description does not

say how the women were selected, so it is difficult to tell. However, selection bias is present because women from most states had no chance of being included in the sample.
2.19 a. Using the list, first number the part-time students 1–3000. Use a random number generator on a calculator or computer to randomly select a whole number between 1 and 3000. The number selected represents the first part-time student to be included in the sample. Repeat the number selection, ignoring repeated numbers, until 10 part-time students have been selected. Then number the full-time students 1–3500 and select 10 full-time students using the same procedure. **b.** No
2.21 a. The pages of the book have already been numbered between 1 and the highest page number in the book. Use a random number generator on a calculator or computer to randomly select a whole number between 1 and the highest page number in the book. The number selected will be the first page to be included in the sample. Repeat the number selection, ignoring repeated numbers, until the required number of pages has been selected. **b.** Pages that include exercises tend to contain more words than pages that do not include exercises. Therefore, it would be sensible to stratify according to this criterion. Assuming that 20 nonexercise pages and 20 exercise pages will be included in the sample, the sample should be selected as follows: Use a random number generator to randomly select a whole number between 1 and the highest page number in the book; the number selected will be the first page to be included in the sample; repeat the number selection, ignoring repeated numbers and keeping track of the number of pages of each type selected, until 20 pages of one type have been selected; then continue in the same way, but ignore numbers corresponding to pages of that type; when 20 pages of the other type have been selected, stop the process. **c.** Randomly select one page from the first 20 pages in the book. Include in your sample that page and every 20th page from that page onward. **d.** Roughly speaking, in terms of the numbers of words per page, each chapter is representative of the book as a whole. It is therefore sensible for the chapters to be used as clusters. Using a random number generator, randomly choose three chapters. Then count the number of words on each page in those three chapters. **e.** Answers will vary. **f.** Answers will vary.
2.23 The researchers should be concerned about nonresponse bias.
2.25 It is not reasonable to consider the participants to be representative of all students with regard to their truthfulness in the various forms of communication. Also, the students knew they were surveying themselves as to the truthfulness of their interactions. This could easily have changed their behavior in particular social contexts and, therefore, could have distorted the results of the study.
2.27 It is quite possible that people who read that newspaper or access this web site differ from the population in some relevant way, particularly considering that they are both New York City-based publications.
2.29 a. Yes. It is possible that students of a given class-standing tend to be similar in the amounts of money they spend on textbooks. **b.** Yes. It is possible that students who pursue a certain field of study tend to be similar in the amounts of money they spend on textbooks. **c.** No. It is unlikely that stratifying in this way will produce groups that are homogeneous in terms of the students' spending on textbooks.
2.31 No, because the people who sent their hair for testing did so voluntarily. It is quite possible that people who would choose to participate in a study of this sort differ in their mercury levels from the population as a whole.
2.33 a. Binding strength **b.** Type of glue **c.** The extraneous variables mentioned are the number of pages in the book and whether the book is bound as a hardback or a paperback. Further extraneous variables that might be considered include the weight of the material used for the cover and the type of paper used.
2.35 Answers will vary. One possible answer: Write the names of the 30 subjects on slips of paper. Mix the slips and then select 10. Assign these 10 subjects to hand-drying method 1. Mix the remaining slips and then select 10. Assign these 10 subjects to hand-drying method 2. Assign the remaining 10 subjects to method 3.
2.39 The figure shows that comparable groups in terms of age have been formed.
2.41 We rely on random assignment to produce comparable experimental groups.
2.45 a. Experiment **b.** no **c.** yes **d.** Yes, because the experiment used random assignment of subjects to treatments. **e.** No, because the subjects were not randomly selected.
2.47 a. Red wine, yellow onions, black tea **b.** Absorption of flavonol into the blood **c.** Gender, amount of flavonols consumed apart from experimental treatment, tolerance of alcohol in wine
2.49 "Blinding" is ensuring that the experimental subjects do not know which treatment they were given and/or ensuring that the people who measure the response variable do not know who was given which treatment.
2.51 a. In order to know that the results of this experiment are valid, it is necessary to know that the assignment of the women to the groups was done randomly. If the women were allowed to choose which groups they went into, it would be impossible to tell whether the stated results were caused by the discussions of art or by the greater social nature of the women in the art discussion group. **b.** Suppose that all the women took part in weekly discussions of art, and that, generally, an improvement in the medical conditions mentioned was observed among the subjects. Then it would be impossible to tell whether these health improvements had been caused by the discussions of art or by some factor that was affecting all the subjects, such as an improvement in the weather over the 4 months. By including a control group, and by observing that the improvements did not take place (generally speaking) for those in the control group, factors such as this can be discounted, and the discussions of art are established as the cause of the improvements.
2.55 Answers will vary. One possible answer: Number the girls from 1 to 700 and then use a random number generator to select 350 of these girls for the book group. Assign the remaining girls to the other group. Then number the boys from 1 to 600 and use a random number generator to select 300 of the boys for the book group Assign the remaining boys to the other group.
2.57 a. If the judges had known which chowder came from which restaurant, then it is unlikely that Denny's chowder would have won the contest, since the judges would probably be conditioned by this knowledge to choose chowders from more expensive restaurants. **b.** In experiments, if the people measuring the response are not blinded, they will often be conditioned to see different responses to some treatments over other treatments, in the same way as the judges would have been conditioned to favor the expensive restaurant chowders. Therefore, it is necessary that the people measuring the response should not know which subject received which treatment, so that the treatments can be compared on their own merits.

2.59 **a.** A placebo group would be necessary if the mere thought of having amalgam fillings could produce kidney disorders. However, since the experimental subjects were sheep, the researchers do not need to be concerned that this would happen. **b.** A resin filling treatment group would be necessary in order to provide evidence that it is the material in the amalgam fillings, rather than the process of filling the teeth, or just the presence of foreign bodies in the teeth, that is the cause of the kidney disorders. If the amalgam filling group developed the kidney disorders and the resin filling group did not, then this would provide evidence that it is some ingredient in the amalgam fillings that is causing the kidney problems. **c.** Since there is concern about the effect of amalgam fillings, it would be considered unethical to use humans in the experiment.
2.61 **a.** This is an observational study. **b.** In order to evaluate the study, we need to know whether the sample was a random sample. **c.** No. Since the sample used in the Healthy Steps study was known to be nationally representative, and since the paper states that, compared with the HS trial, parents in the study sample were disproportionately older, white, more educated, and married, it is clear that it is not reasonable to regard the sample as representative of parents of all children at age 5.5 years. **d.** The potential confounding variable mentioned is what the children watched.
e. The quotation from Kamila Mistry makes a statement about cause and effect and therefore is inconsistent with the statement that the study cannot show that TV was the cause of later problems.
2.63 Answers will vary.
2.65 By randomly selecting the phone numbers, calling back those for which there are no answers, and asking for the adult in the household with the most recent birthday, the researchers are avoiding selection bias. However, selection bias could result from the fact that not all Californians have phones.
2.67 We rely on random assignment to produce comparable experimental groups. If the researchers had hand-picked the treatment groups, they might unconsciously have favored one group over the other in terms of some variable that affects the ability of the people at the centers to respond to the materials provided.
2.69 **a.** Observational study **b.** It is quite possible that the children who watched large amounts of TV in their early years were also those, generally speaking, who received less attention from their parents, and it was the lack of attention from their parents that caused the later attention problems, not the TV-watching.
2.71 For example, it is possible that people who are not married are more likely to go out alone (except for the widowed who are older and therefore tend to stay home). It could then be possible that going out alone is causing the risk of being a victim of violent crime, not the marital status.
2.73 All the participants were women, from Texas, and volunteers. All three of these facts tell us that it is likely to be unreasonable to generalize the results of the study to all college students.
2.75 **a.** The extraneous variables identified are gender, age, weight, lean body mass, and capacity to lift weights. They were dealt with by direct control: all the volunteers were male, about the same age, and similar in weight, lean body mass, and capacity to lift weights. **b.** Yes, it is important that the men were not told which treatment they were receiving; otherwise the effect of giving a placebo would have been removed. If the participants *were* told which treatment they were receiving, then those taking the creatine would have the additional effect of the mere taking of a supplement thought to be helpful (the placebo effect) and those getting the fake preparation would not get this effect. It would then be impossible to distinguish the influence of the placebo effect from the effect of the creatine itself. **c.** Yes, it would have been useful if those measuring the increase in muscle mass had not known who received which treatment. With this knowledge, it is possible that the people would have been unconsciously influenced into exaggerating the increase in muscle mass for those who took the creatine.
2.77 **a.** Answers will vary. **b.** Answers will vary. Some examples include soil type, amount of water, amount of sunlight.
c. experiment
2.79 **a.** Answers will vary. **b.** Answers will vary.

Chapter 3

3.3 **a.** The second and third categories ("Permitted for business purposes only" and "Permitted for limited personal use" were combined into one category ("No, but some limits apply"). **c.** Pie chart, regular bar graph
3.5 **a.**

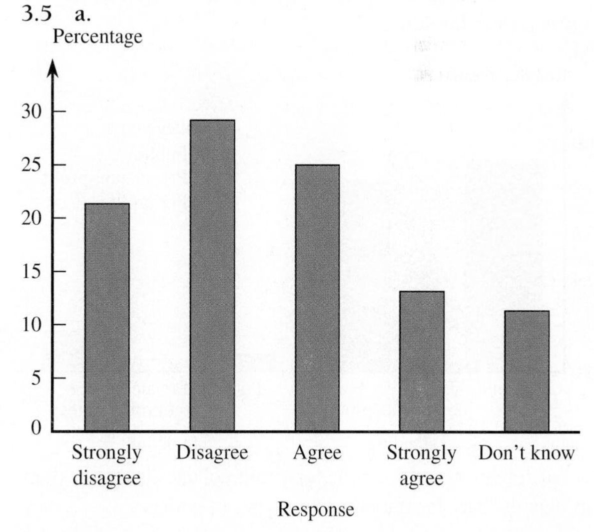

b. Answers will vary. One possible answer: Majority of students not ready for ebooks.
3.7 **b.** Were the surveys carried out on random samples of married women from those countries? How were the questions worded? **c.** In one country, Japan, the percentage of women who say they never get help from their husbands is far higher than the percentages in any of the other four countries included. The percentages in the other four countries are similar, with Canada showing the lowest percentage of women who say they do not get help from their husbands.

3.11 a.

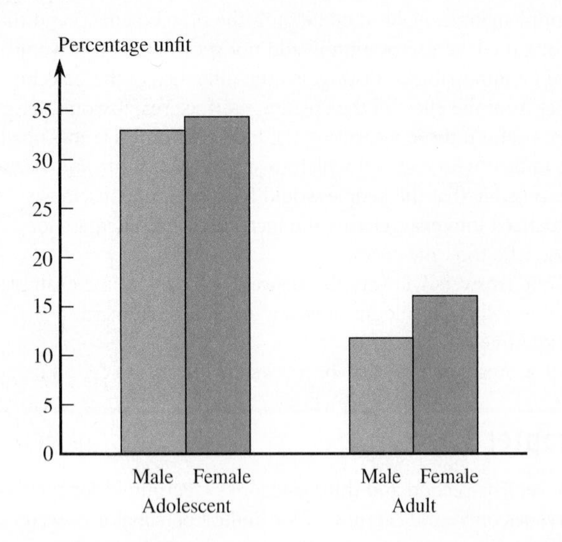

b. The percentage unfit was higher for females in both age categories, but the difference in percentage unfit for males and females was greater for adults.

3.13 a.

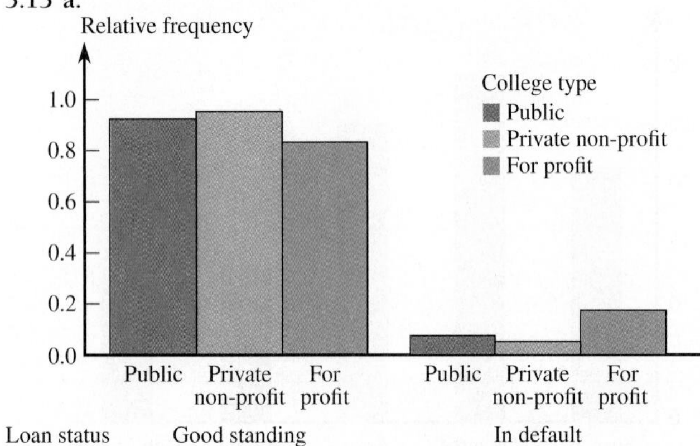

b. The bar for "in default" for the for-profit colleges is higher than the "in default" bars for the other two types of colleges.

3.15

10	578
11	79
12	1114
13	001122478899
14	0011112235669
15	11122445599
16	1227
17	1
18	
19	
20	8

Stem: Ones
Leaf: Tenths

A typical number of births per thousand of the population is around 14, with most birth rates concentrated in the 13.0 to 15.9 range. The distribution has just one peak (at the 14–15 class). There is an extreme value, 20.8, at the high end of the data set, and this is the only birth rate above 17.1. The distribution is not symmetrical, since it has a greater spread to the right of its center than to the left.

3.17 a.

0H	55567889999
1L	0000111113334
1H	556666666667789
2L	00001122233
2H	5

Stem: Tens
Leaf: Ones

A typical percentage of households with only a wireless phone is around 15.

b.

West		East
998	0H	555789
110	1L	00011134
8766	1H	666
21	2L	00
5	2H	

Stem: Tens
Leaf: Ones

A typical percentage of households with only a wireless phone for the West is around 16, which is greater than that for the East (around 11). There is a slightly greater spread of values in the West than in the East, with values in the West ranging from 8 to 25 (a range of 17) and values in the East ranging from 5 to 20 (a range of 15). The distribution for the West is roughly symmetrical, while the distribution in the East shows a slightly greater spread to the right of its center than to the left. Neither distribution has any outliers.

3.19 a.

0	8
1	44
1	67788999
2	00001122233333444
2	5557778
3	012223344
3	57
4	024
4	58

Stem: Tens
Leaf: Ones

b. The data distribution is centered at around 25 cents, but there is a lot of variability from state to state. **c.** None of the data values stand out as outliers.

3.21

0t	4455555
0s	66677777
0*	888888888999999999
1	00001111
1t	222223
1f	445

The stem-and-leaf display shows that the distribution of high school dropout rates is roughly symmetrical. A typical dropout rate is 7%. The great majority of rates are between 4% and 9%, inclusive.

3.23 The distribution of maximum wind speeds is positively skewed and is bimodal, with peaks at the 35–40 and 60–65 intervals.

3.25 b. The typical percentage of workers belonging to a union is around 11, with values ranging from 3.5 to 24.9. There are three states with percentages that stand out as being higher than those of the rest of the states. The distribution is positively skewed. **c.** The dotplot is more informative as it shows where the data points actually lie. For example, in the histogram we can tell that there are three observations in the 20 to 25 interval, but we don't see the actual values and miss the fact that these values are actually considerably higher than the other values in the data set. **d.** The histogram in Part (a) could be taken to imply that there are states with a percent of workers belonging to a union near zero. It is clear from this second histogram that this is not the case. Also, the second histogram shows that there is a gap at the high end and that the three largest values are noticeably higher than those of the other states. This fact is not clear from the histogram in Part (a).

3.27 a.

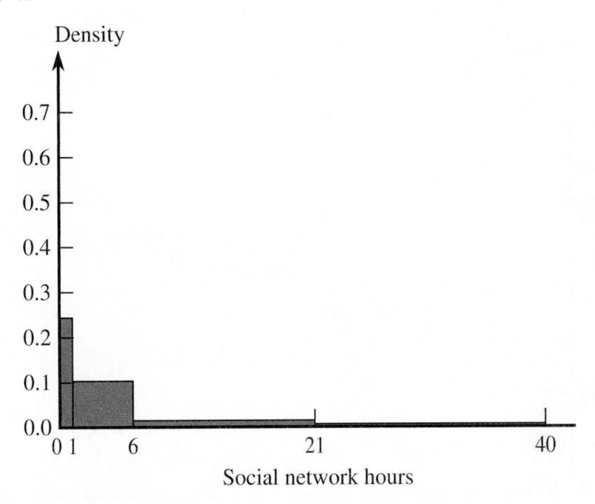

b.

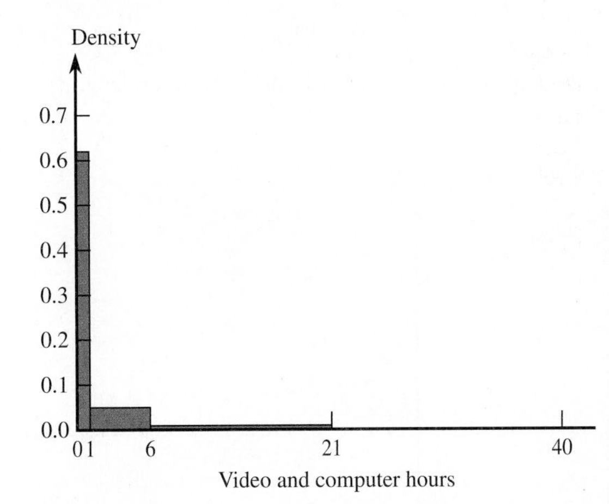

c. Both histograms are positively skewed, but a typical value for social networking hours is greater than the typical value for video and computer hours.

3.29 a.

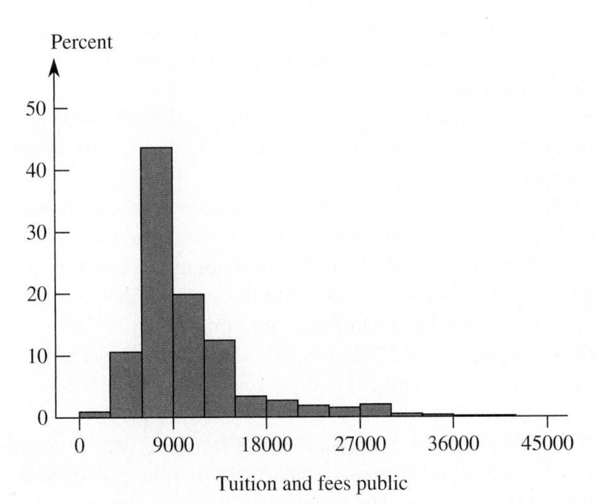

The distribution of tuition and fees at public colleges is positively skewed. A typical value for tuition and fees is around $10,000 and there is a lot of variability from college to college.

b.

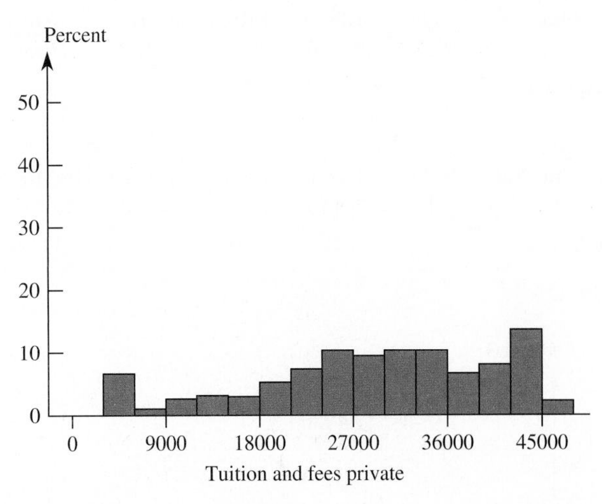

The distribution of tuition and fees at private not for profit colleges is negatively skewed. A typical value for tuition and fees is around $30,000 and there is a lot of variability from college to college. **c.** Tuition and fees at private not for profit colleges tends to be greater than for public colleges. There is also more variability in tuition and fees for the private not for profit colleges than for public colleges.

3.35 a.

Years Survived	Relative Frequency
0 to < 2	.10
2 to < 4	.42
4 to < 6	.02
6 to < 8	.10
8 to < 10	.04
10 to < 12	.02
12 to < 14	.02
14 to < 16	.28

c. The histogram shows a bimodal distribution, with peaks at the 2–4 year and 14–16 year intervals. All the other survival times were considerably less common than these two. **d.** We would need to know that the set of patients used in the study formed a random sample of all patients younger than 50 years old who had been diagnosed with the disease and had received the high dose chemotherapy.

3.37 Answers will vary. One possibility for each part is shown below.

a.

Class Interval	100 to <120	120 to <140	140 to <160	160 to <180	180 to <200
Frequency	5	10	40	10	5

b.

Class Interval	100 to <120	120 to <140	140 to <160	160 to <180	180 to <200
Frequency	20	10	4	25	11

c.

Class Interval	100 to <120	120 to <140	140 to <160	160 to <180	180 to <200
Frequency	33	15	10	7	5

d.

Class Interval	100 to <120	120 to <140	140 to <160	160 to <180	180 to <200
Frequency	5	7	10	15	33

3.39 a.

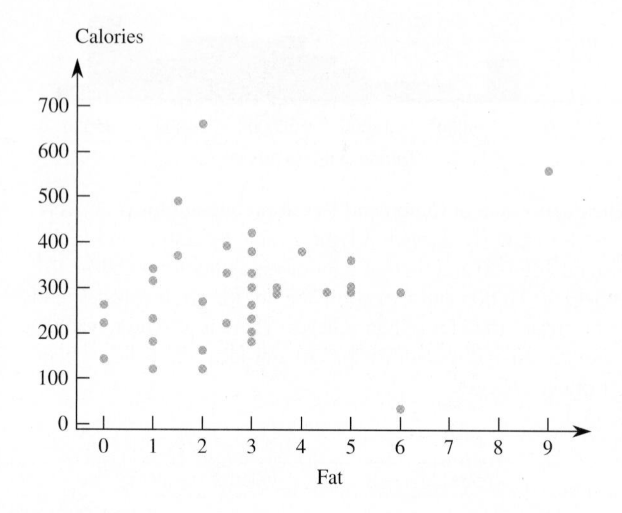

There does appear to be a relationship between fat and calories. As fat increases, calories also tends to increase, which is not surprising.

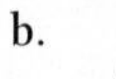
b.

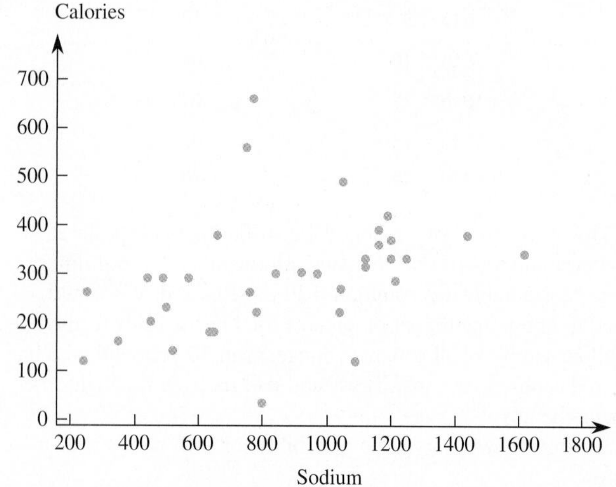

The relationship between sodium and calories is not strong, but it does appear that fast food items with more sodium also tend to have more calories.

c.

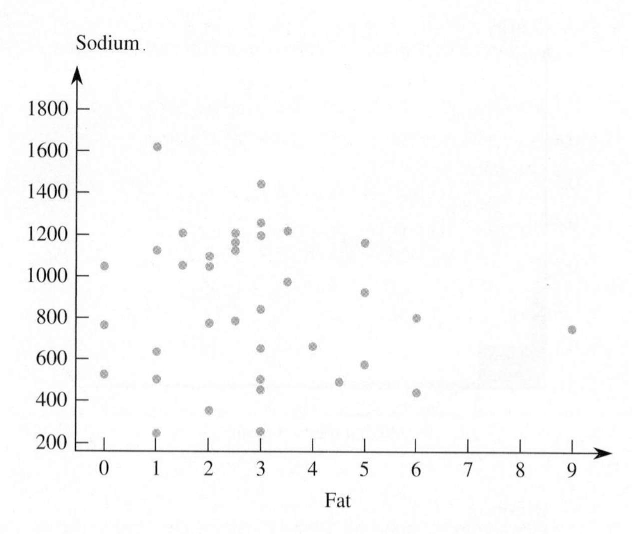

There does not appear to be a relationship between fat and sodium.

d.

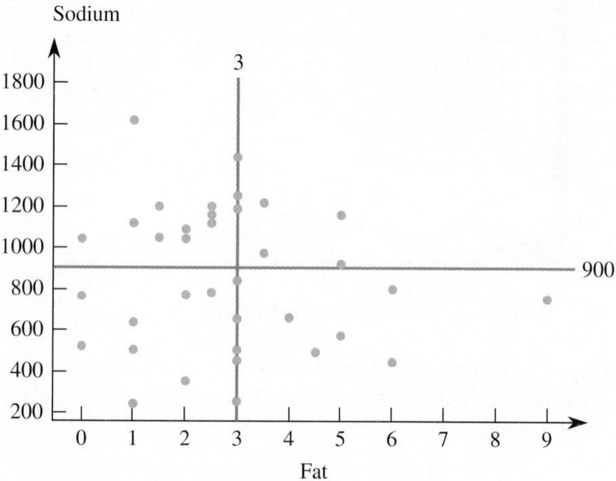

The region in the lower left hand corner corresponds to the healthier fast food choices. These are the choices that are lower in both sodium and fat.

3.41 b. There appears to be a very weak, positive relationship between cost and quality rating. The scatterplot only very weakly supports the statement.

3.43 The plot shows that the amount of waste collected for recycling had grown substantially (not slowly, as is stated in the article) in the years 1990 to 2005. The amount increased from under 30 million tons to nearly 60 million tons in that period, which means that the amount had almost doubled.

3.45 According to the 2001 and 2002 data, there are seasonal peaks at weeks 4, 9, and 14, and seasonal lows at weeks 2, 6, 10–12, and 18.

3.47 b. The graphical display created in Part (a) is more informative, since it gives an accurate representation of the proportions of the ethnic groups. **c.** The people who designed the original display possibly felt that the four ethnic groups shown in the segmented bar section might seem to be underrepresented at the college if they used a single pie chart.

3.49 The first graphical display is not drawn appropriately. The Z's have been drawn so that their heights are in proportion to the percentages shown. However, the widths and the perceived depths are also in proportion to the percentages, and so neither the areas nor the perceived volumes of the Z's are proportional to the percentages. The graph is therefore misleading to the reader. In the second graphical display, however, *only* the heights of the cars are in proportion to the percentages shown. The widths of the cars are all equal. Therefore the areas of the cars are in proportion to the percentages, and this is an appropriately drawn graphical display.

3.51 The piles of cocaine have been drawn so that their heights are in proportion to the percentages shown. However, the widths are also in proportion to the percentages, and therefore neither the areas (nor the perceived volumes) are in proportion to the percentages. The graph is therefore misleading to the reader.

3.55 a.

Stem	Leaf
1	9
2	23788999
3	0011112233459
4	0123

Stem: Tens
Leaf: Ones

b. A typical calorie content for these light beers is 31 calories per 100 ml, with the great majority lying in the 22–39 range. The distribution is negatively skewed, with one peak (in the 30–39 range). There are no gaps in the data.

3.57 a.

Stem	Leaf
0	003334455556888888888999999
1	0001223344567
2	001123689
3	0
4	0
5	
6	6

Stem: Tens
Leaf: Ones

b. A typical percentage population increase is around 10, with the great majority of states in the 0–29 range. The distribution is positively skewed, with one peak (in the 0–9 range). There are two states showing significantly greater increases than the other 48 states: one at 40 (Arizona) and one at 66 (Nevada).

c.

West		East
9988880	0	033344555568889999
432	1	0001234567
982100	2	136
0	3	
0	4	
	5	
6	6	

Stem: Tens
Leaf: Ones

On average, the percentage population increases in the West were greater than those for the East, with a typical value for the West of about 14 and a typical value for the East of about 9. There is a far greater spread in the values in the West, with values ranging from 0 to 66, than in the East where values ranged from 0 to 26. Both distributions are positively skewed, with a single peak for the East data, and two peaks for the West. In the West, there are two states showing significantly greater increases than the remaining states, with values at 40 and 60. There are no such extreme values in the East.

3.59 a. High graft weight ratios are clearly associated with low body weights (and vice versa), and the relationship is not linear. (In fact, roughly speaking, there seems to be an inverse proportionality between the two variables, apart from a small increase in the graft weight ratios for increasing body weights among those recipients with the greater body weights. This is interesting in that an inverse proportionality between the variables would imply that the actual weights of transplanted livers are chosen independently of the recipients' body weights.) **b.** A likely reason for the negative relationship is that the livers to be transplanted are probably chosen according to whatever happens to be available at the time. Therefore, lighter patients are likely to receive livers that are too large and heavier patients are likely to receive livers that are too small.

3.61 b. Continuing the growth trend, we estimate that the average home size in 2010 will be approximately 2500 square feet.

3.63 a.

Disney		Other
975332100	0	0001259
765	1	156
920	2	0
	3	
	4	
4	5	

Stem: Hundreds
Leaf: Tens

b. On average, the total tobacco exposure times for the Disney movies are higher than the others, with a typical value for Disney of about 90 seconds and a typical value for the other companies of about 50 seconds. Both distributions have one peak and are positively skewed. There is one extreme value (548) in the Disney data and no extreme value in the data for the other companies. There is a greater spread in the Disney data, with values ranging from 6 seconds to 540 seconds, than for the other companies, with values ranging from 1 second to 205 seconds.

3.65 c. The segmented bar graph is slightly preferable in that it is a little easier in it than in the pie chart to see that the proportion of children responding "Most of the time" was slightly higher than the proportion responding "Some of the time."

3.67 b. The peaks were probably caused by the incidence of major hurricanes in those years.

3.69 d. In every year the number of related donors was much greater than the number of unrelated donors. In both categories the number of transplants increased every year, but the increases in unrelated donors were greater proportionately than the increases in related donors.

3.71 a.

Skeletal Retention	Frequency
0.15 to <0.20	4
0.20 to <0.25	2
0.25 to <0.30	5
0.30 to <0.35	21
0.35 to <0.40	9
0.40 to <0.45	9
0.45 to <0.50	4
0.50 to <0.55	0
0.55 to <0.60	1

b. The histogram is centered at approximately 0.34, with values ranging from 0.15 to 0.5, plus one extreme value in the 0.55–0.6 range. The distribution has a single peak and is slightly positively skewed.

Cumulative Review 3

CR3.1 No. For example, it is quite possible that men who ate a high proportion of cruciferous vegetables generally speaking also had healthier lifestyles than those who did not, and that it

was the healthier lifestyles that were causing the lower incidence of prostate cancer, not the eating of cruciferous vegetables.
CR3.3 Very often those who choose to respond generally have a different opinion on the subject of the study from those who do not respond. (In particular, those who respond often have strong feelings against the status quo.) This can lead to results that are not representative of the population that is being studied.
CR3.5 Only a small proportion (around 11%) of the doctors responded, and it is quite possible that those who did respond had different opinions regarding managed care than the majority who did not. Therefore, the results could have been very inaccurate for the population of doctors in California.
CR3.7 For example, suppose the women had been allowed to choose whether or not they participated in the program. Then it is quite possible, generally speaking, that those women with more social awareness would have chosen to participate, and those with less social awareness would have chosen not to. Then it would be impossible to tell whether the stated results came about as a result of the program or of the greater social awareness among the women who participated. By randomly assigning the women to participate or not, comparable groups of women would have been obtained.
CR3.9 **b.** Between 2002 and 2003 and between 2003 and 2004, the pass rates rose for both the high school and the state, with a particularly sharp rise between 2003 and 2004 for the state. However, the pass rate for the county fell between 2002 and 2003 and then rose between 2003 and 2004.
CR3.11
a.

0	123334555599	
1	00122234688	
2	1112344477	
3	0113338	
4	37	Stem: Thousands
5	23778	Leaf: Hundreds

The stem-and-leaf display shows a positively skewed distribution with a single peak. There are no extreme values. A typical total length is around 2100 and the great majority of total lengths lie in the 100 to 3800 range.
c. The number of subdivisions that have total lengths less than 2000 is 12 + 11 = 23, and so the proportion of subdivisions that have total lengths less than 2000 is 23/47 = 0.489.
The number of subdivisions that have total lengths between 2000 and 4000 is 10 + 7 = 17, and so the proportion of subdivisions that have total lengths between 2000 and 4000 is 17/47 = 0.361.
CR3.13 **a.** The histogram shows a smooth positively skewed distribution with a single peak. **b.** A typical time difference between the two phases of the race is 150 seconds, with the majority of time differences lying between 50 and 350 seconds. There are about three values that could be considered extreme, with those values lying in the 650 to 750 range. **c.** Estimating the frequencies from the histogram we see that approximately 920 runners were included in the study and that approximately eight of those runners ran the late distance more quickly than the early distance (indicated by a negative time difference). Therefore the proportion of runners who ran the late distance more quickly than the early distance is approximately 8/920 = 0.009.
CR3.15 There is a strong negative linear relationship between racket resonance frequency and sum of peak-to-peak accelerations. There are two rackets with data points separated from the remaining data points. Those two rackets have very high resonance frequencies and their peak-to-peak accelerations are lower than those of all the other rackets.

Chapter 4

4.1 **a.** $\bar{x}$ = \$2118.71. Median = \$1688. **b.** The value of the mean is pulled up by the two large values in the data set. **c.** The median is better since it is not influenced by the two extreme values.
4.3 The mean caffeine concentration for the brands of coffee listed = 125.417 mg/cup. Therefore, the mean caffeine concentration of the coffee brands in mg/ounce is 15.677. This is significantly greater than the previous mean caffeine concentration of the energy drinks.
4.5 It tells us that a small number of individuals who donate a large amount of time are greatly increasing the mean.
4.7 **a.** The mean is greater than the median. **b.** The mean is 683,315.2. The median is 433,246.5. **c.** The median **d.** It is not reasonable to generalize this sample of daily newspapers to the population of the United States since the sample consists only of the top 20 newspapers in the country.
4.9 **a.** The mean is 448.3. **b.** The median is 446. **c.** This sample represents the 20 days with the highest number of speeding-related fatalities, and so it is not reasonable to generalize from this sample to the other 345 days of the year.
4.11 Neither statement is correct. Regarding the first statement, we should note that unless the "fairly expensive houses" constitute a majority of the houses selling, these more costly houses will not have an effect on the median. Turning to the second statement, we point out that the small number of very high or very low prices will have no effect on the median, whatever the number of sales. Both statements can be corrected by replacing the median with the mean.
4.13 The two possible solutions are $x_5 = 32$ and $x_5 = 39.5$.
4.15 The median is 680 hours.
4.17 **a.** $\bar{x} = 52.111$. Variance = 279.111. $s = 16.707$.
b. The addition of the very expensive cheese would increase both the mean and the standard deviation.
4.19 **a.** Lower quartile = 4th value = 41. Upper quartile = 12th value = 62. Iqr = 21. **b.** The iqr for cereals rated good (calculated in exercise 4.18) is 24. This is greater than the value calculated in Part (a).
4.21 $\bar{x} = 51.33$ $s = 15.22$. A typical amount poured into tall slender glass is 51.33 ml. A typical deviation from the mean amount poured is 15.22 ml.
4.23 **a.** Variance = 1176027.905, $s = 1084.448$. **b.** The fairly large value of the standard deviation tells us that there is considerable variation between the repair costs.
4.25 **a.** Lower quartile = 0. Upper quartile = 195. Interquartile range = 195. **b.** The lower quartile equals the minimum value for this data set because there is a large number of equal values (zero in this case) at the lower end of the distribution. This is unusual, and therefore, generally speaking, the lower quartile is not equal to the minimum value.
4.27 The volatility of a portfolio is the amount by which its return varies, and the standard deviation of the annual returns measures exactly that. A lower-risk portfolio is one in which the return is less variable. This can be interpreted in terms of a low standard deviation of annual returns.

4.29 **a.** i **b.** iii **c.** iv

4.31 a.

	Mean	Standard Deviation	Coefficient of Variation
Sample 1	7.81	0.398	5.102
Sample 2	49.68	1.739	3.500

b. The values of the coefficient of variation are given in the table in Part (a). The fact that the coefficient of variation is smaller for Sample 2 than for Sample 1 is not surprising since, relative to the actual amount placed in the containers, it is easier to be more accurate when larger amounts are being placed in the containers.
4.33 **a.** Median = 58. Lower quartile = 13th value = 53.5. Upper quartile = 38th value = 64.4. **b.** Lower quartile − 1.5(iqr) = 53.5 −1.5(10.9) = 37.15. Since 28.2 and 35.7 are both less than 37.15, they are both outliers. **c.** The median percentage of population born and still living in the state is 58. There are two outliers at the lower end of the distribution. If they are disregarded the distribution is roughly symmetrical, with values ranging from 40.4 to 75.8.
4.35 **a.**

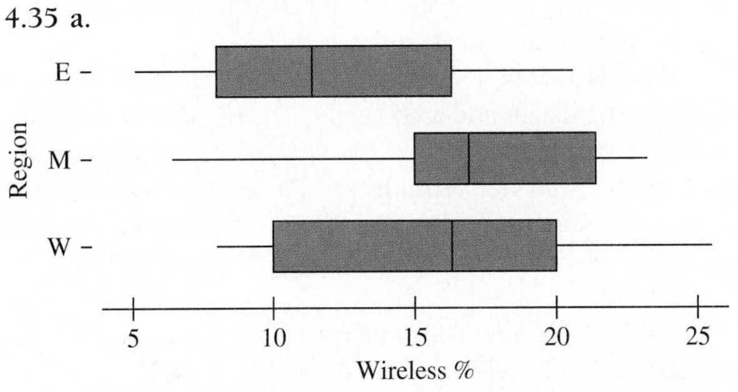

b. The distributions of wireless percent for the Midwest states and West states are centered at a higher value than for states in the East.
4.37 **a.** It would be more appropriate to use the interquartile range than the standard deviation. **b.** Lower quartile = 81.5, upper quartile = 94, iqr = 12.5. (Lower quartile) − 3(iqr) = 81.5 − 3(12.5) = 44. (Lower quartile) − 1.5(iqr) = 81.5 − 1.5(12.5) = 62.75 (Upper quartile) + 1.5(iqr) = 94 + 1.5(12.5) = 112.75. (Upper quartile) + 3(iqr) = 94 + 3(12.5) = 131.5. Since the value for students (152) is greater than 131.5, this is an extreme outlier. Since the value for farmers (43) is less than 44, this is an extreme outlier. There are no mild outliers. **d.** The insurance company might decide only to offer discounts to occupations that are outliers at the lower end of the distribution, in which case only farmers would receive the discount. If the company was willing to offer discounts to the quarter of occupations with the lowest accident rates, then the last 10 occupations on the list should be the ones to receive discounts.
4.39 **a.** Roughly 68% of speeds would have been between those two values. **b.** Roughly 16% of speeds would exceed 57 mph.
4.41 **a.** At least 75% of observations must lie between those two values. **b.** The required interval is (2.90, 70.94). **c.** The distribution cannot be approximately normal.
4.43 For the first test, $z = 1.5$; for the second test, $z = 1.875$. Since the student's z score in the second test is higher, the student did better relative to the other test takers in the second test.
4.45 At least 10% of the students had no debt. At least 25% of the students had no debt. Approximately 50% of the students had a debt of $11,000 or less. Approximately 75% of the students had a debt of $24,600 or less. Approximately 90% of the students had a debt of $39,300 or less.
4.47 The best conclusion we can reach is that at most 16% of weight readings will be between 49.75 and 50.25.
4.49 The distribution is positively skewed. Using Chebyshev's Rule, we can conclude that at most, 18 students changed at least six answers from correct to incorrect.
4.51 **a.**

Per Capita Expenditure	Frequency
0 to <2	13
2 to <4	18
4 to <6	10
6 to <8	5
8 to <10	1
10 to <12	2
12 to <14	0
14 to <16	0
16 to <18	2

b. i. 3.4 **ii.** 5.0 **iii.** 0.8 **iv.** 8.0 **v.** 2.8
4.53 **a.** The minimum value and the lower quartile were both 1. **b.** More than half of the data values were equal to the minimum value. **c.** Between 25% and 50% of patients had unacceptable times to defibrillation.
d. (Upper quartile) + 3(iqr) = 9. Since 7 is less than 9, 7 must be a mild outlier.
4.55 **a.** $\bar{x} = 287.714$. The seven deviations are: 209.286, −94.714, 40.286, −132.714, 38.286, −42.714, −17.714.
b. The sum of the rounded deviations is 0.002.
c. Variance = 12,601.905. Standard deviation = 112.258
4.57 **a.** This is the median, and its value is $4286. The other measure of center is the mean, and its value is $3968.67. **b.** This is smaller than the median and, therefore, less favorable to the supervisors.
4.59 **a.** This is a correct interpretation of the median. **b.** Here the word "range" is being used to describe the interval from the minimum value to the maximum value. The statement claims that the median is defined as the midpoint of this interval, which is not true. **c.** If there is no home below $300,000, then certainly the median will be greater than $300,000 (unless more than half of the homes cost exactly $300,000).
4.61 The new mean is $\bar{x} = 38.364$. The new values and their deviations from the mean are shown in the table below.

Value	Deviation
52	13.636
13	−25.364
17	−21.364
46	7.636
42	3.636
24	−14.364
32	−6.364
30	−8.364
58	19.636
35	−3.364

The value of s^2 for the new values is the same as for the old values.
4.63 a. Lower quartile = 44, upper quartile = 53, iqr = 9. (Lower quartile) − 1.5(iqr) = 44 − 1.5(9) = 30.5. (Upper quartile) + 1.5(iqr) = 53 + 1.5(9) = 66.5. Since there are no data values less than 30.5 and no data values greater than 66.5, there are no outliers. **b.** The median of the distribution is 46. The middle 50% of the data range from 44 to 53 and the whole data set ranges from 33 to 60. There are no outliers. The lower half of the middle 50% of data values shows less spread than the upper half of the middle 50% of data values. The spread of the lowest 25% of data values is slightly greater than the spread of the highest 25% of data values.
4.65 a. $\bar{x} = 192.571$. This is a measure of center that incorporates all the sample values. Median = 189. This is a measure of center that is the "middle value" in the sample. **b.** The mean would decrease and the median would remain the same.
4.67 The median aluminum contamination is 119. There is one (extreme) outlier—a value of 511. If the outlier is disregarded, the data values range from 30 to 291. The middle 50% of data values range from 87 to 182. Even if the outlier is disregarded, the distribution is positively skewed.
4.69 The medians for the three different types of hotel are roughly the same, with the median for the midrange hotels slightly higher than the other two medians. The midrange hotels have two outliers (one extreme) at the lower end of the distribution and the first-class hotels have one (extreme) outlier at the lower end. There are no outliers for the budget hotels. If the outliers are taken into account, the midrange and first-class groups have a greater range than the budget group. If the outliers are disregarded, the budget group has a much greater spread than the other two groups. If the outliers are taken into account, all three distributions are negatively skewed. If the outliers are disregarded, the distribution for the budget group is negatively skewed whereas the distributions for the other two groups are positively skewed.
4.71 The distribution is positively skewed.
4.73 a. The 84th percentile is 120. **b.** The standard deviation is approximately 20. **c.** $z = -0.5$. **d.** 140 is at approximately the 97.5th percentile. **e.** A score of 40 is 3 standard deviations below the mean, and so the proportion of scores below 40 would be approximately (100 − 99.7)/2 = 0.15%. Therefore, there would be very few scores below 40.

Chapter 5

5.1 **Scatterplot 1 (i)** Yes **(ii)** Yes **(iii)** Positive
Scatterplot 2 (i) Yes **(ii)** Yes **(iii)** Negative
Scatterplot 3 (i) Yes **(ii)** No **(iii)** -
Scatterplot 4 (i) Yes **(ii)** Yes **(iii)** Negative
5.3 No. A correlation coefficient of 0 implies that there is no *linear* relationship between the variables. There could still be a nonlinear relationship.
5.5 a.

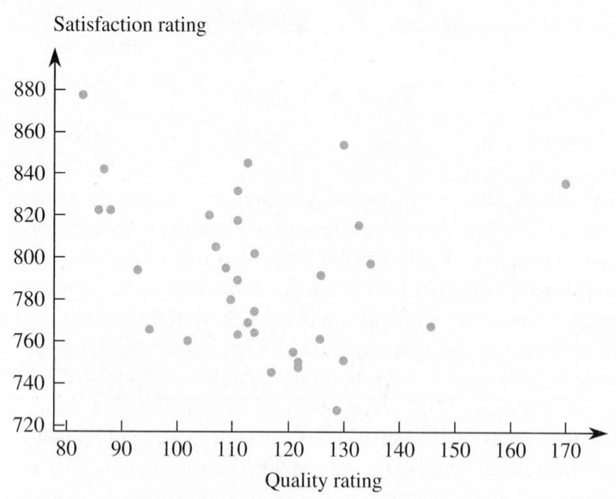

There does not appear to be a relationship between quality rating and satisfaction rating.
b. $r = -0.239$. The linear relationship between quality rating and satisfaction rating is weak and negative.
5.7 a. There is a weak negative linear relationship between average hopping height and arch height. The negative correlation tells us that large arch heights are (weakly) associated with small hopping heights, and vice versa. **b.** Yes. Since all five correlation coefficients are close to zero, the results imply that there is at best a weak linear relationship between arch height and the motor skills measured.
5.9 a. $r = 0.118$ **b.** Yes. The scatterplot does not suggest a strong relationship between the variables.
5.11 $r = 0.935$.
There is a strong positive linear relationship.
5.13 The correlation coefficient is most likely to be close to −0.9.
5.15 a. The equation of the least squares regression line is $\hat{y} = 11.482 - 0.970x$, where x = mini-Wright meter reading and y = Wright meter reading. **b.** When $x = 500$, $\hat{y} = 496.454$. The least squares regression line predicts a Wright meter reading of 496.454 for a person whose mini-Wright meter reading was 500. **c.** When $x = 300$, $\hat{y} = 302.465$. The least squares regression line predicts a Wright meter reading of 302.465 for a person whose mini-Wright meter reading was 500. However, 300 is well outside the range of mini-Wright meter readings in the data set, so this prediction involves extrapolation and is therefore not reliable.
5.17 a. The dependent variable is the number of fruit-and-vegetable servings per day, and the predictor variable is the number of hours of television viewed per day. **b.** Negative. As the number of hours of television viewed per day increases, the number of fruit-and-vegetable servings per day (on average) decreases.
5.19 The equation of the least-squares regression line is $\hat{y} = 0.884 + 0.699x$, where x = passenger complaint rank and y = J.D. Powers quality score. When $x = 3$, $\hat{y} = 2.982$, and when $x = 9$, $\hat{y} = 7.178$.
5.21 a. There is a strong negative relationship between mean call-to-shock time and survival rate. The relationship is close to linear, particularly if the point with the highest x value is disregarded. If that point is included, then there is the suggestion of a curve. **b.** $\hat{y} = 101.32847 - 9.29562x$
5.25 We do not know that the same linear relationship will apply for x values outside this range.

5.27 $b = r(s_y/s_x)$, where s_y and s_x are the standard deviations of the y values and the x values, respectively. Since standard deviations are always positive, b and r must have the same sign.
5.29 a. The equation of the least squares regression line is $\hat{y} = -9.071 + 1.571x$, where x = years of schooling and y = median hourly wage gain. When $x = 15$, $\hat{y} = 14.5$. The least squares regression line predicts a median hourly wage gain of 14.5 percent for the 15th year of schooling. **b.** The actual wage gain percent is very close to the value predicted in Part (a).
5.31 b. The scatterplot for girls shows stronger evidence of a curved relationship.**c.** $\hat{y} = 479.997 + 12.525x$ **e.** The decision to use a curve is supported by the curved pattern in the residual plot.
5.33 a The equation of the least-squares regression line is $\hat{y} = 232.258 - 2.926x$, , where x = rock surface area and y = algae colony density. **b.** $r^2 = 0.300$. Thirty percent of the variation in algae colony density can be attributed to the approximate linear relationship between rock surface area and algae colony density. **c.** $s_e = 63.315$. This is a typical deviation of an algae colony density value from the value predicted by the least-squares regression line. **d.** $r = -0.547$. The linear relationship between rock surface area and algae colony density is moderate and negative.
5.35 a. $\hat{y} = 4788.839 - 29.015x$ **b.** Residuals: −149.964, −44.514, 96.236, 168.886, 97.236, 36.286, −204.164.
c. The clear curve in the residual plot suggests that a linear model is not appropriate. **d.** No, data only included men.
5.37 a. 0.154 **b.** No, since the r^2 value for y = first-year-college GPA and x = SAT II score was 0.16, which is not large. Only 16% of the variation in first-year-college GPA could be attributed to the approximate linear relationship between SAT II score and first-year-college GPA.
5.39 b. $\hat{y} = 85.334 - 0.0000259x$. $r^2 = 0.016$. **c.** The line will not give accurate predictions. **d.** Deleting the point (620,231, 67), the equation of the least-squares line is now $\hat{y} = 83.402 + 0.0000387x$. Removal of the point does greatly affect the equation of the line.
5.41 $r^2 = 0.951$. 95.1% of the variation in hardness is attributable to the approximate linear relationship between time elapsed and hardness.
5.43 a. $r = 0$, $\hat{y} = \bar{y}$. **b.** For values of r close to 1 or −1, s_e will be much smaller than s_y. **c.** $s_e \approx 1.5$. **d.** $\hat{y} = 7.92 + 0.544x$, $s_e \approx 1.02$
5.45 a. The pattern in the scatterplot is curved.
b. The curved pattern is not as pronounced as in the scatterplot from Part (a), but there is still curvature in the plot.
c. Answers will vary.
5.47 a. The relationship appears to be nonlinear.
b. There is an obvious pattern in the residual plot.
c. The second transformation (ln(x) vs. y) seems to be somewhat more successful in straightening the data.
d. $\hat{y} = 62.42 + 43.81x$, where $\hat{y}$ is the predicted success %, and x is the natural logarithm of the energy of shock.
e. 86.94%, 52.64%
5.49 a. The scatterplot shows a positive relationship.
b. The squaring transformation straightens the plot.
c. This transformation straightens the plot. In addition, the variability in log(Agricultural Intensity) does not seem to increase as population density increases, as was the case in Part (b).
d. These transformations were not effective in straightening the plot. There is obvious curvature in the scatterplot.
5.51 a. Any image plotted between the dashed lines would be associated with Cal Poly by approximately the same percentages of enrolling and nonenrolling students. **b.** The images that were more commonly associated with nonenrolling students than with enrolling students were: "Average," "Isolated," and "Back-up school," with "Back-up school" as the most common of these among nonenrolling students. The images that were more commonly associated with enrolling students than with nonenrolling students were (in increasing order of commonality among enrolling students): "Excitingly different," "Personal," "Selective," "Prestigious," "Exciting," "Intellectual," "Challenging," "Comfortable," "Fun," "Career-oriented," "Highly respected," and "Friendly," with this last image marked by over 60% of students who enrolled and over 45% of students who didn't enroll. The most commonly marked image among students who didn't enroll was "Career-oriented."
5.53 a. $r = 0.943$, strong positive linear relationship. No, we cannot conclude that lead exposure causes increased assault rates. A value of r close to 1 tells us that there is a strong linear association but tells us nothing about causation. **b.** $\hat{y} = -24.08 + 327.41x$. When $x = 0.5$, $\hat{y} = 139.625$ assaults per 100,000 people. **c.** 0.89 **d.** The two time-series plots, generally speaking, move together. Thus, high assault rates are associated with high lead exposures 23 years earlier, and low assault rates are associated with low lead exposures 23 years earlier.
5.55 a. $r = -0.981$, strong linear relationship **b.** The word *linear* is not the most effective description of the relationship. A curve would provide a better fit.
5.57 a. One point, (0, 77), is far separated from the other points in the plot. There is a negative relationship. **b.** There appears to be a negative linear relationship between test anxiety and exam score. **c.** $r = -0.912$. This is consistent with the observations given in Part (b). **d.** No, we cannot conclude that test anxiety caused poor exam performance.
5.59 a. There is a positive relationship between the percentages of students who were proficient at the two times. There is the suggestion of a curve in the plot.
b. $\hat{y} = -3.13603 + 1.52206x$
c. When $x = 14$, $\hat{y} = 18.173$. This is slightly lower than the actual value of 20 for Nevada.
5.61 a. $y - \hat{y} = 20.796$ **b.** $r = -0.755$ **c.** $s_e = 11.638$
5.63 a. $r = -0.717$ **b.** $r = -0.835$; the absolute value of this correlation is greater than the absolute value of the correlation calculated in Part (a). This suggests that the transformation was successful in straightening the plot.
5.65 b. Using log(y) and log(x), 15.007 parts per million
5.67 a. $r = 0$ **b.** For example, adding the point (6, 1) gives $r = 0.510$. (Any y-coordinate greater than 0.973 will work.)
c. For example, adding the point (6, −1) gives $r = -0.510$. (Any y-coordinate less than −0.973 will work.)

Cumulative Review 5

CR5.3 The peaks in rainfall do seem to be followed by peaks in the number of *E. coli* cases, with rainfall peaks around May 12, May 17, and May 23 followed by peaks in the number of cases on May 17, May 23, and May 28. (The incubation period seems to be more like 5 days than the 3 to 4 days mentioned in the caption.) Thus, the graph does show a close connection between unusually heavy rainfall and the incidence of the infection. The storms may not be *responsible* for the increased illness levels, however, since the graph can only show us association, not causation.

CR5.5 The random pattern in the scatterplot shows there is little relationship between the weight of the mare and the weight of her foal. This is supported by the value of the correlation coefficient.
CR5.7 **b.** Mean = 3.654%, median = 3.35%.
CR5.9 **b.** Approximately equal **c.** Mean = \$9.459, median = \$9.48 **e.** Mean = \$72.85, median = \$68.61
CR5.11 **a.** $\bar{x} = 2965.2$, $s^2 = 294416.622$, $s = 542.602$, Lower quartile = 2510, Upper quartile = 3112, Interquartile range = 602 **b.** less
CR5.13 **a.** $\bar{x} = 4.93$, Median = 3.6. The mean is greater than the median. This is explained by the fact that the distribution of blood lead levels is positively skewed. **b.** The median blood lead level for the African Americans (3.6) is slightly higher than for the Whites (3.1). Both distributions are positively skewed. There are two outliers in the data set for the African Americans. The distribution for the African Americans shows a greater range than the distribution for the Whites, even disregarding the two outliers.
CR5.15 **a.** Yes **b.** Strong positive linear relationship **c.** Perfect correlation would result in the points lying exactly on some straight line, but not necessarily on the line described.
CR5.17 **a.** 76.64% of the variability in clutch size can be attributed to the approximate linear relationship between snout-vent length and clutch size. **b.** $s_e = 29.250$. This is a typical deviation of an observed clutch size from the clutch size predicted by the least-squares line.
CR5.19 **a.** Yes, the scatterplot shows a strong positive association. **b.** The plot seems to be straight, particularly if you disregard the point with the greatest *x* value. **c.** This transformation is successful in straightening the plot. Also, unlike the plot in Part (b), the variability of the quantity measured on the vertical axis does not seem to increase as *x* increases. **d.** No, this transformation has not been successful in producing a linear relationship. There is a curved pattern in the plot.

Chapter 6

6.1 A chance experiment is any activity or situation in which there is uncertainty about which of two or more possible outcomes will result.
6.3 **a.** {AA, AM, MA, MM} **c. i.** B = {AA, AM, MA} **ii.** C = {AM, MA} **iii.** D = {MM}. D is a simple event. **d.** *B and C* = {AM, MA}; *B or C* = {AA, AM, MA}.
6.5 **b. i.** $A^C = \{(15, 50), (15, 100), (15, 150), (15, 200)\}$

$$\text{ii. } A \cup B = \begin{Bmatrix} (10, 50), (10, 100), (10, 150), (10, 200), (12, 50) \\ (12, 100), (12, 150), (12, 200), (15, 50), (15, 100) \end{Bmatrix}$$

iii. $A \cap B = \{(10, 50), (10, 100), (12, 50), (12, 100)\}$
c. A and C are not disjoint. B and C are disjoint.
6.7 **b.** A = {3, 4, 5} **c.** C = {125, 15, 215, 25, 5}
6.9 **a.** A = {NN, DNN, NDN} **b.** B = {DDNN, DNDN, NDDN} **c.** The number of outcomes is infinite.
6.13 **a.** {expedited overnight delivery, expedited second-business-day delivery, standard delivery, delivery to the nearest store for customer pick-up} **b. i.** 0.2 **ii.** 0.4 **iii.** 0.6
6.15 **a.** {fiction hardcover, fiction paperback, fiction digital, fiction audio, nonfiction hardcover, nonfiction paperback, nonfiction digital, nonfiction audio} **b.** No **c.** 0.72 **d.** 0.28; 0.28 **e.** 0.8
6.17 **a.** P(Red) = 18/38 = 0.474. **b.** No. If the segments were rearranged, there would still be 18 red segments out of a total of 38 segments, and so the probability of landing in a red segment would still be 18/38 = 0.474. **c.** 1000(0.474) = 474. So, if the wheel is balanced you would expect the resulting color to be red on approximately 474 of the 1000 spins. If the number of spins resulting in red is very different from 474, then you would suspect that the wheel is not balanced.
6.19 **a.** 0.3 **b.** No. The problems to be graded will be selected at random, so the probability of being able to turn in both of the selected problems is the same whatever your choice of three problems to complete. **c.** 0.6
6.21 P(credit card used) = 37100/80500 = 0.461.
6.23 **a.** P(female) = 488142/787325 = 0.620. **b.** P(male) = 1 − 0.620 = 0.380.
6.25 **a.** 0.000495, 0.00198 **b.** 0.024 **c.** 0.146
6.27 **a.** BC, BM, BP, BS, CM, CP, CS, MP, MS, PS **b.** 0.1 **c.** 0.4 **d.** 0.3
6.29 **a.** $P(O_1) = P(O_3) = P(O_5) = 1/9$ and $P(O_2) = P(O_4) = P(O_6) = 2/9$ **b.** 1/3, 4/9 **c.** 9/21, 2/7
6.31 **a.** 0.24 **b.** 0.36 **c.** 0.12 **d.** 0.75
6.33 **a.** 0.72 **b.** The value 0.45 is the conditional probability that the selected individual drinks 2 or more cups a day given that he or she drinks coffee. We know this because the percentages given in the display total 100, and yet we know that only 72% of Americans drink coffee. So, the percentages given in the table must be the proportions *of coffee drinkers* who drink the given amounts.
6.37 $P(A|B)$ is larger. $P(A|B)$ is the probability that a randomly chosen professional basketball player is over 6 feet tall—a reasonably large probability. $P(B|A)$ is the probability that a randomly chosen person over 6 feet tall is a professional basketball player—a very small probability.
6.39 **a.** 0.337 **b.** 0.762 **c.** 0.625 **d.** 0.788 **e.** 0.869 **f.** Current smokers are the least likely to believe that smoking is very harmful, and those who have never smoked are the most likely to think that smoking is very harmful. This is not surprising since you would expect those who smoke to be the most confident about the health prospects of a smoker, with those who formerly smoked being a little more concerned, and those who have never smoked being the most concerned.
6.41 **a. i.** 0.167 **ii.** 0.068 **iii.** 0.238 **iv.** 0.833 **v.** 0.41 **vi.** 0.18 **b.** 18- to 24-year-olds are more likely than seniors to regularly wear seat belts.
6.43 **a.** 85% **b.** 0.15 **c.** 0.7225 **d.** 0.1275 **e.** 0.255 **f.** It is not reasonable.
6.45 $P(L)P(F) = 0.29$, not independent
6.47 Not independent
6.49 **a.** 0.001. We have to assume that she deals with the three errands independently. **b.** 0.999 **c.** 0.009
6.51 **a.** 0.81 **b.** P(1–2 subsystem doesn't work) = 0.19, P(3–4 subsystem doesn't work) = 0.19 **c.** P(system won't work) = 0.0361, P(system will work) = 0.9639 **d.** 0.9931, increases **e.** 0.9266, decreases
6.53 **a.** The expert was assuming that there was a 1 in 12 chance of a valve being in any one of the 12 clock positions and that the positions of the two air valves were independent. **b.** The positions of the two air valves are *not* independent, and 1/144 is smaller than the correct probability.
6.55 **a.** No **b.** 0.992 **c.** $P(E_2|E_1) = 39/4999 = 0.00780$. $P(E_2|\textit{not } E_1) = 40/4999 = 0.00800$. If the first board is defective, then it is slightly less likely that the second board will be defective than if the first board is not defective. **d.** Yes
6.57 **a.** 0.00391 **b.** 0.00383
6.59 **a.** 0.6, slightly smaller **b.** 0.556 **c.** 0.333

6.61 a. 0.55 **b.** 0.45 **c.** 0.4 **d.** 0.25
6.63 a P(at least one food allergy and severe reaction) = (0.08)(0.39) = 0.0312. **b.** P(allergic to multiple foods) = (0.08)(0.3) = 0.024.
6.65 a 0.49 **b.** 0.3125 **c.** 0.24 **d.** 0.1
6.67 a. 0.645 **b.** 0.676, higher
6.69 a. i. 0.307 **ii.** 0.693 **iii.** 0.399 **iv.** 0.275 **v.** 0.122 **b.** 30.7% of faculty members use Twitter; 69.3% of faculty members do not use Twitter; 39.9% of faculty members who use Twitter also use it to communicate with students; 27.5% of faculty members who use Twitter also use it as a learning tool in the classroom; 12.2% of faculty members use Twitter and use it to communicate with students **c.** 0.122 **d.** 0.084
6.71 There are different numbers of people driving the three different types of vehicle (and also that there are some drivers who are driving vehicles not included in those three types).
6.73

	Radiologist 1		
	Predicted Male	Predicted Female	Total
Baby Is Male	74	12	**86**
Baby Is Female	14	59	**73**
Total	**88**	**71**	**159**

a. P(prediction is male | baby is male) = 74/86 = 0.860.
b. P(prediction is female | baby is female) = 59/73 = 0.808.
c. Yes. Since the answer to (a) is greater than the answer to (b), the prediction is more likely to be correct when the baby is male.
d. The radiologist was correct for 74 + 59 = 133 of the 159 babies in the study. So P(correct) = 133/159 = 0.836.
6.75 a. i. 0.99 **ii.** 0.01 **iii.** 0.99 **iv.** 0.01 **b.** 0.02
c. 0.5, it is consistent with the quote.
6.77 b. 0.00095 **c.** 0.10085
d. P(has disease|positive test) = 0.00942 This means that in less than 1% of positive tests the person actually has the disease. This happens because so many more people do not have the disease than have the disease, and the minority of people who do not have the disease but still test positive greatly outnumbers the people who *have* the disease and test positive.
6.79 b. 0.27 **c.** 0.43 **d.** Not independent **e.** 0.526 **f.** 0.140
6.81 a. 0.425 **b.** 0.314 **c.** 0.4375 **d.** 0.005 **e.** 0.201 **f.** 0.196 **g.** 0.335 **h.** 0.177 **i.** 0.488 **j.** 0.295 **k.** 0.566
6.83 b. This is not a fair way to distribute licenses, since any given company applying for multiple licenses has approximately the same chance of obtaining all of its licenses as an individual applying for a single license has of obtaining his/her license. It might be fairer for companies who require two licenses to submit two separate applications and companies who require three licenses to submit three separate applications (with individuals/companies applying for single licenses submitting applications as before). Then, 10 of the applications would be randomly selected and licenses would be awarded accordingly.
6.85 Results of the simulation will vary. The probability should be approximately 0.8468.
6.87 b. 0.025 **c.** 0.069
6.89 a. 0.140 **b.** 0.218 **c.** 0.098 **d.** 0.452 **e.** 0.704
6.91 Dependent events
6.93 a. 0.018 **b.** $P(A_1|L) = 0.444$ **c.** $P(A_2|L) = 0.278$
d. $P(A_3|L) = 0.278$
6.97 0.6
6.99 a. 0.106 **b.** 0.446 **c.** 0.547
6.101 a. 4/5 **b.** 19/24 **c.** 18/23 **d.** 0.496
6.103 a. 0.512 **b.** 0.608

Chapter 7

7.1 a. Discrete **b.** Continuous **c.** Discrete **d.** Discrete **e.** Continuous
7.3 a. Positive integers **b.** For example, S with y = 1, RS, LS with y = 2, RLS, RRS,LES,LLS with = 2, RLS, RRS,LES,LLS with y = 3.
7.5 a. The possible values are the real numbers between 0 and 100, inclusive; **b.** y is continuous.
7.7 a. 3, 4, 5, 6, 7 **b.** −3, −2, −1, 0, 1, 2, 3 **c.** 0, 1, 2 **d.** 0, 1
7.9 a. 0.01 **b.** Over a large number of cartons, about 20% will contain one broken egg. **c.** 0.95. Over a large number of cartons, 95% will contain at most two broken eggs. **d.** 0.85. It does not include exactly two broken eggs. **e.** 0.1 **f.** 0.95
7.11 a. 0.82 **b.** 0.18 **c.** 0.65, 0.27
7.15 a.

x	0	1	2	3	4
p(x)	0.4096	0.4096	0.1536	0.0256	0.0016

b. 0 and 1 **c.** 0.1808
7.17 a. The smallest possible y value is 1, and the corresponding outcome is S. The second smallest y value is 2, and the corresponding outcome is FS. **b.** The set of positive integers. **c.** $p(y) = (0.3)^{y-1}(0.7)$, for $y = 1, 2, 3, \ldots$
7.19

y	0	1	2	3
p(y)	0.16	0.33	0.32	0.19

7.21 a. 0.5 **b.** 0.2
7.23 b. 0.08 **c.** 0.36 **d.** 0.4 **e.** 0.4 **f.** equal because associated areas are equal
7.25 a. 0.5, 0.25 **b.** 0.25 **c.** 18
7.27 3.306
7.29 a. 0.56 **b.** 65% **c.** The mean is not (0 + 1 + 2 + 3 + 4)/5 since, for example, far more cartons have 0 broken eggs than 4 broken eggs, and so 0 needs a much greater weighting than 4 in the calculation of the mean.
7.31 a. 0.14 **b.** $\mu - 2\sigma = 4.66 - 2(1.2) = 2.26$. $\mu + 2\sigma = 4.66 + 2(1.2) = 7.06$. The values of x more than 2 standard deviations away from the mean are 1 and 2. **c.** The probability that x is more than 2 standard deviations away from its mean is 0.05.
7.33 a. 2.8, 1.288 **b.** 0.7, 0.781 **c.** 8.4, 14.94 **d.** 7, 61 **e.** 3.5, 2.27 **f.** 15.4, 75.94
7.35 a. 46.5 **b.** $890
7.39 a. Whether y is positive or negative tells us whether or not the peg will fit into the hole. **b.** 0.003 **c.** 0.00632 **d.** Yes. Since there is no reason to believe that the pegs are being specially selected to match the holes (or vice versa), it is reasonable to think that x_1 and x_2 are independent. **e.** Since 0 is less than half a standard deviation from the mean in the distribution of y, it is relatively likely that a value of y will be negative, and, therefore, that the peg will be too big to fit the hole.
7.41 a 3.5, 2.9167, 1.7078 **b** 3.5, 2.9167, 1.7078 **c** 0, 2.094 **d** 0, 5.124 **e** Depends on how much risk you are willing to take. The variability in winnings would be much greater for game 2,

which means the potential for greater winnings is with game 2, but there is also the potential for greater losses.
7.43 **a** 0.988 **b** 0.999 **c** 0.001 **d** 0.400
7.45 **a.** 0.246. Over a large number of random selections of six passengers, 24.6% will have exactly four people resting or sleeping. **b.** 0.262 **c.** 0.901
7.47 **a.** 0.264 **b.** 0.633 **c.** 0.367 **d.** 0.736
7.49 **a.** 0.630 **b.** 0.000 **c.** More likely to be .8.
7.51 **a.** 0.197 **b.** 0.180 **c.** 75, 7.984 **d.** The value 25 is more than 6 standard deviations below the mean. It is surprising that fewer than 25 are said to have failed the test.
7.53 **a.** 200 **b.** 13.416
7.55 **a.** Binomial distribution with $n = 100$ and $p = 0.2$ **b.** 20 **c.** 16, 4 **d.** A score of 50 is $(50 - 20)/4 = 7.5$ standard deviations from the mean in the distribution of x. So, a score of over 50 is very unlikely.
7.57 **a.** 0.017 **b.** 0.811, **c.** 0.425 **d.** The error probability when $p = 0.7$ is now 0.902. The error probability when $p = 0.6$ is now 0.586.
7.59 **a.** There is not a fixed number of trials. **b. i.** 0.062 **ii.** 0.284 **iii.** 0.716 **iv.** 0.779
7.61 **a.** 0.0975 **b.** 0.043 **c.** 0.815
7.63 **a.** 0.9599 **b.** 0.2483 **c.** 0.1151 **d.** 0.9976 **e.** 0.6887 **f.** 0.6826 **g.** 1.0000
7.65 **a.** 0.9909 **b.** 0.9909 **c.** 0.1093 **d.** 0.1267 **e.** 0.0706 **f.** 0.0228 **g.** 0.9996 **h.** approximately 1
7.67 **a.** −1.96 **b.** −2.33 **c.** −1.645 **d.** $z^* = 2.05$ **e.** $z^* = 2.33$ **f.** $z^* = 1.28$
7.69 **a.** $z^* = 1.96$ **b.** $z^* = 1.645$ **c.** $z^* = 2.33$ **d.** $z^* = 1.75$
7.71 **a.** 0.5 **b.** 0.9772 **c.** 0.9772 **d.** 0.8185 **e.** 0.9938 **f.** Approximately 1
7.73 Those with emissions > 2.113 parts per billion
7.75 **a.** 0.2843 **b.** 0.0918 **c.** 0.4344 **d.** 29.928
7.77 **a.** 0.4013 **b.** 0.1314
7.79 The second machine
7.81 94.4 seconds or less
7.83 **b.** The clear curve in the normal probability plot tells us that the distribution of fussing times is not normal. **d.** The transformation results in a pattern that is much closer to linear than the pattern in Part (b).
7.85 **b.** No. The distribution of x is positively skewed. **d.** Yes. The histogram shows a distribution that is slightly closer to being symmetric than the distribution of the untransformed data. **f.** Both transformations produce histograms that are closer to being symmetric than the histogram of the untransformed data, but neither transformation produces a distribution that is truly close to being normal.
7.87 Yes. The curve in the normal probability plot suggests that the distribution is not normal.
7.89 Since the pattern in the normal probability plot is very close to linear, it is plausible that disk diameter is normally distributed.
7.91 **a.** The histogram is positively skewed. **b.** No. The transformation has resulted in a histogram which is still clearly positively skewed.
7.93 Yes. In each case, the transformation has resulted in a histogram that is much closer to symmetric than the original histogram.
7.95 **a.** 0.0016 **b.** 0.9946 **c.** 0.7925
7.97 **a.** 0.1114 **b.** 0.0409 **c.** 0.0968 **d.** 0.9429 **e.** 0.9001
7.99 **a.** There is a fixed number of trials, in which the probability of an undetected tumor is the same in each trial, and the outcomes of the trials are independent since the sample is random. Thus, it is reasonable to think that the distribution of x is binomial. **b.** Yes, since $np = 15.5 \geq 10$ and $n(1 - p) = 484.5 \geq 10$. **c. i.** 0.0606 **ii.** 0.9345 **iii.** 0.0985
7.101 **a.** 0.7970 **b.** 0.7016 **c.** 0.0143 **d.** 0.05
7.103 **a.** No, since $np = 50(0.05) = 2.5 < 10$. **b.** Now $n = 500$ and $p = 0.05$, so $np = 500(0.05) = 25 \geq 10$. So the techniques of this section can be used $P(\text{at least 20 are defective}) \approx 0.8708$.
7.105 **a** 0.8247 **b** 0.0516 **c** 0.68826 **d** Yes, the probability of a pregnancy having a duration of at least 310 days is only 0.0030.
7.107 **a.** 0.7 **b.** 0.45 **c.** 0.55 **d.** 0.71 **e.** 0.65 **f.** 0.45
7.109 **a.** 0.64 **b.** 0.256 **c.** 0.02048 **d.** In order for it to take y selections to find two acceptable batteries, the first $y - 1$ batteries must include exactly 1 acceptable battery (and, therefore, $y - 2$ unacceptable batteries), and the yth battery must be acceptable. There are $y - 1$ ways in which the first $y - 1$ batteries can include exactly 1 acceptable battery. So $p(y) = (y - 1)(0.8)^1(0.2)^{y-2} \cdot (0.8) = (y - 1)(0.2)^{y-2}(0.8)^2$.
7.111 **a.** 0.5934 **b.** $P(x < 25) = 0.0000$. Yes, you are very unlikely to select a car of this model with a fuel efficiency of less than 25 mpg. **c.** $P(x > 32) = 0.0475$. So, if three cars are selected, the probability that they all have fuel efficiencies more than 32 mpg is $(0.0475)^3 = 0.0001$. **d.** 28.026
7.113 $\sigma_x^2 = 2.52$, $\sigma_x = 1.587$
7.115 **a.** 0.9876 **b.** 0.9888 **c.** 6.608
7.117 **a.** No, since 5 feet 7 inches is 67 inches, and if x = height of a randomly chosen woman, then $P(x < 67) = 0.6915$, which is not more than 94%. **b.** About 69%
7.119 **a.**

w	0	1	2	3	4	5
$p(w)$	1/6	5/18	2/9	1/6	1/9	1/18

b. 1.944 hours
7.121 **a.** 0.1552 **b.** 0.2688
c.

x	4	5	6	7
$p(x)$	0.1552	0.2688	0.29952	0.27648

d. 5.697
7.123 The lowest score to be designated an A is 85.28. Since 89 > 85.28, yes, I received an A.
7.125 658 hours

Cumulative Review 7

CR7.3 No. The percentages given in the graph are said to be, for each year, the "percent increase in the number of communities installing" red-light cameras. This presumably means the percentage increase in the number of communities with red-light cameras *installed*, in which case the positive results for all of the years 2003 to 2009 show that a great many more communities had red-light cameras installed in 2009 than in 2002.
CR7.5 0.1243
CR7.7 $P(\text{Service 1}|\text{Late}) = 0.349$ and $P(\text{Service 2}|\text{Late}) = 0.651$. Service 2 is more likely to have been used.
CR7.9 1. $P(M) = 0.021$ 2. $P(M|B) = 0.043$ 3. $P(M|W) = 0.07$
CR7.11 statement iv
CR7.13 **a.** 0.09 **b.** 0.045 **c.** 0.135 **d.** 0.667
CR7.15 **a.** $\mu_x = 2.3$. **b.** $\sigma_x^2 = 0.81$, $\sigma_x = 0.9$
CR7.17 Let x be the number of correct identifications. Assume that the graphologist was merely guessing—in other words, that the probability of success on each trial was 0.5. Then

$P(x \geq 6) = 0.377$. Since this probability is not particularly small, it would not have been unlikely for him/her to get six or more correct identifications when guessing. No ability to distinguish the handwriting of psychotics is indicated.
CR7.19 **a.** 0.1587 **b.** 51.4 minutes **c.** 41.65 minutes
CR7.21 **b.** Positive **c.** Yes. If a symmetrical distribution were required, then a transformation would be advisable.

Chapter 8

8.1 A population characteristic is a quantity that summarizes the whole population. A statistic is a quantity calculated from the values in a sample.
8.3 **a.** Population characteristic **b.** Statistic **c.** Population characteristic **d.** Statistic **e.** Statistic
8.7 **a.** and **b.**

$\bar{x}$	1.5	2	2.5	3	3.5
$p(\bar{x})$	1/6	1/6	1/3	1/6	1/6

c.

$\bar{x}$	1	1.5	2	2.5	3	3.5	4
$p(\bar{x})$	1/16	1/8	3/16	1/4	3/16	1/8	1/16

d. Both distributions are symmetrical, and their means are equal (2.5). However, the "with replacement" version has a greater spread than the first distribution, with values ranging from 1 to 4 in the "with replacement" distribution and from 1.5 to 3.5 in the "without replacement" distribution. The stepped pattern of the "with replacement" distribution more closely resembles a normal distribution than does the shape of the "without replacement" distribution.
8.9 **a.** and **b.**

$\bar{x}$	$2^2/_3$	3	$3^1/_3$	$3^2/_3$
$p(\bar{x})$	0.1	0.4	0.3	0.2

Sample Median	3	4
p(Sample Median)	0.7	0.3

(Max + Min)/2	2.5	3	3.5
p((Max + Min)/2)	0.1	0.5	0.4

c. Since $\mu = 3.2$ and $\mu_{\bar{x}} = 3.2$, $\bar{x}$ is an unbiased estimator of μ, which is not the case for either of the two other statistics. Since the distribution of the sample mean has less variability than either of the other two sampling distributions, the sample mean will tend to produce values that are *closer* to μ than the values produced by either of the other statistics.
8.11 The sampling distribution of $\bar{x}$ will be approximately normal for the sample sizes in Parts (c)–(f), since those sample sizes are all greater than or equal to 30.
8.13 **a.** $\mu_{\bar{x}} = 40$, $\sigma_{\bar{x}} = 0.625$ approximately normal **b.** 0.5762 **c.** 0.2628
8.15 **a.** $\mu_{\bar{x}} = 2$, $\sigma_{\bar{x}} = 0.267$ **b.** In each case $\mu_{\bar{x}} = 2$. When $n = 20$, $\sigma_{\bar{x}} = 0.179$, and when $n = 100$, $\sigma_{\bar{x}} = 0.008$. **c.** All three centers are the same, and the larger the sample size, the smaller the standard deviation of $\bar{x}$. Since the distribution of $\bar{x}$ when $n = 100$ is the one with the smallest standard deviation of the three, this sample size is most likely to result in a value of $\bar{x}$ close to μ.
8.17 **a.** 0.8185, 0.0013 **b.** 0.9772, 0.0000
8.19 $p(0.49 < \bar{x} < 0.51) = 0.9974$; the probability that the manufacturing line will be shut down unnecessarily is $1 - 0.9974 = 0.0026$.
8.21 Approximately 0.
8.23 **a.** $\mu_{\hat{p}} = 0.65$, $\sigma_{\hat{p}} = 0.151$. **b.** $\mu_{\hat{p}} = 0.65$, $\sigma_{\hat{p}} = 0.107$. **c.** $\mu_{\hat{p}} = 0.65$, $\sigma_{\hat{p}} = 0.087$. **d.** $\mu_{\hat{p}} = 0.65$, $\sigma_{\hat{p}} = 0.067$. **e.** $\mu_{\hat{p}} = 0.65$, $\sigma_{\hat{p}} = 0.048$. **f.** $\mu_{\hat{p}} = 0.65$, $\sigma_{\hat{p}} = 0.034$.
8.25 **a.** $\mu_{\hat{p}} = 0.65$, $\sigma_{\hat{p}} = 0.026$. **b.** No, since $np = 100(0.07) = 7$, which is not greater than or equal to 10. **c.** The mean is unchanged, but the standard deviation changes to $\sigma_{\hat{p}} = 0.018$ **d.** Yes, since $np = 14$ and $n(1 - p) = 186$, which are both greater than or equal to 10. **e.** 0.0485
8.27 **a.** $\mu_{\hat{p}} = 0.005$, $\sigma_{\hat{p}} = 0.007$ **b.** No, since $np = 0.5$, which is not greater than or equal to 10. **c.** We need both np and $n(1 - p)$ to be greater than or equal to 10; need $n \geq 2000$.
8.29 **a.** If $p = 0.5$, $\mu_{\hat{p}} = 0.5$, $\sigma_{\hat{p}} = 0.0333$, approximately normal. If $p = 0.6$, $\mu_{\hat{p}} = 0.6$, $\sigma_{\hat{p}} = 0.0327$, approximately normal. **b.** If $p = 0.5$, $P(\hat{p} \geq 0.6) = 0.0013$. If $p = 0.6$, $P(\hat{p}|0.6) = 0.5$ **c.** For a larger sample size, the value of $\hat{p}$ is likely to be closer to p. So, for $n = 400$, when $p = 0.5$, $P(\hat{p} \geq 0.6)$ will be smaller. When $p = 0.6$, $P(\hat{p} \geq 0.6)$ will still be 0.5, and it will remain the same.
8.31 **a.** 0.9744 **b.** Approximately 0
8.33 **a.** $\bar{x}$ is approximately normally distributed with mean 50 and standard deviation 0.1. **b.** 0.9876 **c.** 0.5
8.35 **a.** 0.8185 **b.** **i.** 0.8357 **ii.** 0.9992
8.37 0.0793

Chapter 9

9.1 Statistics II and III are preferable to Statistic I since they are unbiased (their means are equal to the value of the population characteristic). However, Statistic II is preferable to Statistic III since its standard deviation is smaller. So Statistic II should be recommended.
9.3 $\hat{p} = 0.277$
9.5 $\hat{p} = 0.7$
9.7 **a.** $\bar{x} = 421.429$ **b.** $s^2 = 10414.286$ **c.** $s = 102.050$. No, s is not an unbiased statistic for estimating σ.
9.9 **a.** $\bar{x} = 120.6$ therms **b.** The value of τ is estimated to be $10000(120.6) = 1{,}206{,}000$ therms. **c.** $\hat{p} = 0.8$ **d.** sample median $= 120$ therms
9.11 **a.** 1.96 **b.** 1.645 **c.** 2.58 **d.** 1.28 **e.** 1.44
9.13 **a.** The larger the confidence level, the wider the interval. **b.** The larger the sample size, the narrower the interval. **c.** Values of $\hat{p}$ farther from 0.5 give smaller values of $\hat{p}(1 - \hat{p})$. Therefore, the farther the value of $\hat{p}$ from 0.5, the narrower the interval.
9.15 If a large number of random samples of size 1200 were to be taken, 90% of the resulting confidence intervals would contain the true proportion of all Facebook users who would say it is not OK to "friend" someone who reports to you at work.
9.17 **a.** (0.675, 0.705). **b.** We are 98% confident that the proportion of all coastal residents who would evacuate is between 0.675 and 0.705. If we were to take a large number of random samples of size 5046, about 98% of the resulting confidence intervals would contain the true proportion of all coastal residents who would evacuate.
9.19 **a.** (0.642, 0.677). We are 90% confident that the proportion of all Americans age 8 to 18 who own a cell phone is between 0.642 and 0.677. **b.** (0.745, 0.776). We are 90% confident that the proportion of all Americans age 8 to 18 who own an MP3 player is between 0.745 and 0.776. **c.** The interval in

Part (b) is narrower than the interval in Part (a) because the sample proportion in Part (b) is farther from 0.5.
9.21 a. (0.660, 0.740). We are 95% confident that the proportion of all potential jurors who regularly watch at least one crime-scene investigation series is between 0.660 and 0.740. **b.** Wider
9.23 a. (0.223, 0.298). We are 95% confident that the proportion of all U.S. businesses that have fired workers for misuse of the Internet is between 0.223 and 0.298. **b.** The estimated standard error is smaller and the confidence level is lower.
9.25 0.024. We are 95% confident that the proportion of all adults who believe that the shows are either "totally made up" or "mostly distorted" is within 0.024 of the sample proportion of 0.82.
9.27 We are 95% confident that the proportion of all adult drivers who would say that they often or sometimes talk on a cell phone while driving is within $1.96\sqrt{\hat{p}(1-\hat{p})/n} = 0.030$ (that is, 3.0 percentage points) of the sample proportion of 0.36.
9.29 a. (0.119, 0.286). We are 95% confident that the proportion of all patients under 50 years old who experience a failure within the first 2 years after receiving this type of defibrillator is between 0.119 and 0.286. **b.** (0.011, 0.061). We are 99% confident that the proportion of all patients age 50 or older who experience a failure within the first 2 years after receiving this type of defibrillator is between 0.011 and 0.061. **c.** Using the estimate of p from the study, 18/89, the required sample size is $n = 688.685$. A sample of size at least 689 is required.
9.31 A sample size of 2401 is required.
9.33 Using $p = 0.32$, a sample size of 335 is required. Using the conservative value, $p = 0.5$, a sample size of 385 is required. The conservative estimate of p gives the larger sample size. Since the relevant proportion could have changed significantly since 2011, it would be sensible to use a sample size of 385.
9.35 a. 2.12 **b.** 1.80 **c.** 2.81 **d.** 1.71 **e.** 1.78 **f.** 2.26
9.37 The second interval is based on the larger sample size; the interval is narrower.
9.39 a. (7.411, 8.069). **b.** (41.439, 44.921).
9.41 a. The fact that the mean is much greater than the median suggests that the distribution of times spent volunteering in the sample was positively skewed. **b.** With the sample mean much greater than the sample median, and with the sample regarded as representative of the population, it seems very likely that the population is strongly positively skewed and, therefore, not normally distributed. **c.** Since $n = 1086 \geq 30$, the sample size is large enough for us to use the t confidence interval, even though the population distribution is not approximately normal. **d.** (5.232, 5.968). We are 98% confident that the mean time spent volunteering for the population of parents of school-age children is between 5.232 and 5.968 hours.
9.43 a. Narrower. **b.** The statement is not correct. The population mean, μ, is a constant, and therefore, we cannot talk about the probability that it falls within a certain interval. **c.** The statement is not correct. We can say that *on average* 95 out of every 100 samples will result in confidence intervals that will contain μ, but we cannot say that in 100 such samples, *exactly* 95 will result in confidence intervals that contain μ.
9.45 a. The samples from 12 to 23 month and 24 to 35 month are the ones with the greater variability. **b.** The less-than-12-month sample is the one with the greater sample size. **c.** The new interval has a 99% confidence level.
9.47 a. (179.02, 186.98). We are 95% confident that the mean summer weight is between 179.02 and 186.98 pounds.
b. (185.423, 194.577). We are 95% confident that the mean winter weight is between 185.423 and 194.577 pounds. **c.** Based on the Frontier Airlines data, neither recommendation is likely to be an accurate estimate of the mean passenger weight, since 190 is not contained in the confidence interval for the mean summer weight and 195 is not contained in the confidence interval for the mean winter weight.
9.49 A boxplot shows that the distribution of the sample values is negatively skewed, and so the population may not be not approximately normally distributed. Therefore, since the sample is small, it is not appropriate to use the t confidence interval method of this section.
9.51 A reasonable estimate of σ is given by (sample range)/4 = 162.5. A sample size of 1015 is needed.
9.53 First, we need to know that the information is based on a random sample of middle-income consumers age 65 and older. Second, it would be useful if some sort of margin of error was given for the estimated mean of $10,235.
9.55 a. The paper states that queens flew for an *average of* 24.2 ± 9.21 minutes on their mating flights, and so this interval is a confidence interval for a population mean. **b.** (3.301, 5.899)
9.57 a. (0.217, 0.475) **b.** We are 95% confident that the proportion of all young people with pierced tongues who have receding gums is between 0.217 and 0.475. The assumption we have made is that the sample of 52 is representative of the population of young adults with pierced tongues.
9.59 The standard error for the mean cost for Native Americans is much larger than that for Hispanics since the sample size was much smaller for Native Americans.
9.61 a. (0.409, 0.591)
9.63 The point estimate is .43 and the bound on error is approximately .03.
9.65 (0.671, 0.749)
9.67 (8.571, 9429)
9.69 (17.899, 25.901)

Chapter 10

10.1 $\bar{x}$ is a *sample* statistic.
10.3 H_a: $\mu > 100$ will be used.
10.7 H_0: $p = 0.1$ versus H_a: $p < 0.1$
10.9 H_0: $p = 0.5$ H_a: $p > 0.5$
10.11 H_0: $p = 0.83$ versus H_a: $p \neq 0.83$
10.13 a. Type I error, 0.091 **b.** 0.097
10.15 a. A Type I error would be concluding the man is not the father when in fact he is. A Type II error would be concluding the man is the father when in fact he is not the father.
b. $\alpha = 0.001$, $\beta = 0$ **c.** $\beta = 0.008$
10.17 a. A Type I error concluding that there is evidence that more than 1% of a shipment is defective when in fact (at least) 1% of the shipment is defective. A Type II error is not being convinced that more than 1% of a shipment is defective when in fact more than 1% of the shipment is defective. **b.** Type II **c.** Type I
10.19 The probability of a Type I error is equal to the significance level. Here the aim is to reduce the probability of a Type I error, so a small significance level (such as 0.01) should be used.
10.21 a. The researchers failed to reject H_0. **b.** Type II error **c.** Yes
10.23 a. A P-value of 0.0003 means that it is very unlikely (probability = 0.0003), assuming that H_0 is true, that you would get a sample result at least as inconsistent with H_0 as the one obtained in the study. Thus, H_0 is rejected. **b.** A P-value of 0.350 means that it is not particularly unlikely (probability = 0.350), assuming that H_0 is true, that you would get a sample result at least as in-

consistent with H_0 as the one obtained in the study. Thus, there is no reason to reject H_0.
10.25 a. H_0 is not rejected. **b.** H_0 is not rejected. **c.** H_0 is not rejected. **d.** H_0 is rejected. **e.** H_0 is not rejected. **f.** H_0 is not rejected.
10.27 a. Not appropriate **b.** Is appropriate **c.** Is appropriate **d.** Not appropriate
10.29 a. $z = 2.530$, P-value $= 0.0057$, reject H_0 **b.** No. The survey only included women age 22 to 35.
10.31 $z = 6.647$, P-value ≈ 0, reject H_0
10.33 $z = 0.446$, P-value $= 0.328$, fail to reject H_0
10.35 $z = 15.436$, P-value ≈ 0, reject H_0
10.37 $z = 0.791$, P-value $= 0.2146$, fail to reject H_0
10.39 If the null hypothesis is true, $p(\hat{p} \geq 0.33) \approx 0$. It is very unlikely that you would observe a sample proportion as large as 0.33 if the null hypothesis is true.
10.41 The "38%" value given in the article is a proportion of *all* felons; in other words, it is a *population* proportion. Therefore, we know that the population proportion is less than 0.4, and there is no need for a hypothesis test.
10.43 a. 0.484 **b.** 0.686 **c.** 0.025 **d.** 0.000 **e.** 0.097
10.45 a. H_0 is rejected. **b.** H_0 is not rejected. **c.** H_0 is not rejected.
10.47 a. $t = 0.748$, P-value $= 0.468$, fail to reject H_0 **b.** $t = 8.731$, P-value ≈ 0, reject H_0
10.49 $t = 2.417$, P-value $= 0.010$, reject H_0
10.51 $t = 14.266$, P-value ≈ 0, reject H_0
10.53 $t = -5.001$, P-value ≈ 0, reject H_0
10.55 a. $t = 1.265$, P-value $= 0.103$, fail to reject H_0 **b.** $t = 3.162$, P-value $= 0.001$, reject H_0
10.57 a. Yes. Since the pattern in the normal probability plot is roughly linear, and since the sample was a random sample from the population, the t test is appropriate. **b.** The boxplot shows a median of around 245, and since the distribution is a roughly symmetrical distribution, this tells us that the sample mean is also around 245. This might initially suggest that the population mean differs from 240. But, the sample is relatively small, and the sample values range all the way from 225 to 265; such a sample mean would still be feasible if the population mean were 240. **c.** $t = 1.212$, P-value $= 0.251$, fail to reject H_0
10.59 a. Increasing the sample size increases the power. **b.** Increasing the significance level increases the power.
10.61 a. 0.1003 **b.** 0.2358 **c.** 0.0001 **d.** Power when $\mu = 9.8$ is $1 - 0.2358 = 0.7642$; power when $\mu = 9.5$ is $1 - 0.0001 = 0.9999$.
10.63 a. $t = 0.466$, P-value $= 0.329$, fail to reject H_0 **b.** $\beta \approx 0.75$ **c.** Power $\approx 1 - 0.75 = 0.25$
10.65 a. $\beta \approx 0.04$ **b.** $\beta \approx 0.04$ **c.** $\beta \approx 0.24$ **d.** $\beta \approx 0$ **e.** $\beta \approx 0.04$ **f.** $\beta \approx 0.01$
10.67 a. $z = 3.370$, P-value $= 0.0004$, reject H_0
10.69 $z = -11.671$, P-value ≈ 0, reject H_0
10.71 $t = -5.051$, P-value ≈ 0, reject H_0
10.73 $z = 1.069$, P-value $= 0.143$, fail to reject H_0
10.75 $z = -1.927$, P-value $= 0.027$, reject H_0
10.77 $t = -0.599$, P-value $= 0.277$, fail to reject H_0
10.79 $z = 4.186$, P-value ≈ 0, reject H_0
10.81 $t = -5.324$, P-value ≈ 0, reject H_0

Cumulative Review 10

CR10.3 a. Three airlines stand out from the rest and have large numbers of delayed flights. These airlines are ExpressJet, Delta, and Continental, with 93, 81, and 72 delayed flights, respectively. **b.** A typical number of flights delayed per 100,000 flights is around 1.1, with most rates lying between 0 and 1.6. Four airlines stand out from the rest and have high rates, with two of those four having *particularly* high rates. **c.** The rate per 100,000 flights data should be used, since this measures the likelihood of any given flight being late. An airline could stand out in the number of flights delayed data purely as a result of having a large number of flights.
CR10.5 a. 0.134 **b.** 0.041 **c.** $\mu_x = 25$, $\sigma_x = 4.330$ **d.** 0.102
CR10.7 a. 0.4 **b.** 0.18 **c.** 0.26 **d.** 0.45 **e.** Since anyone who accepts a job offer must have received at least one job offer, $P(O|A) = 1$. **f.** 0.18
CR10.9 a. (0.244, 0.416). We are 95% confident that the proportion of all U.S. medical residents who work moonlighting jobs is between 0.244 and 0.416. **b.** (0.131, 0.252). We are 90% confident that the proportion of all U.S. medical residents who have credit card debt of more than \$3000 is between 0.131 and 0.252. **c.** The interval in Part (a) is wider than the interval in Part (b) because the confidence level in Part (a) (95%) is greater than the confidence level in Part (b) (90%) and because the sample proportion in Part (a) (38/115) is closer to 0.5 than the sample proportion in Part (b) (22/115).
CR10.11 A reasonable estimate of σ is given by (sample range)/4 = 0.1. A sample size of 385 is needed.
CR10.13 $z = 7.557$, P-value ≈ 0, reject H_0
CR10.15 a. With a sample mean of 14.6, the sample standard deviation of 11.6 places zero just over one standard deviation below the mean. Since no teenager can spend a negative time online, to get a typical deviation from the mean of just over 1, there must be values that are substantially more than one standard deviation above the mean. This suggests that the distribution of online times in the sample is positively skewed. **b.** $t = 9.164$, P-value ≈ 0, reject H_0

Chapter 11

11.1 The distribution of $\bar{x}_1 - \bar{x}_2$ is approximately normal with mean 5 and standard deviation 0.529.
11.3 $t = -3.770$, df $= 243$, P-value $= 0.0002$, reject H_0
11.5 a. Since boxplots are roughly symmetrical and since there is no outlier in either sample, the assumption of normality is justified, and it is reasonable to carry out a two-sample t test. **b.** $t = 3.332$, P-value $= 0.001$, reject H_0 **c.** (0.423, 2.910). We are 98% confident that the difference between the mean number of hours per day spent using electronic media in 2009 and 1999 is between 0.423 and 2.910.
11.7 $t = -0.445$, P-value $= 0.660$, fail to reject H_0
11.9 a. If the vertebroplasty group had been compared to a group of patients who did not receive any treatment, and if, for example, the people in the vertebroplasty group experienced a greater pain reduction on average than the people in the "no treatment" group, then it would be impossible to tell whether the observed pain reduction in the vertebroplasty group was caused by the treatment or merely by the subjects' knowledge that some treatment was being applied. By using a placebo group, it is ensured that the subjects in both groups have the knowledge of some "treatment," so that any differences between the pain reduction in the two groups can be attributed to the nature of the vertebroplasty treatment. **b.** (−0.687, 1.287). We are 95% confident that the difference in mean pain intensity 3 days after treatment for the vertebroplasty treatment and the fake treatment is between −0.687 and 1.287. **c.** (−1.186, 0.786). We are 95% confident that the difference in mean pain intensity 14 days after

treatment for the vertebroplasty treatment and the fake treatment is between −1.186 and 0.786. (−1.722, 0.322). We are 95% confident that the difference in mean pain intensity 1 month after treatment for the vertebroplasty treatment and the fake treatment is between −1.722 and 0.322. **d.** The fact that all of the intervals contain zero tells us that we do not have convincing evidence of a difference in the mean pain intensity for the vertebroplasty treatment and the fake treatment at any of the three times.
11.11 $t = 1.065$, df = 497, *P*-value = 0.288, fail to reject H_0
11.13 a. μ_1= mean payment for claims not involving errors; μ_2 = mean payment for claims involving errors; H_0: $\mu_1 - \mu_1 = 0$; H_a: $\mu_1 - \mu_2 < 0$ **b.** Answer: (ii) 2.65. Since the samples are large, we are using a *t* distribution with a large number of degrees of freedom, which can be approximated with the standard normal distribution. $P(z > 2.65) = 0.004$, which is the *P*-value given. None of the other possible values of *t* gives the correct *P*-value.
11.15 (−78.154, −61.846). We are 99% confident that the difference in mean time spent watching time-shifted television for 2010 and 2011 (2010 minus 2011) is between −78.154 and −61.846.
11.17 a. $t = 10.359$, *P*-value ≈ 0, reject H_0 **b.** $t = -16.316$, *P*-value ≈ 0, reject H_0 **c.** $t = 4.690$, *P*-value ≈ 0, reject H_0 **d.** The results do seem to provide convincing evidence of a gender bias in the monkeys' choices of how much time to spend playing with each toy, with the male monkeys spending significantly more time with the "masculine toy" than the female monkeys, and with the female monkeys spending significantly more time with the "feminine toy" than the male monkeys. However, the data also provide convincing evidence of a difference between male and female monkeys in the time they choose to spend playing with a "neutral toy." It is possible that it was some attribute other than masculinity/femininity in the toys that was attracting the different genders of monkey in different ways. **e.** The given mean time playing with the police car and mean time playing with the doll for female monkeys are sample means for the same sample of female monkeys. The two-sample *t* test can only be performed when there are two independent random samples.
11.19 a. Since the samples are small it is necessary to know—or to assume—that the distributions from which the random samples were taken are normal. However, in this case, since both standard deviations are large compared to the means, it seems unlikely that these distributions would have been normal. **b.** Since the samples are large, it is appropriate to carry out the two-sample *t* test. **c.** $t = -2.207$, *P*-value = 0.030, fail to reject H_0
11.21 a. $t = -9.863$, *P*-value ≈ 0, reject H_0 **b.** For the two-sample *t* test, $t = -9.979$, df = 22.566, and *P*-value ≈ 0 Thus, the conclusion is the same.
11.23 $t = 0.639$, *P*-value = 0.267, fail to reject H_0
11.25 $t = -0.515$, *P*-value = 0.612, fail to reject H_0
11.27 $t = 2.684$, *P*-value = 0.010, reject H_0
11.29 a. $t = 4.451$, *P*-value ≈ 0, reject H_0 **b.** (−0.210, 0.270). **c.** $t = 3.094$, *P*-value = 0.001, reject H_0 **d.** In Part (a), the male profile heights and the male actual heights are paired (according to which individual has the actual height and the height stated in the profile), and with paired samples, we use the paired *t* test. In Part (c), we were dealing with two independent samples (the sample of males and the sample of females), and therefore, the two-sample *t* test was appropriate.
11.31 $t = -2.457$, *P*-value = 0.018, reject H_0
11.33 a. $t = 4.321$, *P*-value ≈ 0, reject H_0 **b.** $t = 1.662$, *P*-value = 0.055, fail to reject H_0 **c.** A smaller standard deviation in the sample of differences means that we have a lower estimate of the standard deviation of the population of differences. Assuming that the mean wrist extensions for the two mouse types are the same (in other words, that the mean of the population of differences is zero), a sample mean difference of as much as 8.82 is much less likely when the standard deviation of the population of differences is around 10 than when the standard deviation of the population of differences is around 26.
11.35 $t = 3.17$, *P*-value = 0.001, reject H_0
11.37 $z = -3.94$, *P*-value ≈ 0, reject H_0
11.39 a. $z = 1.172$, *P*-value = 0.121, fail to reject H_0 **b.** (−0.036, 0.096). We are 99% confident that the difference between the proportion of Gen Y and the proportion of Gen X who made a donation via text message is between −0.036 and 0.096. In repeated sampling with random samples of size 400, approximately 99% of the resulting confidence intervals would contain the true difference in proportions who donated via text message.
11.41 a. $z = -0.298$, *P*-value = 0.766, fail to reject H_0 **b.** $z = -2.022$, *P*-value = 0.043, reject H_0 **c.** Assuming that the population proportions are equal, you are much less likely to get a difference in sample proportions as large as the one given when the samples are very large than when the samples are relatively small.
11.43 a. (−0.078, 0.058). **b.** Zero is included in the confidence interval. This tells us that there is not convincing evidence of a difference between the proportions.
11.45 No. It is not appropriate to use the two-sample *z* test because the groups are not large enough. We are not told the sizes of the groups, but we know that each is, at most, 81. The sample proportion for the fish oil group is 0.05, and 81(0.05) = 4.05, which is less than 10. So, the conditions for the two-sample *z* test are not satisfied.
11.47 $z = 25.33$, *P*-value ≈ 0, reject H_0
11.49 (−0.043, 0.003). We are 95% confident that the difference in the proportions of college graduates who were unemployed in these two years (2008 minus 2009) is between −0.043 and 0.003.
11.51 $z = 6.306$, *P*-value ≈ 0, reject H_0
11.53 $z = 3.800$, *P*-value ≈ 0, reject H_0
11.55 a. $z = 9.169$, *P*-value ≈ 0, reject H_0 **b.** No. Since this is an observational study, causation cannot be inferred from the result.
11.57 Since the data given are population characteristics, an inference procedure is not applicable. It is *known* that the rate of Lou Gehrig's disease among soldiers sent to the war is higher than for those not sent to the war.
11.59 b. If we want to know whether the e-mail intervention *reduces* (as opposed to *changes*) adolescents' display of risk behavior in their profiles, then we use one-sided alternative hypotheses and the *P*-values are halved. If that is the case, using a 0.05 significance level, we are convinced that the intervention is effective with regard to reduction of references to sex and that the proportion showing any of the three protective changes is greater for those receiving the e-mail intervention. Each of the other two apparently reduced proportions could have occurred by chance.
11.61 a. $t = -6.565$, *P*-value ≈ 0, reject H_0 **b.** $t = 6.249$, *P*-value ≈ 0, reject H_0 **c.** $t = 0.079$, *P*-value = 0.937, fail to reject H_0. **d.** This does not imply that students and faculty consider it acceptable to talk on a cell phone during class; in fact, the low sample mean ratings for both students and faculty show that both groups, on the whole, feel that the behavior is inappropriate.

11.63 **a.** $t = -17.382$, P-value ≈ 0, reject H_0 **b.** $t = 2.440$, P-value = 0.030, reject H_0 **c.** No, the paired t test would not be appropriate since the treatment and control groups were not paired samples.
11.65 $z = 4.245$, P-value ≈ 0, reject H_0
11.67 **a.** $t = -11.952$, P-value ≈ 0, reject H_0 **b.** $t = -68.803$, P-value ≈ 0, reject H_0 **c.** $t = 0.698$, P-value = 0.494, fail to reject H_0
11.69 $t = 0.856$, P-value = 0.210, fail to reject H_0
11.71 $t = -1.336$, P-value = 0.193, fail to reject H_0
11.73 $z = -1.263$, P-value = 0.103, fail to reject H_0
11.75 (−0.274, −0.082). We are 90% confident that $p_1 - p_2$ lies between −0.274 and −0.082, where p_1 is the proportion of children in the community with fluoridated water who have decayed teeth and p_2 is the proportion of children in the community without fluoridated water who have decayed teeth. The interval does not contain zero, which means that we have evidence at the 0.1 level of a difference between the proportions of children with decayed teeth in the two communities, and evidence at the 0.05 level that the proportion of children with decayed teeth is smaller in the community with fluoridated water.
11.77 **a.** (−4.738, 22.738). **b.** $t = 0.140$, P-value = 0.890, fail to reject H_0 **c.** $t = -0.446$, P-value = 0.330, fail to reject H_0
11.79 **a.** $t = 3.948$, P-value ≈ 0, reject H_0 **b.** $t = -1.165$, P-value = 0.249, fail to reject H_0
11.81 $z = 5.590$, P-value ≈ 0, reject H_0

Chapter 12

12.1 **a.** fail to reject H_0 **b.** reject H_0 **c.** reject H_0
12.3 $X^2 = 20.686$, P-value = 0.0003, reject H_0
12.5 $X^2 = 19.599$, P-value ≈ 0, reject H_0
12.7 Since the P-value is small, there is convincing evidence that the population proportions of people who respond "Monday" and "Friday" are not equal.
12.9 **a.** $X^2 = 166.958$, P-value ≈ 0, reject H_0
b. $X^2=5.052$, P-value = 0.025, reject H_0
12.11 $X^2 = 25.486$, P-value ≈ 0, reject H_0
12.13 $X^2 = 1.469$, P-value = 0.690, fail to reject H_0
12.15 **a.** P-value = 0.844, fail to reject H_0 **b.** P-value = 0.106, fail to reject H_0
12.17 $X^2 = 29.507$, P-value = 0.001, reject H_0
12.19 **a.** $X^2 = 90.853$, P-value ≈ 0, reject H_0
b. The particularly high contributions to the chi-square statistic (in order of importance) come from the field of communication, languages, and cultural studies, in which there was a disproportionately high number of smokers; from the field of mathematics, engineering, and sciences, in which there was a disproportionately low number of smokers; and from the field of social science and human services, in which there was a disproportionately high number of smokers.
12.21 **a.** $X^2 = 2.314$, P-value = 0.128, fail to reject H_0 **b.** Yes
c. Yes. Since P-value = 0.127 > 0.05, we do not reject H_0.
d. The two P-values are almost equal; in fact, the difference between them is only due to rounding errors in the Minitab program.
12.23 **a.** $X^2 = 96.506$, P-value ≈ 0, reject H_0 **b.** The result of Part (a) tells us that the level of the gift seems to make a difference. Looking at the data given, 12% of those receiving no gift made a donation, 14% of those receiving a small gift made a donation, and 21% of those receiving a large gift made a donation. (These percentages can be compared to 16% making donations among the expected counts.) So, it seems that the most effective strategy is to include a large gift, with the small gift making very little difference compared to no gift at all.
12.25 $X^2 = 46.515$, P-value ≈ 0, reject H_0
12.27 $X^2 = 3.030$, P-value = 0.387, fail to reject H_0
12.29 $X^2 = 49.813$, P-value ≈ 0, reject H_0
12.31 $X^2 = 1.978$, P-value = 0.372, fail to reject H_0
12.33 **b.** $X^2 = 8.034$, P-value = 0.005, reject H_0
12.35 $X^2 = 1.08$, P-value = 0.982, fail to reject H_0
12.37 $X^2 = 881.360$, P-value ≈ 0, reject H_0
12.39 $X^2 = 4.035$, P-value = 0.258, fail to reject H_0
12.41 **a.** $X^2 = 10.976$, P-value < 0.001, reject H_0
b. representative sample
12.43 $X^2 = 22.855$, P-value ≈ 0, reject H_0

Chapter 13

13.1 **a.** $y = -5.0 + 0.017x$ **c.** 30.7 **d.** 0.017 **e.** 1.7
f. No, the model should not be used to predict outside the range of the data.
13.3 **a.** When $x = 15$, $\mu_y = 0.18$. When $x = 17$, $\mu_y = 0.186$.
b. When $x = 15$, $P(y > 0.18) = 0.5$. **c.** When $x = 14$, $P(y > 0.175) = 0.655$, $P(y < 0.178) = 0.579$.
13.5 **a.** 47, 4700 **b.** 0.3156, 0.0643
13.7 **a.** 0.121 **b.** $s_e = 0.155$; this is a typical vertical deviation of a bone mineral density value in the sample from the value predicted by the least-squares line. **c.** 0.009 g/cm^2 **d.** 1.098 g/cm^2
13.9 **a.** $r^2 = 0.883$ **b.** $s_e = 13.682$, df = 14
13.11 **a.** The plot shows a linear pattern, and the vertical spread of points does not appear to be changing over the range of x values in the sample. If we assume that the distribution of errors at any given x value is approximately normal, then the simple linear regression model seems appropriate.
b. $\hat{y} = -0.00227 + 1.247x$; when $x = 0.09$, $\hat{y} = 0.110$.
c. $r^2 = 0.436$, 43.6% of the variation in market share can be explained by the linear regression model relating market share and advertising share. **d.** $s_e = 0.0263$, df = 8
13.13 **a.** Yes. P-value = 0.023. Since this value is less than 0.05, we have convincing evidence at the 0.05 significance level of a useful linear relationship between shrimp catch per tow and oxygen concentration density. **b.** Yes. $r = \sqrt{0.496} = 0.704$. So there is a moderate to strong linear relationship for the values in the sample. **c.** (17.363, 177.077) We are 95% confident that the slope of the population regression line relating mean catch per tow to O_2 saturation is between 17.363 and 177.077.
13.15 **a.** 0.1537 **b.** (2.17, 2.83) **c.** Yes, the interval is relatively narrow.
13.17 **a.** $a = 592.1$, $b = 97.26$ **b.** When $x = 2$, $\hat{y} = 786.62$, $y - \hat{y} = -29.62$. **c.** (87.76, 106.76)
13.19 $t = -3.66$, P-value ≈ 0, reject H_0
13.21 **a.** (0.081, 0.199) We are 95% confident that the mean change in pleasantness rating associated with an increase of 1 impulse per second in firing frequency is between 0.081 and 0.199. **b.** $t = 5.451$, P-value = 0.001, reject H_0
13.23 **a.** $t = 6.493$, P-value ≈ 0, reject H_0 **b.** $t = 1.56$, P-value = 0.079, fail to reject H_0
13.25 **a.** 0.253 **b.** 0.179; no **c.** 4
13.27 There is a random scatter of points in the residual plot, implying that a linear model relating squirrel population density to percentage of logging is appropriate. The residual plot shows no tendency for the size (magnitude) of the residuals to either

increase of decrease as percentage of logging increases. So it is justifiable to assume that the vertical deviations from the population regression line have equal standard deviations. The last condition is that the vertical deviations be normally distributed. The fact that the boxplot of the residuals is roughly symmetrical and shows no outliers suggests that this condition is satisfied.
13.29 There is a curved pattern in the residual plot, which suggest that the simple linear regression model is not appropriate.
13.31 No. The pattern in the scatterplot looks linear, but the spread around the line does not appear to be constant. The histogram of the residuals also appears to be positively skewed.

Cumulative Review 13

CR13.1 Randomly assign the 400 students to two groups of equal size, Group A and Group B. Have the 400 students take the same course, attending the same lectures and being given the same homework assignments. The only difference between the two groups should be that the students in Group A should be given daily quizzes and the students in Group B should not. After the final exam the exam scores for the students in Group A should be compared to the exam scores for the students in Group B.
CR13.3 **b.** The two airlines with the highest numbers of fines assessed may not be the worst in terms of maintenance violations since these airlines might have more flights than the other airlines.
CR13.5 **a.** (0.651, 0.709) We are 95% confident that the proportion of all adult Americans who view a landline phone as a necessity is between 0.651 and 0.709. **b.** $z = 1.267$, P-value $=$ 0.103, fail to reject H_0 **c.** $z = 9.513$, P-value ≈ 0, reject H_0
CR13.7 **a.** 0.62 **b.** 0.1216 **c.** 0.19 **d.** 0.0684
CR13.9 **b.** $\hat{y} = -12.887 + 21.126x$ **d.** $t = 21.263$, P-value ≈ 0, reject H_0
CR13.11 $X^2 = 26.175$, P-value ≈ 0, reject H_0
CR13.13 $X^2 = 15.106$, P-value $= 0.002$, reject H_0
CR13.15 $t = -113.17$, df $= 45$, P-value ≈ 0, reject H_0
CR13.17 $X^2 = 4.8$, P-value $= 0.684$, fail to reject H_0

Chapter 14

14.1 **a.** A deterministic model does not have the random deviation component e, while a probabilistic model does contain such a component.
14.3 **a.** (mean y value for fixed values of $x1, x2, x3$) $=$ $30 + 0.90x_1 + 0.08x_2 - 4.5x_3$
b. $\beta_0 = 30$, $\beta_1 = 0.9$, $\beta_2 = 0.08$, $\beta_3 = -4.50$ **c.** The average change in acceptable load associated with a 1-cm increase in left lateral bending, when grip endurance and trunk extension ratio are held fixed, is 0.90 kg. **d.** The average change in acceptable load associated with a 1 N/kg increase in trunk extension ratio, when grip endurance and left lateral bending are held fixed, is -4.5 kg. **e.** 23.5 **f.** 95%
14.5 **a.** 13.552 g. **b.** When length is fixed, the mean increase in weight associated with a 1-mm increase in width is 0.828 g. When width is fixed, the mean increase in weight associated with a 1-mm increase in length is 0.373 g.
14.7 **a.** 103.11 **b.** 96.87 **c.** $\beta_1 = -6.6$; 6.6 is the expected decrease in yield associated with a one-unit increase in mean temperature when the mean percentage of sunshine remains fixed. $\beta_2 = -4.5$; 4.5 is the expected decrease in yield associated with a one-unit increase in mean percentage of sunshine when mean temperature remains fixed.
14.9 **b.** Higher for $x = 10$ **c.** When the degree of delignification increases from 8 to 9 the mean chlorine content increases by 7. Mean chlorine content decreases by 1 when degree of delignification increases from 9 to 10.
14.11 **c.** The parallel lines in each graph are attributable to the lack of interaction between the two independent variables.
14.13
a. $y = \alpha + \beta_1x_1 + \beta_2x_2 + \beta_3x_3 + e$
b. $y = \alpha + \beta_1x_1 + \beta_2x_2 + \beta_3x_3 + \beta_4x_1^2 + \beta_5x_2^2 + \beta_6x_3^2 + e$
c. $y = \alpha + \beta_1x_1 + \beta_2x_2 + \beta_3x_3 + \beta_4x_1x_2 + e$;
$y = \alpha + \beta_1x_1 + \beta_2x_2 + \beta_3x_3 + \beta_4x_1x_3 + e$;
$y = \alpha + \beta_1x_1 + \beta_2x_2 + \beta_3x_3 + \beta_4x_2x_3 + e$
d. $y = \alpha + \beta_1x_1 + \beta_2x_2 + \beta_3x_3 + \beta_4x_1^2 + \beta_5x_2^2 + \beta_6x_3^2 + \beta_7x_1x_2 + \beta_8x_1x_3 + \beta_9x_2x_3 + e$
14.15 **a.** Three dummy variables would be needed to incorporate a nonnumerical variable with four categories. For example, you could define $x_3 = 1$ if the car is a subcompact and 0 otherwise, $x_4 = 1$ if the car is a compact and 0 otherwise, and $x_5 = 1$ if the car is a midsize and 0 otherwise. The model equation is then $y = \alpha + \beta_1x_1 + \beta_2x_2 + \beta_3x_3 + \beta_4x_4 + \beta_5x_5 + e$. **b.** For the variables defined in Part (a), $x_6 = x_1x_3$, $x_7 = x_1x_4$, and $x_8 = x_1x_5$ are the additional predictors needed to incorporate interaction between age and type of car.
14.17 **a.** $0.01 < P\text{-value} < 0.05$ **b.** $P\text{-value} > 0.10$ **c.** $P\text{-value} = 0.01$ **d.** $0.001 < P\text{-value} < 0.01$
14.19 **a.** $F = 12118$, P-value ≈ 0, reject H_0 **b.** Since the P-value is small and r^2 is close to 1, there is strong evidence that the model is useful. **c.** The model in Part (b) should be recommended, since adding the variables x_1 and x_2 to the model [to obtain the model in Part (a)] only increases the value of R^2 a small amount (from 0.994 to 0.996).
14.21 $F = 24.41$, P-value < 0.001, reject H_0 and conclude that the model is useful.
14.23 $F = 3.5$, $0.01 < P\text{-value} < 0.05$, reject H_0 and conclude that the model is useful.
14.25 $F = 7.986$, P-value < 0.001, reject H_0 and conclude that the model is useful.
14.27 **a.** $\hat{y} = 1.44 - 0.0523\,length + 0.00397\,speed$ **b.** 1.3245 **c.** $F = 24.02$, P-value ≈ 0, reject H_0 and conclude that the model is useful. **d.** $\hat{y} = 1.59 - 1.40\left(\frac{\text{length}}{\text{speed}}\right)$ **e.** The model in Part (a) has $R^2 = 0.75$ and R^2 adjusted $= 0.719$, whereas the model in Part (d) has $R^2 = 0.543$ and R^2 adjusted $= 0.516$.
14.29 **a.** SSResid $= 390.4347$, SSTo $= 1618.2093$, SSRegr $=$ 1227.7746 **b.** $R^2 = 0.759$; this means that 75.9% of the variation in the observed shear strength values has been explained by the fitted model. **c.** $F = 5.039$, $0.01 < P\text{-value} < 0.05$, reject H_0, and conclude that the model is useful.
14.31 $F = 96.64$, P-value < 0.001, reject H_0, and conclude that the model is useful.
14.35 $\hat{y} = 35.8 - 0.68x_1 + 1.28x_2$, $F = 18.95$, P-value < 0.001, reject H_0, and conclude that the model is useful.

Chapter 15

15.1 **a.** $0.001 < P\text{-value} < 0.01$ **b.** $P\text{-value} > 0.10$ **c.** $P\text{-value} = 0.01$ **d.** $P\text{-value} < 0.001$ **e.** $0.05 < P\text{-value} < 0.10$ **f.** $0.01 < P\text{-value} < 0.05$ (using $df_1 = 4$ and $df_2 = 60$)

15.3 **a.** H_0: $\mu_1 = \mu_2 = \mu_3 = \mu_4$, H_a: At least two of the four μ_i's are different. **b.** P-value = 0.012, fail to reject H_0 **c.** P-value = 0.012, fail to reject H_0
15.5 $F = 6.687$, P-value = 0.001, reject H_0
15.7 $F = 5.273$, P-value = 0.002, reject H_0
15.9 $F = 53.8$, P-value < 0.001, reject H_0
15.11 $F = 2.62$, $0.05 < P$-value < 0.10, fail to reject H_0
15.13

Source of Variation	df	Sum of Squares	Mean Square	F
Treatments	3	75,081.72	25,027.24	1.70
Error	16	235,419.04	14,713.69	
Total	19	310,500.76		

$F = 1.70$, P-value > 0.10, fail to reject H_0
15.15 Since there is a significant difference in all three of the pairs we need a set of intervals that does not include zero. Set 3 is therefore the required set.
15.17 **a.** In *decreasing* order of the resulting mean numbers of pretzels eaten the treatments were slides with related text, slides with no text, slides with unrelated text, and no slides. There were no significant differences between the results for slides with no text and slides with unrelated text and for slides with unrelated text and no slides. However, there was a significant difference between the results for slides with related text and each one of the other treatments and between the results for no slides and for slides with no text (and for slides with related text). **b.** The results for the women and men are almost exactly the reverse of one another, with, for example, slides with related text (treatment 2) resulting in the smallest mean number of pretzels eaten for the women and the largest mean number of pretzels eaten for the men. For the men, treatment 2 was significantly different from all the other treatments; however, for women treatment 2 was not significantly different from treatment 1. For both women and men there was a significant difference between treatments 1 and 4 and no significant difference between treatments 3 and 4. However, between treatments 1 and 3 there was a significant difference for the women but no significant difference for the men.
15.19 **a.**

Sample mean	**Driving**	**Shooting**	**Fighting**
	.42	4.00	5.30

b.

Sample mean	**Driving**	**Shooting**	**Fighting**
	2.81	3.44	4.01

15.21 **a.** $F = 45.64$, P-value ≈ 0, reject H_0 **b.** Yes; T-K interval is (0.388, 0.912)

Index

1 in *k* systematic sample, 42

A

A and B, 287
A or B, 287
Addition rule
  for mutually exclusive events, 300
  general, 324
Additive multiple regression model, 703–704
Additive probabilistic model, 663
Adjusted R^2, 719
Alternative hypothesis, 506
Analysis of variance (ANOVA), *see* Single-factor analysis of variance
ANOVA table, 741–742

B

Bar chart, 87–88
  comparative, 81–82
Bayes Rule, 330–332
Bell-shaped curve, 109–110
  See also Normal distribution
Bias in sampling, 36–37
Biased statistic, 464
Bimodal histogram, 108–109
Binomial distribution, 381–388
  defined, 381
  formula for, 383
  mean of, 387
  normal approximation to, 421–432
  standard deviation of, 387
  tables of probabilities, 385
Binomial experiment, 381
Binomial random variable, 381
Bivariate data
  defined, 10
  scatterplot of, 119
Blocking, 48, 50–51
  defined, 48
  extraneous variables and, 48
  random assignment and, 57
Bound on error of estimation
  defined, 474
  sample size choice and, 475
Boxplot
  comparative, 172–173
  modified, 173–176
  skeletal, 172
  small sample size and, 190

C

Categorical data, 9
  bar chart for, 13
  chi-square test for, 625, 629
  comparative bar chart for, 81–82
  frequency distribution for, 12
  pie chart for, 82–83
Causation
correlation and, 210
Cause-and-effect, determining, 31–33
Cell count *see* Expected cell count and Observed cell count
Census, 35
Center, measures of, 153–160
Central Limit Theorem, 446–448
Chance experiment, 284–285
Chebyshev's rule, 178–180
Chi-square distribution, 628
Chi-square statistic, 628
Chi square test
  goodness-of-fit, 629
  homogeneity, 639
  independence, 642–643
Class interval, 103, 105
Cluster, 42
Cluster sampling, 42
Coefficient of determination, r^2, 231–235
Coefficient of multiple determination, R^2, 718
Common population proportion, 595
Comparative bar chart, 81–82
Comparative boxplot, 172–173
Comparative stem-and-leaf display, 95
Complement of an event, 287
Complete second order model, 708
Completely randomized design, 57
Conditional probability, 305–315
Confidence interval
  general form of, 474
  for population mean, 484
  for population proportion, 470–472
  for comparing two population or treatment means using independent samples, 573
  for comparing two population or treatment means using paired samples, 587–588
  for comparing two population or treatment proportions, 599
  for the slope of the population regression line, 676
Confidence level, 468–469
Confounded variables, 48
Confounding variable, 31
Contingency table, 636
Continuity correction, 420–421
Continuous numerical data
  defined, 11
  frequency distribution for, 103–104
  histogram for, 105
Continuous random variable, 353, 363
  probability distributions for, 363–367
Control group, 54, 62
Convenience sampling, 43
Correlation, 203 -214
Correlation coefficient, 203–204
  and the coefficient of determination, 234
  formula for, 204
  limitations of, 265–266
Cumulative area, 365
Cumulative relative frequency, 113–114
Cumulative relative frequency plot, 112–114

D

Danger of extrapolation, 218, 265, 670
Data analysis process, 5–7
Degrees of freedom, 167
  ANOVA and, 737

Degrees of freedom (*continued*)
Chi-square distributions and, 628
Simple linear regression and, 670
t distribution and, 481–482
two-sample *t* test and, 571–572
Density, 106
Density curve, 363–364
Density function, 363
Density scale, 106
Dependent events, 316
Dependent variable, 214
Descriptive statistics, 7
Deterministic relationship, 663
Deviations from the mean, 164–165
Diagram of experimental design, 54–55
Direct control, 47–48, 50–51
Discrete data, 11
Discrete probability distribution, 364–365
binomial distribution, 381–388
geometric distribution, 388–390
properties of, 358
Discrete random variable
defined, 353–354
mean value of, 367
probability distribution of, 356–362
standard deviation of, 371–372
variance of, 372
Disjoint events *see* Mutually exclusive events
Dotplot, 14–16
Double-blind experiment, 63–64
Dummy variable, 710

E

Empirical Rule, 180–182
standard normal distribution and, 395, 398
Equally likely outcomes, 291–292
Error sum of squares (SSE), 737, 756
Errors in hypothesis testing, 510–516
Estimated regression line, 716–718, 731
Estimation
choosing a statistic, 464
point estimate, 462
interval estimate *see* Confidence interval
bias in, 464
Event, 285
Expected cell count, 638
Expected value, *see* mean
Experiment, 31, 46
completely randomized, 57
double-blind, 64
randomized block design, 57
replicating, 50–51
single-blind, 64
using a control group, 54, 62
using volunteers, 43
Experimental conditions, 46
Experimental design, principles of, 50
Experimental unit, 64
Explanatory variable, 214
Extraneous variable, 47–48, 51
Extrapolation, danger of, 218, 265, 670
Extreme outlier, 173

F

F distribution, 719–720
F test, ANOVA, 733–746
F test for model utility, 721–722
Factor, 31, 46, 733
Finite population correction factor, 495
Fitted value, *see* Predicted value
Five-number summary, 172
Frequency, 12
Frequency distribution, 12–13, 138
for categorical data, 12
for continuous numerical data, 103–112
for discrete numerical data, 99–103
Full quadratic model, 708
Fundamental identity for single-factor ANOVA, 741

G

General addition rule, 324
General additive multiple regression model, 703–704
General multiplication rule, 326
Geometric probability distribution, 388–390
Geometric random variable, 388
Goodness of fit statistic, 628
Goodness of fit test, 629–632

H

Heavy-tailed curve, 110
Histogram
for continuous numerical data with equal width intervals, 105–106
for continuous numerical data with unequal width intervals, 106–108
for discrete numerical data, 100–103
shapes, 108–110
using density, 106–108
Homogeneity, chi-square test of, 635–641
Hypergeometric distribution, 386–387
Hypotheses, 506
alternative hypothesis, 506
null hypothesis, 506
Hypothesis test
errors in, 510–516
for population mean, 528–536
for population proportion, 516–526
for comparing two population or treatment means using independent samples, 562–572
for comparing two population or treatment means using paired samples, 582–587
for comparing two population or treatment proportions, 594–599
for goodness-of-fit, 629
for homogeneity, 639
for independence, 642–643
for the slope of the population regression line, 678–683
power of, *see* Power, hypothesis testing
significance level of, 511
steps for, 523–524

I

Independence, 315–324
testing for, 642–643
Independent events
defined, 316–319
multiplication rule for two, 317
multiplication rule for more than two, 318
Independent samples, 562
Independent variable, 214
Indicator variable, 710–712
Inferential statistics, 7
Influential observation, 229–231, 689
Interaction, 706–710
predictor in multiple regression, 708
Intercept, 214
Interquartile range (iqr), 167–168
Intersection of two events, 287
Interval estimate, *see* Confidence interval

J

Joint probability table, 315

L

Large sample confidence interval
for a population proportion, 470–472
for the difference in two population or treatment proportions, 599

Large-sample hypothesis test
for a population proportion, 516–536
for the difference in two population or treatment proportions, 594–599
Law of large numbers, 295, 337
Law of total probability, 328–330
Least squares regression line
assessing fit of, 224–244
defined, 215–221
deviations from, 215–216
influential observations and, 231
population regression line and, 664
making predictions with, 216
outliers and, 231
slope of, 216
standard deviation about, 235–239
y intercept of, 216
Least-squares principle, 215–216
Line
assessing fit of, 224–244
equation of, 214
least squares principle and, 215–216
Linear combination of random variables, 375
mean and standard deviation of, 375
Linear function of random variable, 374
mean and standard deviation of, 374
Linear regression, 237
model, 663–675
Linear relationship, strength of, 204–205, 208
Lower quartile, 167
Lower-tailed test, 522, 531

M

Margin of error, 493 *see also* Bound on error of estimation
Marginal total, 636
Mean
of a binomial distribution, 387
of a continuous random variable, 373
of a data set, 153–157
deviations from, 164–165, 192
of a discrete random variable, 367–371
of a linear combination, 375
of a linear function, 374
vs. median, 158–159
of a normal distribution, 393–394
outliers and, 156–157
population mean, 155
Mean square, 737
for error (MSE), 737–738
for treatments (MSTr), 737–738
Measurement bias, 36–37
Measure of center, 153–160
Measure of relative standing, 182–183
Measure of spread, 163–168
Median
defined, 157–158
vs. mean, 158–159
outliers and, 158
Mild outlier, 173
Mode, 108
Model utility test
in simple linear regression, 680–682
in multiple regression, 721–722
Modified boxplot, 173–176
Multiple comparison procedure, 746–754
Multiple regression
general additive model, 703–704
interaction variables in, 706–710
model utility *F* test, 721–722
qualitative predictors in, 710–712
Multiplication rule
general, 326
for independent events, 317
Multivariate data, 10
Mutually exclusive events, 288–289
addition rule for, 300–301

N

Negative skew, 109
Nonlinear regression, 244–262
Nonresponse bias, 36
Normal approximation to the binomial distribution, 421–422
Normal curve, 109–110
Normal distribution, 393–409
extreme values in, 404–406
mean value of, 393
standard normal distribution, 394–398
standard deviation of, 393
Normal probability plot, 409–410
Normal score, 409
Null hypothesis, 506
Numerical data, 11
displaying bivariate, 119
dotplot for, 14–16
frequency distribution for continuous numerical data, 103–112
frequency distribution for discrete numerical, 99–103
histogram for continuous numerical data, 105–108
histogram for discrete numerical data, 100–103
stem-and-leaf display for, 92–95
types of, 9

O

Observational study, 30–32
Observed cell count, 636
Observed significance level, 518
see also *P*-value
One-sample *t* confidence interval, 484
One-sample *t* test, 531
One sample *z* confidence interval, 479
One-way frequency table, 625
Outlier, 92, 156, 173
Overcoverage, 66–67

P

Paired samples, 562–563
paired *t* test and, 584–587
Paired *t* confidence interval, 587–588
Paired *t* test, 584–587
Pearson's sample correlation coefficient, *see* Correlation coefficient
Percentile, 183–184
Pie chart
for categorical data, 82–85
Placebo, 62–63, 73
Plotting residuals, 227–231, 688–693
Point estimation, 462–467
Polynomial regression, 704–706
Pooled *t* test, 572
Population, 6
Population correlation coefficient, 209–210
Population mean, 155
comparing two using independent samples, 562–573
comparing two using paired samples, 582–588
confidence interval for, 484
hypothesis tests for, 528–536
one-sample *t* test for, 531
Population proportion
comparing two using large samples, 594–599
large-sample confidence interval for, 470–472
large-sample hypothesis test for, 516–526
Population regression function, 704–705
Population regression line, 664–666
Population standard deviation, 166–167

Population variance, 166–167
Positive skew, 109
Potentially influential observation, 689
Power
  calculating, 541–547
  defined, 540
  effect of sample size and significance level on, 540
  Type II error probabilities and, 541
Practical significance, 535–536, 550
Predicted value, 225–227
Predictor variable, 214
Principle of least-squares, 215–220
Probabilistic model, 663
Probability
  of an event, 292, 295, 297–298
  classical approach for equally likely outcomes, 291–293
  of the complement of an event, 318
  conditional, 305–315
  estimating empirically, 337–338
  estimating using simulation, 339–343
  fundamental properties of, 297–305
  histogram, 359, 363
  of the intersection of two events, 317, 326
  relative frequency approach to, 293–296
  subjective approach to, 296–297
  of the union of two events, 300, 324
Probability distribution
  for continuous random variable, 363–367
  for discrete random variable, 356–362
Probability histogram, 359. 363, 387
P-value, 518
  and alternative hypotheses, 522
  determining for a *t* test, 530–531
  determining for a *z* test, 520–526
  determining for a chi-square test, 628–629
  drawing conclusions based on, 520

Q

Quadratic regression, 247–249, 705, 708
Qualitative data, 9
Qualitative predictor variables, 710–712
Quantitative data, 9
Quartile, 167
  lower, 167
  upper, 167

R

r see Correlation coefficient
Random assignment, 47–53
Random deviation, 664
Random number, 39
Random sampling, 38–40
Random selection, 40
Random variable, 353–356
Randomized block design, 57
Range, 163
Regression, *see* Least-squares regression line and Linear regression
Regression sum of squares (SSRegr), 721
Relative frequency, 12
Relative frequency distribution, 12
Replacement, sampling with and without, 39, 319–321, 385–387
Replication, 50–51
Residual
  in multiple regression, 718
  in linear regression, 225–227
  plotting, 227–231, 688–693
Residual analysis, 686–688
Residual plot, 227–231, 688–693
  standardized residual plot, 688–689
Residual sum of squares (SSResid), 232–233, 718
Response bias, 36–37
Response variable, 214
Right skew, 109

S

Sample
  defined, 6
  independent samples, 562–563
  paired samples, 562–563
Sample mean, 154–155
Sample median, 157–158
Sample proportion, 159–160
Sample regression line *see* Least-squares regression line
Sample size, 40–41
  determining to achieve a given bound on error, 475, 487–488
  and power, 540–541
Sample space, 284–285
Sample standard deviation, 165
Sample variance, 165
Sampling, 35–46
  bias in, 36–37
  cluster, 42
  convenience, 42
  random, 38–40
  stratified, 41–42
  systematic, 42–43
  with and without replacement, 39, 319–321, 385–387
Sampling distribution, 437, 441, 459
  of the difference in sample means (independent samples), 563
  of the difference in sample proportions, 594
  of the sample mean, 442–452
  of the sample proportion, 452–457
  of the slope of the least squares regression line, 675
Sampling frame, 38
Sampling variability, 438–442
Scatterplot, 119
Segmented bar graph, 85–87
Selection bias, 36, 37
Significance level, 511, 520
  and power, 540–541
Simple comparative experiment, 46–61
  design strategies for, 47
Simple event, 286
Simple linear regression, *see* Least-squares regression line and Linear regression
Simple random sample, 38
  selecting, 38–40
Simulation to estimate probabilities, 336–346
Simultaneous confidence level, 749
Single-blind experiment, 63–64
Single factor analysis of variance (ANOVA), 733
  F test for, 733–746
  fundamental identity for, 741, 756
Skeletal boxplot, 172
Skewed histogram, 109
Slope
  defined, 214
  of least-squares regression line, 216, 265–266, 269
  estimate for population regression line, 675–685
Smoothed histogram, 108
Spread, measure of, 163–168
Stacked bar graph *see* Segmented bar graph
Standard deviation
  about the least-squares line, 235–239
  of a data set, 165–167
  of a binomial random variable, 387
  of continuous random variable, 373
  of difference in means, 563, 610
  of a discrete random variable, 371–373

of a linear combination, 375
of a linear function, 374
of the slope of the least-squares line, 675–676
Standard error, 474, 493
Standard normal distribution, 394–398
Standardized residual, 686
Standardized residual plot, 688–689
Statistic, 438
Statistical significance, 535–536
Stem-and-leaf display, 91–99
Comparative, 95
repeated stems, 94
Strata, 41
Stratified random sampling, 41–42
Studentized range distribution, 747
Symmetric histogram, 109
Systematic sampling, 42–43

T

t confidence interval
for a difference in population or treatment means, 573
for a population mean, 484
t critical value, 483–484
t distribution, 481–483
degrees of freedom and, 481
properties of, 482
t test
finding *P*-values for, 530–531
for the slope of the population regression line, 678–679
one-sample *t* test for a population mean, 528–536
paired *t* test, 584–587
pooled *t* test, 572
power of and Type II error probabilities, 544–547
two-sample *t* test for difference in population or treatment means, 562–572
Test procedure, 506
Test statistic, 518
Test of hypotheses *see* Hypothesis test
Time series plot, 123–124
unequal spacing in, 133
Total sum of squares (SSTo), 232–233
Transformations, 251
models transforming *x*, 252–254
models transforming *y*, 254–256
nonlinear regression models, 244–262
Transforming data to normal distribution, 411–415
Treatment, 46
Treatment means
comparing using independent samples, 569–572
comparing using paired samples, 582–593
confidence interval for comparing using independent samples, 573
confidence interval for comparing using paired samples, 587–588
Treatment proportions
confidence interval for difference in, 599
large sample test for difference in, 594–599
Treatment sum of squares (SSTr), 737
Tree diagram, 285
Tukey-Kramer (T-K) multiple comparison procedure, 746–752
simultaneous confidence level and, 749
Two-sample *t* confidence interval, 573
Two-sample *t* test
for comparing two population means, 565
for comparing two treatment means, 569–570
Two-sample *z* confidence interval, 599
Two-sample *z* test
for comparing two proportion, 594–599
Two-way frequency table, 636
Type I error, 510–513
probability of, 511
Type II error, 510–513
probability of, 511, 541–547
probability of and power, 541

U

Unbiased statistic, 464–466
Unconditional probability
compared to conditional probability, 307–309, 329
Undercoverage, 36
Unexplained variability, 233
Uniform distribution, 364
Unimodal histogram, 108–109
Union of two events, 287
Univariate data, 9
Upper quartile, 167

V

Variability
in data, 156
interpreting, 178–186
measuring, 163–168, 178–184
nature and role of, 3–5
sampling, 438–442
Variable, 9
dependent, 214
confounding, 31
explanatory, 214
extraneous, 47–48, 51
independent, 214
interaction between, 706–710
predictor, 214
response, 214
Variance, 165–167 *see also* standard deviation
Venn diagram, 288
Vertical intercept, 214
Voluntary response sampling, 43
Volunteers in experiments, 43, 64

W

Whiskers in a boxplot, 172–173

Y

y intercept, 214
of the least-squares regression line, 216, 265–266
of the population regression line, 665

Z

z confidence interval
for a difference in population or treatment proportions, 599
for a population proportion, 470–472
z critical value, 473, 477
z curve, 394 *see also* standard normal distribution
z-score, 182, 192, 404
and the correlation coefficient, 204
z test
for a difference in population proportions, 594–599
for a population proportion, 516–526